Hydroelectric Power

Hydroelectric Power

Editors

Zhengwei Wang
Yongguang Cheng

Basel • Beijing • Wuhan • Barcelona • Belgrade • Novi Sad • Cluj • Manchester

Editors
Zhengwei Wang
Tsinghua University
Beijing
China

Yongguang Cheng
Wuhan University
Wuhan
China

Editorial Office
MDPI
St. Alban-Anlage 66
4052 Basel, Switzerland

This is a reprint of articles from the Topic published online in the open access journals *Energies* (ISSN 1996-1073), *Machines* (ISSN 2075-1702), and *Water* (ISSN 2073-4441) (available at: https://www.mdpi.com/topics/hydroelectric_power).

For citation purposes, cite each article independently as indicated on the article page online and as indicated below:

Lastname, A.A.; Lastname, B.B. Article Title. *Journal Name* **Year**, *Volume Number*, Page Range.

ISBN 978-3-7258-1071-0 (Hbk)
ISBN 978-3-7258-1072-7 (PDF)
doi.org/10.3390/books978-3-7258-1072-7

Contents

MDPI

Article

An Efficient Method for Computing the Power Potential of Bypass Hydropower Installations

Olivier Cleynen *, Dennis Powalla, Stefan Hoerner and Dominique Thévenin

Laboratory of Fluid Dynamics and Technical Flows, Institute of Fluid Dynamics and Thermodynamics, University Otto von Guericke of Magdeburg, Universitätsplatz 2, 39106 Magdeburg, Germany; dennis.powalla@ovgu.de (D.P.); hoerner@ovgu.de (S.H.); thevenin@ovgu.de (D.T.)
* Correspondence: olivier.cleynen@ovgu.de

Abstract: Small-scale hydropower installations make possible a transition towards decentralized electrical power production with very low ecological footprint. However, the prediction of their power potential is difficult, because the incoming flow velocity and the inlet and outlet water heights are often outside of the control of the operator. This leads to a need for a method capable of calculating an installation's power potential and efficiency rapidly, in order to cover for many possible load cases. In this article, the use of a previously-published theoretical framework is demonstrated with the case of a mid-scale hydropower device, a 26 m long water vortex power plant. It is shown that a simplified CFD simulation with a single output (the mass flow rate) is sufficient to obtain values for the two coefficients in the model. Once this is done, it becomes possible to evaluate the device's real-life performance, benchmarking it against reference values anchored in physical principles. The method can be used to provide design guidance and rapidly compare different load cases, providing answers that are not easily obtained using intuition or even experiments. These results are obtained for a computing cost several orders of magnitude smaller than those associated with a full description of the flow using CFD methods.

Keywords: micro-hydro; pico-hydro; efficiency; computational fluid dynamics

Citation: Cleynen, O.; Powalla, D.; Hoerner, S.; Thévenin, D. An Efficient Method for Computing the Power Potential of Bypass Hydropower Installations. *Energies* **2022**, *15*, 3228. https://doi.org/10.3390/en15093228

Academic Editors: Zhengwei Wang and Yongguang Cheng

Received: 25 March 2022
Accepted: 24 April 2022
Published: 28 April 2022

Publisher's Note: MDPI stays neutral with regard to jurisdictional claims in published maps and institutional affiliations.

1. Introduction

1.1. Scientific Context

Ecological considerations are driving new interest in the development of small-scale hydropower devices. These make possible a transition towards electrical power production that is not only decentralized, bringing significant socioeconomic benefits [1,2], but also better able to abide by increasingly constraining ecological regulations regarding river fauna.

In this way, recent work has gone towards optimizing devices such as Savonius turbines, Darrieus turbines, water wheels, and water vortex power plants, the last of which are further described below. In traditional installations, fish passage is often impeded (thus breaking river continuity [3–5]), there is high risk of lethal fish impact with turbomachine parts [6–9], as well as high risk of barotrauma [10–12]. All of those problematic aspects add up to a high ecological footprint, and all are reduced or avoided with these new types of small-scale machines.

From a turbomachine engineering perspective however, the development of these devices is challenging because they operate in environments that are much less controlled than those of typical large-scale hydropower turbines [13,14]. In particular, the incoming flow velocity, as well as the inlet and outlet water heights, are not under full control and continuously vary along with the natural discharge over the year [15]. This makes the prediction of their power potential difficult, since many different cases have to be studied, as, e.g., in [16].

In this article, the case of a water vortex power plant (WVPP) will be studied. The WVPP is a device that aims to combine both ecology and hydropower, providing river continuity for fish migration while exploiting the discharge of this bypass facility. Here we show why the quantification of its power potential is so challenging. The practical problem encountered on a full-scale laboratory installation will be described. A simple theoretical model will then be applied to the case, demonstrating the use of an efficient method for computing the power potential of bypass hydropower installations.

1.2. Operation of a Water Vortex Power Plant

Water vortex power plants are hydropower devices which use a type of Francis turbine (radial inlet, vertical outlet), but without the use of any guide vanes; instead, their turbine inlet consists of an open-air, spiral-shaped basin. This type of device is suited for combinations of modest flow rates (in the order of $1\,\mathrm{m^3\,s^{-1}}$) and modest hydraulic head (in the order of $1\,\mathrm{m}$). Recent research interest in this type of device has focused on their geometry [17,18] and their compatibility with fish river migration [19].

As part of the research project Fluss-Strom, the dynamics of a 26 m long WVPP were investigated. The constructor of the device markets it as the "Fisch-Freundliches Wehr", for use as fish migration corridors bypassing dams and large weirs. Work was conducted towards erecting a full-scale laboratory installation at the Technische Universität Dresden (Figure 1, for the purpose of carrying out biological investigations with live fish [20–23]. Numerical investigations have also been carried out, as presented further below.

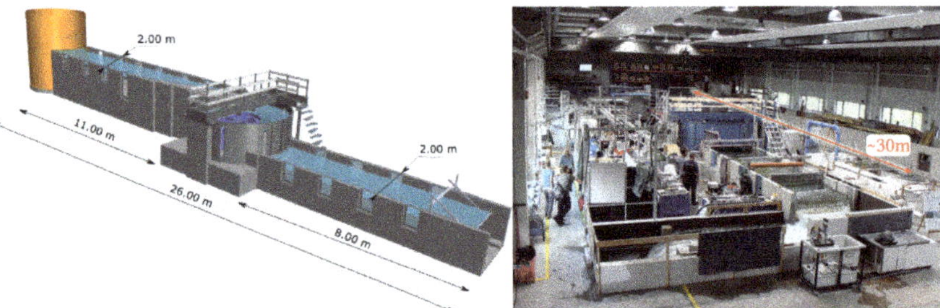

Figure 1. (**Left**): computer drawing of the device with its main dimensions. In both images, the flow is from top left to bottom right, with the turbine basin in the center. (**Right**): photograph of the WVPP installation in the laboratory of the Technische Universität Dresden. (drawing reproduced from Powalla et al. 2021 [24], with geometry provided by Ecoligent GmbH).

In the Dresden laboratory, water is picked up by pumps whose discharge goes through a settling tank before being fed to the WVPP. Water then flows through the free-surface installation, before overflowing an outlet weir, back into the inlet of the pumps. A casual, bystander observation of the dynamics of the installation in Dresden during testing reveals that its behavior is not simple. The system is controlled by prescribing the volume flow (provided by the pumps) and turbine rotation speed; the water levels in the inlet and outlet channels are part of the response.

The need for a model by which to analyze the installation's behavior can be prompted with the following observation. The powerplant is first operated with a given volume flow and turbine rotation speed. Then, the turbine speed is reduced. This increased restriction to the flow causes the water in the upstream channel to "back up", and within the following minute, the water level there has increased significantly. As a result, the turbine's load increases, which, together with the change in rotation speed, causes the power output to change. From the point of view of the experimental scientist, and of the device's future operator, is this change desirable? Is more hydraulic power now available to the turbine?

1.3. Purpose of Article

The present article answers the question above by providing a framework with which to analyze the situation. Here, the former power output (which occurred for a higher turbine speed and identical volume flow) must be meaningfully compared with the new one. Is the turbine more efficient in one of the two cases? Should the expression for efficiency take the increase in inlet cross-sectional area into account?

In the literature, the turbine power of such devices is sometimes non-dimensionalized as an "efficiency", formulated as $\dot{W}_{\text{shaft}}/(\dot{m}g\Delta h)$, as would be done with a high-head cross-flow installation. In this case, however, the device is intended for use as a *bypass* to the main river flow (see Figure 2); the mass flow that is not captured by the WVPP is lost and will pass through the main river flow instead. The denominator in the expression above (the power that corresponds to an "efficiency" of 100 %) is therefore a fleeting amount that is affected by the device's operation, making comparison of different operating points difficult.

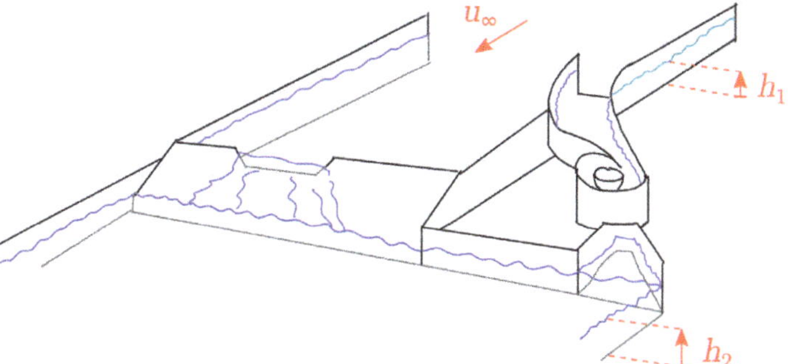

Figure 2. Drawing of the WVPP in the type of installation it has been designed for: as the bypass to the main flow of a river. The heights h_1 and h_2, formally described in ref. [25], remain unaffected by its operation.

These introductory questions prompted by observation of the Dresden laboratory installation can be addressed by defining a maximum by which the actual measured performance of the installation can be compared. To this purpose, we must express power and efficiency in a meaningful way, so they may be compared across different scales and operating conditions.

First, a review of the key parameters that describe the physical processes at hand will be carried out. In a second step, a demonstrative computational fluid dynamics (CFD) simulation will be presented, showing how numerical values for these parameters may be obtained at a relatively low computational cost. Finally, the questions above will be answered with a quantitative example.

2. Full Flow Simulations and Their Limitations

An important tool for fluid flow analysis is of course numerical simulation through computational fluid dynamics (CFD). As part of the same *Fluss-Strom* research project, a series of numerical investigations of the Dresden WVPP installation have been carried out at the University of Magdeburg [24,26–28]. The main focus of those investigations is the device's compatibility with fish migration, with further work currently underway in the laboratory to develop increasingly capable fish behavior models using these simulations [19], but a secondary objective, of concern in the present work, is the quantification of its hydropower potential.

A family of numerical fluid flow simulations is now available to reproduce the flow in the WVPP (Figure 3). These simulations are based on a Reynolds-averaged Navier-Stokes (RANS) approach and have been validated with experimental measurements in the Dresden laboratory. The properties of the simulations are detailed in [24,28]: an unsteady solver is used with volume-of-fluid and k-ω SST models on a structured, 4.7-million-cell grid which accounts for the turbine movement. These are not of concern for the present work; instead, the emphasis is here placed on the computational costs involved in running those simulations: obtaining a single reading for the turbine power in [24] uses up 19,000 CPU-hours. The reasons for this high cost are, firstly, that the inherent physics of the flow are challenging to describe numerically, and secondly, that the simulations must be run for long periods of simulated time (of the order of 45 s) in order to account for the device's long response times. Indeed, the large open-air channels upstream and downstream of the turbine basin act as large mass buffers whenever boundary conditions or turbine speed is changed, while large amounts of momentum are stored in the rotating motion of the water in the basin itself.

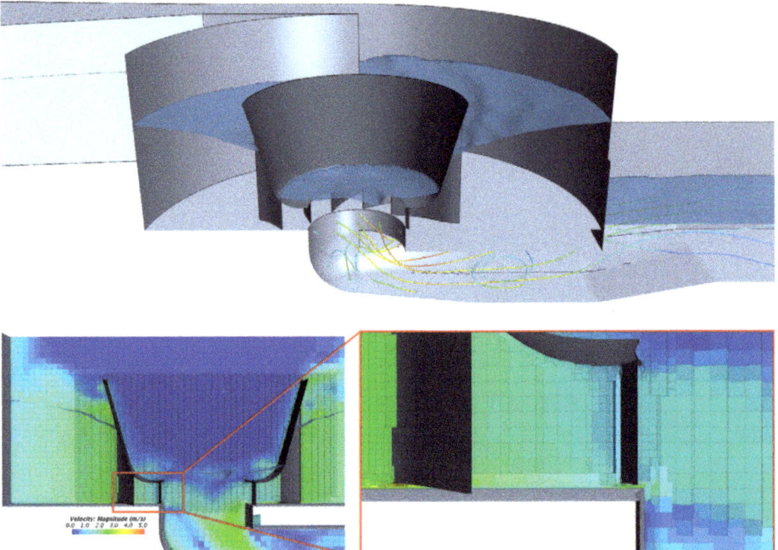

Figure 3. Views from complete simulations of the flow in the WVPP. Top: the device has been sectioned longitudinally so as to display the water level and turbine position in the WVPP. A few streamlines colored according to velocity are visible in the outlet of the turbine. Bottom: view of the mesh structure in the turbine basin, colored according to velocity. Full details of this simulation are published in Powalla et al. 2021 [24].

It therefore follows that a three-dimensional, two-phase CFD simulation accounting for the plant's complete geometry cannot currently be used to map out the behavior of the WVPP across a large range of volume flows, inlet heights, and outlet heights. Instead, a simpler model is needed in order to evaluate the potential power available to the plant, by which its efficiency can be quantified across many conditions.

3. Model for the Energy Budget of a Water Vortex Power Plant

In order to obtain a fast, flexible quantification of the available power across various regimes, the WVPP is now analyzed with the lens of the model presented in Cleynen et al. 2017 [25]. The underlying hypothesis is that the machine is installed as a bypass to a weir or traditional hydraulic dam, as depicted in Figure 2; it is assumed that the inlet and outlet water heights h_1 and h_2 are unaffected by its operation. The mass flow rate $\dot{m}$ is a priori unknown.

In order to quantify and non-dimensionalize power, reference values are chosen. The cross-section of the WVPP's inlet is used as a reference area $A_\mathrm{f} = L_{\text{width inlet}} h_1$, and a representative upstream river velocity U_∞ is used as a reference velocity. The turbine power is non-dimensionalized as the power coefficient from [25], becoming:

$$C_{\text{P hydraulic}} = \frac{\dot{W}_{\text{hydraulic}}}{\frac{1}{2}\rho A_f U_\infty^3} \tag{1}$$

$$C_{\text{P shaft}} = \frac{\dot{W}_{\text{shaft}}}{\frac{1}{2}\rho A_f U_\infty^3} \tag{2}$$

In these Equations (1) and (2), the reference power in the denominator, $\frac{1}{2}\rho A_f U_\infty^3$, is a partly arbitrary quantity, so that C_{P} is not expected to reach any value in particular. Nevertheless, for any given inlet boundary condition, higher power coefficient values unambiguously indicate higher power. When h_1 is increased, the reference power grows in proportion, reflecting the device's increased ability to capture mass flow $\dot{m}$ in the inlet.

The device's *performance* is quantified separately, using the load efficiency η_{load} (the fraction of available hydraulic power that is actually being provided to the turbine) and $\eta_{\text{hydraulic}}$ (the fraction of the hydraulic power provided to the turbine that is actually being converted to shaft power), as per ref. [25], so that the power coefficients can be rewritten as:

$$C_{\text{P hydraulic}} = \frac{1}{\frac{1}{2}\rho A_f U_\infty^3}\, \eta_{\text{load}}\, \dot{W}_{\text{hydraulic, max}} \tag{3}$$

$$C_{\text{P shaft}} = \frac{1}{\frac{1}{2}\rho A_f U_\infty^3}\, \eta_{\text{hydraulic}}\, \eta_{\text{load}}\, \dot{W}_{\text{hydraulic, max}} \tag{4}$$

In [25], a model for the hydraulic power available to the device was developed, based on the actuator theories from Betz, Lanchester and Zhukovsky, and accounting for the altitude drop $\Delta(z + h)$. This power is quantified using a single, relatively simple equation, where the only variable is the velocity through the actuator u_{A}:

$$C_{\text{P hydraulic}} = \left[4 + \frac{K_{\text{D2}}}{R^2}\right]\left(R\frac{u_{\text{A}}}{U_\infty}\right)^3 - 4\left(R\frac{u_{\text{A}}}{U_\infty}\right)^2 - K_{\text{D0}}\left(R\frac{u_{\text{A}}}{U_\infty}\right) \tag{5}$$

In this equation, the size ratio $R \equiv A_{\text{A}}/A_f$ compares the actuator area with the inlet area, as per ref. [25]. The two parameters which must be quantified to obtain a power curve are the static drop coefficient K_{D0} (determined entirely by the device's environment), and the loss coefficient K_{D2}, which quantifies all of the hydraulic losses occurring through the device. These two parameters are defined as per ref. [25] as:

$$K_{\text{D0}} \equiv \frac{-\rho g \Delta(z + h)}{\frac{1}{2}\rho U_\infty^2} \tag{6}$$

$$K_{\text{D2}} \equiv \frac{-F_{\text{loss}}}{\frac{1}{2}\rho u_{\text{A}}^2 A_f} \tag{7}$$

In the present case, the static drop coefficient K_{D0} is already known, since it is determined entirely by the device's installation settings. The loss coefficient K_{D2}, however, is the result of the flow (it is expressed as a function of F_{loss}, the force representing all momentum losses occurring in the flow): some initial reference measurement or numerical simulation is needed in order to estimate its value, and thus predict the device's internal losses during operation. This is achieved subsequently, using a strongly-reduced CFD model.

4. Low-Resource CFD Model of the Power Plant

For the purpose of rapidly quantifying the drop coefficient of the device, a CFD simulation is prepared, based on the simulations presented in Powalla et al. 2021 [24]. The full details of the flow, for example the intricacies of the water movement within the turbine, are not of interest here. Instead, what is needed is only a measure of the device's internal energy losses: dissipation incurred through movement of the water in the inlet and outlet channels, as well as through the rotation occurring in the turbine basin. Those losses can be correctly evaluated when the turbine is abstracted away, i.e., replaced by a pair of linked surfaces. To achieve this, the simulation is therefore greatly simplified, as shown in Figure 4. The global inlet and outlet are set to total pressure boundaries with a static pressure distribution. The modeling of two-phase flow is abandoned, prescribing instead a perfectly flat slip wall for each of the upstream and downstream channel ceilings. The turbine is removed and is replaced with a cylindrical outlet in the upstream channel, and a disc-shaped inlet in the downstream channel. A coupling of the mass flow and total pressure between these two surfaces is implemented. Water exits the top part of the weir in the yellow cylinder-shaped "top outlet", a surface with a prescribed mass flow boundary condition. The mass-flow-averaged total pressure is read out from this surface, and in turn, this value is prescribed as a boundary condition for the "bottom inlet" (purple disc in Figure 4). The resulting mass flow in this "bottom inlet" is read out and serves to prescribe the mass flow in the "top outlet", with some under-relaxation for increased stability.

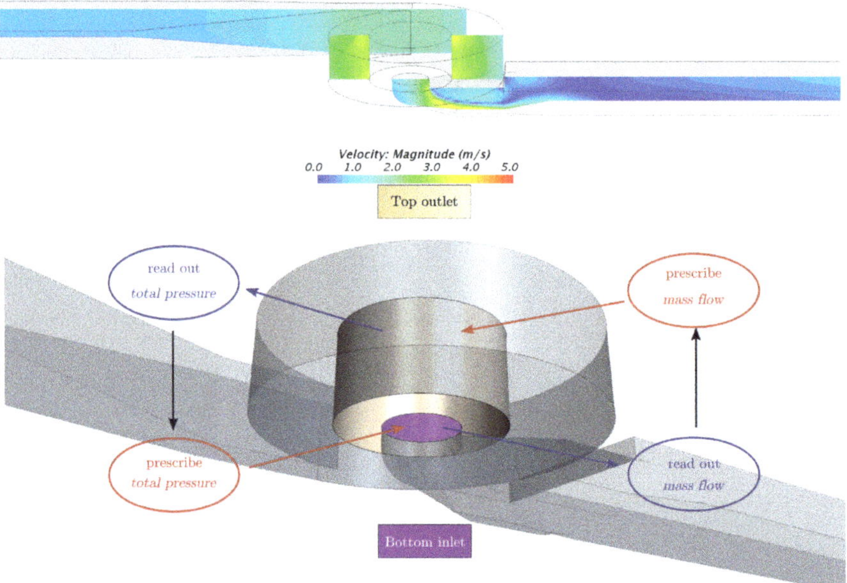

Figure 4. Main features of a simplified simulation for the WVPP, based on the family of simulations depicted in Figure 3, but with much of the sophistication removed. (**Top**): a cross-section of the weir is shown, with plane sections colored according to velocity. (**Bottom**): coupling mechanism between the upper and lower parts of the weir, seen from below the turbine basin.

The resulting, highly-simplified simulation features 2.6 million cells; after a crude initialization, a stable flow field is obtained after 52 s of simulated time, marching with a time step of 0.02 s. This is obtained at the expense of only 550 CPU-hours (less than 3 % of the computational costs of the reference two-phase CFD simulation with moving rotor), making the computation well within reach of an ordinary desktop computer.

In this simulation, since there is no turbine to obstruct the flow, the mass flow is governed by the dissipation losses associated with the transit of water through the complete installation. The mass flow $\dot{m}$ is the only output from the simulation required to quantify F_{losses}, the force representative of all such losses in the model, calculated from the difference between the fluid's momentum at inlet and outlet (see [25]). With F_{losses}, the loss coefficient K_{D2} is quantified, and with it, the power curve of the installation can be drawn, quantifying the hydraulic power available to the turbine as a function of the mass flow.

For the case of the WVPP installed in the laboratory in Dresden, the performance is quantified as follows. The machine features $\Delta z = -0.875\,\text{m}$ and $L_{\text{width inlet}} = L_{\text{width outlet}} = 2\,\text{m}$.

The upstream reference velocity is chosen as $U_\infty = 1.2\,\text{m s}^{-1}$. In the outlet of the turbine chamber leading to the outlet channel, a cross-section is selected arbitrarily to represent an actuator surface, with area $A_A = 0.817\,\text{m}^2$. The choice of that "abstracted turbine" cross-section is a simple matter of convenience and does not affect results. Here, it is expected that this section will likely remain unaffected by modifications of interest in later studies, such as changes to the turbine basin geometry or to the diameter of the throat.

The operating boundary conditions are chosen as $h_1 = 0.825\,\text{m}$ and $h_2 = 0.7\,\text{m}$. These values result in an inlet area $A_f = 1.65\,\text{m}^2$ and a corresponding actuator-to-inlet area ratio $R = 0.495$. The static drop coefficient is then given, before the simulation is run, as $K_{D0} = 13.625$.

The simulation is run until the mass flow has stabilized to a satisfactory level (changing by less than $1\,\text{kg s}^{-2}$). The mass flow then reaches a value of $953\,\text{kg s}^{-1}$ (this is the sole output of the simulation).

The effect of all momentum losses in the system is summed up as the single force $F_{\text{loss}} = -16.09\,\text{kN}$. In this manner, the loss coefficient (Equation (7)) is finally obtained as $K_{D2} = 14.329$. Using the mass flow, the relative actuator velocity (adjusted for relative area) is computed as $R\,u_A/U_\infty = 0.481$ (the fastest the water can ever flow through the device given these boundary conditions). The available power curve, plotted using Equation (5) with the obtained K_{D0} and K_{D2} values, is plotted in Figure 5.

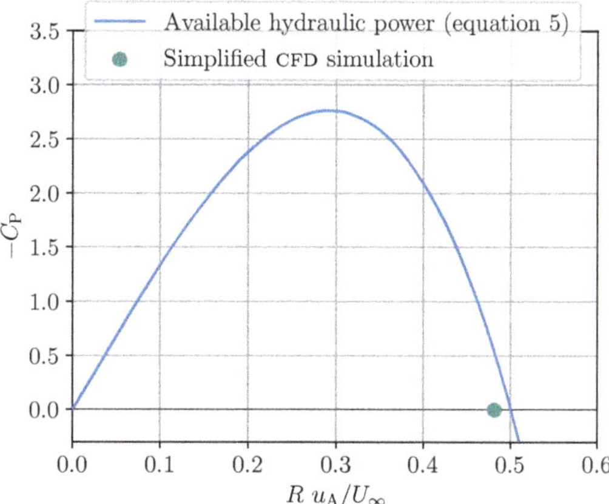

Figure 5. Power coefficient of the WVPP when $h_1 = 0.83\,\text{m}$ & $h_2 = 0.7\,\text{m}$, as predicted using a single, simplified CFD simulation. The vertical axis is the power coefficient C_P based on the frontal area A_f (Equation (1)), and the horizontal axis is the adjusted actuator velocity (non-dimensional inlet velocity, directly proportional to the mass flow). The large dot on the zero-y axis stands for the value read out from the simplified CFD simulation, while the power curve is the solution to Equation (5), calibrated with the K_{D2} value obtained from the simulation.

In this Figure 5, a single point is shown for the single simulation used to generate the curve. A discrepancy between that point and the corresponding prediction according to the power curve is observed. This difference is attributed to non-uniformities in the flow (particularly at the outlet), which are neglected in the intentionally simple post-processing of the simulation. The power curve in Figure 5 features a maximum of $C_{\text{P hydraulic}} = 2.76$ at an adjusted speed $(R\,u_A/U_\infty)_{\text{opt}} = 0.29$ (corresponding to $\dot{m}_{C_{\text{P hydraulic max}}} = 574\,\text{kg s}^{-1}$). This information, available before any detail about the turbine is specified, already provides information useful for its design, for example in determining velocity triangles or sizing mechanical components.

In order to check the validity of the model, further simplified simulations are run, in which the coupling between the upper and lower regions of the simulations is modified: each time, only a specified fraction of the total pressure read out in the top outlet is prescribed in the bottom outlet. The mass flow is therefore reduced by the (fictive) force exerted by the actuator. The result of these simulations are displayed in Figure 6 together with the previously-obtained power curve. Additionally, the turbine shaft power coefficient obtained from a complete, moving-turbine simulation of a corresponding case is displayed as a single, orange-colored point.

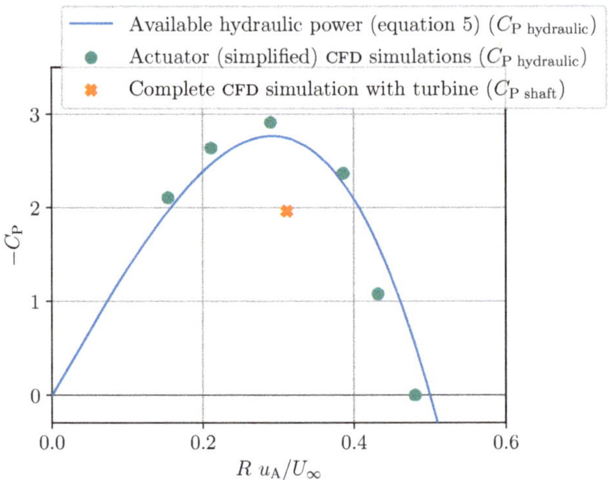

Figure 6. Modeled power curve for the WVPP. The right-most data point and the blue curve are these from Figure 5. The other green points are generated using further simplified simulations from which energy is extracted summarily in the main basin. The orange cross indicates the turbine shaft power coefficient obtained in a complete CFD simulation for conditions very close ($h_1 = 0.85$ m, $h_2 = 0.74$ m) to those used when plotting the power curve.

In Figure 6, the agreement between the values obtained in the actuator simulations and the curve prediction built on a single, unobstructed-flow simulation is, for the purposes of this work, deemed excellent. In the top portion of the power curve, the model underpredicts power by 6 % on average. The curve plotted by is not a best-fit model, but indeed the solution of Equation (5), built on the assumption that the losses internal to the WVPP can be well-described using a single, constant loss coefficient.

In this same figure, the single data point corresponding to a complete simulation (two-phase CFD with rotating turbine) falls well below the curve, at 71 % of $C_{\text{P hydraulic max}}$. In this case, the product of the load and hydraulic efficiencies, which account together for the not-quite-optimal mass flow, free-surface effects in the device, and dissipation losses within and around the turbine, is $\eta_{\text{load}}\eta_{\text{hydraulic}} = 71\,\%$.

5. Results

The model presented above can be used to evaluate performance in different situations. For example, in recent work, a series of four complete simulations of the WVPP was run, with boundary conditions adjusted so that the inlet and outlet heights would be set to respectively $h_1 = 0.64$ m and $h_2 = 0.56$ m. Each time, the rotation speed ω of the turbine was varied, and the power of the turbine was extracted. These two properties are presented together in Figure 7.

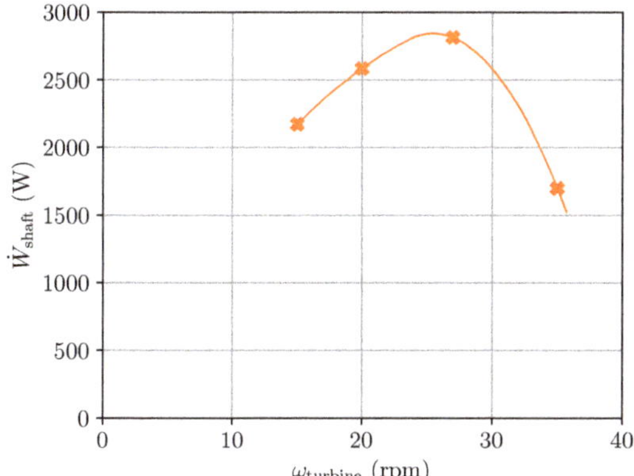

Figure 7. Four turbine shaft power readings carried out in full CFD calculations of the WVPP (individual crosses with a trend curve added as a thin line), with $h_1 = 0.64$ m and $h_2 = 0.56$ m. Both axes are dimensional (power in W and rotation speed in revolutions per minute). In these four simulations, the mass flow decreased steadily as the turbine speed was increased.

The model presented in this article allows for the non-dimensionalization of these values, and their comparison to a maximum theoretical reference point. For this, a new, simplified (single-phase, turbine-less) simulation is run. Compared to the first case studied, the changed inlet height modifies the value of R to 0.641 and the boundary conditions to $K_{D0} = 13.03$. Using the mass flow obtained in the simulation, the loss coefficient is quantified as $K_{D2} = 16.23$. These results make it possible to transform the Figure 7 into the two Figures 8 and 9, which present the power coefficient and product of efficiencies for each of the simulations, together with the theoretical limit for these.

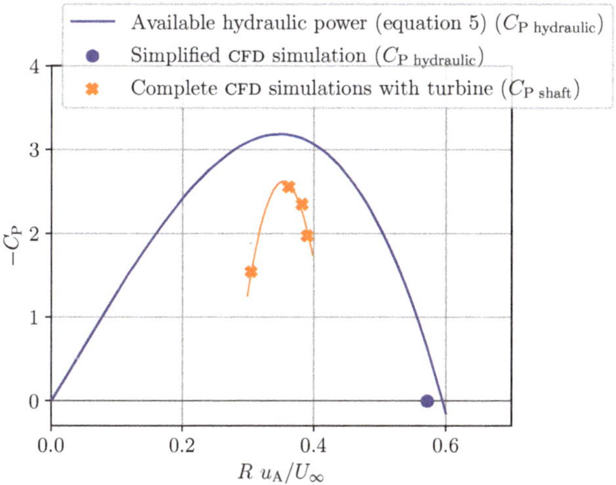

Figure 8. The turbine shaft power readings from Figure 7, this time non-dimensionalized as per the model developed in this chapter (orange crosses). In addition, a single simplified simulation has been run (rightmost circle data point), and the corresponding power curve has been drawn.

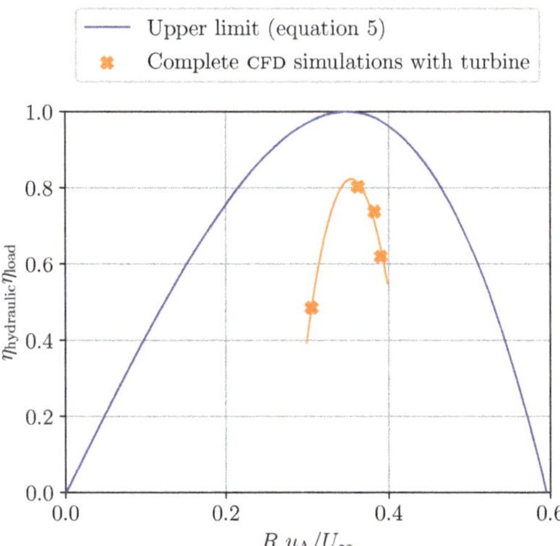

Figure 9. The data from Figure 8, this time presented so that the efficiency of the power extraction (the product of the load efficiency η_{load} and hydraulic efficiency $\eta_{\text{hydraulic}}$) is quantified; the reference power for this is the maximum of the power curve.

Finally, the model can be used to answer the questions formulated in the introduction in a quantitative manner. The simulations in Figure 8 were subjected to an inlet height of 0.825 m; what happens if this height is raised to 1.2 m? The answer is obtained by running one additional simplified CFD simulation. Once the property fields have been initialized with values from the previous case, a converged flow state is obtained after 12 s of simulated time and an expense of 190 CPU-hours (which is 1 % of the computational cost of the reference two-phase CFD simulation with moving turbine). The resulting available power curve is plotted together with that from Figure 8 in Figure 10.

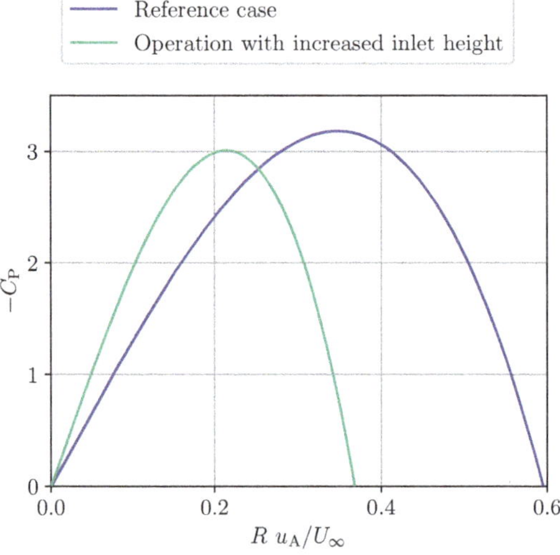

Figure 10. Power curves of the WVPP for two cases: the blue curve, already plotted in Figure 8, is for when $h_1 = 0.825$ m; while the green curve is for the same installation when $h_1 = 1.2$ m. Each curve is based on a single reading from a simplified CFD simulation.

Using the information in Figure 10, the answers are summarized as follows:

- The inlet height has been increased by 45 %, and the available power has increased almost in proportion by 38 % (the maximum power coefficient is decreased by 5 %).
- In order to have access to this additional power, the turbine must operate with a strongly reduced inlet velocity (from $0.35U_\infty$ down to $0.21U_\infty$).

A quick re-dimensionalization of these results indicates that the maximum available power $\dot{W}_{\text{hydraulic max}}$ goes from 4.54 to 6.25 kW, while the corresponding mass flow $\dot{m}_{\text{opt}}$ reduces from 571 to 512 kg s^{-1}; these new values become the reference point by which to quantify the installation's efficiency.

It is therefore seen that the model developed in the first part of this chapter can be used to answer questions that, although simple, have no obvious immediate answer. In this case, increasing the inlet height increases the load on the turbine, but also the internal dissipative losses in the plant; if the inlet velocity were to be kept constant, these losses and the increased momentum abandoned in the outlet (where velocity would increase by virtue of mass conservation) would reduce the power available to the turbine. From the point of view of the operator, this change in operating conditions is only desirable if the turbine and generator are able to operate efficiently at a reduced flow rate.

If the power characteristics of the WVPP were of further interest, models better suited to predicting its losses could naturally be developed, using more sophistication. Here, the approach was voluntarily kept as simple as possible, focusing on illustrating the capabilities of a model which was developed to cover a much more general class of devices, including turbomachines operating in floating installations.

6. Conclusions

This article demonstrates that a theoretical analysis of the achievable performance in floating or bypass hydropower installations allows designers and operators to answer questions of practical importance in a computationally-efficient way.

A one-dimensional model describing the fluid flow through hydraulic devices had formerly been presented: a tool able to characterize the performance of machines operated in conditions where the mass flow rate is a control variable and the outlet water height cannot be controlled. This corresponds for instance to machines operating in a cascading flow alongside a dam. In the model, the device operating speed required to attain full load efficiency, and the corresponding maximum hydraulic power, can be quantified independently of the hydraulic efficiency.

Usage of the model was here demonstrated with the case of a mid-scale hydropower device, a 26 m long water vortex power plant (WVPP). In order to characterize the installation's power budget, only two properties need to be quantified: the static drop coefficient K_{D0}, a function of the boundary conditions only, and the loss coefficient K_{D2}, a measure of the device's hydraulic flow resistance. It is shown that a simplified CFD simulation with a single output (the mass flow rate) is sufficient to obtain a useful value for K_{D2}.

In this manner, the power available to the machine can be quantified given any set of boundary conditions. It becomes possible to evaluate the device's real-life performance, benchmarking it against reference values anchored in physical principles. Our model can be used to provide design guidance, and compare sets of boundary conditions one against the other easily, providing answers that are not easily obtained using intuition or even experiments. These results are obtained with a computing cost reduction of 97–99 % compared to those associated with a full description of the flow using CFD methods.

Author Contributions: Conceptualization, methodology, writing—original draft preparation, O.C.; software, investigation, validation, visualization, O.C. and D.P.; writing—review and editing, O.C, D.P., S.H. and D.T.; supervision, project administration, funding acquisition, S.H. and D.T. All authors have read and agreed to the published version of the manuscript.

Funding: The authors are grateful for the financial support provided by the Fluss-Strom "VP4: Kaskade Fischfreundliches Wehr" project financed by the *Bundesministerium für Bildung und Forschung* (German Federal Ministry of Education and Research) under the project number 03WKC04.

Institutional Review Board Statement: Not applicable.

Informed Consent Statement: Not applicable.

Data Availability Statement: Not applicable.

Acknowledgments: The work of Sergei Sukhorukov in running CFD simulations on the *Neumann* cluster of the University of Magdeburg, as part of his Master's Thesis under the mentorship of the first author, as well as the experimental work of the Technische Universität Dresden, in particular of Nadine Müller, are gratefully acknowledged.

Conflicts of Interest: The authors declare no conflict of interest. The funders had no role in the design of the study; in the collection, analyses, or interpretation of data; in the writing of the manuscript, or in the decision to publish the results.

Nomenclature

A_A actuator frontal area [m^2]
A_f device frontal area [m^2]
C_P power coefficient [—]
g gravitational acceleration [m s^{-2}]
h height from bed to water surface, positive upwards [m]
K_{D0} static drop coefficient (def. (6) as per ref. [25]) [—]
K_{D2} loss coefficient (def. (7) as per ref. [25]) [—]
$\dot{m}$ mass flow [kg s^{-1}]
R size ratio $R \equiv A_A / A_f$, as per ref. [25] [—]
u_A fluid velocity at the actuator [m s^{-1}]
U_∞ free-stream velocity [m s^{-1}]
$\dot{W}$ mechanical or electrical power [W]
z altitude of bed, positive upwards [m]
Δ net difference
η efficiency (from 0 to 1) [—]
ρ density [kg m^{-3}]

References

1. Isa, M.A.; Sudjono, P.; Sato, T.; Onda, N.; Endo, I.; Takada, A.; Muntalif, B.S.; Ide, J. Assessing the Sustainable Development of Micro-Hydro Power Plants in an Isolated Traditional Village West Java, Indonesia. *Energies* **2021**, *14*, 6456. [CrossRef]
2. Alam, Z.; Watanabe, Y.; Hanif, S.; Sato, T.; Fujimoto, T. Social Enterprise in Small Hydropower (SHP) Owned by a Limited Liability Partnership (LLP) between a Food Cooperative and a Social Venture Company; a Case Study of the 20 kW Shiraito (Step3) SHP in Itoshima City, Fukuoka (Japan). *Energies* **2021**, *14*, 6727. [CrossRef]
3. Pavlov, D. *Structures Assisting the Migrations of Non-Salmonid Fish: USSR*; Food & Agriculture Org.: Rome, Italy, 1989; Volume 308.
4. Katopodis, C. *Introduction to Fishway Design*; Freshwater Institute, Central and Arctic Region, Department of Fisheries and Oceans: Winnipeg, MB, Canada, 1992.
5. Silva, A.T.; Lucas, M.C.; Castro-Santos, T.; Katopodis, C.; Baumgartner, L.J.; Thiem, J.D.; Aarestrup, K.; Pompeu, P.S.; O'Brien, G.C.; Braun, D.C.; et al. The Future of Fish Passage Science, Engineering, and Practice. *Fish Fish.* **2018**, *19*, 340–362. [CrossRef]
6. Ebel, G. *Fischschutz und Fischabstieg an Wasserkraftanlagen: Handbuch Rechen- und Bypasssysteme*; Büro für Gewässerökologie und Fischereibiologie Dr. Ebel: Halle, Germany, 2013; Volume 4.
7. Richmond, M.C.; Romero-Gomez, P. Fish Passage through Hydropower Turbines: Simulating Blade Strike Using the Discrete Element Method. *IOP Conf. Ser. Earth Environ. Sci.* **2014**, *22*, 062010. [CrossRef]
8. Romero-Gomez, P.; Lang, M.; Michelcic, J.; Weissenberger, S. Particle-based evaluations of fish-friendliness in Kaplan turbine operations. *IOP Conf. Ser. Earth Environ. Sci.* **2019**, *240*, 042016. [CrossRef]
9. Quaranta, E.; Pérez-Díaz, J.I.; Romero-Gomez, P.; Pistocchi, A. Environmentally enhanced turbines for hydropower plants: current technology and future perspective. *Front. Energy Res.* **2021**, *9*, 592. [CrossRef]
10. Zangiabadi, E.; Masters, I.; Williams, A.J.; Croft, T.; Malki, R.; Edmunds, M.; Mason-Jones, A.; Horsfall, I. Computational prediction of pressure change in the vicinity of tidal stream turbines and the consequences for fish survival rate. *Renew. Energy* **2017**, *101*, 1141–1156. [CrossRef]
11. Klopries, E.M.; Schüttrumpf, H. Mortality assessment for adult European eels (Anguilla Anguilla) during turbine passage using CFD modelling. *Renew. Energy* **2020**, *147*, 1481–1490. [CrossRef]

12. Klopries, E.M.; Schüttrumpf, H. Verbesserung des Prozessverständnisses der Kraftwerkspassage von Aalen durch neuartige Ansätze. *Wasserwirtschaft* **2020**, *110*, 28–32. [CrossRef]
13. Borkowski, D.; Majdak, M. Small hydropower plants with variable speed operation—An optimal operation curve determination. *Energies* **2020**, *13*, 6230. [CrossRef]
14. Tarife, R.; Nakanishi, Y.; Chen, Y.; Zhou, Y.; Estoperez, N.; Tahud, A. Optimization of Hybrid Renewable Energy Microgrid for Rural Agricultural Area in Southern Philippines. *Energies* **2022**, *15*, 2251. [CrossRef]
15. Jung, J.; Han, H.; Kim, K.; Kim, H.S. Machine learning-based small hydropower potential prediction under climate change. *Energies* **2021**, *14*, 3643. [CrossRef]
16. Liszka, D.; Krzemianowski, Z.; Węgiel, T.; Borkowski, D.; Polniak, A.; Wawrzykowski, K.; Cebula, A. Alternative Solutions for Small Hydropower Plants. *Energies* **2022**, *15*, 1275. [CrossRef]
17. Choi, I.H.; Kim, J.W.; Chung, G.S. Effect analysis of pulley on performance of micro hydropower in free surface vortex. *J. Wetl. Res.* **2021**, *23*, 234–241. [CrossRef]
18. Jiang, Y.; Raji, A.P.; Raja, V.; Wang, F.; AL-bonsrulah, H.A.; Murugesan, R.; Ranganathan, S. Multi–Disciplinary Optimizations of Small-Scale Gravitational Vortex Hydropower (SGVHP) System through Computational Hydrodynamic and Hydro–Structural Analyses. *Sustainability* **2022**, *14*, 727. [CrossRef]
19. Powalla, D.; Hoerner, S.; Cleynen, O.; Thévenin, D. A numerical approach for active fish behaviour modelling with a view toward hydropower plant assessment. *Renew. Energy* **2022**, *188*, 957–966. [CrossRef]
20. Müller, N.; Stamm, J. Errichtung eines 1:1 Labormodells für ethohydraulische Untersuchungen an einem Wasserwirbelkraftwerk. In *Dresdner Wasserbauliche Mitteilungen (Heft 60)*; Technische Universität Dresden, Institut für Wasserbau und Technische Hydromechanik: Dresden, Germany, 2018; pp. 123–132.
21. Müller, N.; Stamm, J.; Wagner, F. A water vortex power plant as ethohydraulic test site. In Proceedings of the 5th IAHR Europe Congress, Trento, Italy, 12–14 June 2018. [CrossRef]
22. Müller, N.; Jähnel, C.; Stamm, J.; Wagner, F. Analyse der Strömung in einem Wasserwirbelkraftwerk hinsichtlich des Fischabstiegs. *Wasserwirtschaft* **2019**, *109*, 60–63. [CrossRef]
23. Wagner, F.; Warth, P.; Royan, M.; Lindig, A.; Müller, N.; Stamm, J. Laboruntersuchungen zum Fischabstieg über ein Wasserwirbelkraftwerk. *Wasserwirtschaft* **2019**, *109*, 64–67. [CrossRef]
24. Powalla, D.; Hoerner, S.; Cleynen, O.; Müller, N.; Stamm, J.; Thévenin, D. A computational fluid dynamics model for a water vortex power plant as platform for etho- and ecohydraulic research. *Energies* **2021**, *14*, 639. [CrossRef]
25. Cleynen, O.; Hoerner, S.; Thévenin, D. Characterization of hydraulic power in free-stream installations. *Int. J. Rotating Mach.* **2017**, *2017*. [CrossRef]
26. Lichtenberg, N.; Cleynen, O.; Thévenin, D. Numerical investigations of a water vortex hydropower plant implemented as fish ladder – Part I: The water vortex. In Proceedings of the 4th IAHR Europe Congress, Liege, Belgium, 27–29 July 2016; p. 277.
27. Müller, S.; Cleynen, O.; Thévenin, D. Numerical investigation of the influence of a guide wall in a fish-friendly weir. In Proceedings of the International Conference on Engineering and Ecohydrology for Fish Passage, Corvallis, OR, USA, 19–21 June 2017; p. 15.
28. Müller, S.; Cleynen, O.; Hoerner, S.; Lichtenberg, N.; Thévenin, D. Numerical analysis of the compromise between power output and fish-friendliness in a vortex power plant. *J. Ecohydraul.* **2018**, *3*, 86–98. [CrossRef]

Article

Fluid–Structure Coupling Analysis of the Stationary Structures of a Prototype Pump Turbine during Load Rejection

Qilian He [1,†], Xingxing Huang [2,†], Mengqi Yang [3], Haixia Yang [3], Huili Bi [1] and Zhengwei Wang [1,*]

[1]   Department of Energy and Power Engineering, Tsinghua University, Beijing 100084, China; hql19@mails.tsinghua.edu.cn (Q.H.); bihuili2014@mail.tsinghua.edu.cn (H.B.)
[2]   S.C.I.Energy, Future Energy Research Institute, Seidengasse 17, 8706 Zurich, Switzerland; xingxing.huang@hotmail.com
[3]   Branch Company of Maintenance & Test, CSG Power Generation Co., Ltd., Guangzhou 511400, China; ymq9273@163.com (M.Y.); yanghx13@163.com (H.Y.)
*   Correspondence: wzw@mail.tsinghua.edu.cn
†   These authors contributed equally to this work.

Abstract: During the load rejection transient process of the prototype pump turbine units, the pressure fluctuations of the entire flow passage change drastically due to the rapid closing of guide vanes. The extremely unsteady pressure distribution in the flow domains including the crown chamber and the band chamber may cause a strong vibration on the stationary structures of the unit and result in large dynamic stress on the head cover, stay ring and bottom ring. In this paper, the numerical fluid dynamic analysis of the entire flow passage of a reversible prototype pump turbine during load rejection was performed. The flow characteristics in the runner passage, crown chamber, band chamber, seal labyrinths and balance tubes are analysed. The corresponding unsteady flow-induced dynamic behaviour of the head cover, stay vanes and bottom ring was investigated in detail. The analysed results show that the total deformation of the inner edge of the head cover closed to the main shaft is larger than that of other stationary structures of the unit during the load rejection. The maximum stress of the stay ring is larger than that of the head cover and the bottom ring and the maximum equivalent stress is located at the fillet of the stay vane trailing edge. The fluid–structure coupling calculation method and the analysed results can provide guidance for the design of stationary components of hydraulic machinery such as pump turbines, Francis turbines and centrifugal pumps with different heads.

Keywords: pump turbine; head cover; load rejection; fluid–structure coupling; stress concentration

Citation: He, Q.; Huang, X.; Yang, M.; Yang, H.; Bi, H.; Wang, Z. Fluid–Structure Coupling Analysis of the Stationary Structures of a Prototype Pump Turbine during Load Rejection. *Energies* **2022**, *15*, 3764. https://doi.org/10.3390/en15103764

Academic Editor: Phillip Ligrani

Received: 16 April 2022
Accepted: 18 May 2022
Published: 20 May 2022

Publisher's Note: MDPI stays neutral with regard to jurisdictional claims in published maps and institutional affiliations.

1. Introduction

With the gradual increase in power demand in recent decades, the capacity of traditional power generation such as thermal power and nuclear power has been increasing and the peak-to-valley difference of the power grid has been expanding, which has put forward higher requirements on the regulation capability and flexibility of the grid. As an important source of clean and renewable energy, hydropower has increasingly been developed for power generation worldwide because of its flexible features such as fast start/stop and load adjustment. Hydropower has indeed improved the power grid's ability to resist fluctuations in power demand and has gradually occupied an important share of the grid load. The hydraulic pumped storage power plant (HPSPP) mainly consists of an upstream reservoir, water diversion pipelines, a pumped storage unit and a downstream reservoir. The pump turbine (PT) is the core working equipment of an HPSPP, which can freely choose the direction of energy conversion. When the unit operates under the turbine condition, water flow enters the unit from the spiral case and generates power through the runner. When the unit operates under the pump condition, the flow direction is the opposite and the flow from the draft tube is pumped upstream by the rotating runner. The HPSPP can flexibly select operating conditions according to the demand of the power grid and has functions of peak shaving, valley shaving, frequency regulation, emergency reserve and optimal power supply, which has been vigorously developed and widely used.

The high water head of HPSPP will lead to high pressure excitation on the pump turbine unit and may cause a series of safety problems such as unexpected severe vibration and even fatigue failure on the PT structures. Several cases of unit failure or damage were reported in studies [1–4]. The main reasons for the failure and damage revealed in the reports include the drastic change of hydraulic characteristics under certain operating conditions, abnormal vibration of the unit caused by unreasonable design and fatigue failure of stationary parts such as head cover bolts, etc. According to the demand of the grid, PT units in HPSPPs often experience unstable operation processes such as startup and shutdown, load increase and decrease and even load rejection. Load rejection is one of the transient processes with the most drastic changes in the hydraulic excitation of the PT unit and it has the greatest impact on the safe operation of the unit. Numerous studies [5–9] have shown that the drastic changes of velocity and flow rate during load rejection generate an extremely unstable flow in the flow channels of the PT unit and the pressure excitation on the structural components are much higher than that at rated operation conditions; therefore, it is of great engineering significance to study the load rejection process of the PT unit and the vibration of the stationary structures during load rejection.

Due to the high cost or inconvenience of model tests and prototype tests of PT units, numerical simulations have become an alternative method to study the internal flow characteristics of PT units. The most important issue of numerical calculations is the accuracy compared to tests. A considerable number of studies [9–13] have verified the reliability of numerical calculations of hydraulic machinery including PT units by comparing the calculated results with the ones of field measurements. The comparisons show that the errors are within the acceptable range for engineering projects.

In recent years, the joint 1D/3D simulation method combining 1D pipeline calculation with 3D unsteady flow calculation of turbines [14–17] has been mostly used to study the transient processes of hydraulic turbine units. The 1D/3D joint simulation method can more accurately calculate key parameters such as rotational speed, flow rate and guide vane opening of the turbine unit and achieve valuable results for hydropower engineering applications. The researchers [18–20] have analysed the reliability of 1D pipeline calculation by comparison with measurement results. Regarding the 3D unsteady simulation, some studies [15,21,22] have used the dynamic mesh technology to simulate the transient process of the unit with the time-varying curves of guide vane opening, flow and head as the boundary conditions. The simulation results are in rather good agreement with the experimental data and have the advantage of capturing the pressure pulsation of the whole flow field with high accuracy; however, the computational resource consumption is enormous and the computation time often exceeds several months, which cannot provide rapid feedback to the PT designers in a given short time during practical engineering projects. Some scholars [7,17,23,24] have proposed a new method to calculate the fluid flow at various critical time points during transient processes and studied the main stress/strain changes of the PT structures caused by the fluid flow. The error of the pressure between the simulation results and the experimental results is within 7% in large opening condition and within 12% in minimum opening [24]. In this study, the same approach was adopted to investigate a prototype PT head cover during turbine load rejection.

The research on the characteristics of structural components such as head cover vibration of PT unit is a complex fluid–structure coupling (FSC) problem. One-way FSC and two-way FSC are two typical methods to solve the FSC problem. The characteristic length of the entire flow field can be up to ten meters, while the actual head cover deformation of the PT unit is only several millimetres, so the effect of the head cover deformation on fluid flow is negligible; therefore, the one-way FSC method is reasonable for this study considering the computational resources and the accuracy of the results. It is more reasonable to adopt the one-way FSC method for this study considering the computing resources and accuracy of the results. The studies [25,26] have verified the accuracy of the FSC calculation method, compared the effects of different FSC methods on the calculation results of hydraulic machinery and confirmed the feasibility of the one-way method. Some researchers [27,28] have calculated the stresses and strains of the stationary parts of the unit with the one-way FSC method. The difference between the calculated results and the measured ones is within the acceptable range. In summary, the previous research achievements and the proven methodology provide a solid foundation for FSC analysis of the prototype PT head cover during turbine load rejection.

In this paper, the full 3D fluid domain model of the PT unit and the corresponding 3D structural model were established first. The 1D pipeline calculations were performed to select the key time points during the load rejection process for the full 3D fluid simulations. After that, the pressure of the PT unit at each time point was calculated using the 3D CFD simulation method. The obtained pressures files were sequentially exported, transferred and mapped to the corresponding structural components of the PT unit via the one-way FSC calculation method. Finally, the structural characteristics of stationary components such as head cover and head cover bolts were analysed and discussed in detail to draw conclusions. The achievements in this study can provide an important reference for the optimal design of hydraulic turbine units for power stations.

2. Methods of Numerical Calculation

The numerical calculation methods in this paper include the 1D pressurized pipeline calculation method, 3D turbulence numerical simulation method and FSC method for 3D structure analysis.

2.1. Governing Equations of Unsteady Flow in 1D Pipeline Calculation

The calculation models of the water conveyance system of HPSPP include pipelines, PT unit, upper and lower reservoir, valves and surge shaft, which constitute the mathematical model of the unsteady flow in the pipeline. The basic equations of unsteady flow in the closed pipeline consist of the equation of motion and the equation of continuity. The calculation method of pipeline flow can refer to studies [7,29].

$$g\frac{\partial H}{\partial x} + \bar{v}\frac{\partial \bar{v}}{\partial x} + \frac{\partial \bar{v}}{\partial t} + \frac{f\bar{v}|\bar{v}|}{2D} = 0 \tag{1}$$

$$\bar{v}\frac{\partial H}{\partial x} + \frac{\partial H}{\partial t} + \frac{a^2}{g}\frac{\partial \bar{v}}{\partial x} - \bar{v}\sin\alpha = 0 \tag{2}$$

where $h = Z + P/\rho g$, Z is the elevation of the pipeline axis, p is the pressure, ρ is the liquid density and g is the local gravity acceleration, $\bar{v}$ is the average velocity of the fluid in the pipeline, F is Darcy Westbach friction coefficient, D is the inner diameter of the pipe, X is the length of the pipeline along the axis, A is the wave velocity in the pipeline, T is the transient time and α is the angle between the pipe axis and the horizontal plane.

On the characteristic line $\partial x/\partial t = \pm a$, the original equation can be transformed into a differential equation with constant coefficients; therefore, the process of solving the partial differential equation can be transformed into the discretization of the difference equation on the characteristic line. The compatibility equations are:

$$C^+ : \begin{cases} \frac{dx}{dt} = +a \\ A^2\frac{dH}{dt} + A\frac{a}{g}\frac{dQ}{dt} - QA\sin\alpha + \frac{faQ|Q|}{2gD} = 0 \end{cases} \tag{3}$$

$$C^- : \begin{cases} \frac{dx}{dt} = -a \\ A^2\frac{dH}{dt} - A\frac{a}{g}\frac{dQ}{dt} - QA\sin\alpha - \frac{faQ|Q|}{2gD} = 0 \end{cases} \tag{4}$$

Equations (3) and (4) are called the MoC equations and the compatibility equation established along the characteristic lines. Two characteristic line directions are C^+ and C^-.

The equation can be integrated by the variation of flow rate and water head between two points in the form of difference and then the 1D pipeline calculation can be completed by iteration solution.

2.2. Governing Equations of the 3D Flow Simulation

The conservative differential form Navier–Stokes Equation for fluid flow is:

$$\frac{\partial \rho}{\partial t} + \frac{\partial}{\partial x_i}(\rho u_i) = 0 \tag{5}$$

$$\frac{\partial}{\partial t}\rho u_i + \frac{\partial}{\partial x_j}(\rho u_i u_j) = \frac{\partial \tau_{ij}}{\partial x_j} + \rho f_i \tag{6}$$

The Reynolds averaging (RANS) method regards turbulent motion as the combination of time-averaged flow and instantaneous fluctuating flow, ignoring the influence of density fluctuation, but taking into account the change of average density, averaging the time by the turbulent control equation. The tensor expressions of the continuity equation and the momentum equation are as follows:

$$\frac{\partial \rho}{\partial t} + \frac{\partial}{\partial x_j}(\rho u_j) = 0 \tag{7}$$

$$\frac{\partial}{\partial t}(\rho u_i) + \frac{\partial}{\partial x_j}(\rho u_i u_j) = -\frac{\partial p}{\partial x_i} + \frac{\partial}{\partial x_j}\left(\mu \frac{\partial u_i}{\partial t} - \rho \overline{u'_i u'_j}\right) + S_M \tag{8}$$

where the underlined symbol of the time averages is removed except for the time averages of the fluctuation value. In the equation, ρ is density, p is pressure, i and j range of indicators is (1, 2, 3), u_i represents velocity components in the x, y and z directions, μ is dynamic viscosity and S_M is the generalized source term for the momentum conservation equation. A new unknown quantity appears in the equation, which is defined as Reynolds stress:

$$\tau_{ij} = -\rho \overline{u'_i u'_j} \tag{9}$$

In order to set up a closed-form equation system, the Reynolds stress needs to be treated. Considering the similarities between Reynolds stress and viscous stress, the eddy-viscous model and the corresponding correction method provide a relatively simple method to solve Reynolds stress. Based on the eddy viscosity assumption proposed by Boussinesq, turbulent kinetic energy is introduced:

$$k = \frac{1}{2}\overline{u'_i u'_j} \tag{10}$$

The turbulent Reynolds stress is analogous to the physical viscous stress and is considered to be related to the average velocity gradient and the eddy viscosity:

$$-\overline{u'_i u'_j} = v_t\left(\frac{\partial \overline{u_i}}{\partial x_j} + \frac{\partial \overline{u_j}}{\partial x_i}\right) - \frac{2}{3}k\delta_{ij} \tag{11}$$

Commonly used two-equation eddy viscosity models mainly include $k-\varepsilon$ model, $k-\omega$ model and corresponding improved models. A good coupling model is the $SSTk-\omega$ model, which is standard $k-\varepsilon$ model in the near wall area and standard $k-\omega$ model in the free turbulent flow (turbulent core area):

$$\frac{\partial k}{\partial t} + U_j \frac{\partial k}{\partial x_j} = P_k - \beta^* k\omega + \frac{\partial}{\partial x_j}\left[(v + \sigma_k v_T)\frac{\partial k}{\partial x_j}\right] \tag{12}$$

$$\frac{\partial \omega}{\partial t} + U_j \frac{\partial \omega}{\partial x_j} = \alpha S^2 - \beta \omega^2 + \frac{\partial}{\partial x_j}\left[(\nu + \sigma_\omega \nu_T)\frac{\partial \omega}{\partial x_j}\right] + 2(1 - F_1)\sigma_{\omega 2}\frac{1}{\omega}\frac{\partial k}{\partial x_i}\frac{\partial \omega}{\partial x_i} \tag{13}$$

where the closure coefficients and auxiliary relations are:

$$F_2 = \tanh\left[\left[\max\left(\frac{2\sqrt{k}}{\beta^* \omega y}, \frac{500\nu}{y^2\omega}\right)\right]^2\right] \tag{14}$$

$$P_k = \min\left(\tau_{ij}\frac{\partial U_i}{\partial x_j}, 10\beta^* k\omega\right) \tag{15}$$

$$F_1 = \tanh\left\{\left\{\min\left[\max\left(\frac{\sqrt{k}}{\beta^* \omega y}, \frac{500\nu}{y^2\omega}\right), \frac{4\sigma_{\omega 2}k}{CD_{k\omega}y^2}\right]\right\}^4\right\} \tag{16}$$

$$CD_{k\omega} = \max\left(2\rho\sigma_{\omega 2}\frac{1}{\omega}\frac{\partial k}{\partial x_i}\frac{\partial \omega}{\partial x_i}, 10^{-10}\right) \tag{17}$$

$$\phi = \phi_1 F_1 + \phi_2(1 - F_1) \tag{18}$$

$$\alpha_1 = \frac{5}{9}, \alpha_2 = 0.44 \tag{19}$$

$$\beta_1 = \frac{3}{40}, \beta_2 = 0.0828 \tag{20}$$

$$\beta^* = \frac{9}{100} \tag{21}$$

$$\sigma_{k1} = 0.85, \sigma_{k2} = 1 \tag{22}$$

$$\sigma_{\omega 1} = 0.5, \sigma_{\omega 2} = 0.856 \tag{23}$$

where y is the distance to the nearest wall. Due to the coupling characteristics, the $SST k - \omega$ model has good predictability of flow separation in the reverse pressure gradient region, therefore this model will be used for calculation in this paper.

2.3. Governing Equations of the Fluid–Structure Coupling Analysis

Considering the fluid pressure acting on the structures, the governing equation of structural dynamics based on the finite element method can be described as

$$[M_s]\{\ddot{u}\} + [C_s]\{\dot{u}\} + [K_s]\{u\} = \{F_s(t)\} + \left\{F_{fs}(t)\right\} \tag{24}$$

where $[M_s]$, $[C_s]$ and $[K_s]$ represent the mass, damping matrix and stiffness matrices of the structure, respectively; $\{u\}$, $\{\dot{u}\}$ and $\{\ddot{u}\}$ are the node displacement, velocity and acceleration vectors, respectively; $\{F_s(t)\}$ refers to the external excitation load vector acting on structures, $\left\{F_{fs}(t)\right\}$ represents to the fluid pressure load vector at the interface and t is time.

The fluid pressure distribution in the PT unit can be obtained by 3D computational fluid dynamics (CFD) analysis, as described in Section 2.2. Applying fluid pressure to the fluid–structure coupling surfaces of the structures, the flow-induced structural stress $\{\sigma\}$ (Equation (25)) of the PT unit can be calculated with the displacement $\{u\}$ solved by Equation (24).

$$\{\sigma\} = [D][B]\{u\} \tag{25}$$

where $[D]$ and $[B]$ are the elasticity matrix and the strain-displacement matrix, respectively.

The von Mises stress σ_{vM} (Equation (26)) is adopted to evaluate the stress characters of structures of the PT unit.

$$\sigma_{vM} = \sqrt{\frac{1}{2}[(\sigma_1 - \sigma_2)^2 + (\sigma_2 - \sigma_3)^2 + (\sigma_3 - \sigma_1)^2]} \tag{26}$$

where σ_i ($i = 1, 2, 3$) are the principal stresses.

3. Numerical Simulation of the Fluid Flow

3.1. 1D Pipeline Calculation

The calculation object in this paper is a prototype pumped storage unit. The model of 1D pipeline calculation is shown in Figure 1, including upper and lower reservoir, diversion surge shaft and tailrace surge shaft, turbines and tail gate chamber. The basic parameters of the unit are shown in Table 1. The scope of 1D calculation includes the upstream reservoir to the downstream reservoir and the scope of 3D CFD calculation is the turbine part of the system, including flow domains of the spiral case, stay vane, guide vane, runner, draft tube, upper and lower labyrinth seals and balance tubes. The further FEM calculation includes the head cover, stay ring and bottom ring of the unit.

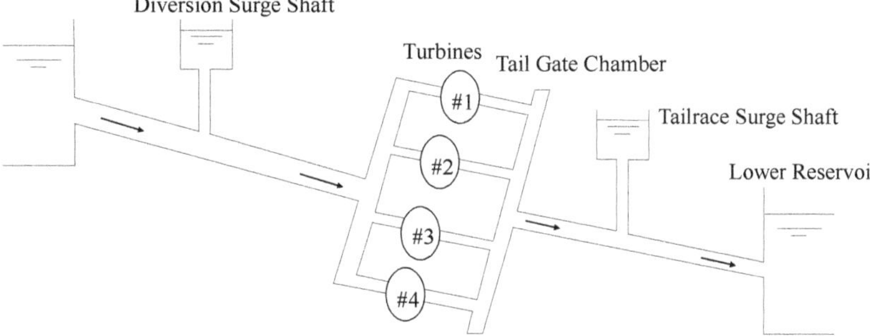

Figure 1. The 1D pipeline calculation model with 4 PT units (#1 to #4).

Table 1. Parameters of the unit.

Parameter	Unit	Value
Unit capacity	MW	300
Rated speed N_r	rpm	500
Number of stay vanes		20
Number of guide vanes		20
Number of runner blades		7

The relative flow rate, relative rotational speed and relative opening curves with time are obtained by 1D calculation as shown in Figure 2 (drawn by the Origin software). Since flow rate and rotational speed are the main factors that affect the intense changes in the flow field and pressure characteristics during load rejection [5,6], the representative extreme value points in the curves of flow rate and rotational speed are selected as key time points for 3D flow field simulations and the study [10] has adopted a similar method to perform the flow field simulation.

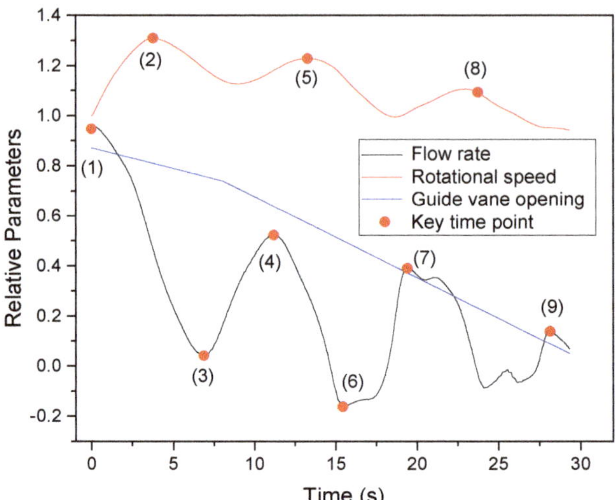

Figure 2. The operating parameters and 9 key time points of the PT unit during load rejection.

The pressures at the spiral case inlet and the draft tube outlet calculated by 1D simulation are shown in Figure 3. The pressure values at different moments are set as the inlet and outlet boundary conditions for the 3D flow calculation.

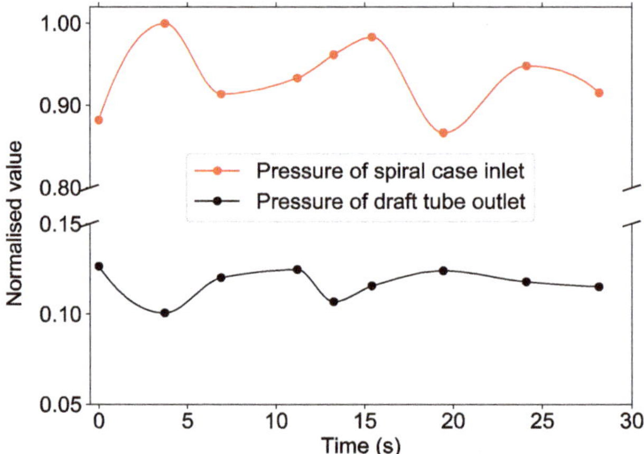

Figure 3. Normalized pressure at the spiral case inlet and draft tube outlet.

3.2. 3D Flow Calculation Model

The 3D model of the PT unit is shown in Figure 4, including spiral case, stay vane, guide vane, runner, draft tube, upper and lower labyrinth seals and balance tubes. The size of labyrinth seal clearance is only 1.5 mm, which has a significant impact on the flow of the unit; therefore, the upper and lower clearances are considered in the model of the fluid domain.

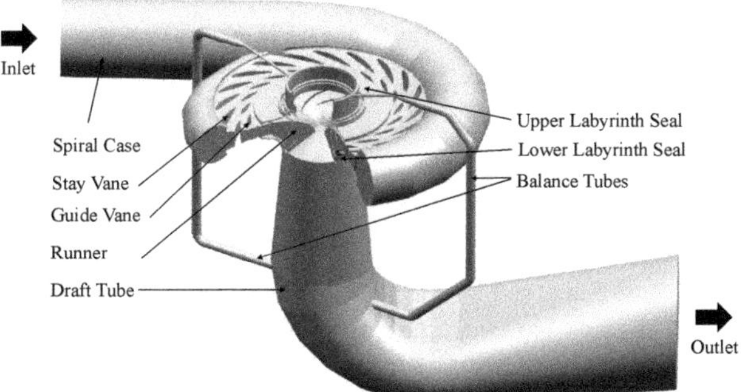

Figure 4. The 3D modeling of flow passage of the unit.

The details of the section from the unit spiral case to the draft tube are shown in Figure 5. Pressure monitoring points (C1, C2, V1, V2, B1, B2) are set at the upper and lower labyrinth seals and the vaneless area.

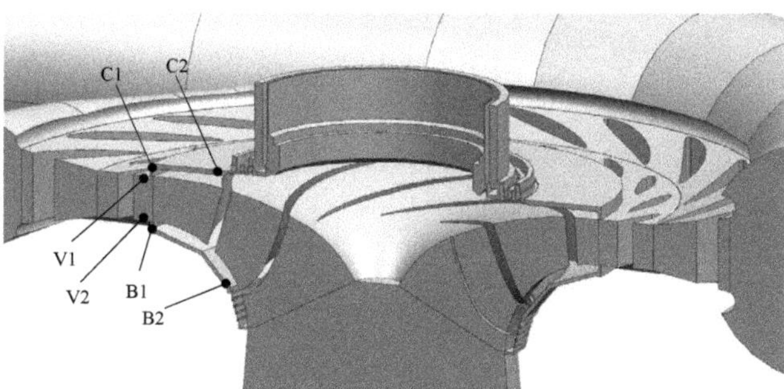

Figure 5. Setup of the monitoring points.

3.3. Mesh Independence Analysis of the 3D Flow Calculation

ANSYS CFX was used to carry out the 3D CFD analysis. The analysis type was set to stable and the discrete format and the solver were set to high resolution. The runner domain was set to rotate at an angular velocity of 500 rpm. The wall condition was set as non-slip wall. The interfaces between guide vanes and runner, runner and draft tube were set to Frozen Rotor and the others are set to General Connection. The turbulence model was set to shear stress transport, the turbulence model was set to shear stress transfer and the turbulence mathematical accuracy was set to first order with a convergence residual of 10^{-5}.

The mesh quality is very important for 3D turbulence calculation. In this paper, the tetrahedron–hexahedron hybrid mesh was used to discrete the fluid domain. In ANSYS Workbench, an automatic mesh generating method according to different guide vane openings was set up to ensure the uniformity of mesh size of guide vane under a given opening. The fluid domain mesh of the PT unit is shown in Figure 6.

In this paper, four sets of meshes with different element sizes were created to calculate the steady-state of full-load condition (point (1) in Figure 3). The mesh independence analysis was carried out by comparing the calculated efficiencies with different set of meshes (Figure 7). The final adopted mesh (3) of the PT unit for load rejection calculation is shown in Table 2.

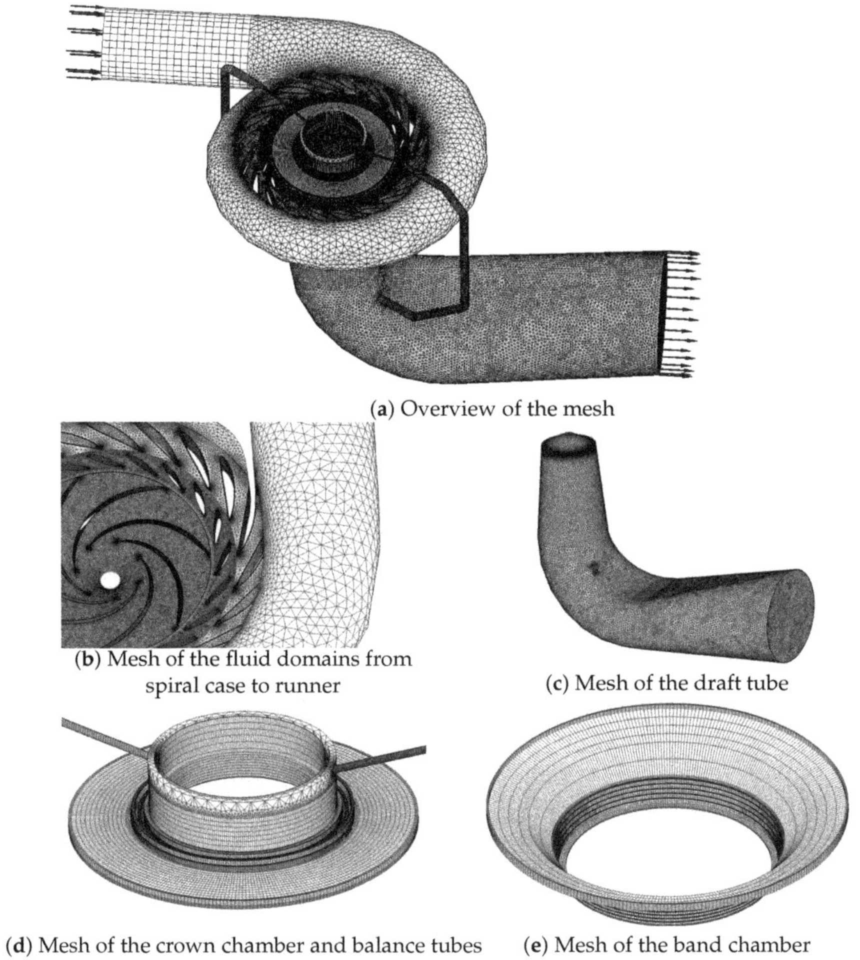

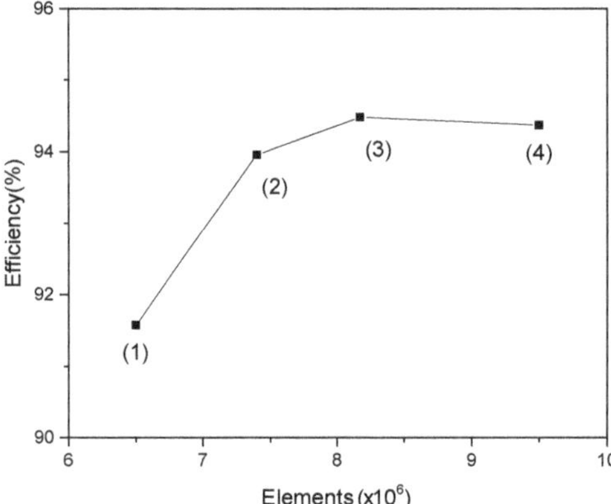

(**a**) Overview of the mesh

(**b**) Mesh of the fluid domains from spiral case to runner

(**c**) Mesh of the draft tube

(**d**) Mesh of the crown chamber and balance tubes

(**e**) Mesh of the band chamber

Figure 6. Mesh of the fluid domains.

Figure 7. Mesh independence analysis using four sets of meshes with different element sizes.

Table 2. Element number of flow domains.

Flow Domain	Elements ($\times 10^6$)
Spiral case & Stay vane	3.26
Guide Vane	0.27
Runner	3.63
Draft tube	0.60
Labyrinth seal & Balance tubes	0.40
Total	8.16

3.4. Results and Discussion of the 3D Flow Calculation

3.4.1. Pressure Change at the Monitoring Points

Figure 8 shows the pressure change of the monitoring points in the calculation. The pressures at various monitoring points fluctuate violently during load rejection, which is closely related to the flow rate and rotational speed of the unit. High rotational speed enhances the shear effect of the runner on the non-rotating flow region and increases the water flow velocities at the runner inlet and the upstream, resulting in more wake cutting on the guide vane flow domain. It also intensifies rotor–stator interaction phenomena and increases pressure pulsation in the vaneless area (t = 3.74 s, t = 13.24 s and t = 24.10 s). In addition, at the moment t = 6.88 s, the opening of the guide vane is large, the rotational speed is high and the flow rate is low, the flow separation and vortices of the incoming flow from the guide vane increase and a large number of blade passage vortices appear in the runner, which are extremely unstable. The irregular flow leads to increased energy dissipation and increased pressure near the vaneless area. In the case of a high flow rate and low rotational speed (t = 11.19 s and t = 19.43 s), the direction of the outflow velocity of the guide vane deviates less from the design operating condition, so the flow inside the unit is more adequate and uniform and the pressure shows a lower level.

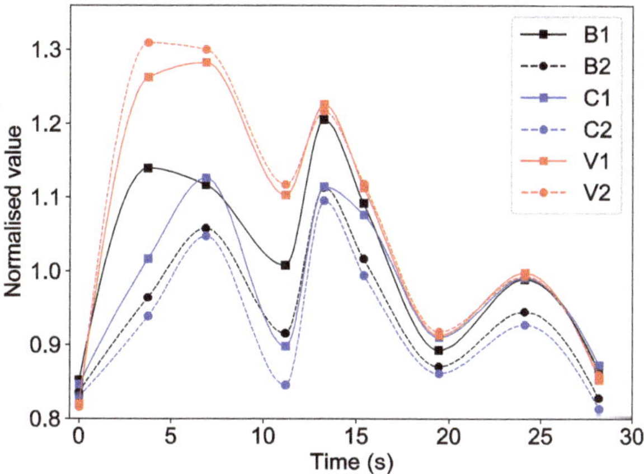

Figure 8. Pressure change at the monitoring points.

3.4.2. Flow Pattern Change in the Runner Passage

In the process of load rejection, the flow from the guide vane to the runner outlet shows extremely unstable turbulent characteristics. Figure 9 shows the streamline distribution in the runner in the whole load rejection process. When the load rejection does not begin (t = 0 s), the flow in the runner is uniform. Afterward, the flow is greatly affected by the changes in flow rate, head and speed.

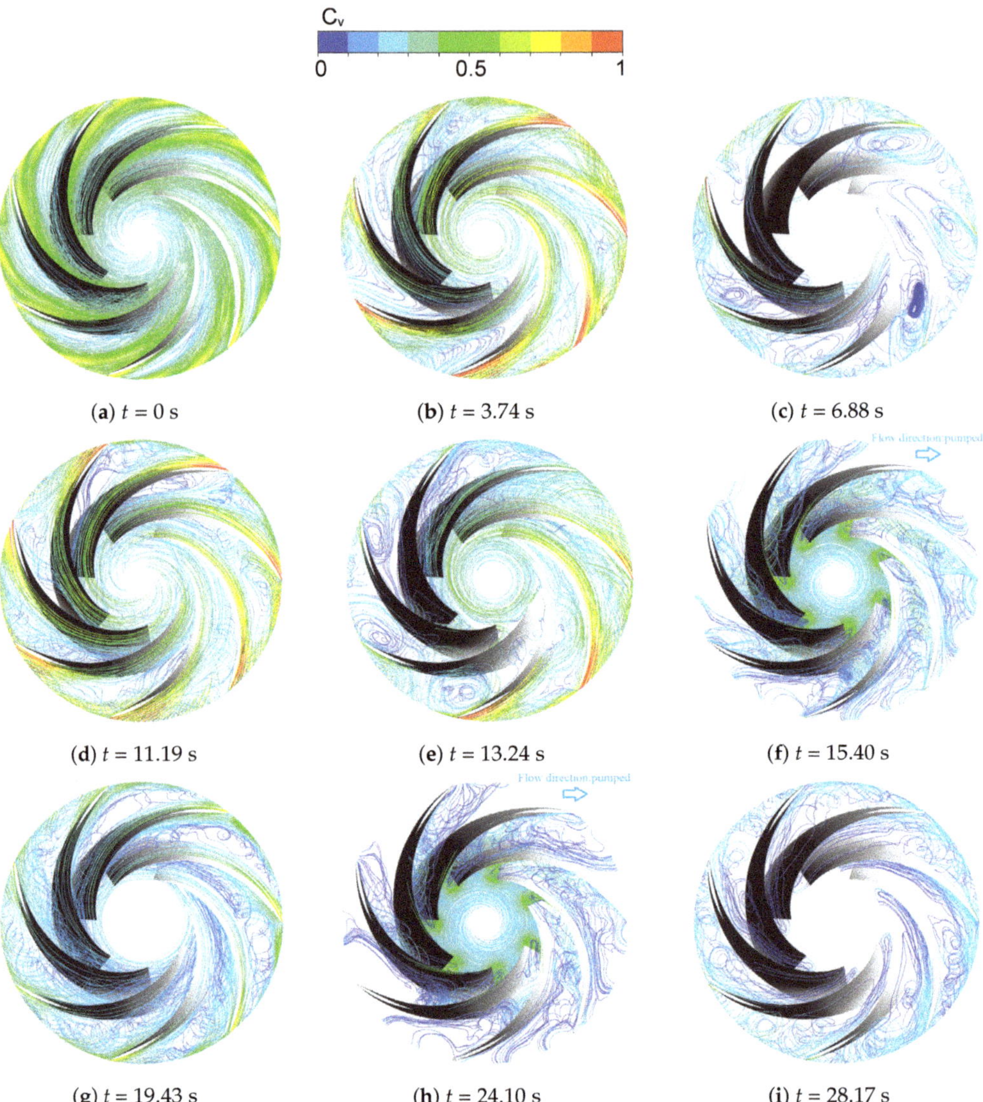

(a) $t = 0$ s   **(b)** $t = 3.74$ s   **(c)** $t = 6.88$ s

(d) $t = 11.19$ s   **(e)** $t = 13.24$ s   **(f)** $t = 15.40$ s

(g) $t = 19.43$ s   **(h)** $t = 24.10$ s   **(i)** $t = 28.17$ s

Figure 9. Flow pattern in the runner passage.

At the moments when the flow rate is high ($t = 3.74$ s, 11.19 s, 13.24 s, 19.43 s), large-scale blade duct vortices appear on the suction surface of the blade due to the peak or high rotational speed. The size of the blade passage vortices corresponds to the size of the flow duct, which disturbs the flow of the whole unit and makes the hydraulic torque of the runner fluctuate violently with the pressure of the unit.

At the moments with negative flow ($t = 15.40$ s and 24.10 s), the unit enters the reverse pump operating condition. Due to the imbalance of speed and flow, the flow between blade passages rotates abnormally and separated flow occurs on the pressure surface of the runner. Both factors cause the wake falling off upstream of the runner blades and a large number of swirl eddies appear in the vaneless area. At the same time, wake falling off also occurs at the leading edge of the guide vanes and stay vanes. The unstable flow of the unit extends to the spiral case area. It is shown in research [15] that the vortex core in the vaneless area is impacted instantaneously by a large number of reverse refluxes into the guide vane area and the stay vane area when the unit experiences the reverse pump operating condition, which is also in accordance with the calculation results in this paper.

3.4.3. Pressure Change in the Flow Passages

Figure 10 shows the pressure change of head cover during load rejection. At the starting moment ($t = 0$ s) and the moments($t = 3.73$ s, 11.19 s, 13.24 s, 19.43 s) when the hydraulic moment is positive or equal to zero, the pressure distribution of the unit is relatively uniform. The pressure difference between the pressure surface of the runner and the suction surface is obvious without pressure concentration.

At the time($t = 6.88$ s, 15.40 s, 24.10 s) when the hydraulic torque is negative, the pressure difference between the pressure surface and suction surface disappears and obvious pressure concentration points appear at guide vanes near the flow clearances.

Figure 11 shows the streamline distribution from the runner to the guide vane, with obvious flow separation at the pressure concentration point and strong circulation behind the guide vane. Pressure concentration is caused by the separation of high-speed annular flow through a small flow gap, resulting in a sudden drop in velocity and local pressure rise as the flow energy is converted into impact on the edge of the vanes.

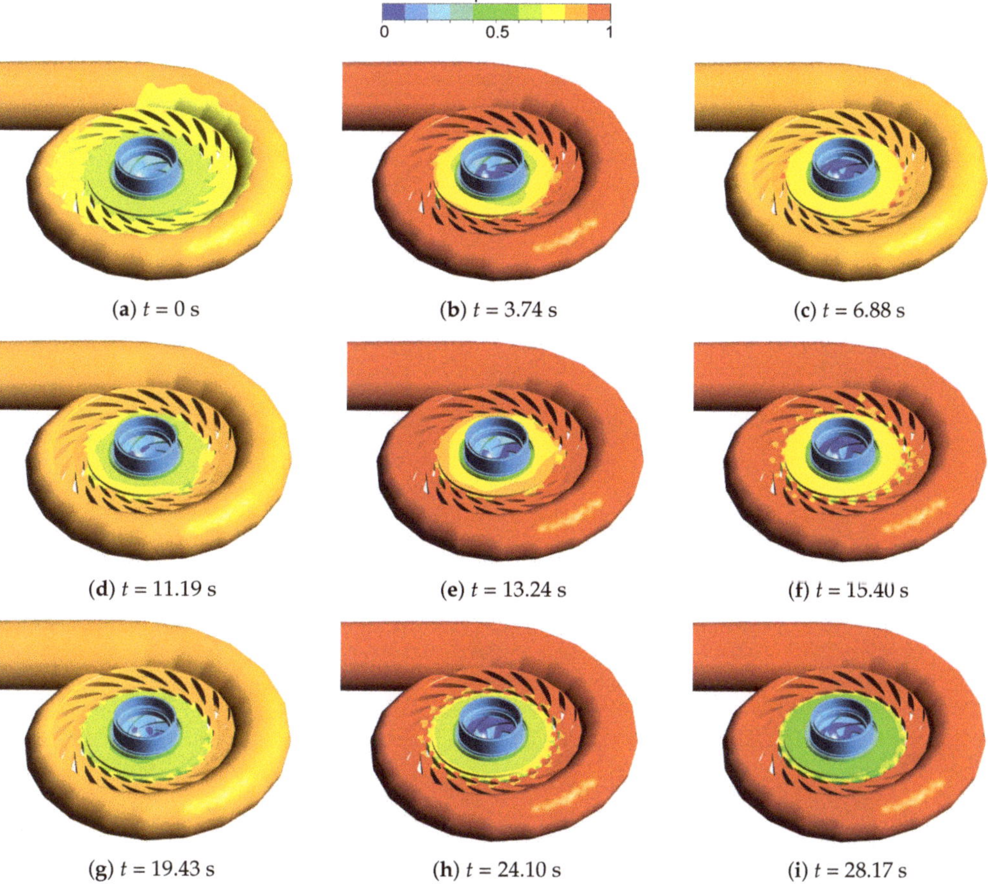

Figure 10. Pressure distribution in the flow passages at different key time moments.

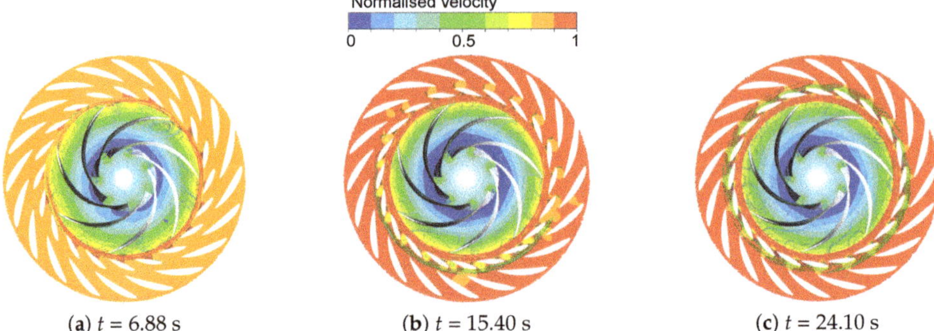

(**a**) $t = 6.88$ s   (**b**) $t = 15.40$ s   (**c**) $t = 24.10$ s

Figure 11. The streamline distribution of the unit at three key time moments.

4. Fluid–Structure Coupling Analysis of the Stationary Structures

4.1. Simulation Model and Boundary Conditions

Figure 12 demonstrates the CAD model and boundary conditions of the stationary structure of the PT unit. The head cover, the stay ring and the bottom ring were bolted together as a single structure. The equivalent stiffness coefficient of the turbine guide bearing was set as 1×10^9 N·m^{-1} and the standard gravity was taken into consideration. Since the thrust bearing structure supporting the entire rotating system was directly placed on the head cover, the equivalent mass of them was applied to the head cover. The upper and lower surfaces of the stay ring were fixed in concrete and the bottom surface of the bottom ring was welded to the draft tube cone. ANSYS Mechanical was used to carry out the fluid–structure coupling analysis.

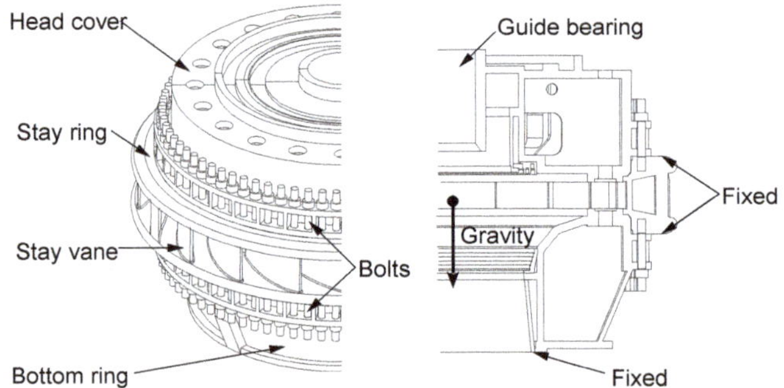

Figure 12. CAD model and boundary conditions of the stationary structures of the PT unit.

The material and mechanical properties of the research object were configured according to Table 3.

Table 3. Material properties of the structures.

	Structures
Density [kg·m^{-3}]	7850
Elastic modulus [MPa]	2.1×10^5
Poisson's ratio [-]	0.3

4.2. Mesh Sensitivity Study

As shown in Figure 13, the structural models were meshed using high-quality tetrahedral elements, better suited for complex geometries. The meshes were refined for the typical stress concentration areas of the head cover, stay ring and bottom ring (P1, P2 and P3).

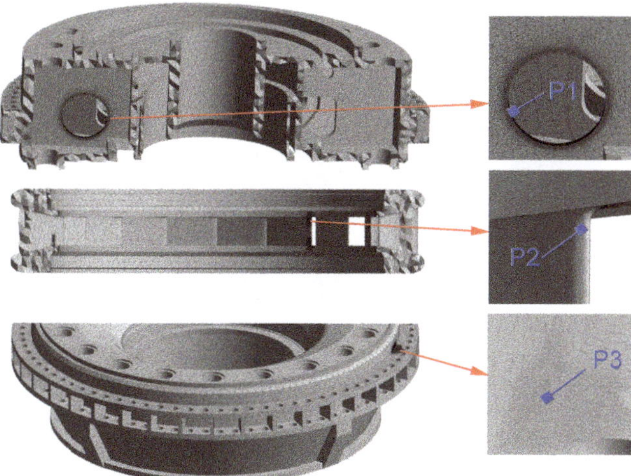

Figure 13. Structural mesh of the stationary structure of the PT unit.

The element size has a significant effect on the FEA results, so it is reasonable to conduct a mesh sensitivity study to obtain a set of mesh that provides a good balance between computational time and the accuracy of the results. Figure 14 presents the simulation duration and convergence of the simulation results for three sets of meshes with different sizes of elements. The element numbers of three sets of meshes are 7.86×10^5, 1.13×10^6 and 2.97×10^6, respectively. Compared to the third set mesh where the element size was significantly reduced, the simulation results of the second set mesh show negligible changes, but the computational time was only 1/9 of the third one; therefore, we conclude that the second set mesh has converged and can be used for further analysis to save calculation time.

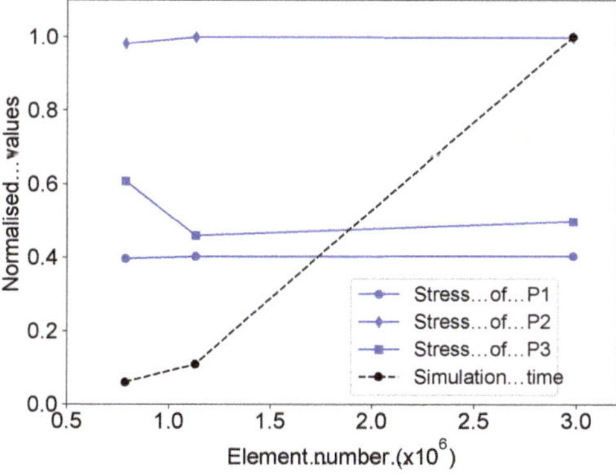

Figure 14. Mesh sensitivity study.

4.3. Fluid Pressure Mapping and Axial Thrust Analysis

To perform the fluid–structure coupling analysis, the pressure distributions acting on the stationary structures at different time points during load rejection were exported from the 3D CFD analysis described earlier. Each pressure file includes the 3D coordinates and the pressure values of the nodes on the fluid–structure coupling interface in the form of (x, y, z, Pressure). Since the structural mesh is different from the fluid mesh on the fluid–structure coupling interface, the fluid nodal pressure values need to be interpolated and mapped onto the nearest nodes of the structures according to the distance between the fluid and structure nodes. The ANSYS built-in tool can complete this task precisely and flawlessly.

Figure 15 shows the mapped pressure distribution on the stationary structures of the PT unit at a given moment in time. By applying the pressure files sequentially to the structural model during load rejection, the flow-induced deformation and stress of the stationary components including the head cover, stay ring and bottom ring can be calculated.

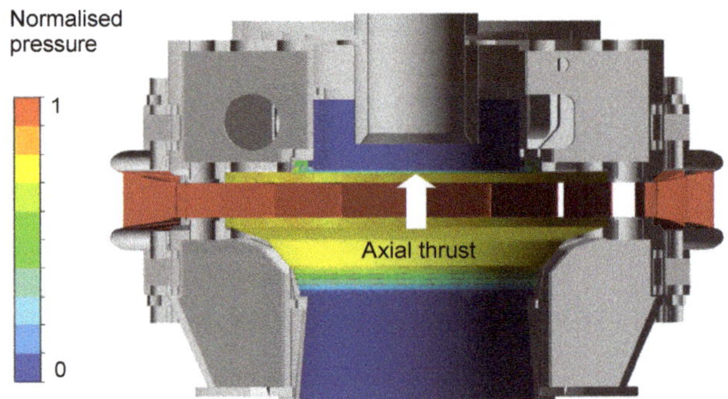

Figure 15. Fluid pressure mapping on the structural structures.

The huge hydraulic pressures acting on the stationary structures of the PT unit lifted the head cover and the upper part of the stay ring and pressed down the bottom ring and the lower part of the stay ring. In addition, rapidly changing pressure loads during load rejection generated large stresses on the head cover, stay vanes and bottom ring.

The axial thrust on the head cover during load rejection needs to be carefully investigated as it causes axial vibration of the head cover, the thrust bearing structure and the entire rotating system. Figure 16 illustrates the relative axial thrust of the head cover during load rejection, which was calculated by dividing the head cover axial thrust by the total mass of the head cover, the thrust bearing structure and the rotating system. The positive direction of the head cover axial thrust is axially upward and the value of the axial thrust varies greatly with time during load rejection, reaching a maximum value at $t = 13.24$ s, which is more than 10 times the total gravity of the head cover, the thrust bearing structure and the rotating system.

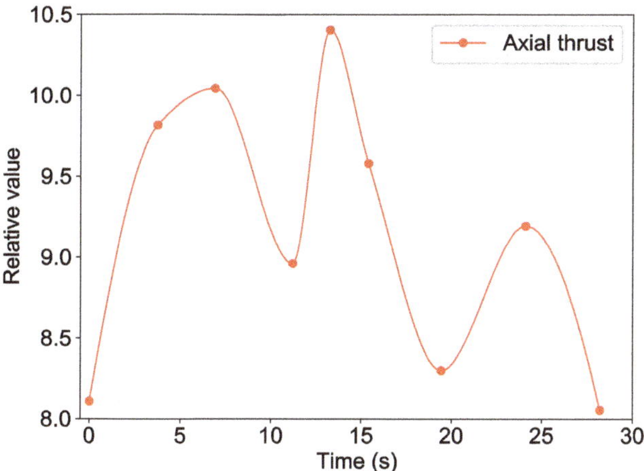

Figure 16. Axial thrust of the head cover during load rejection.

4.4. Results and Discussion

Figure 17 shows the total deformation distribution of the stationary structures caused by the fluid flow in the PT unit during load rejection. The head cover consists of an outer and an inner cover, 4 connecting flanges and 18 stiffener plates with different shaped openings. The huge hydraulic pressure of the fluid acting on the inner surface of the head cover resulted in a large axial thrust, which caused relatively large deformation on the head cover and the largest deformation appeared on the inner head cover where there is no support in the axial direction. The bottom ring is reinforced by 18 stiffener plates and its axial force area is much smaller than that of the head cover, so the total deformation of the bottom ring is at a low level. The stay ring is shorter and stiffer than the head cover and the bottom ring and has a much smaller axial force area, so the flow-induced deformation of the stay ring is also small.

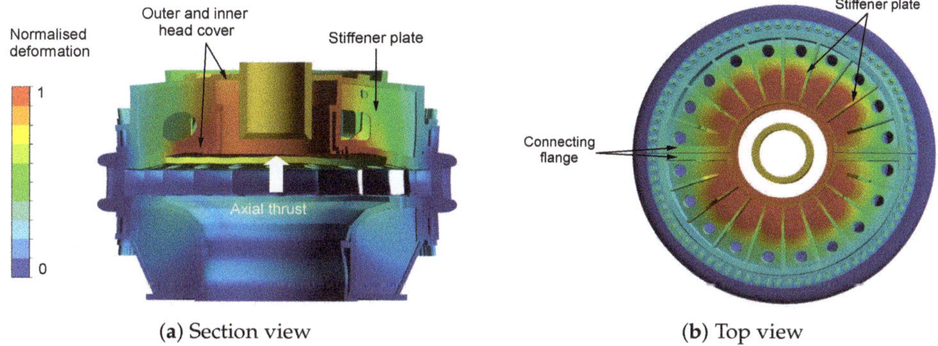

(**a**) Section view   (**b**) Top view

Figure 17. Total deformation distribution of the stationary structures.

Figure 18 shows the flow-induced stress distribution of the stationary structures.

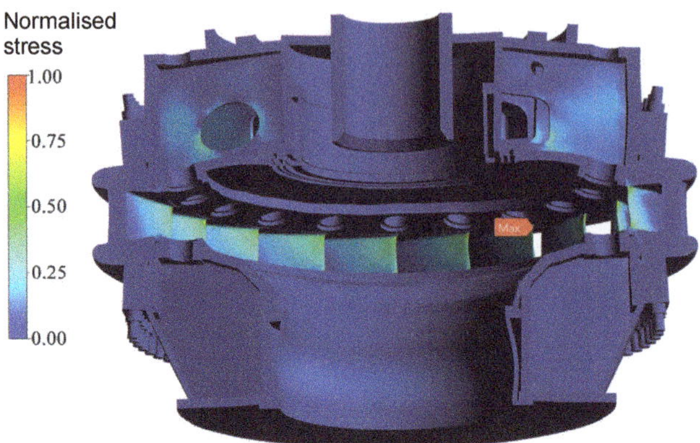

Figure 18. The stress distribution of the stationary structures.

The fluid pressure on the inner surface of the stationary structures pushes the head cover upward and the bottom ring downward. The stay ring between the head cover and the bottom ring is stretched and the maximum stress occurs on the trailing edge fillet of a guide vane of the stay ring (Figure 19).

Figure 19. The maximum stress of the stay ring.

The head cover is only supported by the stay ring in the axial direction but has a large area facing the high pressure water. The stiffener plates are welded between the inner and outer head cover to increase the stiffness and effectively reduce the deformation of the head cover caused by the pressure; however, different shaped openings and holes are made in the stiffener plates for installing various pipes such as pressure balance pipes, cooling water pipes and pressure measurement pipes. The normalized stress distribution of the head cove relative to its maximum stress value is shown in Figure 20, which shows that the hot spots of head cover during load rejection appear on the different shaped openings of the stiffener plates and the maximum stress occurs on the bottom-right side of the circular opening.

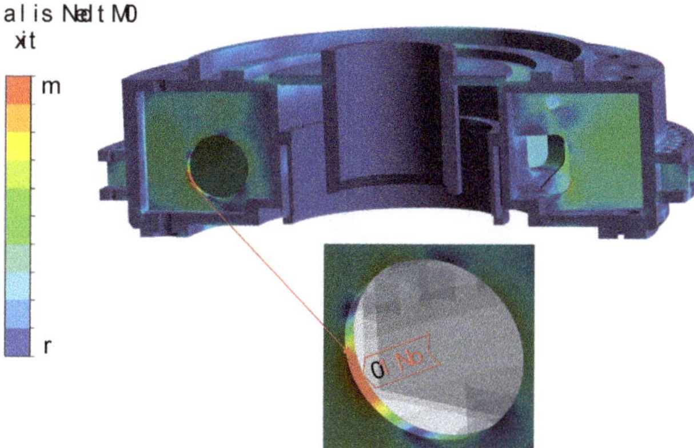

Figure 20. The stress distribution of the head cover.

Figure 21 illustrates the stress distribution of the bottom ring during load rejection. The stresses are normalized with reference to the maximum stress value of the bottom ring and the maximum stress of the bottom ring occurs on the corner of the bottom ring flange.

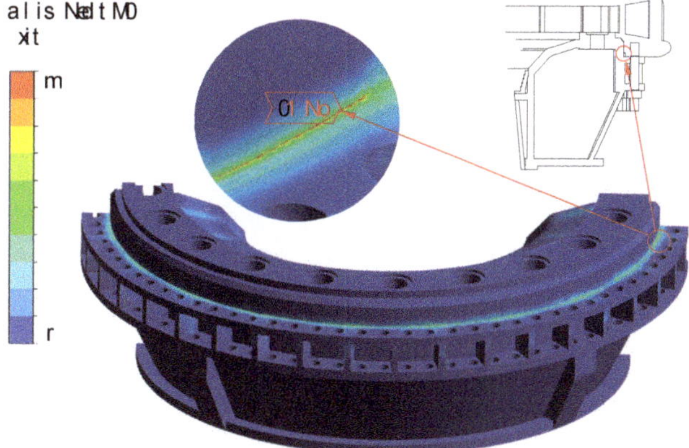

Figure 21. The stress distribution of the bottom ring.

The von Mises stress distributions of the stationary structures at different time points during load rejection are similar and the hot spots of the stationary structures appear in the same locations of the head cover, stay ring and bottom ring, but the maximum stress value change with time. Figure 22 shows the stress comparison of the normalized stresses of the stationary structures during load rejection. All stresses are normalized with reference to the maximum stress of the stay ring, which is larger than the maximum stresses of the head cover and the bottom ring.

The stresses are normalized with reference to the maximum stress value of the bottom ring.

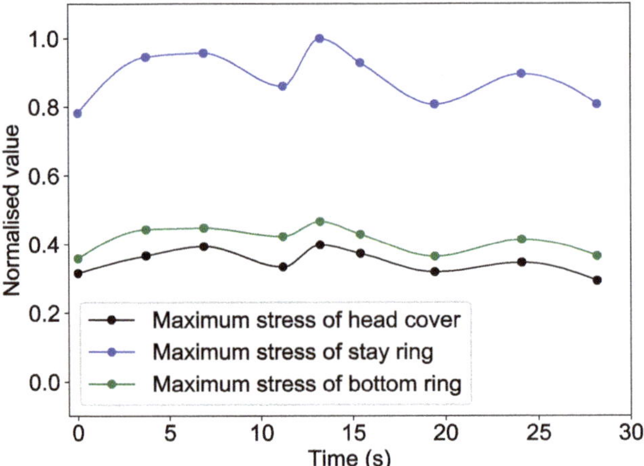

Figure 22. Comparison of the maximum stresses of the stationary structures during load rejection.

The comparison of the axial thrust force on the head cover and the normalized maximum stresses of the stationary structures during load rejection is demonstrated in Figure 23. The tendency of these there parameters follows the same pattern. The maximum stress values of the head cover, stay ring and bottom ring change with the axial thrust force on the head cover and reached the respective maximum value at $t = 13.24$ s.

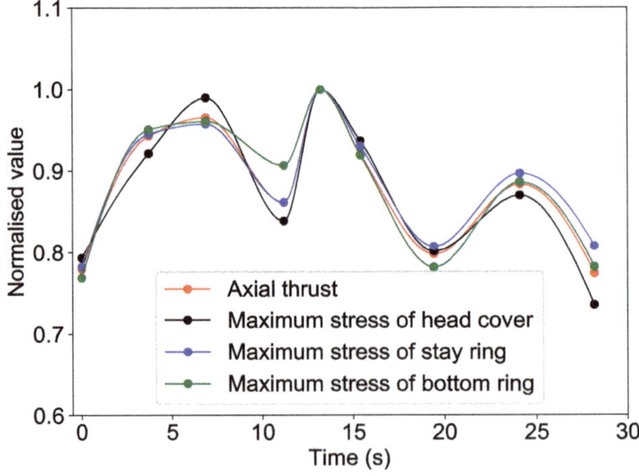

Figure 23. Results comparison of the PT unit during load rejection.

5. Conclusions

This study has carried out the fluid dynamic analysis and the flow-induced structural behaviour analysis of a prototype reversible pump turbine unit during the load rejection based on the measured data.

During the load rejection, the internal flow of the unit changes drastically and the impact of the runner on the upstream is intensified due to the rapid increase in the rotational speed and the flow variation from the design condition leads to more flow separation and energy dissipation at the guide vane outlet. The high rotational speed and the sudden change of flow rate are the main reasons for the unstable flow and pressure characteristics during the load rejection process and the pressure reaches local peaks at $t = 3.74$ s, 6.88 s, 13.24 s and 24.10 s.

When the unit enters the reverse pump working condition ($t = 15.40$ s and 24.10 s), the drastic changes in flow rate, speed and guide vane opening lead to a strong circulation in front of the guide vanes. The water flow through the guide vanes is separated from the circulating flow, resulting in strong local pressures at the backside of the guide vanes.

The pressure change in the flow domains of the spiral case and guide vane during load rejection has the same trend as the pressure at the spiral case outlet. The pressure acting on head cover changed largely during load rejection and reached the maximum value at $t = 13.24$ s.

The huge hydraulic forces of the pump turbine unit during load rejection cause large time-varying stresses on the stationary structures of the pump turbine unit. The maximum stress of the stationary structures including the head cover, stay ring and bottom ring at different time points during load rejection are strongly correlated with that of axial thrust force on the head cover and reached the maximum values at $t = 13.24$ s.

The maximum deformation of the stationary structures during the load rejection occurs on the inner edge of the head cover closed to the main shaft and the maximum equivalent stress is located at the fillet of the stay vane trailing edge.

These findings contribute to an improved physical understanding of the flow characteristics and structural behaviour of the stationary structures of similar pump turbines, Francis turbines and centrifugal pumps and help engineers to better design the turbine units that can operate safely and stably.

Author Contributions: Conceptualization, Z.W. and X.H.; methodology, Q.H. and X.H.; software, Q.H., X.H., M.Y., H.Y. and H.B.; validation, M.Y., H.Y. and H.B.; investigation, Q.H., X.H. and H.B.; writing—original draft preparation, Q.H. and X.H.; writing—review and editing, Q.H., X.H., M.Y., H.Y. and H.B.; supervision, Z.W. All authors have read and agreed to the published version of the manuscript.

Funding: This work was supported by National Natural Science Foundation of China (No.: 51876099).

Institutional Review Board Statement: Not applicable.

Informed Consent Statement: Not applicable.

Data Availability Statement: Not applicable.

Acknowledgments: The authors would like to express their sincere thanks for the financial support of the project: *Research on Lifetime Prediction of Non-rotating Parts of Pump Turbine Unit Based on Rotor-Stator Interaction (RSI), Fluid–Structure Coupling and Fracture Mechanics—Research project on RSI mechanism and its Influence on Non-rotating parts of Pump Turbine Unit* of Branch Company of Maintenance & Test, CSG Power Generation co., LTD.

Conflicts of Interest: The authors declare no conflict of interest.

Abbreviations

The following abbreviations are used in this manuscript.

CFD	computational fluid dynamics
FEM	finite element method
FSC	fluid—structure coupling
FVM	finite volume method
HPSPP	hydraulic pumped storage power plant
PT	pump turbine

References

1. Liu, X.; Luo, Y.; Wang, Z. A review on fatigue damage mechanism in hydro turbines. *Renew. Sustain. Energy Rev.* **2016**, *54*, 1–14. [CrossRef]
2. Casanova, F.; Mantilla, C. Fatigue failure of the bolts connecting a Francis turbine with the shaft. *Eng. Fail. Anal.* **2018**, *90*, 1–13. [CrossRef]

3. Peltier, R.; Boyko, A.; Popov, S.; Krajisnik, N. Investigating the Sayano-Shushenskaya Hydro Power Plant Disaster. *Power* **2010**, *154*, 48.
4. Egusquiza, E.; Valero, C.; Huang, X.; Jou, E.; Guardo, A.; Rodriguez, C. Failure investigation of a large pump-turbine runner. *Eng. Fail. Anal.* **2012**, *23*, 27–34. [CrossRef]
5. Mandair, S.; Morissette, J.F.; Magnan, R.; Karney, B. MOC-CFD coupled model of load rejection in hydropower station. *IOP Conf. Ser. Earth Environ. Sci.* **2021**, *774*, 012021. [CrossRef]
6. Zhou, D.; Chen, H.; Kan, K.; Yu, A.; Binama, M.; Chen, Y. Experimental study on load rejection process of a model tubular turbine. *IOP Conf. Ser. Earth Environ. Sci.* **2021**, *774*, 012036. [CrossRef]
7. Bi, H.; Chen, F.; Wang, C.; Wang, Z.; Fan, H.; Luo, Y. Analysis of dynamic performance in a pump-turbine during the successive load rejection. *IOP Conf. Ser. Earth Environ. Sci.* **2021**, *774*, 012152. [CrossRef]
8. Zhang, H.; Su, D.; Guo, P.; Zhang, B.; Mao, Z. Stochastic dynamic modeling and simulation of a pump-turbine in load-rejection process. *J. Energy Storage* **2021**, *35*, 102196. [CrossRef]
9. He, L.Y.; Wang, Z.W.; Kurosawa, S.; Nakahara, Y. Resonance investigation of pump-turbine during startup process. *IOP Conf. Ser. Earth Environ. Sci.* **2014**, *22*, 32024. [CrossRef]
10. Kolšek, T.; Duhovnik, J.; Bergant, A. Simulation of unsteady flow and runner rotation during shut-down of an axial water turbine. *J. Hydraul. Res.* **2006**, *44*, 129–137. [CrossRef]
11. Ciocan, G.D.; Iliescu, M.S.; Vu, T.C.; Nennemann, B.; Avellan, F. Experimental Study and Numerical Simulation of the FLINDT Draft Tube Rotating Vortex. *J. Fluids Eng.* **2006**, *129*, 146–158. [CrossRef]
12. Huang, X.; Oram, C.; Sick, M. Static and dynamic stress analyses of the prototype high head Francis runner based on site measurement. *IOP Conf. Ser. Earth Environ. Sci.* **2014**, *22*, 032052.:10.1088/1755-1315/22/3/032052. [CrossRef]
13. Goyal, R.; Cervantes, M.J.; Gandhi, B.K. Characteristics of Synchronous and Asynchronous modes of fluctuations in Francis turbine draft tube during load variation. *Int. J. Fluid Mach. Syst.* **2017**, *10*, 164–175. [CrossRef]
14. Nicolet, C.; Alligne, S.; Kawkabani, B.; Simond, J.-J.; Avellan, F. Unstable Operation of Francis Pump-Turbine at Runaway: Rigid and Elastic Water Column Oscillation Modes. *Int. J. Fluid Mach. Syst.* **2009**, *2*, 324–333. [CrossRef]
15. Fu, X.; Li, D.; Wang, H.; Zhang, G.; Li, Z.; Wei, X. Dynamic instability of a pump-turbine in load rejection transient process. *Sci. China Technol. Sci.* **2018**, *61*, 1765–1775. [CrossRef]
16. Mao, Z.; Tao, R.; Bi, H.; Luo, Y.; Wang, Z. Numerical study of hydraulic axial force of prototype pump-turbine pump mode's stop with power down. *IOP Conf. Ser. Earth Environ. Sci.* **2021**, *774*, 012094. [CrossRef]
17. Li, X.; Mao, Z.; Lin, W.; Bi, H.; Tao, R.; Wang, Z. Prediction and Analysis of the Axial Force of Pump-Turbine during Load-Rejection Process. *IOP Conf. Ser. Earth Environ. Sci.* **2020**, *440*, 052081. [CrossRef]
18. Walseth, E.C.; Nielsen, T.K.; Svingen, B. Measuring the Dynamic Characteristics of a Low Specific Speed Pump—Turbine Model. *Energies* **2016**, *9*, 199. [CrossRef]
19. Avdyushenko, A.Y.; Cherny, S.G.; Chirkov, D.V.; Skorospelov, V.A.; Turuk, P.A. Numerical simulation of transient processes in hydroturbines. *Thermophys. Aeromech.* **2013**, *20*, 577–593. [CrossRef]
20. Liu, X.; Liu, C. Eigenanalysis of Oscillatory Instability of a Hydropower Plant Including Water Conduit Dynamics. *IEEE Trans. Power Syst.* **2007**, *22*, 675–681. [CrossRef]
21. Avdyushenko, A.; Chernyi, S.; Chirkov, D. Numerical algorithm for modelling three-dimensional flows of an incompressible fluid using moving grids. *Comput. Technol.* **2012**, *17*, 3–25.
22. Widmer, C.; Staubli, T.; Ledergerber, N. Unstable Characteristics and Rotating Stall in Turbine Brake Operation of Pump-Turbines. *J. Fluids Eng.* **2011**, *133*, 041101. [CrossRef]
23. Nicolle, J.; Giroux, A.M.; Morissette, J.F. CFD configurations for hydraulic turbine startup. *IOP Conf. Ser. Earth Environ. Sci.* **2014**, *22*, 032021. [CrossRef]
24. Mao, Z.; Tao, R.; Chen, F.; Bi, H.; Cao, J.; Luo, Y.; Fan, H.; Wang, Z. Investigation of the Starting-Up Axial Hydraulic Force and Structure Characteristics of Pump Turbine in Pump Mode. *J. Mar. Sci. Eng.* **2021**, *9*, 158. [CrossRef]
25. Münch, C.; Ausoni, P.; Braun, O.; Farhat, M.; Avellan, F. Fluid–structure coupling for an oscillating hydrofoil. *J. Fluids Struct.* **2010**, *26*, 1018–1033. [CrossRef]
26. Benra, F.-K.; Dohmen, H.J. Comparison of Pump Impeller Orbit Curves Obtained by Measurement and FSI Simulation. In Proceedings of the ASME Pressure Vessels and Piping Conference, San Antonio, TX, USA, 22–26 July 2007; pp. 41–48. [CrossRef]
27. Kato, C.; Yoshimura, S.; Yamade, Y.; Jiang, Y.Y.; Wang, H.; Imai, R.; Katsura, H.; Yoshida, T.; Takano, Y. Prediction of the Noise From a Multi-Stage Centrifugal Pump. In Proceedings of the Fluids Engineering Division Summer Meeting, Houston, TX, USA, 19–23 June 2005; pp. 1273–1280. [CrossRef]
28. Jiang, Y.; Yoshimura, S.; Imai, R.; Katsura, H.; Yoshida, T.; Kato, C. Quantitative evaluation of flow-induced structural vibration and noise in turbomachinery by full-scale weakly coupled simulation. *J. Fluids Struct.* **2007**, *23*, 531–544. [CrossRef]
29. Abdelsalam, S.I.; Zaher, A.Z. Leveraging Elasticity to Uncover the Role of Rabinowitsch Suspension through a Wavelike Conduit: Consolidated Blood Suspension Application. *Mathematics* **2021**, *9*, 2008. [CrossRef]

Article

Numerical Simulation and Experimental Validation of a Kaplan Prototype Turbine Operating on a Cam Curve

Raluca Gabriela Iovănel [1],*, Arash Soltani Dehkharqani [2], Diana Maria Bucur [3] and Michel Jose Cervantes [1]

[1]   Division of Fluid and Experimental Mechanics, Luleå University of Technology, 971 87 Luleå, Sweden; michel.cervantes@ltu.se
[2]   R&D Engineer, Svenska Rotor Maskiner, Svarvarvägen 2, 132 38 Saltsjö-Boo, Sweden; arash.soltani@rotor.se
[3]   Department of Hydraulics, Hydraulic Equipment and Environmental Engineering, Politehnica University of Bucharest, 060042 Bucharest, Romania; dmbucur@yahoo.com
*   Correspondence: raluca.iovanel@gmail.com; Tel.: +40-733-433-365

Abstract: The role of hydropower has become increasingly essential following the introduction of intermittent renewable energies. Quickly regulating power is needed, and the transient operations of hydropower plants have consequently become more frequent. Large pressure fluctuations occur during transient operations, leading to the premature fatigue and wear of hydraulic turbines. Investigations of the transient flow phenomena developed in small-scale turbine models are useful and accessible but limited. On the other hand, experimental and numerical studies of full-scale large turbines are challenging due to production losses, large scales, high Reynolds numbers, and computational demands. In the present work, the operation of a 10 MW Kaplan prototype turbine was modelled for two operating points on a propeller curve corresponding to the best efficiency point and part-load conditions. First, an analysis of the possible means of reducing the model complexity is presented. The influence of the boundary conditions, runner blade clearance, blade geometry and mesh size on the numerical results is discussed. Secondly, the results of the numerical simulations are presented and compared to experimental measurements performed on the prototype in order to validate the numerical model. The mean torque and pressure values were reasonably predicted at both operating points with the simplified model. An analysis of the pressure fluctuations at part load demonstrated that the numerical simulation captured the rotating vortex rope developed in the draft tube. The frequencies of the rotating and plunging components of the rotating vortex were accurately captured, but the amplitudes were underestimated compared to the experimental data.

Keywords: hydropower; Kaplan turbine; prototype simulation; CFD; rotating vortex rope

Citation: Iovănel, R.G.; Dehkharqani, A.S.; Bucur, D.M.; Cervantes, M.J. Numerical Simulation and Experimental Validation of a Kaplan Prototype Turbine Operating on a Cam Curve. *Energies* **2022**, *15*, 4121. https://doi.org/10.3390/en15114121

Academic Editors: Zhengwei Wang and Yongguang CHENG

Received: 19 April 2022
Accepted: 1 June 2022
Published: 3 June 2022

Publisher's Note: MDPI stays neutral with regard to jurisdictional claims in published maps and institutional affiliations.

1. Introduction

The introduction of intermittent renewable energy resources such as wind and solar power requires additional power to regulate grid frequency. Hydropower is often used to regulate grid frequency because hydraulic turbines can operate in a relatively wide range of power outputs and load changes and they are able to start and stop in a matter of seconds and minutes, respectively. Such use of hydropower is relatively new and imposes constraints on the hydraulic turbines for which they were not initially designed [1,2]. This may lead to some premature wear and sometimes failures because of large pressure fluctuations mainly arising in the draft tube. A lot of research is now dedicated to minimizing the effect of operating regimes away from the best efficiency and transient operation. The flow phenomena leading to large pressure pulsations need to be well-understood to allow for the development of technology capable of mitigating them. The investigations of such new operating conditions and mitigation techniques are challenging, as the consequences may be both hydrodynamic and structural.

Most studies related to the effect of off-design operations and transients have been performed with model testing and numerical simulations of these model turbines. Model testing allows for the investigation of turbines in a wide range of precise and repeatable operating conditions. Detailed pressure, strain and velocity measurements can now be performed on such turbine models. Extensive measurement campaigns have been carried out, predominantly focusing on Francis turbine models [3–6] but also a bulb model [7,8] and a Kaplan turbine model [9,10].

Ideally, the investigation of hydraulic turbines should be performed on the prototype itself. However, such experiments are challenging because of both the large scales involved and the costs, mainly because of the production losses. There are only a handful of studies concerning full-scale turbines available in the literature [11,12]. None of them offer the same level of detail as experiments performed on models where optical methods may be used for velocity measurements. Usually, full-scale measurements are limited to pressure and strain measurements in the stationary domain [13,14]. Performing measurements on a runner [15,16], presents another level of difficulty available in few studies. Numerical simulations can provide additional information otherwise difficult to obtain from measurements. Together, both types of investigation can contribute to developing data-driven control solutions that optimize the operation and diagnosis of hydraulic machine failures [17,18]. However numerical studies concerning full-scale hydraulic turbines are also sparse. Such simulations are challenging because of the large dimensions, high Reynolds numbers, and lack of experimental data to validate the results [19,20]. There have been no numerical studies referring to full-scale Kaplan prototype turbines.

The advancement in computer hardware, hydraulic numerical simulations and turbulence modelling may currently allow for the easier handling of prototype simulations. Nonetheless, there is a growing need to simplify the numerical models to obtain results in a reasonable amount of time with a reasonable accuracy.

The present work introduces the numerical simulation of a full-scale 10 MW Kaplan turbine prototype at two operating points on a propeller curve. The chosen operating conditions are equivalent to the best efficiency point and part-load operating condition of a single regulated turbine, such as Francis and propeller types. The potential to decrease model complexity was addressed by investigating several issues such as boundary conditions, runner blade clearance, blade geometry and mesh size. The different parameters were systematically investigated to assess their impact on the numerical accuracy of the simulation. The aim was to optimize the numerical accuracy, computational demands, and simulation time. The reduced model was validated using experimental data measured with the corresponding Kaplan prototype.

2. Experimental Measurements

2.1. The Porjus U9 Prototype

An experimental investigation was performed on the Porjus U9 Kaplan turbine prototype by Soltani et al. [1]. This unit is part of the Porjus Hydropower Center, situated in the north of Sweden, and it is primarily used for educational and research purposes. The Porjus U9 Kaplan turbine presented in Figure 1 has 6 runner blades, 20 equally spaced guide vanes, and 18 unequally distributed stay vanes. The runner has a diameter of 1.55 m and is located approximately 7 m below the tailwater. Table 1 presents the rated parameters of the turbine.

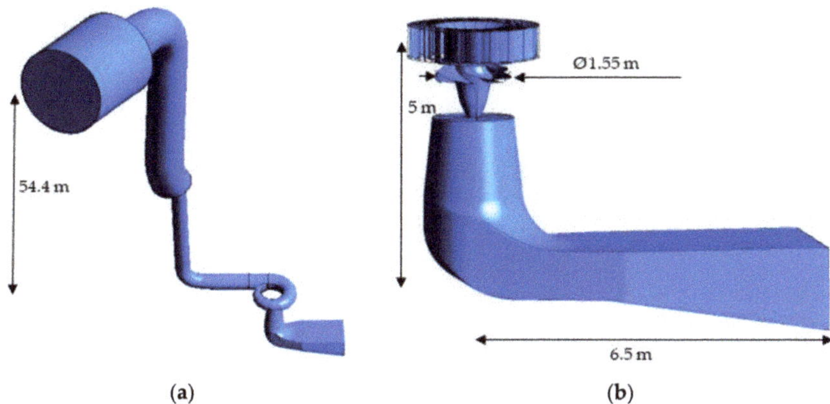

Figure 1. The main dimensions of the Porjus U9 Kaplan turbine prototype: (**a**) complete turbine geometry; (**b**) stay vanes, guide vanes, runner and draft tube.

Table 1. Rated parameters of the Porjus U9 prototype Kaplan turbine.

Parameter [Unit]	Value
Head [m]	55.5
Power [MW]	10
Discharge $[\mathrm{m^3s^{-1}}]$	20
Rotational speed [rpm]	600

2.2. Experimental Setup

The operating points investigated in this paper were part of a measurement campaign performed on the Porjus U9 Kaplan turbine to study different load variations and steady-state operations. Twelve miniature piezoresistive pressure transducers (Kulite LL-080 series) were mounted on small capsules (3 mm in diameter and 5 mm in height) glued on the pressure side (from PS1 to PS6) and on the suction side (from SS1 to SS6) of one runner blade (Figure 2a). An epoxy layer was then added to the blade pressure and suction sides to create a smooth surface (Figure 2b). A similar epoxy layer was applied to the opposite blade to avoid the hydraulic and mass unbalance resulting from the weight of the applied epoxy on the instrumented blade. The pressure transducers were located at the intersection of imaginary circles that passed through 1/3 and 2/3 of the blade's span and 1/4, 1/2, and 3/4 of the blade's chord lines. The operating range of the pressure transducers was 0–700 kPa, and their natural frequency was 380 kHz. Two torsion strain gages (K-XY41-6/350-3-2M manufactured by HBM) were installed on the turbine shaft between the turbine guide bearing and the generator guide bearing to measure the shaft torsional strain. The resistance and gage factor of the torsional strain gages were $350 \pm 0.35\%$ and $2.08 \pm 1\%$, respectively. A detailed description of the data-acquisition system, additional sensors, and operating conditions used in this measurement campaign was provided by Soltani et al. [1].

The uncertainty analysis for the pressure transducers and strain gages was conducted using six nonconsecutive repeated measurements at the best efficiency point. The uncertainties of the pressure transducers and torsion strain gage installed on the shaft are presented in Soltani et al. [1].

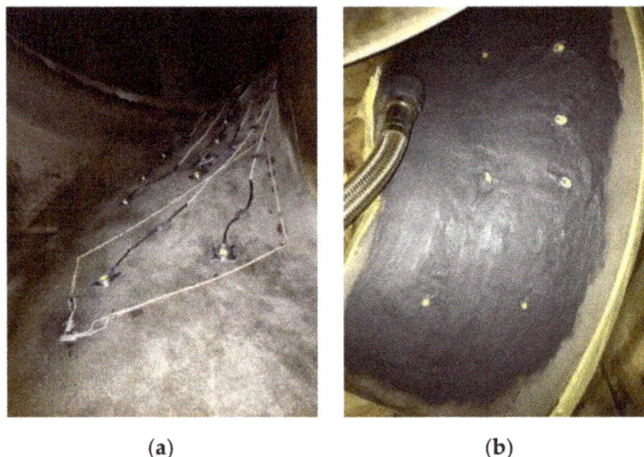

(a)   (b)

Figure 2. (**a**) Pressure transducers mounted on the runner blade; (**b**) epoxy layer added to the runner blade. The hydraulic hose was used to protect the cables from the blade to the runner cone.

2.3. Measuring Program

In this study, two experimentally investigated operating points, located on a propeller curve for a fixed blade angle $\beta = -4.2°$, were selected. The first operating point (OP1) is located on the left side of the propeller curve. It was characterized by a low discharge and the presence of a rotating vortex rope (RVR). The second operating point (OP2) was closer to the best efficiency point and presented a larger discharge. OP1 and OP2 are considered to be the part-load operating point and best efficiency point, respectively. The operating parameters are listed in Table 2.

Table 2. Operating condition parameters.

Operating Point	P_{out} (MW) *	Torque (kN·m)
OP1	4.9	77.9
OP2	6.9	109.8

* Values recorded in the power plant control room.

The experimental power values recorded during the measurement campaign presented in [1] were adjusted according to the values recorded in the power plant control room. Furthermore, the control room recordings were validated by index measurements.

3. Numerical Simulations Setup

Several numerical models were evaluated before performing the final simulations at OP1 and OP2. The object was to minimize the computational cost associated with the prototype dimensions and high Reynolds number and to address the uncertainty associated with the experimental data used as boundary conditions. This chapter follows the investigations performed in this study to reduce the model complexity instead of the classical sequence of geometry, mesh, boundary conditions and flow modelling. Therefore, the flow modelling is firstly presented, followed by a description of the main boundary conditions, different geometries used, and the mesh sensitivity analysis.

The different issues considered in these numerical models were the boundary conditions, mesh sensitivity, and numerical error and the influence of the runner blade clearance, epoxy layer added to two blades, and runner blade angle.

The influence of the boundary conditions was proven significant in previous studies [21–23] showing that specifying a constant total pressure at the inlet of the numerical domain improved the accuracy of the numerical results. To have a better estimation of the total pressure at the inlet of the investigated domain, a steady simulation was first carried out to evaluate the pressure losses in the upstream geometry comprising the spiral casing and penstock.

A mesh sensitivity analysis was performed to find a good compromise between the spatial discretization, the numerical error, and the computational resources. The same arguments led to the investigation of the runner blade clearance modelling, as the discretization of a very small volume considerably increases the mesh size and, consequently, the simulation time.

Another investigation focused on the importance of modelling the epoxy layer added to two runner blades during the experimental investigation. Re-building the geometry of the runner blade with an approximation of the epoxy layer is usually time-consuming, and the potential impact on the numerical results had to be assessed.

Finally, as the experimental uncertainty of the measured runner blade angle could not be quantified because the runner could not be scanned during the measurements, a sensitivity analysis was performed on this parameter as well.

Table 3 summarizes the different performed investigations. The following section presents and discusses the equations, boundary conditions, geometry and mesh used in the different setups.

Table 3. Numerical studies setup summary. The Rayleigh–Plesset cavitation model was employed in all unsteady simulations.

Study	Simulation Type	Numerical Domain	Domains Interfaces	Blade Clearances	Blade Epoxy	Blade Angle Adjustment [°]
Boundary condition	Steady	Penstock, Spiral Casing	Stage	-	-	-
Clearance	Steady	1 Stay vane, 1 Guide vane, 1 Runner blade, Draft tube	Stage	No / Yes (constant) / Yes (variable)	No	+0
Epoxy	Steady	1 Stay vane, 1 Guide vane, 1 Runner blade, Draft tube	Stage	No	No / Yes	+0
Runner blade angle	Steady	1 Stay vane, 1 Guide vane, 1 Runner blade, Draft tube	Stage	No	No	+0 / +2 / +4 / +6 / +8
	Unsteady	1 Stay vane, 1 Guide vane, 1 Runner blade, Draft tube	Stage	No	No	+0 / +6
	Unsteady	1 Stay vane, 1 Guide vane, 6 Runner blades, Draft tube	Stage, Transient Rotor–Stator	No	No	+0 / +6
Mesh sensitivity	Steady	1 Stay vane, 1 Guide vane, 1 Runner blade, Draft tube	Stage	No	No	+0 ($\beta = -4.2°$)

3.1. Flow Modelling

The numerical simulations presented in this paper were carried out using Ansys CFX Solver v16.2. The unsteady Reynolds-averaged Navier–Stokes equations were solved with an element-based finite volume method; therefore, the spatial domain was discretized to build finite volumes. The computational domain was modelled using a hexahedral mesh built in ICEM v16.2 for the guide vane and draft tube domains. The runner was discretized using TurboGrid v16.2, software optimized for the discretization of turbine blade passages. The software allows for the generation of good-quality hexahedral mesh for complex blade geometry.

Combining the advantages of the k-epsilon model in the free stream and k-omega model for the near-wall flow, the k-omega Shear Stress Transport (SST) turbulence model provides a good prediction of flow separation and has been proven to perform well under cavitating conditions [24–26]. The k-omega SST model was employed in all numerical simulations presented in this paper. The k-omega SST models the transport of the turbulent shear stress by solving two transport equations (Equations (1) and (2)): one for the turbulent kinetic energy (k) and one for the turbulent frequency (ω). The turbulent viscosity (ν_t) is modelled according to Equation (3).

$$\frac{\partial(\rho k)}{\partial t} + \frac{\partial}{\partial x_j}(\rho k V_j) = \frac{\partial}{\partial x_j}\left[\left(\mu + \frac{\mu_t}{\sigma_k}\right)\frac{\partial k}{\partial x_j}\right] + P_k + P_{kb} - \beta'\rho k\omega \tag{1}$$

$$\frac{\partial(\rho\omega)}{\partial t} + \frac{\partial}{\partial x_j}(\rho\omega V_j) = \frac{\partial}{\partial x_j}\left[\left(\mu + \frac{\mu_t}{\sigma_{\omega 1}}\right)\frac{\partial\omega}{\partial x_j}\right] + (1-F_1)2\rho\frac{1}{\sigma_{\omega 2}\omega}\frac{\partial k}{\partial x_j}\frac{\partial\omega}{\partial x_j} + \alpha\frac{\omega}{k}P_k - \beta\rho\omega^2 + P_{\omega b} \tag{2}$$

$$\nu_t = \frac{a_1\omega}{\max(a_1,\omega,SF_2)} \tag{3}$$

where P_k is the production rate of turbulence; P_{kb} and $P_{\omega b}$ represent the influence of the buoyancy forces; β, β', α, σ_k, $\sigma_{\omega 1}$ and $\sigma_{\omega 2}$ are coefficients of the turbulence model; and F_1 and F_2 are blending functions that switch from the k-epsilon model to k-omega [27].

The large domain size and high Reynolds number made the resolution of y^+ near the value of 1, impossible. However, the average y^+ values were kept under 25 because the automatic wall function used for the simulations provides reasonable results when compared to the experimental data [28]. In the unsteady simulations, the advection term was first discretized using the upwind advection scheme to achieve convergence faster. These results were provided as initial conditions to an unsteady simulation employing the high-resolution advection scheme. The high-resolution advection scheme automatically maintained the order of the discretization scheme as close to second order as possible to keep the solution robust and bounded. The second-order backward Euler transient scheme was used in all unsteady simulations. The transient time step was set to correspond to one degree of the runner rotation. The influence of the time step size was previously investigated [23], and other studies showed that time steps corresponding to a runner rotation of 0.5–5° is recommended [2]. Finally, the Rayleigh–Plesset cavitation model [24,29] was employed in consideration of the saturation pressure for a temperature of 12.5 °C, as recorded during the measurement campaign. The numerical modelling of the cavitation was introduced after obtaining large negative pressure values near the leading edge of the runner blade and was expected to improve the modelling of the flow [24].

All above-mentioned numerical simulations were validated against experimental data measured at the best efficiency point, OP2 (see Table 2).

3.2. Boundary Conditions

The constant total pressure was used as the inlet boundary condition of the computational domain based on previous studies [22,23]. The outlet was modelled as an opening with specified pressure and direction. The opening boundary condition allows for flow recirculation when the pressure is imposed. When the flow is leaving the domain, the pressure is considered as the relative static pressure as opposed to the total pressure that is considered when the flow enters the domain. Such boundary conditions allow for a shorter computational domain when the flow direction is not known.

To decrease the computational time, the domain of the turbine included a single stay vane-guide vane passage, the runner, and the draft tube. For a better approximation of the total pressure defined at the inlet of the stay vanes, a steady-state simulation of the flow inside the penstock and spiral casing was performed. A mass flow rate of 13 m^3/s was specified at the inlet of the penstock (P), and an opening with 0 Pa relative pressure was defined at the outlet of the spiral casing (SC). A pressure drop of 7874 Pa was obtained between the inlet and the outlet of the domain (Δp_{P-SC}), corresponding to an approximate head difference of 0.8 m.

Henceforward, the gross head of the turbine, H_{gross}, was adjusted in the simulations to consider the static pressure difference between the inlet and the outlet of the computational domain and the pressure losses in the penstock and spiral casing. The total head was calculated as:

$$H = H_{gross} - \frac{\Delta p_{P-SC}}{\rho g} - \left(z_{inlet} - z_{outlet}\right) \tag{4}$$

where z is the elevation.

The boundary conditions for the turbine simulation were defined according to the experimental measurements performed in [1]. The total pressure of 6.9 bar was fixed at the inlet of the computational domain, and the outlet pressure was set as 2.2 bar.

3.3. Geometry

The Porjus U9 Kaplan turbine (Figure 1) includes a penstock, spiral casing, 18 stay vanes, 20 guide vanes, a runner composed of 6 runner blades, and a draft tube. The geometry was obtained from the scaled model geometry. Several aspects had to be clarified before building the definitive numerical model, each concerning a different geometry setup.

To define the total pressure at the inlet of the stay vanes, a steady-state simulation of the flow inside the penstock and spiral casing was first carried out, and the pressure losses throughout this domain were obtained. The computational domain for this investigation is presented in Figure 1a. The large inlet pipe corresponds to the penstock of an older turbine that had a larger flow rate. The smaller pipe was installed with the Porjus U9 turbine.

The modelling of the runner blade clearance required a larger mesh size and, therefore, higher computational demands because of the very small scales involved. The influence of the runner blade clearance size on the numerical results was investigated. Table 4 presents the relative dimensions of the three clearance sizes considered: no clearance, constant clearance near the hub and shroud, and variable clearance from the leading edge (LE) towards the trailing edge (TE).

Table 4. Simulated clearance sizes representing percentages of the runner diameter.

	Mesh Size [10^6 Element]	Hub Clearance [%D]		Shroud Clearance [%D]	
		LE	TE	LE	TE
Geometry *	-	0.02%	0.55%	0.09%	0.004%
No clearance (NoClr)	1.12	no clearance			
Constant clearance (CtClr)	3.25	0.08%		0.05%	
Variable clearance (VarClr)	3.96	0.05% **	0.55%	0.09%	0.05% **

* The geometry was obtained from the scaled model geometry; ** the smallest modelled clearance was 0.05% because the mesh quality criteria could not be satisfied for lower values.

A steady numerical simulation was performed to verify whether the layers of epoxy added for the blade instrumentation had a significant impact on the numerical results. The blade geometry was modified to approximate the real blade covered by the epoxy layer (Figure 3). Considering the thickness of the epoxy layer, the blade thickness in the numerical model was accordingly increased by 5 mm with a decreasing thickness toward the edges of the blade. The estimated average width of the inter-blade channel was 200 mm at $\beta = -4.2°$. Therefore, the added epoxy narrowed the flow channel by a maximum of 2.5% on each side of the blades.

(a)   (b)

Figure 3. (**a**) Runner blade original geometry; (**b**) geometrical representation of the epoxy layer added for blade instrumentation during the measurements.

As the blade angle (Figure 4) used during the measurements contained a certain amount of uncertainty, a sensitivity analysis was performed to assess its influence on the numerical results. The obtained results were compared to the measurements.

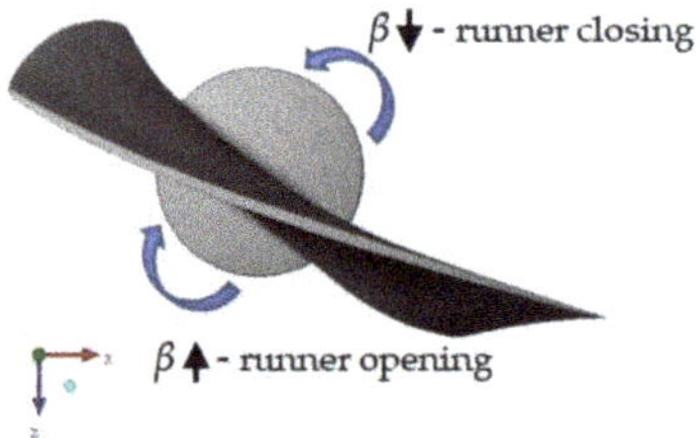

Figure 4. Runner blade angle variation.

The results of the steady simulations carried out for the different blade angle values provided the initial conditions for the unsteady simulations performed for the two extreme runner blade angles, i.e., the unaltered experimental blade angle ($\beta = -4.2°$) and $+6°$; see Table 5. In order to accelerate the convergence of the numerical solutions, the upwind scheme was first used for the discretization of the advection term before running the unsteady simulations with the high-resolution advection scheme (CFX Pre Modelling Guide).

Table 5. Simulated runner blade angles. Experimental value $\beta = -4.2°$.

		Single Runner Blade Passage			Full Runner
		Steady		Unsteady	
$\Delta\beta$ [°]	Advection Scheme	High Resolution	Upwind	High Resolution	High Resolution
0		yes	yes	yes	yes
+2		yes	no	no	no
+4		yes	no	no	no
+6		yes	yes	yes	yes

Two configurations of the computational domain were used in the numerical studies, one using a single runner blade passage and the second using the entire turbine runner with all six blade passages. The three domains (Figure 5)—one stay vane and guide vane passage, one or six runner blade passages, and the complete draft tube—were connected using stage interfaces, except for the case including the full runner domain where a transient rotor–stator interface was used.

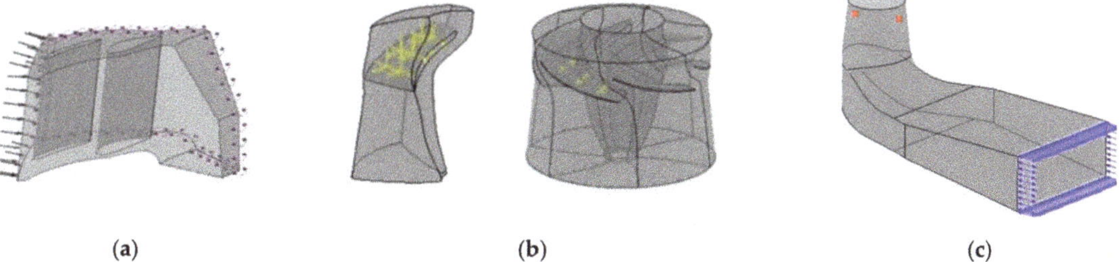

(a)   (b)   (c)

Figure 5. (**a**) Stay vane and guide vane domain; (**b**) runner domain, single blade passage (left), complete runner (right); (**c**) draft tube domain.

Furthermore, the entire runner domain was modelled, thus allowing for the use of a transient rotor–stator interface between the runner and draft tube domains. This type of interface takes the transient behavior of the flow between domains into account.

Twelve pressure monitor points, corresponding to the locations of the pressure sensors mounted on the blade during the experimental measurements [1], were defined on the runner blade: from PS1 to PS6 on the pressure side and from PS1 to PS6 on the suction side (Figure 5b). Additionally, two diametrically opposed monitor points were defined in the draft tube, DT_{in} and DT_{out} (Figure 5c), to capture the pressure fluctuations due to the formation of the RVR at the part-load operating point (OP1 in Table 2) in the numerical studies.

3.4. Mesh Sensitivity Analysis

Figures 6–8 present the meshes created for each computational domain. The hexahedral meshes were built in ICEM v16.2 for all the domains, except for the runner blade passage domain generated with TurboGrid v16.2. The stay vane and guide vane meshes are presented in Figure 6, along with a detail of the boundary layer mesh around the trailing edge of the stay vane and the leading edge of the guide vane.

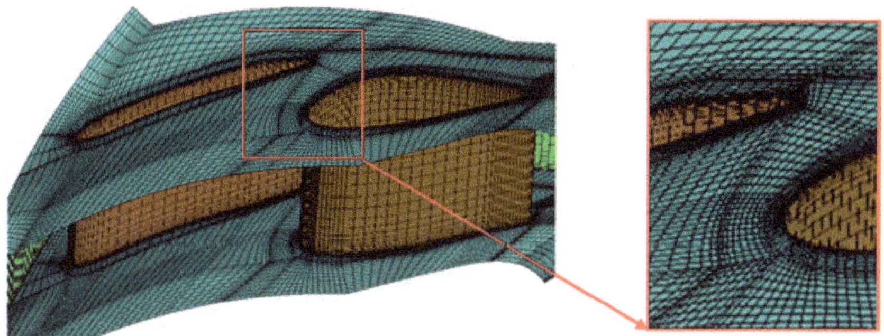

Figure 6. Stay vane and guide vane mesh (**left**) and zoom at the trailing edge of the stay vane and the leading edge of the guide vane (**right**).

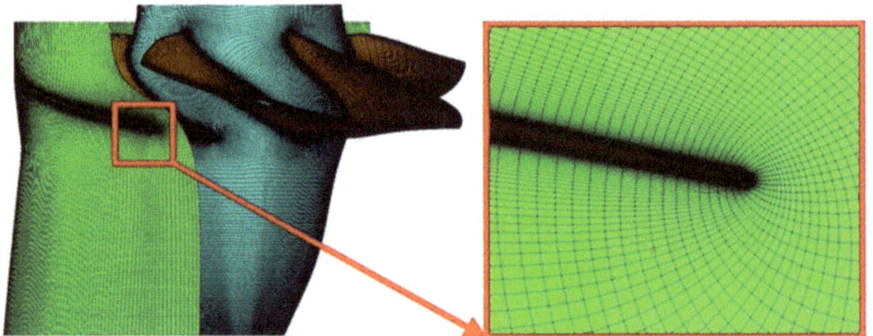

Figure 7. Runner mesh (**left**) and zoom at the trailing edge of the stay vane and the leading edge of the guide vane (**right**).

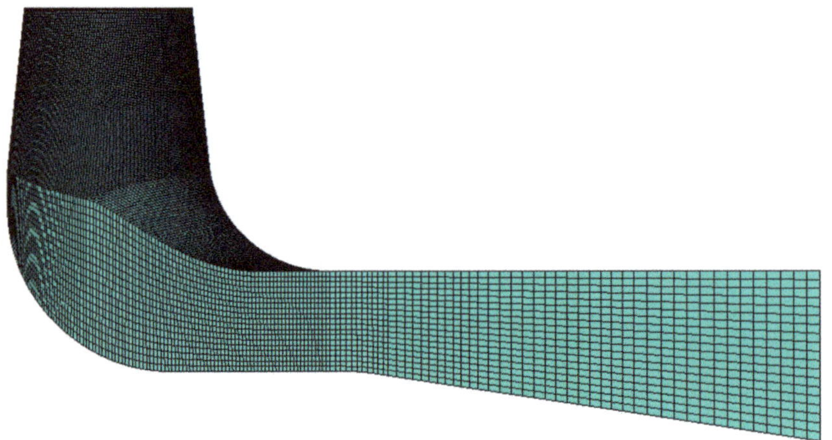

Figure 8. Draft tube mesh.

Similarly, the discretization of the runner domain and the boundary layer mesh are presented in Figure 7. The runner (Figure 7) and draft tube mesh (Figure 8) underwent mesh sensitivity analyses. The meshes were dimensioned according to the conclusions presented further in this section.

A synthesis of the mesh quality parameters evaluated in Ansys CFX Solver for each computational domain mentioned in the previous section is presented in Table 6. The mesh orthogonality angle represents the minimum angle between two adjacent element faces. The mesh expansion factor is the ratio of the smallest element volume to the largest element volume sharing a node. The mesh aspect ratio is the maximum ratio of the largest to the smallest face area of all the elements sharing a node.

Table 6. Mesh size and quality parameters of the different meshes used in the simulations.

Domain	Element Type	Size [$\times 10^6$]	Min. Orthogonality Angle	Max. Expansion Factor	Aspect Ratio
Penstock and spiral casing	hexa	3	20	22	2124
Stay vane and guide vane	hexa	0.34	18.8	35	629
Runner	hexa	1.16	48.7	10	3923
Draft tube	hexa	3.18	30.5	9	7393

Steady-state simulations were performed to estimate the numerical error and evaluate the sensitivity of the numerical model to the discretization of the runner and draft tube domains. The investigated model consisted of a single guide vane passage including a stay vane, one runner blade passage and the draft tube. The guide vane was considered to have an insignificant influence on the investigated flow; therefore, a mesh sensitivity analysis was not performed on this domain, as previously employed in a study of the U9 Kaplan model [23].

In the current study, the runner blade angle was modified to fit the prototype experimental value, and the discretization was entirely redone. No clearances were modelled for the mesh sensitivity analysis. Four levels of mesh refinement were investigated, and these are henceforth referred to as R1–R4. The main size and quality parameters of the runner mesh are presented in Table 7. The global size factor defines the overall mesh size and controls the resolution of the mesh in the entire domain, and the edge refinement factor adjusts the resolution of the mesh in the boundary layer.

Table 7. Runner mesh size and quality parameters.

Simulation	R1	R2	R3	R4
Mesh size [$\times 10^6$] *	0.16	0.43	1.12	2.36
Global size factor	1.5	2	2.5	3
Edge refinement factor	2.5	3	4	5
Min. orthogonality angle	48.7	48.9	48.7	43.9
Max. expansion factor	13	19	10	13
Aspect ratio	727	1255	3923	11,677
y^+ (avg/max)	97/2415	51/2234	25/1970	14/1484

* Hexahedral elements.

The torque value obtained with an 'infinite size' mesh (T_∞) was determined using the Richardson extrapolation. Figure 9 presents the numerical values obtained from the simulations with different levels of mesh refinement of the runner domain. The torque was normalized with the 'infinite size' mesh value. The deviation in numerical pressure values was expected to be less than 4% from coarse to fine discretization [14]. Considering that there was no significant difference between the torque results obtained with the two most refined levels of discretization, all the simulations presented following this section employed runner mesh R3, with 1.12×10^6 elements per blade passage.

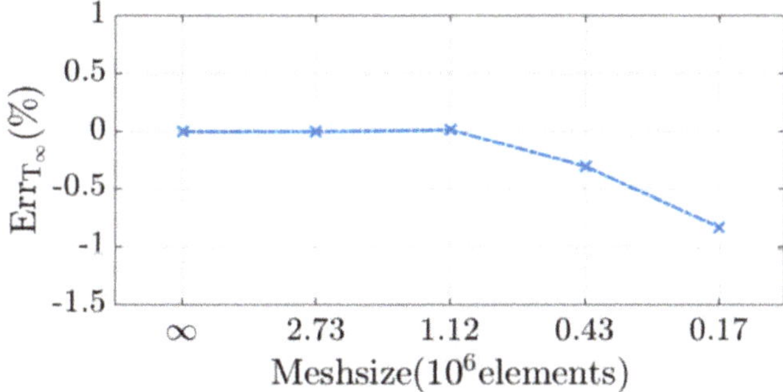

Figure 9. Variation of the normalized torque with the runner mesh size.

The discretization level of the draft tube domain could influence the numerical results obtained in the runner domain, therefore justifying a mesh sensitivity study and the need to estimate the numerical error induced by the draft tube mesh. Three levels of mesh refinement were investigated, and there are henceforth referred to as DT1–DT3. The draft tube's primary size and quality parameters are presented in Table 8.

Table 8. Draft tube mesh size and quality parameters.

Simulation	DT1	DT2	DT3
Mesh size [$\times 10^6$] *	1.01	1.89	3.18
Min. orthogonality angle	30.8	30.6	30.5
Max. expansion factor	11	10	9
Aspect ratio	10,346	8624	7393

* Hexahedral elements.

Similar to the runner mesh sensitivity study, the torque value obtained with an infinite mesh (T_∞) was determined using the Richardson extrapolation. Figure 10 presents the numerical values obtained from the simulations with different levels of mesh refinement of the draft tube domain. The torque was normalized with the infinite mesh value. All the simulations presented after this section employed the draft tube mesh DT3, with 3.18×10^6 elements per blade passage. A 5% discretization error was assumed to be reasonable considering the size of the draft tube mesh.

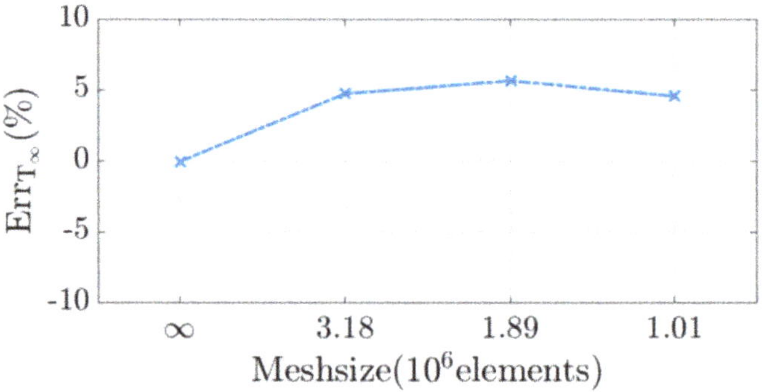

Figure 10. Variation of the normalized torque with the draft tube mesh size.

4. Results and Discussion

4.1. Runner Clearances

Using the selected runner mesh, R3, steady-state simulations were first performed to investigate the influence of the runner blade clearance.

The values were made dimensionless with the result of the 'no clearance' simulation. Figure 11 presents the variation of the torque obtained from the numerical simulations with different types of clearances (Table 4). The 'constant clearance' simulation underestimated the torque compared to the other two cases because the 'constant clearance' values were the largest, thus allowing for a larger flow rate. The 'variable clearance' simulation provided very similar results to the 'no clearance' simulation because the real and/or approximated dimensions of the variable clearances were very small and did not impact the torque results. Modelling the variable clearance did not have enough of a significant impact on the numerical results to justify the added computation time.

A parameter found to be significantly influenced by modelling the runner blade clearances was the pressure monitored on the suction side of the blade (SS3), located closer to the shroud and near the trailing edge. The numerical and experimental pressure values were scaled with the reference pressure measured by Soltani et al. [1] with an empty turbine chamber. The results were consistent to the values presented in Table 4. The shroud clearance at the trailing edge was very small and could not be modelled, so the smallest modelled clearance was 0.05% D. This clearance size was used in both the 'constant clearance' and the' variable clearance' simulation, thus explaining the similar pressure values obtained from the two simulations at SS3 (Figure 12).

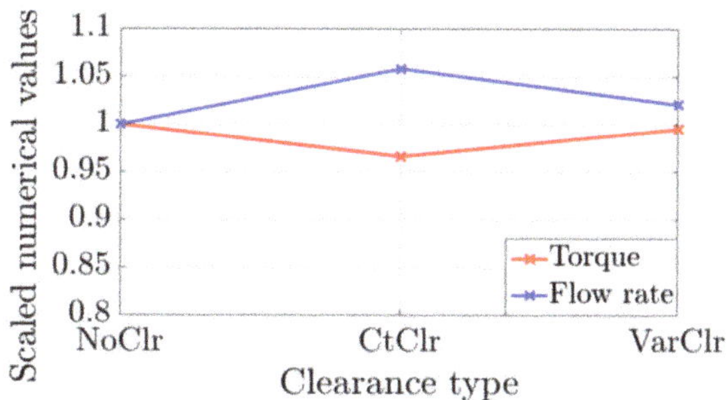

Figure 11. Variation of the scaled torque and flow rate with the runner blade clearance size.

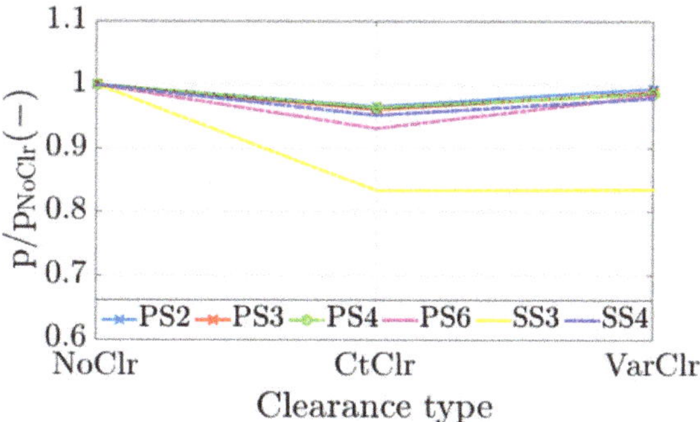

Figure 12. Variation of the scaled pressure with the runner blade clearance size.

Considering the mesh size, the computational resources required to model the small variable clearances, and the influence of the runner blade clearance on the numerical results, the runner blade clearance was not modelled in further simulations.

4.2. Epoxy Study

An epoxy layer was added to one runner blade to hold the pressure sensors and to a second (opposite) runner blade to balance the runner. The influence of the thickness of the epoxy layer was analyzed. The adjustment of the geometry did not have a significant impact on the numerical results. Regarding the pressure values provided by the two simulations, the change in the geometry was noticeable at the SS3 and SS6 monitor points located on the suction side closest to the trailing edge (Figure 13). The torque decreased by 1.8% in the simulation with the modified blade geometry (Figure 14), and the flow rate decreased by 2.5%. The numerical simulations were continued with the original blade geometry without considering the epoxy layer.

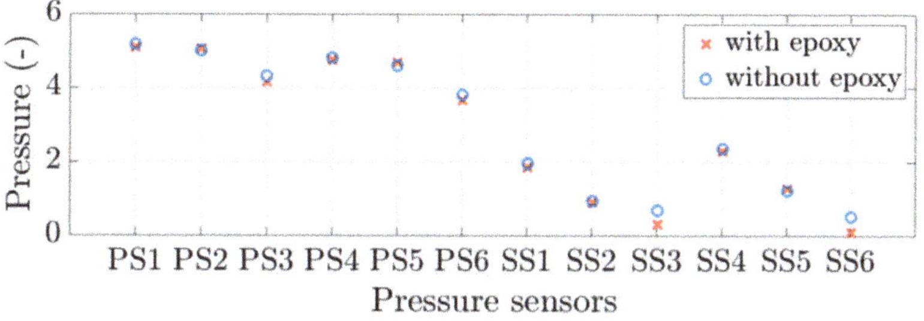

Figure 13. Numerical pressure on the blade with and without epoxy.

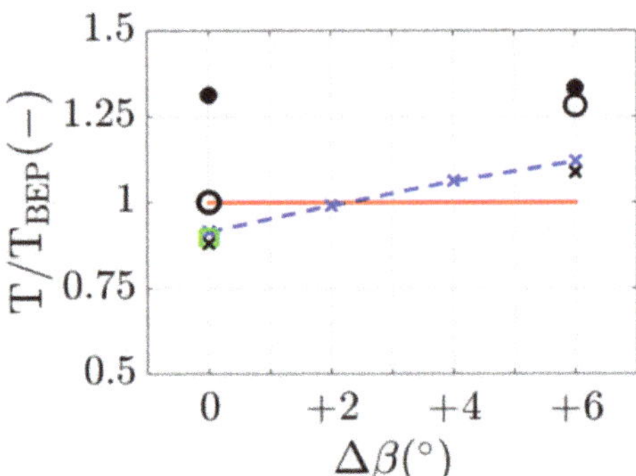

Figure 14. Numerical scaled mean torque values at different runner blade positions and the experimental value. Results of the simulations at OP2. — experimental results; -x- steady-state simulations results; □ added epoxy simulation result; x unsteady upwind simulation results; • unsteady high-resolution simulation results; and o unsteady high-resolution simulation results (complete runner domain).

4.3. Runner Blade Angle

The uncertainty of the experimental measurement on the runner blade angle could not be assessed. Therefore, a sensitivity analysis was performed to quantify the influence of the angle on the accuracy of the numerical simulations.

Figure 14 presents the variation of the non-dimensionalized torque values as a function of the runner blade angle increment. The steady simulations showed a linear increase of the torque as the runner was opened. The blade angle increased (see Figure 4), allowing for a larger lift and thus torque. The results of the unsteady simulations that included a single runner blade passage and employed the upwind advection scheme were similar to the results provided by the steady-state simulations. However, when the high-resolution advection scheme was used, the simulations showed a poor convergence of the monitored variables and the numerical torque values were strongly overestimated. The reason for the poor convergence is unclear but may be attributed to the periodic interfaces defined between the runner blade passages not accurately modelling the influence of the runner blade wakes in the turbulent flow. Including the complete runner geometry in the simulations lead to an improvement of the results of the simulation employing the experimental runner blade angle (Figure 14), and the numerical torque value matched surprisingly well the experimental value. However, at the larger runner blade angle, 6° more than the experimental value, the convergence of the monitored variables was impossible to achieve. The use of a very large blade angle while keeping the same total pressure at the inlet led to incoherent flow structures, large recirculation areas, and a negative flow rate.

Figure 15 presents the non-dimensionalized mean absolute pressure variation function of the runner blade angle. The pressure was monitored on both the pressure (from PS1 to PS6) and the suction side (from PS1 to PS6) of the same runner blade. The numerical and experimental pressure values were scaled with the reference pressure that was measured by Soltani et al. [1]. As the runner blade angle increased, i.e., the runner was opened, the absolute pressure monitored on the blade decreased. At the largest blade angle values, +4° and +6°, the pressure was strongly underestimated near the leading edge (PS4) and in the middle section (PS2) of the blade pressure side. The differences between the numerical results and the experimental data were reduced towards the trailing edge (PS3 and PS6). On the other hand, when using the experimental runner blade angle ($\Delta\beta = 0°$), the numerical pressure values were in better agreement with the experimental values. The pressure was underestimated in the center area of the pressure side and near the leading edge (PS2 and PS4) but reasonably predicted near the trailing edge (PS3 and PS6). On the suction side of the blade, at the middle section, and near the trailing edge, the absolute pressure was overpredicted as the numerical simulation underestimated the hydraulic energy transferred to the runner.

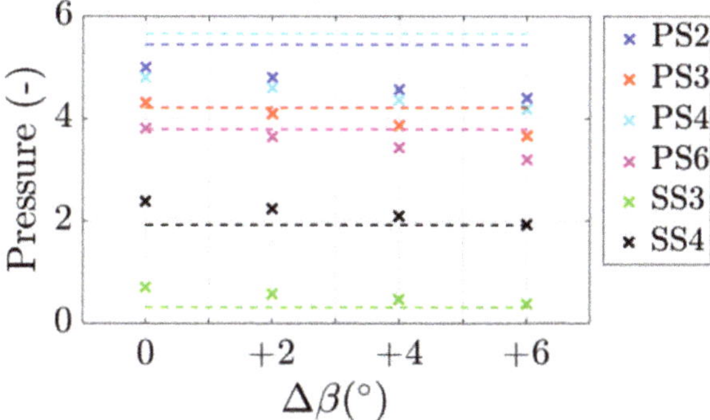

Figure 15. Numerical scaled mean absolute pressure values at different runner blade positions for the steady-state simulations at OP2. The dashed lines represent the experimental values measured at $\beta = -4.2°$ ($\Delta\beta = 0°$).

The experimentally determined runner blade angle, $-4.2°$, was employed in the definitive studies presented in the paper because the pressure measurements and the torque measurements best matched the experimental results.

4.4. Best Efficiency Point (OP2)

The numerical simulations presented in the following sections were unsteady simulations that included a single stay vane-guide vane passage, the complete runner domain and the draft tube. The cavitation model was employed. The runner blade angle was set at the experimental value, $-4.2°$, no clearances were considered, and the epoxy layer was not modelled.

4.4.1. Mean Values

The numerical mean torque value obtained was 109.7 kN·m, only 0.1% smaller than the experimental mean value of 109.8 kN·m. This result was presented in the previous section with the sensitivity analysis of the runner blade angle.

Figure 16 presents the experimental mean absolute pressure data measured on the runner blade and the pressure values obtained from the unsteady numerical simulation. The results were non-dimensionalized with the pressure value experimentally measured in the empty turbine chamber. The pressure variation was similar to the results provided by the steady-state simulations (Figure 15).

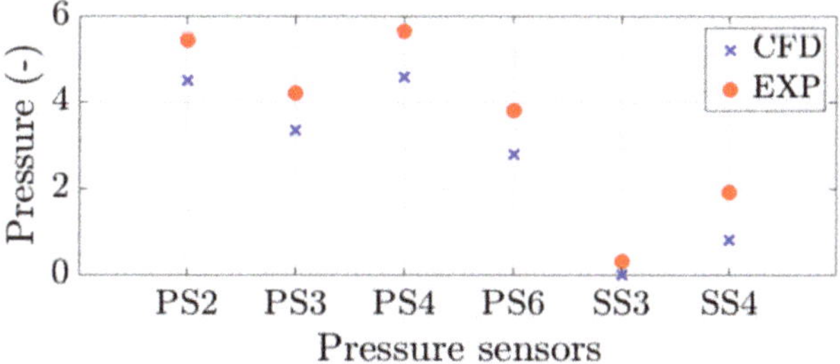

Figure 16. Scaled mean absolute pressure values at OP2.

On the pressure side of the blade, the pressure was underestimated near the leading edge (PS4) and in the middle area (PS2). Towards the trailing edge, the numerical results showed a better match to the experimental data (PS3 and PS6), underestimating the pressure by less than 6%. The simulation overpredicted the pressure on the suction side near the trailing edge (SS3). Figures 17 and 18 present iso-surfaces of the 20% vapor volume fraction captured at the end of the unsteady simulation after approximately three runner rotations. Towards the periphery of the runner blade suction side, the flow was accelerated and cavitation developed as pressure decreases. The flow was not axisymmetric, leading to larger cavitation areas near some blades and not others; see Figure 18.

Figure 17. Iso-surface of 20% vapor volume fraction at OP2 in the runner domain.

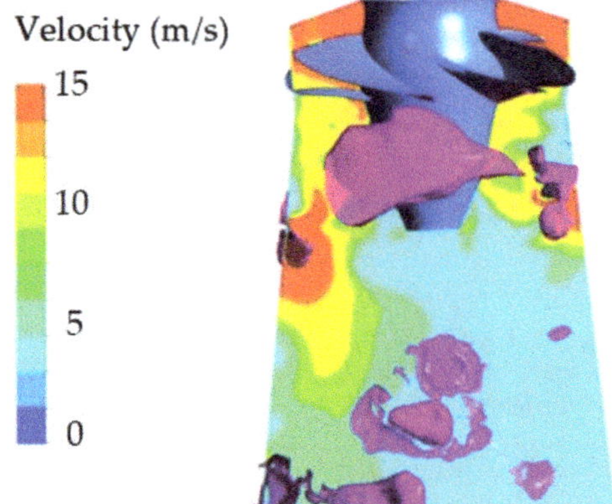

Figure 18. Liquid water velocity in stationary frame and iso-surface of 20% vapor volume fraction at OP2 in the runner domain.

Figure 19 presents the wall shear stress variation on the suction side of the runner blades at OP2. The asymmetric flow was once again confirmed, leading to non-uniform flow separation on the suction side of the runner blades. The OP2 operating point was close to but not exactly the best efficiency point; therefore, it is possible that a vortex was beginning to form in the draft tube cone.

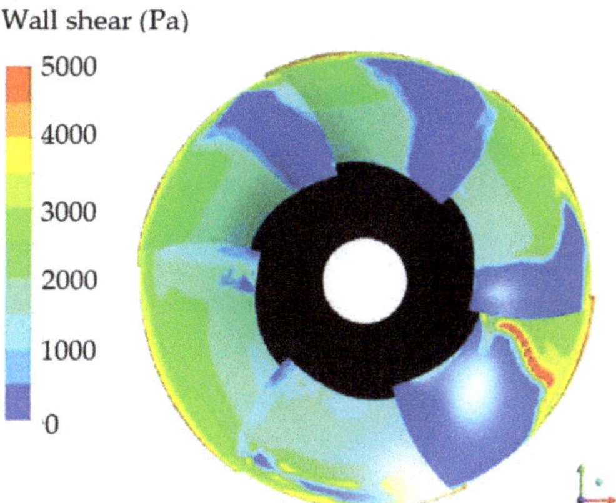

Figure 19. Wall shear stress on the suction side of the runner blades at OP2.

4.4.2. Amplitude Spectrum Analysis

A fast Fourier transform (FFT) analysis was carried out to identify the frequency content of the numerical torque and pressure signals and to compare them to the experimental results. A single smooth area of the signal corresponding to one runner rotation was selected and added to itself to obtain a better resolution of the FFT. The torque was scaled by the mean torque value recorded at OP2 during the experimental measurements. The pressure was scaled by the reference pressure that was experimentally measured [1]. The mean values were subtracted from the numerical and the experimental signals. The lengths of the original signal and the analyzed signal were 0.1 and 110 s, respectively. The frequencies were non-dimensionalized (f^*) using the runner frequency (f_r) of 10 Hz.

Figure 20 presents the amplitude spectrum of the scaled torque for the numerical and experimental data. A factor of 5 was obtained between the amplitudes. The dimensionless frequency of 1, visible in the experimental data and corresponding to the runner frequency, was considered to be caused by the wake of the spiral casing lip junction [30]. In the numerical simulation, the spiral casing was not modelled but the same frequency was captured due to the multiplication of the signal corresponding to a single runner rotation. A second frequency peak at $f^* = 2$ was captured in the experimental signal, though with a larger amplitude than the amplitude of the runner frequency, because two of the six runner blades were covered in epoxy. However, the peak at $f^* = 2$ was visible in the simulations despite the epoxy layers not being modelled due to the selection of the signal to be analyzed (one runner rotation added to itself). The frequency peak at $f^* = 3.16$ found in the experimental signals could represent a shaft torsional natural frequency. This may explain why the frequency was not present in the pressure measurements or the numerical torque.

The dimensionless frequency of 2 was also visible in the amplitude spectrum of the experimental pressure signal (Figure 21). An additional peak was captured in the experiment at $f^* = 20$, corresponding to the number of guide vanes, and not in the numerical results because only one guide vane-stay vane passage was modelled (Figure 5) with a stage interface.

The numerical simulation could not capture all the experimental fluctuations, mainly because of the axial symmetry of the computational domain that did not include the spiral casing lip junction and the epoxy on the blade. However, the runner frequency and its harmonics were visible in the amplitude spectrum of the numerical torque and pressure, most likely because of the selection of a single runner rotation from the signal.

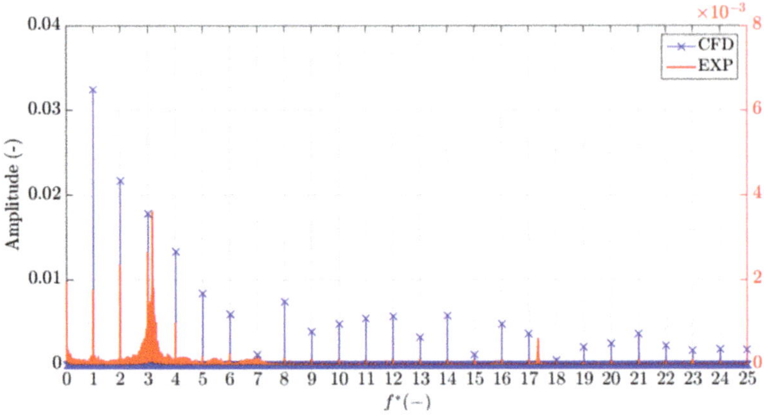

Figure 20. Amplitude spectrum of the scaled torque at OP2. Two vertical axis are used, one for the experiments and one for the numerical results.

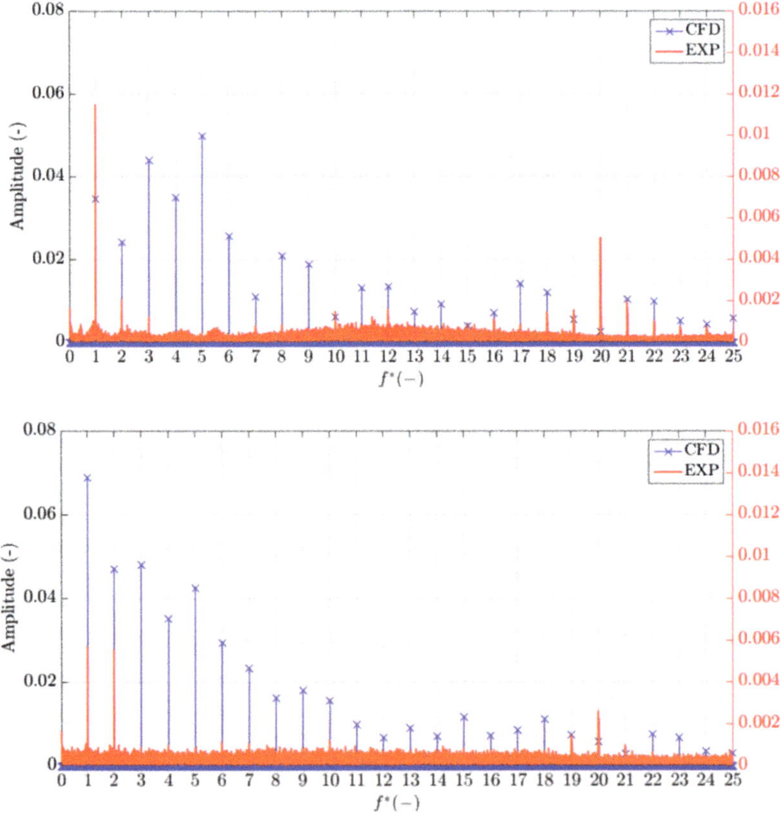

Figure 21. Amplitude spectrum of the scaled pressure PS2 (**top**) and SS4 (**bottom**) at OP2. Two vertical axis are used, one for the experiments and one for the numerical results.

4.5. Part-Load Operation (OP1)

4.5.1. Mean Values

The numerical mean torque value obtained at OP1 was 71.5 kN·m. The numerical value was 8.3% smaller than the experimental mean value of 77.9 kN·m.

The dimensionless mean absolute pressure values obtained from the part-load (OP1) unsteady numerical simulation were compared to the experimental values shown in Figure 22. The numerical results followed the trend of the measured values and presented a good prediction of the pressure on the pressure side of the runner blade, with from 2.9 to 6.1% discrepancy compared with the experiment. The pressure was overestimated on the suction side along the blade span (SS4 and SS3) because the numerical simulation underestimated the fluid angular momentum transformed into torque.

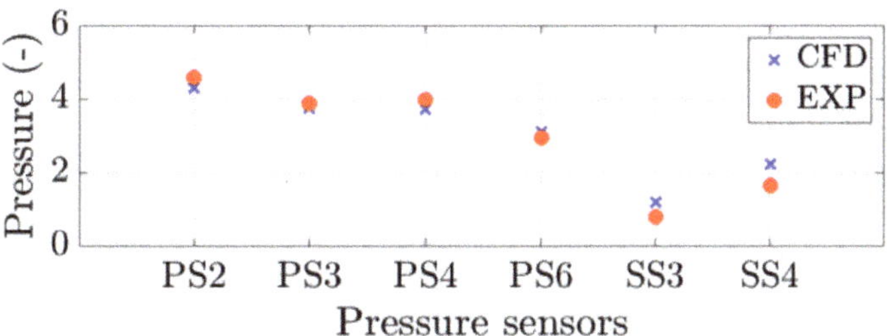

Figure 22. Scaled mean absolute pressure values at OP1.

4.5.2. Amplitude Spectrum Analysis

At the part-load operation (OP1), the flow provided by the guide vanes has a high swirl. As a consequence, the tangential component of the velocity at the outlet of the Kaplan runner is larger, while the axial velocity is reduced [31]. The flow is pulled towards the draft tube walls, creating a low-pressure region below the runner cone in the center of the draft tube and leading to the formation of the RVR [32]. One complete RVR rotation corresponded to about six runner rotations in the present case. The dimensionless frequency of the RVR was found to be 0.17.

The pressure distribution and the velocity vectors in the runner and draft tube domains are presented in Figure 23. Low-pressure regions are visible below the runner cone (Figure 23a), and the corresponding recirculation zones were captured as the flow was concentrated near the draft tube walls (Figure 23b).

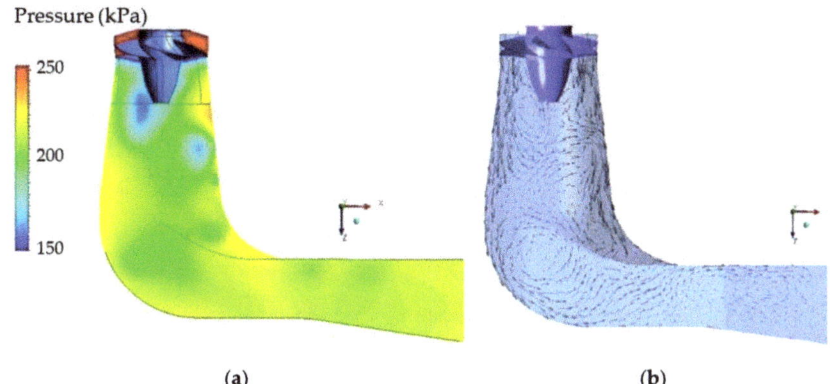

Figure 23. (**a**) Mid-section pressure contour at OP1. (**b**) Velocity vectors.

Figure 24 presents iso-surfaces of the pressure in the draft tube. A double-helix vortex rope (left) was initially formed 3 s before the end of the numerical simulation. After 1.5 s, or 15 runner rotations, a single RVR was captured (middle) and remained so until the last time step (right).

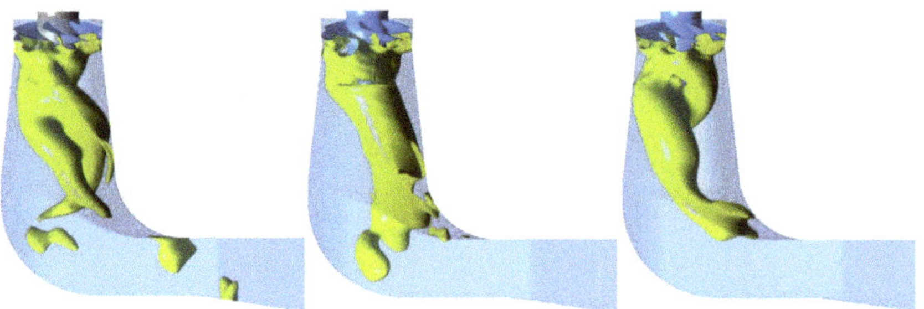

Figure 24. Iso-surface at a constant pressure of 2 bar at OP1. Evolution of the RVR at 1.5 s intervals.

In the stationary reference frame of the draft tube, the RVR can be decomposed into two components [33] from two pressure measurements obtained at the same height and opposite to each other: the plunging and rotating modes. The plunging mode is associated with synchronous pressure pulsations recorded throughout the turbine. The rotating mode leads to pressure oscillations with the same frequency as the plunging mode.

Figure 25 presents the frequency spectrum of the numerical and experimental torque signals. The numerical data analyzed in the following section were obtained from the last 30 simulated runner rotations (10,800 time steps after convergence, or 3 s), i.e., 6 RVR rotations. The numerical simulation captured a frequency peak at $f^* = 0.17$ matching the RVR dimensionless experimental frequency value of 0.17. Similar values were obtained during the measurement campaign of the Porjus U9 model and numerical studies investigating the Kaplan turbine model [10,23]. As the experimental and numerical data were obtained in the rotating domain, the amplitude at this frequency represents the plunging component. The RVR frequency is expected to show a slight variation at different part-load operating regimes [1]. The dimensionless frequency value of 0.83 corresponded to the rotating component of the RVR. It is visible in the experimental data because the blades were not identical and it was not captured by the numerical simulation.

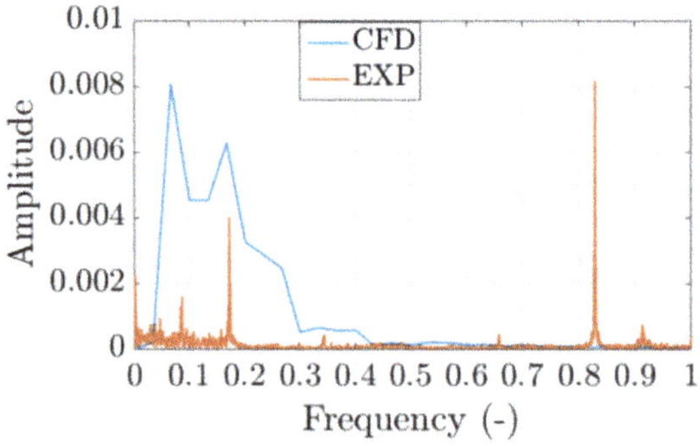

Figure 25. Amplitude spectrum of the scaled torque at OP1.

On the pressure side of the runner blade, at PS4 near the leading edge and PS3 near the trailing edge, the plunging mode of the RVR represented by the peak at $f^* = 0.17$ in the experimental data could also be found in the numerical results at the dimensionless frequency of 0.17 (Figure 26), similar to the torque FFT analysis. The amplitude was underestimated by the numerical model, especially closer to the leading edge (PS4). The rotating component of the RVR, 0.83 dimensionless frequency in the experiments, was captured in the numerical simulation at both locations of the pressure monitor points.

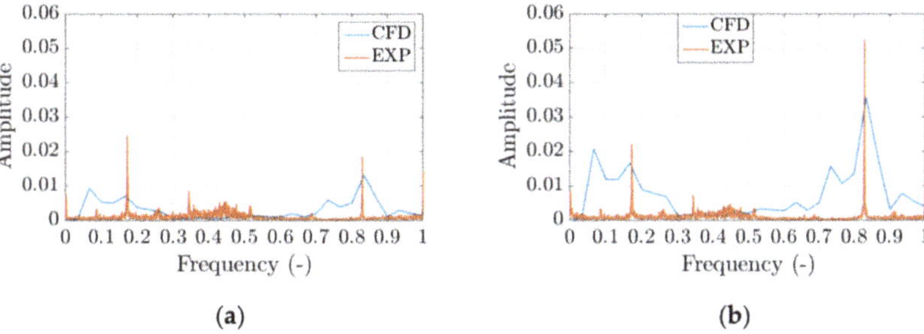

Figure 26. Amplitude spectrum of the scaled pressure at OP1: (**a**) PS4 and (**b**) PS3.

Despite the overestimation of the mean absolute pressure values in the numerical studies on the suction side of the blade (Figure 22), the amplitude of the pressure fluctuations was in a better agreement with the experimental data (Figure 27). The numerical dimensionless frequencies of the plunging and rotating components of the RVR on the blade suction side were coherent with the values obtained in the experiment.

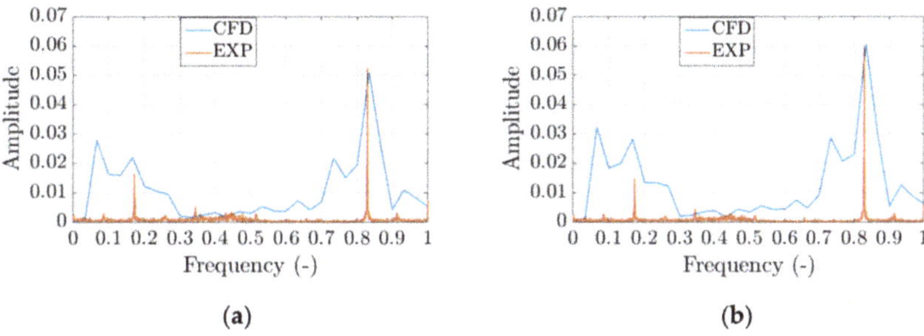

Figure 27. Amplitude spectrum of the scaled pressure at OP1: (**a**) SS4 and (**b**) SS3.

4.6. RVR Decomposition

Two pressure monitor points diametrically opposed (Figure 5), DT_{in} on the inner radius of the draft tube and DT_{out} on the outer radius of the draft tube defined just below the runner cone in the numerical model, were selected for the decomposition of the RVR [33]. The amplitude of the plunging (P_{pl}) and rotating (P_{rot}) modes was calculated as:

$$P_{pl} = \frac{(DT_{in} + DT_{out})}{2} \tag{5}$$

$$P_{rot} = \frac{(DT_{in} - DT_{out})}{2} \tag{6}$$

The two peaks were identified at the dimensionless frequency of 0.17, which was in good agreement with the FFT analysis of the torque and pressure signals (Figure 28). The amplitude of the pressure fluctuations due to the rotating component of the RVR was 30% smaller than the amplitude of the plunging mode. In the frequency spectrum of the plunging component, there were two additional frequency peaks at 0.1 and 0.23 m while in the frequency spectrum of the rotating component there are two peaks at 0.23 and 0.4 dimensionless frequencies. A possible explanation for these frequency peaks could be spectral leaking [34]. As the first and last values of the signal chosen for the analysis were not identical, high frequencies that were aliased between 0 and half of the sampling frequency may have "leaked" their energy into the other frequencies.

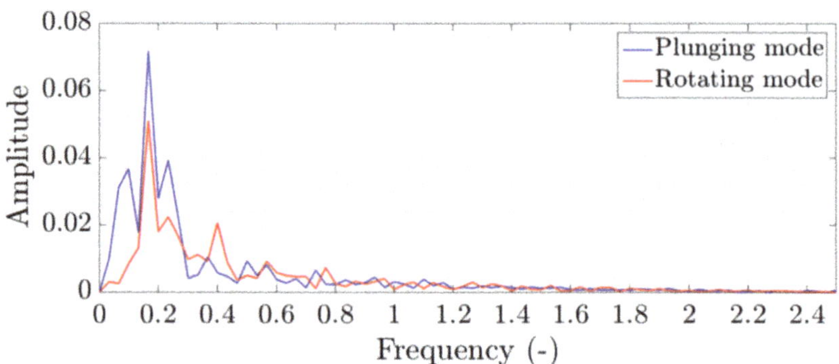

Figure 28. RVR decomposition: plunging and rotating mode.

5. Conclusions

A numerical study of the Porjus U9 Kaplan prototype was carried out. Two operating points on a propeller curve, corresponding to the best efficiency point and part-load operating conditions, were modelled. The effects of the boundary conditions, runner blade clearance, blade geometry and mesh size on the numerical results were first investigated at the best efficiency point to minimize the computational efforts. Then, experimental pressure and torque measurements performed on the prototype were used to validate the numerical results.

Modelling the runner blade clearances and the epoxy layer added to the instrumented runner blade had no significant influence on the numerical results considering the increased mesh size and simulation time, so these parameters could be neglected.

The runner blade angle is a parameter that is challenging to measure on a turbine prototype. However, a parametric study showed that the runner blade angle strongly influences both the convergence and accuracy of numerical results.

The simplified model reasonably predicted the mean torque and pressure values at both operating points.

The steady simulations with a single runner passage and unsteady simulations at BEP provided similar results. Modelling the complete runner domain and including the cavitation model improved the convergence of the simulations and the mean torque prediction. However, the mean unsteady numerical pressure values were similar to the values obtained from the steady simulations. Therefore, steady single-passage simulations at BEP are an excellent first approximation.

At part-load operation, the numerical simulations captured the variation in the pressure due to the RVR observed in the experiment. The frequency of the pressure fluctuations was well-predicted both in the runner and the draft tube considering the simplified, axisymmetric computational domain. In the runner domain, the frequencies of the plunging and rotating components of the RVR were accurately determined by the simulation, except for the torque missing the RVR rotating component. The amplitudes of these pressure fluctuations were also reasonably predicted, especially on the suction side of the runner blade.

The spectral analysis of the numerical signals was strongly influenced by the convergence of the monitored numerical values and the signal selection to be analyzed. Due to the signal selection, the FFT analysis could have been altered by spectral leakage, leading to pressure peaks with no experimental correspondents.

Author Contributions: Conceptualization, M.J.C.; Formal analysis, M.J.C.; Investigation, R.G.I., A.S.D. and M.J.C.; Methodology, M.J.C.; Supervision, D.M.B. and M.J.C.; Writing—original draft, R.G.I.; Writing—review and editing, A.S.D., D.M.B. and M.J.C. All authors have read and agreed to the published version of the manuscript.

Funding: The European Union's Horizon 2020 research and innovation program under grant agreement No. 814958 partially financed the research.

Institutional Review Board Statement: Not applicable.

Informed Consent Statement: Not applicable.

Data Availability Statement: Not applicable.

Acknowledgments: The presented research was carried out as a part of the Swedish Hydropower Centre (SVC), which was established by the Swedish Energy Agency, Elforsk and Svenska Kraftnät, together with the Luleå University of Technology, The Royal Institute of Technology, Chalmers University of Technology, and Uppsala University.

Conflicts of Interest: The authors declare no conflict of interest.

Abbreviations

CtClr	constant clearance
DT_{in}	numerical pressure recorded on the draft tube wall (inner radius)
DT_{out}	numerical pressure recorded on the draft tube wall (outer radius)
f^*	dimensionless frequency
FFT	Fast Fourier transform
f_r	runner frequency
H_{gross}	gross head of the turbine
k	kinetic energy
LE	leading edge
NoClr	no clearance
OP1 and OP2	operating points
P_{out}	power output
P_{pl}	plunging mode pressure
P_{rot}	rotating mode pressure
RVR	rotating vortex rope
SST	Shear Stress Transport
TE	trailing edge
VarClr	variable clearance
y^+	dimensionless distance from the wall
z	elevation
β	runner blade angle
Δp_{P-SC}	pressure drop obtained between the inlet and the outlet of the numerical domain
ν_t	turbulent viscosity
ω	turbulent frequency

References

1. Soltani Dehkharqani, A.; Engstrom, F.; Aidanpää, J.O.; Cervantes, M.J. Experimental investigation of a 10 MW prototype axial turbine runner: Vortex rope formation and mitigation. *J. Fluids Eng.* **2020**, *142*, 101212. [CrossRef]
2. Trivedi, C.H.; Gandhi, B.; Cervantes, M.J. Effect of transients on Francis turbine runner life: A review. *J. Hydraul. Res.* **2013**, *51*, 121–132. [CrossRef]
3. Farhat, M.; Natal, S.; Avellan, F.; Paquet, F.; Couston, M. Onboard measurements of pressure and strain fluctuations in a model of low head Francis turbine—Part 1: Instrumentation. In Proceedings of the 21st IAHR Symposium on Hydraulic Machinery and Systems, Lausanne, Switzerland, 9–12 September 2002.
4. Lowys, P.; Paquet, F.; Couston, M.; Farhat, M.; Natal, S.; Avellan, F. Onboard measurements of pressure and strain fluctuations in a model of low head Francis turbine—Part 2: Measurements and preliminary analysis results. In Proceedings of the 21st IAHR Symposium on Hydraulic Machinery and Systems, Lausanne, Switzerland, 9–12 September 2002.
5. Cervantes, M.J.; Trivedi, C.H.; Dahlhaug, O.G.; Nielsen, T.K. Francis-99 workshop 1: Steady operation of Francis turbines. *J. Phys. Conf. Ser.* **2015**, *579*, 011001. [CrossRef]

6. Trivedi, C.H.; Dahlhaug, O.G.; Selbo Storli, P.T.; Nielsen, T.K. Francis-99 workshop 3: Fluid structure interaction. *J. Phys. Conf. Ser.* **2019**, *1296*, 011001. [CrossRef]

7. Vuillemard, J.; Aeschlimann, V.; Fraser, R.; Lemay, S.; Deschênes, C. Experimental investigation of the draft tube inlet flow of a bulb turbine. *IOP Conf. Ser. Earth Environ. Sci.* **2014**, *22*, 032010. [CrossRef]

8. Fraser, R.; Coulaud, M.; Aeschlimann, V.; Lemay, J.; Deschênes, C. Method for experimental investigation of transient operation on Laval test stand for model size turbines. *IOP Conf. Ser. Earth Environ. Sci.* **2016**, *49*, 062006. [CrossRef]

9. Mulu, B.G.; Cervantes, M.J. Experimental investigation of a Kaplan model with LDA. In Proceedings of the Water Engineering for Sustainable Environment: 33rd IAHR Congress, Vancouver, BC, Canada, 9–14 August 2009.

10. Amiri, K.; Mulu, B.G.; Raisee, M.; Cervantes, M.J. Unsteady pressure measurements on the runner of a Kaplan turbine during load acceptance and load rejection. *J. Hydraul. Res.* **2016**, *54*, 56–73. [CrossRef]

11. Trivedi, C.H.; Gogstad, P.J.; Dahlhaug, O.G. Investigation of the unsteady pressure pulsations in the prototype Francis turbines during load variation and startup. *Renew. Sustain. Energy Rev.* **2017**, *9*, 064502. [CrossRef]

12. Unterluggauer, J.; Sulzgruber, V.; Doujak, E.; Bauer, C. Experimental and numerical study of a prototype Francis turbine startup. *Renew. Energy* **2020**, *157*, 1212–1221. [CrossRef]

13. Wang, F.; Li, X.; Min, Y.; Ma, J. Experimental investigation of characteristic frequency in unsteady hydraulic behavior of a large hydraulic turbine. *J. Hydrodyn.* **2009**, *21*, 12–19. [CrossRef]

14. Trivedi, C.H.; Gogstad, P.J.; Dahlhaug, O.G. Investigation of the unsteady pressure pulsations in the prototype Francis turbines–part 1: Steady state operating conditions. *Mech. Syst. Signal Process.* **2018**, *108*, 188–202. [CrossRef]

15. Gagnon, M.; Jobidon, N.; Lawrence, M. Optimization of turbine startup: Some experimental results from a propeller runner. *IOP Conf. Ser. Earth Environ. Sci.* **2014**, *22*, 032022. [CrossRef]

16. Presas, A.; Luo, Y.; Guo, B.; Wang, Z. Fatigue life estimation of Francis turbines based on experimental strain measurements: Review of the actual data and future trends. *Renew. Sustain. Energy Rev.* **2019**, *102*, 96–110. [CrossRef]

17. You, F.; Sun, L. Machine learning and data-driven techniques for the control of smart power generation systems: An uncertainty handling perspective. *Engineering* **2021**, *7*, 1239–1247. [CrossRef]

18. Bordin, C.; Skjelbred, H.I.; Kong, J.; Yang, Z. Machine learning for hydropower scheduling: State of the art and future research directions. *Procedia Comput. Sci.* **2020**, *176*, 1659–1668. [CrossRef]

19. Huang, X.; Chamberland-Lauzon, J.; Oram, C.; Klopfer, A.; Ruchonnet, N. Fatigue analyses of the prototype Francis runners based on site measurements and simulations. *IOP Conf. Ser. Earth Environ. Sci.* **2014**, *22*, 012014. [CrossRef]

20. Mössinger, P.; Jung, A. Transient two-phase CFD simulation of overload operating conditions and load rejection in a prototype sized Francis turbine. *IOP Conf. Ser. Earth Environ. Sci.* **2016**, *49*, 092003. [CrossRef]

21. Karlsson, M.; Nilsson, H.; Aidanpää, J.O. Influence of inlet boundary conditions in the prediction of rotor dynamic forces and moments for a hydraulic turbine using CFD. In Proceedings of the 12th International Symposium on Transport Phenomena and Dynamics of Rotating Machinery, Honolulu, HI, USA, 17–22 February 2008.

22. Mössinger, P.; Jester-Zürker, R.; Jung, A. Francis-99: Transient CFD simulation of load changes and turbine shutdown in a model sized high-head Francis turbine. *J. Phys. Conf. Ser.* **2017**, *782*, 012001. [CrossRef]

23. Iovănel, R.G.; Dunca, G.; Bucur, D.M.; Cervantes, M.J. Numerical simulation of the flow in a Kaplan turbine model during transient operation from the best efficiency point to part load. *Energies* **2020**, *13*, 3129. [CrossRef]

24. Tiwari, G.; Kumar, J.; Prasad, V.; Patel, V.K. Derivation of cavitation characteristics of a 3 MW prototype Francis turbine through numerical hydrodynamic analysis. *Mater. Today Proc.* **2020**, *26*, 1439–1448. [CrossRef]

25. Li, D.; Song, Y.; Lin, S.; Wang, H.; Qin, Y.; Wei, X. Effect mechanism of cavitation on the hump characteristic of a pump-turbine. *Renew. Energy* **2021**, *167*, 369–383. [CrossRef]

26. Hao, Y.; Tan, L. Symmetrical and unsymmetrical tip clearances on cavitation performance and radial force of a mixed flow pump as turbine at pump mode. *Renew. Energy* **2018**, *127*, 368–376. [CrossRef]

27. *Ansys®CFX, Release 16.2, Help System, Theory Guide*; ANSYS, Inc.: Canonsburg, PA, USA, 2015.

28. Joshi, A.; Assam, A.; Nived, M.R.; Eswaran, V. A Generalised wall function including compressibility and pressure-gradient terms for the Spalart–Allmaras turbulence model. *J. Turbul.* **2019**, *20*, 626–660. [CrossRef]

29. *Ansys®CFX, Release 16.2, Help System, Modelling Guide*; ANSYS, Inc.: Canonsburg, PA, USA, 2015.

30. Amiri, K.; Cervantes, M.J.; Mulu, B.G. Experimental investigation of the hydraulic loads on the runner of a Kaplan turbine model and the corresponding prototype. *J. Hydraul. Res.* **2015**, *53*, 452–465. [CrossRef]

31. Wannassi, M.; Monnoyer, F. Numerical simulation of the flow through the blades of a swirl generator. *Appl. Math. Model.* **2016**, *40*, 1247–1259. [CrossRef]

32. Sotoudeh, N.; Maddahian, R.; Cervantes, M.J. Investigation of rotating vortex rope formation during load variation in a Francis turbine draft tube. *Renew. Energy* **2020**, *151*, 238–254. [CrossRef]

33. Bosioc, A.L.; Susan-Resiga, R.; Muntean, S.; Tanasa, C. Unsteady pressure analysis of a swirling flow with vortex rope and axial water injection in a discharge cone. *J. Fluids Eng.* **2012**, *134*, 1–11. [CrossRef]

34. Lyon, D. The Discrete Fourier Transform, Part 4: Spectral Leakage. *J. Object Technol.* **2009**, *8*, 23–34. [CrossRef]

Article

Experimental and Numerical Study on Vortical Structures and Their Dynamics in a Pump Sump

Václav Uruba [1,2,*], **Pavel Procházka** [1], **Milan Sedlář** [3], **Martin Komárek** [3] and **Daniel Duda** [2]

[1] Institute of Thermomechanics, Czech Academy of Sciences, Dolejškova 1402/5, 182 00 Praha, Czech Republic; prochap@it.cas.cz
[2] Department of Power System Engineering, Faculty of Mechanical Engineering, University of West Bohemia, Universitní 22, 306 14 Plzeň, Czech Republic; dudad@kke.zcu.cz
[3] Centre of Hydraulic Research, Jana Sigmunda 313, 783 49 Lutín, Czech Republic; m.sedlar@sigma.cz (M.S.); m.komarek@sigma.cz (M.K.)
* Correspondence: uruba@it.cas.cz

Abstract: Research on water flow in a pump inlet sump is presented. The main effort has been devoted to the study of the vortical structures' appearance and their behavior. The study was conducted in a dedicated model of the pump sump consisting of a rectangular tank $1272 \times 542 \times 550$ mm^3 with a vertical bellmouth inlet 240 mm in diameter and a close-circuit water loop. Both Computational Fluid Dynamics (CFD) and experimental research methods have been applied. The advanced unsteady approach has been used for mathematical modeling to capture the flow-field dynamics. For experiments, the time-resolved Particle Image Velocimetry (PIV) method has been utilized. The mathematical modeling has been validated against the obtained experimental data; the main vortex core circulation is captured within 3%, while the overall flow topology is validated qualitatively. Three types of vortical structures have been detected: surface vortices, wall-attached vortices and bottom vortex. The most intense and stable is the bottom vortex; the surface and wall-attached vortices are found to be of random nature, both in their appearance and topology; they appear intermittently in time with various topologies. The dominant bottom vortex is relatively steady with weak, low-frequency dynamics; typical frequencies are up to 1 Hz. The origin of the vorticity of all large vortical structures is identified in the pump propeller rotation.

Keywords: pump sump; suction pipe; vortical structure; particle image velocimetry; mathematical modeling

Citation: Uruba, V.; Procházka, P.; Sedlář, M.; Komárek, M.; Duda, D. Experimental and Numerical Study on Vortical Structures and Their Dynamics in a Pump Sump. *Water* **2022**, *14*, 2039. https://doi.org/10.3390/w14132039

Academic Editor: Aonghus McNabola

Received: 26 April 2022
Accepted: 22 June 2022
Published: 25 June 2022

Publisher's Note: MDPI stays neutral with regard to jurisdictional claims in published maps and institutional affiliations.

1. Introduction

Nowadays, hydropower systems are getting much more important since modern society feels the need for sustainable and reliable sources of energy. For example, a hydroelectric power station generates two orders of magnitude less greenhouse gases in comparison with classical power sources ([1]). The hydropower share of the global electricity market is about 17%. On the other hand, the pump, which can be used to maintain the desirable flood level, should also operate with maximal achievable efficiency. This task is quite hard to fulfill, especially for off-design operational conditions. The efficiency of hydraulic machines is very sensitive to the existence of vortex cores entering the propeller. The free surface vortices are unacceptable phenomena, in particular, their existence is proportionally dependent on the value of submergence, and they bring many unwanted aspects, such as rotational flow, cavitation effect and, worse, the air entrainment from the free surface. Changing the flow rate or increasing the submergence value is the typical measure to avoid these vortices. The critical value of submergence depends on various parameters such as intake velocity V, intake diameter D, geometry of the intake bellmouth, circulation value Γ, gravity acceleration g, density of medium ρ, surface tension σ, viscosity of medium ν, etc. The submerged vortices occur as a consequence of the pressure drop from the pump suction. Unlike the free surface vortices, the presence of the submerged vortices is not directly related to the submergence value; still, submergence can influence the cavitation state in the submerged vortex core. Anti Vortex Devices (AVD) could be utilized as a passive control method [2,3]; they can disrupt the angular momentum of the flow and thus suppress the vortices.

This paper deals with the flow phenomena in a pump sump using experimental and numerical methods. The three types of vortices that can enter the impeller inlet are under study: free surface vortices, wall-attached vortices and bottom vortex.

The free surface vortices, also known as air-entrained vortices [4], are studied very often. A total of 1% air entering the pump impeller can cause a rapid decrease in the hydraulic efficiency, and even higher amounts of gaseous components can lead to high bearing loading and could result in serious damage to the pump [5]. There are six levels of classification of the surface vortices [1]: (1) coherent surface swirl; (2) surface dimple; (3) dye core to intake; (4) vortex pulling floating trash; (5) vortex pulling air bubbles; (6) full air core to intake. The higher the level, the more dangerous.

Wall-attached vortices, also known as submerged vortices, are the side vortices and the bottom vortex. Both wall-attached and bottom vortices enter the impeller hub, the first one is attached to the side wall, and the second one is connected to the bottom just below the pump inlet. The connection with the wall can be modeled by a singular point.

It is highly useful to know the operating conditions when these phenomena take place, and then we could try to prevent their formation. For this reason, the flow topology in space and time are examined numerically and experimentally. Due to financial costs, CFD methods predominate, see [6–14]. Experimental studies of suction pumps in a sump can be found seldom in the literature, and if so, the experimental methods are based on the visualization technique [5,6] or the measurement of the swirl angle of a pump-approaching flow [7]. Planar velocity fields can be inspected using the Particle Image Velocimetry (PIV) method effectively [3,4]. To the best of our knowledge, almost no relevant literature has been published on the free surface or bottom vortex dynamics; the time-averaged data are presented as a rule.

Obviously, the published information relative to the intake objects of a pump is very limited. We are going to use the knowledge gained during the project on discharge objects [8]. However, there are numerous substantial differences between these two types of flow. In the intake, vortices are primarily driven due to the presence of the rotating impeller, which causes the swirl. This driving mechanism is common for the vortical structures emerging in this case.

The main goal of the presented paper is the investigation of the vortex structures typical for the pump intake configuration, the surface and bottom vortices, in particular. The main method to study the flow topology and flow dynamics are PIV in the experimental part and the CFD solver ANSYS CFX in mathematical modeling. The numerical tool is to be validated and verified. We will put emphasis on dominant flow structures, their topology and time characteristics in those processes. Hence, the physical model was designed and fabricated to allow optical measurement with maximal possible accuracy. To study the flow-field dynamics, advanced decomposition methods such as Proper Orthogonal Decomposition (POD) and Oscillation Pattern Decomposition (OPD) are to be used to extract the dominant dynamic modes and to determine distinct frequencies presented in the flow. For more details on POD and OPD methods, see [15,16].

2. Test Case Setup

The experiments were carried out in the closed water circuit with a rectangular tank with the dimensions (L × W × H) 1272 × 542 × 550 mm^3. The coordinate system is introduced so that the x-axis is oriented along the tank, the y-axis is identical to the pump axis of revolution/rotation and the z-axis is oriented according to the right-hand rule (Figure 1). The origin is placed on the bottom wall just below the impeller.

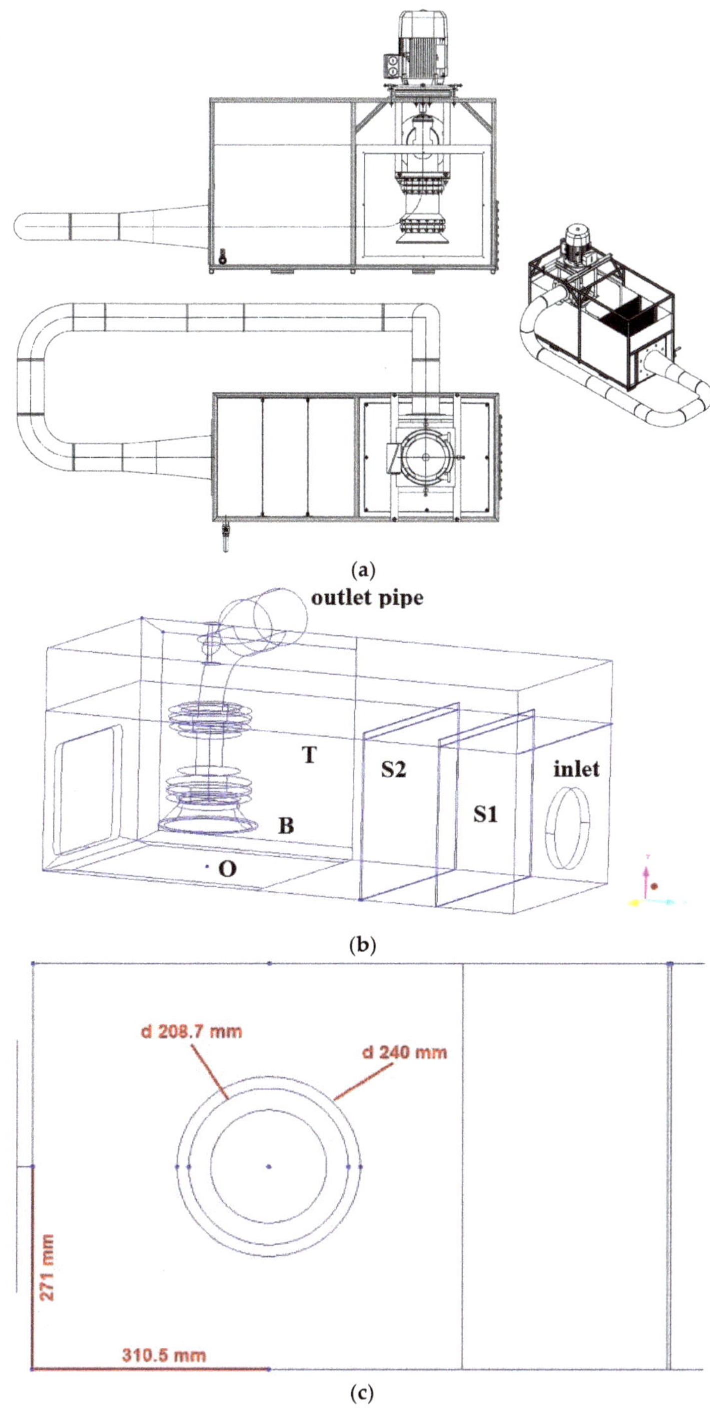

Figure 1. Construction drawings of test stand (**a**), description of the test section, coordinates definition (**b**), top view with dimensions (**c**).

The water enters the tank by pipe from the right in the inlet, passing through two screens, S1 and S2, into the rectangular test section T. The water is sucked out from the tank by a suction bellmouth B and passes out to the outlet pipe. The coordinate system origin is labeled as O. Some key dimensions are shown in Figure 1c.

Both the front and the bottom sides are equipped with transparent windows, which enable lighting the space with a laser sheet and simultaneously recording the PIV images. The axial flow pump is located in the plane of symmetry and 310 mm from the left wall (see Figure 2).

Figure 2. Experimental stand with rectangular tank and transparent windows. Three-dimensional model used in CFD simulations.

The impeller has three blades and is 150 mm in diameter (Figure 3).

Figure 3. Detail of axial flow pump input with bellmouth.

The impeller revolution was set to 1500 min^{-1} counter-clockwise. The lower edge of the suction bellmouth has a diameter of 240 mm and is placed 100 mm above the tank bottom. The water circulates from the pump outlet through the piping back to the tank, which enables the water to be filled easily with seeding particles. On the right-hand side of the tank, two screens are installed (200 and 400 mm from the right-hand wall). They help to create sufficiently uniform inlet flow with fine-grained homogeneous turbulence (Figure 4).

Figure 4. Settling chamber. View of the screens and initial water level.

The screen mesh sizes are 10–12 and 5–8 mm, respectively. The flow rate during all experiments was set to the constant value of 52 L/s. The initial water level has been set to 398 mm above the tank bottom. The volume flow rate was measured with the induction flowmeter (accuracy 0.05 L/s), and the mean water level was controlled by the ultrasonic level transmitter (accuracy 1 mm).

Our experimental research included both the visualizations of unsteady flow phenomena with the high-speed camera and the PIV measurements in selected vertical and horizontal planes.

3. Visualizations

To enable better visualizations of flow phenomena with the high-speed camera, the water has been seeded with particles with a density only slightly higher than water density and illuminated with intensive dispersed light. The sedimentation time of particles was in the order of several hours. These particles followed the water flow and, when lightened, gave a simple picture of the velocity vector field. At the same time, particles already sedimented at the test section bottom enabled visualizing the position and movement of the singular point linked with the bottom vortex. Figure 5 shows three basic vortices below the pump suction bell. The bottom vortex, which is very stable (in the sense of permanent presence), can be easily identified because of the vapor trace in the vortex core. Only one bottom vortex has been identified between the bottom wall and the pump suction. It has been moving in the range of several centimeters from the pump axis of revolution, as can be seen in Figure 5. Its singular point moves on the bottom wall at a speed of several centimeters per second. Much less stable are the vortices rising on the side walls; they could be observed only part of the time. Only very rarely more than one side vortex could be identified at the same time. Different types of vortices could be found between the water level and the pump suction. Both the case with only one free-surface vortex and the case with two counter-rotating ones could be observed (Figure 6), though most of the time, only one vortex could be identified. Free-level vortices have been changing their character from the surface dimples up to the full air core ones.

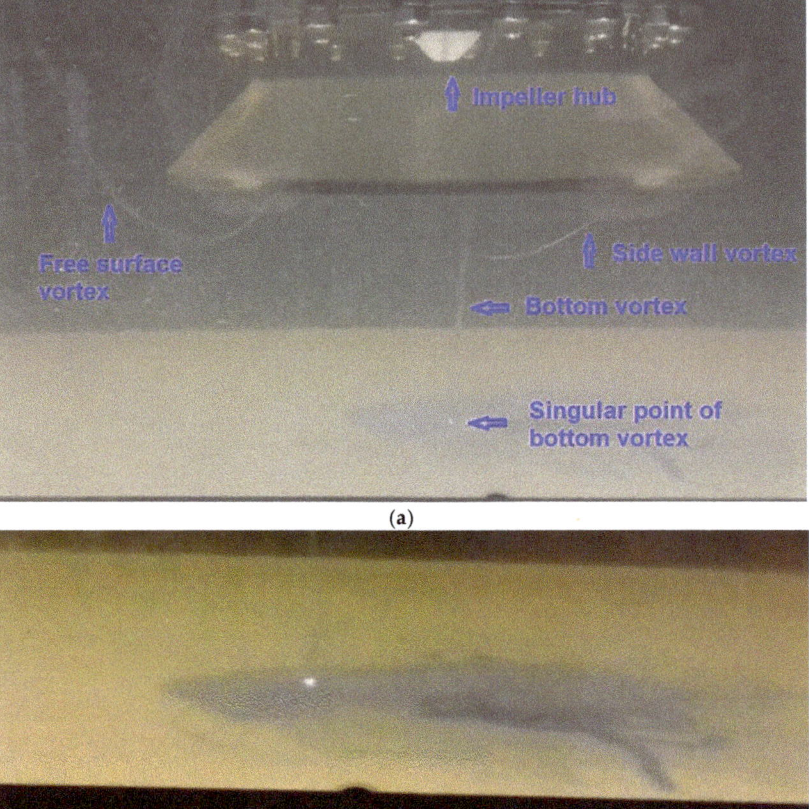

(a)

(b)

Figure 5. View of vortices close to pump suction (**a**) and singular point movement trail (**b**).

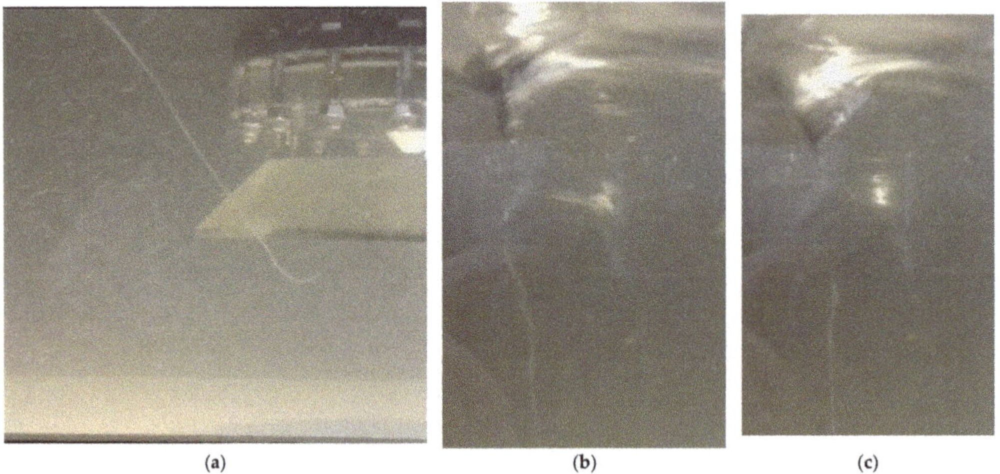

(a)
(b)
(c)

Figure 6. Pair of free-surface vortices, (**a**) side view, (**b**,**c**) bottom view.

4. Methods

The results of the application of both experimental methods and numerical simulations are to be presented. The numerical results were validated.

4.1. Experimental Techniques

For the steady-state regimes and the vortex dynamics, the PIV measurements with one camera (see Figure 7) have been applied as in the case of siphon performance [8]. The PIV results are represented by distributions of the two components of velocity vectors in the plane of the laser sheet, shown in green in Figure 7. The measurement apparatus consisted of the CMOS camera and the laser by Dantec Company. The laser was New Wave Pegasus, Nd:YLF, double-head pulse type with the light wavelength of 527 nm, the maximal frequency is 10 kHz, the single-shot energy is 10 mJ (for 1 kHz). and the corresponding power is 10 W per one head. The camera Phantom V611, with a resolution of 1280×800 pixels, was able to acquire double snaps with a frequency of up to 3000 Hz (for full spatial resolution), and it used the internal memory of 8 GB for the data storage within a single experiment. Hollow glass spheres coated by silver layer HGS-10 were used as tracking particles. Their typical diameter was 10 microns, and they have a very similar density to water, so they can follow the flow very faithfully. The data have been acquired and post-processed using the Dynamic Studio, Matlab and Tecplot software tools. To get a very complex view of the flow structures, we utilized a combination of measurements in both vertical and horizontal planes. However, it is possible to combine only data processed by time-averaging. The vertical planes have been located in the positions $z = 0, 70$ and 130 mm, where z is the distance from the sump symmetry plane. Horizontal planes have been located at the distance of $y = 50, 70, 105, 125, 155, 205, 255$ and 305 mm above the bottom.

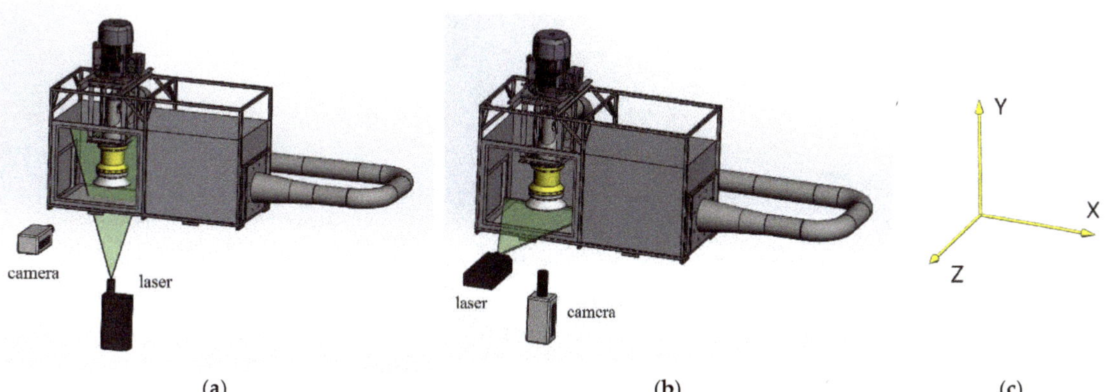

Figure 7. PIV setup; vertical (**a**) and horizontal (**b**) PoMs, Cartesian coordinate system (**c**).

Calibration was performed in one vertical plane ($z = 0$ mm) and also for the horizontal plane at $y = 50$ mm. Horizontal measurements were performed below the impeller and also above it. The traversing system enabled the synchronized displacement of the laser and camera, which allowed the use of a single calibration for more parallel Planes of Measurements (PoMs). Both approaches resulted in dimensions of Field of View (FoV) equal to approx. 330×210 mm^2. The adaptive correlation algorithm using Interrogation Area (IA) 32×32 pixels was utilized, resulting in 79×49 valid vectors in each PoM. Thus, the spatial resolution is about 4 mm.

Two acquisition strategies have been applied. For the time-averaged measurements, 1000 measurements with an acquisition frequency of 10 Hz have been acquired, representing 100 s. However, to study the dynamics of vortical structures, the data were acquired with a frequency of 100 Hz using 4000 snapshots covering 40 s in physical time. The accuracy of PIV velocity measurement is about 2% of the maximal velocity value within the PoM.

4.2. Numerical Simulations

The CFD software ANSYS CFX release 19.2 [17] has been applied to solve the three-dimensional Unsteady Reynolds-Averaged Navier–Stokes (URANS) equation. Because of the rotating flow inside and under the pump, the experimental device misses any symmetry plane, and the computational domain (see Figures 2–4) must represent the full three-dimensional suction object, including all pump components and piping. Moreover, full unsteady interactions between the pump impeller and the stationary parts are kept with the time step representing two degrees of pump impeller revolution. The subdomain of the pump impeller is rotating with a prescribed rotational speed. All the other subdomains inside the pump, rectangular test section and piping, are stationary. There are two interfaces between the stationary and rotating subdomains, as can be seen in Figure 8. Because the impeller nut and the shaft are rotating in the stationary subdomains, their surfaces are treated as rotating walls. On the other hand, the impeller shroud, which is, in fact, stationary, has to be treated as the counter-rotating wall. All the hydraulic surfaces are considered to be smooth, without prescribed roughness. As the computational domain represents a closed loop, there are no inlet and outlet boundaries and the mass flow is generated by the impeller rotation and controlled by the pressure losses in the computational domain, including screens. The screens have been described as porous layers with the permeability derived from the numerical tests with small (100×100 mm^2) parts of the screens modeled in full detail. This strategy has been adopted to keep a reasonable size for the computational grids. The computed mass flow rate has been compared with the mass flow rate measured during the experiments; the detected differences were a few percent. Concerning the boundary condition on the top of the rectangular test section, the opening boundary has been used to enable the air to flow both inside and outside the computational domain.

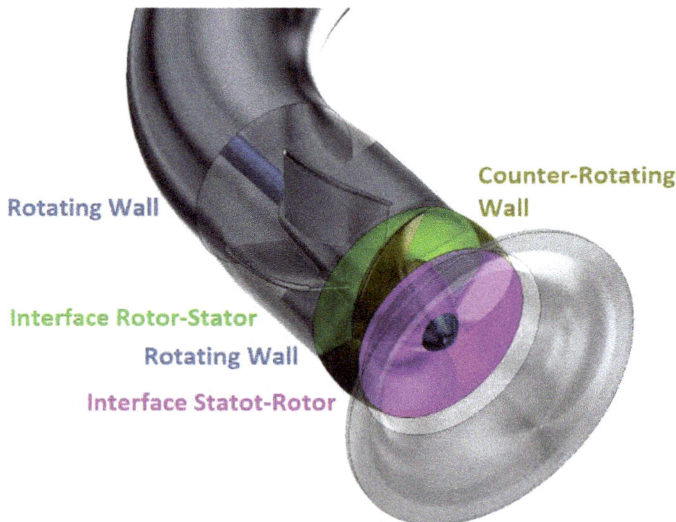

Figure 8. Boundary conditions. CFD.

All simulations are based on constant property fluids. It means that the water, water vapor and air are considered as incompressible fluids with constant densities and constant temperature. The free surface flow, including the gravity effects, is based on the Volume of Fluid (VOF) method evaluating the volume fraction of each fluid. A non-homogeneous model of multiphase flow has been applied with different velocities for the water, vapor and air fractions [17]. The Zwart cavitation model [18] has been employed to describe the interphase mass transfer during the cavitation effects. To capture all possible factors acting on the free surface vortices, the Coriolis forces resulting from the earth rotation have been added into the computational model, though they are much less important than the geometry and other physical factors [19].

The structured computational grids have been used inside the rotating and stationary parts of the pump, pipes and screens. To keep the grid inside the tank sufficiently isotropic with acceptable values of $y+$ monitored in time (close to 1 on the majority of solid surfaces), the unstructured computational grid with prismatic elements inside the boundary layers has been applied. The overall number of computational grid points is approximately 17 million. The maximum distance of the neighboring grid points in the tank is about 3.3 mm, which enables relatively thin vortices to be captured. As it will be described in the next chapter, the calculated vortices have been very unstable, with the random nature of their appearance and topology. For that reason, it has not been possible to provide a standard grid independence test in which the results could be compared and verified directly by the experimental data. Consequently, the computational grids (especially inside the rectangular tank) have been tested and optimized in several steps to keep acceptable values of $y+$ and to capture sufficiently small flow structures. The resulting computational grid represents a reasonable compromise between these requirements and the possibility of using a 16-core-based workstation with 128 GB of memory.

The initial calculations employed the URANS equations with the standard SST turbulence model and then switched to the SAS scale resolving simulations ([20,21]), which are more suitable for predicting small vortical structures. The high-resolution scheme of the second order has been used for the momentum equations, and the first-order scheme has been used for the turbulence modeling. The time discretization is based on the second-order backward Euler scheme.

5. Results

Now, the results are to be presented in terms of vortices description using both experiments and CFD.

5.1. Surface Vortices

The surface vortices were found to be very unstable. They were studied by visualizations (see Section 3) and partly using CFD.

Similar to the visualizations, the CFD simulations also show that the only vortex that is stable and present permanently is the bottom vortex, which moves around the pump axis revolution. Further, one side vortex and one or two free-surface vortices have been kept with the simulations. Unfortunately, they have been very unstable, and the computational capacities do not allow any rule on their formation or frequency analysis to be found. Figures 9 and 10 show a complex system of vortices in the testing tank, including the bottom vortex, one free-surface vortex and one side-wall vortex. The iso-surfaces of constant air volume fraction of 5% (in cyan) and constant vapor volume fraction of 5% (in grey) have been used to visualize the vortex cores in Figure 9. The Q criterion has been applied in Figure 10. A pair of counter-rotating free-surface vortices can be seen in Figure 11, together with the bottom vortex.

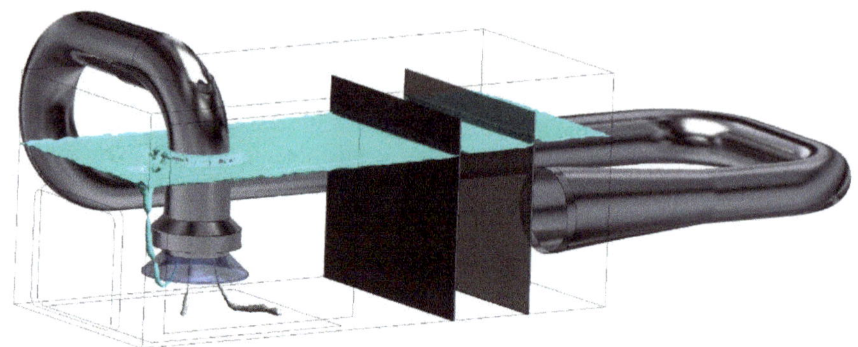

Figure 9. Complex system of vortices in the testing tank. Iso-surfaces of constant air volume fraction 5% (in cyan) and constant vapor volume fraction 5% (in grey). CFD.

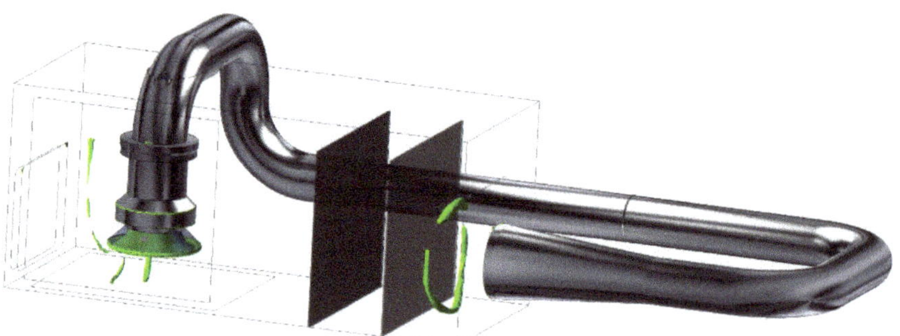

Figure 10. Visualization of vortices in the testing tank. Q criterion (in green). CFD.

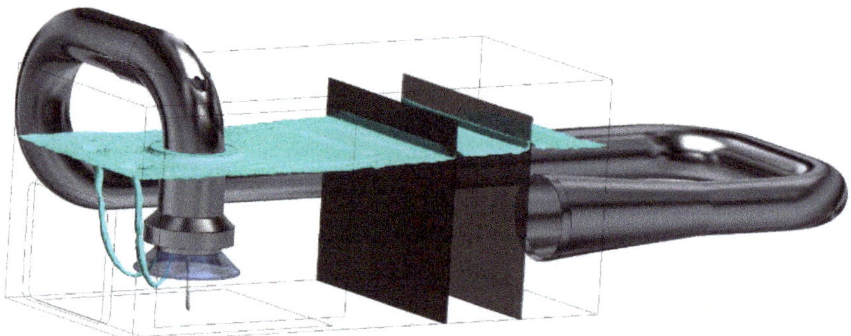

Figure 11. Pair of counter-rotating free-surface vortices and bottom vortices. Iso-surfaces of constant air volume fraction 5% (in cyan) and constant vapor volume fraction 5% (in grey). CFD.

5.2. Bottom Vortex

Because of the unsteady character of flow structures, which have been changing their position as well as the topology, it has been a little bit problematic to present time-averaged results of the numerical simulations. That is why the simulations have been averaged just for a short time when the topology and position of the vortices remained similar (time scale of order 10^{-1} s). Figures 12 and 13 show the position and character of the bottom vortex and one single free-surface vortex in the horizontal planes spaced from 50 to 255 mm from the tank bottom. Both the vector lines and vorticity distributions are shown for every plane. The general term "vector lines" is used in the description of results as "streamlines" are defined as "vector lines of the instantaneous velocity fields in 3D", which is not the case for the time-averaged velocity fields in 2D.

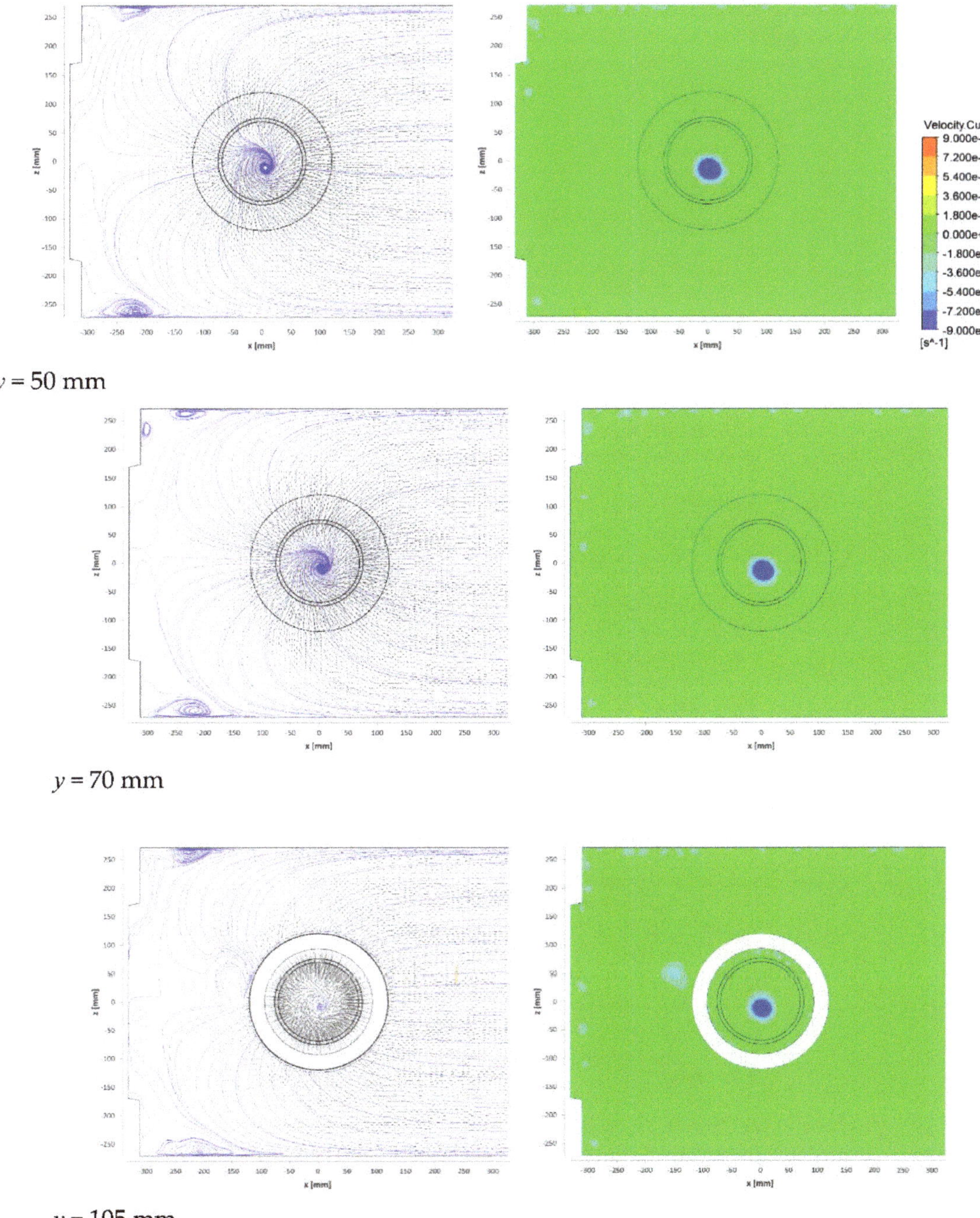

$y = 50$ mm

$y = 70$ mm

$y = 105$ mm

Figure 12. *Cont.*

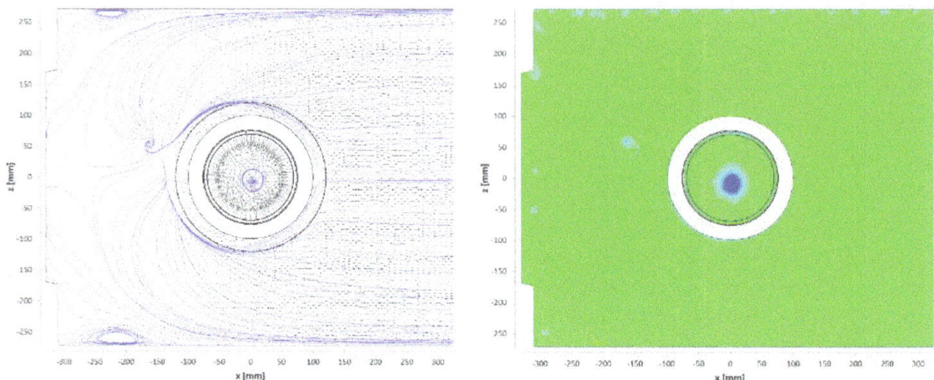

Figure 12. Vector lines (**left**) and vorticity distribution (**right**) in horizontal planes y = 50, 70, 105 and 125 mm. CFD.

In section y = 50–105 mm, the bottom vortex could be identified with its center close to the coordinate system origin. For the planes y = 125 mm and above, the surface vortex located near position [−200;70 mm] could be detected.

5.3. Intake Flow

Figure 14 shows the vector lines and longitudinal velocity distribution in vertical planes z = 0, 62 and 130 mm. Here the vector lines and x-velocity component distributions are shown. The bellmouth section by the projection plane is white, while the object silhouette is shown in black lines.

It can be seen in the vertical plane z = 62 mm that the flow is influenced by the free-surface vortex situated left of the bellmouth.

The numerical simulation shows a good agreement of the basic behavior of the flow phenomena close to the pump suction. Specifically, the bottom vortex is captured in a very good way. The comparison of free vortices obtained by the experiments and by the CFD tools is much more difficult, as these phenomena are highly unsteady and unstable. Their description depends on the way of time averaging, but the most important factor is the fact that during the experimental research, every horizontal or vertical plane is captured in a different instant.

The inflow into the bellmouth from both sides of the x-direction could be detected from the pictures in Figure 14.

The experimental results and velocity fields in vertical planes z = 0, 70 and 130 mm are shown in Figure 15. The flow enters the sump volume from the right-hand side. The black vectors represent in-plane velocity components, and the color is a scalar value (horizontal velocity component); vector lines are added in blue color. The velocity distributions in the vicinity of the bellmouth show very good agreement with CFD data, especially for planar data acquired in the plane of symmetry (z = 0). The vector lines below the suction pipe do not rise up as vertically in comparison with the calculated results. The other two PoMs reveal the flow topology closer to the side wall of the sump. The flow is attracted from above, and the flow is bent 180° in the vicinity of the bellmouth edges. The singular point (zero in-plane velocity) is situated 20 mm to the right of the center. The solid black line indicates the cross-section of the bell. The dashed line is its projection in the other PoM. Note, the bellmouth is not perfectly oriented perpendicularly with the bottom wall in Figure 15 as the physical model was inserted with a small deviation, and so the camera was slightly tilted.

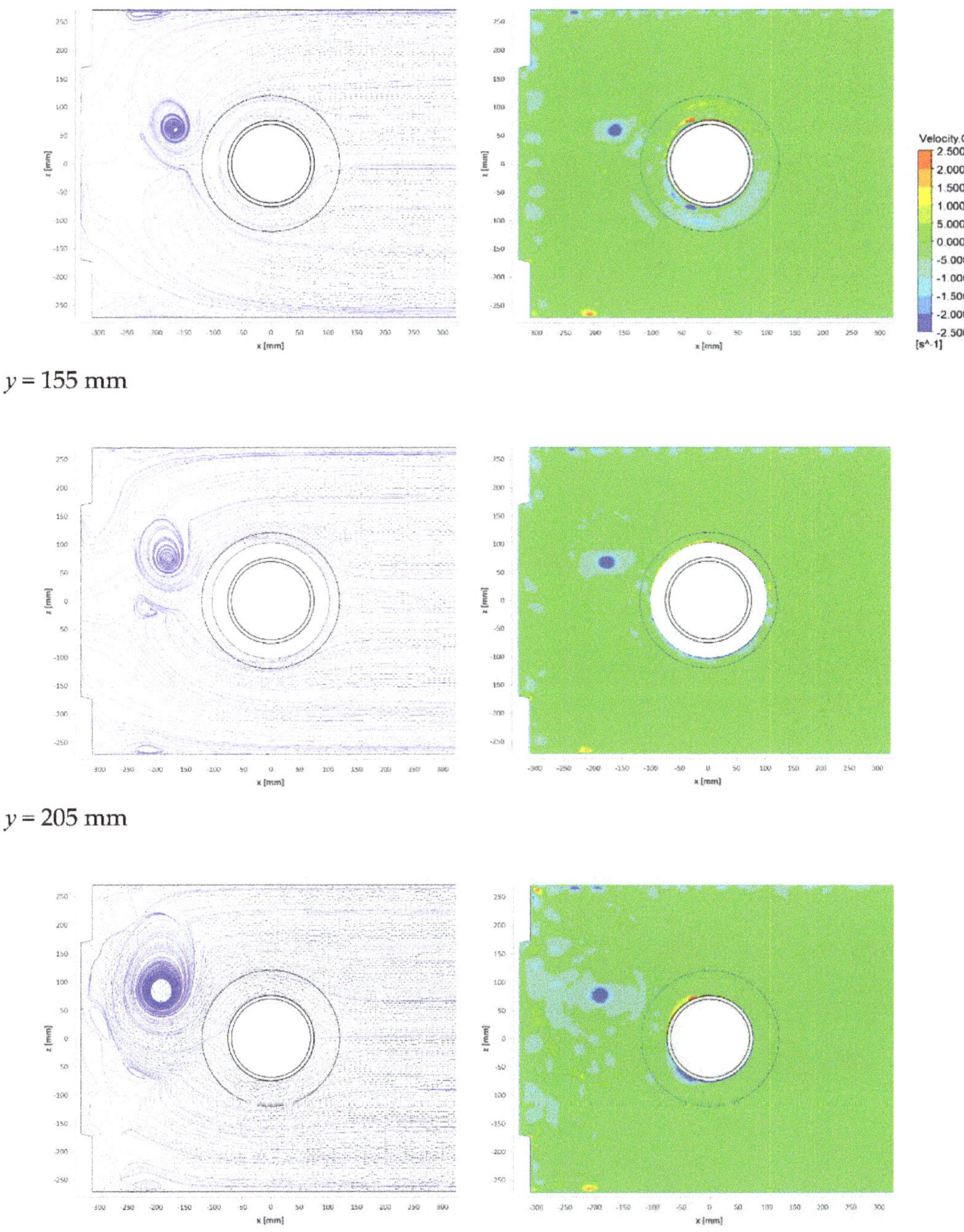

$y = 155$ mm

$y = 205$ mm

$y = 255$ mm

Figure 13. Vector lines (**left**) and vorticity distributions (**right**) in horizontal planes $y = 155, 205$ and 255 mm. CFD.

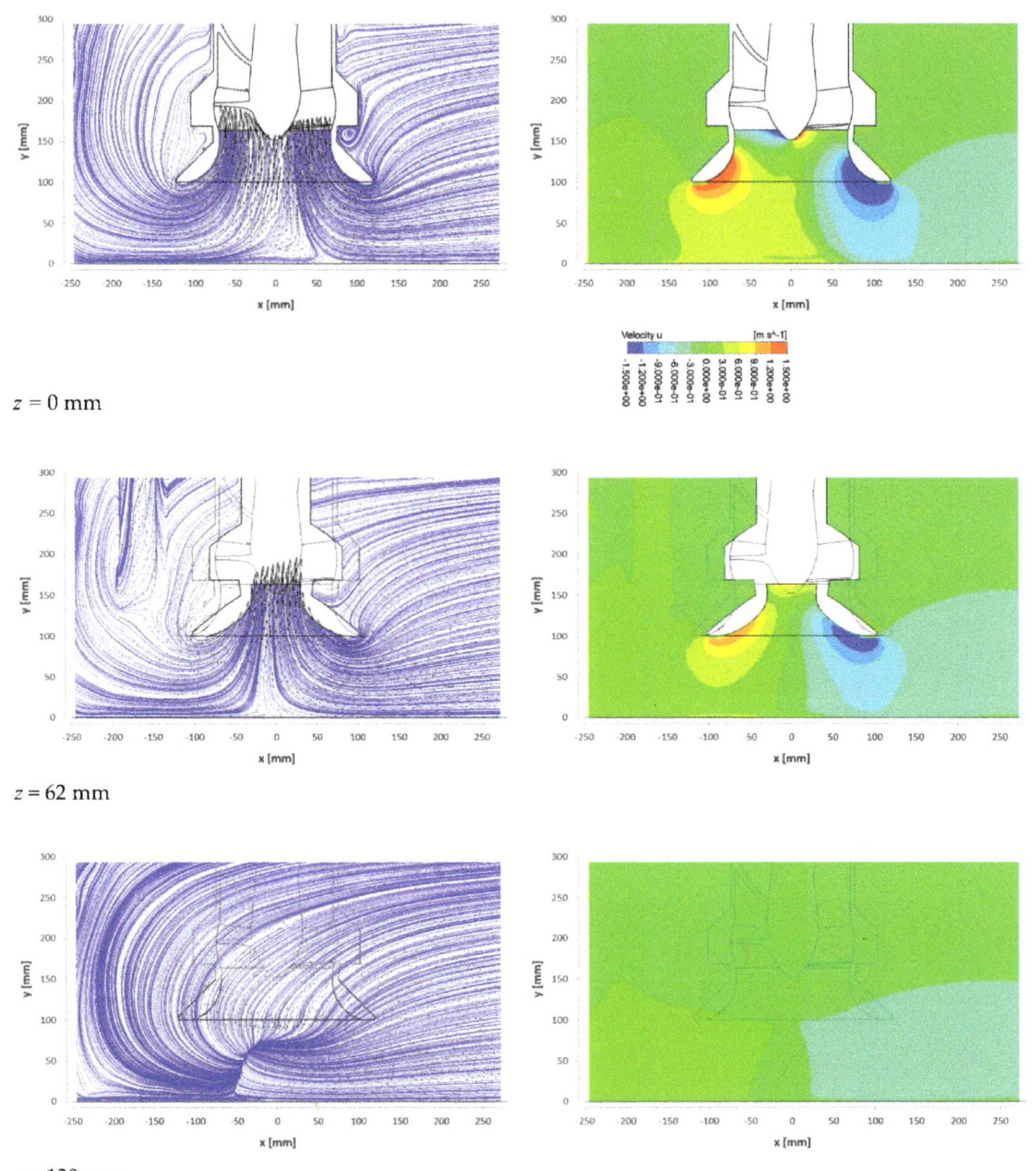

Figure 14. Vector lines (**left**) and longitudinal velocity distribution (**right**) in vertical planes $z = 0, 62$ and 130 mm. CFD.

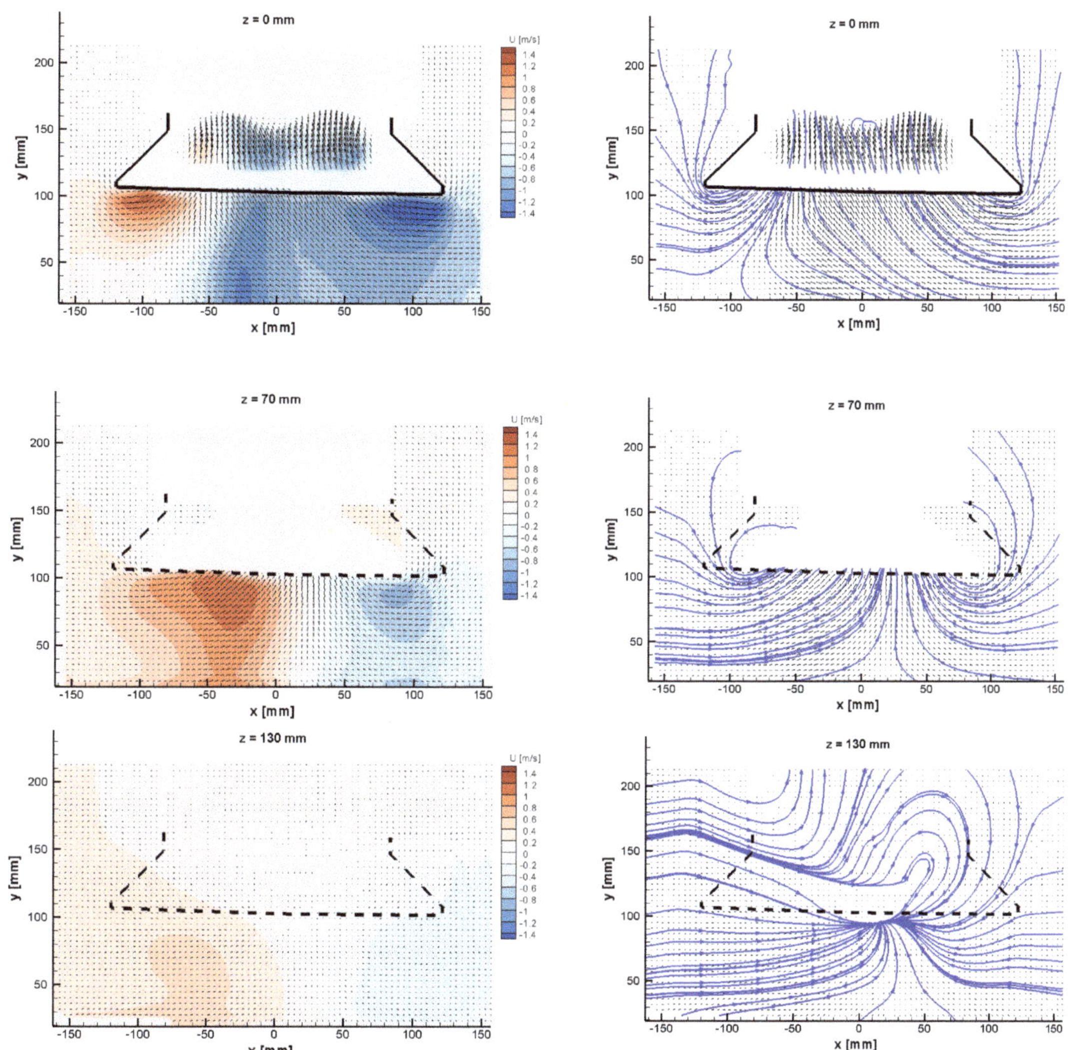

Figure 15. Vertical planes, distribution of horizontal velocity component (**left**) and vector lines (**right**). PIV.

The experimental results in Figure 15 show a not-so-regular input flow as the CFD in Figure 14. While the velocity distribution in the plane of symmetry $z = 0$ is deflected towards the left-hand side, in the position $z = 70$ mm, the situation is opposite for the experiment. The inflow in the case of CFD for both $z = 0$ and 62 mm is more or less symmetrical; compare Figures 14 and 15.

Further, horizontal planes will be introduced. In Figure 16, the positions of various PoMs are plotted to illustrate the flow under and above the bell edge.

There is an intersection of the bottom vortex at $y = 50$ mm, shown in Figure 17. The mean bottom vortex core position is situated close to the coordinate system origin in the geometrical center. The sense of rotation corresponds to the impeller revolution. The flow is entering the pipeline in a helix topology from distant surroundings.

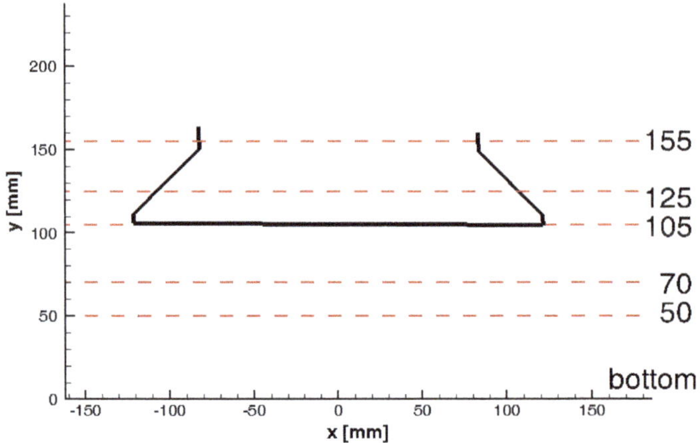

Figure 16. Horizontal plane positions, y = 50, 70, 105, 125 and 155 mm.

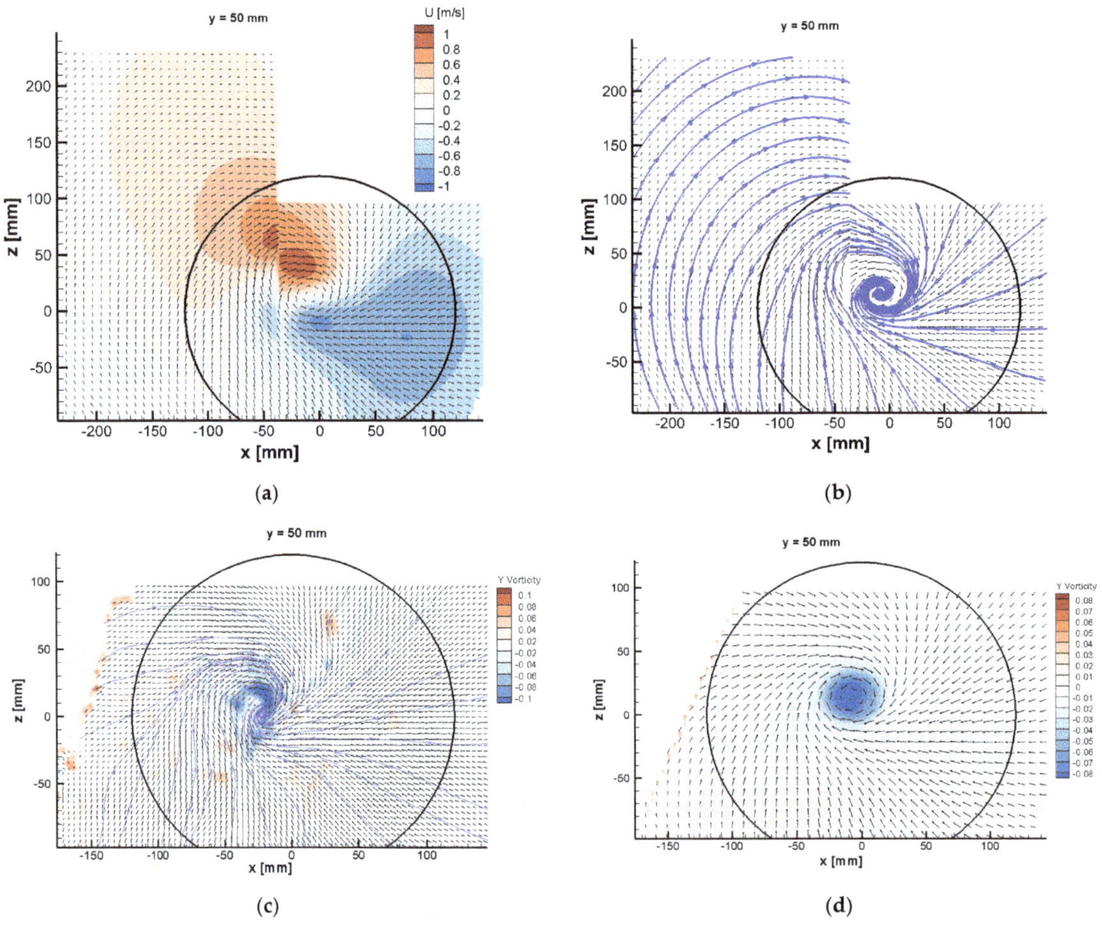

Figure 17. Bottom vortex, y = 50 mm, (**a**) mean velocity distribution, (**b**) vector lines, (**c**) instantaneous vorticity, (**d**) mean vorticity distributions. PIV.

The bottom vortex topology is developing along the y-axis towards the impeller, as it is illustrated in Figure 18. The position of the vortex core does not change very much in the y-direction; however, in position $y = 105$ mm, the inner flow was not resolved because of laser light scatter on the bell edge.

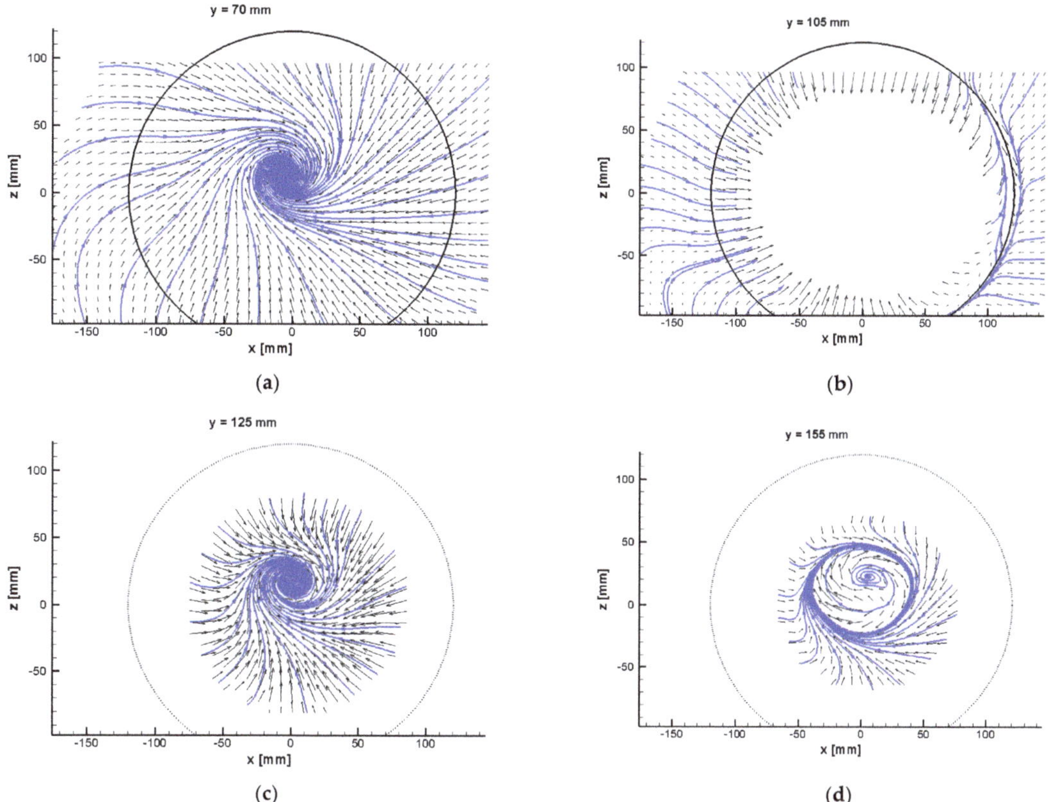

Figure 18. Bottom vortex evolution towards the impeller in the y-direction, horizonral planes $y = 70$ (**a**), 105 (**b**), 125 (**c**) and 155 (**d**) mm. PIV.

The behavior of the free surface vortex is much more chaotic. This vortex can be observed on the left-hand side from the pipe inlet. The topology is developing according to Figure 19. Its center for time-averaged data is reaching the inner corner of the sump as the vortex approaches the water surface. The vortex is even out of PoM for $y = 305$ mm. Nevertheless, its averaged position agrees very well with the computed flow field.

However, the surface vortex behavior is very dynamic with random elements. To demonstrate this fact, the two examples of instantaneous velocity fields in positions $y = 205$ and 305 mm are shown in Figure 20.

There are two instantaneous snapshots of the surface vortex for 205 and 305 mm above the bottom in Figure 20. The left-hand side of this plot reveals standard flow topology, which is changing its position and strength over time (see the next chapter). There is also rarely a situation when two vortices can be observed (this time is a pair with the same orientation).

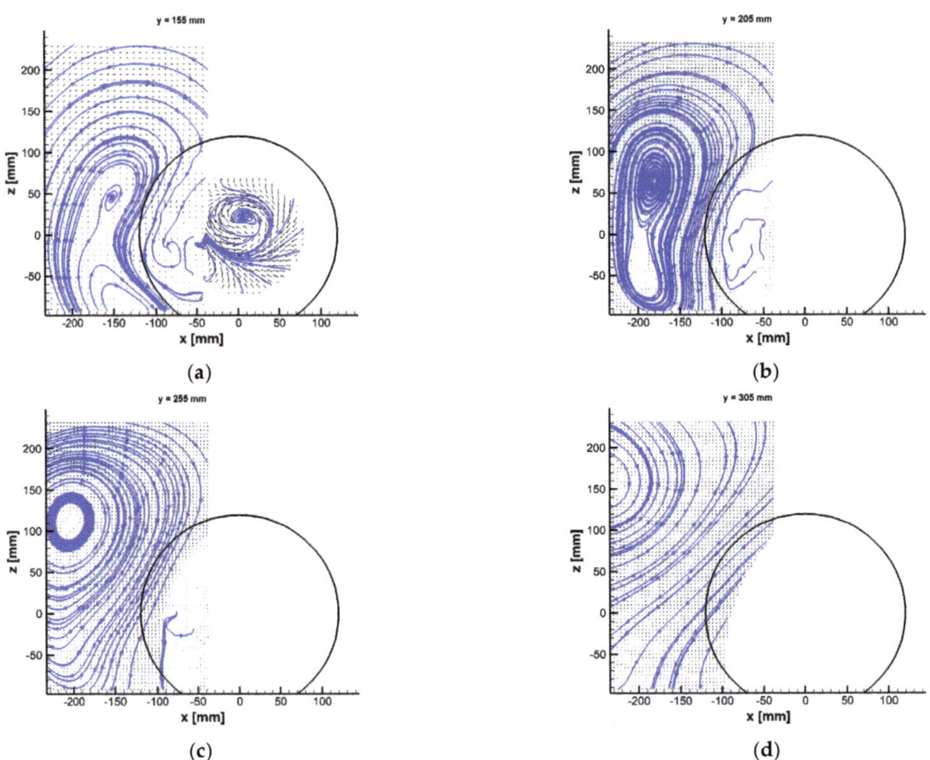

Figure 19. The averaged flow topology of the free-surface vortex at $y = 155$ (**a**), 205 (**b**), 255 (**c**) and 305 (**d**) mm. PIV.

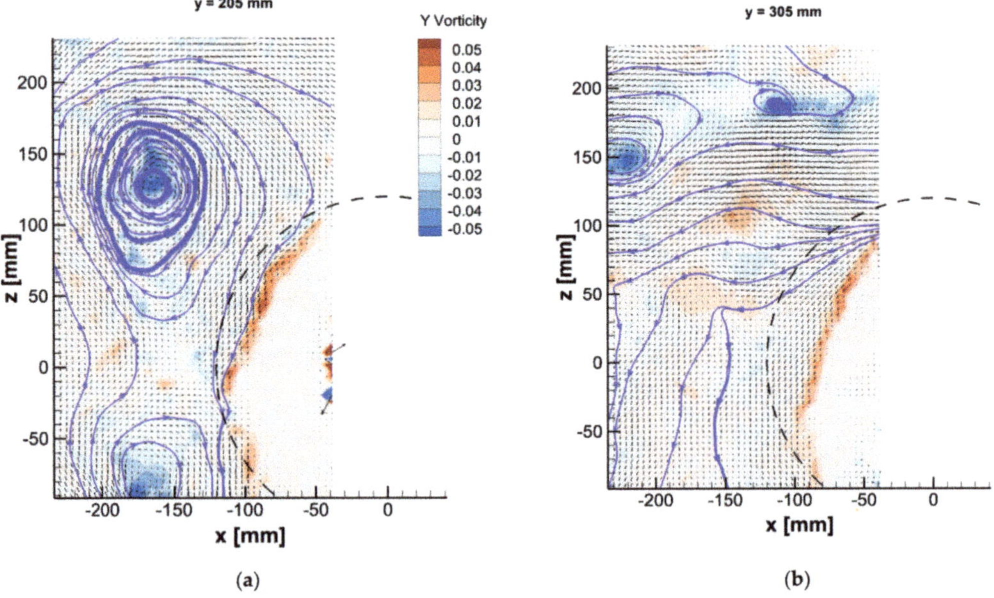

Figure 20. Instantaneous snapshots of the surface vortex at $y = 205$ (**a**) and 305 (**b**) mm. PIV.

5.4. Vortex Dynamics

It was found that both the bottom vortex and the free surface vortices behave dynamically with no distinct periodicity in motion. It was proven that the vortex path is quite chaotic, although the vortex position was always detected inside some circle area. With a few exceptions, the bottom vortex occurs in the second quadrant. The precise position was determined from PIV images via image processing, the vortex core path for the horizontal plane $y = 50$ mm is given in Figure 21 for the time interval 6 s. The bottom vortex core motion is limited by the zone of the size of approximately 20 mm in the x and z-directions. The approximate speed of the core motion was detected up to 10 cm/s.

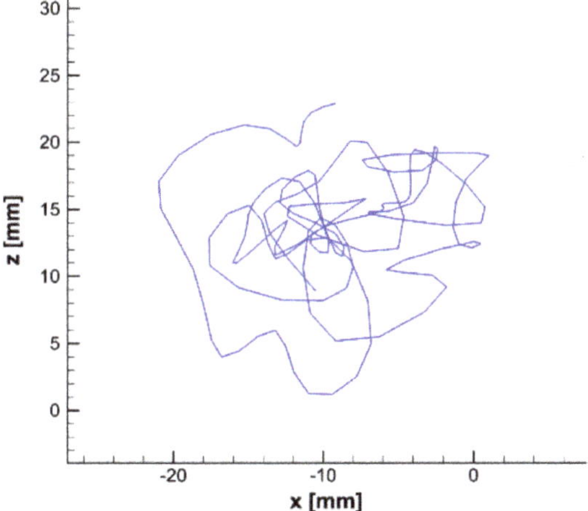

Figure 21. Bottom vortex core path at $z = 50$ mm. PIV.

The first POD mode is shown in Figure 22 as an example. It is formed by a pair of counter-rotating vortices. This mode contains about 16% kinetic energy. Other higher POD modes consist of more vortices.

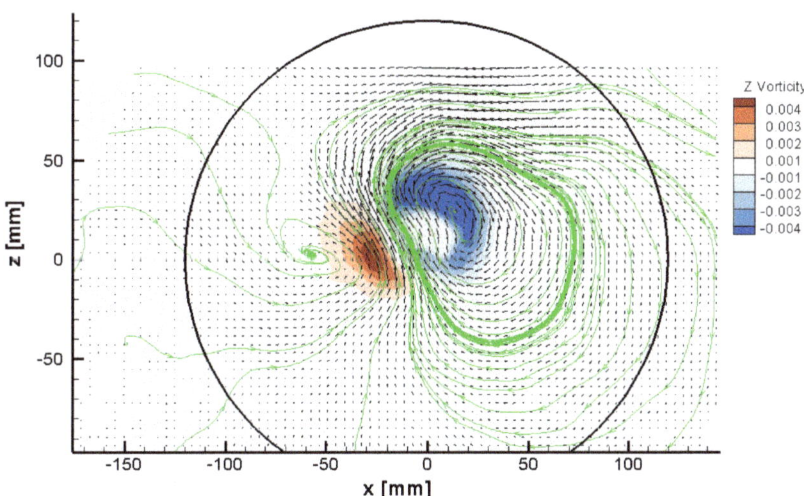

Figure 22. 1. POD mode at $y = 70$ mm, vorticity distribution and vector lines. PIV.

The behavior of the free-surface vortex was studied at plane $y = 205$ mm. The path of its core was also observed using PIV images. The free-surface vortex core was moving inside the area with a size of more than 100 mm, and the core velocity was about 30 cm/s, i.e., three times higher than the velocity of the bottom vortex, see Figure 23.

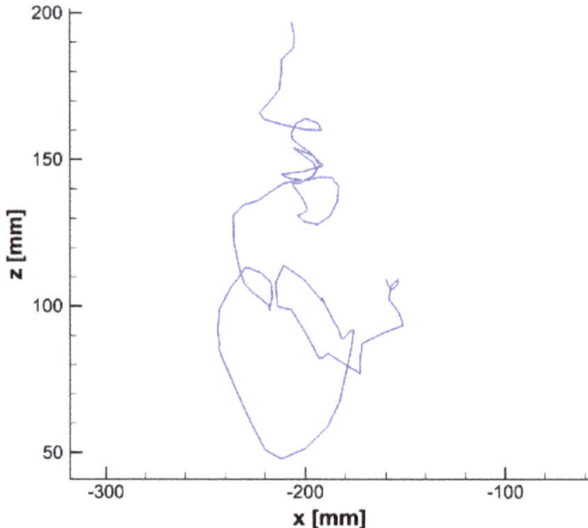

Figure 23. Surface vortex core path at $z = 205$ mm. PIV.

The POD method has been applied to the time-resolved data. Figure 24 reveals the three POD modes for this plane. First, the mode topology can be described as a single vortical structure occupying the space of the time-averaged vortex. The second POD mode consists of two vortices with opposite orientations, while the third mode topology is formed by a single vortex of negative vorticity with strong shear. The content of kinetic energy of the first 10 modes is given in Figure 25.

The dynamics of the bottom vortex have been studied on the horizontal plane at the position $y = 70$ mm in more detail. In Figure 26, there are the mean velocity field and Turbulent Kinetic Energy (TKE) distributions from the PIV measurement.

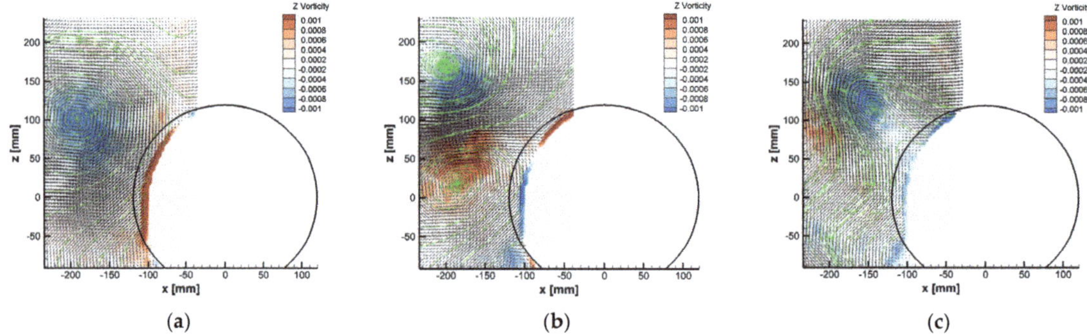

(a)   (b)   (c)

Figure 24. POD modes 1 (**a**), 2 (**b**) and 3 (**c**) at $y = 205$ mm, vorticity distributions and vector lines. PIV.

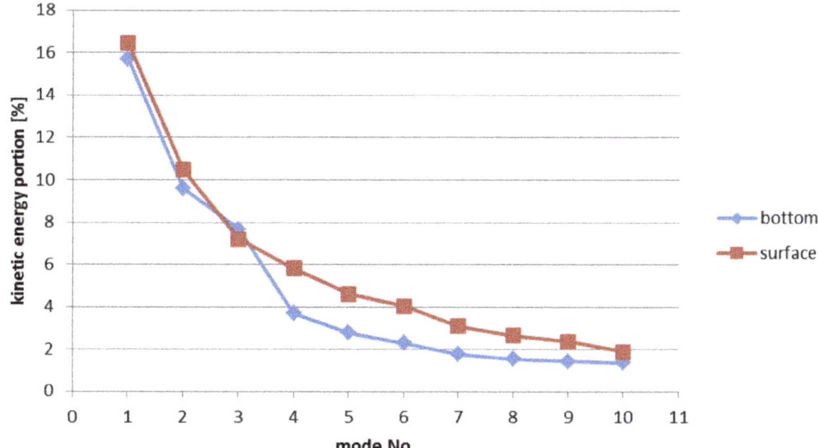

Figure 25. Kinetic energy fraction of the first 10 POD modes for both types of vortices.

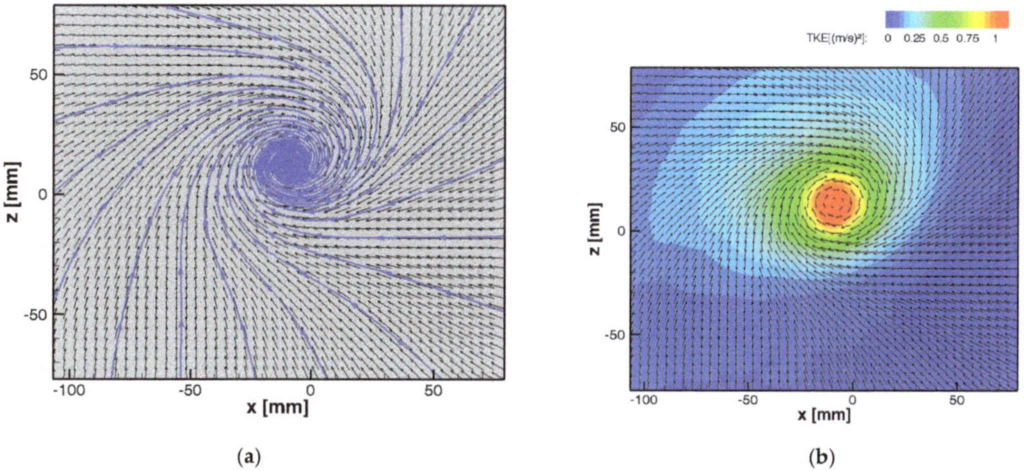

(a)   (b)

Figure 26. Mean fields at $y = 70$ mm, (**a**) velocity field, (**b**) TKE distribution. PIV.

The Oscillation Pattern Decomposition (OPD) analysis has been applied to the velocity field time-resolved data acquired with a frequency of 100 Hz, and 1000 snapshots. The analysis results in OPD modes representing cyclostationary components of the topology dynamics with statistically limited time of life. Each OPD mode is defined by its topology, real and imaginary parts, frequency and mean lifetime. More information on the OPD method and results interpretation can be found in [16].

In Figure 27, there is the OPD spectrum shown, representing the nine most important cyclostationary modes in the plane defined by the frequency and periodicity. The frequency is the typical frequency of pseudo-periodical behavior, while periodicity is the lifetime expressed in multiples of the mode period.

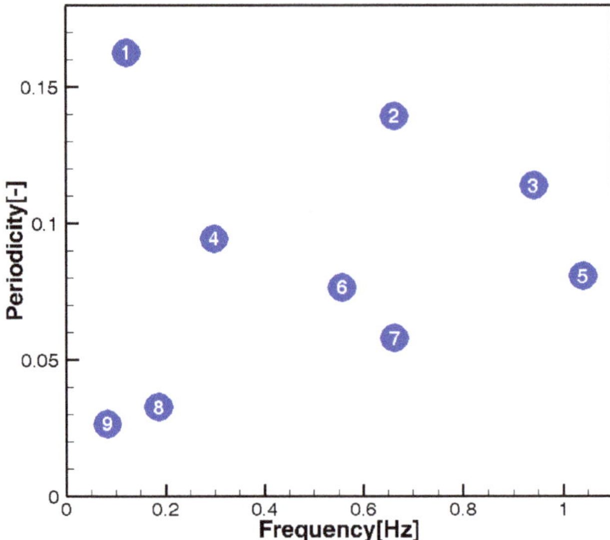

Figure 27. OPD spectrum, the nine cyclostationary OPD modes.

The OPD mode's importance could be quantified by the periodicity value; the higher periodicity, the higher importance. However, even in OPD mode 1, the highest periodicity value is about 0.163; this means that the mode appears randomly and disappears very quickly. The frequency of the first mode is about 0.124 Hz, the real and imaginary topologies are in Figure 28. The mode is represented by the counter-rotating vortex pair rotating around the common center. The color represents vorticity, positive in red and negative in blue.

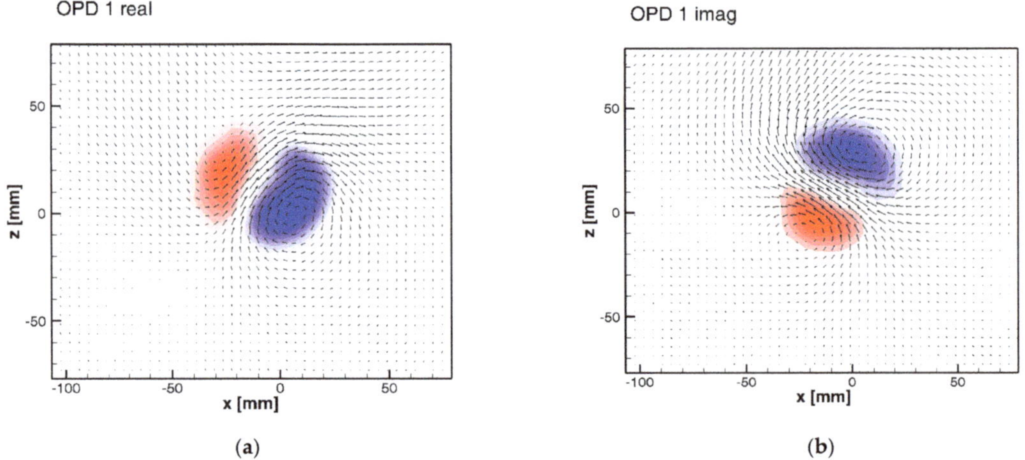

Figure 28. The first OPD mode topology, (**a**) real part, (**b**) imaginary part.

In Figure 29, some higher-order OPD modes topologies are shown; the real parts only. The topology shows a combination of positive and negative vorticity concentrations; the most complex is OPD mode 5, with the highest frequency of 1.040 Hz.

The higher modes consist of several vortices of positive and negative orientations rotating in the plane.

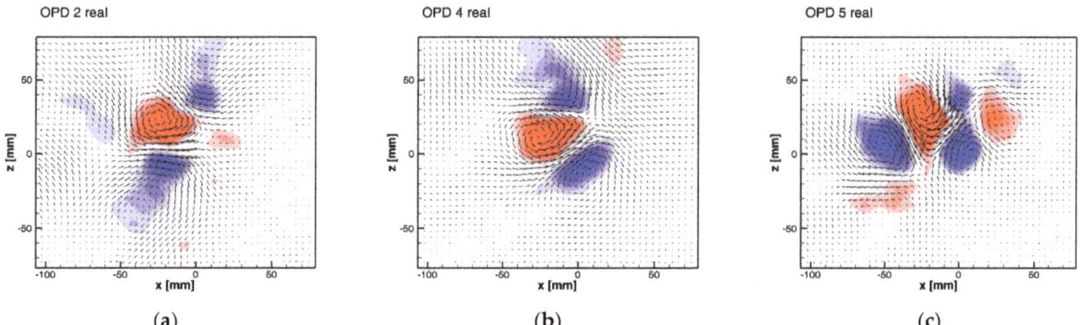

(**a**) (**b**) (**c**)

Figure 29. Real parts of higher OPD modes topology, (**a**) OPD 2, (**b**) OPD 4 and (**c**) OPD 5.

5.5. Validation

The validation of the numerical simulations is performed with the help of flow in the neighborhood of the bellmouth intake. The flow topology is governed by the central vortex touching the bed just below the intake. The vortex is driven by the rotating pump rotor and is characterized by strong axial movement into the intake pipe. The vortex parameters were evaluated on the horizontal plane $y = 50$ mm above the bottom. The (x,z) coordinate system has its origin on the pump axis.

The topology of the vortex is shown in Figure 30; on the left, the result of the PIV measurement, and the CFD result is on the right. The result is shown in the form of mean velocity vector field, and the vector lines are added for better clarity.

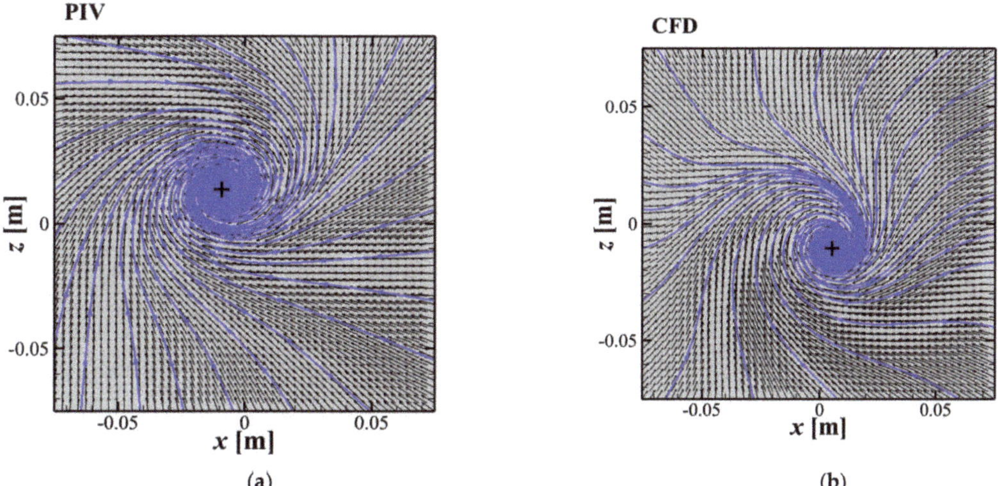

(**a**) (**b**)

Figure 30. Topology of the bottom vortex. (**a**) PIV, (**b**) CFD.

The topologies of the PIV result and the CFD are not exactly the same. The vortex center was detected using the vector lines topology in the spiral focus. The position of the vortex center was detected $[-0.0094;+0.0137]$ m for the PIV data and $[+0.0053;-0.0106]$ m for the CFD data. They are denoted by the cross. The vorticity is concentrated close to the vortex center. The vorticity y-component ω was evaluated:

$$\omega = \frac{\partial w}{\partial x} - \frac{\partial u}{\partial z}. \tag{1}$$

The vorticity distributions are shown in Figure 31 for the PIV and CFD results.

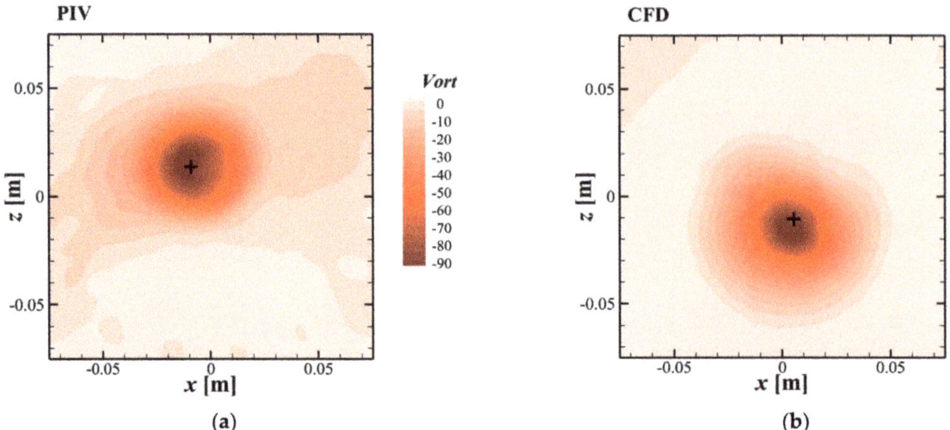

Figure 31. Vorticity distribution, (**a**) PIV, (**b**) CFD.

Please note that the vorticity is negative everywhere, as the clockwise rotation predominates.

To study the vortex differences, the z = const. sections have been evaluated when the section intersects the vortex center and the distance from the center r is considered. In Figure 32, the graphs are shown for both experimental and mathematical modeling.

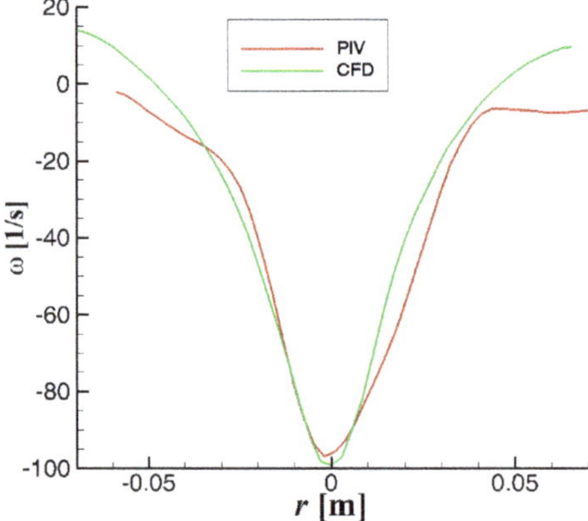

Figure 32. Vorticity profiles intersecting the vortex, PIV and CFD.

The comparison shows a relatively good coincidence of the vorticity in the vortex core; however, outside the vortex core, for $|r| > 0.04$ m, the difference becomes important.

To quantify the difference between the distributions, the integral over the considered zone (rectangular) was evaluated I_ω, which corresponds to the equivalent circulation. For the PIV data, $I_{\omega PIV} = -0.1783$ m^2/s, and for the CFD data, $I_{\omega CFD} = -0.0434$ m^2/s. The difference is rather important; it is clear that CFD underestimates the vorticity value, especially outside the vortex core. Within the vortex core, ± 0.03 m around the vortex center, the agreement of the mathematical modeling results with experiments is much better; the difference is below 3%.

To better estimate the differences between the flow-field topologies, the divergence D has been evaluated for both PIV and CFD data. The 2D definition of the divergence is taken into account:

$$D = \frac{\partial u}{\partial x} + \frac{\partial w}{\partial z}. \tag{2}$$

The flow is incompressible, so the divergence in 3D representation should vanish. Then, the following should hold:

$$\frac{\partial v}{\partial y} = -D \tag{3}$$

Thus, the evaluated divergence provides information about the out-of-plane velocity component v derivative in its direction y. The topologies of Divergence D for PIV and CFD results are shown in Figure 33.

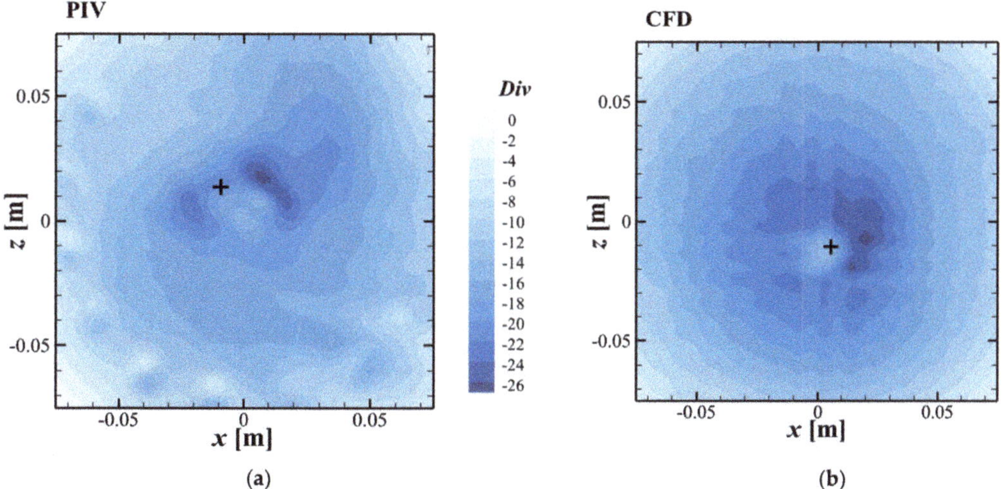

Figure 33. Divergence distribution (**a**) PIV, (**b**) CFD.

As the values of divergence are negative everywhere, this result indicates accelerating flow in the y-direction.

The comparison of divergence in the z = const. section has been evaluated, and the distance from the center r is considered in Figure 34. The difference in the divergence distributions is rather important and visible both in the distribution topology and central section.

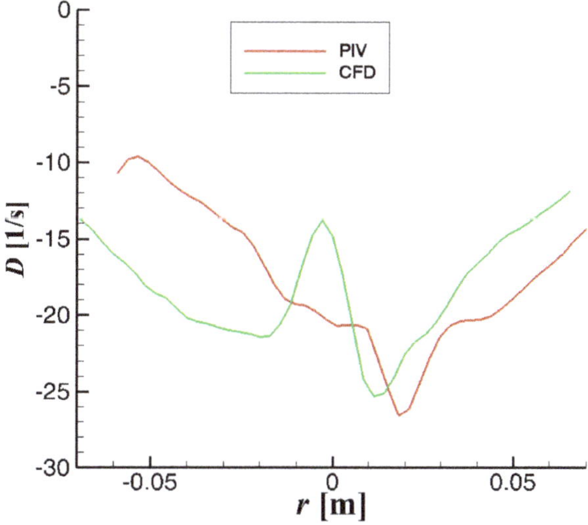

Figure 34. Divergence distribution, PIV (red), CFD (green).

The integral of divergence over the considered zone was evaluated I_D, which corresponds to the equivalent circulation. For the PIV data, $I_{DPIV} = -0.3500$ m^2/s, and for the CFD data, $I_{DCFD} = -0.3034$ m^2/s. The integral values do not differ very much.

6. Discussion

The main task of this article was to validate the mathematical model adjusted to perform CFD simulation of inlet objects of the pump. The second goal was to study the vortex dynamics using both the visualization technique and PIV measurements.

Generally, it can be said that the bottom (submerged) vortex is present throughout the running of the pump. The detected position of the time-averaged vortex core slightly differs from the PIV and CFD data. However, the topology and vortex dimension were determined very much the same for both approaches.

The free-surface vortex is very hard to study quantitatively (and also visualize) as the existence of such a structure is random both in appearance and behavior, and it is limited in time. Very often, only the swirl of the flow is present. The vortex with a full air core is present less than half the time of the device's operation. Sometimes even two distinct free-surface vortices were observed. The occurrence of such structures seems to be completely random. The extent, orientation and location determined from the CFD data matches very well with the PIV data in terms of the averaged dataset, at least for time-mean structures.

The path of the vortex core does not show any regular or even pseudo-regular trajectory for the surface vortex and for the submerged one. The free-surface vortex moves with a velocity about three or four times higher than the bottom vortex.

7. Conclusions

The appearance and behavior study on flow in a pump inlet sump is presented.

The vortical structures were studied in detail, and both mathematical modeling and experimental research methods were applied. The advanced unsteady approach has been applied for mathematical modeling to model the flow-field dynamics. For experiments, the time-resolved PIV method has been used.

Three types of dominant vortical structures have been detected: surface vortices, wall-attached vortices and the bottom vortex. The most intense and stable is the bottom vortex. The surface and wall-attached vortices are found to be of random nature as for their appearance and topology; they appear intermittently in time with various topologies. The dominant bottom vortex is relatively steady with weak dynamics. The frequencies of pseudo-periodical behavior were detected up to 1 Hz.

The mathematical modeling has been validated against the experimental data. The bottom vortex core circulation was modeled with an error below 3%. The rest of the flow field was validated qualitatively only. The flow patterns, especially the vortices detected in experiments, are well captured by mathematical modeling.

The origin of the vorticity of all big vortical structures was identified in the pump propeller rotation.

Author Contributions: Conceptualization, M.S. and V.U.; experimental setup design, M.S. and P.P.; validation, P.P., V.U. and M.S.; CFD computation, M.S.; experiments execution, P.P., V.U., M.S. and M.K.; vortex analysis, D.D. and V.U.; writing—original draft preparation, P.P. and V.U.; writing—review and editing, P.P., V.U., M.S. and M.K.; visualization, M.S.; project administration, M.S. and V.U.; funding acquisition, V.U. All authors have read and agreed to the published version of the manuscript.

Funding: The authors thank the Institute of Thermomechanics, Academy of Sciences of the Czech Republic, for funding by the institutional support RVO: 61388998.

Data Availability Statement: The experimental and CFD datasets can be provided by authors on demand.

Acknowledgments: This research was supported by the Czech Ministry of Education, Youth and Sports, grant number CZ.02.1.01/0.0/0.0/17_049/0008408.

Conflicts of Interest: The authors declare no conflict of interest.

Abbreviations

CFD	Computational Fluid Dynamics
FoV	Field of View
IA	Interrogation Area
POD	Proper Orthogonal Decomposition
PoM	Plane of Measurement
PIV	Particle Image Velocimetry
OPD	Oscillation Pattern Decomposition
SST	Shear Stress Transport
TKE	Turbulent Kinetic Energy
URANS	Unsteady Reynolds-Averaged Navier-Stokes equation
VoF	Volume of Fluid

References

1. Domfeh, M.K.; Gyamfi, S.; Amo-Boateng, M.; Andoh, R.; Ofosu, E.A.; Tabor, G. Free surface vortices at hydropower intakes: A state-of-the-art review. *Sci. Afr.* **2020**, *8*, e00355. [CrossRef]
2. Kim, C.G.; Kim, B.H.; Bang, B.H.; Lee, Y.H. Experimental and CFD analysis for prediction of vortex and swirl angle in the pump sump station model. *IOP Conf. Ser. Mater. Sci. Eng.* **2015**, *72*, 42044. [CrossRef]
3. Park, I.; Kim, H.-J.; Seong, H.; Rhee, D.S. Experimental Studies on Surface Vortex Mitigation Using the Floating Anti-Vortex Device in Sump Pumps. *Water* **2018**, *10*, 441. [CrossRef]
4. Okamura, T.; Kamemoto, K.; Matsui, J. CFD prediction and model experiment on suction vortices in pump sump. In Proceedings of the 9th Asian International Conference on Fluid Machinery, Jeju, Korea, 16–19 October 2007.
5. Gupta, S.; Panda, J.P.; Nandi, N. A model study of free vortex flow. In Proceedings of the ICTACEM, Kharagpur, India, 29–31 December 2014.
6. Nagahara, T.; Sato, T.; Okamura, T. Effect of the Submerged Vortex Cavitation Occurred in Pump Suction Intake on Hydraulic Forces of Mixed Flow Pump Impeller. In Proceedings of the Fourth International Symposium on Cavitation, Pasedena, CA, USA, 20–23 June 2001.
7. Amin, A.; Kim, B.H.; Kim, C.G.; Lee, Y.H. Numerical Analysis of Vortices Behavior in a Pump Sump. *IOP Conf. Ser. Earth Environ. Sci.* **2019**, *240*, 32020. [CrossRef]
8. Sedlář, M.; Procházka, P.; Komárek, M.; Uruba, V.; Skála, V. Experimental Research and Numerical Analysis of Flow Phenomena in Discharge Object with Siphon. *Water* **2020**, *12*, 3330. [CrossRef]
9. Papierski, A.; Błaszczyk, A.; Kunicki, R.; Susik, M. Surface Vortices and Pressure in Suction Intakes of Vertical Axial-Flow Pumps. *Mech. Mech. Eng.* **2012**, *16*, 51–71.
10. Bayeul-Lainé, A.C.; Bois, G.; Issa, A. Numerical simulation of flow field in water-pump sump and inlet suction pipe. *IOP Conf. Ser. Earth Environ. Sci.* **2010**, *12*, 12083. [CrossRef]
11. Long, N.I.; Shin, B.R.; Doh, D.-H. Study on Surface Vortices in Pump Sump. *J. Fluid Mach.* **2012**, *74*, 60–66. [CrossRef]
12. Shin, B. Numerical Study of Effect of Flow Rate on Free Surface Vortex in Suction Sump. *Trans. Jpn. Soc. Comput. Eng. Sci.* **2018**, *2018*, 20180010.
13. Tokyay, T.; Constantinescu, G. Coherent structures in pump-intake flows: A large eddy simulation (LES) study. In Proceedings of the Korea Water Resources Association Conference, Seoul, Korea, 11–16 September 2005; pp. 231–232.
14. Sokolovskiy, M.A.; Carton, X.J.; Filyushkin, B.N. Mathematical Modeling of Vortex Interaction Using a Three-Layer Quasi-geostrophic Model. Part 2: Finite-Core-Vortex Approach and Oceanographic Application. *Mathematics* **2020**, *8*, 1267. [CrossRef]
15. Uruba, V. Decomposition methods in turbulent research. *Eur. Phys. J. Conf.* **2012**, *25*, 1095. [CrossRef]
16. Uruba, V. Near Wake Dynamics around a Vibrating Airfoil by Means of PIV and Oscillation Pattern Decomposition at Reynolds Number of 65,000. *J. Fluids Struct.* **2015**, *55*, 372–383. [CrossRef]
17. ANSYS Inc. *ANSYS CFX-Solver Theory Guide*; Release 19.2; ANSYS Inc.: Canonsburg, PA, USA, 2019.
18. Zwart, P.J.; Gerber, A.G.; Belamri, T. A Two-Phase Flow Model for Predicting Cavitation Dynamics. In Proceedings of the ICMF 2004 International Conference on Multiphase Flow, Yokohama, Japan, 30 May–3 June 2004.
19. Joa, J.C.; Kanga, D.G.; Kima, H.J.; Roha, K.W.; Yunea, Y.G. The Effect of Coriolis Force on the Formation of Dip on the Free Surface of Water Draining from a Tank. In Proceedings of the Transactions of the Korean Nuclear Society Autumn Meeting, Pyeongchang, Korea, 25–26 October 2007.
20. Menter, F.R.; Egorov, Y. A scale-adaptive simulation model using two-equation models. In Proceedings of the 43rd AIAA Aerospace Sciences Meeting and Exhibit, Reno, NV, USA, 10–13 January 2005. [CrossRef]
21. Menter, F.R.; Schutze, J.; Kurbatskii, K.A. Scale-Resolving Simulation Techniques in Industrial CFD. In Proceedings of the 6th AIAA Theoretical Fluid Mechanics Conference, Honolulu, HI, USA, 27–30 June 2011. [CrossRef]

Article

Optimal Decomposition for the Monthly Contracted Electricity of Cascade Hydropower Plants Considering the Bidding Space in the Day-Ahead Spot Market

Yang Wu [1], Chengguo Su [2], Shuangquan Liu [1,*], Hangtian Guo [2], Yingyi Sun [2], Yan Jiang [1] and Qizhuan Shao [1]

[1] Yunnan Power Dispatching & Control Center, Yunnan Power Grid Co., Ltd., Kunming 650011, China; wuy@yn.csg.cn (Y.W.); jiangyan@yn.csg.cn (Y.J.); shaoqizhuan@126.com (Q.S.)
[2] School of Hydraulic Science and Engineering, Zhengzhou University, Zhengzhou 450001, China; suchguo@163.com (C.S.); ght52969@163.com (H.G.); yingyi_sun@163.com (Y.S.)
* Correspondence: liushuangquan@yn.csg.cn

Abstract: With the gradual opening of China's electricity market, it is effective for cascade hydropower plants to simultaneously participate in both the monthly contract market and the day-ahead spot market to obtain higher power generation benefits. Hence, this paper studies the optimal decomposition model for the monthly contracted electricity of cascade hydropower plants considering the bidding space in the day-ahead spot market. The close hydraulic and electric connection between cascade hydropower plants, the implementation requirements of contracted electricity, and the uncertainty of the day-ahead market clearing price are all well considered. Several linearization techniques are proposed to address the nonlinear factors, including the objective function and the power generation function. A successive approximation (SA) approach, along with a mixed-integer linear programming (MILP) approach, is then developed to solve the proposed model. The presented model is verified by taking the decomposition of the monthly contracted electricity of cascade hydropower plants in China as an example. The results indicate that the developed model has high computational efficiency and can increase the power generation benefits compared with the conventional deterministic model. The effect of the penalty coefficient for imbalanced monthly contracted electricity is also evaluated, which provides a practical reference for market managers.

Keywords: cascade hydropower plants; decomposition of the monthly contracted electricity; day-ahead spot market; uncertainty of the clearing price; successive approximation; mixed-integer linear programming

Citation: Wu, Y.; Su, C.; Liu, S.; Guo, H.; Sun, Y.; Jiang, Y.; Shao, Q. Optimal Decomposition for the Monthly Contracted Electricity of Cascade Hydropower Plants Considering the Bidding Space in the Day-Ahead Spot Market. *Water* **2022**, *14*, 2347. https://doi.org/10.3390/w14152347

Academic Editors: Zhengwei Wang and Yongguang CHENG

Received: 26 May 2022
Accepted: 26 July 2022
Published: 29 July 2022

Publisher's Note: MDPI stays neutral with regard to jurisdictional claims in published maps and institutional affiliations.

1. Introduction

Hydropower is a type of renewable energy with flexible operation and mature technology, which is clean, low carbon, and has been favored by countries all over the world [1,2]. With more than 9000 hydropower dams registered across all continents, it supplies almost 70% of all renewable energy globally [3–5]. As of the end of 2019, the world's total installed hydropower capacity stood at about 1308 GW. Clean electricity generation from hydropower achieved a record 4306 TWh in 2019, the single greatest contribution from a renewable energy source in history. Fifty countries added hydropower capacity in 2019. Those with the highest individual increases in installed capacity were Brazil (4.92 GW), China (4.17 GW), and Laos (1.89 GW). The 11,233 MW Belo Monte project in Brazil became fully operational in 2019, while other major projects include the 1285 MW Xayaburi project in Laos, followed by the 990 MW Wunonglong and 920 MW Dahuaqiao projects in China [6].

China's total installed hydropower capacity climbed to 300 GW, accounting for 27% of the world's installed hydropower capacity and 17% of China's total installed power capacity [7]. In 2015, China implemented a new round of power industry reform and established a medium- and long-term electricity market [8]. Since 2017, China has organized and promoted the construction of spot markets in eight regions, including Southern China (starting from Guangdong Province) and Sichuan Province as the first batch of pilot projects [9,10]. It has been a developing trend for hydropower plants to participate in both the contract market and the day-ahead market [11,12]. Medium- and long-term contracts can lock the electricity price to avoid the risk of price fluctuation, but the price is generally low. The day-ahead market has created high profit opportunities for hydropower plants involved in the transaction, but the price volatility and uncertainty are strong, so there is a large profit risk. Because of the limitations of runoff, storage capacity, installed capacity, and other factors of hydropower plants, as well as the complex hydraulic and electric coupling relationship between cascade hydropower plants [13], there is a close internal relationship between the contract market and the day-ahead market of power distribution. Medium- and long-term contracts signed by the hydropower generation company (HGenCo) are required be physically settled, which may affect the total revenue of the HGenCo and bidding space of the day-ahead market. Hence, how to determine the power generation of the cascade hydropower plants that participates in the contract market and day-ahead market to obtain the maximum benefit is an extremely difficult problem for the HGenCo.

Because the power output of hydropower plants is limited by reservoir inflow and storage capacity [14], it is difficult to directly apply research on the participation of other types of power plants in the electricity market [15–17] to the present problem. Scholars have carried out some research on the participation of hydropower plants in the electricity market, and most of them can be broadly divided into two categories: (1) The first category is the optimal decomposition of yearly or monthly contracted electricity. For example, Lu et al. [18] proposed a long-term optimal operation method for cascade hydropower plants considering the allocation of power generation in multiple markets and the uncertainty of multiple variables. Li et al. [19] developed an information gap decision-making theory-based method for optimal medium-term stochastic cascade hydropower operation in a multimarket environment, which considers the hydrological and economic uncertainties. Shrestha et al. [20] studied the optimal management of hydropower resources in the medium term to maximize the expected revenue of a Nordic hydropower producer. Luo et al. [21] developed an optimal scheduling model for long-term generation schedules of a cascade hydropower plant, which takes into account the uncertainty of multiple market prices. Chen et al. [12] proposed an integrated solution methodology based on a multi-core parallel tabu genetic algorithm to provide the optimal assignment of bilateral contracts, considering the simulation of a hydro-dominated market. (2) The second category is the trading and dispatching strategy of hydropower plants in the day-ahead spot market. Yuan et al. [22] proposed an efficient method to solve the benefit-based optimal self-scheduling of several cascaded hydro plants in a pool-based day-ahead electricity market. Conejo et al. [23] addressed the self-scheduling of a hydro generating company to maximize the benefit from selling energy in the day-ahead market. Pousinho et al. [24] established a stochastic mixed-integer linear programming (MILP) model to maximize the expected total benefit of a HGenCo in the pool-based day-ahead market. Kongelf et al. [25] proposed a stochastic MILP approach to formulate a coordinated planning problem for a hydropower producer, considering the uncertainty of portfolio size in multiple electricity markets. However, there are few studies on the decomposition of monthly contracted electricity for cascade hydropower plants connected with the bidding in the day-ahead spot market.

The optimal decomposition of monthly contract electricity is part of the mid- and long-term optimal operation problem of cascade hydropower plants, which is a typical nonlinear programming (NLP) problem with multiple variables, high dimensionality, and complex constraints. At present, the methods of solving such an NLP problem can be divided into three categories: dynamic programming (DP) and its improved algorithm [26], intelligent algorithms represented by particle swarm optimization algorithm [27], and the mathematical programming methods [28,29]. The model constructed in this paper struggles to meet the requirement of having no aftereffect because it takes into account the uncertainty of the day-ahead market clearing price and involves the coupling relationship between monthly contract electricity and day-ahead market trading electricity. Thus, DP would be a hindrance in formulating the present problem as a multi-stage optimization problem. The intelligent algorithms would struggle to effectively handle the complex operation constraints and cannot guarantee a global optimal solution within finite iterations. The computational efficiency of NLP algorithm is affected by the initial solution and easily falls into the local optimal solution, and therefore are also unsuitable for solving the present problem. Among mathematical programming methods, the MILP approach becomes more and more mature and has been widely used in in the field of optimal reservoir operation due to its good performance in addressing complex constraints and producing stable calculation results [30–32].

Aiming at the above problems, this paper emphasizes on the optimal decomposition model for the monthly contract electricity of cascade hydropower plants considering the bidding space in the day-ahead spot market. In this model, the generation benefits of cascade hydropower plant participating in both the monthly contract market and the day-ahead spot market are maximized. The uncertainty of the day-ahead market clearing price is modeled by a scenario analysis technique. An efficient and novel method, coupling a successive approximation (SA) approach and a MILP approach is then proposed to solve the proposed model. The rationality and validity of the proposed model and method are verified by an example of monthly contracted electricity decomposition plan of a cascade hydropower plant in China. The major contributions of this paper are clarified as follows: (1) An optimal scheduling model for cascade hydropower plants participating in both the monthly contract market and the day-ahead spot market is established, considering the uncertainty of the day-ahead market electricity price. This model mirrors the real-life situation of cascade hydropower plants participating in China's electricity market. (2) SA along with a MILP approach is developed to solve this complex issue, which guarantees an optimal or near-optimal solution and significantly improves the solving efficiency.

The organization of the remaining parts of this paper is shown as follows. The mathematical formulation of this problem is established in Section 2. Section 3 provides the solution technique for the optimization model. The optimization results of the case study are presented and discussed in Section 3. Finally, conclusions are drawn in Section 4.

2. Mathematical Formulation

2.1. Problem Description

According to the trading rules of the electricity market, each hydropower plant signs the monthly electricity contract for the next month with the power purchasing user in the current month, and the contract stipulates the amount of electricity to be traded and the settlement price. In order to improve the consumption of renewable energy sources like hydropower, hydropower plants are allowed to participate in the day-ahead market according to the market situation and their own power generation capacity by self-scheduling during their operation within a month. In this case, the hydropower plant only declares the electricity quantity, not the price, and uses the market clearing price as the settlement price for the electricity quantity traded in the day-ahead market (i.e., as the price-taker). According to the settlement rules, the trading electricity in the day-ahead market will be settled the next day, while the monthly contract electricity will be settled at the beginning of the next month. Therefore, the monthly contracted electricity that has been agreed in advance by the hydropower plants can still bide in the day-ahead market, but the imbalanced portion of the monthly contracted electricity will be compensated to the power purchasing user according to the market rules or the contract agreement. At present, the medium- and long-term market is the energy market, and the power generation curves of the power plants are not agreed. Therefore, the power market operation institution or the power grid dispatching center will require the hydropower plants to report the water level process for the next month, the daily electricity generation, and the monthly and day-ahead market combined transaction electricity at the end of the month. To get more benefits, it is necessary for cascade hydropower plants to optimize the decomposition scheme for monthly contract electricity and the trading electricity in the day-ahead market based on their power generation capacity and predicted day-ahead market clearing price. Due to the inevitable deviation between the predicted day-ahead market clearing price and the actual value, it is challenging to develop a decomposition scheme for the monthly contracted electricity. Therefore, the uncertainty of the day-ahead market clearing price need to be well considered when the decomposition scheme for the monthly contract electricity of cascade hydropower plants is formulated.

In this paper, it is assumed that the day-ahead market clearing price is given. Moreover, cascade hydropower plants belong to the same HGenCo, and each hydropower plant participates in the market transaction independently. However, in order to give full play to the compensation and regulation function between cascade plants, all hydropower plants are uniformly operated by their owners.

2.2. Uncertainty Treatment of Day-Ahead Market Clearing Price

This section uses the scenario analysis technique to generate a series of scenarios to model and analyze the randomness of day-ahead market clearing price, in order to transform the stochastic optimization problem into an equivalent deterministic one.

(1)　It is assumed that the forecasted error of the clearing price series $\left\{p_1^{df}, p_2^{df}, \ldots, p_T^{df}\right\}$ obeys the normal distribution $N(\mu, \sigma^2)$ at any time period, where μ and σ are the mean and standard deviation of the forecasted error, respectively, and they satisfy that $\mu = 0, \sigma = 0.1 \times p_t^{df}$.

(2)　The Latin Hypercube Sampling (LHS) method [31] is adopted for generating the scenarios for the forecasted error of the day-ahead market price. In this method, the sampling probability distribution is stratified first, and then samples are randomly selected from each layer in turn, which can effectively improve the coverage degree of sampling samples to the distribution space of random variables.

(3)　To adequately reflect the stochastic characteristics of the day-ahead market clearing price, more price scenarios will be generated through LHS, which mainly aims to avoid a low calculation accuracy due to a small number of scenarios. In this paper, the fast backward/forward method [33] is adopted to balance the solving accuracy and efficiency, that is, to minimize the number of scenarios while maintaining the main characteristics of price scenarios.

2.3. Objective Function

The goal of optimal monthly contract electricity decomposition is to maximize the expected monthly income, whilst meeting various operational constraints. The cost of cascade hydropower plants is mainly construction cost; thus, power generation cost is not considered in this model. According to the electricity market trading rules, the monthly contract is a physical contract. This implies that although part of the monthly contracted electricity quantity of from hydropower plants can participate in the day-ahead market bidding, it needs to accept a certain penalty. The mathematical expression of the objective function is presented as follows:

$$\max F = F_1 + F_2 - F_3 + F_4 \tag{1}$$

$$F_1 = \sum_{i=1}^{I} P_i^m \times E_i^{mb} \tag{2}$$

$$F_2 = \sum_{i=1}^{I} P_i^{pm} \times \max(E_i^{ma} - E_i^{mb}, 0) \tag{3}$$

$$F_3 = \sum_{i=1}^{I} P_i^{nm} \times \max(E_i^{mb} - E_i^{ma}, 0) \tag{4}$$

$$F_4 = \sum_{s=1}^{S} \sum_{i=1}^{I} \sum_{t=1}^{T} p_s \times E_{i,t}^{d} \times P_{s,t}^{d} \tag{5}$$

where F denotes the target total monthly benefits (in CNY); F_1 denotes the basic income according to the monthly electricity contract (in CNY); F_2 denotes the extra income obtained from the excess settlement of the monthly contracted electricity (in CNY); F_3 denotes the penalty cost for an insufficient amount of monthly contracted electricity (in CNY); F_4 denotes the expected benefits of cascade power plants from participating in the day-ahead market (in CNY); I and i are the total number and index of plants, respectively; T and t denote the dispatching cycle and dispatching period, respectively; E_i^{mb} and P_i^{m} are the monthly contract electricity and electricity price, respectively, of the hydropower plant i (in MWh and CNY/MWh); E_i^{ma} denotes monthly contract electricity that actually completed (in MWh); P_i^{pm} and P_i^{nm} are the electricity prices corresponding to the excess and deficiency of monthly contracted electric quantity, respectively (in CNY/MWh). They satisfy that $P_i^{pm} = (1 - \tau) \times P_i^{m}$, $P_i^{nm} = (1 + \tau) \times P_i^{m}$, and τ is the penalty coefficient of imbalanced contracted electricity; S and s are the total number and index of scenarios for day-ahead market clearing price; p_s denotes the probability of price scenario s; $P_{s,t}^{d}$ denotes the day-ahead market clearing price in scenario s; $E_{i,t}^{d}$ denotes the trading electricity of plant i in the day-ahead market in time period t (in MWh).

2.4. Constraints

(1)   Water balance constraints

$$V_{i,t} = V_{i,t-1} + 3600 \cdot (Q_{i,t}^{in} - Q_{i,t}) \times \Delta t \tag{6}$$

where $Q_{i,t}^{in}$ denote the total inflow of plant i in period t (in m^3/s); $Q_{i,t}$ denotes the total water discharge of plant i in period t (in m^3/s); Δt is the duration of period t (in h); $V_{i,t}$ is the water storage at the end of period t (in m^3).

(2)   Hydraulic connection constraints

$$Q_{i,t}^{in} = Q_{i-1,t} + Q_{i,t}^{lin} = Q_{i-1,t}^{g} + Q_{i-1,t}^{s} + Q_{i,t}^{lin} \tag{7}$$

where $Q_{i,t}^{lin}$ denote the local inflow of plant i in period t (in m^3/s); $Q_{i-1,t}^{g}$ and $Q_{i-1,t}^{s}$ denote the generating water flow and water spillage, respectively (in m^3/s). Note that $Q_{i,t}^{s}$ is set to 0 in this paper because hydropower curtailment is generally not allowed according to China's clean energy consumption policy.

(3)   Water level constraints

$$Z_{i,\min} \leq Z_{i,t} \leq Z_{i,\max} \tag{8}$$

$$Z_{i,0} = Z_{i,begin} \tag{9}$$

$$Z_{i,T} = Z_{i,end} \tag{10}$$

where $Z_{i,\max}$ and $Z_{i,\min}$ denote the upper and lower bounds of the reservoir water level (in m); $Z_{i,t}$ is the reservoir water level in period t (in m); $Z_{i,begin}$ is the initial water level and $Z_{i,end}$ is the target water level at the end of the planning horizon (in m).

(4) Water discharge constraints

$$Q_{i,\min} \leq Q_{i,t} \leq Q_{i,\max} \tag{11}$$

where $Q_{i,\min}$ and $Q_{i,\max}$ denote the lower and upper bounds of the total water discharge of plant i, respectively (in m^3/s).

(5) Power output constraint

$$P_{i,\min} \leq P_{i,t} \leq P_{i,\max} \tag{12}$$

where $P_{i,\max}$ and $P_{i,\min}$ represent the upper and lower bounds of the power output of plant i, respectively (in MW); $P_{i,t}$ represents the power output of plant i in period t (in MW), which satisfies Equation (13).

$$P_{i,t} = k_{i,t} \cdot Q^g_{i,t} \times H_{i,t}/1000 \tag{13}$$

$$k_{i,t} = f_{i,kh}(H_{i,t}) \tag{14}$$

where $k_{i,t}$ is the output factor of plant i, and is related to the water head [34,35]; $H_{i,t}$ is the water head of plant i in period t (in m).

(6) Water head constraints

$$H_{i,t} = (Z_{i,t-1} + Z_{i,t})/2 - zd_{i,t} \tag{15}$$

where $zd_{i,t}$ denotes the tailwater level in period t (in m).

(7) Forebay water level–water storage relationship

$$V_{i,t} = f_{i,zv}(Z_{i,t}) \tag{16}$$

$f_{i,zv}(\cdot)$ represents the forebay water level–water storage relationship function of plant i.

(8) Tailwater level–water discharge relationship

$$zd_{i,t} = f_{i,zq}(Q_{i,t}) \tag{17}$$

$f_{i,zq}(\cdot)$ represents the tailwater level–water discharge relationship function of plant i.

(9) Constraints on trading electricity in the day-ahead market

$$E^d_{i,t} \leq P_{i,t} \times \Delta t \tag{18}$$

The above formula indicates that the trading electricity of plant i in the day-ahead market in time period t shall not exceed the power generation of the day.

(10) Trading electricity constraints

$$E^{ma}_i + \sum_{t=1}^{T} E^d_{i,t} = \sum_{t=1}^{T} P_{i,t} \times \Delta t \tag{19}$$

The above formula indicates that the sum of the actually completed monthly contracted electricity and the trading electricity in the day-ahead market of the current month shall be equal to the total monthly power generation of the power plant.

3. Solving Technique

According to the analysis in Section 1, the optimized model is nonlinear and non-convex, making it difficult to solve by conventional algorithms. To solve such a complex problem, a good idea is to approximate each nonlinear function by a series of linear functions. Hence, in this model, the nonlinear functions are linearized through several linearization techniques, so that the original NLP model is transformed into an MILP formulation. Based on this, a mature optimization solver can be adopted to solve the MILP model.

In the established model, Equations (3) and (4) contain a $\max(\cdot)$ function. The water head function (i.e., Equation (15)) and power generation function (i.e., Equation (13)) are also nonlinear functions. Considering that the linear approximation of the water head constraint has been well established by many studies [29,30], this work aims to study the linearization of the objective function (Equations (3) and (4)) and the power production function (Equation (13)).

3.1. Linear Approximation of the Objective Function

We introduce a binary variable $\mu_i \in \{0, 1\}$, and Equations (3) and (4) can be converted into the following constraints.

$$F_2 = \sum_{i=1}^{I} \mu_i \times P_i^{pm} \times (E_i^{ma} - E_i^{mb}) \tag{20}$$

$$F_3 = \sum_{i=1}^{I} (1 - \mu_i) \times P_i^{nm} \times (E_i^{mb} - E_i^{ma}) \tag{21}$$

$$E_i^{ma} - E_i^{mb} \leq M \times \mu_i \tag{22}$$

$$E_i^{mb} - E_i^{ma} \leq M \times (1 - \mu_i) \tag{23}$$

where M is a very large number; μ_i is a binary indicator variable, which is equal to 1 if the amount of monthly contracted electricity that is actually settled is higher than the amount signed off, and otherwise, it is equal to 0.

Equations (20) and (21) are still nonlinear because there is still multiplication of integer variables and continuous variables (i.e., $\mu_i \times E_i^{ma}$). Hence, we introduce two auxiliary variables, X_i and K_i, to convert Equations (20) and (21) into the following linear constraints.

$$F_2 = \sum_{i=1}^{I} P_i^{pm} \times (X_i - \mu_i \times E_i^{mb}) \tag{24}$$

$$F_3 = \sum_{i=1}^{I} P_i^{pm} \times \left[(1 - \mu_i) \times E_i^{mb} - K_i \right] \tag{25}$$

$$X_i = E_i^{ma} - K_i \tag{26}$$

$$X_i \leq M \times \mu_i \tag{27}$$

$$X_i \geq -M \times \mu_i \tag{28}$$

$$K_i \leq M \times (1 - \mu_i) \tag{29}$$

$$K_i \geq -M \times (1 - \mu_i) \tag{30}$$

When $\mu_i = 1$, i.e., the monthly contracted electricity that is actually settled is higher than the amount signed for in the contract, K_i is equal to 0 because of Equations (29) and (30). F_3 is then equal to 0 according to Equation (25), which is consistent with the fact that the penalty cost is 0 when then monthly contracted electricity has been settled. Equations (24) and (20) are equivalent at this time due to the constraint of Equation (32). Similarly, when $\mu_i = 0$, F_2 is equal to 0 and Equations (25) and (21) are equivalent. To sum up, the combination of linear constraints in Equations (22)–(30) is the equivalent transformation of Equations (3) and (4).

3.2. Linear Approximation of the Power Generation Function

The output factor $k_{i,t}$ is closely related to the water head $H_{i,t}$, while $H_{i,t}$ depends on the generation water flow [34,35]. Hence, the power generation function is a very complicated nonlinear function for the generation water flow. If the generation water flow is chosen as a decision variable and the net water head is fixed, the power generation function will become a linear one. Based on the above analysis, the successive approximation (SA) approach is introduced, which is an efficient and practical method to deal with complex nonlinear optimization problems [36–38]. The basic idea of this method is to solve a series of fixed water head subproblems using the MILP approach, and the solution for the original complex problem can be obtained after successive calculations. A flow chart for the proposed SA approach is illustrated in Figure 1, and the specific procedures are as follows.

(1) Set the convergence precision δ of the SA approach and let the index of iterations $n = 1$.

(2) The initial solution has a great influence on the computational efficiency of the SA approach. Thus, to enhance the convergence speed, the initial water head of all the hydropower plants from upstream to downstream $\left\{ H_{i,0}^0, H_{i,1}^0, \cdots, H_{i,T}^0 \right\}$ is generated using Equations (31)–(33). The initial output factor of each hydropower plant $\left\{ k_{i,0}^0, k_{i,1}^0, \cdots, k_{i,T}^0 \right\}$ is then calculated using Equation (14).

$$\Delta W_i = V_{i,begin} - V_{i,end} + \sum_{t=1}^{T} Q_{i,t}^{in} \times \Delta t \tag{31}$$

$$\overline{Q_i} = \Delta W_i / \sum_{t=1}^{T} \Delta t \tag{32}$$

$$\overline{H_{i,t}} = (Z_{i,begin} + Z_{i,end})/2 - f_{i,zq}(\overline{Q_i}) \forall t \in [1, T] \tag{33}$$

where ΔW_i and $\overline{Q_i}$ denotes the total discharge volume and average water discharge, respectively, during the scheduling horizon. $\overline{H_{i,t}}$ is the average water head of hydropower plant i during the scheduling horizon.

(3) Based on the given water head and the output factor of each hydropower plant, the MILP based model for the optimal decomposition of monthly contracted electricity for cascade hydropower plants is established by using the linearization techniques presented in Section 3.1 and Ref. [30].

(4) An efficient optimization solver is adopted to solve the MILP model, and the dispatching schemes, including the water discharge, forebay water level, and power output of each hydropower plant, can be obtained.

(5) Calculate the new water head $H_{i,t}^n$ and corresponding output factor $k_{i,t}^n$ after the nth iteration using Equations (14)–(17). Judge if $max\left|H_{i,t}^n - H_{i,t}^{n-1}\right|/H_{i,t}^n \leq \delta, \forall i, \forall t$. If true, end the iteration and output the latest solution as the optimal dispatching scheme, otherwise let $n = n + 1$ and repeat steps (3)–(5).

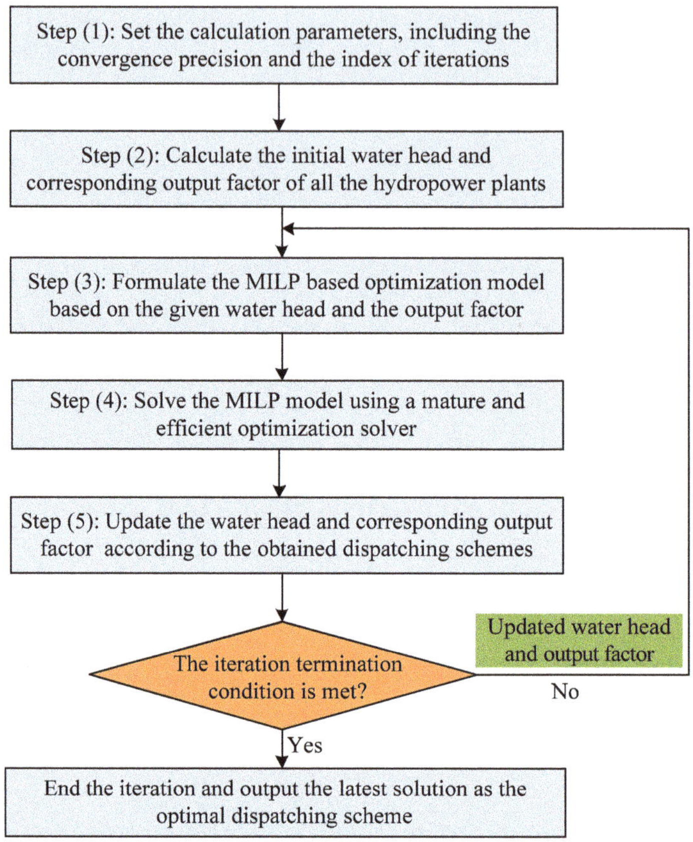

Figure 1. Flow chart of the successive approximation.

4. Case Studies

In this paper, the monthly contracted electricity decomposition for two cascade hydropower plants was taken as an example to test the proposed model and method. The cascade power plants involved in the calculation include two hydropower plants, upstream plant A and downstream plant B, whose main operating parameters are shown in Table 1. The actual operational conditions of the cascade hydropower plants during March 2020, including local inflow, monthly contract electricity, contract price, water level at the beginning of month, and control water level at the end of month, were considered as references for the case study. The day-ahead market clearing price is forecast by each hydropower plant at the end of February 2020. The operational control conditions of each hydropower plant are shown in Table 2. The local inflow of each plant and the forecasted day-ahead market clearing price are shown in Figures 2 and 3, respectively. The allowable deviation of control water level at the end of the month was set to 0.01 m.

Table 1. Characteristic parameters of the cascade hydropower plants.

Plant	Regulation Performance	Normal Water Level/m	Dead Water Level/m	Installed Capacity /MW	Maximum Generating Water Flow/(m³/s)	Minimum Total Water Discharge /(m³/s)
A	Seasonal	835	818	2 × 60	260	5
B	Weekly	756	740	2 × 65	209	5

Table 2. Operation control conditions of each plant.

Plant	Monthly Contract Electricity/kWh	Contract Price/(CNY/kWh)	Water Level at the Beginning of Month/m	Control Water Level at the End of Month/m
A	673×104	0.19062	822.29	821.86
B	832×104	0.19039	751.04	752.76

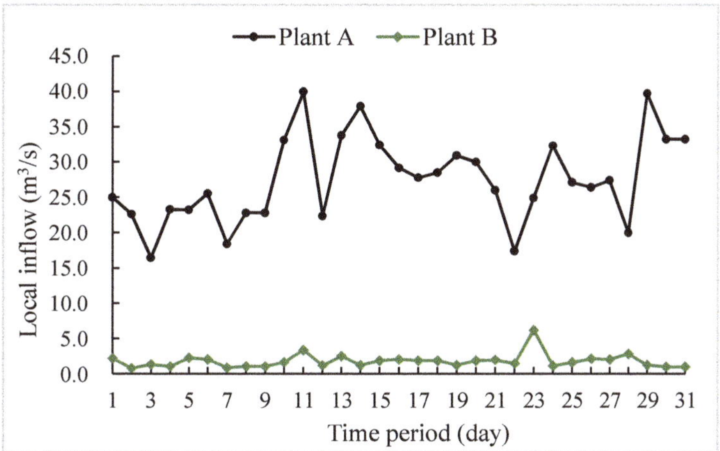

Figure 2. Local flow of each hydropower plant.

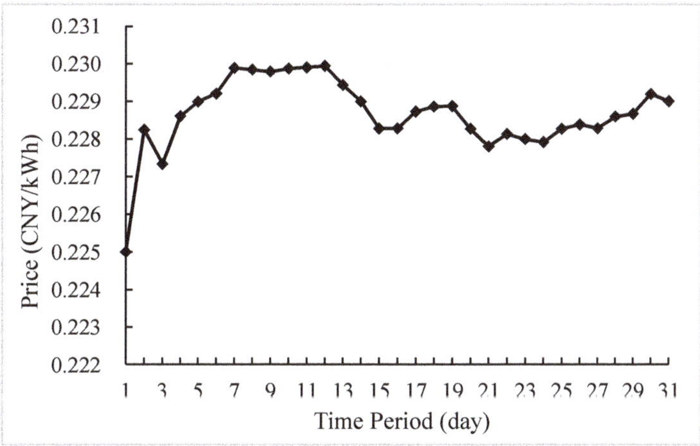

Figure 3. Forecasted day-ahead market clearing price.

In total, 300 scenarios were generated using the LHS method, and the fast backward/forward method was used to reduce the number of scenarios to 50. In the calculation, the scheduling cycle was 1 month, and the dispatching period was 1 day. The penalty coefficient for imbalanced contracted electricity τ was set as 0.3. A commercial optimization software package, LINGO solver, was used for solving the proposed model, and the computing environment was a ThinkPad PC with quad-core CPU and 16 GB memory.

The total revenue of the cascade hydropower plants is 578.92×10^4 CNY, including 286.62×10^4 CNY from the monthly contracted electricity revenue and 291.3×10^4 CNY from the day-ahead market trading electricity revenue. The calculation time of the model is 137 s, which fully meets the timeliness requirements of medium- and long-term scheduling of cascade hydropower plants, reflecting the high solving efficiency of the optimization model established in this paper.

The power generation and water level process of each hydropower plant obtained by the optimized calculation are shown in Figure 4. As can be seen, the upstream plant A can make full use of its own seasonal regulating storage capacity to realize the temporal and spatial redistribution of runoff, so as to respond to the change of day-ahead market clearing price and improve its own income. During periods with high electricity prices, plant A uses its limited power generation capacity to participate in day-ahead market trading and increase its power output as much as possible to generate more electricity. Conversely, in periods with low electricity prices, plant A only generates electricity with the minimum generating water flow, and almost all the generated electricity goes towards the settlement of the monthly contracted electricity due to the penalty that can be incurred for an imbalance from the contracted electricity. Downstream plant B needs the upstream plant for flow compensation due to its small, regulated storage capacity and almost no local inflow. In periods with low electricity prices, plant B only generates electricity with the minimum flow, continuously raises the water level for water storage, and increases the water head. When electricity prices are high, the highest possible water level is maintained at plant B in order to participate in day-ahead market trading and maximize power generation before the water level gradually falls back to the control water level at the end of the month.

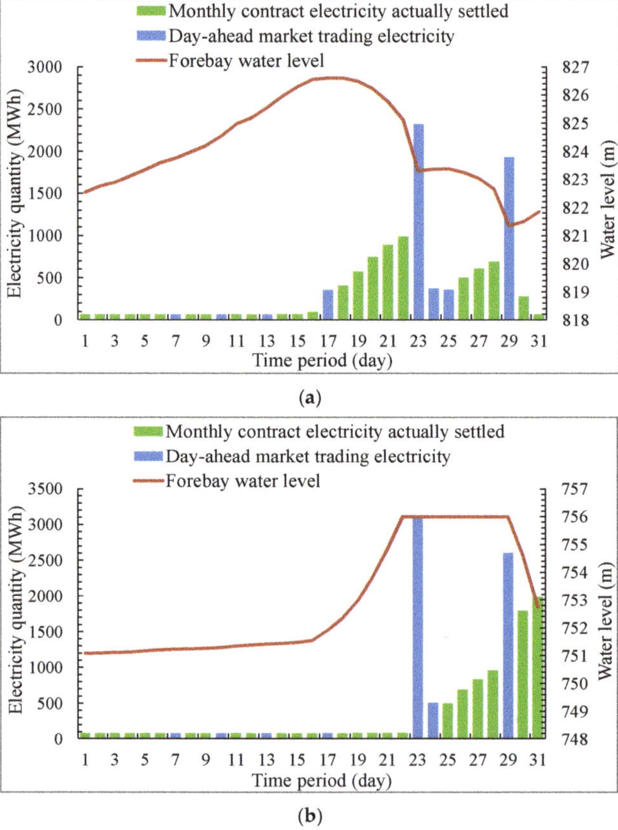

Figure 4. The power generation and water level process of each plant. (**a**) Plant A; (**b**) Plant B.

The benefits of the stochastic scheduling model established in this paper (hereinafter referred to as Model (1)), are compared with the traditional deterministic scheduling model based on predicted values (hereinafter referred to as Model (2)). In Model 2, the predicted value of day-ahead market electricity price is taken as the determined input value for scheduling. The two models adopt the same control objectives and constraints, and a comparison of the results is shown in Table 3. It can be seen that, on the basis of guaranteeing the monthly contracted electricity, the total power generation and the total generation revenue of the cascade hydropower plants optimized by Model 1 is 2720×10^4 kWh and 578.92×10^4 CNY, respectively, while under Model 2, the values are 2731×10^4 kWh and 567.49×10^4 CNY, respectively. Compared to the deterministic model, the total revenue of the proposed model increases by 2% when power generation is reduced. This shows that when making a monthly contract electricity decomposition plan, taking into account the uncertainty of the day-ahead market clearing price can significantly increase the expected benefits of the HGenCo.

Table 3. The comparison results of the Model 1 and Model 2.

Model	$F_1/\times 10^4$ **CNY**	$F_2/\times 10^4$ **CNY**	$F_3/\times 10^4$ **CNY**	$F_4/\times 10^4$ **CNY**	$F/\times 10^4$ **CNY**
Model 1	286.62	0	0	292.3	578.92
Model 2	286.62	0	0	280.87	567.49

To verify the impact of penalty coefficient for imbalanced monthly contracted electricity (τ) on the optimal scheduling results, optimization results with different τ values are compared and analyzed, which are presented in Table 4. When τ is small ($\tau = 0.1$ or 0.2), the negative deviation penalty price of monthly contracted electricity of plant A and B is generally lower than the day-ahead market clearing price. In this case, cascade hydropower plants will choose to violate the monthly electricity transaction contract and compensate the contract buyer and allow for more generation to participate in the day-ahead market transaction to obtain higher profits. However, when $\tau = 0.3$ or 0.4, the negative deviation penalty price of the monthly contracted electricity of plant A and B is relatively high. In this case, the cascade hydropower plants will fulfill the monthly contract, and only the remaining generation will participate in the day-ahead market. Therefore, the dispatching agencies should analyze all possible situations when making power trading rules and formulate reasonable penalty coefficients for imbalances of the monthly contracted electricity so as to avoid a large number of defaults and ensure the long-term stable operation of the electricity market.

Table 4. Optimization results with different τ values.

τ	Completed Monthly Contracted Electricity/$\times 10^4$ **kWh**	Day-Ahead Market Trading Electricity/$\times 10^4$ **kWh**	Total Power Generation /$\times 10^4$ **kWh**	Total Revenue/$\times 10^4$ **CNY**
0.1	0	2711	2711	609.45
0.2	202	2430	2712	581.90
0.3	1505	1215	2720	578.92
0.4	1505	1215	2720	578.92

5. Discussion

As shown in Figure 3, the periods with the highest forecasted prices are days 7–13. However, there is always an inevitable deviation between the predicted day-ahead market clearing price and the actual value. In this paper, 50 scenarios are generated to represent all possible actual day-ahead market clearing prices. After careful analysis, we found that in most of the scenarios, the high electricity price is mainly concentrated in days 22–30. To obtain higher expected power generation benefits, it is recommended to participate in day-ahead market trading and increase its power output as much as possible to generate more electricity during days 22–30.

To the best of our knowledge, there are very few studies on the decomposition of monthly contracted electricity for cascade hydropower plants connected with the bidding in the day-ahead spot market. Thus, we cannot find other relevant optimization models and compare them. The deterministic scheduling model based on predicted values is usually adopted by the operators of the hydropower plants in practical engineering applications, and the deterministic scheduling model is compared with the proposed optimization model as benchmark model in many literatures [39–41]. For deterministic operation, we considered the day-ahead market clearing prices to be a known input of the operation model, assuming that the market clearing prices could be forecast accurately. However, there is always an inevitable deviation between the predicted day-ahead market clearing price and the actual value. To consider the impact of forecast errors on the decision-making process, the stochastic model was adopted to produce operational decisions using multiple day-ahead market clearing price processes to characterize forecast uncertainty. Since the operational decisions were required to satisfy all the scenarios, the total generation profit of the stochastic model is inevitably affected and is only 2% higher than those of deterministic model. However, as the forecast errors of the day-ahead market clearing price are inevitable in actual operation, the proposed model can avoid solutions that imply small profits or major costs, hedging against risk and uncertainty.

As a consequence of the market power of some producers, two types of generation companies can be listed: price-takers [42] and price-makers [39]. Price-takers accept market clearing prices without being able to affect them. Instead, price-makers have market power, thus being able to influence market prices to increase profit. In competitive electricity markets, the profit of the generation companies depends not only on their own decisions, but also on the decisions of the other companies. Under perfect competition, the market share of every generation company is small, and no company can affect the market price. In this case, every company takes market prices for granted when devising its offering strategy, acting as a price-taker. However, some generation companies may have a relatively high market share and are capable of exercising their market power, influencing market prices for their own benefit. It means that a perfect competition model cannot be used, since the companies act as price-makers. When devising its offering strategy, a price-maker hydro producer takes into account the fact that it can affect market prices with its offers. Our focus is to obtain the mid-term scheduling strategy of a price-taker hydropower generation company (HGenCo). The case studies in this paper include two hydropower plants with a total installed capacity of 250 MW. The market share of this HGenCo is small and cannot affect the market price.

The penalty coefficient for imbalanced monthly contracted electricity (τ) is very important for the smooth settlement of the monthly contracted electricity. When τ is small ($\tau = 0.1$ or 0.2), cascade hydropower plants will choose to violate the monthly electricity transaction contract and allow for more generation to participate in the day-ahead market transaction to obtain higher profits. While when $\tau = 0.3$ or 0.4, the cascade hydropower plants will fulfill the monthly contract. The value of penalty coefficient depends on the interest tendency of market managers. If the market managers hope to avoid large defaults on medium-and long-term contracts, they can set the penalty coefficient to 0.3 or 0.4. If they want to encourage power plants to participate freely in various markets to reap greater benefits, they can set the penalty coefficient very low, such as 0.1.

The establishment of an optimal scheduling model is based on the operation mechanism of medium- and long-term market and day-ahead spot market. China's power supply structure and power consumption characteristics have certain uniqueness, and the construction of China's electricity market is still in the initial stage, so China's electricity market mechanism has certain differences from other countries. For example, the Nordic electricity market has developed a model of financial contracts plus spot trading. The time span of trading in the financial contract market includes weeks, quarters, and years. Financial contracts do not require physical delivery and are primarily used as a means of price hedging and risk management. This means that, after signing a medium- and long-term contract, the power plants can participate in the spot market without default penalty. Hence this model cannot be generalized to other countries' markets in our opinion. However, it can provide some reference for other countries that are in the initial stage of electricity market construction, such as India and Brazil.

National and foreign direct investments are needed for economic developments and national projects, including hydropower works, mainly in relation to less-favored economic areas where social risk could appear [43]. Such hydropower plants are useful to alleviate areas with poverty and solve urban and rural issues of electricity for poor communities [44,45]. However, when more cascade hydropower plants are included in the proposed model, the computational efficiency will be reduced to some extent, and the calculation may take several hours due to the consideration of the day-ahead market clearing price uncertainty and multiple iterations of the SA approach. Hence, model decomposition techniques and parallel techniques should be integrated into the solution method in the future research to further improve the solution efficiency and increase the engineering application value of the proposed model.

6. Conclusions

In view of the coexistence of the medium- and long-term contract markets and the day-ahead spot market in China's electricity market, an optimal decomposition model for the monthly contracted electricity of cascade hydropower plants is established, considering the bidding space of day-ahead market. The validation of the proposal model is applied to the decomposition of the monthly contracted electricity of China's cascade hydropower plants as an example. The following conclusions are drawn:

(1) A scenario analysis technique and several effective linearization strategies are put forward to address the uncertain and nonlinear factors in the optimization model, including the uncertain day-ahead market clearing price, the nonlinear objective function, and the nonlinear power generation function of each hydropower plant. For such a complex research problem, the combination of the SA approach and MILP approach is computationally efficient with a calculation time of 137 s.

(2) The total revenue obtained from the proposed stochastic optimization model is 578.92×10^4 CNY. Compared to the deterministic model, the total revenue of the proposed model increases by 2% when power generation is reduced. Furthermore, as the forecast errors of the day-ahead market clearing price are inevitable in actual operation, the proposed model can avoid solutions that imply small profits or major costs, hedging against risk and uncertainty.

(3) The penalty coefficient for imbalanced monthly contracted electricity (τ) is very important for the smooth settlement of the monthly contracted electricity. When τ is small ($\tau = 0.1$ or 0.2), cascade hydropower plants will choose to violate the monthly electricity transaction contract and allow for more generation to participate in the day-ahead market transaction to obtain higher profits. While, when $\tau = 0.3$ or 0.4, the cascade hydropower plants will fulfill the monthly contract. Therefore, market managers need to formulate a reasonable penalty coefficient to avoid a large number of defaults and ensure the long-term stable operation of the electricity market.

At present, China's electricity market construction is still in the preliminary stage, and many factors are still to be studied and tested in practice. Due to the complex operation constraints of cascade hydropower plants and multiple uncertainties, it is challenging to develop a decomposition scheme for the monthly contracted electricity of cascade hydropower plants. This paper can provide guidance for China's hydropower participation in the electricity market, and also provide reference for other countries that are in the initial stage of electricity market construction, such as India and Brazil. In addition, the uncertainty of runoff and the accelerated solution method of the model will be further considered in the future study.

Author Contributions: Data curation, Y.S.; Formal analysis, Q.S.; Funding acquisition, C.S.; Investigation, Y.S.; Methodology, Y.W. and H.G.; Project administration, S.L.; Software, C.S.; Supervision, S.L.; Validation, Y.J.; Writing—original draft, Y.W.; Writing—review and editing, C.S. and S.L. All authors have read and agreed to the published version of the manuscript.

Funding: The research presented in this paper was supported by the National Natural Science Foundation of China (No. 52109041), China Postdoctoral Science Foundation (No. 2021M690139) and the Science and Technology Project of Yunnan Power Grid Co., LTD (No. YNKJXM20200167).

Institutional Review Board Statement: Not applicable.

Informed Consent Statement: Not applicable.

Data Availability Statement: Not applicable.

Conflicts of Interest: The authors declare no conflict of interest.

Abbreviations

HGenCo	Hydropower generation company
NLP	Nonlinear programming
MILP	Mixed integer linear programming
DP	Dynamic programming
SA	Successive approximation

References

1. Yang, W.; Norrlund, P.; Saarinen, L.; Witt, A.; Smith, B.; Yang, J.; Lundin, U. Burden on hydropower units for short-term balancing of renewable power systems. *Nat. Commun.* **2018**, *9*, 2633. [CrossRef]
2. Cheng, C.; Yan, L.; Mirchi, A.; Madani, K. China's booming hydropower: Systems modeling challenges and opportunities. *J. Water Resour. Plan. Manag.* **2017**, *143*, 02516002. [CrossRef]
3. Llamosas, C.; Sovacool, B.K. The future of hydropower? A systematic review of the drivers, benefits and governance dynamics of transboundary dams. *Renew. Sustain. Energy Rev.* **2021**, *137*, 110495. [CrossRef]
4. Dukpa, R.D.; Joshi, D.; Boelens, R. Contesting hydropower dams in the Eastern Himalaya: The cultural politics of identity, territory and self-governance institutions in Sikkim, India. *Water* **2019**, *11*, 412. [CrossRef]
5. Crețan, R.; Vesalon, L. The political economy of hydropower in the communist space: Iron Gates revisited. *Tijdschr. Econ. Soc. Geogr.* **2017**, *108*, 688–701. [CrossRef]
6. 2020 Hydropower Status Report. Available online: https://www.hydropower.org/publications/2020-hydropower-status-report (accessed on 10 May 2022).
7. Statistics of the National Power Industry in 2020. Available online: http://www.nea.gov.cn/2021-01/20/c_139683739.htm (accessed on 10 May 2022).
8. Zeng, M.; Yang, Y.; Wang, L.; Sun, J. The power industry reform in China 2015: Policies, evaluations and solutions. *Renew. Sustain. Energy Rev.* **2016**, *57*, 94–110. [CrossRef]
9. Guo, H.; Davidson, M.R.; Chen, Q.; Zhang, D.; Jiang, N.; Xia, Q.; Kang, C.; Zhang, X. Power market reform in China: Motivations, progress, and recommendations. *Energy Policy* **2020**, *145*, 111717. [CrossRef]
10. Bao, M.; Guo, C.; Wu, Z.; Wu, J.; Li, X.; Ding, Y. Review of electricity spot market reform in China: Current status and future development. In Proceedings of the 2019 IEEE Sustainable Power and Energy Conference (iSPEC), Beijing, China, 21–23 November 2019.
11. Cheng, C.; Chen, F.; Li, G.; Ristic, B.; Mirchi, A.; Qiyu, T.; Madani, K. Reform and renewables in China: The architecture of Yunnan's hydropower dominated electricity market. *Renew. Sustain. Energy Rev.* **2018**, *94*, 682–693. [CrossRef]
12. Chen, F.; Liu, B.; Cheng, C.; Mirchi, A. Simulation and regulation of market operation in hydro-dominated environment: The Yunnan case. *Water* **2017**, *9*, 623. [CrossRef]
13. Kuriqi, A.; Pinheiro, A.N.; Sordo-Ward, A.; Garrote, L. Water-energy-ecosystem nexus: Balancing competing interests at a run-of-river hydropower plant coupling a hydrologic-ecohydraulic approach. *Energy Convers. Manag.* **2020**, *223*, 113267. [CrossRef]
14. Kuriqi, A.; Pinheiro, A.N.; Sordo-Ward, A.; Garrote, L. Influence of hydrologically based environmental flow methods on flow alteration and energy production in a run-of-river hydropower plant. *J. Clean. Prod.* **2019**, *232*, 1028–1042. [CrossRef]
15. Khaloie, H.; Abdollahi, A.; Shafie-khah, M.; Anvari-Moghaddam, A.; Nojavan, S.; Siano, P.; Catalao, J.P.S. Coordinated wind-thermal-energy storage offering strategy in energy and spinning reserve markets using a multi-stage model. *Appl. Energy* **2020**, *259*, 114168. [CrossRef]

16. Khaloie, H.; Mollahassani-Pour, M.; Anvari-Moghaddam, A. Optimal behavior of a hybrid power producer in day-ahead and intraday markets: A bi-objective CVaR-based approach. *IEEE Trans. Sustain. Energy* **2021**, *12*, 931–943. [CrossRef]
17. Sanchez de la Nieta, A.A.; Paterakis, N.G.; Gibescu, M. Participation of photovoltaic power producers in short-term electricity markets based on rescheduling and risk-hedging mapping. *Appl. Energy* **2020**, *266*, 114741. [CrossRef]
18. Lu, J.; Li, G.; Cheng, C.; Liu, B. A long-term intelligent operation and management model of cascade hydropower stations based on chance constrained programming under multi-market coupling. *Environ. Res. Lett.* **2021**, *16*, 055034. [CrossRef]
19. Li, G.; Lu, J.; Yang, R.; Cheng, C. IGDT-based medium-term optimal cascade hydropower operation in multimarket with hydrologic and economic uncertainties. *J. Water Res. Plan. Manag.* **2021**, *147*, 5021015. [CrossRef]
20. Shrestha, G.B.; Pokharel, B.K.; Lie, T.T.; Fleten, S.E. Medium term power planning with bilateral contracts. *IEEE Trans. Power. Syst.* **2005**, *20*, 627–633. [CrossRef]
21. Luo, B.; Miao, S.; Cheng, C.; Lei, Y.; Chen, G.; Gao, L. Long-term generation scheduling for cascade hydropower plants considering price correlation between multiple markets. *Energy* **2019**, *12*, 2239. [CrossRef]
22. Yuan, X.; Wang, Y.; Xie, J.; Qi, X.; Nie, H.; Su, A. Optimal self-scheduling of hydro producer in the electricity market. *Energy Convers. Manag.* **2010**, *51*, 2523–2530. [CrossRef]
23. Conejo, A.; Arroyo, J.; Contreras, J.; Villamor, F. Self-scheduling of a hydro producer in a pool-based electricity market. *IEEE Trans. Power Syst.* **2002**, *17*, 1265–1272. [CrossRef]
24. Pousinho, H.; Contreras, J.; Catalão, J. Short-term optimal scheduling of a price-maker hydro producer in a pool-based day-ahead market. *Iet. Gener. Transm. Dis.* **2012**, *6*, 1243–1251. [CrossRef]
25. Kongelf, H.; Overrein, K.; Klaeboe, G.; Fleten, S. Portfolio size's effects on gains from coordinated bidding in electricity markets: A case study of a Norwegian hydropower producer. *Energy Syst.* **2019**, *10*, 567–591. [CrossRef]
26. Wu, X.; Cheng, C.; Lund, J.R.; Niu, W.; Miao, S. Stochastic dynamic programming for hydropower reservoir operations with multiple local optima. *J. Hydrol.* **2018**, *564*, 712–722. [CrossRef]
27. Fu, X.; Li, A.; Wang, L.; Ji, C. Short-term scheduling of cascade reservoirs using an immune algorithm-based particle swarm optimization. *Comput. Math. Appl.* **2011**, *62*, 2463–2471. [CrossRef]
28. Wang, J.; Cheng, C.; Shen, J.; Cao, R.; Yeh, W.W.G. Optimization of large-scale daily hydrothermal system operations with multiple objectives. *Water Resour. Res.* **2018**, *54*, 2834–2850. [CrossRef]
29. Catalao, J.P.S.; Mariano, S.J.P.S.; Mendes, V.M.F.; Ferreira, L.A.F.M. Scheduling of head-sensitive cascaded hydro systems: A nonlinear approach. *IEEE Trans. Power Syst.* **2009**, *24*, 337–346. [CrossRef]
30. Borghetti, A.; D'Ambrosio, C.; Lodi, A.; Martello, S. An MILP approach for short-term hydro scheduling and unit commitment with head-dependent reservoir. *IEEE Trans. Power Syst.* **2008**, *23*, 1115–1124. [CrossRef]
31. Su, C.; Cheng, C.; Wang, P.; Shen, J.; Wu, X. Optimization model for long-distance integrated transmission of wind farms and pumped-storage hydropower plants. *Appl. Energy* **2019**, *242*, 285–293. [CrossRef]
32. Su, C.; Yuan, W.; Cheng, C.; Wang, P.; Sun, L.; Zhang, T. Short-term generation scheduling of cascade hydropower plants with strong hydraulic coupling and head-dependent prohibited operating zones. *J. Hydrol.* **2020**, *591*, 125556. [CrossRef]
33. Dupacova, J.; Growe-Kuska, N.; Romisch, W. Scenario reduction in stochastic programming: An approach using probability metrics. *Math. Program* **2003**, *95*, 493–511. [CrossRef]
34. Niu, W.; Feng, Z.; Cheng, C. Optimization of variable-head hydropower system operation considering power shortage aspect with quadratic programming and successive approximation. *Energy* **2018**, *143*, 1020–1028. [CrossRef]
35. Lu, P.; Zhou, J.; Wang, C.; Qiao, Q.; Mo, L. Short-term hydro generation scheduling of Xiluodu and Xiangjiaba cascade hydropower stations using improved binary-real coded bee colony optimization algorithm. *Energy Convers. Manag.* **2015**, *91*, 19–31. [CrossRef]
36. Ge, X.; Xia, S.; Lee, W.; Chung, C.Y. A successive approximation approach for short-term cascaded hydro scheduling with variable water flow delay. *Electr. Power. Syst. Res.* **2018**, *154*, 213–222. [CrossRef]
37. Zhou, Y.; Guo, S.; Chang, F.; Xu, C. Boosting hydropower output of mega cascade reservoirs using an evolutionary algorithm with successive approximation. *Appl. Energy* **2018**, *228*, 1726–1739. [CrossRef]
38. He, Z.; Wang, C.; Wang, Y.; Wei, B.; Zhou, J.; Zhang, H.; Qin, H. Dynamic programming with successive approximation and relaxation strategy for long-term joint power generation scheduling of large-scale hydropower station group. *Energy* **2021**, *222*, 119960. [CrossRef]
39. Pousinho, H.M.I.; Contreras, J.; Bakirtzis, A.G.; Catalão, J.P.S. Risk-constrained scheduling and offering strategies of a price-maker hydro producer under uncertainty. *IEEE Trans. Power Syst.* **2013**, *28*, 1879–1887. [CrossRef]
40. Yuan, W.; Wang, X.; Su, C.; Cheng, C.; Liu, Z.; Wu, Z. Stochastic optimization model for the short-term joint operation of photovoltaic power and hydropower plants based on chance constrained programming. *Energy* **2021**, *222*, 119996. [CrossRef]
41. Ming, B.; Liu, P.; Guo, S.; Cheng, L.; Zhou, Y.; Gao, S.; Li, H. Robust hydroelectric unit commitment considering integration of large-scale photovoltaic power: A case study in China. *Appl. Energy* **2018**, *228*, 1341–1352. [CrossRef]
42. Díaz, F.J.; Contreras, J.; Muñoz, J.I.; Pozo, D. Optimal scheduling of a price-taker cascaded reservoir system in a pool-based electricity market. *IEEE Trans. Power Syst.* **2011**, *26*, 604–615. [CrossRef]
43. Virtanen, M. Foreign direct investment and hydropower in Lao PDR: The Theun-Hinboun hydropower project. *Corp. Soc. Responsib. Environ. Manag.* **2006**, *13*, 183–193. [CrossRef]

44. Fan, J.; Liang, Y.; Tao, A.; Sheng, K.; Ma, H.; Xu, Y.; Wang, C.; Sun, W. Energy policies for sustainable livelihoods and sustainable development of poor areas in China. *Energy Policy* **2011**, *39*, 1200–1212. [CrossRef]
45. Méreiné-Berki, B.; Málovics, G.; Creţanc, R. You become one with the place: Social mixing, social capital, and the lived experience of urban desegregation in the Roma community. *Cities* **2021**, *117*, 103302. [CrossRef]

Article

Convolutional Neural Network Identification of Stall Flow Patterns in Pump–Turbine Runners

Junjie Wu and Xiaoxi Zhang *

School of Environmental Science and Engineering, Xiamen University of Technology, Xiamen 361024, China
* Correspondence: zhangxiaoxi@xmut.edu.cn

Abstract: Stall flow patterns occur frequently in pump turbines under off-design operating conditions. These flow patterns may cause intensive pressure pulsations, sudden increases in the hydraulic forces of the runner, or other adverse consequences, and are some of the most notable subjects in the study of pump turbines. Existing methods for identifying stall flow patterns are not, however, sufficiently objective and accurate. In this study, a convolutional neural network (CNN) is built to identify and analyze stall flow patterns. The CNN consists of input, convolutional, downsampling, fully connected, and output layers. The runner flow field data from a model pump–turbine are simulated with three-dimensional computational fluid dynamics and part of the classifiable data are used to train and test the CNN. The testing results show that the CNN can predict whether or not a blade channel is stalled with an accuracy of 100%. Finally, the CNN is used to predict the flow status of the unclassifiable part of the simulated data, and the correlation between the flow status and the relative flow rate in the runner blade channel is analyzed and discussed. The results show that the CNN is more reliable in identifying stall flow patterns than using the existing methods.

Keywords: pump–turbine; stall flow patterns; flow identification; convolutional neural network

Citation: Wu, J.; Zhang, X. Convolutional Neural Network Identification of Stall Flow Patterns in Pump–Turbine Runners. *Energies* **2022**, *15*, 5719. https://doi.org/10.3390/en15155719

Academic Editors: Zhengwei Wang and Yongguang CHENG

Received: 5 July 2022
Accepted: 2 August 2022
Published: 5 August 2022

Publisher's Note: MDPI stays neutral with regard to jurisdictional claims in published maps and institutional affiliations.

1. Introduction

Pumped storage power plants serve several purposes, including peak control, valley filling, energy storage, frequency regulation, phase regulation, and power backup. They are not only important complements for existing power systems dominated by thermal electricity but also fundamental prerequisites for developing power generation from new energy resources, such as wind and solar power. In the global drive to reduce carbon emissions, the capacity of pumped storage power plants will certainly increase significantly. In China, for example, after over 60 years of construction, the total installed capacity of pumped storage power plants has surpassed 30 million kilowatts (kW), but this figure is predicted to reach 120 million kW by 2030 [1]. Given this rapid growth, the development of reliable and efficient pumped storage equipment is increasingly important.

Pump turbines, which can operate in both pumping and generating modes, are one of the most important pieces of equipment in pumped storage power plants. To properly capitalize on the benefits of pumped storage power plants, pump turbines must frequently run among different operating modes. This increases the likelihood of running into unstable operating regions, such as the S-shaped characteristic region in the generating mode [2] and the hump instability region of the pumping mode [3]. When running in these regions, pump turbines may suffer from intensive pressure pulsations, unit vibrations, and even structural damage [4,5]. These unfavorable phenomena are usually caused by special flow patterns in the pump-turbine, such as vortex ropes in the draft-tube [6], inter-blade vortices [7], and rotating stalls in the runner [8].

A rotating stall is caused by flow separations between the runner blades when the flow rate of the pump-turbine falls. The separating flows then develop into large-scale vortices (i.e., stall cells) that block the flow of the blade channel and rotate sub-synchronously with

---

the runner [9]. This flow structure is always accompanied by violent high-amplitude and low-frequency pressure pulsations [10], which impose enormous alternating stresses on the runner and shaft systems of the pump-turbine [11], and is one of the leading causes of structural failure [12]. Identifying rotating stalls and evaluating the resulting damage is thus crucial in the research and development of new pump turbines.

Currently, there are three primary ways to identify rotating stalls. The first is to estimate whether stall cells exist by observing the distribution of streamlines or velocity vectors in the runner artificially. For example, Widmer et al. [13] and Botero et al. [14] identify stall cells in the guide-vane channels of pump turbines through visualized flow fields using three-dimensional numerical simulations and model tests, respectively. This is a straightforward method based on the flow field features; however, the identification process is quite subjective and lacks a unified standard.

The second way is to analyze the abnormal pressure pulsations induced by the rotating stall. According to Hasmatuchi et al. [15], the rotating stall causes significant low-frequency pressure pulsations in the pump-turbine, and the dominant frequency is at about 70% of the runner rotational frequency. The propagating stall cells in the runner can thus be identified by observing the alternations of peaks and troughs in the pressure signals at various locations around the runner. However, Li et al. [16] note that low-frequency pressure pulsations can also be caused by the draft-tube vortex of the pump-turbine. This indicates that identifying rotating stalls by abnormal pressure pulsations may not be very appropriate but should be used as a complement to other ways, such as flow visualization. Yang et al. [17], for instance, use high-speed flow visualization to identify stall cells and spectrum analysis of the pressure signals to determine the frequency of the rotating stall.

The final method requires determining the blade channel flow rates in the runner. The basic idea is that the flow rates of the stalled blade channels are considerably reduced due to the blockage created by the stall cells, while the flow rates of the un-stalled channels are relatively high. Thus, by comparing the flow rates of different runner blade channels, researchers can determine whether stall cells exist. For example, Cavazzini et al. [18] simulate the transient flow patterns of a pump-turbine during the load rejection scenario at a constant guide-vane opening and investigate how the rotating stall forms and develops in the runner and guide-vane zones. However, although this method looks more persuasive than the other two, it is merely a semi-quantitative approach because there is not a certain critical flow rate for determining whether or not there is a stall cell in the blade channel. Therefore, all current methods are imperfect, and developing a more objective and accurate method for identifying the stall flow patterns is necessary to more fully understand rotating stalls in pump turbines.

The convolutional neural network (CNN), an important algorithm in the field of deep learning, shows satisfactory performance in processing two-dimensional data with grid-like topology [19]. Now, it has been widely used in pattern identification, such as face [20] and image [21] identification. Due to its powerful ability to extract features and its high identification accuracy, it has also been used successfully in other fields, such as medicine [22] and botany [23]. CNN may thus also be used in identifying special flow patterns in pump turbines, and there have indeed been studies attempting to apply CNN to flow pattern identification. Zhang et al. [24] developed a fast-response transient flow pattern identification algorithm based on a long short-term memory neural network (LSTM) and CNN, and this model can instantly identify separated flows with an accuracy of 94%. Strofer et al. [25] proposed a region-CNN algorithm for identifying the recirculation flow regions and boundary layers in two-dimensional flows. This model can also be used in identifying horseshoe vortices in three-dimensional flows. Franz et al. [26] identified and tracked ocean eddies based on an LSTM and CNN; the results demonstrate that the model is capable of identifying numerous eddies concurrently. Tang et al. [27] proposed a CNN-based flow field feature identification algorithm that can accurately identify clockwise, counterclockwise, and saddle-like vortex structures in two-dimensional flow fields. Deng et al. [28] developed a vortex identification CNN model that incorporates both local and global flow field information, and the identification accuracy of the model can exceed 98%. Nie et al. [29] proposed a flow pattern classification method for gas-liquid two-phase flow based on the principle of CNN image recognition, which can identify five different flow patterns and achieve an accuracy of over 90%. Brantson et al. [30] proposed a 4-layer CNN to identify the flow patterns of simultaneous gas-liquid flow in a pipeline, and the results indicate that the model has an accuracy of 99.90% used in recognizing bubbly, slug, and churn flow patterns. Seal et al. [31] combined CNN to develop a visualization tool for classifying in-tube condensation flow patterns with an accuracy of up to 98%. In the above applications, CNN can automatically extract and generalize features from flow field data, which means that the identifications do not require predetermined mathematical models and are rarely influenced by the researchers' subjective experience. This precisely compensates for the shortcomings of the above existing methods for identifying stall flow patterns. In addition, all of the identifying accuracies of the CNN-based models are high enough. Therefore, although there are not many related studies, CNN has shown great potentialities in flow pattern identification.

In this study, we construct a CNN-based rotating stall identification model for pump turbines using three-dimensional simulated flow field data from a model pump-turbine. The contribution of the work is to provide an objective and accurate tool for identifying stall flow patterns in pump–turbine runners. The remainder of this paper is structured as follows. Section 2 describes the generation of the datasets used for the construction and evaluation of the CNN. Section 3 demonstrates the architecture of the CNN, as well as its training and evaluation results. In Section 4, the correlation between the relative flow rate and the un-stalled probability of the blade channel is analyzed, and the results are discussed in Section 5. The main conclusions are presented in Section 6.

2. Data Preparation

2.1. Flow Field Data

Stall flow patterns indicate a type of special velocity field with strong flow separations and large-scale vortices in the blade channels of the runner. The data for velocity vectors in the runner are thus essential elements for CNN feature extraction. These data are obtained by numerical simulations through three-dimensional computational fluid dynamics. The computational domain is the full flow passage of a model pump-turbine, including the spiral-casing, stay-vane and guide-vane zones, runner, and draft-tube, with a runner diameter of 0.55 m (Figure 1). The simulations are conducted by the ANSYS FLUENT 18.0 software (ANSYS, Canonsburg, PA, USA) and the flow field data in the pump-turbine are solved by three-dimensional unsteady Navier–Stokes equations with a time step size of 0.001 s. To solve the unsteady turbulent flow with strong separations in the pump-turbine, the scale adaptive turbulence model is used in ANSYS FLUENT software [32]. The above models are solved by the finite volume method, and the computational domain is discretized using a hybrid mesh scheme with multiple mesh topologies. The total number of meshes chosen is 9.54 million based on a series of trial calculations among cases with different total cell numbers and the meshes were generated using the ANSYS ICEM CFD 18.0 software (ANSYS, Canonsburg, PA, USA); details about the mesh can be found in previous studies [33].

Figure 2 displays the comparisons between the simulated pump-turbine characteristics (defined by the unit flow rate Q_{11} and unit speed n_{11}; see Equations (1) and (2)) and the measured characteristic curve supplied by the manufacturer. There appears to be a high degree of agreement between the two curves, indicating that the simulated results are sufficiently accurate and can be used for further flow analysis.

$$n_{11} = \frac{nD}{\sqrt{H_{\mathrm{w}}}} \tag{1}$$

$$Q_{11} = \frac{Q}{\left(D^2\sqrt{H_{\mathrm{w}}}\right)} \tag{2}$$

where n_{11} is the unit speed, n is the rotational speed, D is the runner inlet diameter, H_w is the working head of the pump-turbine, Q_{11} is the unit flow rate, and Q is the flow rate.

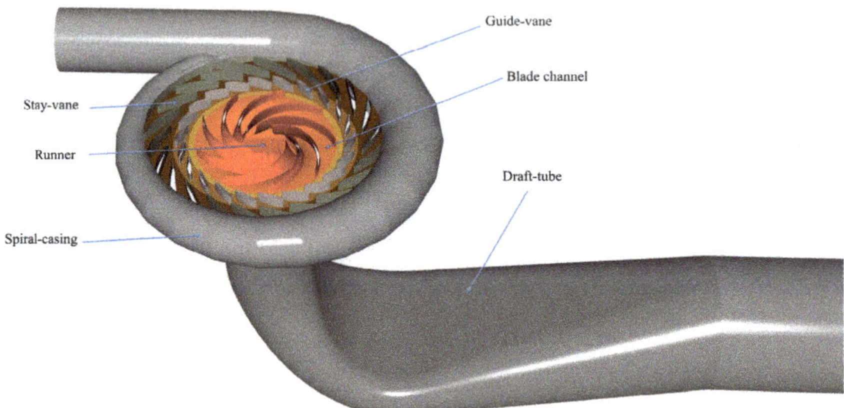

Figure 1. Computational domain of the model pump–turbine.

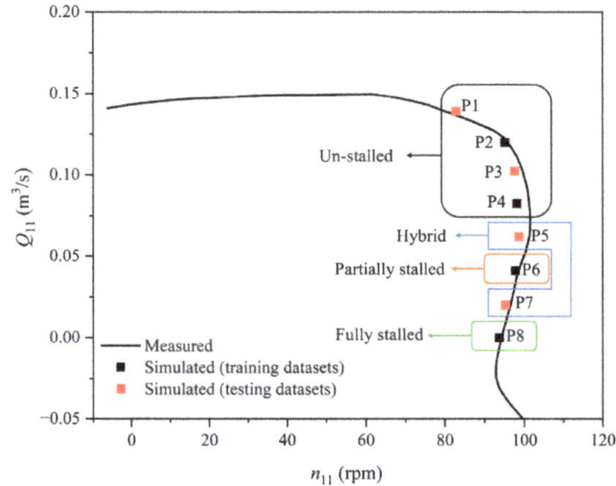

Figure 2. Comparison of simulated and measured pump-turbine characteristics.

2.2. Samples

Following an initial flow field analysis, the runner velocity field data for eight representative operating points (P1–P8 in Figure 2) are chosen to create samples for constructing the CNN. These eight points can be preliminarily classified as the un-stalled condition (P1–P4 in Figure 2), the partially stalled condition in which partial channels in the runner are blocked by stall cells (P6 in Figure 2), the fully stalled condition in which all channels are blocked by stall cells (P8 in Figure 2), and the hybrid condition (P5 and P7 in Figure 2), based on the flow field features in the runner blade channels. To extract as many samples as possible from the limited number of simulated operating points, the velocity fields at different points, different moments, and different blade channels are taken as different samples in this study because a rotating stall in the pump–turbine runner is a typical unsteady flow structure that varies with time and space, so stall flow patterns are different at different moments, as well as at different blade channels, of the same operating point.

2.3. Classification

For the un-stalled and fully stalled conditions (P1–P4 and P8 in Figure 2), the blade channel can easily be classified as un-stalled or stalled, respectively. It is difficult to classify the partially stalled (P6 in Figure 2) and the hybrid conditions (P5, P7 in Figure 2), which is the motivation for this study. However, the data for these two conditions should be used in constructing the CNN to improve its generalization. Here, we use the identification method based on the blade channel flow rate to partially solve this problem.

The basic idea behind this method is that stall cells block the blade channels, resulting in significant reductions in the flow rates at the corresponding blade channels, while the flow patterns in the un-stalled blade channels are relatively smooth (i.e., the flow rates of the corresponding blade channels are relatively high). Although we cannot identify the flow status of all blade channels, the flow status of channels with significantly high or low flow rates can easily be identified. To compare the blade channel flow rates at different operating points, the relative form of this variable is used, as defined in Equation (3):

$$q_{ir} = \frac{q_i}{\frac{Q_{BEP}}{9}} \tag{3}$$

where q_{ir} is the relative flow rate of blade channel i, q_i is the absolute flow rate of blade channel i, Q_{BEP} is the total absolute flow rate of the runner in the best efficiency point (i.e., P1 in Figure 2), and 9 is the total channel number of the runner.

Figure 3 compares the distributions of relative flow rates and streamlines among different runner blade channels at a given moment in the partially stalled condition (P6 in Figure 2). We can see that the relative flow rate of channel 6 (C6) is significantly below the average level, indicating that C6 is stalled; the relative flow rate of C4 is significantly above the average level, indicating that C4 is un-stalled. The flow statuses of the other blade channels are difficult to determine because the relative flow rates of these blade channels are between the maximum and minimum values. Consequently, only data for two runner channels at one moment of P6 can be used in constructing the CNN.

For the hybrid condition (P5 and P7 in Figure 2), the runner may present a fully stalled, partially stalled, or un-stalled status at different moments during the rotation, so only the most typical blade channel velocity fields are used. For example, P5 is an un-stalled channel dominating operating point because the general runner flow patterns are smoother than the partially stalled condition P6. Therefore, only the blade channel with the largest relative flow rate (i.e., the indubitably un-stalled channel) at every moment is used. For the same reason, P7 is a stalled channel dominating operating point, and only the blade channel with the smallest relative flow rate (i.e., the indubitably stalled blade channel) at every moment is used.

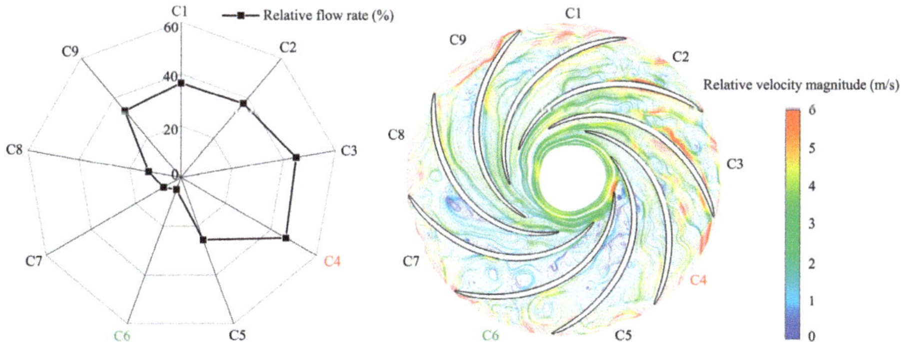

Figure 3. Distributions of the blade channel relative flow rates and streamlines at a given moment in the partially stalled condition.

According to the above procedure, 3200 classifiable samples are selected from P1 to P8, with a total of 1600 each for stalled and un-stalled samples. The samples from P2, P4, P6, and P8 form the training dataset used to train the CNN. The samples from P1, P3, P5, and P7 form the testing dataset, which is used to evaluate the generalization of the CNN. To ensure that the training and testing datasets are completely independent and able to present the operating range of the pump-turbine as widely as possible, the samples of these two datasets are taken from cross-distributed operating points along the characteristic curve, as shown in Figure 2. The detailed compositions of the datasets are shown in Table 1. It is worth mentioning that, before the initial training process, 20% of the samples from the training dataset are randomly selected as validation samples for use in cross-validating the training results and adjusting the structure of the CNN.

Table 1. Compositions of the datasets.

Datasets	Conditions	Operating Points	Un-Stalled Samples	Stalled Samples
Training	Un-stalled	P2	450	0
		P4	450	0
	Fully stalled	P8	0	900
	Partially stalled	P6	600	600
Testing	Un-stalled	P1	34	0
		P3	33	0
	Hybrid	P5	33	0
		P7	0	100
	Total		1600	1600

2.4. Data Format

The last important treatment regarding the dataset is transforming the runner's classified velocity field data to an acceptable input format for the CNN. This treatment can be divided into two steps: coordinate and data transformation, as shown in Figure 4.

The purpose of coordinate transformation is to transform the three-dimensional and irregularly shaped runner flow passage into a two-dimensional rectangular area (i.e., into matrix format). To make the rectangular area representative, the middle span of the runner, as shown by the blue axisymmetric spatial surface in Figure 4, is chosen as the object of the transformation. All computational mesh nodes in this spatial surface are projected onto a rectangular plane, and this plane is then re-meshed according to an appropriate aspect ratio. This new mesh system represents a matrixed area with 88 rows and 700 columns.

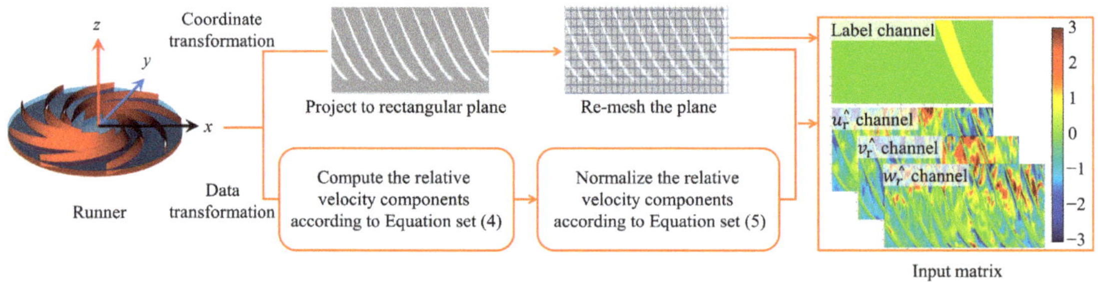

Figure 4. Sample format transformation process.

The purpose of data transformation is to normalize the velocity field data to remove the impact of data magnitude across samples and enhance the convergence speed and accuracy of the CNN. First, the absolute velocity vectors are represented by three components, u, v, and w, along the x, y, and z axes. These components are then transformed to relative forms, u_r, v_r, and w_r, to remove the influence of runner rotational speed on the velocity field, as defined by Equation set (4):

$$\begin{cases} u_r = u - (n \times 2\pi \times y) \\ v_r = v + (n \times 2\pi \times x) \\ w_r = w \end{cases} \tag{4}$$

where u, v and w are the components of the absolute velocity vector along the x, y and z axes, respectively; u_r, v_r and w_r are the relative forms of u, v and w, respectively.

The standard deviation is used to normalize the three relative velocity components, as shown in Equation set (5):

$$\begin{cases} \hat{u}_r = (u_r - u_{\text{mean}}) / u_{\text{std}} \\ \hat{v}_r = (v_r - v_{\text{mean}}) / v_{\text{std}} \\ \hat{w}_r = (w_r - w_{\text{mean}}) / w_{\text{std}} \end{cases} \tag{5}$$

where $\hat{u}_r$, $\hat{v}_r$, and $\hat{w}_r$ are the normalized forms of u_r, v_r, and w_r, respectively; u_{mean}, v_{mean}, and w_{mean} are the mean values of u_r, v_r, and w_r, respectively; and u_{std}, v_{std}, and w_{std} are the standard deviations of u_r, v_r, and w_r, respectively. The normalized relative velocity components on the new mesh nodes (i.e., the matrixed area) can thus be interpolations of the data in the original mesh nodes. If the mesh nodes are located on the blade, the velocity components are defined as 0.

The blade channels also have to be labeled because we treat different blade channels at different moments as different samples. That is, although a sample in the input dataset contains the data for all blade channels in the runner, only the data for one blade channel are used. To mark the required channel, we must label the nodes of the new mesh system: all nodes of the required blade channel are assigned a value of 1, whereas all nodes from other blade channels and the void regions representing the blades are assigned a value of 0.

After processing, each sample can be represented by a matrix with 4 channels, 88 rows, and 700 columns (size: $4 \times 88 \times 700$). These four channels represent the normalized relative velocity components $\hat{u}_r$, $\hat{v}_r$, $\hat{w}_r$, and the blade channel markers, respectively, as shown by the color maps on the right side of Figure 4 (the color bar indicates the normalized values for all the flour channels), and make up the input data for the CNN. The whole data preparation process can be summarized by Figure 5.

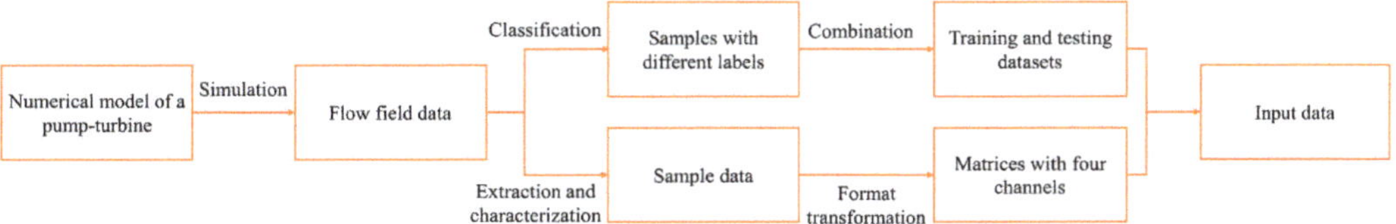

Figure 5. Flow chart of the data preparation.

3. Construction of the CNN

3.1. Architecture

To extract features from the flow field data that are as deep as possible, we design a multi-layer CNN that consists of an input layer, five convolutional and downsampling layers, two fully connected layers, and an output layer, as depicted in Figure 6. Among these layers, the input layer receives samples (i.e., the four-channel matrices mentioned above), and the output layer provides the probability of whether the received samples are stalled or not. The functions of the remaining layers are described below.

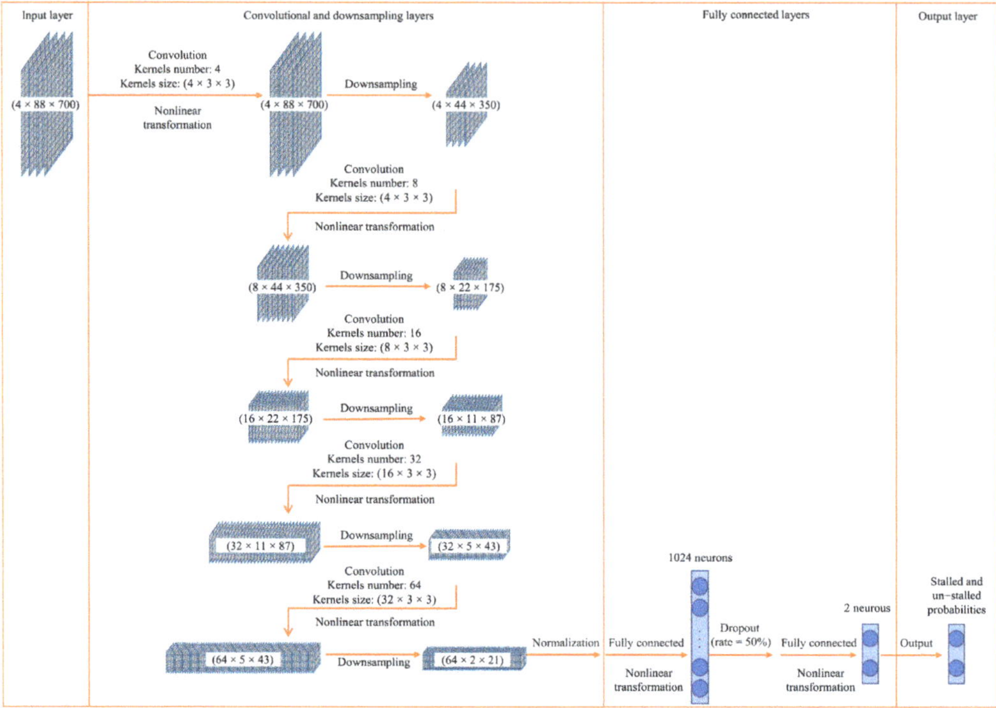

Figure 6. CNN architecture.

Following the input layer, the convolutional and downsampling layers alternate. The convolutional operation, which is essentially the process of extracting different features from the input matrix using different convolution kernels, is the foundation of the CNN. After a sufficient number of convolutional layers, underlying features from the shallow to deep levels of the input matrix can be extracted gradually. The convolution is described by Equation (6):

$$Y_i = \sum_{j=0}^{M} X_j * W_j^i + b^i \; i \in 1, 2, \cdots, N,$$ (6)

where X is the input matrix with M channels, and j is the channel index; W is the convolution kernel, a matrix with the same number of channels as the input matrix X, and its superscript i and subscript j indicate the indexes of the convolution kernels and the channel, respectively; b is the bias vector, and its superscript indicates that the current convolution is performed using the ith element of the bias vector; Y is the output matrix with N channels, and its subscript i indicates the channel index. It should be noted that the total channel number of the convolution kernel corresponds to that of the input matrix; the total convolution kernel number corresponds to the total channel number of the output matrix. The more channel numbers the output matrix has, the deeper the underlying features that can be extracted. Details about the size (channel $\times$ row $\times$ column) of the matrix at different layers of the CNN are shown in Figure 6.

Figure 7 depicts the convolutional operation for extracting matrix features from the convolution kernel, which functions like a filter; the orange box shows the region where the current convolutional operation is conducted by the convolution kernel, and the black box shows the region where the next convolutional operation is conducted by the convolution kernel. First, the ith convolution kernel determines a region in the input matrix, as shown by the orange box in Figure 7. Then, the elements of each channel in the orange box are convolved with the corresponding channel of the present convolution kernel. Finally, the element of the corresponding bias vector is added, as shown in Equation (6). The result of the operation is element y_{11} of the ith channel of the output matrix Y (i.e., Y_i). This operation can be understood as the convolution kernel filtering and linearly superimposing the local features among different channels of the input matrix and providing a more comprehensive and deeper underlying local feature than the input matrix does. The next element, y_{12}, of the output matrix Y can be computed in the same way after changing the operating region, such as the region indicated by the black box. After computing all the elements of Y_i, the above procedure is repeated with the $i + 1$th convolution kernel to obtain the elements of the $i + 1$th channel of the output matrix Y.

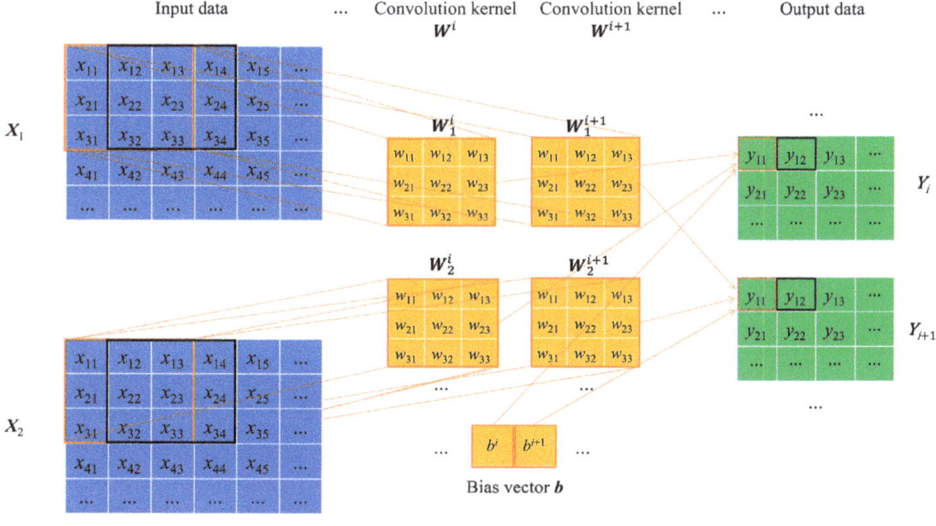

Figure 7. Convolutional operation procedure.

The downsampling layer, or pooling layer, follows each of the convolutional layers and is used to extract the primary features of the matrix obtained by the convolutional layer and prevent overfitting. The max-pooling downsampling method is employed in this study, wherein only the maximum value of the local (2 × 2) area of the input matrix is recovered as an element of the output matrix, as shown in Figure 8, in which the orange box shows the region where the current max-pooling operation is conducted. After max-pooling, the matrix retains the same number of channels, but its length and width are halved, so the number of elements is decreased to one-fourth of the input matrix.

The purpose of the fully connected layers is to project the features into the label space of the sample. The fully connected layer can synthesize the extracted features and construct connections between the features and the labels. There are two fully connected layers in the present CNN (see Figure 6): the first one transforms the two-dimensional features passed through the convolutional and downsampling layers into a one-dimensional vector for synthesizing features; the second one constructs connections between the synthesized features and the labels indicating the flow status.

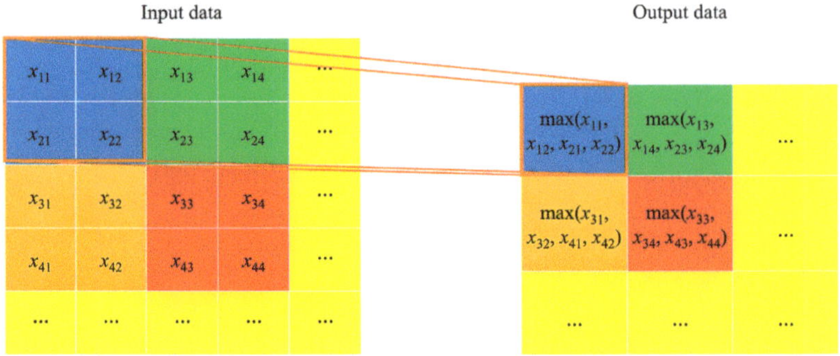

Figure 8. Max-pooling procedure.

All of the above operations are linear transformations, so nonlinear transformations should be introduced to enable the CNN to deal with complex nonlinear problems. This operation is also known as activation. Activations in this study are conducted through two types of nonlinear functions—the LeakyReLu function and the Softmax function—and the result of every convolution or full connection should be activated by one of these two functions (see Figure 6). The LeakyReLu function, as defined by Equation (7), can resolve, to the greatest possible extent, the problem of the occurrence of gradient explosion and gradient disappearance [34] during CNN training and is used after most operations. The Softmax function, which is effective at solving multi-classification issues, is only used before the output layer to transform the classification results into probability values in the range 0–1 as illustrated by Equation (8):

$$f(x) = \begin{cases} x, \ x > 0 \\ \alpha x, \ x \leq 0 \end{cases} \tag{7}$$

$$f(z_i) = \frac{e^{z_i}}{\sum_{k=1}^{N} e^{z_k}} \tag{8}$$

In Equation (7), x is the element of the convolutional layer output matrix; α is a constant between 0 and 1; and $f(x)$ is the activation result. In Equation (8), z_i is the output value of the ith node of the fully connected layer, N is number of output nodes (or the number of classification categories), and e is the base of the natural logarithm.

3.2. Training and Testing

The CNN training process consists of forward and backward propagations. During the forward propagations, the loss values, which measure the difference between the identified results of the CNN and the actual classification, are calculated from the input to the output layers; during the backward propagations, the elements of the convolution kernels and bias vectors are updated by iteration from the output to input layers to minimize the loss value. If the loss values remain low enough after several forward and backward propagations, the training process might be finished. After training, the resulting CNN should be tested with the test dataset to evaluate its generalization.

Figure 9 illustrates the training results, including the loss values and accuracies. It can be seen that both of the curves tend to be steady after 20 training epochs, and the loss values and accuracies then remain lower than 0.02 and higher than 99.5%, respectively. This indicates that there are minimal divergences between the predicted results of the CNN and the actual classifications. Finally, the CNN training process is terminated after 30 epochs and the obtained CNN is tested with the testing dataset, which has no intersections with the training datasets. The resulting loss value is 0.0193 with a 100% accuracy, suggesting that the network has excellent generalization and can be used as an accurate and credible research tool for identifying stall flow patterns.

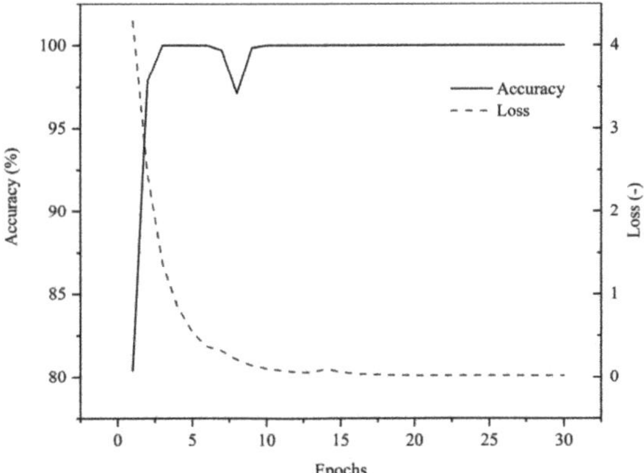

Figure 9. CNN training results.

4. Correlation between Flow Status and Flow Rate

As mentioned above, analyzing the flow rate of the channel is an easy way to identify whether a blade channel is stalled. However, this method can only be used to identify the flow status of blade channels with relatively larger or smaller flow rates than the average levels under the operating points with significant flow rate differences among blade channels of the runner and is invalid for other cases. In this section, we use the CNN classification results as benchmarks to determine a critical relative flow rate for identifying blade channel flow status to extend the applications of the flow rate assessment method.

To conduct the study, a new dataset called the predicting dataset is generated. To ensure the numbers of stalled and un-stalled samples are as equal as possible in this dataset, we select samples from low flow rate operating points (P4–P8 in Figure 2), which are more prone to form stall flow patterns than large flow rate points. The predicting dataset consists of 8550 samples, including 2683 classifiable samples that have been used in the training and testing process (Table 1), and 5867 uncertain samples. It should be noted that the purpose of this section is to use the CNN rather than test its accuracy, so the training and testing samples can be used to collect as many samples as possible.

The stalled probabilities of the samples and the relative flow rates of the corresponding blade channels predicted by the CNN are depicted in a scatter plot, as shown in Figure 10, in which a scatter represents a sample in the predicting dataset. The background colors represent the kernel density reflecting the distribution characteristics of these scatters: red-colored regions denote densely distributed samples, and blue-colored regions denote sparsely distributed samples. It can be seen that most of the samples are located in two distinct regions in the figure. One region is in the upper-left side (shown by the red dashed curve in Figure 10) and represents un-stalled samples, as the stalled probability of these samples is significantly less than 50%. There are about 35.5% of the samples gathering in this region. Another region is at the lower-right side representing stalled samples (shown by the yellow dashed curve in Figure 10), with a stalled probability much higher than 50%, and possesses about 41.7% of the samples. Obviously, the total percentage of samples in these two regions (about 77.2%) is much larger than in other regions (about 22.8%), indicating that most of the samples can be definitely classified by the CNN.

The critical relative flow rate can be sought with the help of the distributing characteristics of the samples in Figure 10. The search process consists of three steps: (1) draw a vertical line in Figure 10 through the 50% stall probability (as shown by the black dash line in Figure 10) and divide the samples into stalled and un-stalled; (2) draw a horizontal line randomly and divide Figure 10 into four sub-regions with the vertical line; and (3) slide the horizontal line vertically until the total number of samples in sub-region 1 (stalled region) and sub-region 3 (un-stalled region) is greatest, as shown by the black solid line in

Figure 10. The ordinate indicated by the horizontal line can be determined as the critical relative flow rate. This approach is based on the idea that, if the crucial flow rate exists, the classified results for most of the samples are the same for the critical relative flow rate and the CNN—that is, the total number of samples in sub-regions 1 and 3, in which the samples are classified as stalled or un-stalled, respectively, by both methods, is the largest.

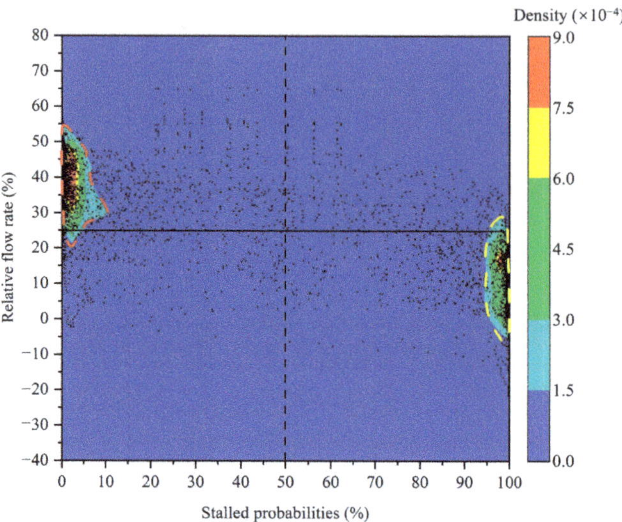

Figure 10. Scatter plot of the classified samples.

The critical relative flow rate for identifying stalled or un-stalled blade channels is found to be 24.8%. Using this value, 92.2% of the classified results are exactly the same as those determined by the CNN, including 47.6% of stalled samples (in sub-region 1) and 44.6% of un-stalled ones (in sub-region 3). Only 7.8% of the results differ for classification by the two methods (relative flow rate and CNN). The critical relative flow rate of 24.8% can thus be used to identify the stall flow patterns in the pump–turbine runner blade channel with an accuracy of more than 90% when using CNN classification results as the benchmark.

5. Discussion

This section discusses the reason for the different results identified by the methods based on the relative flow rate and the CNN. The representative runner flow patterns at a specific moment of P5 are selected for the discussion. The relative flow rates and the un-stalled probabilities computed by the CNN for the nine blade channels are illustrated in Figure 11, in which the x-axis represents the channel index, the left y-axis represents the relative flow rate, and the right y-axis represents the un-stalled probability. The hollow and solid scatters represent the relative flow rate and un-stalled probability of the corresponding blade channels, respectively. The red line in Figure 11 indicates both the relative flow rate of 24.8% and the un-stalled probability of 50%, and is the line of demarcation between the stalled and un-stalled status for both methods. As can be seen, although the identified results (stalled or un-stalled status) for both methods appear nearly identical for all blade channels, the degree of stalled or un-stalled status obtained by the two methods may differ for a certain channel. To be specific, the flow patterns in C3 and C4 show similar un-stalled probabilities, 48.22% and 53.72%, respectively, but the relative flow rate of C4 is nearly four times larger than that of C3. Additionally, the relative flow rates of C4 and C6 (26.27% and 26.30%, respectively) are nearly equal, but the un-stalled probability of C6 is almost double that for C4.

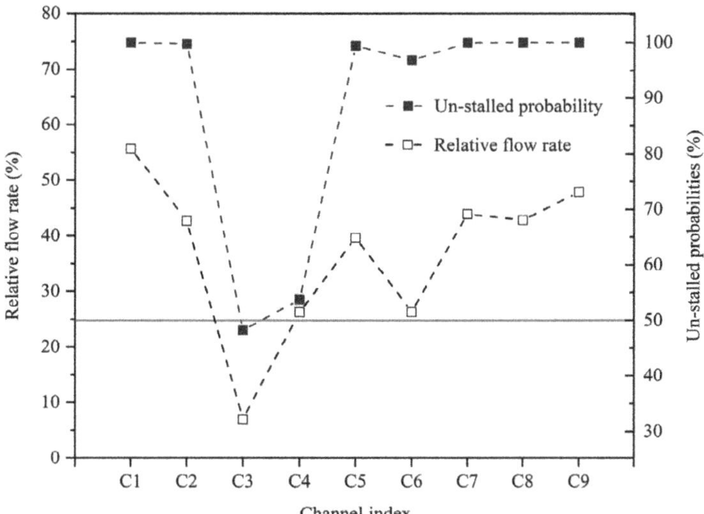

Figure 11. Relative flow and un-stalled probability for a given sample.

These differences can be explained by the real flow patterns in the blade channels. Figure 12 depicts the runner streamlines (drawn by the relative velocity vectors) of the same samples shown in Figure 11. The red dashed circles in Figure 12 indicate the vortices and the arrows indicate the rotational directions of the vortices. There are significant vortices in C3, C4, and C6, but the sizes and locations of the vortices differ. In C3, there are two large vortices (vortices A and B) near the leading and tail edges of the blades, respectively; in C4, there is only one large vortex (vortex C) near the leading edge; and, in C6, there are two small leading-edge vortices (vortices D and E). First, these vortices reduce the effective sectional area of the flow passage. As a result, the flow rate of C3, which is fully blocked by two large vortices, is considerably smaller than that of C4 and C6, both of which are not fully blocked. The blockage degree for C4, with a large vortex, is similar to that of C6, with two small vortices, so the flow rates of these two channels are nearly equal. It is also rational to deduce that the criterion for identifying whether there are stall flow patterns in a blade channel for the CNN is whether there are large vortices near the leading edge of the blades. For example, the un-stalled probabilities for C3 and C4, which have similar leading-edge vortices, are nearly identical. The data for C6 are much larger than those for C3 and C4 because the leading-edge vortices are smaller than those in C3 and C4. This CNN criterion agrees with most previous studies about the relationship between rotating stalls and vortices distribution in pump turbines, as summarized by Zhang et al. [9].

According to the above analysis, the existence of vortices is a sufficient but not necessary condition for flow rate reduction, and a necessary but not sufficient condition for the existence of stall flow patterns. This is the main reason for the different identified results between using the relative flow rate and the CNN. Considering that the CNN is constructed according to the pre-classified flow patterns in the pump–turbine blade channels and performs excellently in predicting the testing samples, it is more reliable than using the relative flow rate to identify stall flow patterns.

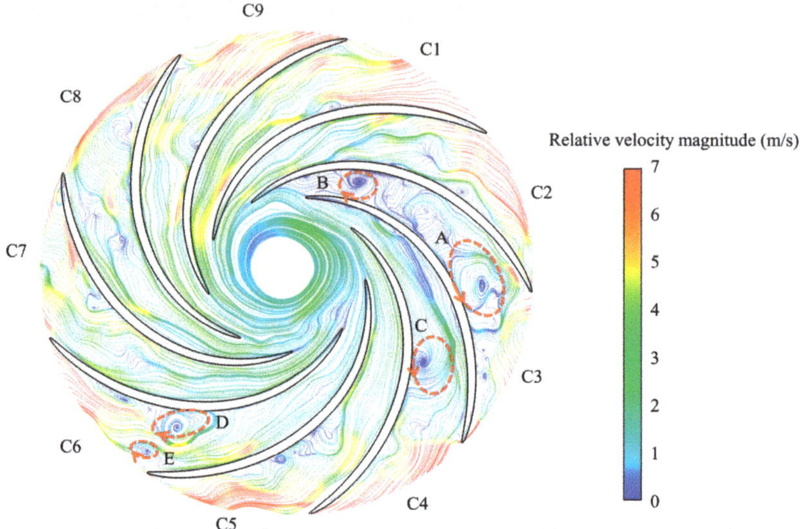

Figure 12. Streamline distribution of a given sample.

6. Conclusions

In this study, a CNN is built to identify stall flow patterns in pump–turbine runners using simulated flow field data, and the correlation between the flow status and the relative flow rate in the runner blade channel is analyzed. The main conclusions are as follows:

(1) The CNN designed can predict whether or not the runner blade channels are stalled with an accuracy of 100% for the samples in the testing dataset, indicating that the CNN has excellent generalization and can be used as a standard tool for the identification of stall flow patterns in further research.

(2) Using the classifications provided by the CNN as a benchmark, a critical relative flow rate of 24.8% can be estimated for identifying whether or not the runner blade channels are stalled, and the probability of giving the same identified results by this critical value and the CNN is 92.2%.

(3) Through analysis of runner flow patterns, we find that the criterion for the CNN to identify whether or not a blade channel is stalled is whether there are large vortices near the leading edge of the blades. This criterion agrees with most previous studies about the correlation between rotating stalls and vortices, and is further proof of the reliability of the CNN.

To sum up, this study provides a new tool for investigating rotating stall flow patterns in pump–turbine runners, and this tool is sufficiently objective and accurate comparing to the existing methods. One limitation of this study is that the usability and validity of the proposed CNN model in other types of pump turbines have not been demonstrated. Further investigations should be emphasized on verifying the model more comprehensively.

Author Contributions: Conceptualization, X.Z.; methodology, X.Z.; software, J.W.; investigation, J.W.; writing—original draft preparation, J.W.; writing—review and editing, X.Z.; funding acquisition, X.Z. All authors have read and agreed to the published version of the manuscript.

Funding: This research was funded by the National Natural Science Foundation of China (51909226) and the Xiamen Science and Technology Planning Project of China (3502Z20206075).

Institutional Review Board Statement: Not applicable.

Informed Consent Statement: Not applicable.

Data Availability Statement: Not applicable.

Conflicts of Interest: The authors declare no conflict of interest.

Nomenclature

C	Channel index
CNN	Convolutional neural network
LSTM	Long short-term memory neural network
P	Operating point
b	Bias vector
D	Runner inlet diameter
H_w	Working head of the pump-turbine
n	Rotational speed
n_{11}	Unit speed
Q	Flow rate
Q_{11}	Unit flow rate
Q_BEP	Total absolute flow rate of the runner in the best efficiency point
q_i	Absolute flow rate of blade channel i
$q_{i\mathrm{r}}$	Relative flow rate of blade channel i
u, v and w	Absolute velocity components on the x-axis, y-axis, and z-axis, respectively
$u_\mathrm{mean}, v_\mathrm{mean}$ and w_mean	Mean values of $u_\mathrm{r}, v_\mathrm{r}$ and w_r, respectively
$u_\mathrm{r}, v_\mathrm{r}$ and w_r	Relative velocity components on the x-axis, y-axis, and z-axis, respectively
$\hat{u}_\mathrm{r}, \hat{v}_\mathrm{r}$ and $\hat{w}_\mathrm{r}$	Normalized forms of $u_\mathrm{r}, v_\mathrm{r}$, and w_r, respectively
$u_\mathrm{std}, v_\mathrm{std}$ and w_std	Standard deviations of $u_\mathrm{r}, v_\mathrm{r}$, and w_r, respectively
W	Convolution kernel
X	Input matrix
Y	Output matrix

References

1. National Energy Administration. *Medium and Long-Term Development Plan for Pumped Storage Energy (2021–2035)*; National Energy Administration: Beijing, China, 2021; p. 5. (In Chinese)
2. Zuo, Z.; Fan, H.; Liu, S.; Wu, Y. S-shaped characteristics on the performance curves of pump-turbines in turbine mode—A review. *Renew. Sustain. Energy Rev.* **2016**, *60*, 836–851. [CrossRef]
3. Cavazzini, G.; Houdeline, J.; Pavesi, G.; Teller, O.; Ardizzon, G. Unstable behaviour of pump-turbines and its effects on power regulation capacity of pumped-hydro energy storage plants. *Renew. Sustain. Energy Rev.* **2018**, *94*, 399–409. [CrossRef]
4. Zuo, Z.; Liu, S. Flow-Induced Instabilities in Pump-Turbines in China. *Engineering* **2017**, *3*, 504–511. [CrossRef]
5. Wang, W.; Tai, G.; Pei, J.; Pavesi, G.; Yuan, S. Numerical investigation of the effect of the closure law of wicket gates on the transient characteristics of pump-turbine in pump mode. *Renew. Energy* **2022**, *194*, 719–733. [CrossRef]
6. Kim, S.; Suh, J.; Yang, H.; Park, J.; Kim, J. Internal flow phenomena of a Pump-Turbine model in turbine mode with different Thoma numbers. *Renew. Energy* **2022**, *184*, 510–525. [CrossRef]
7. Li, D.; Fu, X.; Wang, H.; Zhao, R.; Wei, X. Evolution mechanism of a prototype pump turbine after pump power-off. *Phys. Fluids* **2021**, *33*, 106109. [CrossRef]
8. Suh, J.; Yang, H.; Kim, J.; Joo, W.; Park, J.; Choi, Y. Unstable S-shaped characteristics of a pump-turbine unit in a lab-scale model. *Renew. Energy* **2021**, *171*, 1395–1417. [CrossRef]
9. Zhang, Y.; Zhang, Y.; Wu, Y. A review of rotating stall in reversible pump turbine. *Inst. Mech. Eng. Part C J. Mech. Eng. Sci.* **2017**, *231*, 1181–1204. [CrossRef]
10. Wang, W.; Xu, L.; Li, Z.; Tang, W.; Wang, X.; Shang, L.; Liu, D.; Liu, X. Stability Analysis of Vaneless Space in High-Head Pump-Turbine under Turbine Mode: Computational Fluid Dynamics Simulation and Particle Imaging Velocimetry Measurement. *Machines* **2022**, *10*, 143. [CrossRef]
11. Zhang, X.; Chen, Y.; Yang, Z.; Chen, Q.; Liu, D. Influence of rotational inertia on the runner radial forces of a model pump-turbine running away through the S-shaped characteristic region. *IET Renew. Power Gen.* **2020**, *14*, 1883–1893. [CrossRef]
12. Tanaka, H. Vibration Behavior and Dynamic Stress of Runners of Very High Head Reversible Pump-turbines. *Int. J. Fluid Mach. Syst.* **2011**, *4*, 289–306. [CrossRef]
13. Widmer, C.; Staubli, T.; Ledergerber, N. Unstable Characteristics and Rotating Stall in Turbine Brake Operation of Pump-Turbines. *J. Fluids Eng.* **2011**, *133*, 041101. [CrossRef]
14. Botero, F.; Hasmatuchi, V.; Roth, S.; Farhat, M. Non-intrusive detection of rotating stall in pump-turbines. *Mech. Syst. Signal Process.* **2014**, *48*, 162–173. [CrossRef]
15. Hasmatuchi, V.; Farhat, M.; Roth, S.; Botero, F.; Avellan, F. Experimental Evidence of Rotating Stall in a Pump-Turbine at Off-Design Conditions in Generating Mode. *J. Fluids Eng.* **2011**, *133*, 051104. [CrossRef]

16. Li, D.; Sun, Y.; Zuo, Z.; Liu, S.; Wang, H.; Li, Z. Analysis of Pressure Fluctuations in a Prototype Pump-Turbine with Different Numbers of Runner Blades in Turbine Mode. *Energies* **2018**, *11*, 1474. [CrossRef]
17. Yang, J.; Pavesi, G.; Yuan, S.; Cavazzini, G.; Ardizzon, G. Experimental Characterization of a Pump-Turbine in Pump Mode at Hump Instability Region. *J. Fluids Eng.* **2015**, *137*, 051109. [CrossRef]
18. Cavazzini, G.; Covi, A.; Pavesi, G.; Ardizzon, G. Analysis of the Unstable Behavior of a Pump-Turbine in Turbine Mode: Fluid-Dynamical and Spectral Characterization of the S-shape Characteristic. *J. Fluids Eng.* **2016**, *138*, 21105. [CrossRef]
19. Liu, W.; Wang, Z.; Liu, X.; Zeng, N.; Liu, Y.; Alsaadi, F.E. A survey of deep neural network architectures and their applications. *Neurocomputing* **2017**, *234*, 11–26. [CrossRef]
20. Li, L.; Mu, X.; Li, S.; Peng, H. A Review of Face Recognition Technology. *IEEE Access* **2020**, *8*, 139110–139120. [CrossRef]
21. Chai, Q. Research on the Application of Computer CNN in Image Recognition. *J. Phys. Conf. Ser.* **2021**, *1915*, 32041. [CrossRef]
22. Sarvamangala, D.R.; Kulkarni, R.V. Convolutional neural networks in medical image understanding: A survey. *Evol. Intell.* **2022**, *15*, 1–22. [CrossRef] [PubMed]
23. Abade, A.; Ferreira, P.A.; de Barros Vidal, F. Plant diseases recognition on images using convolutional neural networks: A systematic review. *Comput. Electron. Agric.* **2021**, *185*, 106125. [CrossRef]
24. Zhang, Y.; Azman, A.N.; Xu, K.; Kang, C.; Kim, H. Two-phase flow regime identification based on the liquid-phase velocity information and machine learning. *Exp. Fluids* **2020**, *61*, 212. [CrossRef]
25. Ströfer, C.M.; Wu, J.; Xiao, H.; Paterson, E. Data-Driven, Physics-Based Feature Extraction from Fluid Flow Fields. *Commun. Comput. Phys.* **2018**, *25*, 625–650. [CrossRef]
26. Franz, K.; Roscher, R.; Milioto, A.; Wenzel, S.; Kusche, J. In Ocean Eddy Identification and Tracking Using Neural Networks. In Proceedings of the IGARSS 2018—2018 IEEE International Geoscience and Remote Sensing Symposium, Bonn, Germany, 22–27 July 2018; pp. 6887–6890. [CrossRef]
27. Tang, B.; Li, Y. CNN-based Flow Field Feature Visualization Method. *Int. J. Perform. Eng.* **2018**, *14*, 434–444. [CrossRef]
28. Deng, L.; Wang, Y.; Liu, Y.; Wang, F.; Li, S.; Liu, J. A CNN-based vortex identification method. *J. Vis.* **2019**, *22*, 65–78. [CrossRef]
29. Nie, F.; Wang, H.; Song, Q.; Zhao, Y.; Shen, J.; Gong, M. Image identification for two-phase flow patterns based on CNN algorithms. *Int. J. Multiph. Flow* **2022**, *152*, 104067. [CrossRef]
30. Brantson, E.T.; Abdulkadir, M.; Akwensi, P.H.; Osei, H.; Appiah, T.F.; Assie, K.R.; Samuel, S. Gas-liquid vertical pipe flow patterns convolutional neural network classification using experimental advanced wire mesh sensor images. *J. Nat. Gas Sci. Eng.* **2022**, *99*, 104406. [CrossRef]
31. Seal, M.K.; Noori Rahim Abadi, S.M.A.; Mehrabi, M.; Meyer, J.P. Machine learning classification of in-tube condensation flow patterns using visualization. *Int. J. Multiph. Flow* **2021**, *143*, 103755. [CrossRef]
32. Menter, F.R.; Egorov, Y. The Scale-Adaptive Simulation Method for Unsteady Turbulent Flow Predictions. Part 1: Theory and Model Description. *Flow Turbul. Combust.* **2010**, *85*, 113–138. [CrossRef]
33. Zhang, X.; Chen, Q. Dynamic evolution of rotating stall in a model pump-turbine during runaway transient scenario: Three-dimensional simulation. *IOP Conf. Ser. Earth Environ. Sci.* **2021**, *774*, 12134. [CrossRef]
34. Yang, G.; Schoenholz, S.S. Mean Field Residual Networks: On the Edge of Chaos. In Proceedings of the 31st International Conference on Neural Information Processing Systems, Long Beach, CA, USA, 4–9 December 2017; pp. 2865–2873.

*energies*

Article

Parameter Identification of a Governing System in a Pumped Storage Unit Based on an Improved Artificial Hummingbird Algorithm

Liying Wang, Luyao Zhang, Weiguo Zhao * and Xiyuan Liu

School of Water Conservancy and Hydropower, Hebei University of Engineering, Handan 056038, China; wangliying@hebeu.edu.cn (L.W.); zly01221214@163.com (L.Z.); 15690103970@163.com (X.L.)
* Correspondence: zhaoweiguo@hebeu.edu.cn

Abstract: Parameter identification is an important method to establish the governing system of a pumped storage unit. Appropriate parameters will make the governing system obtain better control performance. Therefore, in this study, an improved artificial hummingbird algorithm (IAHA) is proposed for the parameter identification of the governing system in a pumped storage unit. The algorithm integrates two key strategies to improve the optimization ability of the algorithm. First, the Chebyshev chaotic map is employed to initialize the artificial hummingbirds, which in turn increases and enhances the global search capability of the initial population. Second, the Levy flight is introduced in the guided foraging phase to expand the search space and avoid premature convergence. The performance of the proposed IAHA algorithm is compared with that of four other algorithms on 23 standard test functions, and the results show that IAHA has higher accuracy and faster convergence than the other four algorithms. Finally, IAHA was applied to the parameter identification of the governing system of a pumped storage unit to verify the effectiveness of the algorithm in tracking real-world problems.

Keywords: artificial hummingbird algorithm; Chebyshev chaotic map; levy flight; pumped storage; parameter identification; governing system

Citation: Wang, L.; Zhang, L.; Zhao, W.; Liu, X. Parameter Identification of a Governing System in a Pumped Storage Unit Based on an Improved Artificial Hummingbird Algorithm. *Energies* **2022**, *15*, 6966. https://doi.org/10.3390/en15196966

Academic Editor: Dimitrios Katsaprakakis

Received: 28 August 2022
Accepted: 19 September 2022
Published: 23 September 2022

Publisher's Note: MDPI stays neutral with regard to jurisdictional claims in published maps and institutional affiliations.

1. Introduction

A pumped storage power station is an important energy storage technology. It can effectively alleviate the impact of intermittent fluctuation energy, such as scenery on the power system, and improve the absorption capacity of clean energy. To solve the security and stability of clean energy entering the power grid on a large scale [1]. Currently, the pumped storage power stations built or being built in China are developing in an intelligent direction. The complex water diversion system of a pumped storage power station brings great challenges to unit frequency governing and power grid security and stability. Therefore, it is necessary to carry out parameter identification of the regulation system of pumped storage units to ensure the safe and stable operation of units.

In modeling studies of governing systems, due to the diversity of unit and governor characteristics and their operating conditions, it is often difficult to derive accurate model parameters directly from their basic operating principles [2,3]; it is difficult to establish a complete system simulation model that can be used practically in power system simulation software or in the performance evaluation of governing systems. In order to solve this problem, the system identification method was widely used [4].

System identification is the determination of a mathematical model that describes the behavior of a system based on its input and output time functions. It consists of two basic parts: structure identification and parameter identification. The governing system is a complex dynamic system; the mathematical model structure can be determined by applying the mechanism analysis method, so the parameters of the model are determined by the parameter identification method [5]. In recent decades, the traditional methods for parameter identification were the least squares method [6], input response method [7], and maximum likelihood estimation method [8], but these methods have many limitations, for example, the least squares method requires sufficient system inputs, and the maximum likelihood estimation method can easily fall into local optimality, etc. In recent years, identification methods based on meta-heuristic algorithms have been developed that treat the parameter identification problem as an optimization problem. Since metaheuristic algorithms are a global optimization method, they can establish an objective function for parameter identification by optimizing the output error between the original system and the identification system, comparing with traditional identification methods, metaheuristic algorithms are more suitable for the parameter identification of complex systems, and the identification performances of these algorithms depend on optimization capability.

Many metaheuristic algorithms have been presented and successfully applied to different areas: particle swarm optimization algorithm ((PSO) simulates the foraging behavior of birds in a group [9,10]; the chimpanzee optimization algorithm (ChOA) simulates the cooperative hunting behaviors of attack, drive, intercept, and chase chimpanzees [11]; the artificial rabbit optimization (ARO) algorithm is inspired by the survival strategies of rabbits found in nature, including detour foraging and random hiding [12]; the genetic algorithm (GA) simulates biological evolutionary mechanisms in natural environments [13]; the artificial ecosystem optimization algorithm (AEO) simulates the energy flow process in the earth ecosystem [14]; the gravitational search algorithm (GSA) involves the optimization of populations based on the law of gravity and Newton's second law [15]; the ant colony optimization (ACO) algorithm is a simulation of the way ants find paths in nature [16]; the Black Widow algorithm (BWO) simulates its entire life cycle [17]; the Tom search algorithm (ASO) simulates the displacement of atoms in a molecular system composed of atoms due to their mutual force and system constraint [18]; the gray wolf optimization (GWA) algorithm simulates gray wolf prey predation activities [19]; the artificial fish swarming algorithm (AFSA) simulates the foraging, clustering, and tail-chasing behaviors of fish [20]. The algorithm for the ant lion optimization algorithm (ALO) simulates the hunting mechanism of an ant lion hunting ants [21]; the whale optimization algorithm (WOA) is based on encircling prey, bubble netting prey, and searching for whale prey [22]; the manta ray foraging optimization algorithm (MRFO) simulates the foraging process of manta rays in the ocean [23]. Strong robustness, adaptability, and randomness are characteristics of these optimization algorithms. Although these metaheuristics outperform conventional numerical approaches in handling challenging engineering problems, they also have some drawbacks and nevertheless hold enormous promise for performance in optimization.

Some authors have made improvements to metaheuristic algorithms, for example, Zhongqiang Wu et al. [24] proposed an improved ant lion optimization algorithm to identify the parameters of the solar cell model by adding chaotic sequences. M Ali et al. proposed an algorithm to identify the parameters of a polymer electrolyte membrane fuel cell model using the gray wolf optimization algorithm [25]. Xiao Zhang et al. applied an elite backward learning particle swarm algorithm for the identification of PV cell parameters [26]. The metaheuristic algorithms have some limitations due to local optimum and premature phenomena, despite the fact that these algorithms have been successfully used in a variety of fields for parameter identification problems.

In 2022, Weiguo Zhao et al. proposed the artificial hummingbird algorithm (AHA) [27], which was inspired by simulating hummingbirds, special flight abilities, and their intelligent foraging strategies. The method's advantages—it has few parameters, a fast speed, and performs well in solving optimization problems. This paper proposes an improved metaheuristic algorithm, named the improved artificial hummingbird algorithm (IAHA). In order to make the initialization more uniform and rich, IAHA added the Chebyshev chaotic map to initialize the artificial hummingbird and the Levy flight to improve the search efficiency when guiding foraging.

2. Artificial Hummingbird Algorithm (AHA)

2.1. Brief Introduction of AHA

AHA is a population-based metaheuristic algorithm that mainly simulates three foraging behaviors of hummingbirds: guided foraging, territorial foraging, and migratory foraging. During the foraging process, three flight skills are modeled—omnidirectional, diagonal, and axial flight. At the same time, an access table simulating the hummingbird's extraordinary memory ability is constructed to guide the hummingbird to perform global optimization in the algorithm.

The three flying skills are defined as follows:

The flight skill simulation is extended to the d-D space with axial flight defined as follows:

$$D^{(i)} = \begin{cases} 1 & \text{if } i = randi([1,d]) \, i = 1, \cdots, d \\ 0 & \text{else} \end{cases} \tag{1}$$

Diagonal flight is defined as follows:

$$D^{(i)} = \begin{cases} 1, & \text{if } i = p(j) \, P = randperm(k), k \in [2, \lceil r_1(d-2) \rceil + 1] \\ 0, & \text{else.} \end{cases} \tag{2}$$

Omnidirectional flight is defined as follows:

$$D^{(i)} = 1 \quad i = 1, \cdots, d \tag{3}$$

where $randi([1,d])$ generates a random integer from 1 to d, $randperm(k)$ creates a random permutation of integers from 1 to k, and r_1 is a random number in (0, 1]. The diagonal flight in a d-D space is inside a hyperrectangle.

The AHA first initialization a set of random solutions and a visit table. In each iteration, guided or territorial foraging is performed 50% of the time. Hummingbirds can migrate toward their intended food sources using guided foraging, which is based on nectar filling rates and a visit table. Territorial foraging allows hummingbirds to easily move to neighboring regions within their own territory and find new food sources as candidates. Migration foraging is performed every two iterations. Up until the stop rule is reached, all operations and calculations are performed interactively. Finally, the food source with the highest rate of nectar-refilling replenishment is returned as an approximate global optimum.

(1) A population of n hummingbirds are randomly initialized to n food sources as follows:

$$x_i = Low + r \times (Up - Low) \qquad i = 1, \cdots \cdots n \tag{4}$$

where Low and Up are the lower and upper boundaries for a d-dimensional problem, respectively, r is a random vector in $[0, 1]$, and x_i represents the position of the ith food source.

$$VT_{i,j} = \begin{cases} 0 & \text{if } i \neq j \\ null & i = j \end{cases} \tag{5}$$

where $i = j, VT_{i,j} = null$ indicates that a hummingbird is taking food at its specific food source; $i \neq j, VT_{i,j} = 0$ indicates that the jth food source has just been visited by the ith hummingbird in the current iteration.

(2) Guided foraging: With the aforementioned flight capabilities, the hummingbird can access its target food source to obtain candidate food sources, so the mathematical equation for simulating guiding foraging behavior and candidate food sources is as follows:

$$v_i(t+1) = x_{i,tar}(t) + a \times D \times (x_i(t) - x_{i,tar}(t)) \tag{6}$$

$$a \sim N(0,1) \tag{7}$$

where $x_i(t)$ is the position of the ith hummingbird food source in time t, $x_{i,tar}(t)$ is the position of the ith hummingbird target food source in time t, a is normally distributed, with mean = 0 and a standard deviation of 1.

The position update of the ith food source is as follows:

$$x_i(t+1) = \begin{cases} x_i(t) & f(x_i(t)) \leq f(v_i(t+1)) \\ v_i(t+1) & f(x_i(t)) > f(v_i(t+1)) \end{cases} \tag{8}$$

where $f(\cdot)$ indicates the function fitness value. Equation (8) shows that if the nectar refilling rate of the candidate food source is better than that of the current one, the hummingbird abandons the current food source and stays at the candidate one resulting from Equation (6) for feeding.

(3) Territorial foraging: After reaching a target food source where nectar is eaten, hummingbirds may seek new food sources. Therefore, a hummingbird can easily move to a neighboring region within its territory, where new food sources can be found that may be better candidate solutions. The mathematical equation for simulating the local search of hummingbirds in territorial foraging strategies and candidate food sources is as follows:

$$v_i(t+1) = x_i(t) + b \times D \times x_i(t) \tag{9}$$

$$b \sim N(0,1) \tag{10}$$

where b is normally distributed, with mean = 0 and a standard deviation of 1.

(4) When food is often scarce in a territory frequented by hummingbirds, the bird often migrates to more distant food sources to forage. In the AHA algorithm, a migration coefficient is defined. The hummingbirds in the food source with the weakest filling rate will, at random, migrate to another new food source throughout the entire search space when the number of iterations exceeds the predefined value of the migration coefficient. At this point, the hummingbird will abandon the original source and stay at the new source foraging, and then the migration foraging of the hummingbird from the source with the worst nectar-filling rate to the randomly generated new source can be given as follows:

$$x_{wor}(t+1) = Low + r \times (Up - Low) \tag{11}$$

where x_{wor} is the food source with the worst nectar refilling rate in the population.

2.2. Improved Artificial Hummingbird Algorithm Based on Chebyshev Chaotic Map and Levy Flight (IAHA)

2.2.1. Chebyshev Chaotic Map

Chaos is a random state in a deterministic system, a phenomenon of the evolution of a nonlinear system. It is caused by deterministic rules and is very sensitive to the initial conditions, which is long-term behavior without a fixed period. Chaos has an ergodic, and searching using chaotic variables is obviously superior to an unordered search [28].

The Chebyshev chaotic map has a wide and more uniform distribution range, and it can be distributed in the interval $[-1, 1]$. When $k \geq 2$ (k is the order), no matter how close the initial value is chosen, the iterated sequence is uncorrelated and chaotic and ergodic within this range. The equation is shown as follows:

$$x_{n+1} = \cos(k \arccos x_n) \quad x_n \in [-1, 1] \tag{12}$$

This paper uses the equation to generate uniformly distributed points to initialize the position of artificial hummingbirds, improve the global search ability of the initial population, and improve the solution accuracy of the algorithm.

2.2.2. Levy Flight

In the 1930s, the Levy distribution was a probability distribution proposed by Levy, a French mathematician. After that, many researchers have carried out many studies on the Levy distribution. So far, it has been proven that the foraging trajectory of many animals in nature follows the Levy distribution. The Levy flight follows the principle of the Levy distribution of many random phenomena, such as Brownian motion, random walk, etc. [29] At present, Levy flight is widely used in intelligent optimization. For example, the Cuckoo algorithm adopts Levy flight to update the position [30]. Levy flight can expand the search space, so it is easier to avoid premature convergence by introducing Levy flight into the AHA algorithm.

Levy flight position updated to:

$$x_i(t+1) = \begin{cases} x_i(t) + \alpha \oplus Levy(\lambda) & f(x_i(t)) \le f(v_i(t+1)) \\ v_i(t+1) & f(x_i(t)) > f(v_i(t+1)) \end{cases} \tag{13}$$

where x_i^t is the tth generation position of x_i, $\oplus$ is the dot multiplication, α is the step size control parameter, and $Levy(\lambda)$ is the random search path, which satisfies:

$$Levy \sim u = t^{-\lambda}, \quad 1 < \lambda \le 3 \tag{14}$$

Its step size obeys the Levy distribution, and step size s is calculated as [29]:

$$s = \frac{\mu}{|v|^{1/\beta}} \tag{15}$$

where μ, v are normally distributed, defined as:

$$\mu \sim N(0, \sigma_\mu^2)$$

$$v \sim N(0, \sigma_v^2)$$

where

$$\sigma_\mu = \frac{(1+\beta)(\sin\frac{\pi\beta}{2})}{\frac{1+\beta}{2}\beta^2\frac{\beta-1}{2}} \tag{16}$$

$$\sigma_v = 1$$

where β is usually a constant of 1.5.

According to the above, the flow chart of the IAHA algorithm is shown in Figure 1. Its basic steps are as follows:

Step 1: Set the parameters: a maximum number of iterations of T, a number of artificial hummingbirds of Pop, and a dimension of the fitness function of *Dim*.

Step 2: The Chebyshev chaotic map was used to initialize the food source location of artificial hummingbirds; the corresponding fitness function was calculated and the optimal value was recorded.

Step 3: Introducing Levy flight-guided foraging or territorial foraging to update the food source location of artificial hummingbirds. The probability of both forages was 50%.

Step 4: In the worst-case scenario of a food source, hummingbirds may visit the same food source as their target source after two repetitions, in which case, migratory foraging is performed.

Step 5: Determine if the result has reached the maximum number of iterations of the algorithm, if it has reached the maximum accuracy, then the optimal food source location is obtained, otherwise, transpose step 3.

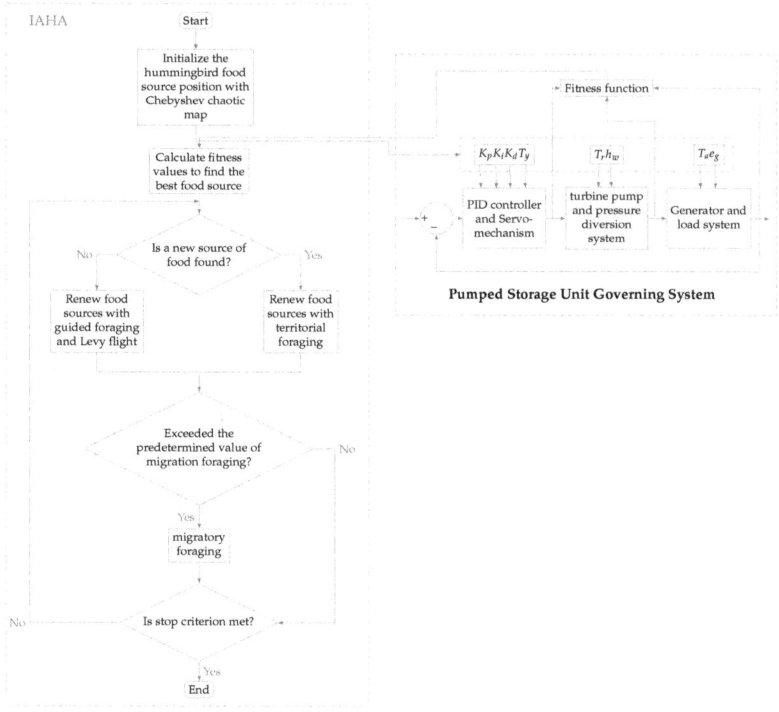

Figure 1. Flowchart of IAHA.

3. Performance Testing and Analysis

In order to verify the effectiveness of the IAHA algorithm. We compare the IAHA method with four well-established optimizers, including the artificial hummingbird algorithm (AHA), particle swarm optimization algorithm (PSO), ant lion optimization algorithm (ALO), and gravitational search algorithm (GSA). In addition, the number of iterations of all test functions in the five algorithms was 500, and the population size was 30. The rest of the parameters are shown in Table 1.

Table 1. Parameter settings for each algorithm in the test function.

Algorithm	Parameter Settings
PSO	$c_1 = 2; c_2 = 2; \omega_{max} = 0.8; \omega_{min} = 0.2$ ω decreases linearly from 0.8 to 0.2
GSA	$G_0 = 100; a = 20$

The metaheuristic algorithm is a kind of random search algorithm. For the same optimization problem, the results are usually not identical if the same algorithm is used many times. To avoid excessive errors arising from randomness in a single result, the algorithm was repeated 20 times and the statistical results for each algorithm are shown in Tables 2–4. The tables include the mean (mean), standard deviation (Std), best (best), and worst (worst) values of the 23 test functions (see Table A1 in "Appendix A"), in which the bold are the best values of the mean, standard deviation, best, and worst values of the different test functions for the five algorithms.

3.1. Unimodal Test Functions

F1–F7 are unimodal functions, which mainly test the capability of the development stage. Moreover, these types of functions do not have local extrema, but only global optimal values, which are easy to optimize. Based on this, the convergence speed may be more important than the global optimal value of the algorithm. The statistical results of the IAHA algorithm proposed compared with the other four algorithms are shown in Table 2 and Figure 2. Among the five intelligent optimization algorithms, the IAHA algorithm has a significant advantage in terms of mean and standard deviation, except for F6, where both the IAHA algorithm and the AHA algorithm achieve optimal values. As a result, the IAHA algorithm is superior in development ability, convergence, and stability.

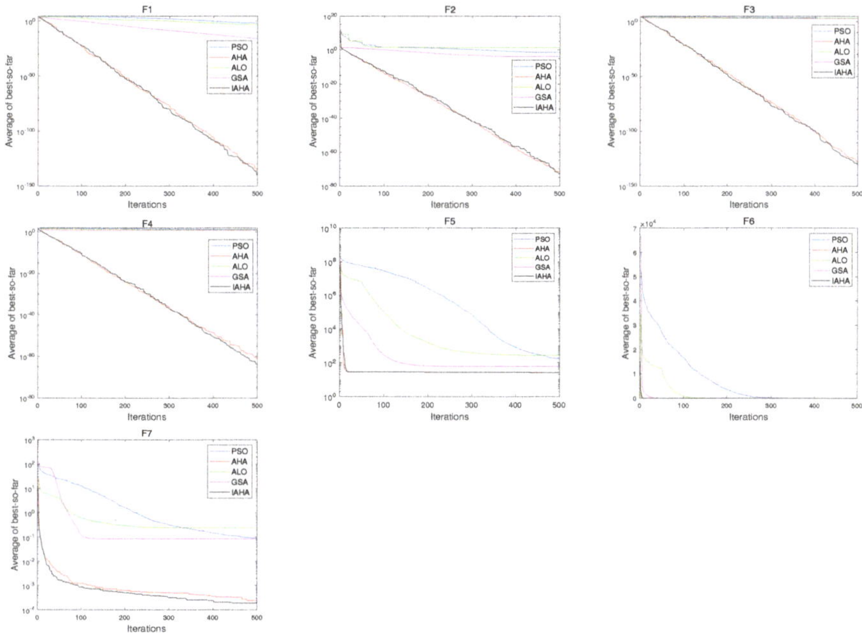

Figure 2. The average convergence curve of different optimization algorithms on a unimodal test function.

Table 2. Statistical results of unimodal test functions for each algorithm.

Function	Index	IAHA	AHA	PSO	ALO	GSA
F1	Mean	9.48×10^{-141}	1.72×10^{-136}	3.05×10^{-2}	1.34×10^{-3}	2.27×10^{-16}
	Std	4.08×10^{-140}	7.67×10^{-136}	7.57×10^{-2}	7.96×10^{-4}	9.07×10^{-17}
	Best	8.48×10^{-163}	1.19×10^{-158}	3.50×10^{-4}	2.38×10^{-4}	7.15×10^{-17}
	Worst	1.83×10^{-139}	3.43×10^{-135}	3.31×10^{-1}	3.06×10^{-3}	4.85×10^{-16}
F2	Mean	9.65×10^{-74}	1.85×10^{-72}	4.28×10^{-2}	38.00	1.83×10^{-4}
	Std	2.96×10^{-73}	8.18×10^{-72}	7.35×10^{-2}	40.06	4.97×10^{-4}
	Best	1.36×10^{-83}	4.7×10^{-82}	3.44×10^{-3}	4.77	3.66×10^{-8}
	Worst	1.07×10^{-72}	1.16×10^{-71}	2.59×10^{-1}	1.19×10^2	1.70×10^{-3}
F3	Mean	1.05×10^{-129}	6.04×10^{-128}	4.10×10^4	4.80×10^4	9.90×10^3
	Std	4.70×10^{-129}	2.19×10^{-127}	1.36×10^3	2.72×10^3	3.37×10^2
	Best	2.99×10^{-145}	2.26×10^{-148}	1.67×10^3	8.02×10^2	4.31×10^2
	Worst	2.10×10^{-128}	9.63×10^{-127}	7.66×10^3	1.05×10^4	1.78×10^3
F4	Mean	7.15×10^{-65}	4.80×10^{-63}	19.51	17.64	7.13
	Std	1.95×10^{-64}	1.94×10^{-62}	4.18	4.58	1.91
	Best	1.31×10^{-73}	1.45×10^{-70}	12.50	11.90	3.35
	Worst	7.98×10^{-64}	8.67×10^{-62}	31.18	24.64	10.76

Table 2. *Cont.*

Function	Index	IAHA	AHA	PSO	ALO	GSA
F5	Mean	26.64	26.80	1.80×10^2	2.93×10^2	62.28
	Std	3.17×10^{-1}	3.69×10^{-1}	1.26×10^2	3.85×10^2	56.67
	Best	26.04	26.10	52.86	27.15	27.09
	Worst	27.29	27.46	4.91×10^2	1.22×10^3	2.25×10^2
F6	Mean	0	0	1.8	1.84×10^{-3}	6.7
	Std	0	0	1.77	1.55×10^{-3}	8.77
	Best	0	0	0	3.80×10^{-4}	0
	Worst	0	0	6	5.27×10^{-3}	30
F7	Mean	1.90×10^{-4}	2.48×10^{-4}	9.16×10^{-2}	2.53×10^{-1}	8.68×10^{-2}
	Std	1.61×10^{-4}	1.84×10^{-4}	3.21×10^{-2}	6.79×10^{-2}	3.82×10^{-2}
	Best	1.27×10^{-5}	3.79×10^{-5}	2.39×10^{-2}	1.15×10^{-1}	2.95×10^{-2}
	Worst	5.23×10^{-4}	6.66×10^{-4}	1.56×10^{-1}	3.51×10^{-1}	1.65×10^{-1}

3.2. High-Dimensional Multi-Peak Test Functions

F8–F23 are multi-peak functions, which mainly demonstrate the ability of the exploration phase. Among these, F8–F13 are high-dimensional multi-peak functions and F14–F23 are low-dimensional multi-peak functions. The multi-peak test functions have multiple local extremes and global optima, which are more difficult to optimize compared to unimodal functions. As a result, the multi-peak function algorithm's capability of global optimization is given additional focus. The statistical results of the high-dimensional multi-peak function test for each algorithm are shown in Table 3 and Figure 3. In F8, the mean value and standard deviation of IAHA rank second, second only to AHA. In F9–F11, the mean and standard deviation of IAHA and AHA are the same. The mean value of IAHA (in F12 and F13) is the best. Although the standard deviation of F13 is not the smallest, its best value and the worst value are optimal compared with other algorithms. Thus, it can be seen that IAHA has a better advantage in the global search.

Table 3. Statistical results of high-dimensional multi-peak test functions for each algorithm.

Function	Index	IAHA	AHA	PSO	ALO	GSA
F8	Mean	-1.15×10^4	-1.16×10^4	-5.78×10^3	-5.66×10^3	-2.61×10^3
	Std	3.73×10^2	3.50×10^2	9.04×10^2	8.60×10^2	4.75×10^2
	Best	-1.23×10^4	-1.22×10^4	-7.35×10^3	-9.29×10^3	-4.06×10^3
	Worst	-1.07×10^4	-1.08×10^4	-4.30×10^3	-5.42×10^3	-1.87×10^3
F9	Mean	0	0	41.99	75.67	31.44
	Std	0	0	11.33	21.79	6.73
	Best	0	0	25.87	30.84	18.90
	Worst	0	0	72.75	1.16×10^2	43.78
F10	Mean	8.88×10^{-16}	8.88×10^{-16}	1.41	5.68	1.22×10^{-8}
	Std	0	0	8.32×10^{-1}	3.48	3.71×10^{-9}
	Best	8.88×10^{-16}	8.88×10^{-16}	1.24×10^{-2}	2.32	4.95×10^{-9}
	Worst	8.88×10^{-16}	8.88×10^{-16}	2.50	12.67	1.88×10^{-8}
F11	Mean	0	0	6.77×10^{-2}	6.83×10^{-2}	25.62
	Std	0	0	1.23×10^{-1}	4.03×10^{-2}	5.87
	Best	0	0	2.37×10^{-3}	1.85×10^{-2}	15.83
	Worst	0	0	5.73×10^{-1}	1.87×10^{-1}	36.56
F12	Mean	6.78×10^{-4}	8.67×10^{-4}	2.21	12.36	1.97
	Std	1.54×10^{-3}	1.99^{-3}	1.42	4.69	1.07
	Best	3.71×10^{-5}	7.02×10^{-5}	1.99×10^{-1}	5.59	1.91×10^{-1}
	Worst	6.67×10^{-3}	7.34×10^{-3}	5.63	23.23	4.21
F13	Mean	1.82	1.96	3.19	30.74	7.37
	Std	5.16×10^{-1}	3.39×10^{-1}	4.33	17.89	5.53
	Best	5.25×10^{-1}	1.40	2.74×10^{-1}	2.60×10^{-2}	1.08×10^{-1}
	Worst	2.48	2.53	16.02	62.73	22.86

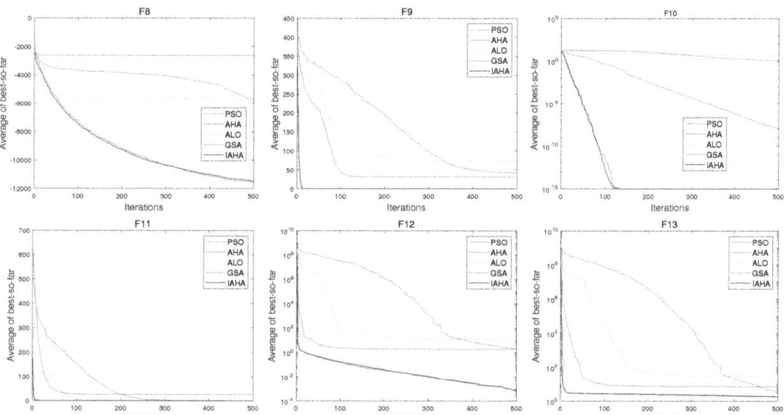

Figure 3. The average convergence curve of different optimization algorithms on high-dimensional multi-peak test function.

3.3. Low-Dimensional Multi-Peak Test Functions

F14–F23 are low-dimensional multi-peaked functions, which have fewer local extremes than higher-dimensional multi-peaked functions and are relatively easier to optimize. Table 4 and Figure 4 show the statistical results of the low-dimensional multi-peak test functions of each algorithm. The mean values of the functions in F16–F19 tested by the IAHA algorithm and other algorithms are the same and reach optimal values. In addition, the standard deviation of the IAHA algorithm in F16 and F19 is second only to the GSA algorithm. In F20, the result of the IAHA algorithm is slightly better than that of other algorithms. The IAHA algorithm has a minimum standard deviation in F15, F18, F19, F21–F23. However, regarding the mean values of the AHA algorithm and IAHA algorithm, eight functions are the same and do not show better advantages. Therefore, because of the particularity of the function, the results obtained by the proposed IAHA algorithm and other algorithms are basically close to the theoretical global optimum.

Table 4. Statistical results of low-dimensional multi-peak test functions of each algorithm.

Function	Index	IAHA	AHA	PSO	ALO	GSA
F14	Mean	9.98×10^{-1}	9.98×10^{-1}	9.98×10^{-1}	3.017	6.38
	Std	0	0	5.09×10^{-17}	2.76	3.84
	Best	9.98×10^{-1}	9.98×10^{-1}	9.98×10^{-1}	9.98×10^{-1}	1.06
	Worst	9.98×10^{-1}	9.98×10^{-1}	9.98×10^{-1}	10.76	13.32
F15	Mean	3.07×10^{-4}	3.07×10^{-4}	4.97×10^{-4}	2.27×10^{-3}	6.03×10^{-3}
	Std	4.21×10^{-11}	1.54×10^{-9}	1.32×10^{-4}	4.47×10^{-3}	4.21×10^{-3}
	Best	3.07×10^{-4}	3.07×10^{-4}	3.07×10^{-4}	6.14×10^{-4}	8.10×10^{-4}
	Worst	3.07×10^{-4}	3.07×10^{-4}	6.84×10^{-4}	2.06×10^{-2}	1.68×10^{-2}
F16	Mean	-1.03	-1.03	-1.03	-1.03	-1.03
	Std	1.76×10^{-16}	1.69×10^{-16}	2.22×10^{-16}	1.04×10^{-13}	1.02×10^{-16}
	Best	-1.03	-1.03	-1.03	-1.03	-1.03
	Worst	-1.03	-1.03	-1.03	-1.03	-1.03
F17	Mean	3.98×10^{-1}	3.98×10^{-1}	3.98×10^{-1}	3.98×10^{-1}	3.98×10^{-1}
	Std	0	0	6.06×10^{-10}	5.26×10^{-14}	0
	Best	3.98×10^{-1}	3.98×10^{-1}	3.98×10^{-1}	3.98×10^{-1}	3.98×10^{-1}
	Worst	3.98×10^{-1}	3.98×10^{-1}	3.98×10^{-1}	3.98×10^{-1}	3.98×10^{-1}
F18	Mean	3	3	3	3	3
	Std	2.04×10^{-16}	3.95×10^{-16}	1.28×10^{-15}	6.49×10^{-13}	4.34×10^{-15}
	Best	3	3	3	3	3
	Worst	3	3	3	3	3

Table 4. *Cont.*

Function	Index	IAHA	AHA	PSO	ALO	GSA
F19	Mean	-3.86	-3.86	-3.86	-3.86	-3.86
	Std	2.24×10^{-15}	2.26×10^{-15}	2.22×10^{-15}	1.06×10^{-12}	1.87×10^{-15}
	Best	-3.86	-3.86	-3.86	-3.86	-3.86
	Worst	-3.86	-3.86	-3.86	-3.86	-3.86
F20	Mean	-3.32	-3.30	-3.26	-3.27	-3.32
	Std	4.20×10^{-12}	4.36×10^{-2}	6.10×10^{-2}	6.02×10^{-2}	2.67×10^{-2}
	Best	-3.32	-3.32	-3.32	-3.32	-3.32
	Worst	-3.32	-3.30	-3.30	-3.30	-3.30
F21	Mean	-10.15	-10.15	-7.50	-6.49	-6.37
	Std	2.31×10^{-6}	6.22×10^{-6}	3.19	2.88	3.66
	Best	-10.15	-10.15	-10.15	-10.15	-10.15
	Worst	-10.15	-10.15	-2.68	-2.63	-2.68
F22	Mean	-10.40	-10.40	-7.54	-6.65	-9.86
	Std	1.19×10^{-9}	1.55×10^{-5}	3.58	3.57	1.82
	Best	-10.40	-10.40	-10.40	-10.40	-10.40
	Worst	-10.40	-10.40	-2.75	-2.75	-2.75
F23	Mean	-10.54	-10.54	-9.05	-7.07	-10.13
	Std	6.20×10^{-9}	9.08×10^{-4}	3.06	3.62	1.81
	Best	-10.54	-10.54	-10.54	-10.54	-10.54
	Worst	-10.54	-10.54	-2.81	-2.42	-2.43

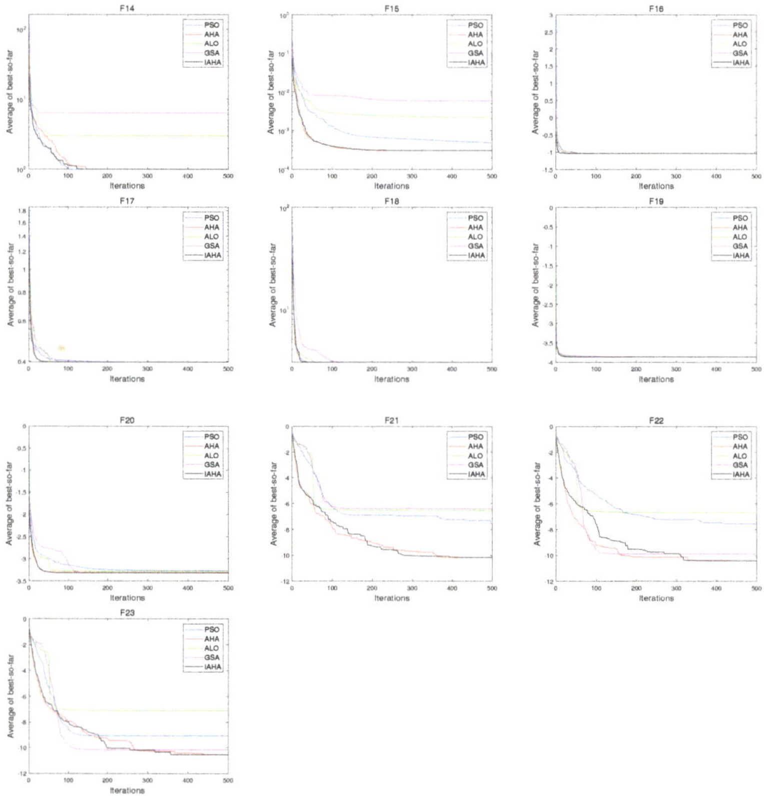

Figure 4. Average convergence curve of different optimization algorithms on low—dimensional multi—peak test function.

4. Analysis of Parameter Identification under Frequency Disturbance Conditions

4.1. Mathematical Model of Governor

4.1.1. PID Controller Model

PID controllers (due to their simple structures, high stability, and convenient adjustments) are the main technologies of industrial control. At present, PID controllers are used in most hydropower stations in China, and their transfer functions are as follows [31]:

$$G_{PID}(s) = \frac{K_p + \frac{K_i}{s} + \frac{K_d}{(T_{1v}s+1)}}{\frac{b_p K_i}{s} + 1} \tag{17}$$

where K_p is the proportional parameter, K_i is the integral parameter, K_d is the derivative parameter, T_{1v} is the derivative time constant, b_p is the permanent speed droop, and s is the Laplace operator.

4.1.2. Electrohydraulic Servo System Model

The electrohydraulic servo system is the executive agency of the governor, which consists of the comprehensive amplification element, electrohydraulic converter, the main control valve, and the main servomotor [32]. Considering that the time constant of the main servomotor is much larger than that of the auxiliary servomotor, the time constant of the auxiliary servomotor is ignored. In order to prevent the guide vane from moving too fast, the main servomotor usually has a limited amplitude (in regard to achieving the speed limit of the main servomotor) [33]:

$$G_y(s) = \frac{1}{T_y s + 1} \tag{18}$$

where T_y is the servomotor time constant.

4.2. Mathematical Model of the Pressure Water Diversion System

The pressure diversion system is an important hydraulic construction of the pumped storage power station; it mainly consists of the upper reservoir, headrace tunnel, upper surge chamber, penstock, casing, turbine pump, draft tube, downstream surge chamber, tailwater tunnel, and lower reservoir [34]. The two types of water hammer models for the water diversion system are elastic and rigid. In this paper, the elastic water hammer model was chosen, and the hyperbolic tangent function was used as the mathematical model as follows:

$$G_h(s) = \frac{H(s)}{Q(s)} = -2h_w \frac{e^{\frac{T_r}{2}s} - e^{-\frac{T_r}{2}s}}{e^{\frac{T_r}{2}s} + e^{-\frac{T_r}{2}s}}$$

$$= -2h_w \frac{sh(\frac{T_r}{2}s)}{ch(\frac{T_r}{2}s)} \tag{19}$$

where T_r is the water hammer pressure wave time constant, h_w is the pipeline characteristic coefficient, Q is the flow rate, and H is the head.

The rigid water hammer model is unable to accurately represent the dynamic properties of water flow in the pipe because it ignores the frictional resistance of water flow as well as the elasticity of the water body and pipe wall. Therefore, this paper makes use of the fourth-order elastic water hammer model. The transfer function is as follows:

$$G_h(s) = \frac{H(s)}{Q(s)} = -h_w \frac{T_r s + \frac{1}{24} T_r^3 s^3}{1 + \frac{1}{8} T_r^2 s^2 + \frac{1}{384} T_r^4 s^4} \tag{20}$$

4.3. Turbine Pump Mathematical Model

Under some operating conditions, the turbine pump operates with small fluctuations around the specified operating point, and the variation in the nonlinear characteristics of the turbine pump is not particularly significant within a small range. In this case, the full characteristic model of the turbine pump is simplified at a certain operating point to obtain a linear model of the turbine pump, which is calculated as follows [35]:

$$\begin{cases} m_t(t) = e_x x(t) + e_y y(t) + e_h\, h(t) \\ q(t) = e_{qx} x(t) + e_{qy} y(t) + e_{qh} h(t) \end{cases} \tag{21}$$

where x is speed, y is guide vane opening, h is working head, $e_x, e_y, e_h, e_{qx}, e_{qy}, e_{qh}$ are turbine transfer coefficient at a certain operating point.

4.4. Generator and Load Mathematical Model

The first-order model (reflecting the dynamic characteristics and self-regulating performance of the generator rotor) is usually used in research on the modeling and optimization control of the governing system of the pumped storage unit. Considering the balance between the torque of the turbine pump and the torque of the generator, the transfer function of the generator and load is [36]:

$$G_g(s) = \frac{1}{T_a s + e_g} \tag{22}$$

where T_a is the inertial time constant of the generator and e_g is the adjustment coefficient of the generator.

A six-parameter model was used for the turbine pump in the pumped storage unit, and the water diversion system adopted the four-order elastic water hammer model, where the model took into account the nonlinear link of the electrohydraulic follower system with limiting amplitude; the model structure of the governing system of the pumped storage unit is shown in Figure 5.

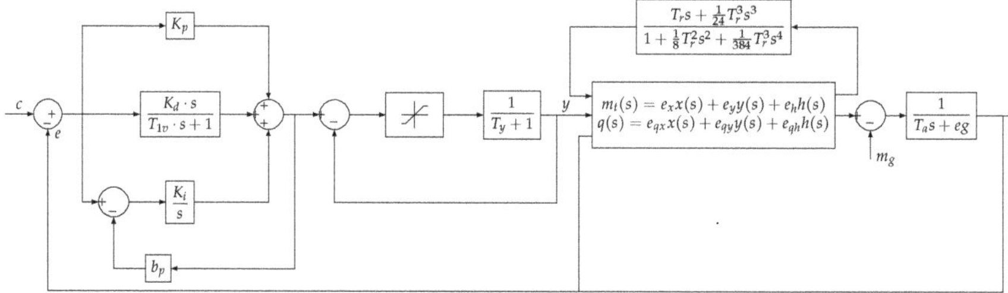

Figure 5. Simulation model of the governing system in the pumped storage unit.

4.5. Parameter Identification of the Governing System

The parameter identification with a known model structure is usually transformed into an optimization problem [37]. The optimization objective function is the deviation between the system output variable and the identified system output variable, and then the integration is performed. Under the assumption that the identification system can reflect the real system, the optimal value can be found through unknown parameters to achieve the identification effect and make the output consistent with the real system.

4.5.1. Objective Function

In the governing system of the pumped storage units, although the PID controller parameters can be set on the turbine pump governor to verify the integrity and effectiveness of the proposed identification strategy, the parameter set to be identified includes the PID control parameters. Therefore, eight parameters are identified as follows: proportional parameter K_p, integral parameter K_i, derivative parameter K_d, servomotor time constant T_y, pipeline characteristic coefficient h_w, time constant of water hammer pressure wave T_r, generator inertia time constant T_a, and generator regulation coefficient e_g. In the model of the system, the square of the difference between the measurable output selected for the objective function and the reintegration within 50 s after the disturbance starts. The objective function is as follows:

$$C(\theta) = \sum_{i=1}^{N} (x_i - \widehat{x}_i)^2 + \sum_{i=1}^{N} (y_i - \widehat{y}_i)^2 + \sum_{i=1}^{N} (m_t - \widehat{m}_t)^2 \tag{23}$$

where N is the sample capacity, x_i, y_i, and m_t are the speed, guide vane opening, and torque of the real system output; $\widehat{x}_i$, $\widehat{y}_i$, and $\widehat{m}_t$ are the speed, guide vane opening, and torque of the identification system output; $\theta = [K_p \ K_i \ K_d \ T_y \ h_w \ T_r \ T_a \ e_g]$ are the parameters to be identified.

4.5.2. Identification Strategy

The process of identification is illustrated in Figure 6. Firstly, the system is given an excitation signal to obtain the respective dynamic responses $[x, y, m_t]$. Secondly, the intelligent optimization algorithm is used to initialize the parameters and obtain the dynamic response of the identification system $[\widehat{x}, \widehat{y}, \widehat{m}_t]$, and then the fitness function is calculated. Finally, the fitness function is optimized according to the intelligent optimization algorithm to obtain the minimum value of the fitness function.

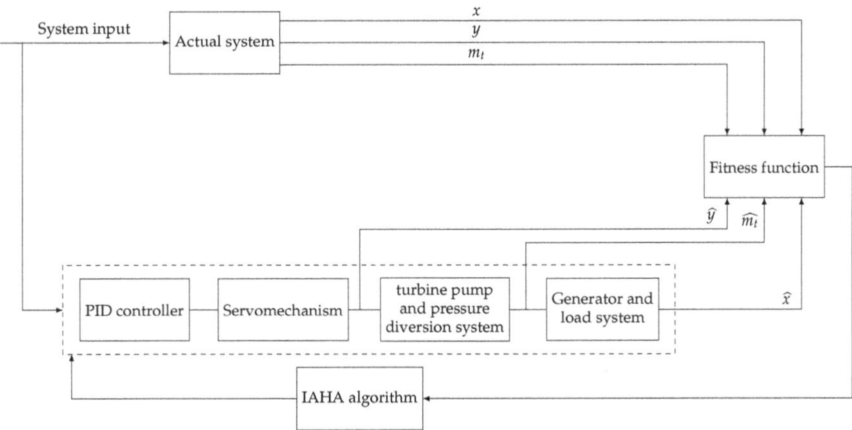

Figure 6. Identification strategy of the governing system.

4.5.3. Analysis of the Simulation Test Results

To verify the accuracy of the improved artificial hummingbird algorithm for identifying the governing system in the pumped storage unit, this paper compares the following four optimization algorithms—the particle swarm optimization algorithm (PSO), ant lion optimization algorithm (ALO), gravitational search algorithm (GSA), artificial hummingbird algorithm (AHA). The governing system in the pumped storage unit established in the previous section was used as the object of study and parameter values $[K_p \ K_i \ K_d \ T_y \ h_w \ T_r \ T_a \ e_g]$ were set as the parameter values of the system, as shown in Table 5. The output state variables for simulation under certain operating conditions were used as the output variables. In order to evaluate the accuracy of parameter identification, the parameter error (PE) and average parameter error (APE) were used to measure the accuracy of the model parameters. The calculation formulas of PE and APE are as follows:

$$PE = \frac{|\theta_i - \widehat{\theta}_i|}{\theta_i} \times 100\% \quad i = 1, 2, 3, \cdots, m \tag{24}$$

$$APE = \frac{1}{m} \sum_{i=1}^{m} \frac{|\theta_i - \widehat{\theta}_i|}{\theta_i} \times 100\% \tag{25}$$

The experiment was carried out under the condition of frequency disturbance; the frequency of the unit step was applied in the stable state of the unit. The simulation duration was 50 s, the sampling period was 0.01 s, and the number of identification tests was 20. Because the identification of the optimization algorithm was random, the average value of the results was taken. The six parameters of the hydraulic turbine can be obtained as follows: $e_x = -1.925, e_y = 0.7133, e_h = 1.413, e_{qx} = -0.7, e_{qy} = 0.5833, e_{qh} = 0.8555$, according to the rotational speed–flow characteristic curve and rotational speed–torque characteristic curve [38].

Tables 5–7 show the results of the parameter identification under 5%, 10%, and 15% frequency disturbances by different optimization algorithms. From the PE index of the parameter error of each algorithm, the identification of individual parameters was not optimal compared to PSO, ALO, GSA, and AHA, but only next to the best. From the APE index of the average parameter identification accuracy, it can be seen that under the three frequency disturbance conditions, 15% of the APE was only 1.28%, and the IAHA algorithm was better than the other four algorithms. IAHA improves the identification accuracy of the AHA algorithm accordingly. This shows that the proposed Chebyshev chaotic map and Levy flight improvement strategy effectively improve the search and convergence of the algorithm.

Table 5. Algorithm identification results of 5% frequency disturbance.

θ	Real Value	PSO		ALO		GSA		AHA		IAHA	
		$\widehat{\theta}$	PE	$\widehat{\theta}$	PE	$\widehat{\theta}$	PE	$\widehat{\theta}$	PE	$\widehat{\theta}$	PE
K_p	3.21	3.4725	0.0818	3.4089	0.0620	2.7837	0.1328	**3.1871**	0.0071	3.1822	0.0087
K_i	2.68	2.7284	0.0181	2.6924	0.0046	2.8566	0.0659	2.7007	0.0077	**2.7007**	0.0077
K_d	1.24	1.5805	0.2746	1.5197	0.2256	1.6001	0.2904	1.3074	0.0544	**1.2964**	0.0455
T_y	0.30	0.5283	0.761	0.5040	0.68	0.3615	0.205	0.3152	0.0507	**0.3146**	0.0487
h_w	1.00	1.0480	0.0480	1.0580	0.0580	0.8805	0.1195	0.9518	0.0482	**0.9691**	0.0309
T_r	1.50	1.5566	0.0377	1.5049	0.0033	1.6834	0.1223	1.5536	0.0357	**1.5283**	0.0189
T_a	8.86	9.1299	0.0305	8.8038	0.0063	7.7609	0.1241	8.8523	0.0009	**8.8335**	0.0030
e_g	1.50	1.4974	0.0017	1.5004	0.0003	1.5104	0.0069	1.4996	0.0003	**1.5001**	0.00007
APE		0.1567		0.1300		0.1334		0.0256		**0.0204**	

Table 6. Algorithm identification results of 10% frequency disturbance.

θ	Real Value	PSO		ALO		GSA		AHA		IAHA	
		$\widehat{\theta}$	PE	$\widehat{\theta}$	PE	$\widehat{\theta}$	PE	$\widehat{\theta}$	PE	$\widehat{\theta}$	PE
K_p	3.21	3.6584	0.1397	3.3515	0.0441	2.9985	0.0659	3.1554	0.0170	**3.2077**	0.0007
K_i	2.68	2.6507	0.0109	2.7020	0.0082	2.7975	0.0438	2.7050	0.0093	**2.690**	0.0049
K_d	1.24	1.3532	0.0913	1.5432	0.2445	1.6201	0.3065	1.3087	0.0554	**1.2974**	0.0463
T_y	0.30	0.5116	0.7053	0.4603	0.5343	0.3820	0.2733	**0.3084**	0.0280	0.3155	0.0517
h_w	1.00	0.9294	0.0706	0.9430	0.057	0.9297	0.0703	0.9657	0.0343	**0.9769**	0.0231
T_r	1.50	1.6058	0.0705	1.6508	0.1005	1.7389	0.1593	1.5303	0.0202	**1.5269**	0.0179
T_a	8.86	9.4549	0.0671	8.8340	0.0029	7.6352	0.1382	8.8043	0.0063	**8.8533**	0.0008
e_g	1.50	1.4962	0.0025	1.5001	0.00007	1.5108	0.0072	1.5009	0.0006	**1.5007**	0.0005
APE		0.1447		0.1240		0.1330		0.0214		**0.0182**	

Table 7. Algorithm identification results of 15% frequency disturbance.

θ	Real Value	PSO		ALO		GSA		AHA		IAHA	
		$\hat{\theta}$	PE	$\hat{\theta}$	PE	$\hat{\theta}$	PE	$\hat{\theta}$	PE	$\hat{\theta}$	PE
K_p	3.21	3.4168	0.0644	3.3238	0.0355	2.8265	0.1195	3.1421	0.0241	**3.1651**	0.0140
K_i	2.68	2.7092	0.0109	2.7251	0.0168	2.9182	0.0889	2.7138	0.0126	**2.7067**	0.0100
K_d	1.24	1.7219	0.3886	1.5115	0.2190	1.5179	0.2241	1.3120	0.0581	**1.2893**	0.0398
T_y	0.30	0.4489	0.4963	0.4500	0.500	0.3286	0.0953	0.3075	0.0250	**0.3009**	0.0030
h_w	1.00	1.1061	0.1061	1.0637	0.0637	0.9323	0.0677	0.9701	0.0299	**0.9869**	0.0131
T_r	1.50	1.4625	0.025	1.4911	0.0059	1.6037	0.0691	1.5165	0.0110	**1.5032**	0.0021
T_a	8.86	8.7925	0.0076	8.684	0.0199	7.8136	0.1181	8.7751	0.0096	**8.7994**	0.0068
e_g	1.50	1.4993	0.0005	1.4996	0.0003	1.5038	0.0025	1.5014	0.0009	**1.4997**	0.0002
APE		0.1406		0.1076		0.0982		0.0211		**0.0128**	

Figures 7a–c, 8a–c, and 9a–c, respectively, show the comparison diagram of the dynamic response of the unit speed, torque output, and guide vane opening output of the original parameters and IAHA algorithm optimization identification parameters under different frequency disturbance conditions, and they were locally amplified. It can be seen from the figure that the IAHA identification system is consistent with the original system, with high accuracy. Moreover, d–f are the identification errors of different links. It can be seen that the error finally approaches zero.

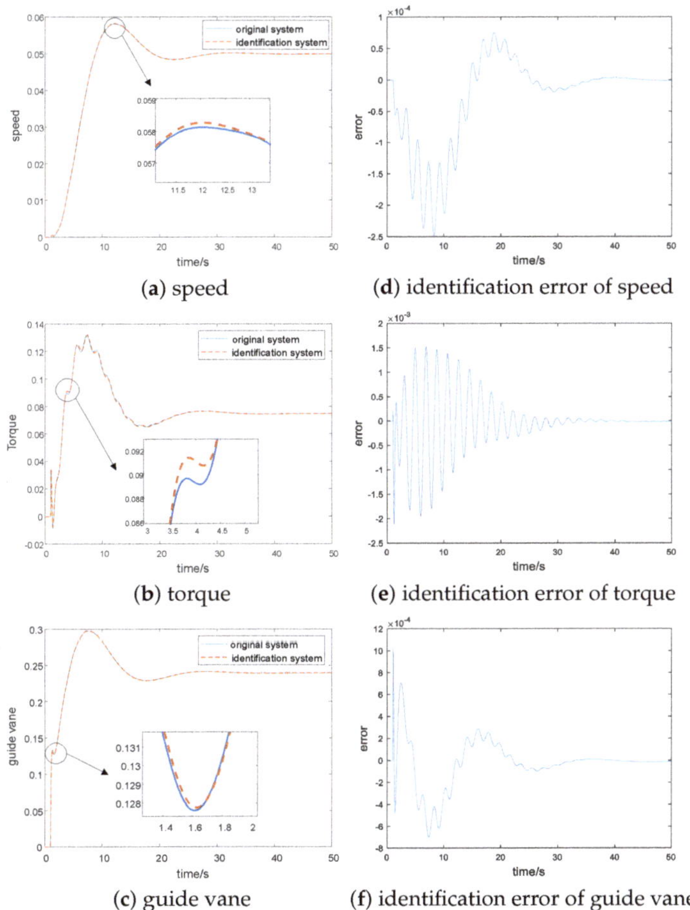

Figure 7. Comparison between the outputs of the three parts of the original and the identification systems under 5% frequency disturbance.

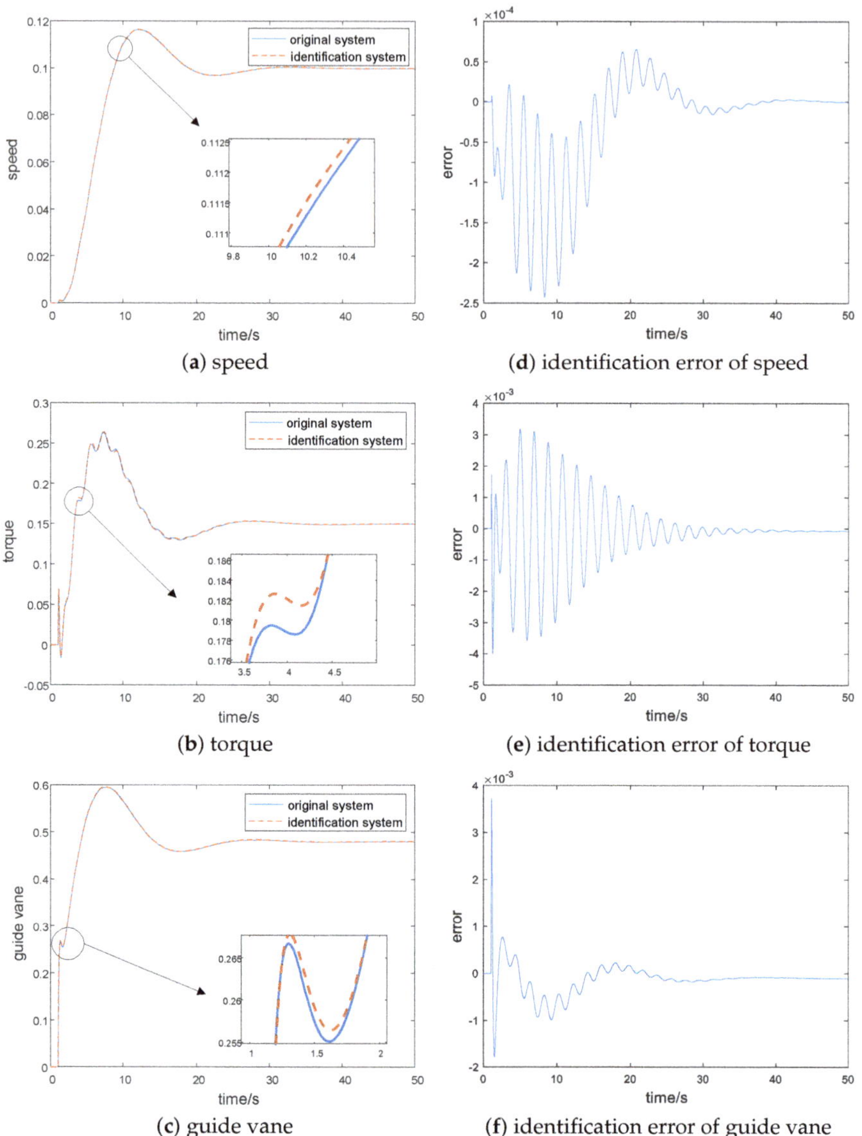

(a) speed   (d) identification error of speed

(b) torque   (e) identification error of torque

(c) guide vane   (f) identification error of guide vane

Figure 8. Comparison between the outputs of the three parts of the original and the identification systems under a 10% frequency disturbance.

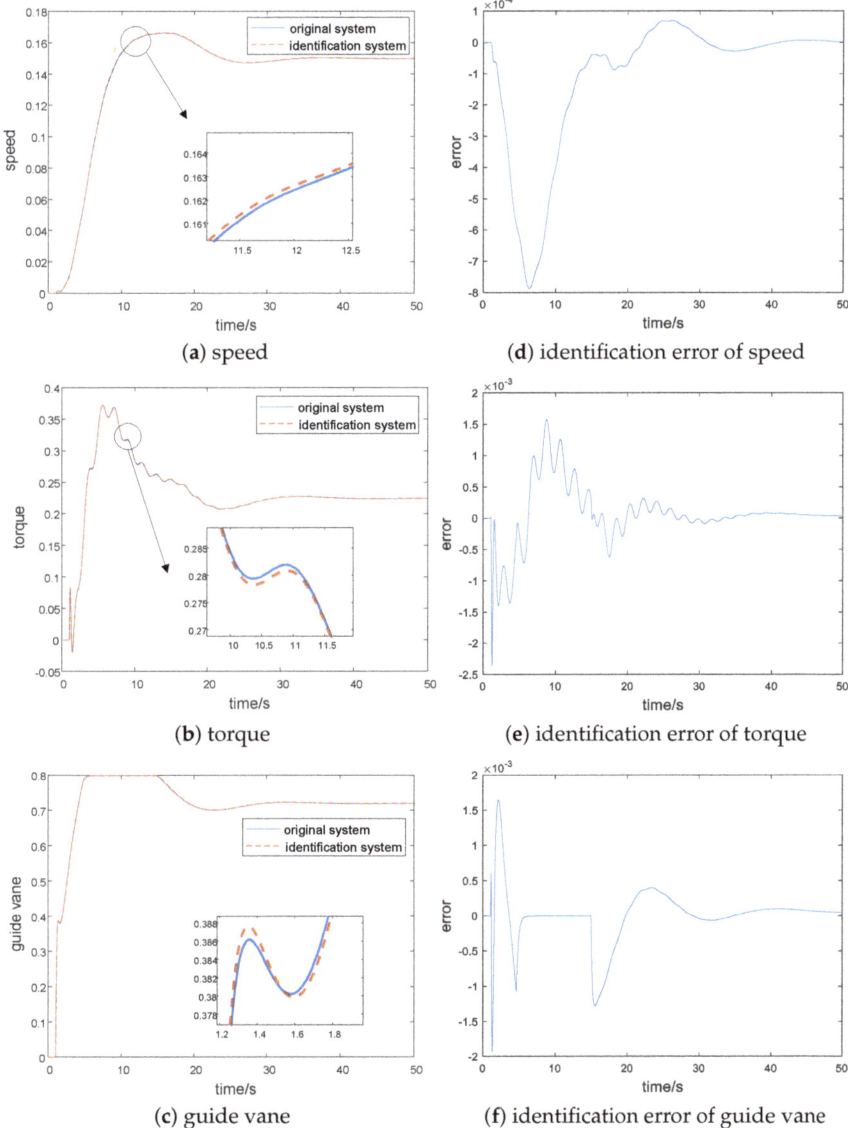

(a) speed

(d) identification error of speed

(b) torque

(e) identification error of torque

(c) guide vane

(f) identification error of guide vane

Figure 9. Comparison between the outputs of the three parts of the original and the identification systems under a 15% frequency disturbance.

Figures 10–12 show the convergence curves of fitness functions of different optimization algorithms under 5%, 10%, and 15% frequency interferences, respectively. The fitness function of the PSO is always the largest in the three graphs. It falls into the local optimum when it iterates 100 times and converges ahead of time. When GSA and ALO iterate 120 times, the fitness function value tends to be stable. The fitness function of the AHA algorithm is second only to IAHA. The IAHA algorithm is always the smallest, and has a rapid downward trend in the end. It reflects that IAHA has the ability of global optimization.

Figures 13–15 show the comparison charts of the dynamic response of speed, torque, and guide vane opening output of different algorithms under different frequency disturbance conditions. It can be seen intuitively from the figure that the consistency between the output curve identified by IAHA and the output curve of the real system is higher than that of the other four algorithms, indicating that the accuracy is also higher.

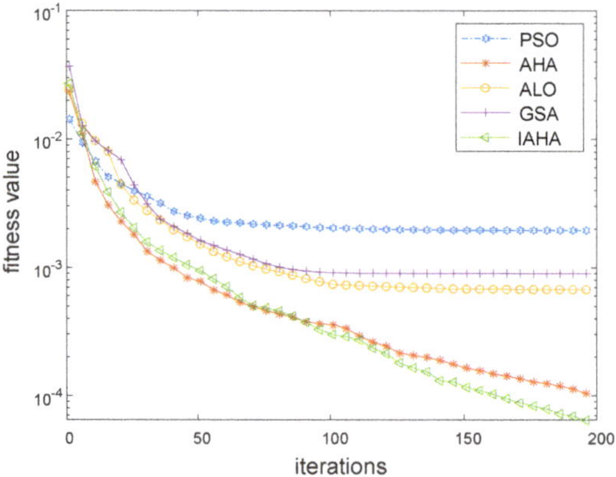

Figure 10. Comparison of the average iteration process under a 5% frequency disturbance condition.

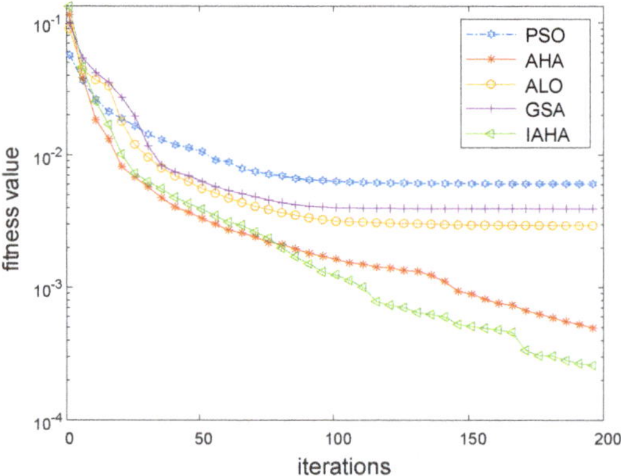

Figure 11. Comparison of the average iteration process under a 10% frequency disturbance condition.

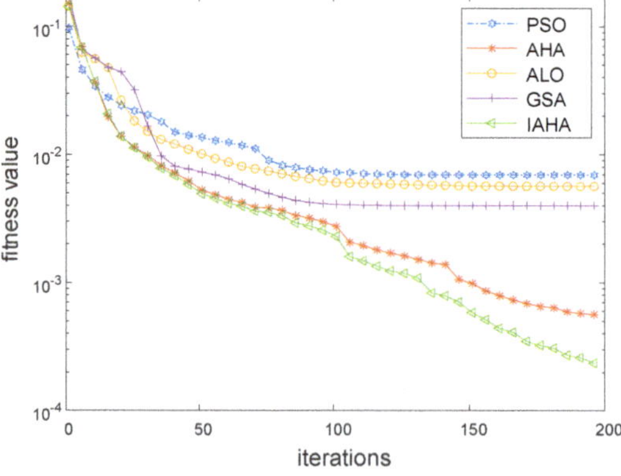

Figure 12. Comparison of the average iteration process under a 15% frequency disturbance condition.

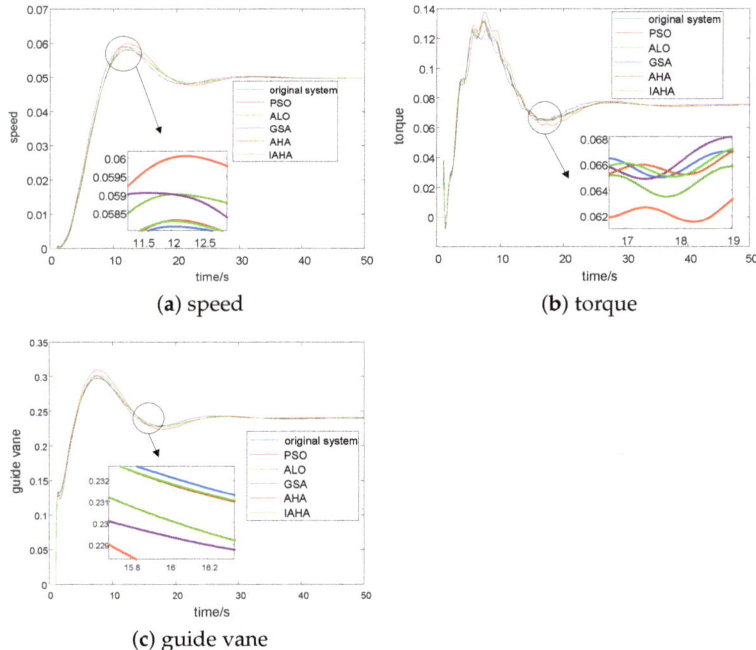

(**a**) speed

(**b**) torque

(**c**) guide vane

Figure 13. Comparison of the output between different optimization algorithms for 5% frequency disturbance conditions.

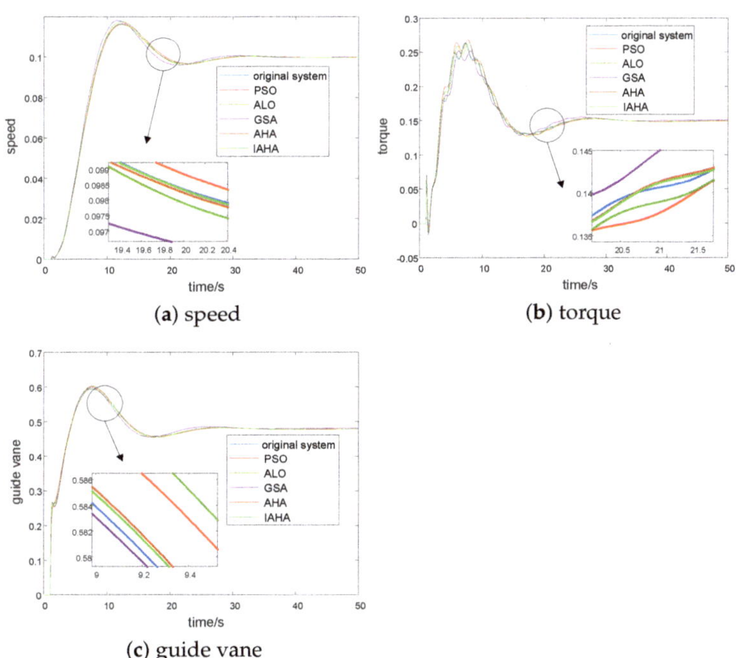

(**a**) speed

(**b**) torque

(**c**) guide vane

Figure 14. Comparison of the output between different optimization algorithms for 10% frequency disturbance conditions.

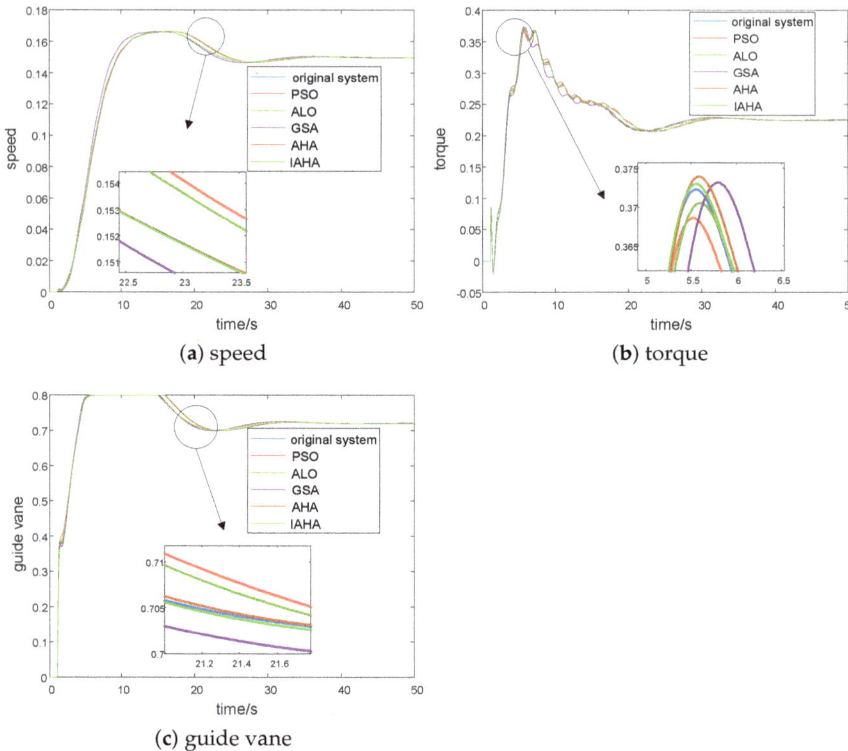

(a) speed

(b) torque

(c) guide vane

Figure 15. Comparison of the outputs between different optimization algorithms for 15% frequency disturbance conditions.

5. Conclusions

In this paper, an improved AHA (IAHA) is proposed. Compared with AHA, there are two improved strategies. First, the population initialization of the Chebyshev chaotic map was used to expand the search range and improve the accuracy. Second, Levy flight was added to guide foraging, which gives the algorithm better convergence and stability. In order to verify the optimization performance of the IAHA algorithm, the 23 standard functions, which include unimodal and multimodal functions, were evaluated. The calculated values in the statistical analyses were the mean value, standard deviation, optimal value, and the worst value. IAHA performs best in most functions. The practical application involves identifying the parameters of the governing system of pumped storage units and calculating the average value through 20 independent operation algorithms. The results show that the errors of the IAHA are only 2.04%, 1.82%, and 1.28% under 5%, 10%, and 15% frequency disturbance conditions. Compared with PSO, GSA, ALO, and AHA, the identification accuracy of the governing system is improved.

Author Contributions: Conceptualization, L.W. and W.Z.; methodology, L.W.; software, L.Z. and X.L.; validation, L.W. and W.Z.; formal analysis, L.W. and W.Z.; resources, L.W.; writing—original draft preparation, L.Z.; writing—review and editing, L.W. and W.Z.; visualization, L.Z.; supervision, L.W.; project administration, L.W.; funding acquisition, L.W. All authors have read and agreed to the published version of the manuscript.

Funding: This research was funded by National Natural Science Foundation of China: 11972144.

Institutional Review Board Statement: Not applicable.

Informed Consent Statement: Not applicable.

Data Availability Statement: Not available.

Acknowledgments: This work was supported by a grant from the National Natural Science Foundation of China (11972144).

Conflicts of Interest: The authors declare no conflict of interest.

Appendix A

Table A1. Standard test function.

Function	D	Range
$f_1(x) = \sum_{i=1}^{n} x_i^2$	30	$[-100, 100]^D$
$f_2(x) = \sum_{i=1}^{n} \lvert x_i \rvert + \prod_{i=1}^{n} \lvert x_i \rvert$	30	$[-100, 100]^D$
$f_3(x) = \sum_{i=1}^{n} \left(\sum_{j=1}^{i} x_j \right)^2$	30	$[-100, 100]^D$
$f_4(x) = max_i \{ \lvert x_i \rvert, 1 \le i \le n \}$	30	$[-100, 100]^D$
$f_5(x) = \sum_{i=1}^{n-1} \left(100(x_{i+1} - x_i^2)^2 + (x_i - 1)^2 \right)$	30	$[-30, 30]^D$
$f_6(x) = \sum_{i=1}^{n} (x_i + 0.5)^2$	30	$[-100, 100]^D$
$f_7(x) = \sum_{i=1}^{n} i x_i^4 + random[0, 1)$	30	$[-1.28, 1.28]^D$
$f_8(x) = -\sum_{i=1}^{n} \left(x_i \sin(\sqrt{\lvert x_i \rvert}) \right)$	30	$[-500, 500]^D$
$f_9(x) = \sum_{i=1}^{n} \left(x_i^2 - 10 \cos(2\pi x_i) + 10 \right)^2$	30	$[-5.12, 5.12]^D$
$f_{10}(x) = -20 \exp\left(-0.2 \sqrt{\frac{1}{n} \sum_{i=1}^{n} x_i^2} \right) - \exp(\frac{1}{n} \sum_{i=1}^{n} cos 2\pi x_i) + 20 + e$	30	$[-32, 32]^D$
$f_{11}(x) = \frac{1}{4000} \sum_{i=1}^{n} (x_i - 100)^2 - \prod_{i=1}^{n} \cos(\frac{x_i - 100}{\sqrt{i}}) + 1$	30	$[-600, 600]^D$
$f_{12}(x) = \frac{\pi}{n} \left\{ 10 sin^2(\pi y_1) + \sum_{i=1}^{n-1} (y_1 - 1)^2 \left[1 + 10 sin^2(\pi y_1 + 1) \right] + (y_n - 1)^2 \right\} + \sum_{i=1}^{30} u(x_i, 10, 100, 4)$	30	$[-50, 50]^D$
$f_{13}(x) = 0.1 \{ sin^2(3\pi x_1) + \sum_{i=1}^{29} (x_i - 1)^2 p[1 + 10 sin^2(3\pi x_{i+1})] + (x_n - 1)^2 [1 + sin^2(2\pi x_{30})] \} + \sum_{i=1}^{30} u(x_i, 5, 10, 4)$	30	$[-50, 50]^D$
$f_{14}(x) = [\frac{1}{500} + \sum_{j=1}^{25} \frac{1}{j + \sum_{j=1}^{2} (x_i - a_i)^6}]^{-1}$	2	$[-65.536, 65.536]^D$
$f_{15}(x) = \sum_{i=1}^{11} \lvert a_i - \frac{x_1(b_i^2 + b_i x_2)}{b_i^2 + b_i x_3 + x_4} \rvert^2$	4	$[-5, 5]^D$
$f_{16}(x) = 4x_1^2 + 2.1x_1^4 + \frac{1}{3}x_1^6 + x_1 x_2 - 4x_2^2 + 4x_2^4$	2	$[-5, 5]^D$
$f_{17}(x) = (x_2 - \frac{5.1}{4\pi^2} x_1^2 + \frac{5}{\pi} x_1 - 6)^2 + 10(1 - \frac{1}{8\pi} cos x_1) + 10$	2	$[-5, 10] \times [0, 15]$
$f_{18}(x) = \left[1 + (x_1 + x_2 + 1)^2 (19 - 14x_1 + 3x_1^2 - 14x_2 + 6x_1 x_2 + 3x_2^2) \right] \times [30 + (2x_1 + 1 - 3x_2)^2 (18 - 32x_1 + 12x_1^2 + 48x_2 - 36x_1 x_2 + 27x_2^2)]$	2	$[-2, 2]^D$
$f_{19}(x) = -\sum_{i=1}^{4} exp[-\sum_{j=1}^{3} a_{ij}(x_j - p_{ij})^2]$	3	$[0, 1]^D$
$f_{20}(x) = -\sum_{i=1}^{4} exp[-\sum_{j=1}^{6} a_{ij}(x_j - p_{ij})^2]$	6	$[0, 1]^D$
$f_{21}(x) = -\sum_{i=1}^{5} \lvert (x_i - a_i)(x_i - a_i)^T + c_i \rvert^{-1}$	4	$[0, 10]^D$
$f_{22}(x) = -\sum_{i=1}^{7} \lvert (x_i - a_i)(x_i - a_i)^T + c_i \rvert^{-1}$	4	$[0, 10]^D$
$f_{23}(x) = -\sum_{i=1}^{10} \lvert (x_i - a_i)(x_i - a_i)^T + c_i \rvert^{-1}$	4	$[0, 10]^D$

References

1. Xu, Y.; Zhou, J.; Zhang, C.; Zhang, Y.; Li, C.; Qian, Z. A parameter adaptive identification method for a pumped storage hydro unit regulation system model using an improved gravitational search algorithm. *Simulation* **2017**, *93*, 679–694. [CrossRef]
2. Jiang, X.; Wang, Z.; Zhu, H.; Wang, W. Hydraulic turbine system identification and predictive control based on gasa-bpnn. *Int. J. Miner. Metall. Mater.* **2021**, *28*, 1240–1247. [CrossRef]
3. Tian, T.; Zhao, W.; Zhen, W.; Liu, C. Application of improved whale optimization algorithm in parameter identification of hydraulic turbine at no-load. *Arab. J. Sci. Eng.* **2020**, *45*, 9913–9924. [CrossRef]
4. Trudnowski, D.; Agee, J. Identifying a hydraulic-turbine model from measured field data. *IEEE Trans. Energy Convers.* **1995**, *10*, 768–773. [CrossRef]
5. Feng, C.; Chang, L.; Li, C.; Ding, T.; Mai, Z. Controller optimization approach using lstm-based identification model for pumped-storage units. *IEEE Access* **2019**, *7*, 32714–32727. [CrossRef]

6. Di Piazza, M.C.; Luna, M.; Vitale, G. Dynamic pv model parameter identification by least-squares regression. *IEEE J. Photovolt.* **2013**, *3*, 799–806. [CrossRef]
7. Zhang, X.; Lian, L.; Zhu, F. Parameter fitting of variogram based on hybrid algorithm of particle swarm and artificial fish swarm. *Future Gener. Comput. Syst.* **2021**, *116*, 265–274. [CrossRef]
8. Stoica, P.; Li, J. On nonexistence of the maximum likelihood estimate in blind multichannel identification. *IEEE Signal Process. Mag.* **2005**, *22*, 99–101. [CrossRef]
9. Sun, W.; Kong, X.Y.; Yang, Q.; Zhang, F. Parameter identification method for turbine speed governor system based on particle swarm optimization. *Appl. Mech. Mater.* **2014**, *448*, 2511–2515. [CrossRef]
10. Liu, L.; Liu, W.; Cartes, D.A. Particle swarm optimization-based parameter identification applied to permanent magnet synchronous motors. *Eng. Appl. Artif. Intell.* **2008**, *21*, 1092–1100. [CrossRef]
11. Hu, G.; Dou, W.; Wang, X.; Abbas, M. An enhanced chimp optimization algorithm for optimal degree reduction of said–ball curves. *Math. Comput. Simul.* **2022**, *197*, 207–252. [CrossRef]
12. Wang, L.; Cao, Q.; Zhang, Z.; Mirjalili, S.; Zhao, W. Artificial rabbits optimization: A new bio-inspired meta-heuristic algorithm for solving engineering optimization problems. *Eng. Appl. Artif. Intell.* **2022**, *114*, 105082. [CrossRef]
13. Mantri, G.; Kulkarni, N. Design and optimization of pid controller using genetic algorithm. *Int. J. Res. Eng. Technol.* **2013**, *2*, 926–930. [CrossRef]
14. Zhao, W.; Wang, L.; Zhang, Z. Artificial ecosystem-based optimization: A novel nature-inspired meta-heuristic algorithm. *Neural Comput. Appl.* **2020**, *32*, 9383–9425. [CrossRef]
15. Li, P.; Duan, H. Path planning of unmanned aerial vehicle based on improved gravitational search algorithm. *Sci. China Technol. Sci.* **2012**, *55*, 2712–2719. [CrossRef]
16. Blum, C. Ant colony optimization: Introduction and recent trends. *Phys. Life Rev.* **2005**, *2*, 353–373. [CrossRef]
17. Hu, G.; Du, B.; Wang, X.; Wei, G. An enhanced black widow optimization algorithm for feature selection. *Knowl.-Based Syst.* **2022**, *235*, 107638. [CrossRef]
18. Zhao, W.; Wang, L.; Zhang, Z. Atom search optimization and its application to solve a hydrogeologic parameter estimation problem. *Knowl.-Based Syst.* **2019**, *163*, 283–304. [CrossRef]
19. Moayedi, H.; Nguyen, H.; Kok Foong, L. Nonlinear evolutionary swarm intelligence of grasshopper optimization algorithm and gray wolf optimization for weight adjustment of neural network. *Eng. Comput.* **2021**, *37*, 1265–1275. [CrossRef]
20. Zhang, C.; Zhang, F.; Li, F.; Wu, H. Improved artificial fish swarm algorithm. In Proceedings of the 2014 9th IEEE Conference on Industrial Electronics and Applications, Hangzhou, China, 9–11 June 2014; pp. 748–753. [CrossRef]
21. Tian, T.; Liu, C.; Guo, Q.; Yuan, Y.; Li, W.; Yan, Q. An improved ant lion optimization algorithm and its application in hydraulic turbine governing system parameter identification. *Energies* **2018**, *11*, 95. [CrossRef]
22. Ding, T.; Chang, L.; Li, C.; Feng, C.; Zhang, N. A mixed-strategy-based whale optimization algorithm for parameter identification of hydraulic turbine governing systems with a delayed water hammer effect. *Energies* **2018**, *11*, 2367. [CrossRef]
23. Hu, G.; Li, M.; Wang, X.; Wei, G.; Chang, C.-T. An enhanced manta ray foraging optimization algorithm for shape optimization of complex CCG-Ball curves. *Knowl.-Based Syst.* **2022**, *240*, 108071. [CrossRef]
24. Wu, Z.; Yu, D.; Kang, X. Parameter identification of solar cell model based on improved ant lion optimization algorithm. *J. Sol. Energy* **2019**, *40*, 3435–3443.
25. Ali, M.; El-Hameed, M.; Farahat, M. Effective parameters' identification for polymer electrolyte membrane fuel cell models using grey wolf optimizer. Renew. *Energy* **2017**, *111*, 455–462. [CrossRef]
26. Zhang, X.; Lin, G.; Hu, H. Parameter identification of photovoltaic cells based on elite reverse learning particle swarm optimization. *J. Hunan Inst. Eng.* **2021**, *31*, 1–7. [CrossRef]
27. Zhao, W.; Wang, L.; Mirjalili, S. Artificial hummingbird algorithm: A new bio-inspired optimizer with its engineering applications. *Comput. Meth. Appl. Mech. Eng.* **2022**, *388*, 114194. [CrossRef]
28. Varol, Altay, E.; Alatas, B. Bird swarm algorithms with chaotic mapping. *Artif. Intell. Rev.* **2020**, *53*, 1373–1414. [CrossRef]
29. Liu, Y.; Cao, B. A novel ant colony optimization algorithm with levy flight. *IEEE Access* **2020**, *8*, 67205–67213. 10.1109/access.2020.2985498. [CrossRef]
30. Roy, S.; Chaudhuri, S.S. Cuckoo search algorithm using levy flight: A review. *Int. J. Mod. Educ. Comput. Sci.* **2013**, *5*, 10. [CrossRef]
31. Zhao, W.; Shi, T.; Wang, L.; Cao, Q.; Zhang, H. An adaptive hybrid atom search optimization with particle swarm optimization and its application to optimal no-load pid design of hydro-turbine governor. *J. Comput. Des. Eng.* **2021**, *8*, 1204–1233. [CrossRef]
32. Cai, W.; Cai, B. Research on fine modeling of hydraulic servo system of hydraulic turbine governor. *Hydropower Pumped Storage* **2021**, *7*, 33–38.
33. Zhou, J.; Zhang, C.; Peng, T.; Xu, Y. Parameter identification of turbine pump governing system using an improved backtracking search algorithm. *Energies* **2019**, *11*, 1668. [CrossRef]
34. Zhang, N.; Li, C.; Li, R.; Lai, X.; Zhang, Y. A mixed-strategy based gravitational search algorithm for parameter identification of hydraulic turbine governing system. *Knowl.-Based Syst.* **2016**, *109*, 218–237. [CrossRef]
35. Guo, W.; Zhu, D. Nonlinear modeling and operation stability of variable speed pumped storage power station. *Energy Sci. Eng.* **2021**, *9*, 1703–1718. [CrossRef]
36. Zhang, J.; Li, Z.; Li, Y.; Liang, X.; Wei, Z.; Zhu, Z. Optimization of pid parameters of hydro-genrator unit governor based on hybrid particle swarm optimization. *China Rural. Water Hydropower* **2019**, *1*, 180–183.

37. Li, C.; Zhou, J. Parameters identification of hydraulic turbine governing system using improved gravitational search algorithm. *Energy Convers. Manag.* **2011**, *52*, 374–381. [CrossRef]
38. Wang, L.; Zhang, K.; Zhao, W. Nonlinear modeling of dynamic characteristics of pump-turbine. *Energies* **2022**, *15*, 297. [CrossRef]

Article

Flow-Induced Vibration of Non-Rotating Structures of a High-Head Pump-Turbine during Start-Up in Turbine Mode

Mengqi Yang [1], Weiqiang Zhao [2], Huili Bi [2], Haixia Yang [1], Qilian He [2], Xingxing Huang [3] and Zhengwei Wang [2,*]

[1] Branch Company of Maintenance & Test, CSG Power Generation Co., Ltd., Guangzhou 511400, China
[2] State Key Laboratory of Hydroscience and Engineering, Tsinghua University, Beijing 100084, China
[3] S.C.I. Energy, Future Energy Research Institute, Seidengasse 17, 8706 Zurich, Switzerland
* Correspondence: wzw@mail.tsinghua.edu.cn

Abstract: Pumped storage-power plants play an extremely important role in the modern smart grid due to their irreplaceable advantages in load peak-valley regulation, frequency modulation, and phase modulation. The number of start-stops per day of pump-turbine units is therefore also increasing. During the start-up transient process in turbine mode, the complex flow in runner passage, crown and band chambers, and seal labyrinth is able to induce severe vibration of non-rotating structures such as head cover, stay-ring, and pose a threat to the safe operation of the pump-turbine unit. In this article, the flow-induced vibration of the structures of a pump-turbine unit during its start-up process in turbine mode is studied. In the first place, this investigation establishes a three-dimensional model of the full flow passage and carries out a full three-dimensional CFD calculation based on one-dimensional pipeline calculation results for the start-up transient process. In the next place, by applying the fluid–structure interaction calculation method, the finite element analysis of non-rotating components of the pump-turbine unit is carried out. The flow-induced stresses and deformations of head cover, stay-ring, etc., are obtained and analyzed. The results reveal that the maximum deformation of the non-rotating structures is located at the inner edge of the head cover while the maximum stress appears at the trailing edge fillet of a stay vane. In summary, the dynamic stress of the non-rotating structures changes largely during the start-up process. The stress is strongly related to the axial thrust caused by the fluid flow. The achieved results can provide guidance for further fatigue life assessment of non-rotating structures and contribute to the structural safety design of pump-turbine units.

Keywords: pump-turbine; turbine mode; start-up; non-rotating structures; flow-induced vibration

Citation: Yang, M.; Zhao, W.; Bi, H.; Yang, H.; He, Q.; Huang, X.; Wang, Z. Flow-Induced Vibration of Non-Rotating Structures of a High-Head Pump-Turbine during Start-Up in Turbine Mode. *Energies* **2022**, *15*, 8743. https://doi.org/10.3390/en15228743

Academic Editor: Helena M. Ramos

Received: 9 October 2022
Accepted: 14 November 2022
Published: 21 November 2022

Publisher's Note: MDPI stays neutral with regard to jurisdictional claims in published maps and institutional affiliations.

1. Introduction

In recent years, with the increasing requirement for clean and green energies, people are paying more and more attention to new renewable energies such as wind energy, solar energy, marine energy, and so on. However, the intermittent output of new renewable energies provides an unstable supply to the power grid, which brings a great challenge to the stability of the energy system. On the other hand, conventional power resources such as coal power are difficult to modify in a short time (several minutes), which is not able to match the consumption fluctuation of electricity in the power grid. More flexibility in the power system has to be developed to deal with this new scenario. Energy storage techniques highlight the importance because of their mature technology and fast response ability. The pumped storage power is a type of energy that consumes the surplus electricity of the power grid when there is an excessive supply and converts the potential energy of water to electricity and supply to the power grid during peak hours [1]. Pumped storage stations effectively play the role of dynamic balance between the supply and consumption in the power grid [2]. Therefore, the operation stability of pump-turbines is of paramount significance to the safety of the power plants and the grid.

In order to keep the balance of the power grid, the pumped-storage power plant has to switch its function between pump mode and turbine mode. Because of the entrance of new renewable energies, pump storage stations have to switch their function more and more frequently. Some of them have to start-up and close-down more than 10 times in a single day [3]. Pump-turbine is a pump as turbine by nature, and it can be flexibly switched between pump and turbine conditions [4]. Under frequent start-up and close-down, load change, load rejection and other transient processes, pump-turbine units have to face a series of unsteady hydraulic excitations such as turbulence, rotor–stator interaction (RSI), vortex rope, etc. The alternate stress induced by the excitations will cause damage to the structure of the turbine units. Unstable flows such as RSI and vortex under transient processes may lead to strong vibrations to the non-rotating structures such as the head cover and stay-ring of the pump-turbine, which may develop stress concentrations [5]. A long-time operation may cause fatigue damage to the non-rotating structures, thus threatening the operation safety of the unit. Generally, hydraulic excitations mainly induce the vibration of the head cover while the vibration of the upper and lower brackets are generated by the rotor [6]. Because of the special design of the pump-turbines, the rotor–stator interaction is more severe than the regular type of hydraulic turbines [7]. Theoretical analysis, numerical simulation and experimental research have been done on the analysis of hydraulic excitation in pump-turbine units [8]. Researchers have carried out numerical simulations on the fluid domain of hydraulic turbine units in order to obtain their vibration behavior [9], modal characteristics [10], etc. A lot of experience has been accumulated in the calculation under steady-state and transient conditions. Valentín, D., et al. [11] researched the dynamic response of a Francis turbine runner by means of both Experimental Modal Analysis and Finite Element Method (FEM). The accuracy of the numerical simulation was proved to be high with the comparison of the experimental results. The behavior of a Francis turbine at part load condition is studied by Computational Fluid Dynamics and the vibration and stress show good accuracy compared with the field test results [12]. Egusquiza, E., et al. calculated the natural frequency, modal shape, and damping of a pump-turbine impeller by means of FEM. The obtained results have been validated with the field test and can be used on similar machines [13].

The transient process of pump-turbines is one of the most critical conditions that need to be researched because of its complicated inner flow and strong vibration [14]. Due to the inconvenience and high-cost of field measurements of hydraulic turbine units, numerical simulation methods have become a substitute to study the internal flow characteristics of machines. In a number of previous research, numerical simulations have been verified to be reliable in the prediction of the inner flow of hydraulic machinery by comparing the simulation result with the field tests. Tomaž, K. proposed a novel method for the prediction of water turbine characteristics during transient operating regimes based on numerical flow simulation with the finite volume method. The method has been validated by the model test [15]. Song, X. carried out a free surface vortex prediction method by means of two-phase flow method. The simulation clearly shows that the predicted results are consistent with the existing problems of the unit on site [16]. Wang, W. established a numerical model of a pump-turbine unit for pressure pulsation prediction. The prediction result shows good agreement with the condition monitoring data [17]. The above research has proven that for engineering problems, the comparisons show that the errors are within an acceptable range. Lingyan He et al. [18] investigated the resonance phenomenon of a pump-turbine unit during its start-up process by means of three-dimensional (3-D) full flow path analysis. The results revealed that the resonance was caused by the matching between the RSI and natural frequencies of the runner mode. Zhongyu Mao et al. studied the axial force of the shaft in the pump-turbine unit during the start-up process. The consequences of axial force during the start-up such as deformation and stress were analyzed in detail [19]. Funan Chen et al. analyzed transient unsteady flow in pump-turbine units and obtained reliable dynamic stress results during the start-up process. The results show that the RSI can be a leading factor of dynamic stresses during the transient process [20]. There are also some investigations that have investigated the stress level of non-rotation components by means of stress measurement through numerical simulation [21] or experiments of stress and strain [22,23]. During the transient process, the flow is mainly influenced by the critical time points, which are the extrema in the pressure curve [24,25]. The critical points can be obtained from one-dimensional pipeline calculation. In literature [19,26], scholars have proposed the simulation method at critical moments during transient processes and calculated the variation of stress/strain on the structures caused by the fluid flow. The above studies on the influence of the start-up process on the non-rotating structures in turbine mode based on the fluid-structure interaction (FSI) method still need to be supplemented.

In this study, the fluid domain of a reversible pump-turbine during the turbine start-up process has been conducted by means of numerical simulation. In Section 2, the basic theory of 1-D and 3-D simulation is introduced. In Section 3, the transient flow in the hydraulic system is calculated by the one-dimensional (1-D) transient flow calculation method. Then, the 3-D models of the fluid domain and the non-rotating structures are built up. After that, a full 3-D computational fluid dynamic (CFD) calculation is conducted on the whole fluid domain, which includes the labyrinth seals of the crown and band for the turbine start-up. In Section 4, Fluid-Structure Interaction (FSI) is used in order to couple the 1-D calculation and 3-D calculation. In FSI, the flow and pressure results obtained by 1-D simulation are transferred to the 3-D model as the boundary conditions. The obtained simulation results are then used as the excitation force of FSI analysis to calculate the fatigue and vibration of the non-rotating structures. The flow-induced stresses and deformations of head cover, stay-ring, etc., are obtained and analyzed. Section 5 gives the main conclusions. The calculated results have shown that the variation of guide vane opening (GVO) and spiral casing water level have a great influence on the vibration of the head cover during the turbine start-up, and the vibration tends to be stable after entering the load-increasing stage. The achieved results can provide data for further fatigue life assessment of non-rotating structures and contribute to the structural safety design of pump-turbine units.

2. Methodology of FSI Calculation

The analysis process in this study consists of 1-D pipeline calculation, fluid-structure interaction calculation and finite element analysis. Firstly, the 1-D pipeline calculation result provides border conditions for the calculation of FSI. Then the FSI will be used for the Finite Element Analysis (FEA) on the 3-D structure of the pump-turbine unit and the result of FEA will be analyzed in detail. The basic governing equations of the above methodologies are introduced below.

2.1. Governing Equations of 1-D Pipeline Calculation

The mathematical unsteady flow models of the pipeline system of the researched power plant consist of the penstock, surge shaft, valves, pump-turbine unit, and upper and lower reservoirs. The basic equations of unsteady flow in the closed pipeline include the kinematic equation of motion and the continuity equation.

$$g\frac{\partial H}{\partial X} + V\frac{\partial V}{\partial X} + \frac{\partial V}{\partial t} + \frac{F_d V|V|}{2D} = 0 \tag{1}$$

$$V\frac{\partial H}{\partial X} + \frac{a^2}{g}\frac{\partial V}{\partial X} + \frac{\partial H}{\partial t} - V\sin\alpha = 0 \tag{2}$$

$$H = Z + \frac{p}{\rho_f g} \tag{3}$$

where Z, D, and X represent the elevation, diameter, and length of the pipeline, respectively. ρ_f and p are the density and pressure of the fluid. V represents the fluid velocity in the pipeline. A is the velocity of the wave in the pipeline, t is the transient time, F_d is the Darcy

Westbach friction coefficient, α represents the angle between the pipeline and the horizontal plane. By combining Equations (1)–(3), the full differential equation can be obtained:

$$C^+ : \begin{cases} \frac{dx}{dt} = +a \\ A^2 \frac{dH}{dt} + A\frac{a}{g}\frac{dQ}{dt} - QA\sin\alpha + \frac{F_d aQ|Q|}{2gD} = 0 \end{cases} \tag{4}$$

$$C^- : \begin{cases} \frac{dx}{dt} = -a \\ A^2 \frac{dH}{dt} - A\frac{a}{g}\frac{dQ}{dt} - QA\sin\alpha - \frac{F_d aQ|Q|}{2gD} = 0 \end{cases} \tag{5}$$

Equations (4) and (5) are called the method of characteristic (MoC) equations and the compatibility equation is established along the characteristic lines. Two characteristic line directions are C^+ and C^-. The equation can be integrated by the variation of flow rate and head of the water in the form of difference, and then the 1-D pipeline calculation can be completed by iteration solution.

2.2. Governing Equations of 3-D FSI Calculation

The laws of conservation of mass and momentum are the governing equations of the 3-D unsteady turbulent flow in the flow passage:

$$\frac{\partial \rho_f u_j}{\partial t} + \frac{\partial \rho_f u u_j - \tau_j}{\partial x_j} = f_j \tag{6}$$

where u_j represents the velocity (on j direction) of the flow field, respectively, f_j is the volume force vector of the fluid on j direction, and τ_j is the shear force tensor:

$$\tau_j = -p + \mu\nabla \cdot u_j + 2\mu e_j \tag{7}$$

where μ and e_j are the dynamic viscosity and the velocity stress tensor of the fluid, respectively:

$$e_j = 1/2((\partial u_j)/(\partial x_j) + (\partial u_j^T)/(\partial x_j^T)) \tag{8}$$

The fluid pressure at the FSI interface is calculated according to the above theory. The force applied on the structure $f_j(t)$ by the fluid can be calculated by summing the pressure of fluid $p_j(t)$ with the FSI interface S_j.

$$f_j(t) = \int p_j(t) \cdot dS_j \tag{9}$$

By applying fluid pressure to the fluid-structure coupling surfaces of the structures, the following equations can be adopted to analyze the flow-induced structural vibration of the non-rotating components of the pump-turbine unit:

$$[M]\ddot{x} + [C]\{\dot{x}\} + [K]\{x\} = \{F(t)\} \tag{10}$$

$$\{\sigma\} = |E||S|\{x\} \tag{11}$$

where $[M]$ is the mass matrix of the structure, $[C]$ is the damping matrix, and $[K]$ is the stiffness matrix of the structure. $\{x\}$, $\{\dot{u}\}$ and $\{\ddot{u}\}$ are the displacement, velocity, and acceleration vectors of the structure, $F(t)$ is the external volume force vector applied to the structure including the pressure load of fluid, $[E]$ is the elasticity matrix, and $[S]$ is the strain-displacement matrix of the structure, respectively.

3. Set-Up of Flow-Induced Vibration Calculation

3.1. Transient Calculation of 1-D Pipeline

The researched pump-turbine unit is a vertical-shaft machine with a rated power of 300 MW and a rated rotating speed of 500 rpm. The turbine unit has 7 blades and 20 guide vanes. The rated head of the researched machine is 520 m. The structure of the flow passage of the pipeline is shown in Figure 1, which is composed of an upper and a lower reservoir, an upper and a tailrace surge shaft, four turbine units, and the penstock.

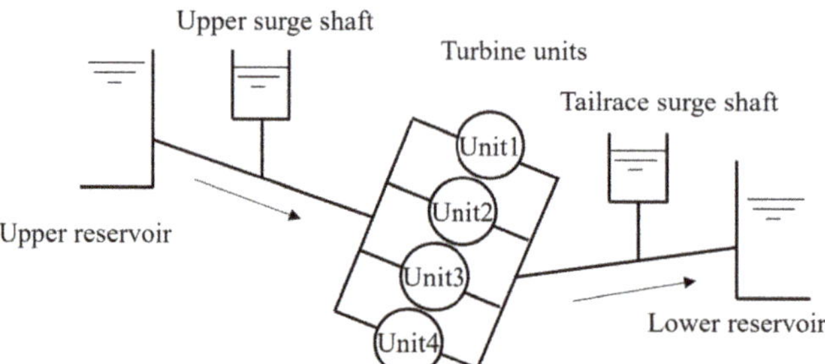

Figure 1. The layout of the pipeline system of the researched pumped storage power plant.

The 1-D pipeline numerical calculation during turbine start-up is carried out on the pipeline in order to determine the boundary conditions for 3-D CFD simulation for the turbine channels. The boundary conditions of the 1-D pipeline calculation are the water level of the reservoirs and the opening angle of the guide vanes. In this study, the water level between the upper and lower reservoirs is similar to the rated head.

The start-up process in turbine mode starts at 0 s and finishes at 60 s. During this time period, the guide vane opening increases from 0% to 27% and then decreases slowly from its peak value to 19% (Figure 2). By applying these boundary conditions on the 1-D model, the discharge and pressure in the pipeline are calculated, and the results are shown in Figure 2. The pressure at the spiral casing inlet and draft tube outlet calculated by 1-D simulation is displayed in Figure 3. A dimensionless parameter pressure coefficient ϕ described by Equation (12) is used in order to obtain a general rule of the pressure variation p during the turbine start-up process.

$$\phi = p/(\rho g H_r) \tag{12}$$

where H_r is the rated head of the unit.

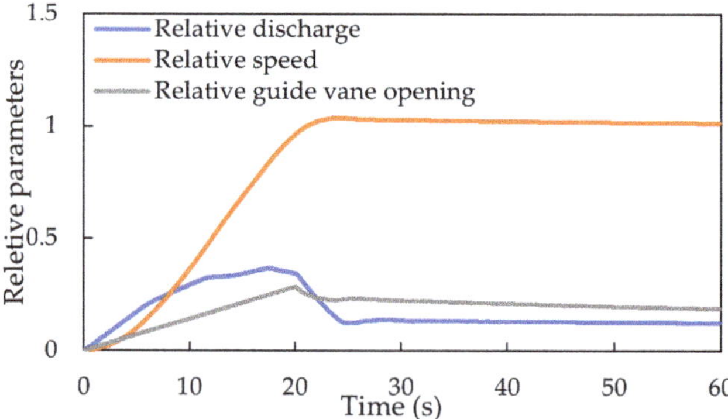

Figure 2. Regulation of the guide vane opening during turbine start-up and discharge and rotating speed calculated.

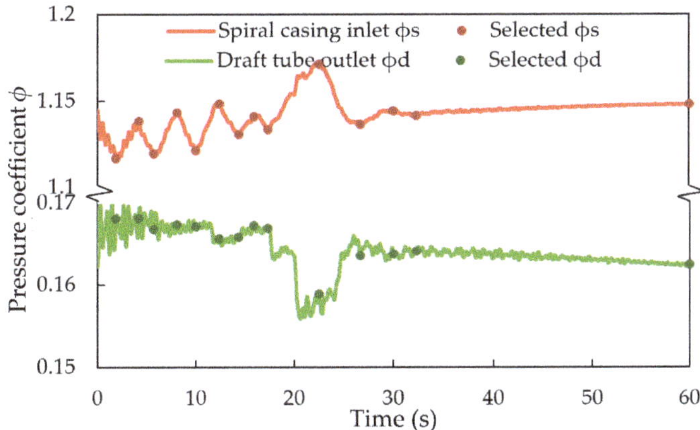

Figure 3. Pressure coefficients at spiral casing inlet and draft tube outlet during the turbine start-up.

The pressure coefficient at the spiral casing inlet oscillates (ϕ_s) around 1.14 while the pressure coefficient at the draft tube outlet (ϕ_d) oscillates around 0.17. A sudden increase in pressure at the time of 20 s is induced by the water hammer effect when the guide vane opening decreases. A total of 14 time points in Figure 3 have been picked according to the extreme values of ϕ_s. The pressure values from the selected points are used as the boundary conditions of the spiral casing inlet and the draft tube outlet for the 3-D CFD simulation. The other surfaces in the CFD model were set as no-slip walls. The runner flow channel was set as the rotation domain with the corresponding rotational speed during turbine start-up, and the other flow channels were set as the stationary domains.

3.2. Setup of the 3-D CFD Numerical Simulation

Based on the measured parameters shown in Figure 3, 14 time points during the turbine start-up process were selected to perform the detailed flow field simulation. The 3-D simulation model of the flow passage of the researched pump-turbine is shown in Figure 4, including the spiral casing, stay and guide vanes, runner, draft tube, labyrinth seals and pressure-balance pipelines. For the following CFD numerical simulation, the flow passage in the labyrinth seal has been taken into consideration since it has a significant impact on the flow in the unit, although the size of the labyrinth seal clearance is only 1.5 mm.

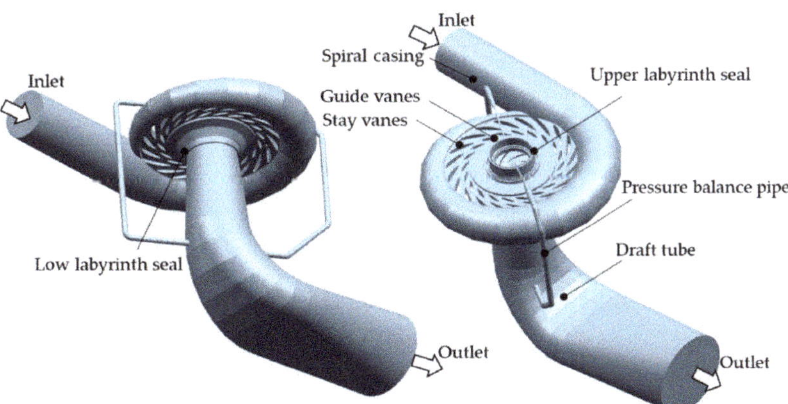

Figure 4. Fluid domain of the researched pump-turbine.

The mesh for the fluid domain of the researched machine is constructed by ANSYS Meshing Tool. The element density on the boundary layers of the flow passage and the gaps between the adjacent guide vanes have been increased in order to refine the mesh quality. The hybrid mesh, which includes both tetrahedral and hexahedral elements, can achieve a good balance between the computational time and simulation quality. By using the hybrid mesh, which combines two types of elements, a rather good balance between calculation time and calculation accuracy can be achieved. The meshes of the flow passage are shown in Figure 5.

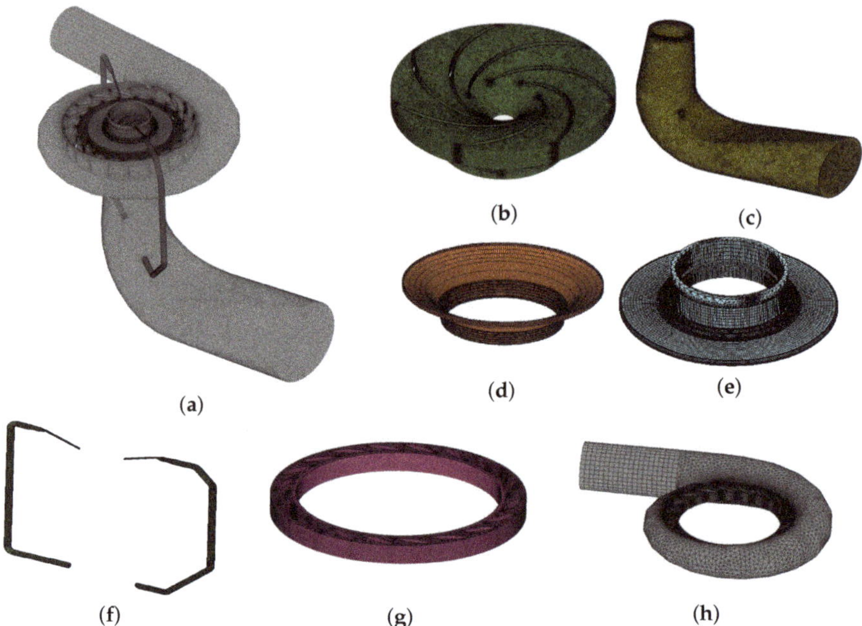

Figure 5. Mesh of different domains of the fluid model. (**a**) Entire simulation model; (**b**) Runner; (**c**) Draft tube; (**d**) Lower chamber with its labyrinth seal; (**e**) Upper chamber with its labyrinth seal; (**f**) Pressure balance pipelines; (**g**) Guide vane flow passage; (**h**) Spiral casing.

The fluid mesh independence test is performed before the calculation. Four sets of mesh (as shown in Table 1) with different numbers of elements have been developed in order to validate the grid of the fluid domain. The efficiency of the machine under the steady-state of the rated power generation condition is calculated with different sets of mesh. As shown in Figure 6, mesh set 3, marked with the green square box, is taken to carry out the CFD analysis.

Table 1. Number of elements of the four sets of meshes.

Fluid Domain	Mesh 1 ($\times 10^6$)	Mesh 2 ($\times 10^6$)	Mesh 3 ($\times 10^6$)	Mesh 4 ($\times 10^6$)
Runner	2.94	3.27	3.63	4.25
Guide vane	0.22	0.24	0.27	0.32
Spiral casing and stay vane	2.64	2.93	3.26	3.81
Draft tube	0.49	0.54	0.60	0.70
Labyrinth seals and pressure balance pipes	0.32	0.36	0.40	0.47
Total model	6.61	7.34	8.16	9.55

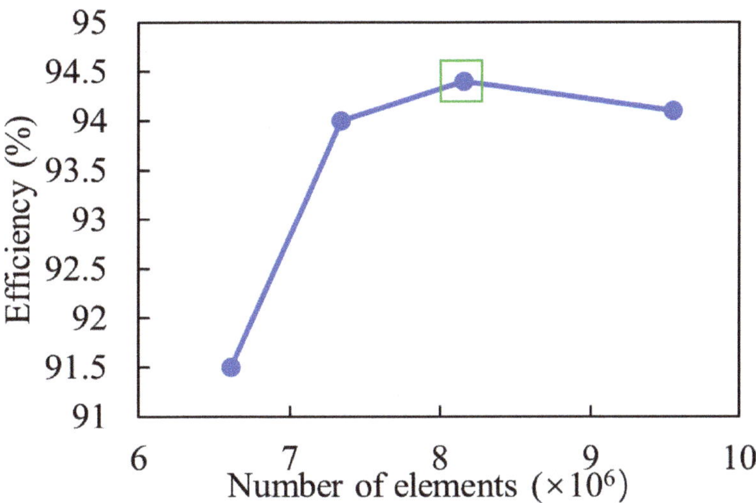

Figure 6. The mesh independence test of the CFD simulation.

3.3. Results of the 3-D CFD Numerical Simulation

The velocity distributions of the flow passage from the spiral casing to the draft tube during the turbine start-up process are shown in Figure 7. The color of the streamlines indicates the velocity of the flow. From the figure it can be seen that the flow velocity in the spiral casing is the lowest (nearly zero) at the beginning of the start-up process. At the start, the gap between the adjacent guide vanes is small, so the fluid velocity before the guide vanes is low. Because of the small discharge, the circumferential asymmetry phenomenon raises the flow passage of the spiral casing. From each figure it can be seen that the red color, which means a higher velocity, always appears in the guide vane and vaneless region. The velocity of the fluid in the stay vane region is between the velocity of the guide vane area and spiral casing. From the spiral casing to the runner, the velocity keeps raising. The velocity increases dramatically after the guide vanes open. It reaches its highest value at the time of 17.32 s. At this moment, the streamline before and after the guide vanes gets more uniform, and the circumference flow in the vaneless region disappears. The velocity of the flow in the spiral casing increases in the beginning but decreases in the middle of the start-up process. The highest flow velocity appears in the middle of the start-up process. At Point 10 (t = 22.48 s), the velocity decreases. According to the Bernoulli principle, the pressure increases, which can be seen from Figure 3. The discharge keeps decreasing until t = 26.67 s and then reaches the velocity of no-load condition.

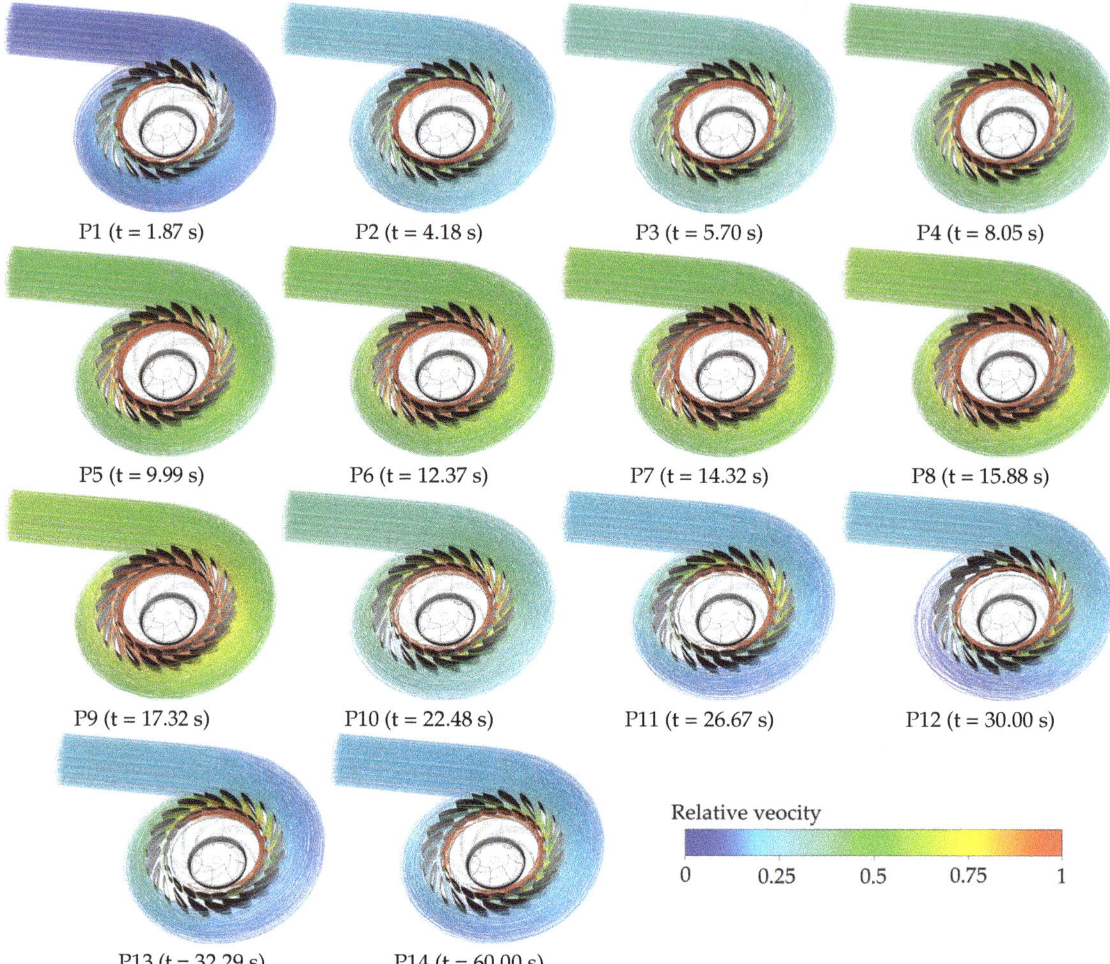

Figure 7. The flow velocity distributions of the pump-turbine unit during turbine start-up.

The pressure distributions in the upper and lower side of the flow passage from stay vanes to draft tube at four typical moments (t = 1.87 s, t = 4.18 s, t = 22.48 s, t = 60.00 s) are displayed in Figure 8. The upper side of the flow passage includes the region of the stay vanes, guide vanes, upper chamber, and upper labyrinth seal, while the lower side includes the stay vanes, guide vanes, upper chamber, and upper labyrinth seal. In each subfigure, the upper side is on the left and the lower side is on the right. From Figure 8 it can be seen that the upper side at the beginning of the start-up process has the largest pressure. At this time, the discharge is low so that the pressure applied on the flow passage equals the static pressure of the water. With the guide vanes open, the flow rate increases. According to the Bernoulli principle, the pressure decreases with the rise of the flow velocity. However, at the time of 22.48 s, the pressure increases in the flow passage, which corresponds to the lowest discharge in the flow passage. The same phenomenon can also be seen in Figure 3. In Figure 3, the pressure in the spiral casing inlet increases at 20 s, which is caused by the water hammer. Figure 8 also indicates that the pressure before the guide vanes is much higher than that in the runner chamber, which is due to the small opening of the guide vanes. In addition, during turbine start-up, the pressure on the upper side of the flow channel is always higher than that on the lower side. The pressure drops sharply from the runner chamber to the upper and lower labyrinth seals.

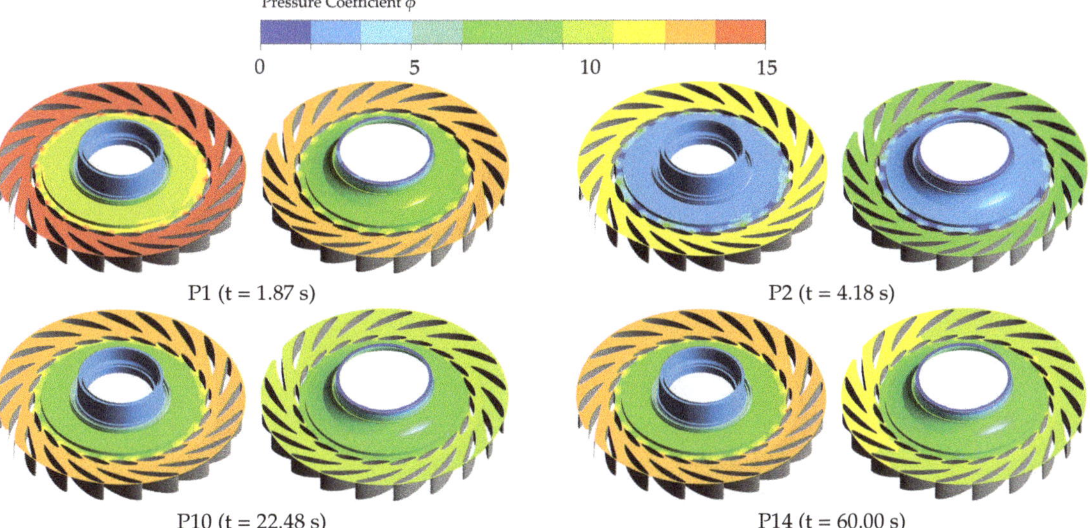

Figure 8. Pressure distribution of the pump-turbine unit during turbine start-up.

4. Flow-Induced Vibration Analysis of Non-Rotating Structures

4.1. Setup of the Flow-Induced Vibration Analysis for Non-Rotating Structures

The geometry model and boundary conditions of the non-rotating structures of the researched pump-turbine unit are demonstrated in Figure 9. The entire non-rotating structure is assembled by the head cover, bottom ring, stay ring, stay vanes, and high-strength bolts. The density, elastic modulus, and Poisson's ratio of the non-rotating structures are 7850 $\text{kg} \cdot \text{m}^{-3}$, 2.1×10^{11} Pa, and 0.3, respectively.

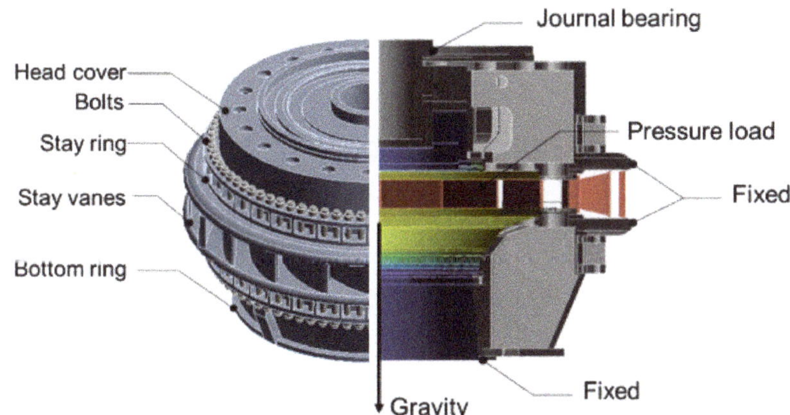

Figure 9. Setup of the non-rotating structures for flow-induced vibration simulation.

Since the non-rotating structures of the unit have been embedded in concrete during construction, the surfaces of the structure connected to the concrete are set as a fixed constraint boundary condition for the flow-induced vibration simulation. The turbine journal bearing stiffness is 1×10^6 $\text{N} \cdot \text{mm}^{-1}$ and the earth gravity is 9.8 $\text{m} \cdot \text{s}^{-2}$. The pressure profiles calculated by 3-D CFD are applied to the inner surface of the non-rotating structures for flow-induced vibration simulation using ANSYS Mechanical.

Three groups of meshes with 0.8, 1.1, and 3.0 million high-quality tetrahedral elements are created for the non-rotating structures (Figure 10) for the mesh independence study. The local meshes for the stress concentration locations of the head cover, stay vane, and bottom ring are also refined.

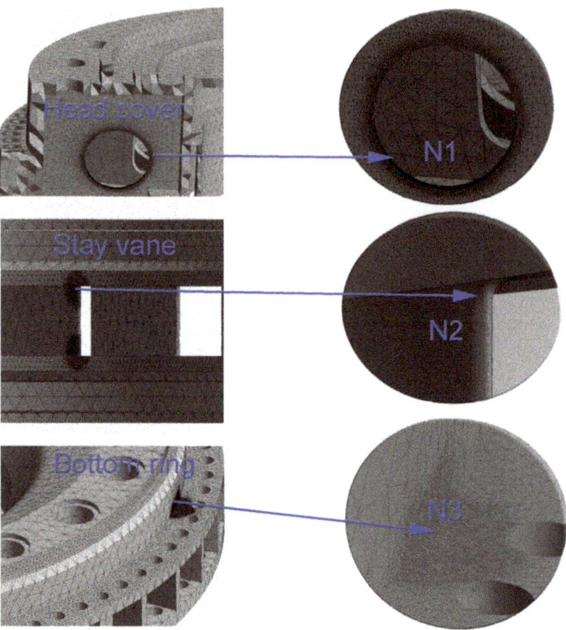

Figure 10. The meshes of the non-rotating structures of the pump-turbine unit.

Figure 11 compares the stresses at nodes N1, N2, and N3 for non-rotating structures with different amounts of elements under designed operating conditions, and the corresponding computational time consumption. The mesh independence study shows that the results of the second group of mesh are similar to those of the third group of mesh, but the computation time is only about 0.11% of that of the third one. Therefore it is reasonable to use the second group of mesh for following flow-induced vibration simulations.

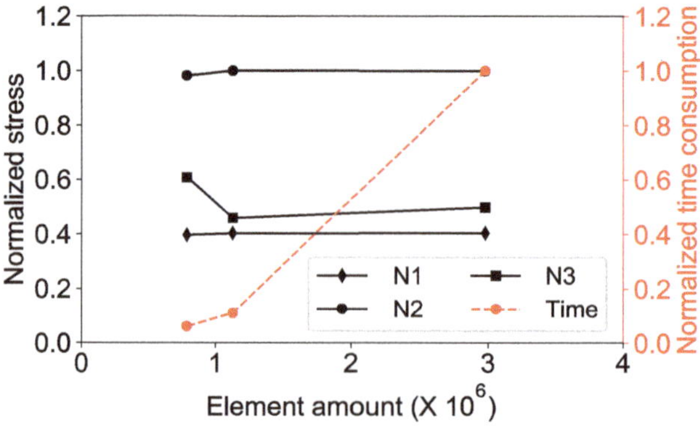

Figure 11. The mesh independence study of the flow-induced vibration simulation.

4.2. Results of the Flow-Induced Vibration Analysis for Non-Rotating Structures

Pressure profile data contains coordinates and pressure values for all nodes of the CFD mesh. With ANSYS built-in tools, the pressure data can be accurately interpolated and mapped at the nearest node at the fluid-structure interface of non-rotating structures. This allows the flow-induced vibration in non-rotating structures of the pump-turbine unit to be analyzed. The flow-induced deformation distribution of the non-rotating structures at different time moments during turbine start-up has a similar pattern. The maximum deformation is located at the same location (Figure 12), but the maximum deformation value is changing with time (Figure 13).

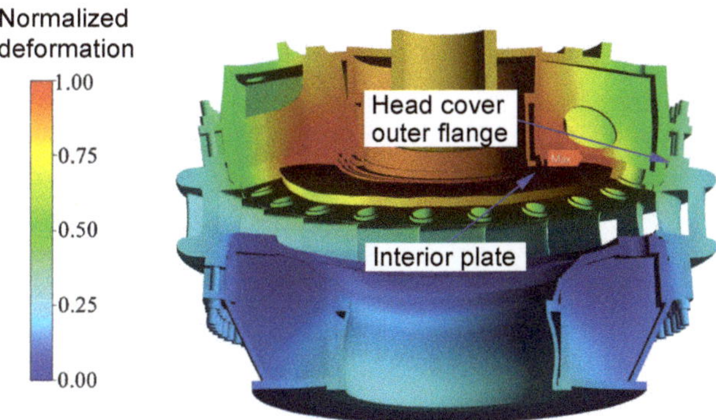

Figure 12. The deformation distribution of the non-rotating structures (longitudinal section).

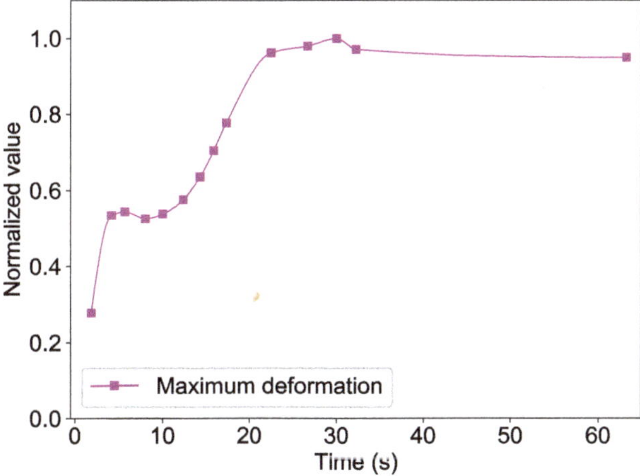

Figure 13. The maximum deformation of the non-rotating structures with time.

The head cover outer flange is assembled to the stay ring with high-strength bolts, but the interior plates are suspended without support in the axial direction. Due to the large hydraulic pressure loads applied on the inner surfaces of the head cover, the interior plates of the head cover have larger axial deformation than other plates.

The inner surface area of the stay ring is much smaller than that of the head cover, so the hydraulic loads applied on the stay ring are consequently lower than the head cover. Furthermore, the stay vanes of the stay ring are very stiff, so the flow-induced deformation of the stay ring is very small.

The height of the bottom ring is larger and stiffener than the head cover. In addition, the effective area of the inner surface of the bottom ring to bear the axial hydraulic load is smaller than that of the head cover, so the deformation of the bottom ring is also smaller than that of the head cover.

For ease of manufacture and installation, the head cover of the investigated pump-turbine unit is of split structures, which are assembled into an integral structure through the bolts of the connecting flanges. It can be seen from Figure 14 that the deformation distribution of the head cover is also symmetrical. Since the thickness of the connecting flange is thicker than that of the reinforcing plates of the head cover, the deformation of the head cover near the connecting flanges is smaller than the deformation in other locations of the head cover.

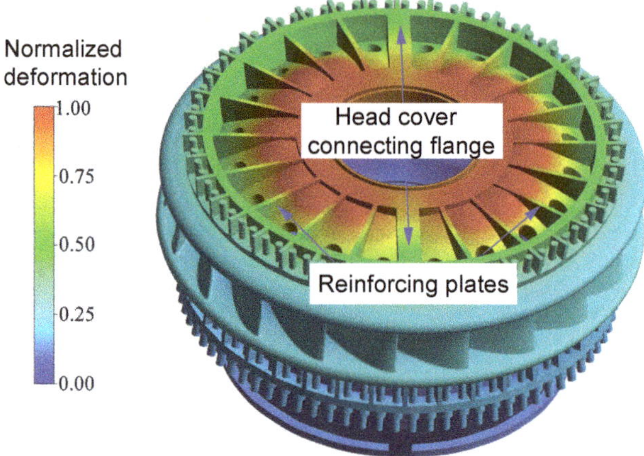

Figure 14. The deformation distribution of the non-rotating structures (cross-section).

The flow-induced stresses of the non-rotating structures of the pump-turbine unit at different moments during turbine start-up also present a similar distribution, with the maximum stresses concentrated at the stay ring. The flow-induced stress distribution during turbine start-up normalized with the maximum stress value of the stay vanes (Figure 15).

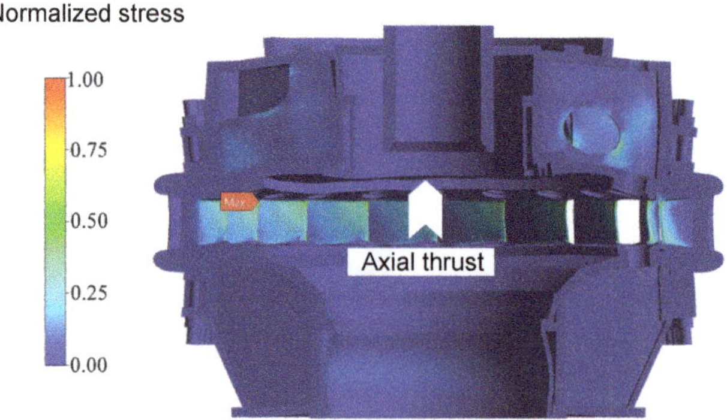

Figure 15. The flow-induced stress of the non-rotating structures.

The stress distributions of the stay ring, bottom ring and head cover shown in Figures 16–18 are normalized by their respective maximum stress values.

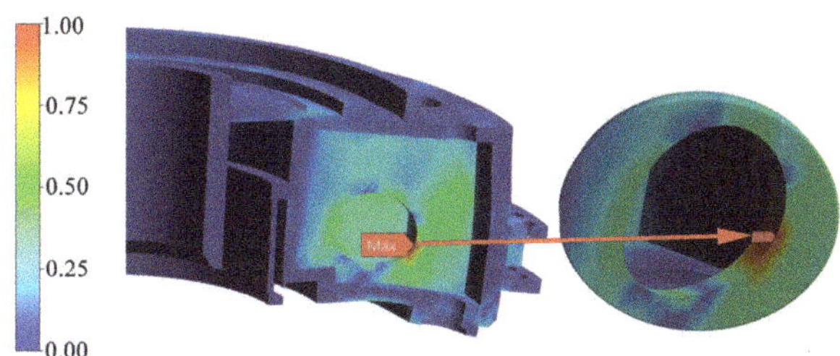

Figure 16. The maximum stress of the head cover during turbine start-up.

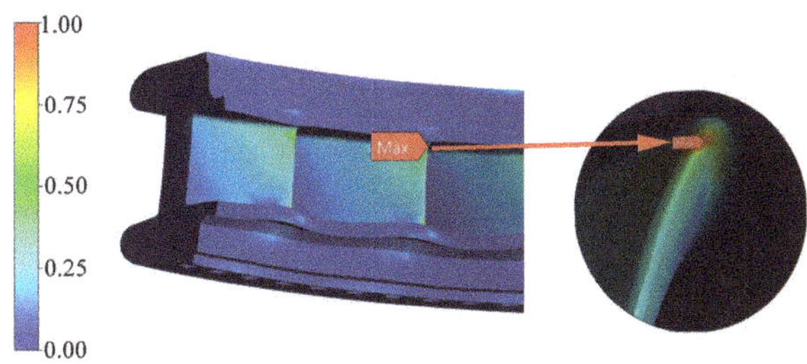

Figure 17. The maximum stress of the stay ring during turbine start-up.

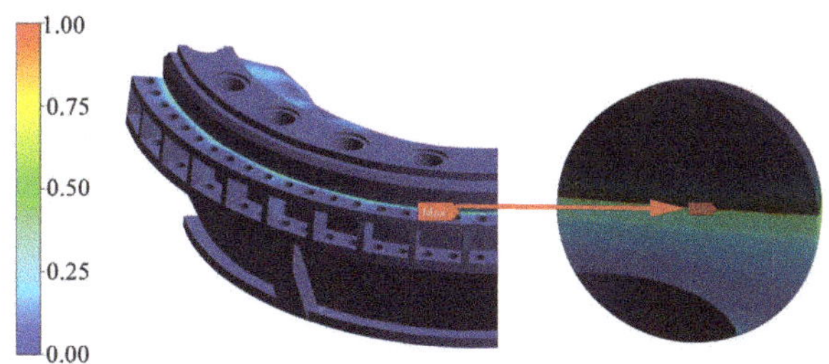

Figure 18. The maximum stress of the bottom ring during turbine start-up.

During the start-up process in turbine mode, the pressure load applied to the inner surface of the non-rotating structure lifts the head cover, presses down the bottom ring, and stretches all the stay vanes of the stay ring, and the maximum stress occurs on the trailing edge fillet of a stay vane of the stay ring. The maximum stress of the head cover is located at the round hole of a reinforcing plate, which has a lower stiffness than other reinforcing plates. and the maximum stress of the bottom ring is on the connecting flange of the bottom ring. To analyze the variation of the maximum stresses of the non-rotating structures with time, Figure 19 compares the normalized maximum stresses during turbine startup.

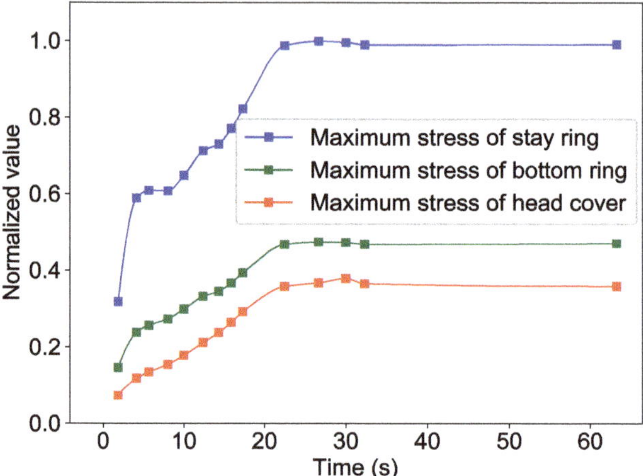

Figure 19. Comparison of maximum stresses in non-rotating structures during turbine start-up.

It can be seen that the maximum stress of the stay ring is always larger than the maximum stresses of the bottom ring and head cover during the whole turbine start-up process. However, the maximum stresses of different components have a similar variation law with time.

Figure 20 shows the comparison of normalized maximum deformation, maximum stresses, and axial thrust of the non-rotating structures during turbine start-up. During turbine start-up, the maximum deformation, maximum stresses, and axial thrust of the non-rotating structures increase rapidly with increasing hydraulic parameters such as GVO, discharge, and rotational speed. After t = 25 s, the hydraulic parameters remain almost constant and the maximum stresses in the non-rotating structures become also stable. The parameter comparison in Figure 20 also reveals that the variation of the flow-excited vibration behavior including structural deformation and stress is consistent with the variation of the axial thrust generated by the pressure load on the structure during turbine start-up.

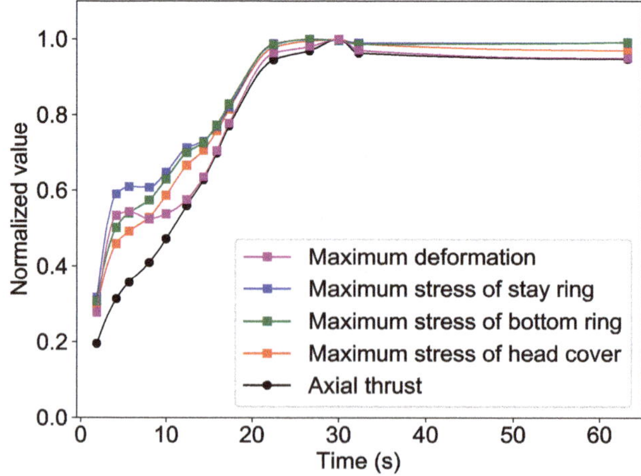

Figure 20. Comparison of maximum deformation, maximum stresses, and axial thrust during turbine start-up.

5. Conclusions

In this study, the flow-induced vibration of non-rotating structures of a high-head pump-turbine during start-up in turbine mode has been researched. A 1-D simulation is carried out in order to obtain the variation of pressure and discharge during the start-up process. The results are used as the border condition of the fluid-structure interaction analysis on the 3-D model of the pump-turbine unit.

The numerical simulation results show that at the beginning of the start-up, the discharge is low and the pressure is the highest. With the guide vane opening increasing, the pressure decreases with the rise of the flow velocity. At the time of 22 s, the pressure increases in the flow passage, which is caused by the water hammer. The flow field calculation results are used as the boundary conditions to calculate the behavior of the non-rotating structures for further fluid-structure interaction analysis.

The changing hydraulic pressure loads during turbine start-up cause changes in the distribution of the deformation and stresses on the non-rotating structures of the pump-turbine unit. However, the deformation distribution and stress distribution at different moments during the turbine start-up process follow similar patterns, respectively.

The maximum deformation of the non-rotating structures is located at the inner edge of the head cover, which is without support in the axial direction. The maximum stress of the non-rotating structures appears at the trailing edge fillet of a stay vane of the stay ring, which is stretched by the large hydraulic pressure. The flow-induced deformation and stresses of the non-rotating structures increase rapidly with the increasing hydraulic parameters at the beginning of the turbine start-up process. When the unit reaches the rated state, the hydraulic parameters remain almost constant and the stresses of the non-rotating structures also become stable. The flow-induced stresses of the non-rotating structures change largely during turbine start-up, and the variation of the deformation and stresses have a strong relationship with the axial thrust caused by the large pressure caused by the fluid flow. It is recommended to avoid too many frequent start-stops of the pump-turbine unit to reduce the risk of structural fatigue damage.

Author Contributions: Conceptualization, M.Y. and W.Z.; methodology, Q.H.; software, H.B.; Q.H.; X.H.; validation, H.B.; H.Y., formal analysis, W.Z.; writing—original draft preparation, W.Z.; writing—review and editing, X.H.; visualization, W.Z.; supervision, Z.W. All authors have read and agreed to the published version of the manuscript.

Funding: This research was funded by National Natural Science Foundation of China grant number 51876099.

Data Availability Statement: Not applicable.

Acknowledgments: The authors gratefully acknowledge the financial support of the project: Research on Lifetime Prediction of Non-rotating Parts of Pump–Turbine Unit Based on Rotor–Stator Interaction (RSI), Fluid–Structure Coupling and Fracture Mechanics—Research project on RSI mechanism and its Influence on Non-rotating Parts of Pump–Turbine Unit of Branch Company of Maintenance and Test, CSG Power Generation Co., Ltd.

Conflicts of Interest: The authors declare no conflict of interest.

Abbreviations

The following abbreviations are used in this manuscript:

RSI	Rotor–Stator Interaction
CFD	Computational Fluid Dynamics
FSI	Fluid-Structure Interaction
GVO	Guide Vane Opening
FEA	Finite Element Analysis

References

1. Zhao, W.; Presas, A.; Egusquiza, M.; Valentín, D.; Egusquiza, E.; Valero, C. On the use of Vibrational Hill Charts for improved condition monitoring and diagnosis of hydraulic turbines. *Struct. Health Monit.* **2022**, *21*, 14759217211072409. [CrossRef]
2. Zhao, W.; Presas, A.; Egusquiza, M.; Valentín, D.; Egusquiza, E.; Valero, C. Increasing the operating range and energy production in Francis turbines by an early detection of the overload instability. *Measurement* **2021**, *181*, 109580. [CrossRef]
3. Wang, H.; Jiang, H. Design of the control sequence for monitoring and control system of Pushihe Pumped Storage Power Station. In Proceedings of the 2012 IEEE International Conference on Computer Science and Automation Engineering (CSAE), Zhangjiajie, China, 25–27 May 2012; Volume 1, pp. 329–331.
4. Ansari, B.; Aligholami, M.; Rostamzadeh Khosroshahi, A. An experimental and numerical investigation into using hydropower plant on oil transmission lines. *Energy Sci. Eng.* **2022**, *in press.* [CrossRef]
5. Zhao, W.; Egusquiza Estévez, E.; Valero Ferrando, M.D.C.; Egusquiza Montagut, M.; Valentín Ruiz, D.; Presas Batlló, A. A novel condition monitoring methodology based on neural network of pump-turbines with extended operating range. In Proceedings of the 16th IMEKO TC10 Conference on Testing, Diagnostics & Inspection as a Comprehensive Value Chain for Quality & Safety, Berlin, Germany, 3–4 September 2019; pp. 154–159.
6. Zhang, Y.; Zheng, X.; Li, J.; Du, X. Experimental study on the vibrational performance and its physical origins of a prototype reversible pump turbine in the pumped hydro energy storage power station. *Renew. Energy* **2019**, *130*, 667–676. [CrossRef]
7. Egusquiza, E.; Valero, C.; Valentin, D.; Presas, A.; Rodriguez, C.G. Condition monitoring of pump-turbines. New challenges. *Measurement* **2015**, *67*, 151–163. [CrossRef]
8. Rodriguez, C.G.; Mateos-Prieto, B.; Egusquiza, E. Monitoring of rotor-stator interaction in pump-turbine using vibrations measured with on-board sensors rotating with shaft. *Shock Vib.* **2014**, *2014*, 276796. [CrossRef]
9. Kaneko, Y.; Iwasaki, Y. Study on Vibration Response of High Head Pump Turbine Runner (Vibration Characteristics of Mistuned Runner). *J. Environ. Eng.* **2008**, *3*, 37–48. [CrossRef]
10. Liang, Q.; Rodríguez, C.; Egusquiza, E.; Escaler, X.; Avellan, F. Modal response of hydraulic turbine runners. In Proceedings of the 23rd IAHR Symposium on Hydraulic Machinery and Systems, Yokohama, Japan, 17–21 October 2006; Volume 1, pp. 1–9.
11. Valentín, D.; Presas, A.; Egusquiza, E.; Valero, C.; Bossio, M. Dynamic response of the MICA runner. Experiment and simulation. In *Proceedings of the Journal of Physics: Conference Series*; IOP Publishing: Bristol, UK, 2017; Volume 813, p. 012036.
12. Valentín, D.; Presas, A.; Egusquiza, M.; Valero, C.; Egusquiza, E. Behavior of Francis turbines at part load. Field assessment in prototype: Effects on power swing. In *Proceedings of the IOP Conference Series: Earth and Environmental Science*; IOP Publishing: Bristol, UK, 2019; Volume 240, p. 062012.
13. Egusquiza, E.; Valero, C.; Presas, A.; Huang, X.; Guardo, A.; Seidel, U. Analysis of the dynamic response of pump-turbine impellers. Influence of the rotor. *Mech. Syst. Signal Process.* **2016**, *68*, 330–341. [CrossRef]
14. He, Q.; Huang, X.; Yang, M.; Yang, H.; Bi, H.; Wang, Z. Fluid–Structure Coupling Analysis of the Stationary Structures of a Prototype Pump Turbine during Load Rejection. *Energies* **2022**, *15*, 3764. [CrossRef]
15. Kolšek, T.; Duhovnik, J.; Bergant, A. Simulation of unsteady flow and runner rotation during shut-down of an axial water turbine. *J. Hydraul. Res.* **2006**, *44*, 129–137. [CrossRef]
16. Song, X.; Luo, Y.; Wang, Z. Numerical prediction of the influence of free surface vortex air-entrainment on pump unit performance. *Ocean. Eng.* **2022**, *256*, 111503. [CrossRef]
17. Wang, W.; Wang, X.; Wang, Z.; Ni, M.; Yang, C. Analysis of Internal Flow Characteristics of a Startup Pump Turbine at the Lowest Head under No-Load Conditions. *J. Mar. Sci. Eng.* **2021**, *9*, 1360. [CrossRef]
18. He, L.; Wang, Z.; Kurosawa, S.; Nakahara, Y. Resonance investigation of pump-turbine during startup process. In *Proceedings of the IOP Conference Series: Earth and Environmental Science*; IOP Publishing: Bristol, UK, 2014; Volume 22, p. 032024.
19. Mao, Z.; Tao, R.; Chen, F.; Bi, H.; Cao, J.; Luo, Y.; Fan, H.; Wang, Z. Investigation of the starting-up axial hydraulic force and structure characteristics of pump turbine in pump mode. *J. Mar. Sci. Eng.* **2021**, *9*, 158. [CrossRef]
20. Chen, F.; Bi, H.; Ahn, S.H.; Mao, Z.; Luo, Y.; Wang, Z. Investigation on dynamic stresses of pump-turbine runner during start up in turbine mode. *Processes* **2021**, *9*, 499. [CrossRef]
21. Casanova, F.; Mantilla, C. Fatigue failure of the bolts connecting a Francis turbine with the shaft. *Eng. Fail. Anal.* **2018**, *90*, 1–13. [CrossRef]
22. Wang, Z.; Yang, J.; Wang, W.; Qu, J.; Huang, X.; Zhao, W. Research on the Flow-Induced Stress Characteristics of Head-Cover Bolts of a Pump-Turbine during Turbine Start-Up. *Energies* **2022**, *15*, 1832. [CrossRef]
23. Chen, F.; Chen, L.; Wang, Z.; Yu, J.; Luo, C.; Zhao, Z.; Ren, S.; Li, J.; Deng, D. Computation of static stresses of the head cover bolts in a pump turbine. In *Proceedings of the IOP Conference Series: Materials Science and Engineering*; IOP Publishing: Bristol, UK, 2019; Volume 493, p. 012143.
24. Bi, H.; Chen, F.; Wang, C.; Wang, Z.; Fan, H.; Luo, Y. Analysis of dynamic performance in a pump-turbine during the successive load rejection. In *Proceedings of the IOP Conference Series: Earth and Environmental Science*; IOP Publishing: Bristol, UK, 2021; Volume 774, p. 012152.

25. Li, X.; Mao, Z.; Lin, W.; Bi, H.; Tao, R.; Wang, Z. Prediction and analysis of the axial force of pump-turbine during load-rejection process. In *Proceedings of the IOP Conference Series: Earth and Environmental Science*; IOP Publishing: Bristol, UK, 2020; Volume 440, p. 052081.
26. Nicolle, J.; Giroux, A.; Morissette, J. CFD configurations for hydraulic turbine startup. In *Proceedings of the IOP Conference Series: Earth and Environmental Science*; IOP Publishing: Bristol, UK, 2014; Volume 22, p. 032021.

 *machines*

Article

Flow-Induced Dynamic Behavior of Head-Cover Bolts in a Prototype Pump-Turbine during Load Rejection

Weiqiang Zhao [1,†], Xingxing Huang [2,†], Mengqi Yang [3], Haixia Yang [3], Huili Bi [1], Qilian He [1] and Zhengwei Wang [1,*]

1   State Key Laboratory of Hydroscience and Engineering, Tsinghua University, Beijing 100084, China
2   S.C.I.Energy GmbH, Future Energy Research Institute, Seidengasse 17, 8706 Zurich, Switzerland
3   Branch Company of Maintenance & Test, CSG Power Generation Co., Ltd., Guangzhou 511400, China
*   Correspondence: wzw@mail.tsinghua.edu.cn
†   These authors contributed equally to this work.

Abstract: In order to ensure stable grid operatiFon and improve power quality, active or passive load rejection of pumped storage power stations (PSPS) inevitably occurs from time to time. The rapid closing of the guide vanes will cause drastic changes in pressure pulsations in the flow channel of the pump-turbine (PT) unit. The high-level pressure pulsations during load rejection transfer to the entire flow passage of the PT unit and generate strong vibrations on the head-cover and the connecting bolts. In this study, the 1D/3D joint simulation of the pipeline in a pumped storage power station and the turbine flow channels including the flow domains of the runner, crown chamber, band chamber, upper and lower labyrinths and pressure balance tubes is carried out first. Then, by applying the calculated pressure loads on the head-cover, stay vanes and bottom ring of the PT unit, the flow-induced dynamic behavior of the structures including the head-cover bolts is analyzed in detail. The results demonstrate that pressure loads on head-cover bolts change dramatically during the load rejection process. The flow-induced deformation of the inner head-cover during the load rejection is larger than that of other structures, and the flow-induced displacement and stress of different head-cover bolts are not uniform. The achieved conclusions in this study can be a useful reference for the design and operation of head-cover bolts for other PT units and high-head Francis turbine units.

Keywords: pump-turbine; head-cover bolts; load rejection; flow-induced dynamic behavior; computational fluid dynamics; deformation and stress

Citation: Zhao, W.; Huang, X.; Yang, M.; Yang, H.; Bi, H.; He, Q.; Wang, Z. Flow-Induced Dynamic Behavior of head-Cover Bolts in a Prototype Pump-Turbine during Load Rejection. *Machines* **2022**, *10*, 1130. https://doi.org/10.3390/machines10121130

Academic Editor: Davide Astolfi

Received: 24 October 2022
Accepted: 23 November 2022
Published: 28 November 2022

Publisher's Note: MDPI stays neutral with regard to jurisdictional claims in published maps and institutional affiliations.

1. Introduction

Due to the fast-increasing demand for electricity in recent decades, traditional power plants such as thermal power and nuclear power have to enlarge their capacity in order to balance the demand and the supply in the power grid. On the other hand, with the target of achieving net-zero emission in the middle of this century, more and more countries are introducing green and clean power, including wind power and solar power to their power system [1]. The above two factors both bring challenges to the stability of the power grid: the output of traditional power can hardly be regulated because the heat is difficult to be dissipated, while the new renewable energies rely on intermittent sunlight and wind, which depend on the changing weather. A type of huge energy that can be fast-regulated is required for this new scenario. As an important part of renewable energies, hydraulic power is being increasingly developed worldwide because of its the large-enough capacity and great flexibility. Pumped storage power is a type of hydropower that can consume energy by converting the electricity to the potential energy of water and generating power with turbine mode when there is a requirement for energy from the grid. It usually consists of a pumped-storage power station (PSPA), an upper reservoir and a lower reservoir. Pump-turbine (PT) is the core component of a PSPA, which can switch between turbine mode and pump mode according to the demand of the power grid [2]. In turbine mode, the water flows through the spiral casing into the runner and flows out through the draft tube. In pump mode, the water flow into the draft tube from the lower reservoir is pumped by the runner to the upper reservoir. By switching the operation mode and adjusting the output or input, PT units are able to achieve peak clipping, valley filling, frequency and phase regulation and emergency reserve, which are becoming more and more necessary to the power system.

Compared with Francis turbine units, PT units usually have a higher head and less number of blades. Because of this special design, PT units have to withstand higher pressure pulsation during the operation [3]. The severe pressure pulsation brings excessive vibration and higher stress concentration to the structure, which causes fatigue and failure to the machine. The short remaining useful life will cause economic loss to the PSPAs and even danger to the operators in the plants. A number of damage or failure cases have been reported in the previous studies [4–7]. According to the research, the main reasons for the failure of the structure include drastic changes in hydraulic excitation under certain operating conditions, excessive vibration induced by the resonance and fatigue failure of the head-cover bolts. The head-cover bolt is a component that is used to connect the head-cover and the stay ring. The design of the head-cover bolts is of paramount importance to the operation safety of PT units. Failures such as loose, fatigued and broken head-cover bolts will cause serious accidents such as component damage, power outages, and even casualties to the operators in the power plants [8,9]. Some accidents caused by the failure of head-cover bolts have been reported in literature [5,6]. A number of researchers have carried out studies on the head-cover bolts and some conclusions can be used as guidance to different stages of the installation of bolts. During the design stage, the variation of the relative stiffness of the bolts and head-cover for PTs have to be considered [10]. In the operation of the PT unit, the head-cover bolts mainly suffer from axial stress. Because of the uneven distribution of the hydraulic pressure, the stress distribution along the circumference is also uneven [11]. The stress and deformation mainly concentrate on the connection positions between the head-cover and the bolts while the maximum stress appears on the bolt at the position where the diameter changed [12]. Generally, the first four turns basically bear 30% of the total load. With the increase in load, the high-stress area extends downward, which helps decrease the area of high stress [13]. Brekke pays special attention to the maintenance of the head-cover bolts in his research and discussed the optimization methodologies of the safety of turbine bolt connections [14]. The previous investigations have discussed the distribution of stress on different bolts under steady operating conditions as well as the stress on certain bolts under transient conditions. However, the stress distribution on different head-cover bolts during the load rejection process of PT units has not been studied in detail. Because of the asymmetric characteristic of hydraulic pressure in the runner chamber during the load rejection process, the circumference distribution of the deformation and stress of head-cover bolts varies greatly. Excessive stress concentration might appear on certain bolts even though the average stress is within the yield strength. The load rejection process is a challenging transient condition for turbine units and is a major fatigue damage contributor for the turbine unit [15,16]. Therefore, a careful stress investigation on the head-cover bolts during the load rejection process is important for the safe design of large hydraulic PT units.

The load rejection process causes extremely high pressure in the flow passage of turbine units, which is harmful to the operation safety of the machine. The field test on the load rejection process is inconvenient for a prototype PT unit. In recent decades, numerical simulations have become an alternative and competitive method to study the internal flow characteristics of PT units. With the development of computer technology, simulation accuracy keeps increasing. Considerable numbers of research [17–21] have proved the reliability of computational fluid dynamics (CFD) of PT by comparing the simulation results with a field test. In literature [22], the simulation error between the CFD results and the field text is less than 7% and 12% in large opening condition and minimum opening, respectively. In this research, numerical simulation is applied to investigate prototype high-head PT head-cover bolts during the load rejection process.

The coupling between the vibration of structures such as the head-cover of PTs and the fluid field is a typical Fluid-Structure Interaction (FSI) problem, which is generally solved by one-way FSI and two-way FSI. In a two-way FSI, the structure is excited by the hydraulic pressure while the effect of the deformation of the structure has an influence on the fluid field reversely. For engineering problems, the scale of the whole flow passage can be more than ten meters, while the scale of the deformation of the PT structure is only several millimeters. The effect of the structure deformation on the fluid field is neglectable. Therefore, the one-way FSI is suitable for the numerical simulation in this research with a reasonable calculation accuracy. The one-way FSI simulation methodology has been verified in literature [23,24], where the result of different FSI methods has been compared. By means of one-way FSI, some researchers tried to calculate the deformation and stress of the structure of a hydraulic machine [25,26]. The difference between the FSI simulation and field test result is acceptable for engineering problems. The previous research has provided sufficient foundation for the efficiency of the FSI method on the prototype PT head-cover during the load rejection process.

The research on the characteristics of structural components such as head-cover vibration of PT units is a complex FSI problem. One-way FSI and two-way FSI are two typical methods to solve the FSI problem. The characteristic length of the entire flow field can be up to ten meters, while the actual head-cover deformation of the PT unit is only several millimeters, so the effect of the head-cover deformation on fluid flow is negligible; therefore, the one-way FSI method is reasonable for this study considering the computational resources and the accuracy of the results. It is more reasonable to adopt the one-way FSI method for this study considering the computing resources and accuracy of the results. The studies [23,24] have verified the accuracy of the FSI calculation method, compared the effects of different FSI methods on the calculation results of hydraulic machinery and confirmed the feasibility of the one-way method. Some researchers [25,26] have calculated the stresses and strains of the stationary parts of the unit with the one-way FSI method. The difference between the calculated results and the measured ones is within the acceptable range.

In this research, the flow-induced dynamic behavior of head-cover bolts in a PT unit during its load rejection process is investigated by means of the 1D and 3D hybrid numerical simulation method. In Section 2, the basic theory is introduced by displaying the equations of 1D and 3D simulation. In Section 3, a 1-dimensional (1D) calculation was performed on the pipeline in order to obtain the pressure and discharge in the machine during the load rejection process. The key time points were selected according to the peak and valley pressure and used as the boundary conditions of the 3-dimensional (3D) calculation. Then, a physical model of both the fluid domain and structure of the researched unit was built for the 3D fluid simulation. Computational fluid dynamics (CFD) was applied to the model to calculate the pressure distribution of each key time point. In Section 4, the obtained pressure distribution was mapped on the structure model of the unit by the one-way FSI method. Finally, the dynamic behavior of the head-cover and the head-cover bolts were investigated and analyzed in detail. The conclusions in Section 5 of this research provide important guidance for the design of head-cover bolts for the PT units.

2. Numerical Simulation Methodology

The numerical simulation methods in this paper are composed of 1D pipeline simulation, 3D CFD calculation and one-way FSI simulation method.

2.1. 1D Pipeline Calculation Methodology

The 1D pipeline model of the researched unit includes the upper reservoir, penstock, surge shaft, valves and lower reservoir. In each of the components of the 1D pipeline model, the mathematical model of the unsteady flow is used. The basic equations of unsteady flow in the model include the kinematic equation of motion and the continuity equation.

$$g\frac{\partial H}{\partial X} + V\frac{\partial V}{\partial X} + \frac{\partial V}{\partial t} + \frac{fV|V|}{2D} = 0 \tag{1}$$

$$V\frac{\partial H}{\partial X} + \frac{\partial H}{\partial t} + \frac{a^2}{g}\frac{\partial V}{\partial X} - V\sin\alpha = 0 \tag{2}$$

$$H = Z + P/\rho g \tag{3}$$

where Z, g and ρ are the axis elevation of the pipeline, local acceleration of gravity and water density, respectively. V is the average velocity of the water. D is the diameter of the flow passage in the pipeline. F is the Darcy Westbach friction coefficient. ρ is the liquid density and g is the local acceleration of gravity and water density, respectively. F is Darcy Westbach friction coefficient, D and X are the inner diameter and the axis length of the pipe, respectively. A and p are the wave velocity and the pressure of the water in the pipeline, T is the transient time and α is the angle between the pipe axis and the horizontal plane.

Equations (1)–(3) can be combined and transformed into method of characteristic (MoC), which is a differential equation with constant coefficients. On the characteristic line, $\partial X / \partial t = \pm a$, the original equation can be transformed into a differential equation with constant coefficients; therefore, the process of solving the partial differential equation can be transformed into the discretization of the differential equation on the characteristic line. The compatibility equations are:

$$C^+ : \begin{cases} \frac{dX}{dt} = +a \\ A^2\frac{dH}{dt} + A\frac{a}{g}\frac{dQ}{dt} - QA\sin\alpha + \frac{faQ|Q|}{2gD} = 0 \end{cases} \tag{4}$$

$$C^- : \begin{cases} \frac{dX}{dt} = -a \\ A^2\frac{dH}{dt} - A\frac{a}{g}\frac{dQ}{dt} - QA\sin\alpha - \frac{faQ|Q|}{2gD} = 0 \end{cases} \tag{5}$$

where C^+ and C^- are the two directions of the characteristic line. The compatibility equation is established along the characteristic line. The equation can be integrated into the form of a difference through the change in flow and head between two points, and then solved iteratively to complete the 1D pipeline calculation.

2.2. 3D Flow Simulation Methodology

The 3D flow simulation methodology is based on the Navier–Stokes Equation for fluid flow in conservative differential form:

$$\frac{\partial \vec{u}}{\partial t} + \vec{u} \cdot \nabla \vec{u} = -\frac{1}{\rho}\nabla p + \nu\nabla^2\vec{u} + \vec{g} \tag{6}$$

$$\nabla \cdot \vec{u} = u\vec{i} + u\vec{j} + u\vec{k} = 0 \tag{7}$$

Equation (6) is the Navier–Stokes Equation and Equestion (7) is the formulas for incompressible fluids. Where ρ and t represent the fluid density and time, $\vec{u}$ is the velocity vector in 3D space. Based on the equation and finite element method, the pressure pulsation and distribution in the flow passage can be obtained. The pressure can be applied on the fluid–structure coupling surfaces of the structures.

2.3. Fluid–Structure Interaction Analysis Methodology

The dynamic characteristics of the structure are influenced by the forces and pressure acting on the structure. Considering the pressure, the finite element methodology can be described as:

$$[M]^e\{\ddot{\delta}(t)\}^e + [C]^e\{\dot{\delta}(t)\}^e + [K]^e\{\delta(t)\}^e = \{F_S(t)\} + \left\{F_{fS}(t)\right\} \tag{8}$$

where $[M]^e$, $[C]^e$ and $[K]^e$ represent the mass matrix, damping matrix and stiffness matrices of the structure, respectively; $\{\ddot{\delta}(t)\}^e$, $\{\dot{\delta}(t)\}^e$ and $\{\delta(t)\}^e$ refers to the acceleration, velocity and displacement of the structure, respectively; $\{F_S(t)\}$ is the excitation force vector applies on structures, while $\{F_{fS}(t)\}$ is to the fluid pressure vector acting on the surface of the structure.

By 3D computational fluid dynamics (CFD) analysis, the fluid pressure distribution in the PT unit can be obtained. As described in Section 2.2, the fluid pressure can be mapped to the fluid–structure interference surfaces and the flow-induced structural stress σ of the machine can be calculated with the displacement $\{\delta(t)\}^e$ by Equation (9).

$$\{\sigma\} = \frac{[B]}{[S]}\{u\} \tag{9}$$

where $[B]$ and $[D]$ are the strain-displacement matrix and stiffness matrix, respectively.

Equation (10) shows the calculation of von Mises stress σ_{vM} yield by stress in three directions, which is used for evaluating the structural stress characters of the turbine unit.

$$\sigma_V = \frac{1}{\sqrt{2}}\sqrt{(\sigma_{11} - \sigma_{22})^2 + (\sigma_{22} - \sigma_{33})^2 + (\sigma_{33} - \sigma_{11})^2} \tag{10}$$

where σ_{ii} ($i = 1, 2, 3$) are the principal stresses.

3. Numerical Simulation of the Fluid Flow

3.1. 1D Simulation on Pipeline Flow

The researched machine is a PT unit with a rated power of 300 MW and a rated speed of 500 rpm. The rated head is 520 m. It has 7 blades in the runner and 20 blades on both the stay vanes and guide vanes. Figure 1 shows the sketch of the water conveyance system of the researched PT unit. The 1D model of the pipeline system includes the upper and lower reservoir, turbine unit and surge shafts.

The 3D CFD calculation domain is the turbine part of the system, including the flow passage of the spiral casing, stay vane, guide vane, runner, draft tube, upper and lower labyrinth seals and pressure-balance pipelines.

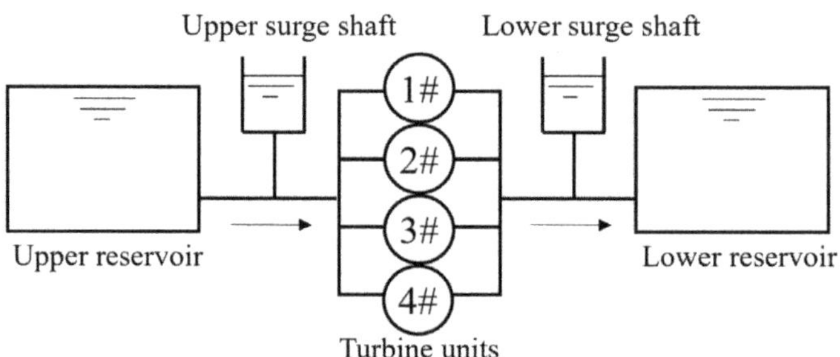

Figure 1. The 1D calculation model of the researched PT unit.

Figure 2 shows the close curve of the relative guide vanes and variation of relative rotating speed and discharge obtained by the 1D simulation. From the figure, it can be observed that the guide vanes close monotonously from the maximum opening to zero after the load rejection. The rotating speed of the runner increases at the beginning of the load rejection process while vibrating due to the close-down of the guide vanes. The extremes are selected from both the rotating speed curve and the discharge curve. The extreme points are used to perform the fluid dynamic calculation [18]; giving the rotating speed and discharge are the key parameters that affect the flow in the fluid domain during the load rejection process [27,28]. Figure 3 shows the relative pressure (rate between the pressure and rated head of the machine) of spiral casing inlet (p_{SC}) and draft tube outlet (p_{DT}) calculated by 1D simulation. The pressure values vibrate with the variation of the discharge and rotating speed. p_{SC} and p_{DT} are used as the boundary conditions of the 3D flow simulation.

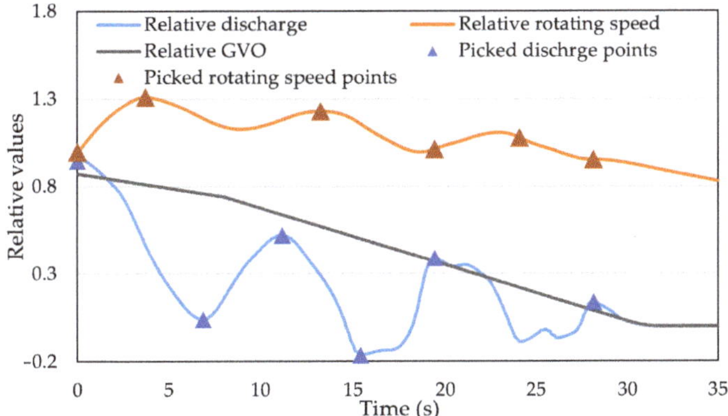

Figure 2. The operating parameters and 9 key time points of the PT unit during load rejection.

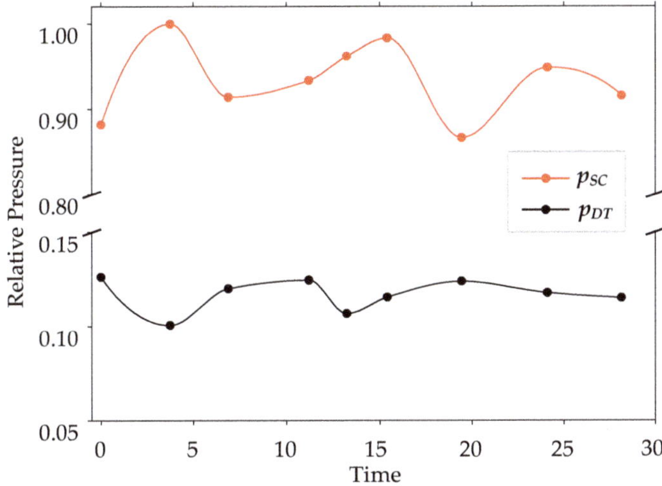

Figure 3. Normalized pressure at the spiral casing inlet and draft tube outlet.

3.2. 3D CFD Simulation

3.2.1. 3D Model of Flow Calculation

The 3D fluid domain model includes the flow passage of the spiral casing, stay vanes, guide vanes, runner and draft tube. As shown in Figure 4 The flow passage of the upper labyrinth seal and lower labyrinth seal (1.5 mm) between the runner and non-rotational structure are also included in the fluid model in order to increase the simulation accuracy.

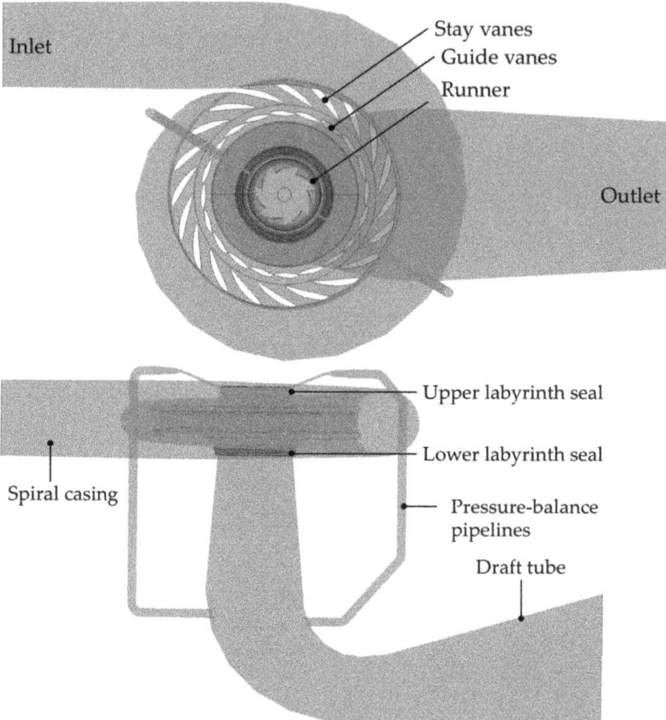

Figure 4. The 3D modeling of flow passage of the unit.

3.2.2. Mesh Independence Analysis of the 3D Flow Calculation

ANSYS CFX was used to carry out the 3D CFD analysis. The analysis type was set to stable and the discrete format and the solver were set to high resolution. The runner domain was set to rotate at an angular velocity of 500 rpm. The wall condition was set as a non-slip wall. The interfaces between guide vanes and runner, runner and draft tube were set to Frozen Rotor and the others are set to General Connection. The turbulence model was set to shear stress transport, the turbulence model was set to shear stress transfer and the turbulence mathematical accuracy was set to first order with a convergence residual of 10^{-5}.

The mesh quality is very important for 3D turbulence calculation. In this paper, the tetrahedron–hexahedron hybrid mesh was used to discrete the fluid domain. In this study, the mesh in the flow passage of guide vanes was generated automatically according to different GVOs in order to ensure the mesh quality of the guide vanes under a given opening. Figure 5 shows the mesh of different fluid domains of the PT unit. Four sets of mesh have been compared in order to determine the best mesh that achieves a balance between the simulation time and accuracy. The comparison of the efficiencies of different sets of mesh and the details about the mesh of each part of the fluid domain is listed in Figure 5 and Table 1 in literature [16]. The final adopted mesh (3) of the PT unit for load rejection calculation is shown in Table 1.

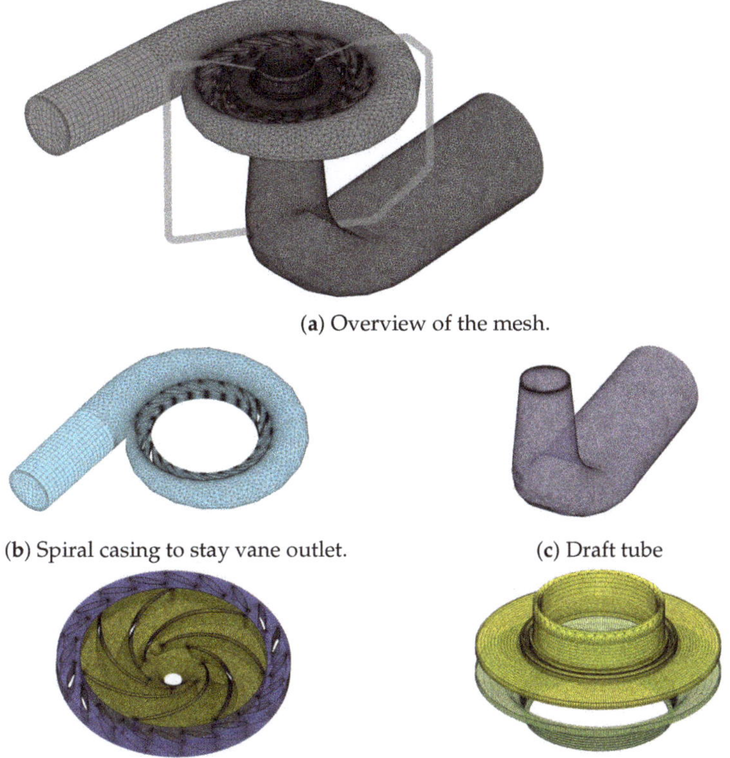

(**a**) Overview of the mesh.

(**b**) Spiral casing to stay vane outlet.

(**c**) Draft tube

(**d**) Guide vanes to runner outlet

(**e**) Crown labyrinth seal & band labyrinth seal.

Figure 5. Mesh of the fluid domains.

Table 1. Element number of flow domains.

Flow Domain	Elements ($\times 10^6$)
Spiral casing and Stay vane	3.26
Guide Vane	0.27
Runner	3.63
Draft tube	0.60
Labyrinth seal and Pressure-balance pipelines	0.40
Total	8.16

3.3. 3D Fluid Simulation Results

3.3.1. Pressure Variation at the Monitoring Points

Four pressure monitoring points (P_S, P_G, P_{U1}, P_{U2}) are set at the stay vanes, guide vanes, crown gap and upper labyrinth seal. The details of the pressure monitoring points and fluid domain from the spiral casing to the draft tube are shown in Figure 6. The variation of pressure at the monitoring points during the load rejection process is shown in Figure 7. During the load rejection process, the pressure of each monitoring point fluctuates with great amplitudes, which is closely related to the discharge and rotating speed of the machine. The high rotating speed enhances the shearing effect of the runner on the non-rotating flow region, increasing the water flow velocity at the inlet and upstream of the runner, resulting in more wake cutting on the vane flow domain. It also enhances the rotor–stator interaction phenomenon and increased the pressure pulsation in the bladeless region. At t = 6.88 s, the GVO and the rotating speed are high and the flow rate is low.

At this moment, the turbulence and vortices of the flow from the guide vane increase, and a large number of blade channel vorticity currents appear in the runner, which is extremely unstable. The turbulent flow induces increased energy dissipation and pressure in the vaneless area. At the moment with high discharge and low speed (t = 11.19 s and t = 19.43 s), the deviation between the outflow speed direction of the guide vane and the design condition is small, so the internal flow of the unit is sufficient and uniform and the pressure shows a low level.

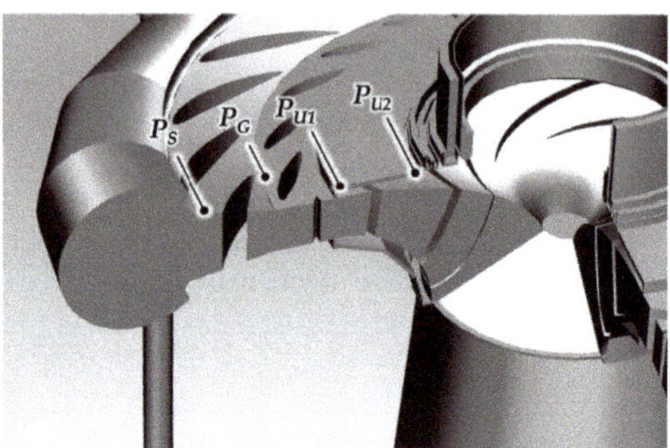

Figure 6. Pressure monitoring points' selection.

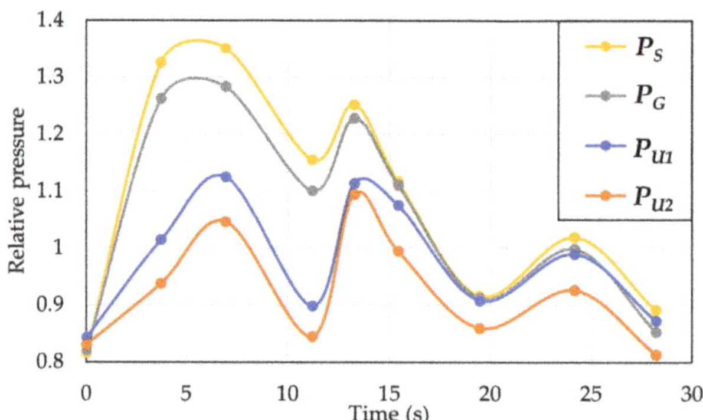

Figure 7. Pressure change at the monitoring points.

3.3.2. Flow Pattern Change in the Runner Passage

In the process of load rejection, the flow from the guide vane to the runner outlet shows extremely unstable turbulent characteristics. Figure 8 shows the streamline distribution in the runner in the whole load rejection process. When the load rejection does not begin (t = 0 s), the flow in the runner is uniform. Afterward, the flow is greatly affected by the changes in flow rate, head and speed.

During the process of load rejection, the fluid in the flow passage shows extremely unstable turbulent characteristics. Figure 8 shows the streamline distribution in one of the sections of the runner during the whole load rejection process. Before the load rejection (t = 0 s), the fluid in the runner flows uniformly. After that, the flow is greatly affected by the change of flow, head and speed. From each figure during the load rejection process, it can be observed that the flow in each flow passage between the blades is different from one another. The flow during load rejection shows apparent asymmetry. At the moments with large discharge (Figure 8b,d,e,g), large vortices appear on the suction surface of the blades. The scale of the vortices implies the size of the flow duct, which disturbs the flow in the machine. The unstable flow induces pressure fluctuation in the runner, which influences the hydraulic torque of the machine.

At the moments with the negative flow (t = 15.40 s and 24.10 s), the unit enters the reverse pump operating condition. Due to the imbalance of speed and flow, the flow between blade passages rotates abnormally and separated flow occurs on the pressure surface of the runner. Both factors cause the wake to fall off upstream of the runner blades and a large number of swirl eddies appear in the vaneless area. At the same time, wake falling off also occurs at the leading edge of the guide vanes and stay vanes. The unstable flow of the unit extends to the spiral casing area. It is shown in research [29] that the vortex core in the vaneless area is impacted instantaneously by a large number of reverse refluxes into the guide vane area and the stay vane area when the unit experiences the reverse pump operating condition, which is also in accordance with the calculation results in this paper.

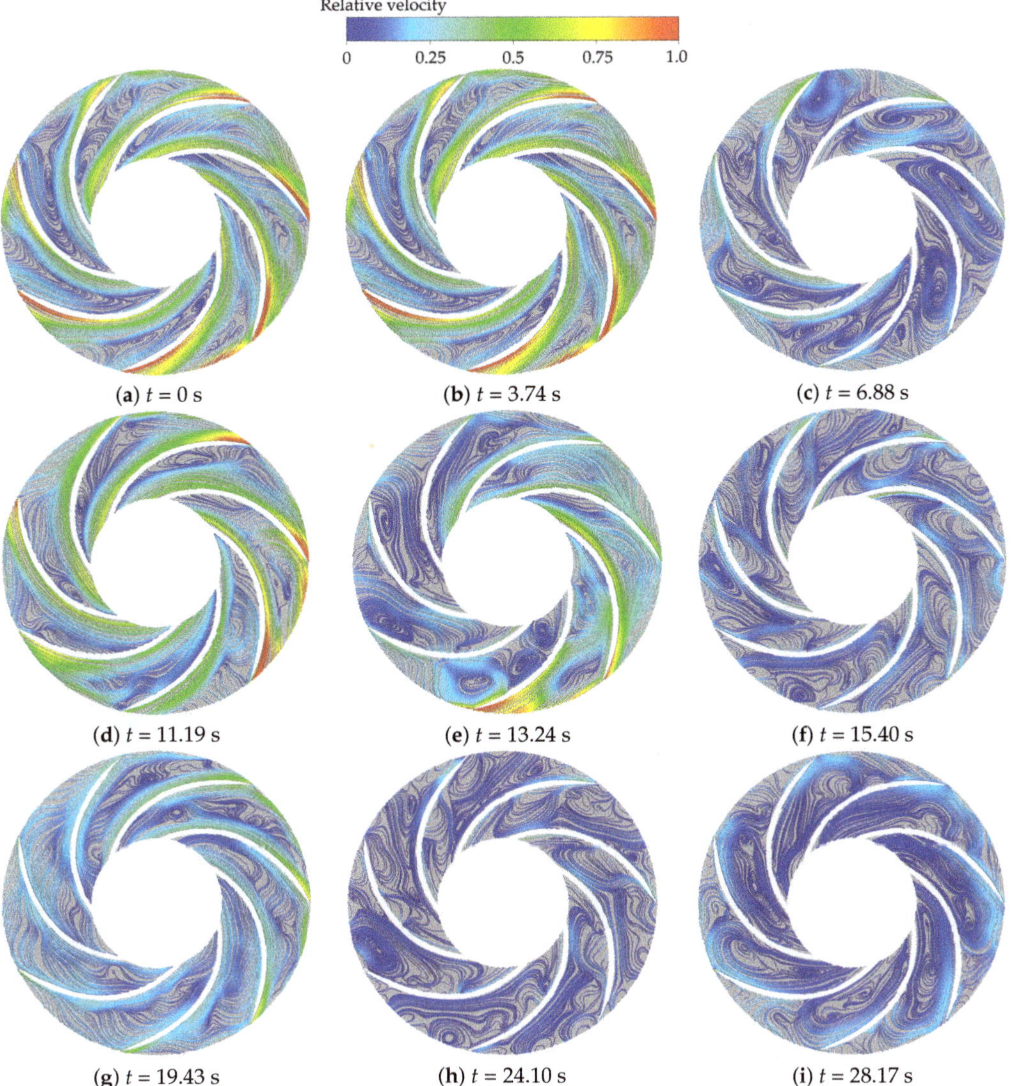

Figure 8. Flow pattern in the runner passage.

At t = 15.40 s and 24.10 s, the discharges are negative while the rotating speed is positive, which means the unit enters the reverse pump condition. Due to the imbalance of pressure, the flow between the blades rotates abnormally, and the separation flow appears on the pressure surface of the runner. The separation causes the wake to shed at the trailing edge of the guide vanes, resulting in a large number of vortices in the bladeless area. At the same time, wake shedding also occurs at the leading edges of the blades.

3.3.3. Result of Pressure Variation

Figure 9 shows the pressure change of the upper side of the flow passage of stay vanes, guide vanes, vaneless area, crown seal and upper labyrinth during load rejection. Before the load rejection (t = 0 s) the distribution of the pressure is uniform on the lower side of the head-cover. For the moments of t = 3.73 s, t = 11.19 s, t = 13.24 s and t = 19.43 s, the hydraulic moment is positive or equal to zero, and the pressure distribution of the unit is relatively uniform, i.e., there is no pressure concentration applied on the head-cover. For the moments with negative hydraulic torque (t = 6.88 s, 13.24 s, 15.40 s, 24.10 s), there appears to be circumferential pressure unbalance and pressure concentration in the flow passage of the guide vanes. The pressure concentration is caused by the separation of high-speed circumferential flow through the clearance. In the clearance, the flow comes across a sudden velocity decrease and local pressure rise, which causes strong impacts on the head-cover.

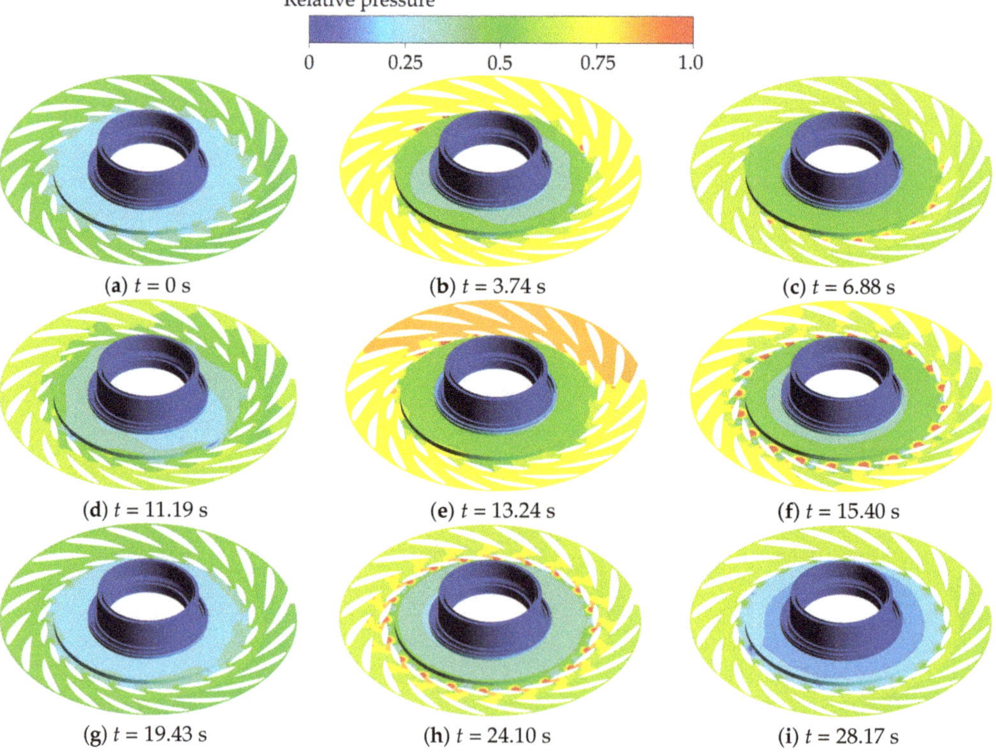

Figure 9. Pressure distribution in the flow passages at different key time moments.

4. Flow-Induced Dynamic Behavior of Structures

4.1. Simulation Setup and Mesh Independence Analysis

The studied stationary structures of the PT unit consist of a head-cover, a stay ring, a bottom ring and connecting bolts (Figure 10). The head-cover and bottom ring are assembled to the stay ring by 80 head-cover bolts and 80 bottom ring bolts. The nominal diameter of the bolts is 110 mm. The material properties of the stationary structures are listed in Table 2.

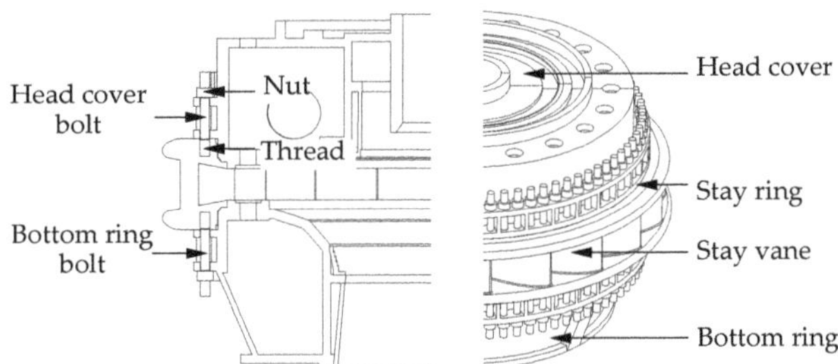

Figure 10. CAD model and boundary conditions of the stationary structures of the PT unit.

Table 2. Material properties of the studied structures.

Property	Value
Elastic modulus [Pa]	2.1×10^{11}
Poisson's ratio [-]	0.3
Density [kg·m^{-3}]	7850

The flow-induced dynamic behavior of the stationary structures of the PT unit is analyzed in detail with the help of the numerical code ANSYS Mechanical. The finite element mesh and the boundary conditions are shown in Figure 11. All nodal degrees of freedom (DOFs) on the stay ring surfaces embedded in concretes are fixed, and the nodal DOFs of the bottom ring surface welded into the draft tube are also fixed. The calculated pressure loads (P_{Load}) are applied on the fluid-structure coupling surfaces and caused large axial force on the head-cover ($F_{axial\ thrust}$). The stiffness of turbine radial guide bearing (1.0×10^9 N·m^{-1}) and the gravity ($g = 9.8$ m · s^{-2}) are considered for the analysis.

After applying a preload force ($F_{Preload}$) to each bolt, the head-cover, the stay ring and the bottom ring are bolted together as a single structural assembly. The nut is considered to be bonded to the bolt shank, and the bolt thread is bonded to the stay ring. In this study, the bolt thread details are not considered. The connection surfaces between the head-cover and the stay ring are in frictional contact. The same setting applies to the attachment surfaces between the bottom ring and the stay ring, the attachment surfaces between the nuts and the head-cover, and the attachment surfaces between the nuts and the bottom ring. The friction coefficient of frictional contacts is set to 0.1. The exported pressure files from CFD simulations at each time point during the load rejection are mapped on the finite element model of the stationary structures sequentially so that the flow-induced dynamic behavior of the studied structures can be analyzed.

The geometrically complex head-cover, stay ring and bottom ring are meshed with high-quality tetrahedral elements, while the bolts are geometrically simple and meshed with high-quality hexahedral elements. The meshes of the head-cover bolts and bottom ring bolts, especially at the typical stress concentration areas, are refined (S1 and S2).

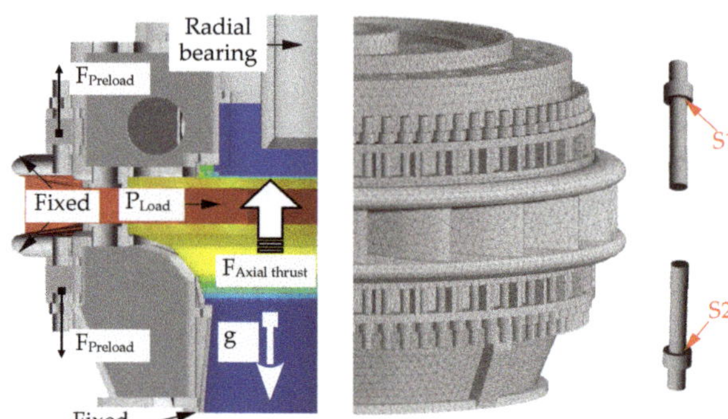

Figure 11. Simulation setup and finite element mesh of the stationary structure of the unit.

To achieve a reasonable balance between the accuracy of the computation results and the simulation time, a mesh independence analysis shall be carried out. Three groups of meshes with different element sizes are established for the studied structures. The node numbers of three sets of meshes are 1.6×10^6, 2.4×10^6 and 5.2×10^6, respectively. Figure 12 shows that the second group mesh marked by the red box consumes only 0.7% of the simulation time of the third group, but the results are similar. Therefore, it is logical to use the second group mesh to complete the investigation.

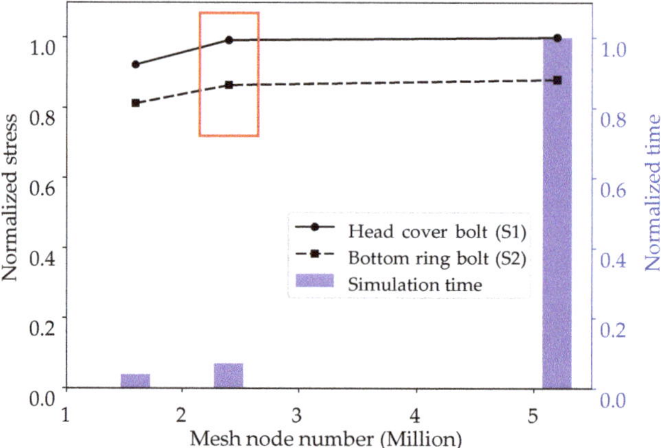

Figure 12. Mesh independence analysis.

4.2. Results and Discussions

During the turbine load rejection process, the hydraulic pressure load acting on the stationary structures of the unit changes significantly. Figure 13 shows the normalized axial thrust on the head-cover. It can be observed that the axial thrust on the head-cover is time-dependent and the force direction is axially upward during the whole load rejection process. The axial thrust has three local maxima during load rejection and reaches its maximum at $t = 13.2$ s. The maximum axial thrust is more than ten times the overall weight of the head-cover plus the rotating shaft line.

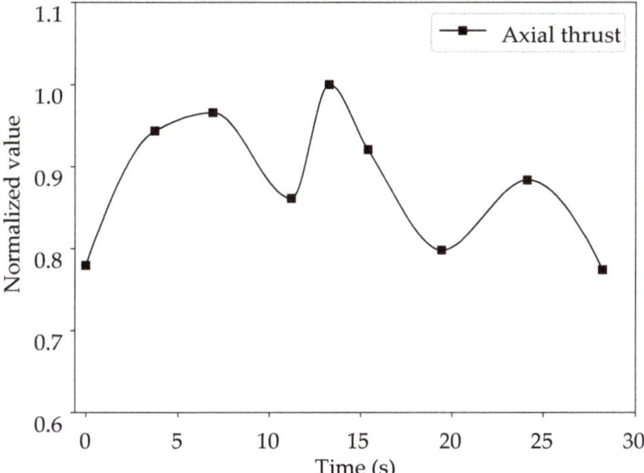

Figure 13. Axial thrust of the head-cover during load rejection.

The maximum deformation of the structures induced by unsteady flow during load rejection is also changing with time, but the deformation distribution at the different moments keeps the same pattern, as shown in Figure 14.

Flange beams and bolt flanges are used to strengthen the connections between the stay ring and the head-cover and the bottom ring. The adjacent head-cover bolts (bolt couple) between the flange beams are deformed close to each other, resulting from the hydraulic forces induced by the unsteady flow. The same deformation characteristics are also present on the bottom ring bolts.

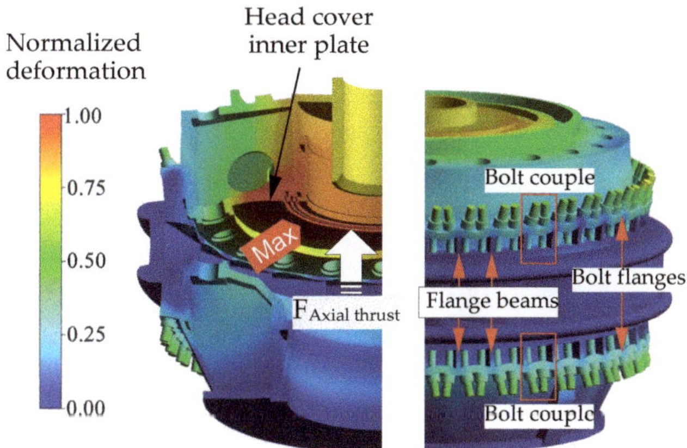

Figure 14. Deformation distribution of the stationary structures in section view.

The vibration behavior of the head-cover and head-cover bolts during load rejection are strongly related to the axial thrust on the head-cover. Since the inner head-cover has no support in the axial direction, the maximum deformation of the structures appears at the inner plate of the head-cover lifted by the large axial thrust. During the load rejection process, the maximum deformation of the head-cover with time follows the variation of the axial thrust and reaches its maximum at $t = 13.2$ s (Figure 15).

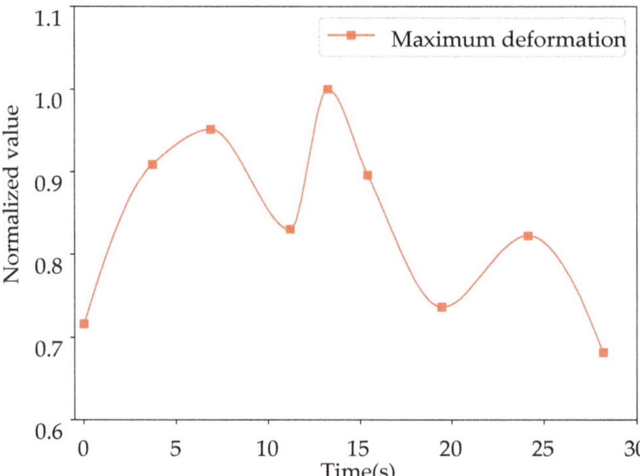

Figure 15. The maximum deformation of the head-cover during load rejection.

In order to facilitate production processing, transportation and installation, the head-cover is usually manufactured as two symmetrical halves, which are assembled into a single structure during the installation in the power plant. As can be seen from Figure 16, the flow-induced deformation of the head-cover is also symmetrical. Since the connecting flanges of the head-cover are thicker than the stiffener plates, the deformation of the head-cover inner plate near the connecting flanges is smaller than the deformation of the inner plate at other locations. Therefore, increasing the thickness of the stiffener plates inside the head-cover can increase the local stiffness and effectively reduce the deformation of the inner plate of the head-cover.

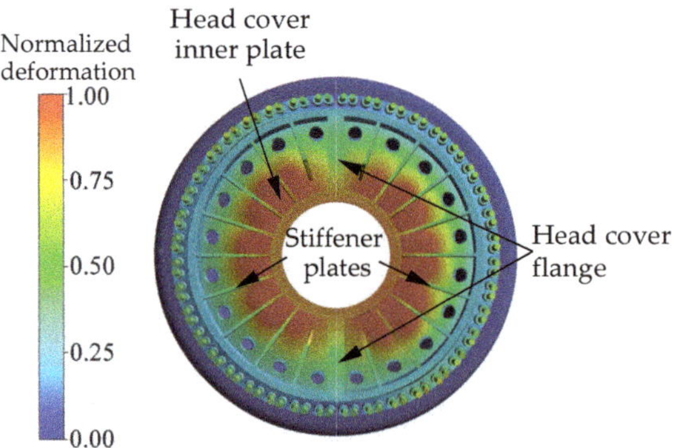

Figure 16. Deformation distribution of the stationary structures in top view.

The flow-induced stress distribution during turbine load rejection in the large stationary structures of the unit including the head-cover, stay ring and bottom ring is shown in Figure 17. The large hydraulic pressure acts on the stationary structures of the unit, pressing the bottom ring downward and lifting the head-cover. As a result, the guide vanes between them are stretched. Since the trailing edge of the stay vane is much thinner than the leading edge, the maximum flow-induced stress is concentrated at the rounded corner of the trailing edge of the stay vane.

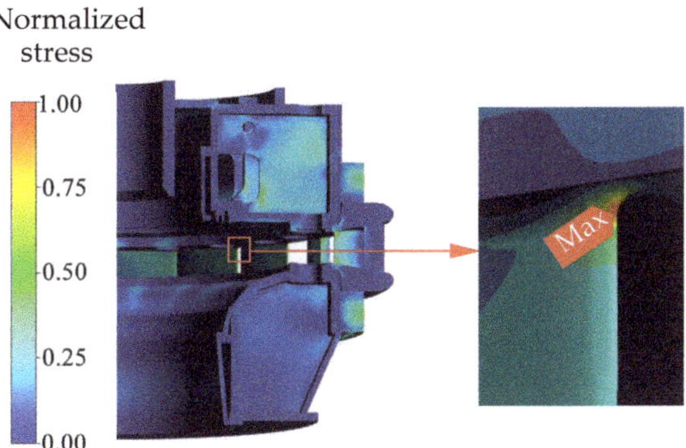

Figure 17. The stress distribution of the large stationary structures.

Since the connecting bolts bear the large pressure from the flow channel and tighten the stationary structures of the unit together, high stresses during the turbine load rejection appear also on the head-cover bolts, bottom ring bolts, the flange beams, and the bolt flanges (Figure 17). The bolt couples are evenly distributed along the circumference in the large stationary structures so that the stress distribution in each bolt couple shows a similar distribution (Figure 18).

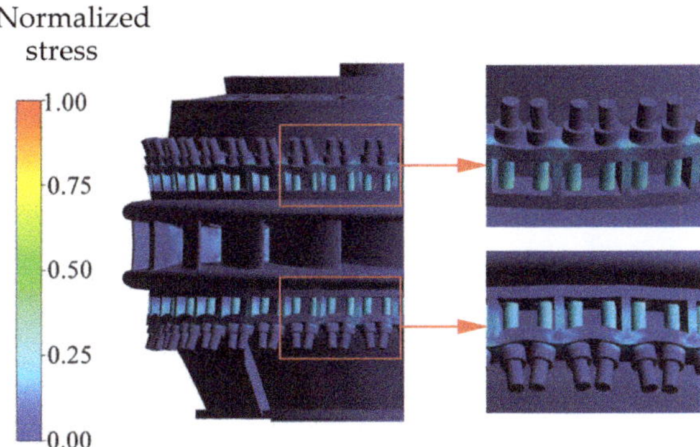

Figure 18. The stress distribution of the large stationary structures and the connection bolts.

The stress distributions of the head-cover bolts and the bottom ring bolts during the process of turbine load rejection are demonstrated in Figure 19. The analysis shows that the stress distribution of the connection bolts is similar in general. The maximum stress of the head-cover bolt is located at the bolt fillet close to the head-cover bolt flange, while the maximum stress of the bottom ring bolt is concentrated at the bolt fillet near the bottom ring bolt flange.

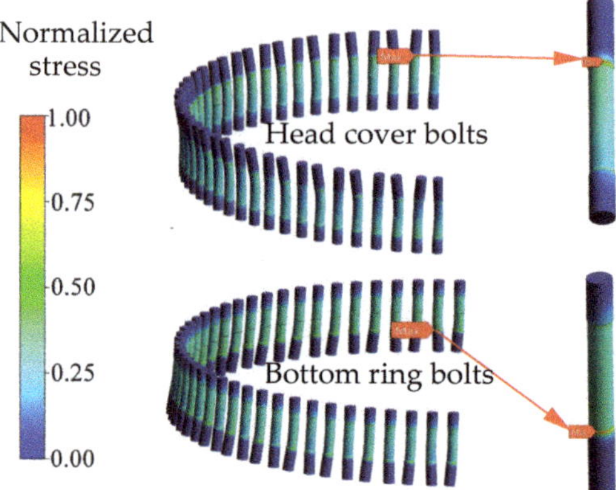

Figure 19. The stress distribution of head-cover bolts and bottom ring bolts.

The stress distribution of the large stationary structures, head-cover bolts and bottom ring bolts during turbine load rejection varies with time and keeps the same pattern at different moments. Figure 20 compares the maximum stresses of the stay ring, head-cover bolts, and bottom ring bolts during turbine load rejection. The stresses are normalized with reference to the maximum stress of the head-cover bolts.

The axial effective area of the head-cover is much larger than that of the bottom ring, so the axial force on the head-cover is larger than that on the bottom ring. In addition, the bottom ring with sold reinforced plates is more rigid than the head-cover. So the maximum stress of head-cover bolts is higher than the maximum stress of bottom ring bolts during turbine load rejection, and the peak-peak value of head-cover bolts is also larger than that of the bottom ring bolts. Figure 20 also shows that the peak-peak value of the stay ring is larger than the connection bolts, but the absolute stress values are much lower than the connection bolts.

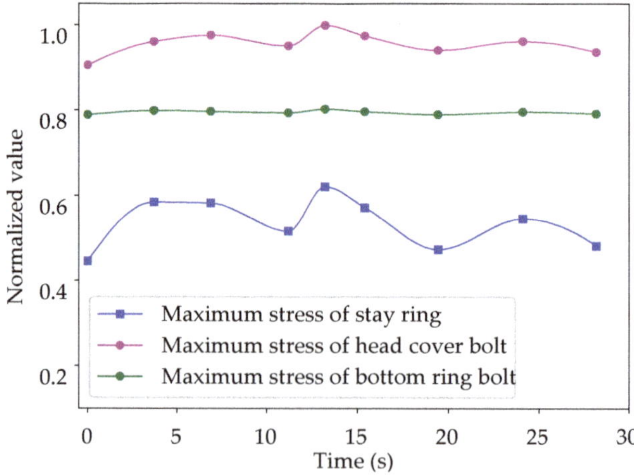

Figure 20. Comparison of the maximum stresses of the stationary structures during load rejection.

In order to study the relation between the axial thrust and the structural dynamic behavior, Figure 21 compares the axial thrust and the maximum deformation and stresses of the structures during the load rejection process. All values are normalized with reference to their maximum values, respectively. The maximum deformation and stresses of the structures follow the same trend of axial thrust with time during the turbine load rejection and reach their respective maximum values at $t = 13.24$ s. However, the variation ranges of these parameters are significantly different.

Figure 21 shows the variation range of the axial thrust, maximum structural deformation and maximum stress of the stay ring. It can be observed that the range is between 0.7 to 1, which is larger than the variation range of the maximum stress of the connection bolts, between 0.9 to 1. For the connection bolts, the variation range of the head-cover bolts (0.93–1) is larger than that of the bottom ring bolts (0.98–1), and the absolute stress value of the head-cover bolts is higher than that of the bottom ring bolts, so under the same operation condition, the head-cover bolts are more prone to fatigue failure with enough alternative load cycles.

For the connection bolts, the variation range of the head-cover bolts (0.93–1) is greater than that of the bottom ring bolts 0.98–1, and the absolute stress value of the head-cover bolts is higher than that of the bottom ring bolts, as shown in Figure 20. Therefore, under the same operating conditions, the head-cover bolts are more prone to fatigue failure than bottom ring bolts under sufficient cycles of alternating loads.

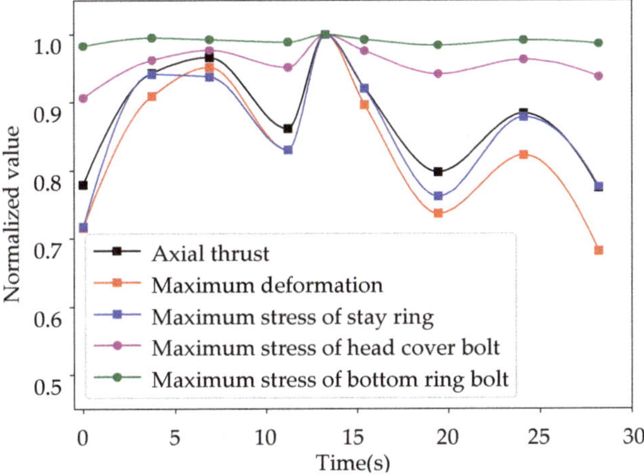

Figure 21. Results' comparison of the PT unit during load rejection.

5. Conclusions

The flow-induced dynamic behavior of the stationary structures including head-cover bolts in a prototype pump-turbine unit during turbine load rejection has been studied in detail.

Firstly, the variation of the inlet and outlet pressures during the load rejection process has been calculated by 1D pipeline simulation. Then, the 3D model of the researched PT unit is built from the spiral casing to the draft tube. The pressure-balance pipelines and labyrinth seals are included in the model. After that, the pressure obtained by 1D simulation is used as the boundary conditions of the CFD simulation carried out on the 3D model. The calculation results show that large-scale vortices appear in the runner during the load rejection process. The pressure distribution analysis shows the pressure concentration on the head-cover.

By applying the pressure loads calculated by CFD simulations, the flow-induced deformation and stress of the structures have been analyzed. The maximum deformation and stresses of the structures during load rejection are changing with time, but the distributions of deformation and stress at the different moments keep the same pattern.

The maximum deformation during turbine load rejection is located on the inner plate of the head-cover. Increasing the thickness of the stiffener plates of the head-cover can improve the local rigidity of the head-cover, thereby effectively reducing the deformation of the inner plate of the head-cover.

The maximum stress of the large stationary structures appears at the rounded corner of the trailing edge of the stretched stay vane, and the maximum stress of the bolts in the head-cover and bottom ring is concentrated at the bolt fillet close to the bolt flange.

The maximum deformation and stresses in the structures during turbine load rejection follow the tendency of axial thrust variation with time and reach their respective maximum values at $t = 13.24$ s.

During turbine load rejection, the variation ranges of axial thrust, maximum deformation, and maximum stress are significantly different. The normalized variation range of the axial thrust, maximum structural deformation, and maximum stress of the stay ring (0.7 to 1) is larger than that of the maximum stress of the connection bolts (0.9 to 1). Both the absolute stress value and the peak-peak value of head-cover bolts are larger than those of the bottom ring bolts, therefore the head-cover bolts are more prone to fatigue failure with enough alternative load cycles than the bottom ring bolts. It is recommended to strengthen the online monitoring and on-site inspection of head-cover bolts and bottom ring bolts and replace the bolts after a long service time.

Author Contributions: Conceptualization, Z.W.; methodology, W.Z. and X.H.; software, H.B., Q.H., W.Z., M.Y., H.Y. and X.H.; validation, H.B., M.Y. and H.Y.; investigation, H.B., Q.H., W.Z. and X.H.; writing—original draft preparation, W.Z., X.H. and Q.H.; writing—review and editing, W.Z., X.H., M.Y., H.Y., H.B. and Q.H.; supervision, Z.W. All authors have read and agreed to the published version of the manuscript.

Funding: The work is supported by Joint Open Research Fund Program of State key Laboratory of Hydroscience and Engineering and Tsinghua—Ningxia Yinchuan Joint Institute of Internet of Waters on Digital Water Governance (sklhse-2022-Iow13). This work is supported by National Natural Science Foundation of China (No.: 51876099).

Institutional Review Board Statement: Not applicable.

Informed Consent Statement: Not applicable.

Data Availability Statement: Not applicable.

Acknowledgments: The authors would like to express their sincere thanks for the financial support of the project: Research on Lifetime Prediction of Non-rotating Parts of pump-turbine Unit Based on Rotor-Stator Interaction (RSI), Fluid–Structure Coupling and Fracture Mechanics—Research project on RSI mechanism and its Influence on Non-rotating parts of pump-turbine Unit of Branch Company of Maintenance & Test, CSG Power Generation Co., LTD.

Conflicts of Interest: The authors declare no conflict of interest.

Abbreviations

The following abbreviations are used in this manuscript:

PSPS	Pumped storage power station
CFD	Computational fluid dynamics
FEM	Finite element method
FSI	Fluid–structure interaction
FVM	Finite volume method
PT	Pump-turbine

References

1. Zhao, W. Improved Condition Monitoring of Hydraulic Turbines Based on Artificial Intelligence Techniques. Ph.D. Thesis, Universitat Politècnica de Catalunya, Barcelona, Spain, 2021.
2. Zhao, W.; Presas, A.; Egusquiza, M.; Valentín, D.; Egusquiza, E.; Valero, C. On the use of Vibrational Hill Charts for improved condition monitoring and diagnosis of hydraulic turbines. *Struct. Health Monit.* **2022**, *21*, 14759217211072409. [CrossRef]
3. Zhao, W.; Egusquiza, M.; Valero, C.; Valentín, D.; Presas, A.; Egusquiza, E. On the use of artificial neural networks for condition monitoring of pump-turbines with extended operation. *Measurement* **2020**, *163*, 107952. [CrossRef]

4. Liu, X.; Luo, Y.; Wang, Z. A review on fatigue damage mechanism in hydro turbines. *Renew. Sustain. Energy Rev.* **2016**, *54*, 1–14. [CrossRef]
5. Casanova, F.; Mantilla, C. Fatigue failure of the bolts connecting a Francis turbine with the shaft. *Eng. Fail. Anal.* **2018**, *90*, 1–13. [CrossRef]
6. Peltier, R.; Boyko, A.; Popov, S.; Krajisnik, N. Investigating the Sayano-Shushenskaya hydro power plant disaster. *Power* **2010**, *154*, 48.
7. Egusquiza, E.; Valero, C.; Huang, X.; Jou, E.; Guardo, A.; Rodriguez, C. Failure investigation of a large pump-turbine runner. *Eng. Fail. Anal.* **2012**, *23*, 27–34. [CrossRef]
8. Wang, Z.; Yang, J.; Wang, W.; Qu, J.; Huang, X.; Zhao, W. Research on the Flow-Induced Stress Characteristics of Head-Cover Bolts of a Pump-Turbine during Turbine Start-Up. *Energies* **2022**, *15*, 1832. [CrossRef]
9. Song, X.j.; Yao, R.; Chao, L.; Wang, Z.w. Study of the formation and dynamic characteristics of the vortex in the pump sump by CFD and experiment. *J. Hydrodyn.* **2021**, *33*, 1202–1215. [CrossRef]
10. Luo, Y.; Chen, F.; Chen, L.; Wang, Z.; Yu, J.; Zhu, X.; Zhao, Z.; Ren, S.; Li, J.; Lu, X. Study on stresses of head cover bolts in a pump turbine based on FSI. In *IOP Conference Series: Earth and Environmental Science*; IOP Publishing: Bristol, UK, 2021; Volume 804, p. 042062.
11. Luo, Y.; Chen, F.; Chen, L.; Wang, Z.; Yu, J.; Luo, C.; Zhao, Z.; Ren, S.; Li, J.; Deng, D. Stresses and relative stiffness of the head cover bolts in a pump turbine. In *IOP Conference Series: Materials Science and Engineering*; IOP Publishing: Bristol, UK, 2019; Volume 493, p. 012113.
12. Chen, F.; Chen, L.; Wang, Z.; Yu, J.; Luo, C.; Zhao, Z.; Ren, S.; Li, J.; Deng, D. Computation of static stresses of the head cover bolts in a pump turbine. In *IOP Conference Series: Materials Science and Engineering*; IOP Publishing: Bristol, UK, 2019; Volume 493, p. 012143.
13. Chen, L.; Li, H.; Yu, J.; Luo, Y.; Wang, Z.; Zhu, X.; Zhao, Z.; Lu, X. Stress analysis of screw connection of key structural components in pump turbine. In *IOP Conference Series: Earth and Environmental Science*; IOP Publishing: Bristol, UK, 2021; Volume 804, p. 032037.
14. Brekke, H. Performance and safety of hydraulic turbines. In *IOP Conference Series: Earth and Environmental Science*; IOP Publishing: Bristol, UK, 2010; Volume 12, p. 012061.
15. Huang, X.; Chamberland-Lauzon, J.; Oram, C.; Klopfer, A.; Ruchonnet, N. Fatigue analyses of the prototype Francis runners based on site measurements and simulations. In *IOP Conference Series: Earth and Environmental Science*; IOP Publishing: Bristol, UK, 2014; Volume 22, p. 012014.
16. He, Q.; Huang, X.; Yang, M.; Yang, H.; Bi, H.; Wang, Z. Fluid–Structure Coupling Analysis of the Stationary Structures of a Prototype Pump Turbine during Load Rejection. *Energies* **2022**, *15*, 3764. [CrossRef]
17. He, L.; Wang, Z.; Kurosawa, S.; Nakahara, Y. Resonance investigation of pump-turbine during startup process. In *IOP Conference Series: Earth and Environmental Science*; IOP Publishing: Bristol, UK, 2014; Volume 22, p. 032024.
18. Kolšek, T.; Duhovnik, J.; Bergant, A. Simulation of unsteady flow and runner rotation during shut-down of an axial water turbine. *J. Hydraul. Res.* **2006**, *44*, 129–137. [CrossRef]
19. Ciocan, G.D.; Iliescu, M.S.; Vu, T.C.; Nennemann, B.; Avellan, F. Experimental study and numerical simulation of the FLINDT draft tube rotating vortex. *J. Fluids Eng.* **2007**, *129*, 146–158. [CrossRef]
20. Huang, X.; Oram, C.; Sick, M. Static and dynamic stress analyses of the prototype high head Francis runner based on site measurement. In *IOP Conference Series: Earth and Environmental Science*; IOP Publishing: Bristol, UK, 2014; Volume 22, p. 032052.
21. Goyal, R.; Cervantes, M.J.; Gandhi, B.K. Characteristics of Synchronous and Asynchronous modes of fluctuations in Francis turbine draft tube during load variation. *Int. J. Fluid Mach. Syst.* **2017**, *10*, 164–175. [CrossRef]
22. Mao, Z.; Tao, R.; Chen, F.; Bi, H.; Cao, J.; Luo, Y.; Fan, H.; Wang, Z. Investigation of the starting-up axial hydraulic force and structure characteristics of pump turbine in pump mode. *J. Mar. Sci. Eng.* **2021**, *9*, 158. [CrossRef]
23. Münch, C.; Ausoni, P.; Braun, O.; Farhat, M.; Avellan, F. Fluid–structure coupling for an oscillating hydrofoil. *J. Fluids Struct.* **2010**, *26*, 1018–1033. [CrossRef]
24. Benra, F.K.; Dohmen, H.J. Comparison of pump impeller orbit curves obtained by measurement and FSI simulation. In *Proceedings of the ASME Pressure Vessels and Piping Conference, San Antonio, TX, USA, 22–26 July 2007*; Volume 42827, pp. 41–48.
25. Kato, C.; Yoshimura, S.; Yamade, Y.; Jiang, Y.Y.; Wang, H.; Imai, R.; Katsura, H.; Yoshida, T.; Takano, Y. Prediction of the noise from a multi-stage centrifugal pump. *Fluids Engineering Division Summer Meeting, Houston, TX, USA, 19–23 June 2005*; Volume 41987, pp. 1273–1280.
26. Jiang, Y.; Yoshimura, S.; Imai, R.; Katsura, H.; Yoshida, T.; Kato, C. Quantitative evaluation of flow-induced structural vibration and noise in turbomachinery by full-scale weakly coupled simulation. *J. Fluids Struct.* **2007**, *23*, 531–544. [CrossRef]
27. Mandair, S.; Morissette, J.; Magnan, R.; Karney, B. MOC-CFD coupled model of load rejection in hydropower station. In *IOP Conference Series: Earth and Environmental Science*; IOP Publishing: Bristol, UK, 2021; Volume 774, p. 012021.
28. Zhou, D.; Chen, H.; Kan, K.; Yu, A.; Binama, M.; Chen, Y. Experimental study on load rejection process of a model tubular turbine. In *IOP Conference Series: Earth and Environmental Science*; IOP Publishing: Bristol, UK, 2021; Volume 774, p. 012036.
29. Fu, X.; Li, D.; Wang, H.; Zhang, G.; Li, Z.; Wei, X. Dynamic instability of a pump-turbine in load rejection transient process. *Sci. China Technol. Sci.* **2018**, *61*, 1765–1775. [CrossRef]

 *energies*

Article

Research on the HHT-Based Analysis Method of Tidal Power Generation Power Distribution Law [†]

Yuanfeng Huang, Yuqi Liu *, Yani Ouyang and Haifeng Wang

Institute of Electrical Engineering, Chinese Academy of Sciences, Beijing 100190, China
* Correspondence: liuyuqi@mail.iee.ac.cn
† This paper is an extended version of our paper published—In Proceedings of the E3S Web of Conferences, Odesa, Ukraine, 16 April 2021.

Abstract: Tidal power generation technology has advanced quickly in recent years. In this study, the Hilbert-Huang transform (HHT) was used to examine the electrical energy distribution law of tidal power generation according to the time periods of days, months, and years based on observed data for tidal power generation. Our analysis summarized the tidal power generation law as follows: in the span of one day, the motor ran for 14 h before shutting off for 10 h. The frequency was 28 Hz, and the maximum voltage was 259 V. The tidal power generation swung twice a month, reaching its peak in the middle of the month and its trough at the beginning. The tidal power output fluctuated twice a year, reaching its peak in August, September, and October and its trough in February, March, and April. In the investigation of tidal power generation patterns, the HHT transformation's precision and potency were confirmed.

Keywords: tidal power generation; the law of electrical energy distribution; Hilbert-Huang Transform (HHT); empirical mode decomposition (EMD); intrinsic mode functions (IMF)

Citation: Huang, Y.; Liu, Y.; Ouyang, Y.; Wang, H. Research on the HHT-Based Analysis Method of Tidal Power Generation Power Distribution Law. *Energies* 2022, 15, 9494. https://doi.org/10.3390/en15249494

Academic Editors: Yongguang Cheng and Zhengwei Wang

Received: 16 November 2022
Revised: 9 December 2022
Accepted: 12 December 2022
Published: 14 December 2022

Publisher's Note: MDPI stays neutral with regard to jurisdictional claims in published maps and institutional affiliations.

1. Introduction

The widespread use of coal, oil, and natural gas in recent years has resulted in numerous issues, including the greenhouse effect, environmental degradation, and ecological devastation, which have presented severe dangers to the existence and advancement of humans worldwide [1,2]. The aforementioned environmental issues can be resolved by substituting renewable energy for fossil energy. Ocean energy, wind energy, and solar energy are currently the three main new energy sources accessible to humans. Wave energy, tidal power, salinity gradient energy, and offshore wind energy are some of them. Ocean energy is currently the focus of new energy research.

Tidal power generation has always been a popular area of research both domestically and internationally as a type of innovative marine energy generation. We can only more effectively direct the development and usage of tidal power generation by explaining the electric energy distribution legislation of its generation. The analysis approach of the tidal power electric energy distribution law is investigated with a focus on the unique electric-energy-generation features of the tidal power generation system. By concentrating on the periodic features of tidal power generation, it is possible to more accurately forecast and analyze the tidal quality and give correct data references to direct the application of tidal power generation equipment testing.

In the contemporary study and analysis of electrical energy laws, the Fourier transform, the Wavelet transform, and the Stockwell transform (ST) are frequently utilized. When analyzing steady signals, the Fourier transform is typically used, and both its fast Fourier transform and extended discrete Fourier transform create phase-detection errors [3,4]. The window function of the ST cannot be a root function. It is adjusted to the actual needs of specific projects, lacking generality. Additionally, it is computationally intensive [5,6].

Although the Wavelet transform is suitable for the analysis of nonstationary signals, the selection of wavelet bases is complicated [7]. The Hilbert-Huang transform (HHT), a novel nonlinear non-smooth signal analysis technique, is adaptive, and its basic function does not need to be predetermined, which provides good data-processing flexibility and circumvents the drawback of the Wavelet transform's preset basis function [8].

HHT has been extensively used in numerous fields as a mature analysis algorithm. In preliminary research, researchers in [9] investigated the application of a novel automatic segmented HHT (SHHT) method for evaluating the SEC of the voltage sag caused by different types of symmetrical and unsymmetrical faults in both transmission and distribution networks. Reference [10] proposes the application of the HHT method for the decomposition of the PQ data into their individual frequency components to separate the nonstationary voltage sag waveform containing the fundamental frequency component. In reference [11], the feasibility of HHT for fault detection in VSC-based high-voltage direct current systems is analyzed. Reference [12] proposes the use of HHT and CNN for the classification of PS alterations. The HHT is used to obtain the time-frequency spectrum of the PSs that are signals characterized by a time-varying trend. Reference [13] proposes a fault-diagnosis method based on the combination of the Hilbert-Huang transform and convolutional neural network for the fault identification of transmission lines in distribution networks containing distributed power sources. It was discovered that HHT is very suitable for the nonlinear and non-smooth stochastic characteristics of the output electric energy of tidal power generation, even though both domestic and foreign scholars have largely ignored its use in the analysis of the electric energy law of tidal power generation.

In this article, we introduce a novel concept for employing HHT to study the distribution law of tidal power generation. Based on the measured tidal power data, the analysis first examines the start-stop, sharp acceleration, and sharp deceleration of the tidal generation system over the course of one day and then summarizes the transformation law of tidal generation over the course of 24 h. To determine the monthly and yearly distribution rules, the Hilbert spectra of the tidal power generation after 30 days and 12 months are then calculated using HHT. The Hilbert spectra shows the trend of the tidal power generating system over one month and one year. Finally, to confirm the precision and efficiency of HHT analysis in the investigation of tidal power generation patterns, the monthly and annual distribution curves are compared with the observed data.

2. Materials and Methods

Tidal power is the kinetic energy created by the seawater's regular flow due to the Sun and Moon's changing gravitational pull on the water. As seen in Figure 1a, when the Sun, Moon, and Earth are in a straight line, the gravitational pulls of the Moon and Sun are overlaid, causing greater tides known as spring tides. A neap tide is generated when the Sun, Moon, and Earth are at an angle, canceling out the gravitational forces that would otherwise result in higher tides Figure 1b. On days with spring tides, the tidal current velocity is higher, and the captured and output power of the system generating tidal power are greater; on days with neap tides, the tidal current velocity is lower, and the system's captured and output power are consequently lower [14].

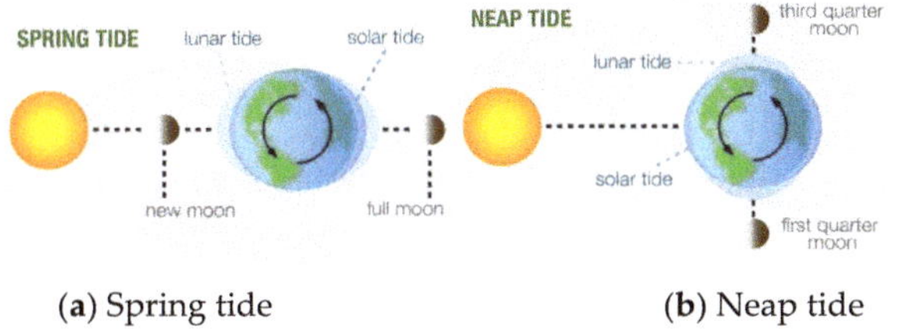

(**a**) Spring tide   (**b**) Neap tide

Figure 1. Relative position of the Earth, Moon and Sun at high and low tides.

The tidal power generation platform "Zhongke Hydropower No. 1" built by the Institute of Electrical Engineering of the Chinese Academy of Sciences in Zhoushan, Zhejiang Province, is shown in Figure 2. A maximum of 20 kW of electricity can be generated by the platform, which uses two tidal power generation devices to harness the tidal power of nearby seas in real-time. The platform for generating tidal power is also outfitted with a full tidal current detection system, which can track and record characteristics, including tidal current speed, flow direction, output voltage, current, and power in real-time.

Figure 2. "Zhongke Hydropower No. 1" tidal power generation platform.

The platform, which costs roughly 200,000 RMB, underwent a yearlong test in the water off Zhoushan, Zhejiang Province, China, between 2016 and 2017. The generated electricity was cleaned and delivered to the islands' electrically dependent machinery. According to estimates, power costs roughly RMB 2 per unit on average.

On spring tide days, the platform records the more typical daily tidal energy generation data obtained, as shown in Figure 3.

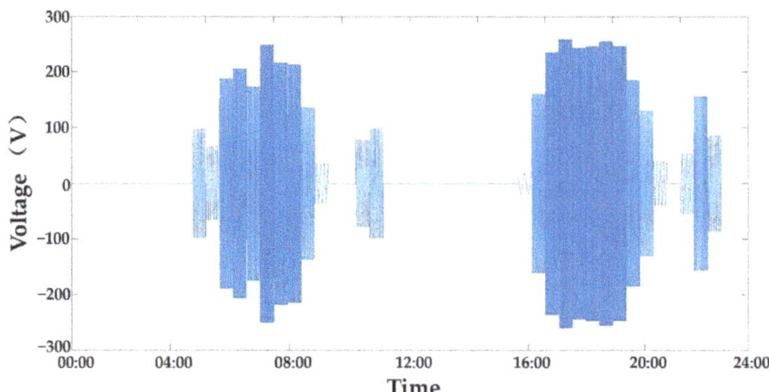

Figure 3. Daily variation pattern of tidal power generation electricity.

It is clear in Figure 3 that over the 24 h of a high-tide day, the output power of tidal power encounters four peak power-generation processes: high tide, low tide, high tide, and low tide. Different seawater flow rate circumstances also cause proportional changes in the motor's output voltage and power. Seawater moves through the tidal power plant at roughly 0.4 m/s. When the seawater velocity reaches its first peak, the motor's highest output voltage is roughly 250 V; when it falls below a particular point, the motor shuts off. After that, the direction of the seawater flow shifts, the tidal velocity steadily increases, and the tidal power generator set also resumes operating from a halt.

In conclusion, the output power of tidal power generation systems similarly exhibits regular periodic fluctuations along with the periodic changes in the seawater flow rate. The efficiency of tidal power generation at the load side or the grid connection are both directly impacted by this transition, which occurs with some regularity. Studying the power distribution law of tidal power generation is crucial for this reason.

3. Hilbert-Huang Transform

On nonlinear and nonstationary data, Hilbert-Huang transform (HHT) is used to perform analysis and processing more effectively [15]. To obtain the intrinsic mode function (IMF) and the Hilbert spectrum and time-frequency energy spectrum of the signal for signal analysis, HHT first applies empirical mode decomposition (EMD) to the signal to generate the IMF.

3.1. Empirical Mode Decomposition (EMD)

Empirical mode decomposition adaptively decomposes any complex signal into a series of intrinsic mode functions (IMF) based on the signal characteristics. It satisfies the following two conditions:

- The number of extreme value points of the signal is equal to zero points or differs by one;
- The local mean of the upper envelope defined by the extreme value and the lower envelope defined by the extreme minima of the signal is zero.

The EMD process is as follows:

- For the input signal $x(t)$, find the maximum value point $x(t_i)$ and the minimum value point $x(t_j)$;
- Construct the upper and lower envelopes of the signal by interpolating the third spline function for the maximum and minimum points, and calculate the mean value function $xl(t)$ for the upper and lower envelopes;
- Examine whether $xl(t)$ satisfies the IMF condition, if it does, go to the next step, otherwise perform the first two steps on xl until the kth step satisfies the IMF condition, and obtain the first IMF component, noted as c_1;
- Get the first residual $r_1 = x - c_1$, operate on r_1 as in the above three steps to get c_2, and so on;
- Until r_n is a monotonic signal or there is only one pole.

Eventually, the original signal [16] is expressed as:

$$x = \sum_{i=1}^{n} c_i + r_n \tag{1}$$

3.2. Hilbert Spectrum Analysis

Suppose $x(t)$ is an arbitrary signal and the Hilbert transform of $x(t)$ is:

$$y(t) = \frac{1}{\pi} \int_{-\infty}^{+\infty} \frac{x(\tau)}{t - \tau} d\tau \tag{2}$$

Its Hilbert inverse transform is:

$$x(t) = \frac{1}{\pi} \int_{-\infty}^{+\infty} \frac{y(\tau)}{\tau - t} d\tau \tag{3}$$

Obtain the resolved signal:

$$z(t) = x(t) + iy(t) = a(t)e^{j\varnothing(t)} \tag{4}$$

where: $a(t)$ is the instantaneous amplitude and $\varnothing(t)$ is the phase.

$$a(t) = \sqrt{x^2(t) + y^2(t)} \tag{5}$$

$$\varnothing(t) = \arctan\frac{y(t)}{x(t)} \tag{6}$$

The instantaneous frequency is:

$$f(t) = \frac{1}{2\pi}\frac{d\varnothing(t)}{dt} \tag{7}$$

Equations (5) and (7) are expressions for instantaneous amplitude and instantaneous frequency, which are obtained from a simple modulation and demodulation process.

The voltage and frequency of the power generation fluctuate from moment to moment due to the energy source's waveform nature. The HHT's measurements of instantaneous voltage and frequency clearly show the variable features.

4. Analysis of Tidal Power Generation Data

4.1. Analysis of Daily Distribution Pattern of Electricity Generated by Tidal Power

The output power data of the tidal energy generator set throughout the day are shown in Figure 3. The output voltage of the motor changes throughout the day in Figure 3. Voltage surges and drops occur throughout the generator operation, and voltage interruptions occur during motor stalls. Next, the EMD decomposition of the signal in Figure 3 is shown. The first three IMFs were taken for analysis, as shown in Figure 4. It can be seen in Figure 4 that the signal's symmetry causes the signal's primary modes to be concentrated in IMF1. A residual function that is monotonically decreasing is the outcome of the EMD decomposition.

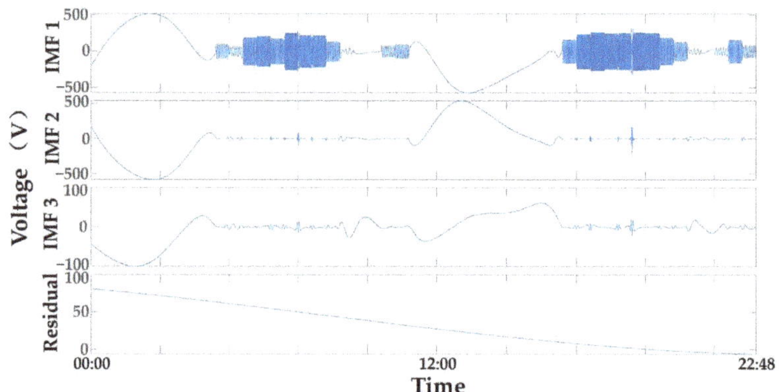

Figure 4. The EMD decomposition of the daily generation waveform of tidal power.

The Hilbert spectrum of the all-day tidal energy generation data is shown in Figure 5.

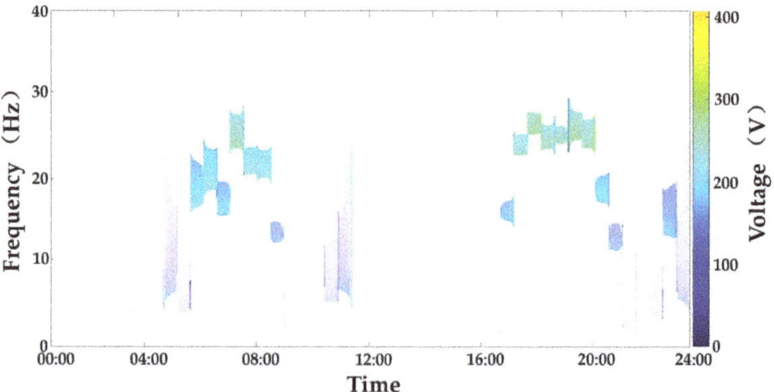

Figure 5. Hilbert spectrum of the daily generation waveform of tidal power.

In the Hilbert spectrum, it can be seen that:

- There is no curve from 0:00–4:30, indicating that the voltage and frequency are zero. The motor is at a standstill, with a standstill time of 4.5 h;
- From 4:30 to 5:00, the curve appears in dark blue color. Compared with the color bar on the right side, the amplitude is small. The curve of the spectrogram is wider, and the middle value is taken as its frequency, which is lower at this time. After that, the curve in the spectrogram of the signal shows an upward trend, indicating that the frequency increases. The color changes from dark blue to light blue, compared with the right color bar, indicating that the amplitude increases;
- From 7:00 to 7:30, the amplitude and frequency reach the first peak of 248.9 V and 26.6 Hz, after which the curve in the spectrum begins to decrease, which indicates a decrease in frequency. The color began to deepen from light blue to dark blue, suggesting a decrease in amplitude;
- From 9:30–10:30, the curve disappears, and the voltage decreases to zero at 9:30, and the motor stops for 1 h;
- From 10:30 to 11:30, the curve reappears and the motor starts again and runs for an hour;
- From 11:30–15:30, the curve disappears, and the voltage decreases to zero at 11:30. Moreover, the motor stops for 4 h;
- From 15:30, the curve appears again in dark blue, with a smaller amplitude and lower frequency. After, the curve height rises rapidly and the color changes from dark blue to light blue again, suggesting that the signal amplitude and frequency increase. The motor speed speeds up, and the output voltage increases;
- From 18:00 to 18:30, the amplitude and frequency reach the second peak with an amplitude of 258.7 V and a frequency of 27.7 Hz. After, the curve in the spectrum is maintained at a high level for a while and then begins to decline, with the color gradually changing from light blue to dark blue and the output voltage gradually decreasing until 22:00, when the voltage reduces to zero and the motor stops;
- From 22:00–22:30, the motor stops for half an hour, and the spectrum curve reappears at 22:30, indicating that the motor starts;
- From 22:30 to 24:00, the curve rises and falls, and the color changes from dark to light, then light to dark. The voltage and frequency rise and then fall.

In conclusion, the motor ran for 14 h and then idled for 10 h. The highest voltage was 259 V, and the frequency was 28 Hz, as can be seen in the Hilbert spectrum, which shows that the duration of the motor operating for a day, the duration of halting, and the duration of the maximum voltage appearing are consistent with Figure 3.

4.2. Analysis of Monthly Tidal Power Generation Electricity-Distribution Pattern

The distribution of electrical energy measured by the tidal energy generation system in September 2016 is shown in Figure 6.

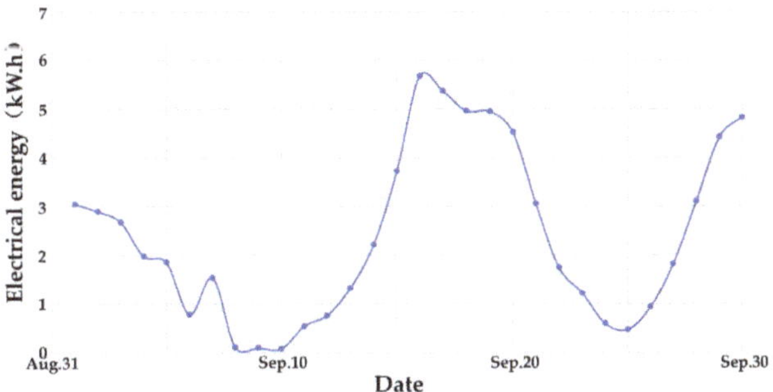

Figure 6. Monthly distribution of tidal power generation.

The maximum power-generating point occurred on 16 September, the big tidal day, when the daily power generation was above 2 kW·h for more than half the month. Power generation increased over a five-day span from 16 September to 20 September, and it also increased at the start and end of the month. Power generation decreased from 8 September to 10 September and on 24–25 September, respectively.

With the results in Figure 6, we generated the graph of the monthly variation of tidal power generation electricity by calculating the daily average voltage of tidal power generation, as shown in Figure 7.

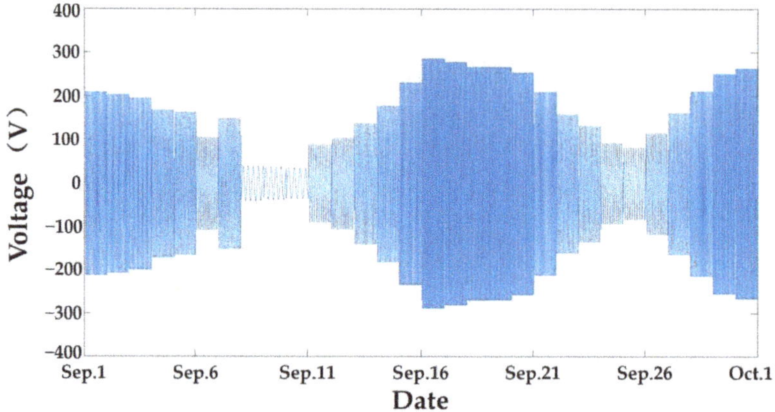

Figure 7. Monthly variation pattern of tidal power generation electricity.

The EMD decomposition waveform of the tidal power monthly generation waveform is shown in Figure 8. The first three IMF signals are taken for analysis. Due to the symmetry of the signal, most of its modes are in IMF1, and the decomposition of the signal's abrupt moments is also mirrored in IMF2 and IMF3.

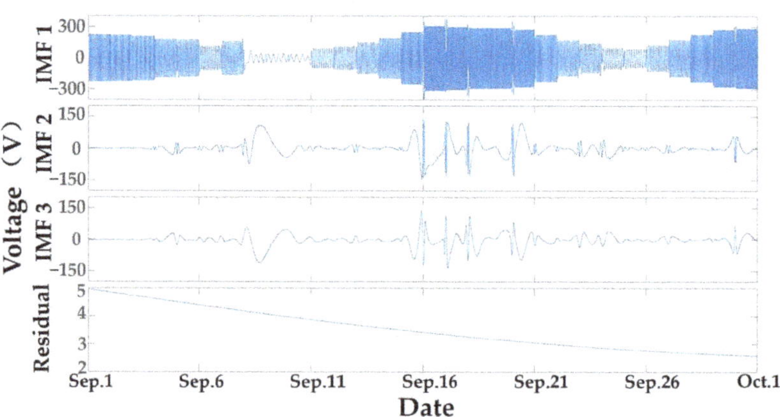

Figure 8. EMD decomposition of the monthly generation waveform of tidal power.

The Hilbert spectrum of the monthly generation waveform of tidal energy is shown in Figure 9.

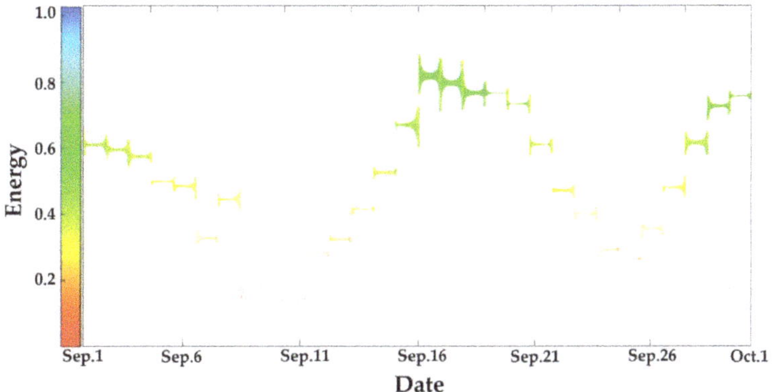

Figure 9. Hilbert spectrum of the monthly generation waveform of tidal power.

After normalizing it, it can be seen that:

- From 1 September to 10 September, the curve gradually moves from light green to yellow to red as it oscillates downward, showing a decline in energy and a subsequent decline in daily power generation;
- From 10 September to 16 September, the color gradually shifts from red to yellow to light green as the curve climbs quickly, signifying an increase in energy and a quickening of daily power generation;
- On 16 September, the energy of the curve reaches its highest point. With a daily power generation of 5.69 kW·h, the 16th is the month's highest power-generating day;
- From 16 September to 24 September, the color gradually shifts from light green to yellow and then from yellow to red as the curve steadily declines, showing a gradual decline in energy before a sudden decline in daily power generation;
- From 24 September to 30 September, the curve rises rapidly, with the color gradually changing from red to yellow and then from yellow to light green, indicating a rise in energy and a rapid increase in power generation.

In summary, the HHT transformation's distribution pattern of tidal power generation is compatible with the observed data shown in Figure 6.

4.3. Analysis of the Annual Distribution Pattern of Tidal Power Generation Electricity

The yearly distribution pattern of the system is shown in Figure 10 based on the measured power generation data of the tidal power generating system in 2016–2017, where the horizontal coordinate corresponds to the current month, and the vertical coordinate represents the system's monthly power generation.

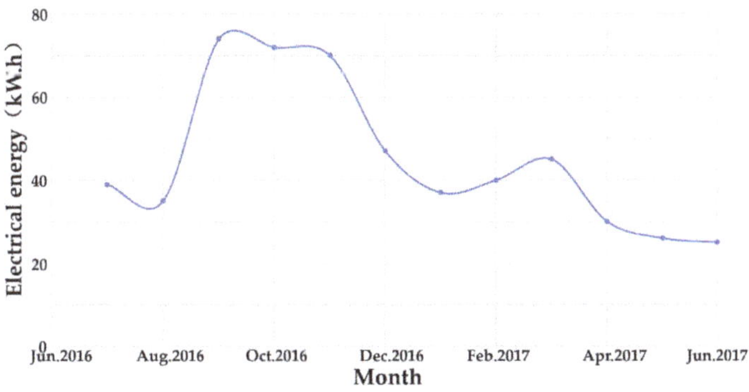

Figure 10. Monthly distribution of tidal power generation (2016–2017).

According to Figure 10, the power generated by tidal current energy generators is higher in September, October, and November, with monthly power generation of over 70 kW·h; it is lower in April, May, and June, with monthly power generation of less than 50 kW·h; and it is essentially between 35 kW·h and 50 kW·h in the remaining months.

The Hilbert-Huang transform and EMD decomposition were then applied to the signal in Figure 11 to generate the corresponding results shown in Figures 12 and 13. The average voltage of tidal power generation was used as the input for this HHT process to obtain its annual voltage distribution pattern (as shown in Figure 11).

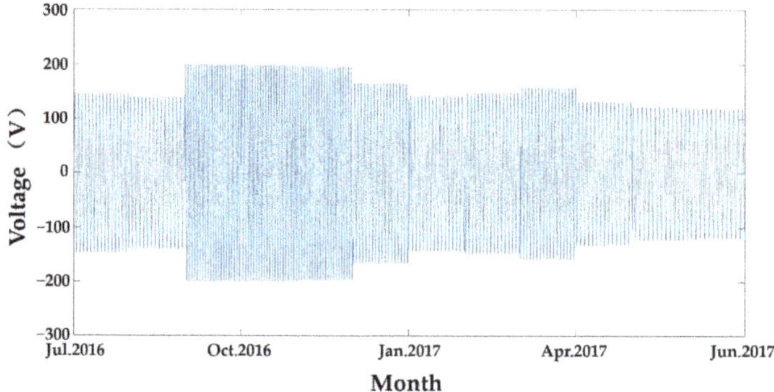

Figure 11. Annual variation pattern of electricity generated by tidal power.

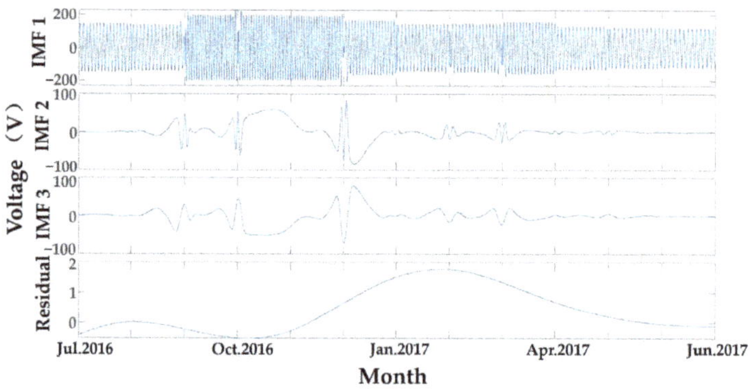

Figure 12. EMD decomposition of the annual generation waveform of tidal power.

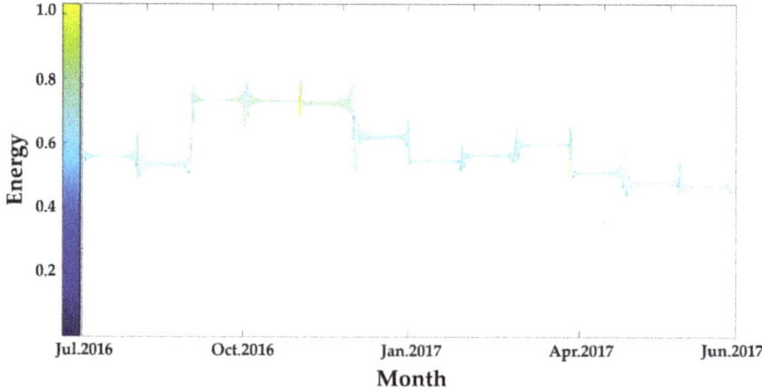

Figure 13. Hilbert spectrum of the annual generation waveform of tidal power.

According to Figure 13:

- From July 2016 to September 2016, the curve oscillates upward and gradually changes color from dark blue to light blue, indicating a sharp increase in energy and a sharp increase in power generation. The curve reaches its highest point in September, indicating the maximum energy and the most power generation in that month;
- From September 2016 to December 2016, the curve decreases slowly and then rapidly, and the color changes from light blue to dark blue, indicating that the energy decreases slowly and then rapidly;
- From January 2017 to March 2017, the curve increases slowly with a dark blue color and gradually becomes lighter, demonstrating an increase in energy and a gradual increase in power generation;
- From March 2017 to June 2017, the curve decreases with a dark blue and gradually deepening color, indicating a decrease in energy and a gradual decrease in power generation.

In conclusion, the distribution law of tidal power generation discovered by the HHT is also compatible with the observed data shown in Figure 10, demonstrating the viability and efficiency of the HHT for studying the distribution law of tidal power generation.

5. Conclusions

This paper proposes a concept for analyzing the electric energy distribution law of tidal power generation by using HHT transformation to process and analyze the measured data of tidal power generation. This study eventually obtained the electric energy distribution law of tidal power generation in 24 h, 30 days, and 12 months as follows:

1. Within 24 h, the tidal power generation started and stopped more rapidly, and the voltage varied greatly within a short period of time. The tidal power generation time was mainly divided into four time periods, corresponding to two high tide and two low tide processes;
2. There were two fluctuations in tidal energy generation within 30 days, with the peak position located in the middle of the month and the trough position at the beginning of the month;
3. There were also two fluctuations in tidal power output during the 12-month period, with peaks in August, September and October and troughs in February, March and April.

A full verification of the accuracy and efficacy of the application of HHT analysis in the study of tidal power generation law was provided by the fact that the distribution law of tidal power generation electricity obtained using the HHT in this paper was essentially consistent with the measured tidal power generation data of the offshore platform. In addition, the following issues were considered in this report and require more study in follow-up work:

1. The spectrum maps obtained in this paper sometimes show curve oscillations. In further studies, the image can be deeply optimized by optimizing data selection by selecting the sampling frequency;
2. This paper focuses on verifying the accuracy and effectiveness of applying HHT analysis in studying the law of tidal power generation. A future step can be to consider the effectiveness and feasibility of other algorithms, such as wavelet and Fourier applied in this field and compare and analyze the results with those of this paper;
3. Additionally, further studies also can apply the ideas of this paper to the analysis of the electrical energy law or power quality of other ocean energy power generation systems, such as wave energy, and summarize the characteristics of different forms of ocean energy power generation to guide the application of ocean energy power generation.

Author Contributions: Conceptualization, Y.H. and Y.L.; methodology, Y.H.; software, Y.O.; validation, Y.H., Y.L. and Y.O.; formal analysis, Y.H.; investigation, Y.L.; resources, Y.H.; data curation, Y.O.; writing—original draft preparation, Y.H.; writing—review and editing, Y.L.; visualization, Y.O.; supervision, H.W.; project administration, H.W.; funding acquisition, H.W. All authors have read and agreed to the published version of the manuscript.

Funding: This research was funded by [National Key Research and Development Project] grant number [2018YFB1501905-4], [the Institute of Electrical Engineering, CAS] grant number [E1551502] and [National Natural Science Foundation of China] grant number [51877204].

Conflicts of Interest: The authors declare no conflict of interest.

References

1. Zhou, L.; Han, Z.; Ren, J.; Li, J.; Zhang, D.; Feng, Z.; Zhang, C. Current situation and changing characteristics of oil and natural gas proved reserves in China. *China Min. Mag.* **2019**, *28*, 6–11.
2. Alvarez, G.E.; Marcovecchio, M.G.; Aguirre, P.A. Optimization of the integration among traditional fossil fuels, clean energies, renewable sources, and energy storages: An MILP model for the coupled electric power, hydraulic, and natural gas systems. *Comp. Ind. Eng.* **2020**, *139*, 106141. [CrossRef]
3. Wen, H.; Teng, Z.; Wang, Y.; Zeng, B.; Zheng, D. Improved Windowed Interpolation FFT Algorithm and Application for Power Harmonic Analysis. *Trans. China Electrotech. Soc.* **2012**, *27*, 270–277.
4. Xu, Y.; Zhao, Y. Identification of Power Quality Disturbance Based on Short-Term Fourier Transform and Disturbance Time Orientation by Singular Value Decomposition. *Power Syst. Technol.* **2011**, *35*, 174–180.
5. Yang, H.; Liu, S.; Xiao, X.; Deng, W.; Chen, D. S-transform-based Expert System for Classification of Voltage Dips. *Proc. CSEE* **2007**, *27*, 98–104.
6. Li-Ping, Q.; Chang-Long, H.; Jie, Z.; Soc, I.C. Research Summary of Power Quality Disturbance Detection and Classification Recognition Method Based on Transform Domain. In Proceedings of the 19th International Symposium on Distributed Computing and Applications for Business Engineering and Science (DCABES), Xuzhou, China, 16–19 October 2020; pp. 50–53.
7. Yan, Z.; Hao, C.; Chen, M.; Yang, T.; Zheng, L. Power quality disturbance detection and localization based on orthogonal wavelet packet algorithm. *Technol. Innov. Appl.* **2019**, *15*, 16–17.
8. Gao, J.; Zhu, S. Detection Method of Power Quality in ADN Based on Improved Hilbert-Huang Transform. *Northeast Electr. Power Technol.* **2019**, *40*, 31–34.
9. Hasan, S.; Muttaqi, K.M.; Sutanto, D. Application of the Automatic Segmented Hilbert Huang Transform Method for the Evaluation of the Single-Event Characteristics of Voltage Sags in Power Systems. *IEEE Trans. Ind. Appl.* **2021**, *57*, 1882–1891. [CrossRef]
10. Hasan, S.; Muttaqi, K.M.; Sutanto, D. Detection and Characterization of Time-Variant Nonstationary Voltage Sag Waveforms Using Segmented Hilbert-Huang Transform. *IEEE Trans. Ind. Appl.* **2020**, *56*, 4563–4574. [CrossRef]
11. Li, D.; Ukil, A.; Satpathi, K.; Yeap, Y.M. HilbertHuang Transform Based Transient Analysis in Voltage Source Converter Interfaced Direct Current System. *IEEE Trans. Ind. Electron.* **2021**, *68*, 11014–11025. [CrossRef]
12. Carni, D.L.; Lamonaca, F. Toward an Automatic Power Quality Measurement System: An Effective Classifier of Power Signal Alterations. *IEEE Trans. Instrum. Meas.* **2022**, *71*, 2514208. [CrossRef]
13. Sun, X.; Guan, H.; Zhao, Y.; Zhang, T.; Wane, Y.; Yang, B. Research on Fault Diagnosis Method of Distributed Power Distribution Network Based on HHT and CNN. In Proceedings of the 33rd Chinese Control and Decision Conference (CCDC), Kunming, China, 22–24 May 2021; pp. 129–133.
14. Huang, Y.; Wang, Y.; Cao, X.; Wang, H. Analysis and research on power distribution law of power generation based on HHT transform. In Proceedings of the E3S Web of Conferences, Odesa, Ukraine, 16 April 2021.
15. Wang, K.; Wang, H.; Huang, Y.; Wang, Y. Power quality analysis of tidal current energy generation based on HHT transform. In Proceedings of the 23rd International Conference on Electrical Machines and Systems (ICEMS), Electr Network, Hamamatsu, Japan, 24–27 November 2020; pp. 997–1001.
16. Wang, Q.; Wang, T.; Sheng, J.; Zhang, X.; Shi, G. The power quality disturbance detection and classification method based on EMD. In Proceedings of the IEEE PES Innovative Smart Grid Technologies, Tianjin, China, 21–24 May 2012; pp. 1–3.

Article

Multi-Objective Optimal Long-Term Operation of Cascade Hydropower for Multi-Market Portfolio and Energy Stored at End of Year

Haojianxiong Yu [1], Jianjian Shen [1,*], Chuntian Cheng [1], Jia Lu [2] and Huaxiang Cai [3]

[1]   Institute of Hydropower & Hydro informatics, Dalian University of Technology, Dalian 116024, China
[2]   China Yangtze Power Co., Ltd., Yichang 443133, China
[3]   Kunming Power Exchange Center, Kunming 650011, China
*   Correspondence: shenjj@dlut.edu.cn

Abstract: Taking into account both market benefits and power grid demand is one of the main challenges for cascade hydropower stations trading on electricity markets and secluding operation plan. This study develops a multi-objective optimal operation model for the long-term operation of cascade hydropower in different markets. Two objectives were formulated for economics benefits and carryover energy storage. One was to maximize the market utility value based on portfolio theory, for which conditional value at risk (CVaR) was applied to measure the risk of multi-markets. Another was the maximization of energy storage at the end of a year. The model was solved efficiently through a multi-objective particle swarm optimization (MOPSO). Under the framework of the MOPSO, the chaotic mutation search mechanism and elite individual retention mechanism were introduced to diversify and accelerate the non-inferior solution sets. Lastly, a dynamic updating of archives was established to collect the non-inferior solution. The proposed model was implemented on the hydropower plants on the Lancang River, which traded on the Yunnan Electricity Market (YEM). The results demonstrated non-inferior solution sets in wet, normal and dry years. A comparison in solution sets revealed an imbalanced mutual restriction between objectives, such that a 2.65 billion CNY increase in market utility costs a 13.96 billion kWh decrease in energy storage. In addition, the non-inferior solution provided various schemes for actual demands based on other evaluating indexes such as the minimum output, power generation and spillage.

Keywords: electricity market; mean-CVaR; cascade dispatching; MOPSO

Citation: Yu, H.; Shen, J.; Cheng, C.; Lu, J.; Cai, H. Multi-Objective Optimal Long-Term Operation of Cascade Hydropower for Multi-Market Portfolio and Energy Stored at End of Year. *Energies* **2023**, *16*, 604. https://doi.org/10.3390/en16020604

Academic Editor: K. T. Chau

Received: 3 November 2022
Revised: 5 December 2022
Accepted: 27 December 2022
Published: 4 January 2023

1. Introduction

Since the new round of electricity from China in 2015, lots of regions and provinces have established electricity markets and exchange centers. At the end of 2021, the market trading of electricity included 3778.74 billion kWh, accounting for 45.5% of the total social electricity consumption. With the gradual liberalization of the power market, a way to promote power generation consumption through market trading was provided for hydropower stations. Especially in Southwest China, the hydropower-rich provinces, such as Yunnan, Sichuan, Guizhou, etc., have successively carried out electric power reform pilot projects. As the main sending sources for the West-to-East Electricity Transmission (WEET) project, these hydropower plants face multi-party demands besides their own market benefits need: flood control, river navigation and water spillage [1–3]. These demands require large carryover storage of cascades. The carryover energy storage at the end of a year is an important state to reflect these demands in the long-term operation of cascade hydropower [4,5]. Under such circumstances, considering both the overall benefits in different markets and the cascade carryover storage in optimizing the generation scheduling for cascades is a practical problem to be solved by the hydropower enterprises. However, this problem becomes much more complicated due to the uncertainty of electricity prices, while market risks should be taken into account. In addition, the hydraulic and market constraints of the cascade hydropower plants increase the complexity further [6].

Despite the benefits of efficiency improvements for power system deregulation, the introduction of electricity markets can also bring about many risks in power trading. Many studies have focused on risk management of the electricity market due to the volatility of market prices. Early studies mainly combined the future market with the node market, and used forward contract with the short-term market to avoid market risks caused by price fluctuations [7,8]. Decision analysis and Monte Carlo simulation have been used to find the optimal combination between markets under economic returns [9,10]. In recent years, with the participation of new energy and increase in power market categories, methods of considering multi-market portfolio optimization have gradually received lots of research attention using two different methods: the mean-variance (MV)-based method, using, for example, the modern portfolio model (MPT) [11–13]; and the conditional value at risk (CVaR)-based method [14–17]. Mean-variance models penalize risk in the objective function, where the measure of risk is the variance of profit, and CVaR models use their own risk definition based on the probability of reaching a minimum profit [12]. Yan et al. [11] a novel portfolio theory-based approach is proposed for optimally managing ES in various markets to maximize benefits and reduce the risk for ES owners. Garcia et al. [12] proposed MPT models which were combined with a generalized autoregressive conditional heteroscedastic (GARCH) prediction technique for a generation company to optimally diversify their energy portfolio. Liu and Wu [13] provided an optimization strategy for power generation companies' energy allocation based on an MPT quantitative analysis and mathematical derivation of risk-free and risky transactions in the PJM market. Zhao et al. [14] proposed a dynamic multi-stage optimization configuration model for electricity assets based on the conditional value at risk (CVaR) and the impacts of different asset reallocation strategies on the electricity assets configuration effect were analyzed. Zhang et al. [15] designed a risk prevention linkage mechanism of credit evaluation and risk measurement for retailers, including a credit risk measurement model based on the CvaR method for accurately describing the retailers' credit risk. Hang developed a decision model based on a CVaR methodology, whose object was to maximize the power supplier's utilities of electricity purchasing and selling. There are other techniques based on the value at risk (VaR) model [18] that have already been implemented in electricity markets [19,20]. However, VaR is not as consistent a risk measurement tool as the CvaR because VaR is neither convex nor sub-additive. The Mean-variance-based and CvaR-based methods mostly focus on risk measurement and management of a single period, and become extremely complicated in multiple periods.

Meanwhile, the operation scheduling for hydropower plants is a typical multi-variable and non-linear problem [21–24], especially for cascade hydropower plants. The most common main objective of hydropower plants' optimal operation is to maximize the economic benefits in the electricity market [24], maximize the power generation during conventional scheduling [21–23] or maximize the carryover energy storage [5], so it is difficult to take into account both the demand of the power grid and hydropower stations at the same time, and it is not suitable for the actual operation demand of the large cascaded stations. Under these circumstances, optimizing the generation scheduling for cascades in multiple markets, improving the overall benefits and avoiding price risks for the market portfolio, and maximizing the carryover energy storage at the end of the year should be among the vital multi-objective problems to be solved. There is an extensive literature which explores the multi-objectives optimization scheduling of reservoirs, which remains a research hotpot [25–28]. There are two kinds of method used to solve multi-objective optimization problems, one is the traditional mathematical analytic method, and the other is the evolutionary algorithm. An evolutionary algorithm is a kind of population-based method with a random search method, which has the ability to approach the real Pareto front through continuous evolution. Therefore, it is widely used in solving multi-objective problems, for example, using the Non-dominated Sorting Genetic Algorithm-II (NSGA-II) and population-based algorithms. Among the population-based algorithms, the multi-objective particle swarm optimization algorithm (MOPSO) has the characteristic of fast convergence.

To our knowledge, few studies have considered both the market benefit and power grid demand, especially in the circumstances of China's electricity market, in which power exchange centers take charge of trading while the power grid takes charge of operations. A recent paper by Lu et al. [29] established a risk analysis method for cascade hydropower to participate in the market, which was coupled with the monthly market and day-ahead market. Different from the work of Liu, we focus on a multi-objective model for a one year operation schedule which suits both the power grid and power exchange center trading, and we use price scenarios to characterize the possible situations of the actual clear price in multiple markets on the premise of knowing the mean and variance of historical price data, which gives the optimal market portfolio with a confidence level. In addition, we analyze the model's performance under different inflow situations, including flood, normal and dry conditions.

In this paper, a multi-objective optimization model for the long-term operation of hydropower plants were developed, in which one objective was the maximization of market benefits considering the price risk and the other was the maximization of carryover energy storage. The model aimed to give non-inferior sets under typical runoff conditions according to historical runoff data. To fully consider the volatility of multi-market prices, a mean-CvaR-based utility function was constructed to calculate the multi-market revenues and risks in whole schedule periods for the market portfolio. A MOPSO-based framework was proposed to solve the model and several mechanisms, including a chaotic mutation search, elite individual retention and dynamic updating archives, were introduced to accelerate convergence. The developed model was evaluated by optimizing the monthly generation schedules of China's Xiaowan-Nuozhadu hydropower system during a one-year period. The methodological framework used to investigate this relationship is illustrated in Figure 1.

The contributions of this paper are threefold. (1) We proposed a utility function for a multi-market portfolio which measured the benefits and risk based on the mean-CVaR method in relation to long-term hydropower scheduling. (2) A multi-objective particle swarm optimization algorithm was established with a chaotic mutation search mechanism, elite individual retention mechanism and dynamic updating of archives to accelerate algorithm convergence. (3) The effectiveness of the proposed model was validated by applying it to perform the long-term scheduling of the Xiaowan-Nuozhadu hydropower system in China. Combined with other operation indicators, the non-inferior frontier solution could provide optimal schemes which were adapted to different demands.

The remainder of this study is organized as follows. Section 2 introduces the objectives and constraints. Section 3 explains the methods used to solve the model including the market relative and multi-objective particle swarm optimization algorithm. Section 4 provides a detailed information of the case study. Section 5 includes an analysis and discussion of the case study results. The conclusions are summarized in Section 6.

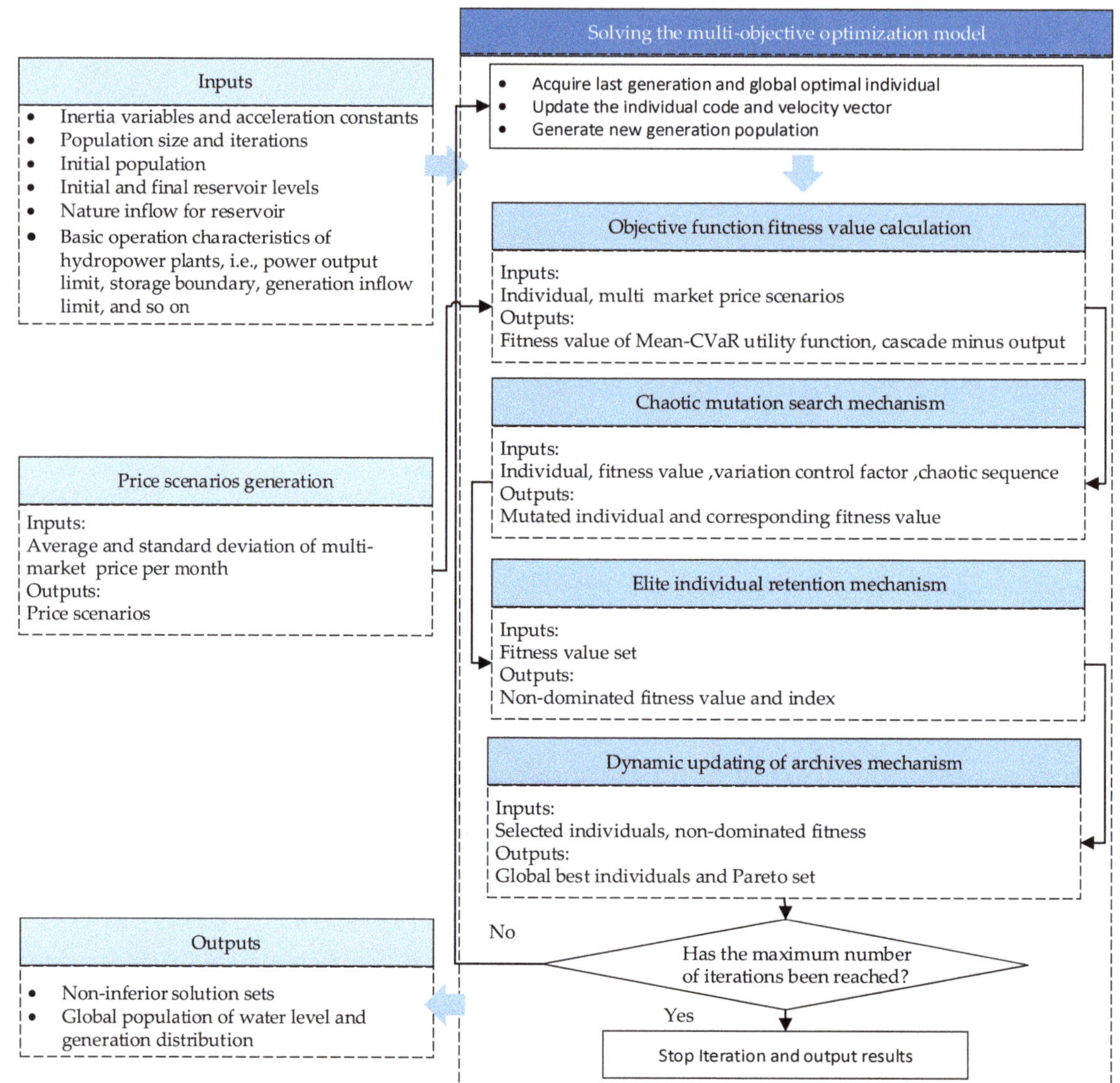

Figure 1. Methodological framework of the proposed optimization model.

2. Objectives and Constraints

2.1. Obejctieve

This paper considers two optimization objectives: the maximum market utility function value and maximum carryover storage of cascade stations.

The former expects that the maximum market benefits of cascade hydropower under a minimum risk of loss, which is represented by the utility function value; the latter expects maximize carryover storage of cascade stations which meets the demand of peak regulation and frequency regulation in the scheduling period. This model is a multi-objective optimization problem, which can be expressed as

$$\max F(x) = [f_1, f_2] \tag{1}$$

where f_1 is the utility function of the cascade stations' multi-market benefits and can be expressed as

$$f_1 = \sum_{t=1}^{T} U(t) = \sum_{t=1}^{T} (E(t) + CVaR(Lost(t), \beta)) \tag{2}$$

where f_2 is the carryover energy storage at the end of the period and can be expressed as

$$f_2 = ES_T \tag{3}$$

where N refers to the total number of cascade hydropower stations, and n refers to the index number of hydropower stations; T is the total number of scheduling periods, and t is the index number period which is on monthly scale; $E(t)$ is the expected multi-market return in period t; $Lost(t)$ is the lost function in period t; $CVaR(Lost(t), \beta)$ is the conditional value at risk under the probability of $1 - \beta$ in period t; and ES_T is the energy storage of hydropower stations at the end of period T.

2.2. Station Constraint

To solve the objective function, the following constraints must be met:

1. Water balance constraint: ensure the water balance of upstream and downstream power stations and adjacent periods, as follows:

$$V_n^{t+1} = V_n^t + \left(I_n^t - q_n^t - r_n^t \right) \cdot \Delta t \cdot 3600 \tag{4}$$

where V_n^t and V_n^{t+1} are the storage capacity of hydropower station n in the period t and $t + 1$, measured in m^3; I_n^t represents the inflow of hydropower station n in the period t, measured in m^3/s; q_n^t represents the generation flow of hydropower station n in the period t, measured in m^3/s; r_n^t represents the spillage water flow of hydropower station n during the period t, measured in m^3/s; and Δt is the number of hours for the period.

2. Water level restriction: the water level restriction of hydropower station dispatching ensures that the reservoir operates within a safe and reasonable water level range:

$$\underline{Z_n^t} \leq Z_n^t \leq \overline{Z_n^t} \tag{5}$$

where Z_n^t represents the water level of hydropower station n in the period t; $\underline{Z_n^t}$ and $\overline{Z_n^t}$ are the lower limit and upper limit of the water level of hydropower station n in the period t, respectively, measured in m.

3. Output constraint: the upper and lower output limits of the hydropower station are determined according to the nameplate output, maintenance plan, minimum technical output limit, etc., of the hydropower unit:

$$\underline{P_n^t} \leq P_n^t \leq \overline{P_n^t} \tag{6}$$

where P_n^t represents the output of hydropower station n in the period t, measured in W; $\underline{P_n^t}$ and $\overline{P_n^t}$ are the lower and upper output limits of hydropower station n in period, respectively, measured in W.

4. The turbine flow constraint mainly depends on the overflow capacity of all water turbine units of the hydropower station:

$$q_n^t \leq q_n^t \leq \overline{q_n^t} \tag{7}$$

where q_n^t represents the turbine flow of hydropower station n in the period t, measured in m^3/s; q_n^t and $\overline{q_n^t}$ are the lower limit and upper limit of the turbine flow of hydropower station n in period t, respectively, measured in m^3/s.

5. Water discharges constraint: the discharge limit of a hydropower station is related to the discharge capacity of the hydropower units and the discharge capacity of the spillway:

$$S_n^t \leq S_n^t \leq \overline{S_n^t} \tag{8}$$

where S_n^t represents the discharge of hydropower station n in period t, measured in m^3/s; S_n^t, $\overline{S_n^t}$ represent the lower limit and upper limit of the discharge of hydropower station n in period t, respectively, measured in m^3/s.

6. Power generation function:

$$P_n^t = q_n^t \cdot wr_n[h_n^t] \tag{9}$$

$$h_n^t = \frac{Z_n^{t-1} + Z_n^t}{2} - Zd_n^t \tag{10}$$

where h_n^t represents the average gross water head of station n in period t; Zd_n^t is the average tailrace level of station n in period t; and $wr_n[\cdot]$ represents the water released-energy conversion function of station n.

7. Carryover energy constraint, based on the expression in [4]:

$$vu_n = v_{n,T} - v_{n,min} + vu_{n-1} \tag{11}$$

$$ES_T = \sum_{n=1}^{N} \frac{vu_n}{\eta_n} \tag{12}$$

$$\eta_n = T^{-1} \cdot \sum_{t=1}^{T} wr_n[h_n^t] \tag{13}$$

where $v_{i,T}$ represents the storage of station n in period T; $v_{n,min}$ is the minimum storage of station; vu_n is the sum of the storage of station n and its upstream station, measured in m^3; and η_i is the average water consumption rate of station n, measured in m^3/kWh.

8. The relationship between the water level and reservoir storage:

$$V_n^t = f_n^{zv}[Z_n^t] \tag{14}$$

where $f_n^{zv}[\cdot]$ represents the water level as a function of station storage for station n.

9. Relationship between the tailrace water level and reservoir discharge:

$$Zd_n^t = f_n^{zq}\left[q_n^t\right] \tag{15}$$

where $f_n^{zq}[\cdot]$ represents the tailrace water level as a function of reservoir discharge for station n.

10. The market electricity of a station is calculated as follows:

$$Q_n^t = 3.6 \cdot P_n^t \cdot \Delta t \tag{16}$$

where Q_n^t represents the total market electricity of station n in period t, measured in kWh.

3. Model Solving Method

3.1. Mean-CVaR Market Model

3.1.1. CVaR Model

Let $Lost(x, y)$ denote the loss associated with decision variables vector x and random variables vector y. x can be represented as a market portfolio, and vector y stands for the uncertainties affecting the loss. If the probability density function of y is given as $p(y)$, the distribution function of loss function will be determined for x, which is given by:

$$\varphi(x, a) = \int_{Lost(x,y)<a} p(y)dy \tag{17}$$

Given the confidence level $\beta(0 < \beta < 1)$, the VaR of the portfolio is given by the following expression:

$$VaR = \min\{a \in R : \varphi(x, \alpha) > \beta\} \tag{18}$$

It is assumed that a is the VaR for the specified portfolio under confidence level β. Nevertheless, even though regulators and fund managers have widely used VaR, it has some drawbacks. The VaR approach has the undesirable property of a lack of sub-additivity, making it difficult to calculate when using scenarios and, if that is the case, it is a non-convex and non-smooth function with multiple local extreme values [30,31]. Therefore, an alternative measure of losses for VaR with more attractive properties is the conditional value at risk (CVaR) measure [32]. Given confidence level β, the CVaR of the portfolio is given by the following expression:

$$CVaR = E(Lost(x, y) > VaR) = (1 - \beta)^{-1} \int_{L(x,y)>VaR} f(x, y)p(y)dy \tag{19}$$

The Monte Carlo method is generally used to solve the definite integral of a function. Its basic idea is to obtain a large number of experimental solutions of the problem through a large number of random variable samples. On this basis, the statistical characteristics (usually the mean value) of the optimization variables are taken as the solution of the problem [33]. The Monte Carlo method is applied to simulate the cumulative distribution function by taking the L sample scene vector $\{P^1(t), P^2(t), \ldots, P^L(t)\}$, in which the vector $P^l(t) = \{p_{m,t}^l\}$, $p_{m,t}^l$ represents the possible price value of the market m in period t. Then, the total market value $R^l(t) = P^l(t) \cdot W(t)$, $W(t) == \{w_{m,t}\}$, $w_{m,t}$ represents the distribution proportion of market m in period t. We consider $Lost(l, t) = \overline{P}(t) \cdot W(t) - R^l(t)$, which means the loss benefits of the forecast and actual possible price in the market, where $\overline{P}(t)$ represents the average electricity price in period t. We denote F_{CVaR} as an approximation of the CVaR obtained by sampling the probability distribution in $Lost(l,t)$. F_{CVaR} is obtained using the following expression:

$$F_{cvar} = a + (L \cdot (1 - \beta))^{-1} \sum_{l=1}^{L} (Lost(l,t) - a)^{+} \tag{20}$$

where a is the VaR when the CVaR is minimized, c represents the cost of the hydropower station, F_{cvar} represents the expected loss of market benefits in the period under confidence level β.

Then, the total market profit of scene l in period t is given with the following expression:

$$profit_t^l = \sum_{n}^{N} \left((\sum_{m}^{M} p_{m,t}^l \cdot w_{m,t} - c) \cdot Q_{n,t} \right) \tag{21}$$

where $Q_{n,t}$ is the power generation of station n in period t, measured in kWh, and the cost of hydropower stations is 0 in this paper.

3.1.2. Mean-CVaR Utility Function

Previous modern portfolio theory (MPT) models based on the portfolio theory of Markowitz [34] have allocated energy between spot and contract markets in real markets. Liu and Wu [35] used MPT to allocate energy of risky and non-risky assets in the PJM market. Although MPT has been a rather popular measure of risk in portfolio selection, it has its limitations. One distinguished limitation is that analysis based on variance considers high returns as equally undesirable as low returns because high returns will also contribute to the extreme of variance. Due to the variance limitation, the CvaR can be used to replace variance for measuring risk. Then, the mean-CVaR utility function can be given with the following expression:

$$E(t) = L^{-1} \cdot \sum_{l=1}^{L} profit_t^l \tag{22}$$

$$Lost(l,t) = \sum_{n=1}^{N} \left(Q_{n,t} \cdot \left(\sum_{m=1}^{M} w_{m,t} \cdot (\overline{p}_{m,t} - p_{m,t}^l) \right) \right) \tag{23}$$

$$a = quan(Lost(l,t), 1 - \beta) \tag{24}$$

$$U(t) = E(t) + CVaR(t, \beta) = L^{-1} \cdot \sum_{l=1}^{L} profit_t^l + F_{cvar} \tag{25}$$

where $quan(seq, \gamma)$ represents the quantile of the seq sequence at confidence level γ, which is the same meaning as the expression of VaR. However, it is hard to consider the risk of whole scheduling periods, as it is much more complicated than calculating the risk of one period. For the sake of simplifying the calculation, we take the minimization of CvaR in period t as the risk measure based on a random price scenario set and take the average value of multi-market profit returns as the expected return. The mean-CVaR utility function obtains the final value by participating in market proportion decision variables and generating a capacity decision.

3.2. Solution of Cascade Multi-Objective Optimization Problem

3.2.1. Standard Particle Swarm Optimization

The particle swarm optimization algorithm is a swarm intelligence optimization algorithm. In the algorithm, particles update their speed and position in the solution space according to the best position of the population and the best position in their history. The speed and position expression formula is as follows:

$$x_{i,j}^{k+1} = x_{i,j}^k + v_{i,j}^{k+1} \tag{26}$$

$$v_{i,j}^{k+1} = \omega v_{i,j}^k + c_1 r_1 \left(p_{i,j}^k - x_{i,j}^k \right) + c_2 r_2 \left(g_{i,j}^k - x_{i,j}^k \right) \tag{27}$$

where, i is the individual number, $i = 1, 2, 3, \ldots, m$; j is the dimension number, $j = 1, 2, 3, \ldots, n$; k is the number of iterations; $x_{i,j}^{k+1}$ and $x_{i,j}^k$ are the individual codes; $v_{i,j}^k$ and $v_{i,j}^{k+1}$ are the velocity vectors; $p_{i,j}^k$ is the single generation optimal individual; $g_{i,j}^k$ is the global optimal individual; ω is the inertial variable; c_1 and c_2 are acceleration constants respectively; r_1 and r_2 are the random number subject to a [0, 1] uniform distribution.

3.2.2. Multi-Objective Particle Swarm Optimization Algorithm

At present, the multi-objective particle swarm optimization has been widely used in multi-objective problem solving [36,37]. For the multi-objective model proposed in this paper, a multi-objective particle swarm optimization (MOPSO) algorithm was constructed. In order to obtain a better Pareto efficient frontier solution set, the local search ability was improved through the use of a chaotic mutation search mechanism. The file mechanism based on an elite retention strategy dynamically updates the globally and historically optimal set of non-dominated solutions. The congestion distance selection mechanism was used to determine the globally and historically optimal individuals of each generation to improve the PSO algorithm. Finally, according to the characteristics of the proposed model, an individual coding method based on mapping mechanism was proposed.

- Chaotic mutation search mechanism

The particle swarm optimization algorithm has a strong global search ability and weak local search ability. Considering that with the increase of the number of iterations, the quality of individuals is greatly improved, and the probability of better individuals nearby is improved, the local search for historical optimal files and global files becomes relatively important. For this reason, the standard uses a relatively mature chaotic cubic map to perform mutation operations on individuals in the file in order to expect to obtain other better solutions or non-inferior solutions. The calculation formula is as follows:

$$X_i' = X_i \cdot (1 + \phi \cdot Z_n) \tag{28}$$

$$Z_n = 4(Z_{n-1})^3 - 3Z_{n-1} \tag{29}$$

$$\phi = 1 - (k - 1)/k_{max} \tag{30}$$

where ϕ is the variation control factor; Z_n is a chaotic sequence, $Z_n \in [-1, 1]$; and k and k_{max} are the iteration number and maximum iteration number, respectively. If the fitness value of the variant individual is the dominant solution of the original individual it will be replaced, the dominated solution will not be replaced and the non-dominated solution will be replaced according to a certain probability. At the same time, the whole chaotic mutation search mechanism also occurs randomly under a given probability.

- Elite individual retention mechanism and dynamic updating of archives

For the individuals of the population in the algorithm, the corresponding individual file is established according to the literature to save the non-dominated solution set of individuals in the historical process and provide choices for the next generation of historical optimal individuals. Similarly, the global file includes the storage of non-dominated individuals obtained in the iterative process. The update mechanism of individual files and global files is as follows:

$$P(k+1) = \begin{cases} P(k) \cup x(k+1) \setminus \{p\}, \exists p, f(x) \prec f(p) \\ P(k), \exists p, f(p) \prec f(x) \\ P(k) \cup x(k+1), f(p) \sim f(x) \end{cases} \tag{31}$$

where $P(k+1)$, $P(k)$ represents the individual files or global files of the k and $k+1$ generation; the individual is p in the file; and $x(k+1)$ is the individual corresponding to the individual file in the generation $k+1$. If individual $x(k+1)$ in the file is controlled, the controlled individual in the file will be deleted and $x(k+1)$ will be added to the file. If it is controlled by an individual in the file, the file will remain unchanged. If $x(k+1)$ is not controlled by individual archives, it will be directly added to the archives.

As the number of iterations increases, the number of individuals in the file increases gradually. To ensure the speed of iterative calculation, we set the number of individuals contained in the file. The strategy of the file update was to perform a non-inferior hierarchical sorting of individuals in the file, and by calculating the individual crowding degree, when the number of files exceeded the limit, the individuals with low crowding degree would be deleted from the file. The crowding degree calculation formula is as follows:

$$\begin{cases} I(d_1) = I(d_N) = \infty \\ I(d_i) = \sum_{o=1}^{O} \frac{I(i+1) \cdot f_o - I(i-1) \cdot f_o}{f_o^{max} - f_o^{min}} \end{cases} \tag{32}$$

where $I(d_1)$ and $I(d_N)$ are the first and last individuals after sorting, respectively, and N is the total number of current individuals in the file; f_o^{max} and f_o^{min} are the maximum and minimum values of the file individual, respectively, in the o objective function o, and O is the total number of objective functions in the fitness function; $I(i) \cdot f_o$ is the value of objective function o corresponding to individual i; and $I(d_i)$ is the crowding degree value of the individual in the sorted file.

At the same time, based on the crowding degree value, the mechanism of global optimization for each generation selection history and the mechanism of historical optimization for each individual are given. The selection probability for any individual in the file is as follows:

$$prob(l) = \begin{cases} \frac{r_s}{2}, l = 1, N \\ (1 - r_s) \frac{I(d_l)}{\sum\limits_{i=2}^{N-1} I(d_i)}, i = 2, 3, \ldots, N-1 \end{cases} \tag{33}$$

where r_s is a larger value, which was set to 0.7 in our model.

3.3. Overall Solution Framework

Coupled with the above sections, the overall solution framework of the multi-objective optimal scheduling problem of cascade hydropower stations is shown in Figure 2.

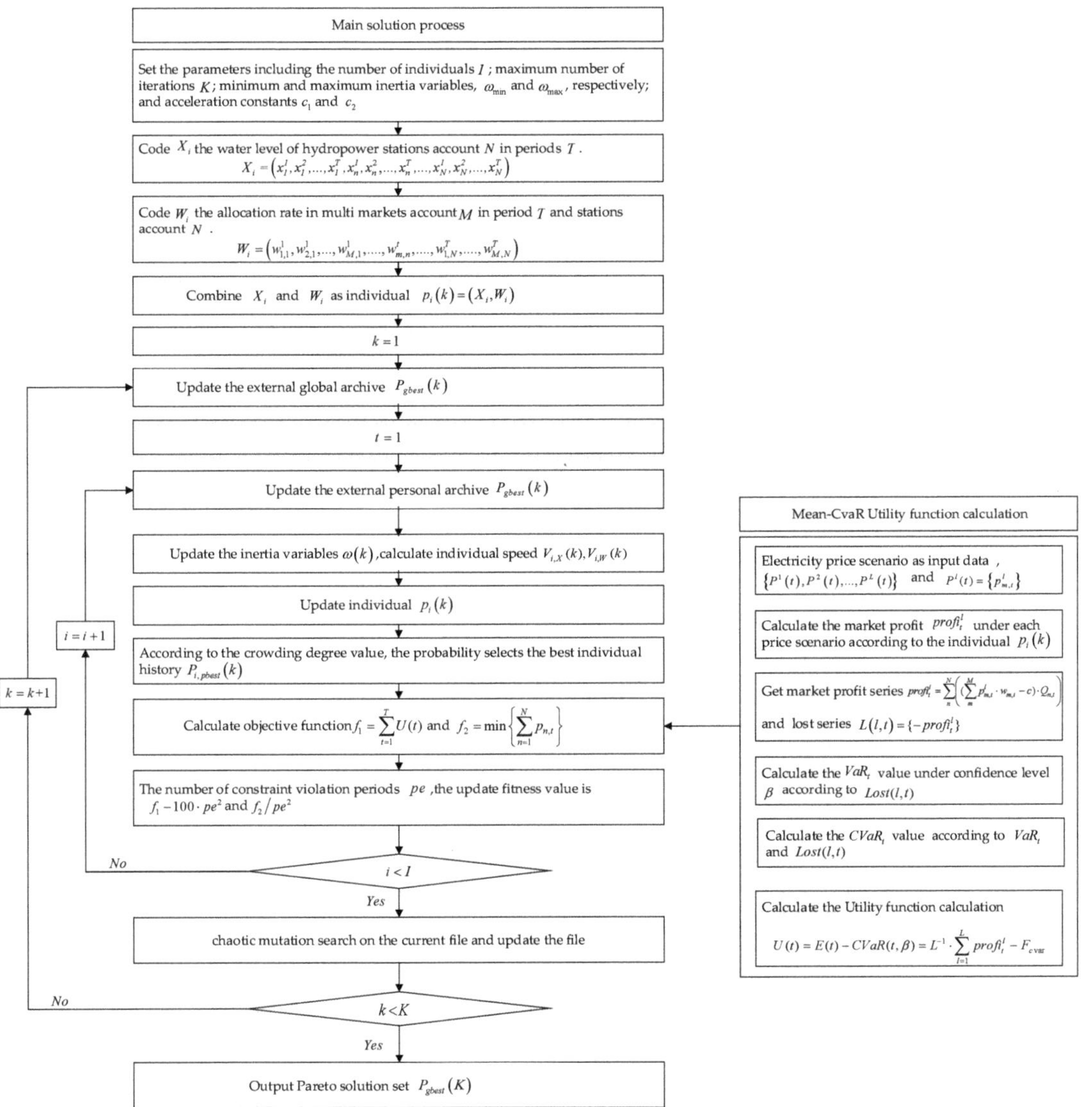

Figure 2. Multi-objective Particle Swarm Optimization Solution Framework.

4. Case Study

4.1. Basin Cascade Data

The proposed model was applied to the cascade hydropower plants of the Lancangjiang River in Yunnan Province, China. Seven hydropower plants have been built on the middle and lower Lancangjiang River, with a total installed capacity of 15,915 MW. Figure 3 shows the distribution chart of the middle and lower Lancangjiang River cascade, with reservoir characteristics in Table 1. From May to October the maximum water level of stations is at flood level, while for the rest of the year it is a normal level. Two hydropower plants have long-term regulation abilities in the studied cascade, and they are Xiaowan and Nuozhadu. The storage capacities of the other five reservoirs were much smaller and three of them did not trade on the Yunnan Electricity Market. The cascade power decisions were distributed through the operation of these two large reservoirs.

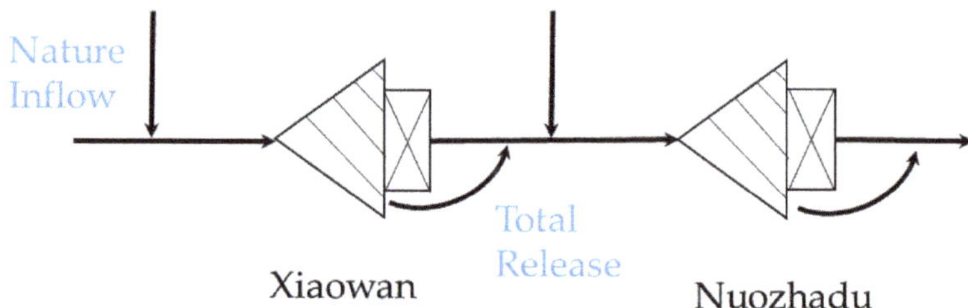

Figure 3. The distribution chart of Xiaowan and Nuozhadu.

Table 1. Characteristics of hydropower plants.

Plant Name	Installed Capacity (MW)	Normal Level (m)	Flood Control Level (m)	Dead Level (m)	Maximum Discharge (m^3/s)	Maximum Generation Flow (m^3/s)
Xiaowan	4200.00	1240.00	1236.00	1166.00	20,683.00	2202.00
Nuozhadu	5850.00	812.00	804.00	765.00	27,418.00	3240.00

4.2. Electricity Market Information

The studied Yunnan Electricity Market (YEM) has three majority trading markets: a year-bilateral trading market, a matchmaking trading market and a month-bilateral trading market in long-term market. Based on the public price information of Kunming Power Exchange Center (KPEC), which was the power exchange center of the YEM, the average and the standard deviation of the market price were treated as known and are given in Table 2. The price scenarios were parameterized using the statistical characteristic value in Table 2. The total number of scenarios was 100.

Table 2. The parameters of market clear price for multi-markets in a different month (CNY/MWh).

Month	Centralized Market		Bilateral Market		Day Ahead Market	
	Avg.	Std.	Avg.	Std.	Avg.	Std.
Jan.	239.82	12.19	234.88	8.53	244.15	3.27
Feb.	242.29	14.3	233.95	2.56	246.3	5.88
Mar.	240.11	13.02	234.85	2.46	237.77	10.25
Apr.	240.07	12.56	246.34	6.52	240.16	12.17
May	236.01	13.01	231.23	2.67	238.81	9.42
Jun.	174.58	16.79	172.04	1.83	186.58	23.12
Jul.	123.61	17.21	120.01	1.86	131.25	29.59
Aug.	114.71	18.26	112.69	1.79	123.54	34.29
Sep.	114.66	18.8	104.57	9.97	126.4	40.79
Oct.	125.64	16.05	132.27	6.46	148.09	32.64
Nov	236.75	15.62	249.27	2.38	250.25	14.23
Dec.	244.96	19.92	275.03	15.67	270.17	18.38

Note: Avg. indicates the average value of the market clear price; Std. indicates the standard deviation of the market clear price.

4.3. Inflow Information

In order to verify the effectiveness of the model method in this paper, three different inflow years were selected: a wet year, normal year and dry year. Figure 4 shows the interval inflow of each cascade hydropower station in three typical years.

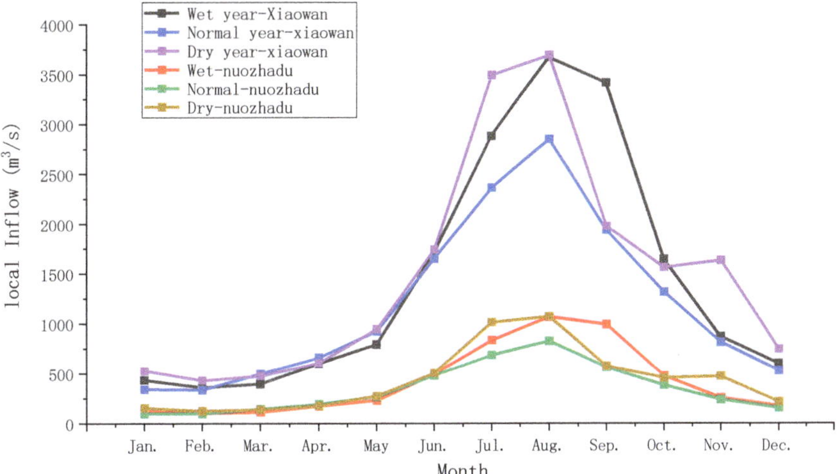

Figure 4. The monthly local inflow of each reservoir in wet, normal and dry years.

5. Results and Discussion Case Study

This analysis involves the following parts: (1) A non-inferior frontier solution set analysis under different incoming water conditions, (2) a solution set for water level and market proportion analysis and (3) a frontier solution set selection strategy.

5.1. Non-Inferior Frontier Solution Set under Different Incoming Water Conditions

In this paper, the population number was 100, the number of iterations was 500 and the inertia variable was adaptively changed with the iteration process. The purpose was to have a better global search capability in the early stage of the iteration, and focus on a local search at the end of the iteration. Therefore, the maximum inertia variable was set to 0.9 and the minimum inertia variable was set to 0.2, resulting in:

$$\omega_k = \omega_{max} - \frac{k}{K}(\omega_{max} - \omega_{min}) \tag{34}$$

where ω_k is the inertia variable of the time, and k and K are the current iteration number and the maximum iteration number, respectively. The local speed factor was 1.2 and the global speed factor was 0.8. The individual contained two kinds of chromosomes, both of which used floating point coding: One was a water level chromosome; the initial water levels were set to 1230 m for Xiaowan and 805 m for Nuozhadu. The other was the proportion chromosome of risk market participation in the period, which was composed of a multi-dimensional array of reservoirs, which generates constraint satisfaction. Figure 5 shows the non-inferior set in a dry, normal and wet year. It can be seen that:

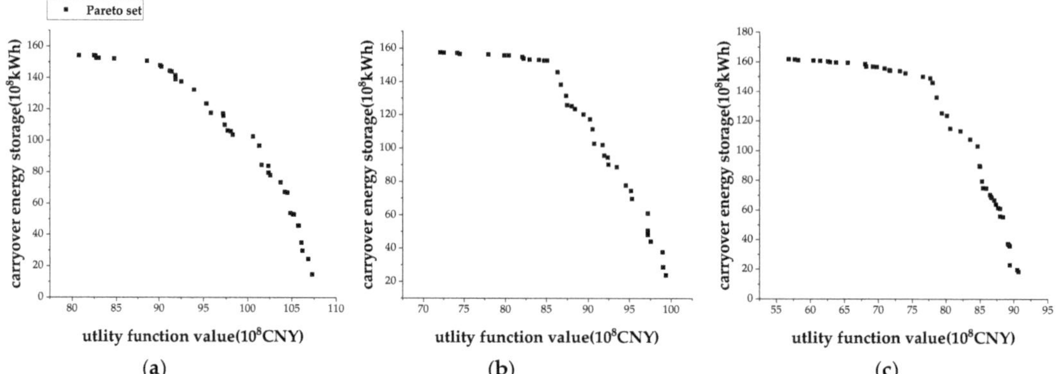

Figure 5. The non-inferior set result under different inflow years: (**a**) wet year, (**b**) normal year, (**c**) dry year.

1. Under the different inflow conditions, the trend of the non-inferior set was similar, and the distribution of the overall points was relatively uniform. Figure 5 obviously demonstrated that the maximum and minimum values of the cascade utility function values changed significantly depending on the inflow conditions, i.e., the maximum market utility values in wet and dry years were 10.73 billion CNY and 9.07 billion CNY, respectively, which represents a reduction of 1.66 billion CNY, while the minimum values were 8.07 billion CNY and 5.67 billion CNY, respectively, representing a reduction of 2.40 billion CNY. It was found that the nature inflow has an obvious influence on market benefits but has little impact on the range of carryover energy storage.
2. The two objectives of market utility function value and carryover energy storage were obviously inversely proportional under the three different inflow conditions. With the increase of market utility value, the rate of decreasing carryover energy storage at the end of year increased, which means that it will take more cost of energy storage reduction to improve the market benefits when the benefits increase to a certain level. On the one hand, when the energy storage at the end of year is kept at a high level, the market benefits can be increased at a small cost.
3. The non-inferior set can cover a wider space. Under the three types of water inflow (flood, normal and dry), the range of market utility function was 2.65 billion CNY, 2.73 billion CNY and 3.39 billion CNY, respectively, while the range of carryover energy storage at end of year was 13.95 billion kWh, 13.36 billion kWh and 14.34 billion kWh, respectively.

5.2. Operation Analysis in Solution Sets

In this sub-section we take five schemes of points from the leftmost, rightmost and middle of the non-inferior solutions sets under the wet year. These fifteen schemes of solution sets are listed in Table 3, ordered by market utility function value. Corresponding to solution groups in Table 3, the water level of cascade stations are shown in Figure 5, in which the more blue colors represent a greater market utility value of the cascade stations, and the more green colors represent a smaller market utility value of the cascade stations.

Table 3. Non-inferior set of two objective functions.

Non-Inferior Set Scheme Number	f_1/Billion CNY	f_2/Billion kWh
1	8.08	15.43
2	8.25	15.42
3	8.26	15.41
4	8.27	15.27
5	8.30	15.27
6	9.72	11.71
7	9.72	11.58
8	9.73	11.00
9	9.77	10.64
10	9.80	10.6
11	10.58	4.59
12	10.61	3.50
13	10.62	2.98
14	10.68	2.45
15	10.73	1.47

It can be seen that the average market utility values was 8.23 billion CNY and the average carryover energy storage was 15.36 billion kWh in schemes 1–5, while in schemes 11–15 the average was 10.64 and 2.99, respectively. It is obvious that the amplitude of market utility values was smaller than that of carryover energy storage. From the perspective of objective values, it was found that the increase in the rate of market benefits was far lower than the decrease in the rate of energy storage, as the increment of market benefits between schemes 5 and 6 was 1.42 billion CNY and the decrement of energy storage was 3.56 billion kWh, while the same values were 0.78 billion CNY and 6.01 billion kWh between schemes 10 and 11. This further explains that the two objectives are in conflict and restrict each other, which indicates that how an operator balances between them will directly affect the overall operation efficiency of the cascade stations now and in the future.

Generally, as shown in Figure 6a, the trend of Xiaowan's water level curves in different schemes were similar from January to April: in the dry season the water level fell and water vacated the reservoir, while from June to September, in the flood season, the water level rose and water was stored in the reservoir. Under the two objectives of market utility value and energy storage, the water level at the end of year decreased in the schemes which pursued more market benefits. However, the trend of Nuozhadu's water level curves were quite different in these schemes, and they are shown in Figure 6b. Take schemes 8–15, which pursued the market utility value, as examples: the water levels in the dry season (February to March) fell rapidly to increase the power generation, while the water levels in the flood season (January to October) increased slowly, to take advantage of the large amount of inflow to increase the power generation. On the other hand, the water level of schemes which pursued energy storage at the end of the year were at higher positions relatively to achieve a lower consumption rate to ensure the market utility value.

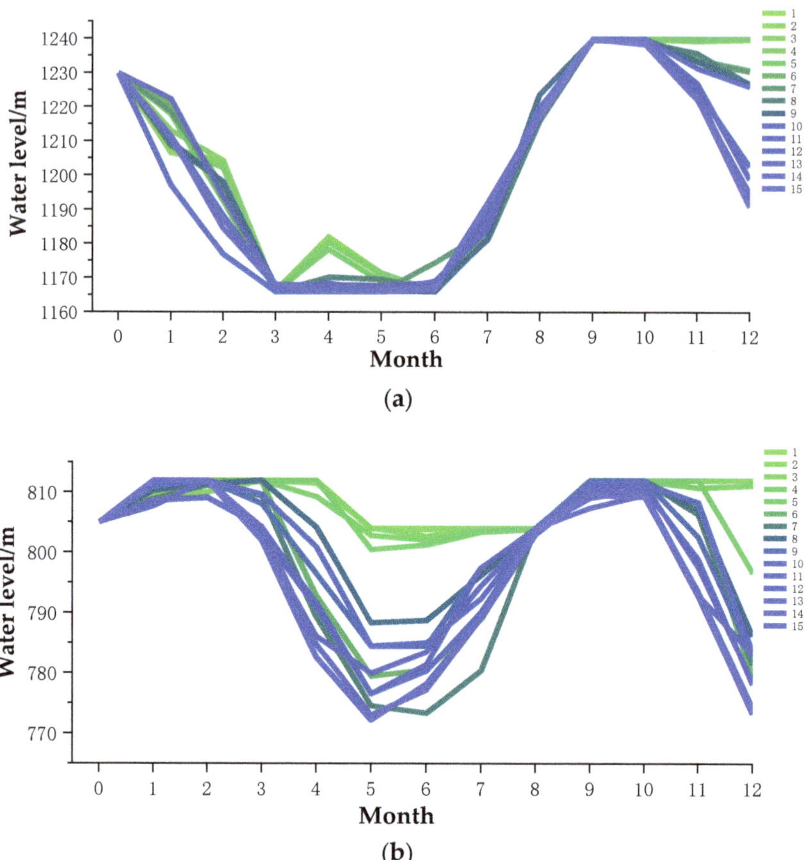

Figure 6. The optimization process of regulating reservoirs in schemes 1 to 15. (**a**) Xiaowan; (**b**) Nuozhadu.

Figure 7 shows the portfolio proportion and generation in whole scheduling periods and Figure 8 shows the expected benefits and CVaR in whole scheduling periods. In terms of the types of markets hydropower stations choose to participate in, most of the months include two or more types which have considered market portfolios to reduce the value at risk under confidence levels. Combined with Figure 8, although the generation of stations in the flood season (June to October) is larger than in the other seasons (dry and normal), the expected benefits are not greater than the others in terms of the growth of utility function value, which is affected by a low average price. On the other hand, the CVaR increased due to the high power generation during the flood season. Thus, this indicates that the accuracy of the market price forecast is more important in the flood season. It is significant that, in November and December in schemes 11–15, the cascade lowered the water level to gain huge benefits instead of maintaining a high level for storage, which, combined with Figure 6, further explained the mutual constraints between objectives.

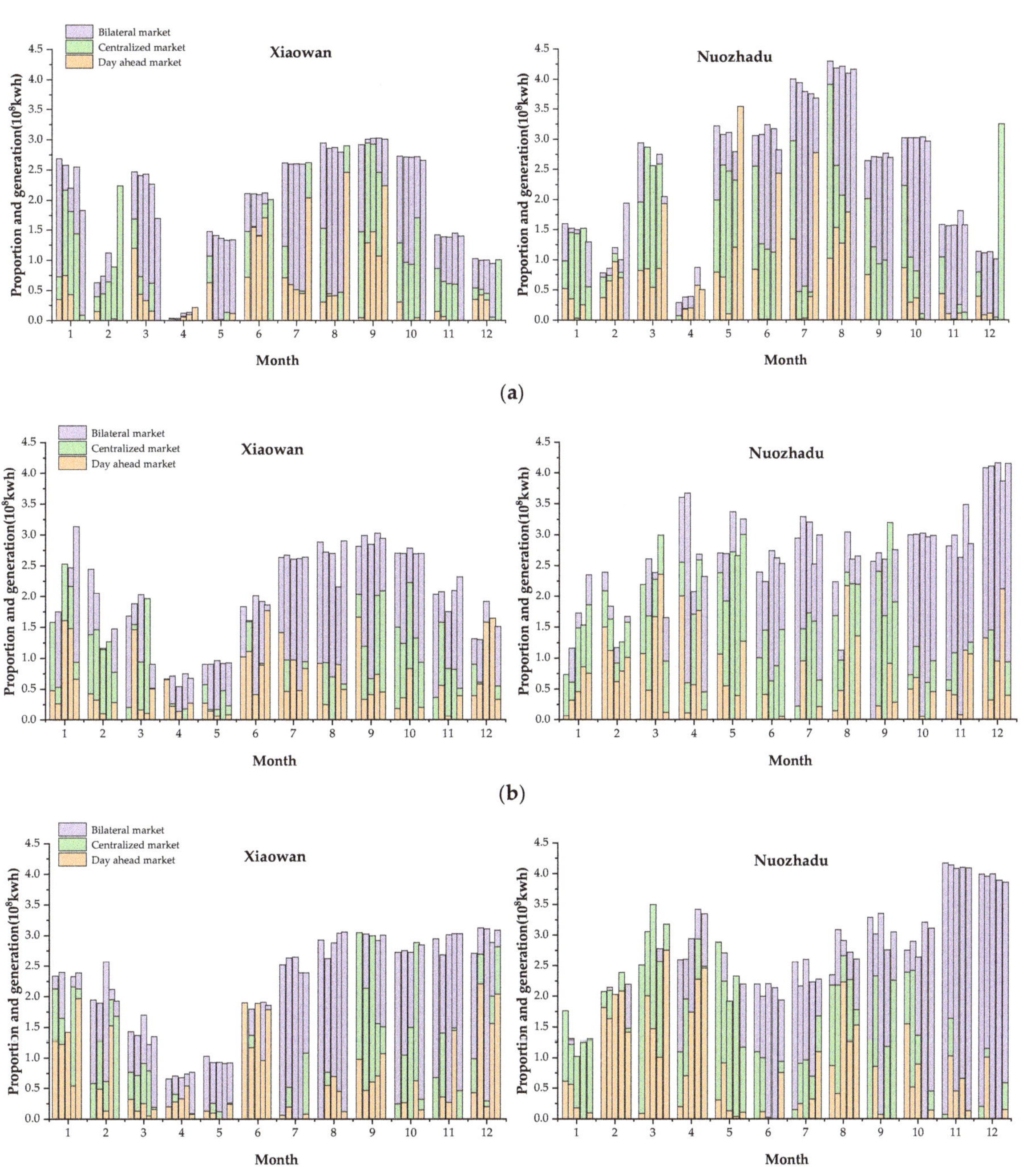

Figure 7. Market portfolio and generation of regulating reservoirs in schemes 1 to 15. (**a**) Schemes 1–5; (**b**) Schemes 6–10; (**c**) Schemes 11–15.

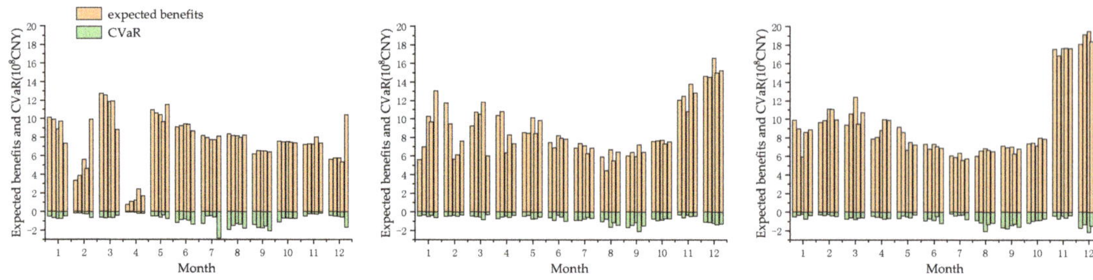

Figure 8. Expected benefits and CVaR of cascade reservoirs in schemes 1 to 15.

5.3. Selection Strategy of Non-Inferior Solution Set

In order to better screen the non-inferior solution for station operators, in this section we selected the power generation, minimum output and spillage energy as the evaluation indexes based on the actual scheduling demand, combined with a non-inferior solution set proposed in Table 3, and analyzed the performance of these schemes according to the previous evaluation indicators, and provided selection suggestions. The evaluation index of the non-inferior solution set is shown in Table 4.

Table 4. Evaluation index of non-inferior solution set.

Non-Inferior Set Number	Evaluation Index		
	Minimum Output /MW	Power Generation /Billion kWh	Spillage Energy /Billion kWh
1	459.29	51.26	0.00
2	670.15	51.31	0.00
3	875.40	51.29	0.00
4	1587.29	51.42	0.04
5	1259.84	53.27	0.00
6	5264.10	55.34	0.62
7	6078.05	55.00	0.57
8	4386.56	55.97	0.43
9	5632.49	55.92	0.77
10	4639.70	56.09	0.55
11	5369.64	58.96	1.20
12	5583.61	59.12	1.60
13	5074.72	59.77	1.02
14	5520.08	59.36	1.84
15	5394.72	59.56	2.15

It can be seen from Table 3 that the maximum market utility value was achieved in scheme 15, which was 10.73 billion CNY; the maximum carryover energy storage was achieved by scheme 1, which was 15.43 billion kWh; the maximum power generation was achieved by scheme 13, which was 59.77 billion kWh; the minimum output was achieved by scheme 8, which was 6078.05 MW; and the minimum spillage of energy was achieved by schemes 1, 2, 3 and 5, which had spillage of 0 billion kWh. According to the actual dispatching and market demand, the following suggestions can be given: (1) Considering the stable operation of the power grid, scheme 7 can be suggested, and the minimum system output values were higher than 6000 MW. (2) Taking into account the reduction of water spillage, schemes 1, 2, 3 and 5 can be suggested. (3) Considering the consumption of cascade power generation, schemes 13–15 schemes can be suggested.

6. Conclusions

In this study, we proposed a multi-objective optimization model considering both market benefits and energy storage at the end of the year for cascades in long-term operation schedules. Based on the mean-CVaR model, a utility function was established to measure the market portfolio benefits and risks in monthly trading. A chaotic mutation search mechanism, elite individual retention mechanism and dynamic updating of archives were incorporated into the MOPSO to guarantee the convergence efficiency of the algorithm and improve the diversity of solutions. To evaluate the proposed model, simulations with different inflow years were conducted to evaluate the rationality of the model. The following conclusions were drawn from the analysis:

1. The proposed multi-objective model can give a reasonable non-inferior solution set which considers both the market benefits and energy storage.
2. The optimal solution set reflected the mutual restriction of the objectives (that as market benefits of cascade stations increased, less energy storage for the cascade stations could be maintained at the end of scheduling).
3. The operation analysis revealed an internal relationship between benefits and risk. In a flood season, the benefits were not great, which was affected by low average prices, while the risk was relatively large due to the high power generation.
4. Based on the non-inferior solution, various schemes can be provided according to the different actual needs.

It should be mentioned that the market reform in China is still at an early stage and the number of available samples is relatively small. This paper was based on price distribution assumptions and the price scenarios were generated stochastically instead of using real historical data from markets. Due to the electricity market reform policy, more and more information might be available to the public in the future, so that the interaction characteristics between the markets could be analyzed in detail. The practical application of the proposed methodology can be extended to other objectives demanding detail such as flood control, irrigation and navigation. It is also recommended that further research be extended to cascade stations crossing provinces and regions.

Author Contributions: Conceptualization, H.Y. and J.S.; methodology, H.Y.; software, J.S.; validation, J.S. and C.C.; formal analysis, H.Y. and J.S.; investigation, H.C. and J.L.; writing—original draft preparation, H.Y.; writing—review and editing, J.S. and C.C. All authors have read and agreed to the published version of the manuscript.

Funding: This work was supported by the Fundamental Research Funds for the Central Universities (Nos.DUT22QN224 and DUT22JC21).

Data Availability Statement: The data presented in this study are available on request from the corresponding author. The data are not publicly available and please contact the correspondence author.

Acknowledgments: The writers are very grateful to the anonymous reviewers and editors for their constructive comments.

Conflicts of Interest: The authors declare no conflict of interest.

References

1. Zhang, Z.; Zhang, S.; Geng, S.; Jiang, Y.; Li, H.; Zhang, D. Application of decision trees to the determination of the year-end level of a carryover storage reservoir based on the iterative dichotomizer 3. *Int. J. Electr. Power Energy Syst.* **2015**, *64*, 375–383. [CrossRef]
2. Tan, Q.-F.; Wen, X.; Fang, G.-H.; Wang, Y.-Q.; Qin, G.-H.; Li, H.-M. Long-term optimal operation of cascade hydropower stations based on the utility function of the carryover potential energy. *J. Hydrol.* **2020**, *580*, 124359. [CrossRef]
3. Shen, J.; Cheng, C.; Wang, S.; Yuan, X.; Sun, L.; Zhang, J. Multiobjective optimal operations for an interprovincial hydropower system considering peak-shaving demands. *Renew. Sustain. Energy Rev.* **2020**, *120*, 109617. [CrossRef]
4. Niu, W.; Wu, X.; Feng, Z.; Shen, J.; Cheng, C. The Optimal Operation Method of Multi-reservoir System Under the Cascade Storage Energy Control. *Proc. Chin. Soc. Electr. Eng.* **2017**, *37*, 3139–3147.

5. Zeng, Y.-H.; Jiang, T.-B.; Zhang, Y.-C. A long-term scheduling model for stored energy maximization of Three Gorges cascade hydroelectric stations and its decomposition algorithm. *Power Syst. Technol.* **2004**, *28*, 5–8.
6. Shen, J.J.; Cheng, C.T.; Jia, Z.B.; Zhang, Y.; Lv, Q.; Cai, H.X.; Wang, B.-C.; Xie, M.-F. Impacts, challenges and suggestions of the electricity market for hydro-dominated power systems in China. *Renew. Energy* **2022**, *187*, 743–759. [CrossRef]
7. Kaye, R.J.; Outhred, H.R.; Bannister, C.H. Forward contracts for the operation of an electricity industry under spot pricing. *IEEE Trans. Power Syst.* **1990**, *5*, 46–52. [CrossRef]
8. Bjorgan, R.; Liu, C.C.; Lawarree, J. Financial risk management in a competitive electricity market. *IEEE Trans. Power Syst.* **1999**, *14*, 1285–1291. [CrossRef]
9. Sheble, G.B. Decision analysis tools for GENCO dispatchers. *IEEE Trans. Power Syst.* **1999**, *14*, 745–750. [CrossRef]
10. Siddiqi, S.N. Project valuation and power portfolio management in a competitive market. *IEEE Trans. Power Syst.* **2000**, *15*, 116–121. [CrossRef]
11. Yan, X.; Gu, C.; Wyman-Pain, H.; Li, F. Capacity Share Optimization for Multiservice Energy Storage Management Under Portfolio Theory. *IEEE Trans. Ind. Electron.* **2019**, *66*, 1598–1607. [CrossRef]
12. Garcia, R.C.; Gonzalez, V.; Contreras, J.; Custodio, J.E. Applying modern portfolio theory for a dynamic energy portfolio allocation in electricity markets. *Electr. Power Syst. Res.* **2017**, *150*, 11–23. [CrossRef]
13. Liu, M.; Wu, F.F. Risk management in a competitive electricity market. *Int. J. Electr. Power Energy Syst.* **2007**, *29*, 690–697. [CrossRef]
14. Zhao, W.-H.; Wang, W.; Shi, Q.-S.; Dao, Q. Dynamic multi-stage optimization configuration model for electricity assets based on conditional value at risk. *Power Syst. Technol.* **2009**, *33*, 77–82.
15. Zhang, Y.; Zhao, H.; Li, B.; Zhao, Y.; Qi, Z. Research on credit rating and risk measurement of electricity retailers based on Bayesian Best Worst Method-Cloud Model and improved Credit Metrics model in China's power market. *Energy* **2022**, *252*, 124088. [CrossRef]
16. Zhang, X.-P.; Chen, L.; Wu, R.-L. Analysis of multi-period combined bidding of power suppliers based on weighted CVaR. *Proc. CSEE* **2008**, *28*, 79–83.
17. Zhang, Q.; Wang, X.; Wang, J. Electricity Purchasing and Selling Risk Decision for Power Supplier Under Real-time Pricing. *Autom. Electr. Power Syst.* **2010**, *34*, 22–27.
18. White, H.; Kim, T.-H.; Manganelli, S. VAR for VaR: Measuring tail dependence using multivariate regression quantiles. *J. Econom.* **2015**, *187*, 169–188. [CrossRef]
19. Denton, M.; Palmer, A.; Masiello, R.; Skantze, P. Managing market risk in energy. *IEEE Trans. Power Syst.* **2003**, *18*, 494–502. [CrossRef]
20. Dahgren, R.; Liu, C.C.; Lawarree, J. Risk assessment in energy trading. *IEEE Trans. Power Syst.* **2003**, *18*, 503–511. [CrossRef]
21. Wu, X.; Cheng, C.; Zeng, Y.; Lund, J.R. Centralized versus Distributed Cooperative Operating Rules for Multiple Cascaded Hydropower Reservoirs. *J. Water Resour. Plan. Manag.* **2016**, *142*, 05016008. [CrossRef]
22. Wen, X.; Zhou, J.; He, Z.; Wang, C. Long-Term Scheduling of Large-Scale Cascade Hydropower Stations Using Improved Differential Evolution Algorithm. *Water* **2018**, *10*, 383. [CrossRef]
23. Cheng, C.-T.; Shen, J.-J.; Wu, X.-Y.; Chau, K.-W. Operation challenges for fast-growing China's hydropower systems and respondence to energy saving and emission reduction. *Renew. Sustain. Energy Rev.* **2012**, *16*, 2386–2393. [CrossRef]
24. Luo, B.; Miao, S.; Cheng, C.; Lei, Y.; Chen, G.; Gao, L. Long-Term Generation Scheduling for Cascade Hydropower Plants Considering Price Correlation between Multiple Markets. *Energies* **2019**, *12*, 2239. [CrossRef]
25. Fang, R.; Popole, Z. Multi-objective optimized scheduling model for hydropower reservoir based on improved particle swarm optimization algorithm. *Environ. Sci. Pollut. Res.* **2020**, *27*, 12842–12850. [CrossRef]
26. Reddy, M.J.; Kumar, D.N. Multi-objective particle swarm optimization for generating optimal trade-offs in reservoir operation. *Hydrol. Process.* **2007**, *21*, 2897–2909. [CrossRef]
27. Reddy, M.J.; Kumar, D.N. Optimal reservoir operation using multi-objective evolutionary algorithm. *Water Resour. Manag.* **2006**, *20*, 861–878. [CrossRef]
28. Feng, Z.-K.; Liu, D.; Niu, W.-J.; Jiang, Z.-Q.; Luo, B.; Miao, S.-M. Multi-Objective Operation of Cascade Hydropower Reservoirs Using TOPSIS and Gravitational Search Algorithm with Opposition Learning and Mutation. *Water* **2019**, *11*, 2040. [CrossRef]
29. Lu, J.; Li, G.; Cheng, C.; Yu, H. Risk Analysis Method of Cascade Plants Operation in Medium Term Based on Mul-ti-Scale Market and Settlement Rules. *IEEE Access* **2020**, *8*, 90730–90740. [CrossRef]
30. Mausser, H.; Rosen, D.; Ieee, I. Beyond VaR: From Measuring Risk to Managing Risk. In Proceedings of the IEEE/IAFE Conference on Computational Intelligence for Financial Engineering (CIFE), New York, NY, USA, 28–30 March 1999; pp. 163–178. [CrossRef]
31. Artzner, P.; Delbaen, F.; Eber, J.M.; Heath, D. Coherent measures of risk. *Math. Financ.* **1999**, *9*, 203–228. [CrossRef]
32. Rockafellar, R.T.; Uryasev, S. Conditional value-at-risk for general loss distributions. *J. Bank. Financ.* **2002**, *26*, 1443–1471. [CrossRef]
33. Duenas, P.; Reneses, J.; Barquin, J. Dealing with multi-factor uncertainty in electricity markets by combining Monte Carlo simulation with spatial interpolation techniques. *Iet Gener. Transm. Distrib.* **2011**, *5*, 323–331. [CrossRef]
34. Markowitz, H.M. Foundations of Portfolio Theory. *J. Financ.* **1991**, *46*, 469–477. [CrossRef]
35. Liu, M.; Wu, F.F. Portfolio optimization in electricity markets. *Electr. Power Syst. Res.* **2007**, *77*, 1000–1009. [CrossRef]

36. Zhao, S.Z.; Suganthan, P.N. Two-lbests based multi-objective particle swarm optimizer. *Eng. Optim.* **2011**, *43*, 1–17. [CrossRef]
37. Abido, M.A. Multiobjective particle swarm optimization for environmental/economic dispatch problem. *Electr. Power Syst. Res.* **2009**, *79*, 1105–1113. [CrossRef]

Article

Simulating Fish Motion through a Diagonal Reversible Turbine

Phoevos (Foivos) Koukouvinis and John Anagnostopoulos *

Laboratory of Hydraulic Turbomachines, Department of Mechanical Engineering, National Technical University of Athens, Heroon Polytechniou 9, Zografou, 157 80 Athens, Greece
* Correspondence: anagno@fluid.mech.ntua.gr; Tel.: +30-210-7721080

Abstract: Utilization of unharnessed hydro-power necessitates designing fish-friendly hydraulic machinery. Towards this effort, the present work investigates various methods for tracking fish motion, ranging from particle tracking methods to accurate, but computationally expensive, body tracking methods, such as immersed boundaries and overset meshes. Moreover, a novel uncoupled 6-Degree of Freedom tracking technique is proposed, based on an approximated pressure field around the tracked body of interest, using steady-state flow field data, and including collision detection to walls. The proposed method shows promising results in terms of accuracy, being comparable to the more computationally expensive fully coupled methods at a tiny fraction of the execution time. The new method reveals location of fish–blade impact, as well as statistics of forces, pressure and flow shear that a passing fish is subjected to, both in the normal and reverse operation of the turbine. The low computational cost of the proposed method renders it attractive for optimization studies.

Keywords: fish-friendly hydropower; body tracking; fluid–structure interaction; collision detection

Citation: Koukouvinis, P.; Anagnostopoulos, J. Simulating Fish Motion through a Diagonal Reversible Turbine. *Energies* **2023**, *16*, 810. https://doi.org/10.3390/en16020810

Academic Editors: Zhengwei Wang and Yongguang Cheng

Received: 25 November 2022
Revised: 5 January 2023
Accepted: 6 January 2023
Published: 10 January 2023

1. Introduction

Hydropower is the largest contributor to the renewable energy market, making up almost 16% of the total produced electrical energy worldwide [1]. Compared to other renewables, it has an estimated contribution in the energy production of almost 60%, based on 2021 data [2]. The role of hydropower is expected to increase dramatically in the future for two main reasons:

1. It can greatly contribute as an energy storage solution, along with batteries, allowing the increased penetration of other renewable sources [3], such as wind and solar, the latter being inherently less predictable due to weather patterns. Estimates indicate that the installed capacity of reversible hydro/pumped storage will increase by 56% till 2026 [4].

2. There is a substantial amount of unharnessed hydropower potential, with estimates indicating that, roughly, 25% of the technically achievable and 50% of the economically feasible (for the current technologies) potential is currently being utilized [5]. Indicatively, significant potential can also be found in existing infrastructure that currently lacks generating units (e.g., existing barrages, weirs, dams, canal fall structures, water supply schemes) by adding new hydropower facilities. To this purpose, the installed hydropower capacity is expected to increase steadily [6], with projections showing a steady 3% increase annually of hydropower contribution to the global energy market until the year 2030, assuming a Net Zero Emissions scenario [7].

As outlined above, the increasing penetration of hydropower, both as a means of electric grid stabilization/energy storage solution and flexible energy production, through the use of new and existing infrastructures, is expected to have an impact on aquatic life [8,9]. In recent years, there has been a steadily increasing interest in fish-friendly turbine design concepts, that, aside from efficiency objectives, also examine aquatic life survivability indicators.

Despite the increasing interest in hydropower, the concept of fish-friendly hydropower and fish-friendly turbines is not new. Since early 2000s there have been efforts to design novel turbine concepts with minimum effects on passing fish. In particular, several critical damage mechanisms have been identified [10,11], namely:

1. The possibility of impact of the fish on the rotating runner and/or stator vanes, due to the surrounding flow. The aspect of mechanical trauma could also involve grinding of fish in small passages/gaps.

2. The local pressure variation over time and in particular sudden pressure decrease, as pressure increase was found to have little effect on passing fish [11]. More extreme pressure drops could even lead to barotrauma [12] and/or cavitation inside the fish tissue, even though these are in general limited, as there are guidelines on minimum pressure for turbine operation.

3. The intensity of shear stresses, vorticity and turbulence which mechanically lead to strain of the fish tissues and/or disorient them. Abrasion can also occur as the fish is forced to the circumference of turbine pipelines due to imparted whirl [11], e.g., inside the draft tube.

Many studies have been conducted with experimental means to understand the mechanisms of fish damage and to extrapolate these findings to fish-friendly turbine concepts. Indicatively, investigations with X-rays and high-resolution photography on simulated decompression of fish tissues have illustrated that the main mechanism of barotrauma is large pressure drops [12], potentially rupturing fish bladders. Similar experimental investigations with water jets revealed shear stress thresholds that fish can experience without sustaining tissue damage (descaling) [13]. More recent experimental investigations involve autonomous sensor devices that replicate fish trajectories through a turbine, at the same time collecting data on local accelerations [14] and even pressure [15].

In an effort to actually predict the mortality and injury of fish passing through hydraulic machinery, correlations for blade strike probability and the resulting mortality or injury have been derived using live fish [16]. More recent works actually involve entirely computational methods, modeling the passing fish with various assumptions. Early approaches involved the simplification of treating a fish trajectory as a flow streamline [13] and deriving pressure or shear stress variation along this path. Newer works involve Lagrangian approaches where the fish is approximated as a particle [17] or a linear arrangement of particles with the Discrete Element Method (DEM) [18]. However, these methods have their shortcomings, mainly related to the simplifications for estimating the fluid-induced forces on the tracked fish. Whereas more accurate methods exist (e.g., overset meshes, immersed boundaries), only recently (see [19]) have they been employed in fish tracking studies and still were associated with severe limitations (lack of collision detection, enormous computational cost), prohibitive for practical application and extraction of statistical data.

The present work aims to contribute towards the goal of predicting fish motion through a hydraulic machine with simulation methods, while also collecting statistical information that could correlate with fish injury. Novel elements of the present work involve:

1. A new method for tracking the motion of rigid, immersed bodies in flow is proposed. The method aims to be placed in between fast, Lagrangian, particle methods which use empirical laws for fluid interaction and computationally expensive, fully coupled, 6-DoF methods. An additional and critical element of the present method is inclusion of wall collisions; it is highlighted that the latter is not a standard option in most CFD tools available.

2. A comparison of the performance of several methods used in the literature for tracking immersed bodies, along with the more accurate fully coupled 6-DoF tracking and the newly proposed method, assessing the importance of the aforementioned assumptions of each method.

3. The study of a reversible, diagonal-type (Deriaz) turbine in both turbine and pumping modes. The model under consideration is a five-bladed turbine previously optimized in its energy recovery characteristics [20]. Diagonal turbines could be considered an advantageous solution compared to relatively low-head Francis turbines in sites with

significant flow rate variation or when variable operation and flexibility are desired, thanks to their double regulation feature, as well as when the survivability of the passing fish is an important environmental issue, thanks to the mixed flow design and the smaller blade numbers of the runner.

In the frame of the present work, hydrodynamic forces acting on the body due to pressure are calculated and recorded, as are any collisions of the body on the rotating blades or on the hub/shroud surfaces. These data are then used to quantify the probability and the severity of injuries to passing fish. The new tool is also used to assess the behavior of the same diagonal turbine in reverse operation as a pump, due to the increasing interest in pumped storage plants at lower static head sites, within the range of this turbine type.

2. Numerical Methods

In this section, a brief overview of popular existing methodologies on body tracking in computational fluid dynamics will be provided, with emphasis on the newly proposed algorithm for uncoupled body tracking, which demonstrates promising results.

2.1. Particle-Based Methods

Particle-based methods, in the form of Discrete Particle Method, Dense Discrete Particle Method or Discrete Element Method, are very popular in computational mechanics and have also been investigated for fish tracking in hydraulic turbines; see particularly [17,18]. The common aspect of all these methods is that the interaction of the tracked body with the surrounding fluid is performed using Newton's Law of motion, by taking into account various forces. A set of equations commonly resolved in a Lagrangian approach for particle tracking are as follows:

Equation	Formulation	
Momentum equation	$m \cdot \frac{du}{dt} = \mathbf{F}_d + \mathbf{F}_p + \mathbf{F}_{vm} + \mathbf{F}_L$	(1)
Drag force	$\mathbf{F}_d = \frac{1}{2} C_d \cdot \rho \cdot A \cdot |\mathbf{u_r}| \mathbf{u_r}$	(2)
Pressure gradient force	$\mathbf{F}_p = -V_p \cdot \nabla p$	(3)
Virtual mass force	$\mathbf{F}_{cm} = C_{vm} \cdot V_p \cdot \rho \cdot \left(\frac{D\mathbf{u}}{Dt} - \frac{d\mathbf{u}_p}{dt} \right)$	(4)

where: C_d stands for the drag coefficient, ρ density of the fluid, A the surface area of the particle, $\mathbf{u_r}$ for the relative velocity vector, i.e., flow velocity minus the particle velocity, $\mathbf{u}$-$\mathbf{u}_p$, V_p the particle volume, p pressure, and C_{vm} the virtual mass coefficient. Additional forces can be included, as lift forces (both Saffman and Magnus), rotationally induced drag, as well as collision forces and non-spherical modifications of drag coefficient. A detailed description of such models is provided in [21].

A basic limitation of this family of methods is that the fluid–body interaction is done assuming particle-derived force laws, which, although they may feature corrections for non-spherical particles, in their core do not take into account body orientation. Even in the case of particle-to-particle or particle-to-boundary interactions it is assumed that the particle shape is point-like, with the only exception being the Discrete Element Method, which can take into account simple approximations of general shapes (e.g., ellipse, cylinder, etc).

2.2. Overset Meshes

Overset meshes are a technique of mesh generation and manipulation that allows relatively easy handling of complex motions/deformations [22]. In its core, it features a background mesh and one or more component meshes, which are attached to, e.g., moving bodies. Interpolations are performed from the component meshes to the background mesh to define connectivity between the otherwise disconnected regions and to form the computational domain. In general, it is a well-established, accurate and versatile method to handle 6-DoF body tracking, following a very similar methodology as the one outlined in Section 2.4. Its main drawbacks are: (a) very computationally expensive, (b) cannot take into account collisions (at least not by default), (c) the method has special requirements on mesh sizing at the overlap region and may require a per-case adjustment of schemes for cell-cutting and grid priorities, to minimize the presence of cells without suitable neighbors to interpolate to and from (also called orphan cells).

2.3. Immersed Boundaries

Immersed boundaries are a technique for mapping the geometry of solid bodies on a background mesh [23]; this is commonly achieved either by introducing appropriate source terms in the momentum equation, or by cell-cutting to conform to the body geometry. Similar to overset meshes, it is an accurate technique, hence it is quite computationally expensive (although less expensive than overset meshes). However, a main requirement is that the mesh resolution cannot be coarser than the smallest dimension of the body to be tracked, or else the solid mapping to the fluid domain will not be accurate (non-smooth surface, artificial gaps/holes).

2.4. Uncoupled 6-DoF Tracking

2.4.1. Fluid Coupling and 6-DoF

The body to be described is represented using a triangular tessellation, forming a surface grid. This grid can be further simplified to a set of points, each located at the barycenter of the respective triangular element, and a surface area dS, together with the local normal vector, $\mathbf{n}$; see also Figure 1a. At each point of the body's surface, the nearest neighbors (cell centers) of the solved flow field are found using a knn-search algorithm [24]. Then, once the closest neighbors are found, the flow variables are interpolated. Various interpolation methods have been tested in the scope of this work; in particular Inverse Distance (ID), Radial Basis Functions (RBF), Nearest Neighbor (NN), Least Squares/Linear Regression (LS) and, as a baseline, the scatteredInterpolant function in Matlab, using localized Delaunay tessellation; all aforementioned methods give similar results, as will be presented in Section 4. A brief formulation of the interpolation methods is presented below in Table 1, apart from the scatteredInterpolant function for which the interested reader is addressed to Matlab documentation and the relevant reference [25].

Table 1. Formulations of the different interpolation methods used in the scope of the present work. The symbol φ is a placeholder that stands for any interpolated quantity, e.g., velocities (u, v, w), or pressure, p. The symbol, d, denotes the Euclidean norm, or distance, of the location where interpolation is to be carried out from the nearby, i.e., neighboring, sampling points.

Interpolation Method	Formulation	
Nearest Neighbor (NN)	$\phi = \phi_i$ if $d_i = \min(d, \forall neighbors)$	(5)
Inverse Distance (ID)	$\phi = \dfrac{\sum\limits_{Neighbors} \frac{\phi}{d}}{\sum\limits_{Neighbors} \frac{1}{d}}$	(6)
Radial Basis Functions (RBF)	$\phi = \sum\limits_{Neighbors} f(d_i) \cdot w_i$ where $f(d_i) = e^{-C_{RBF} \cdot d_i}$ and weights are: $\begin{bmatrix} f(d_{11}) & f(d_{12}) & \cdots & f(d_{1n}) \\ f(d_{21}) & f(d_{22}) & \cdots & f(d_{2n}) \\ \cdots & \cdots & \cdots & \cdots \\ f(d_{n1}) & f(d_{n2}) & \cdots & f(d_{nn}) \end{bmatrix} \begin{bmatrix} w_1 \\ w_2 \\ \cdots \\ w_n \end{bmatrix} = \begin{bmatrix} \phi_1 \\ \phi_2 \\ \cdots \\ \phi_n \end{bmatrix}$	(7)
Linear Regression (Least Squares)	$\phi = \sum\limits_{Neighbors} w_i \cdot x_i$ where the weight vector is calculated as: $W = (X^T X)^{-1} X^T Y$ $X = \begin{bmatrix} x_1 \\ x_2 \\ \cdots \\ x_n \end{bmatrix} Y = \begin{bmatrix} \phi_1 \\ \phi_2 \\ \cdots \\ \phi_n \end{bmatrix}$	(8)

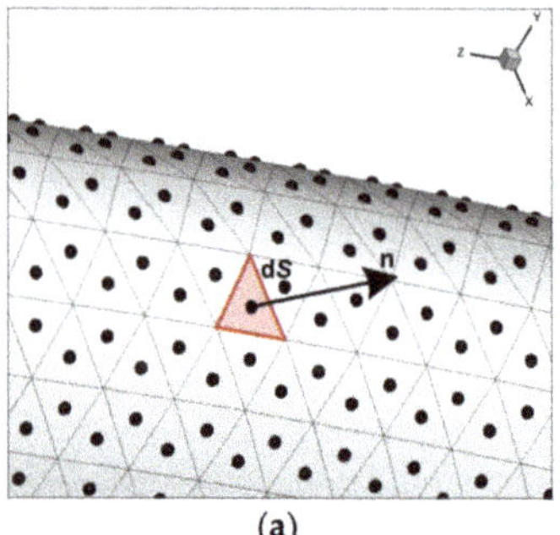
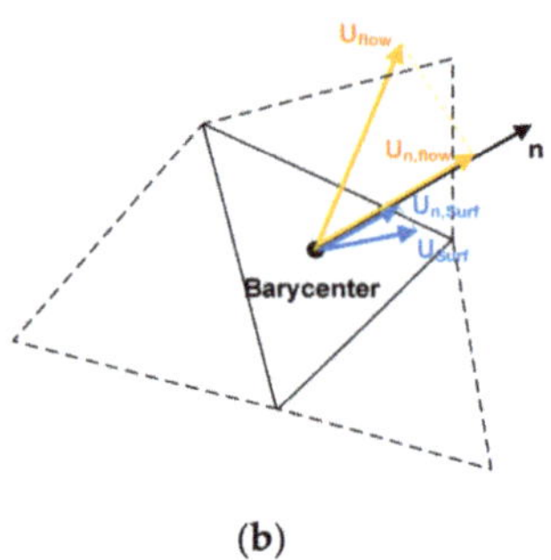
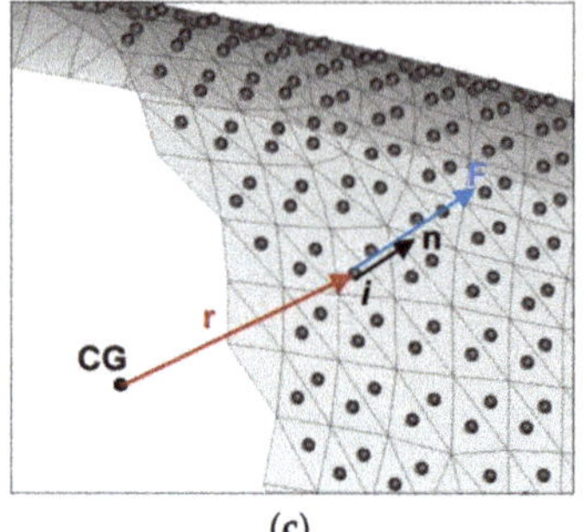

(a) (b) (c)

Figure 1. (**a**) Tessellated surface representation, (**b**) definition of the slip velocity, projected to the wall normal (**c**) Torque calculation, in respect to the center of gravity (CG).

Local pressure at each face element is calculated applying the Bernoulli equation, as the sum of interpolated flow pressure, p_{flow}, plus dynamic pressure due to slip velocity, p_{dyn}; (see also Figure 1b), i.e.,:

$$p_{local} = p_{flow} + p_{dyn} \tag{9}$$

$$p_{dyn} = \frac{1}{2}\rho \cdot \left(u_{n,flow} - u_{n,surf}\right)^2 \cdot sign\left(u_{n,surf} - u_{n,flow}\right) \tag{10}$$

where n index implies the projection on the local surface normal vector, and ϱ is the flow density (taken the same as water here). Moreover, u_{flow} implies the velocity of the interpolated flow field, whereas u_{surf} implies the velocity of the surface element, calculated by the body kinematics. Once local pressure is calculated, then the total force, **F**, and torque, **T**, on the body can be calculated by integrating along the body surface of the contributions on each infinitesimal surface element dS, associated with a normal vector **n**:

$$d\mathbf{F} = -p \cdot \mathbf{n} \cdot dS \tag{11}$$

$$d\mathbf{T} = \mathbf{r} \times d\mathbf{F} \tag{12}$$

Center of gravity motion can be integrated through Newton's law of motion, i.e.,:

$$\mathbf{F} = m \cdot \ddot{\mathbf{x}} \tag{13}$$

However, body rotation is more complicated, as it needs to be expressed in a body–local coordinate system, where the moment of inertia tensor, I, is naturally expressed. For this, it is necessary to introduce the Euler angles, $[\varphi, \theta, \psi]$, which express a sequence of rotations to convert the global coordinate system to the body-local coordinate system, by first rotating around the global Z-axis (ψ), then the transformed Y-axis (θ) and finally around the transformed X-axis (φ) [26]. With these angles, torques expressed in the global coordinate system are transformed to a local coordinate system, using the R-transformation matrix, shown below:

$$R = \begin{bmatrix} \cos\theta \cdot \cos\psi & \cos\theta \cdot \sin\psi & -\sin\theta \\ \sin\varphi \cdot \sin\theta \cdot \cos\psi - \cos\varphi \cdot \sin\psi & \sin\varphi \cdot \sin\theta \cdot \sin\psi + \cos\varphi \cdot \cos\psi & \sin\varphi \cdot \cos\theta \\ \cos\varphi \cdot \sin\theta \cdot \cos\psi + \sin\varphi \cdot \sin\psi & \cos\varphi \cdot \sin\theta \cdot \sin\psi - \sin\varphi \cdot \cos\psi & \cos\varphi \cdot \cos\theta \end{bmatrix} \tag{14}$$

For more details on derivation the reader is addressed to [26]. The transformation matrix is applied to convert global (indicated with the index G) torque vector to the body–local (indicated with index B), as $\mathbf{T_B} = R.\mathbf{T_G}$. Then, the body–local torque vector is used to calculate the angular acceleration at the body frame, keeping in mind that the body–coordinate system is rotating. Hence, the following formula applies:

$$\dot{\boldsymbol{\omega}}_B = I^{-1} \cdot (\mathbf{T}_B - \boldsymbol{\omega}_B \times I\boldsymbol{\omega}_B) \tag{15}$$

Then, body frame angular acceleration can be integrated to update the body-frame angular velocities and then transformed back to the global frame, using the inverse R transformation. At the same time, angular velocities of the Euler angles, $\boldsymbol{\omega}_E$, can be calculated from the body–coordinate angular velocities, $\boldsymbol{\omega}_B$, through a similar transform, denoted with G:

$$\boldsymbol{\omega}_E = G \cdot \boldsymbol{\omega}_B \tag{16}$$

$$G = \begin{bmatrix} 1 & \sin\varphi \cdot \tan\theta & \cos\varphi \cdot \tan\theta \\ 0 & \cos\varphi & -\sin\varphi \\ 0 & \sin\varphi \cdot \sec\theta & \cos\varphi \cdot \sec\theta \end{bmatrix} \tag{17}$$

Integration of linear/rotational motion and Euler angles is implemented using an explicit Euler, corrected Euler (Heun) or implicit Euler method [27], depending on the number of corrections applied. Sub-cycling can also be applied, that is, advancing angles and angular velocities at smaller intervals than linear displacements and velocities; this option greatly helps to conserve the angular momentum, as will be shown later on in Section 4. An indicative calculation process of the algorithm is shown in Figure 2. In any case, tracking is performed until the body exits the outlet boundary of the turbine/pump configuration.

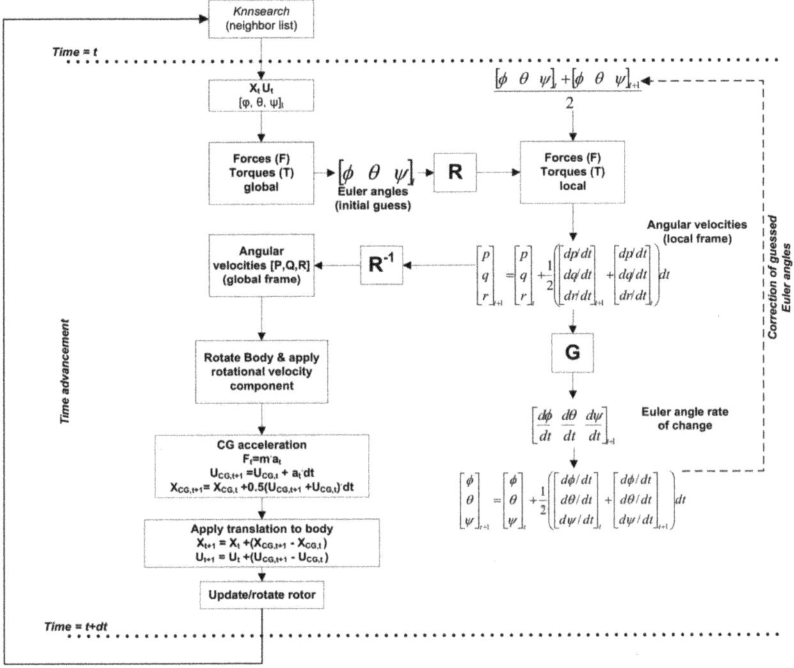

Figure 2. The algorithm used for 6-DoF body tracking.

The fish model is a simplification of a more complex fish shape, obtained from a CAD repository [28]. The fish is approximated as an ellipsoid with dimensions of $0.173 \times 0.04 \times 0.017$ (L $\times$ H $\times$ D), mass of 63.6 g, assuming neutral buoyancy, and moments of inertia of: $I_1 = 1.012 \times 10^{-4}$ kg·m^2, $I_2 = 9.71 \times 10^{-5}$ kg·m^2, $I_3 = 6.099 \times 10^{-6}$ kg·m^2 (off-diagonal moment of inertia terms are zero). While the more accurate methods discussed here can handle the complexity of actual fish geometry, it was preferred to resort to a simpler geometric shape, in the form of an ellipsoid, for replicability and easier reference.

2.4.2. Collision Detection

An important aspect of the uncoupled 6-DoF approach proposed here is the inclusion of a collision detection algorithm that prevents the tracked body from penetrating solid boundaries. The algorithm is based on the application of an external spring type force from each boundary (wall) element, generally represented as:

$$F = k_{spring} \cdot abs(\min(0, d_{proj})) \cdot \mathbf{n} \tag{18}$$

where k_{spring} is the spring constant (tested values in the range of 100–1000 N/m), $\mathbf{n}$ is the local wall normal vector and d_{proj} is the projected distance of each tracked body surface element to the closest wall element. The process for identifying collisions is as follows (see also Figure 3); for each surface element (highlighted with dark blue in Figure 3) of the tracked body,

- use knn-search to find the closest (i.e., min distance, d_{min}) wall boundary element (indicated with red highlight in Figure 3)
- project the minimum distance vector at the local wall normal, $\mathbf{n}$, to calculate d_{proj}
- if the projection is positive, do nothing (no penetration)
- if the projection is negative, apply Equation (18).

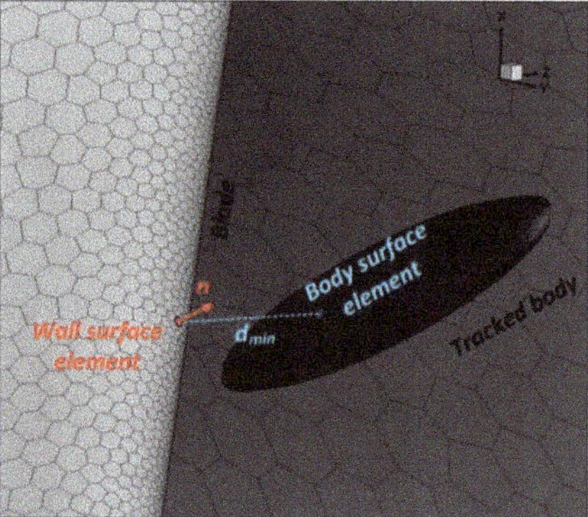

Figure 3. A schematic showing the identification of collisions and the annotations used in Equation (3).

The aforementioned methodology assumes that the surface is represented at an adequate resolution and that curvature is resolved well enough so that large changes in wall normal direction are captured. In the opposite case, no-penetration for all impact angles cannot be ensured. As an indicative demonstration, the collisions of the tracked object are presented in Figure 4, as it passes from the stator blades of the simulated turbine; for a detailed description, see the next section. The point of this demonstration is that the method presented here is capable of tracking efficiently collisions with complex, curved bodies.

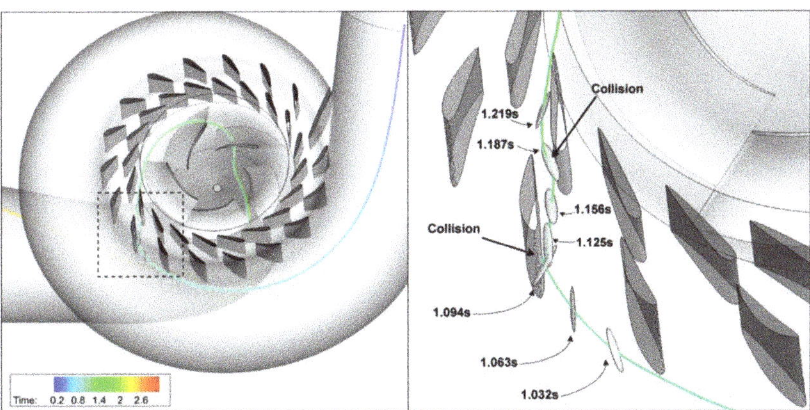

Figure 4. A schematic showing the path of the tracked object through the turbine, colored according to trajectory timing. The magnified inset shows instances before and after the collision at stator blades.

3. Case Description

An optimized reversible diagonal turbine has been the focus of the present investigation. The design has been a result of optimization, as discussed in previous publications by the authors [20]. The design features a five-bladed rotor and two series of stator blades, an outer one with 17 blades and an inner one with 20 blades. The main dimensions of the hydraulic machine are: casing inlet surface area 2.26 m^2, draft tube outlet surface area 4.99 m^2, max. radius of casing 3 m, runner diameter 2 m. An indicative view of the geometry is shown in Figure 5.

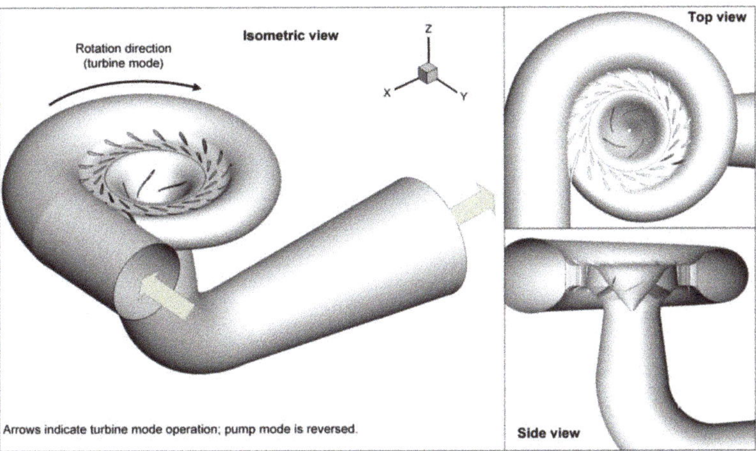

Figure 5. Views of the reversible diagonal turbine, showing details of the design such as the two rows of stator blades, the five-bladed runner and the draft tube.

The conditions of operation of the reversible turbine are for (1) turbine mode: flow rate 12.9 m^3/s, total head 45 m, rotation speed 300 rpm, efficiency 89%, (2) pump mode: flow rate 10.3 m^3/s, total head 28 m, rotation speed 300 rpm, efficiency 86%. An indicative view of the turbine operation, as well as the computational mesh, is shown in Figure 6.

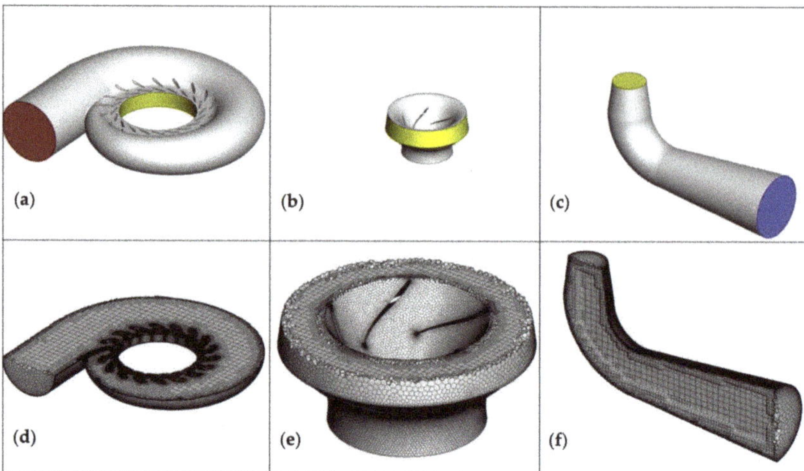

Figure 6. Top row: the regions of the computational domain, (**a,d**) the spiral casing, (**b,e**) the rotor, (**c,f**) the draft tube. Bottom row: details of the computational mesh. All images correspond to configurations for the sliding mesh approach.

Design performance, in both normal and reversed operation, has been examined using a sliding mesh approach; in this approach, non-conformal interfaces are defined at surfaces separating rotating from stationary regions. These interfaces act as interpolation zones over which information passes from one mesh region to the other. The structuring of the computational domain is shown in Figure 6a–c, indicating the different parts involved (casing volume, rotor and duct), along with non-conformal interfaces (indicated in yellow). Views of the mesh are included in Figure 6d–e, showing details of mesh structure; the computational mesh is hybrid, poly-hex core, with refinement towards wall boundaries and especially rotor/stator blades, ensuring a maximum y+ of ~300. Three layers of prism-type cells are included at all wall boundaries to capture near-wall gradients. The total computational mesh consists of 2.3 million cells, as demonstrated by a mesh-independence study conducted through prior work of the authors [20].

Each investigated configuration is adapted to the particularities of the modeling method:

— Particle tracking or the uncoupled 6-DoF tracking are performed with sliding meshes. In this case, the rotor consists of a moving (rotating) mesh in the case of particle tracking, or rotating reference frame in the case of the uncoupled 6-DoF tracking. Particle tracking was done in a transient analysis in ANSYS Fluent v2021, whereas the uncoupled 6-DoF was performed using a steady state flow field, also from ANSYS Fluent v2021. In the frame of the present investigation, particles emulating the fish are approximated as equivalent spheres of radius 24.7 mm, considering drag, lift, added mass force and rotation. Non-spherical effects are taken into account using a non-spherical factor of 0.66 corresponding to the surface area ratio of the fish model and a sphere of the same equivalent volume.

— Overset mesh body tracking was performed using similar meshing methods and resolution as the sliding mesh approach, but with a different structure to accommodate for overset domains; an indicative view of the mesh structure, showing overlapping regions, is in Figure 7. The set-up can be summarized as follows: a rotating mesh was used for the rotor, which was considered as component mesh with the lowest grid priority. The fish was introduced in another component mesh, enclosed within a sphere of diameter 0.74 m, or 4.35 fish lengths; it had intermediate grid priority. The fish domain 6-DoF properties (mass, moments of inertia) are the same as the uncoupled

6-DoF approach (see Section 2.4.1). Finally, the background mesh consisting of the complete flow path (including casing, stator blades and duct) had the highest grid priority. Overset simulation was performed using ANSYS Fluent v2021. The body was tracked with an implicit scheme, to maintain stability while using large time steps, in the order of 1 ms.

— Immersed boundary was set up using CFX v2021, by introducing the flow passage through the turbine and modeling the rotor blades and the fish as immersed bodies. However, due to the large aspect ratio of the fish and blades (effectively very thin bodies) it was impossible to obtain a good representation for reasonable mesh resolutions and a tractable computational cost.

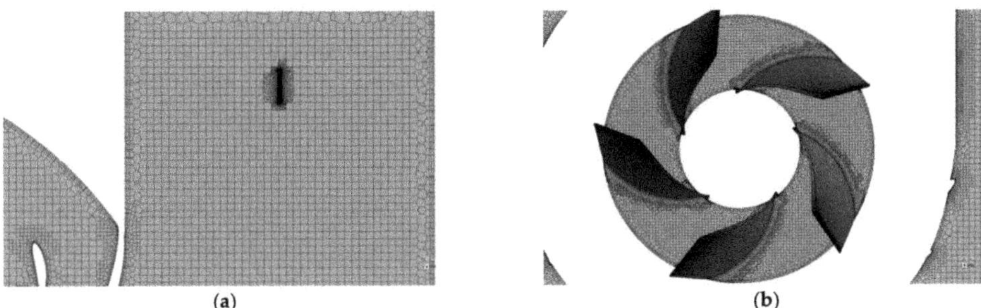

(a)   (b)

Figure 7. Indicative views of the overset mesh: (**a**) near the region of the tracked body (ellipsoid) (**b**) at the rotor blades. The overlap of the mesh zones is visible.

4. Validation

To test the validity of the uncoupled 6-DoF method proposed here, several tests have been performed.

4.1. Influence of the Interpolation Method

Here a comparison of the interpolation methods that can be used to reconstruct the flow field on the surface of the tracked body is presented. The results are shown in Figure 8a, showing a similar trend up to the rotor.

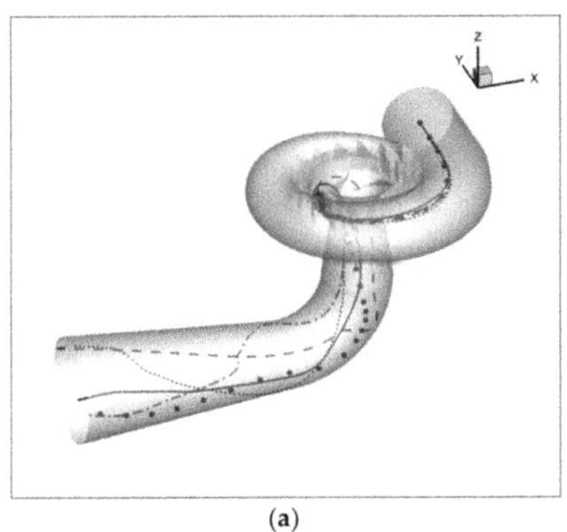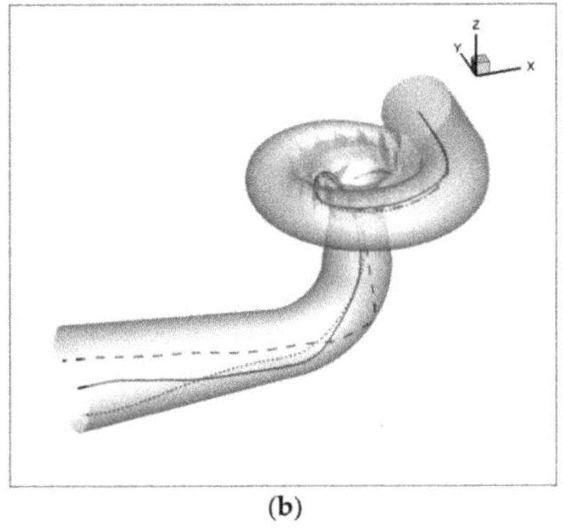

(a)   (b)

Figure 8. Body trajectories in turbine mode (**a**) Effect of interpolation method: Inverse distance (continuous line), scatteredInterpolant (filled circles), linear regression (dashed line), nearest neighbor (dotted line), radial basis functions (dashed-dotted line). (**b**) Effect of resolution: dashed line 500 elements, solid line 1000 elements, dotted line 2000 elements.

When comparing the methods it is notable that the localized Delaunay tessellation is the most computationally intensive, which is reasonable given the cost of constructing tetrahedral elements from the localized point cloud to perform interpolations. The Radial Basis Function interpolation has the drawback of a user-defined term, c_{RBF} (see Table 1), which is case-dependent and can lead to ill-conditioned matrices, potentially causing instabilities due to inaccurate interpolation after inversion. A similar issue, although to a lesser extent, has been observed with the Least Squares/Linear Regression method; even if lacking tuned coefficients, instabilities may arise due to inversion, depending on the number and positions of neighboring sampling points identified. The most robust method tended to be the Inverse Distance interpolation, which is rather stable and comparable to all other methods. The fastest method is the Nearest Neighbor search, which however tends to be sensitive in areas of sparse sampling points (i.e., coarse background mesh).

4.2. Resolution of the Body

Different resolutions were tested for the ellipsoid fish model, ranging from 0.5–2 k surface elements (triangles). All gave a very similar trajectory and body rotations up to the point of the interaction with the runner, as shown in Figure 8b; hence, a resolution of 1 k elements was deemed enough for the remaining investigations. Beyond that point, differences are observed, which are due to the strongly non-linear driving force from the flow field, which effectively represents a swirling flow. Hence, small variations of the position of the body can lead to detrimental changes in the trajectory.

The discrepancies after the rotor interaction, both due to interpolation method and resolution, are intrinsic to the tracking of bodies immersed in fluids. In fact, the motion of a body through a fluid has been proven to be very sensitive to small variations of the exact flow conditions, leading to different results in both the orientation and the exact trajectory of a fixed point on the body; see [29–31]. Furthermore, such chaotic motions have been identified in the case of tumbling ellipsoid objects, such as the one examined in the scope of the present work [32]. Hence, flow velocity variations, generated due to error accumulation from interpolation or resolution, are expected to lead to different trajectories after a point. Nevertheless, the relative convergence of flow paths up to the point of the turbine rotor demonstrates that the fast-tracking method proposed is capable of providing a reasonable estimation of the trajectory of a fish-like object until the interaction with the rotor.

4.3. Prediction of Free Rotation of the Fish Model

Here we test the evolution of angular velocities of the body, without any torques/forces, to assess conservation of angular momentum, comparing against a reference solution from SIMULINK 6-DoF solver, with initial conditions $\omega_x = 10$ rad/s, $\omega_y = 50$ rad/s, and $\omega_z = 100$ rad/s. As shown in Figure 9, reproduction of the reference results is perfect when sub-cycling is used. Note that this case poses an extreme configuration, as any rotation in the simulations to be further examined will effectively be dampened by the surrounding fluid. However, this idealized case serves as validation for testing the numerical integration scheme proposed in Section 2.4.1.

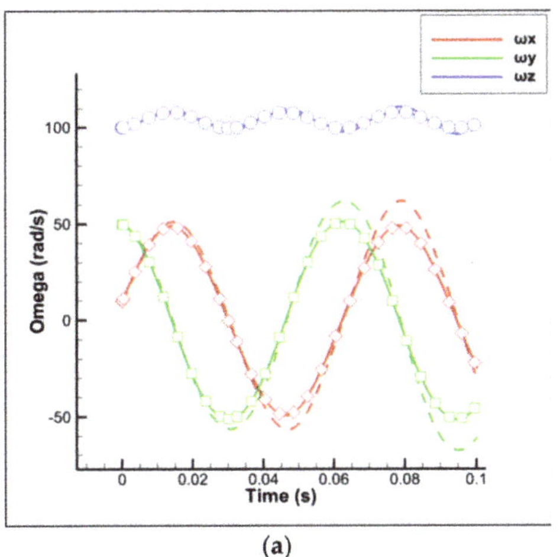

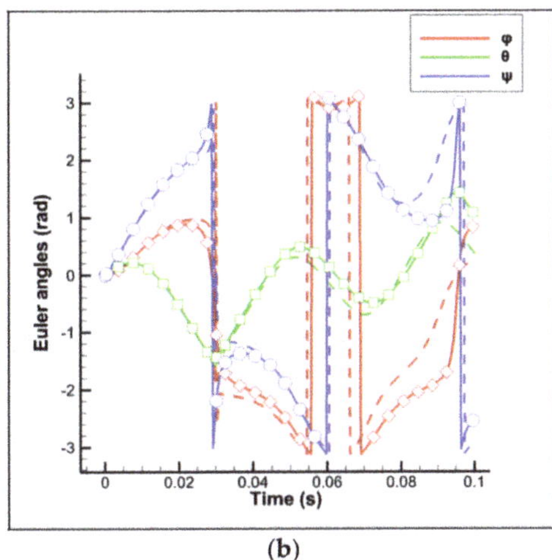

(a) **(b)**

Figure 9. Validation of the 6-DoF solver: (**a**) angular velocities, (**b**) Euler angles. Points are SIMULINK, dashed lines is the present solver without sub-cycling and continuous lines with sub-cycling.

4.4. Prediction of a Fish Track in the Diagonal Turbine Model/Comparison of Different Techniques

Here we test the different methods discussed in Section 2. As shown in Figure 10, the developed method is effectively equivalent to the much more computationally demanding overset 6-DoF technique, whereas also takes into account boundary collisions. Unsteady particle tracking tends to follow the same path as the two other methods, but can deviate substantially; see for example Figure 10, the Turbine mode. This is justified given that both the particle tracking and streamlines effectively perform a point-like interpolation of flow variables at the center of the equivalent sphere.

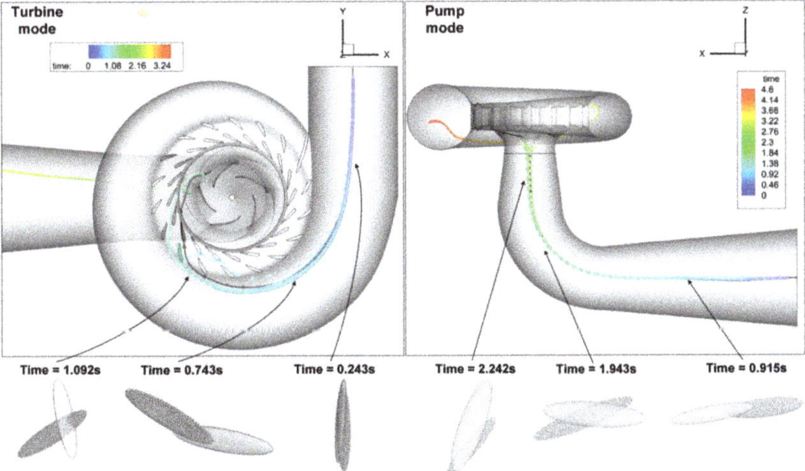

Figure 10. Comparison of fish tracks predicted with different models. Continuous colored line corresponds to the uncoupled 6-DoF. Spheres are positions predicted with overset 6-DoF (spheres are markers, *not* the body shape). Squares are positions predicted with unsteady particle tracking. Black line with vectors is a streamline for reference. The bottom row shows the position and orientation predicted by the fast-tracking 6-DoF approach (grey spheres), compared with the surface from overset 6-DoF.

To give an indication of the computational cost:

- the uncoupled 6-DoF requires ~1–2 h for a single track using a single processor (indicatively Intel i7-3632), while limiting the body motion to 2.5–10% of the max. body dimension per time step for stability, depending on the location in the turbine (lower near the blades, or in vortices, e.g., draft tube in turbine mode).
- the unsteady particle tracking requires 1–2 days on a 6-core machine (indicatively Intel i7-8850), depending on the time step used; higher time steps are faster but more prone to no-clipping.
- the overset method has an immense computational cost, requiring 3–4 days to reach the runner on a 16-core machine (indicatively Intel Xeon Gold 6140). The simulation was not pursued further, due to lack of the boundary collision algorithm implemented in Fluent v2021, which caused the body to effectively pass through the stator blades and rotor wall, resulting in an unrealistic configuration.
- the streamline is trivial to complete, requiring mere seconds; however, it does not consider body motion or rotor rotation.

5. Statistics of Trajectories

The relatively low computational cost of the uncoupled 6-DoF approach allows the observation of multiple tracks, including any boundary collisions. In Figure 11 indicative trajectories of fish at different starting positions are shown, both for turbine and pump operation; the most aggressive collisions are shown, above a threshold force of 100 N (for a reference, hydrodynamic forces are at least 10 times smaller, depending on the location). By analyzing the locations of impact, it was found that:

- in turbine mode, 18.75% of tracked fish undergo a rotor blade impact and 43% undergo a stator blade impact, although this is much milder than the rotor impact. Indicatively, stator blade impact is in the order of 100 N, whereas rotor blade impact may be in excess of 400 N.
- in pumping mode, 10% of tracked fish undergo a rotor blade impact and 40% a stator blade impact. Here however, the magnitude of impacts is rather comparable, in the order of 100 N, most likely due to the fact that at the inlet section of the runner at pump mode the circumferential velocity is smaller than the turbine mode. As a measure of comparison, the hydrodynamic forces are much smaller, in the order of 1–10 N, depending on the location (higher near/at the runner).

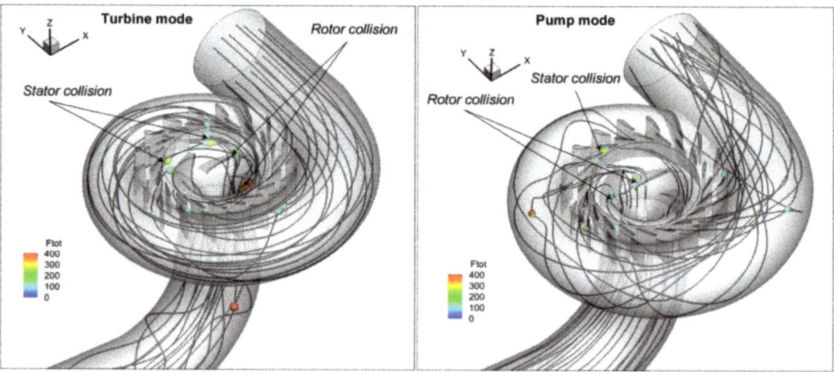

Figure 11. Trajectories and observed collisions of fish for different starting positions, at turbine mode (**left**) and pumping mode (**right**).

An alternative way to illustrate the above is through a histogram comparing the forces that a passing fish is subjected to, as shown in Figure 12. In this histogram, a comparison of turbine mode and pumping mode operation is shown. Forces between [20, 400] N are much more probable in pumping mode, whereas forces between [400, 1000] N are somewhat more pronounced in turbine mode. However, in the latter case they represent a tiny fraction of samples, nearly 0.5%.

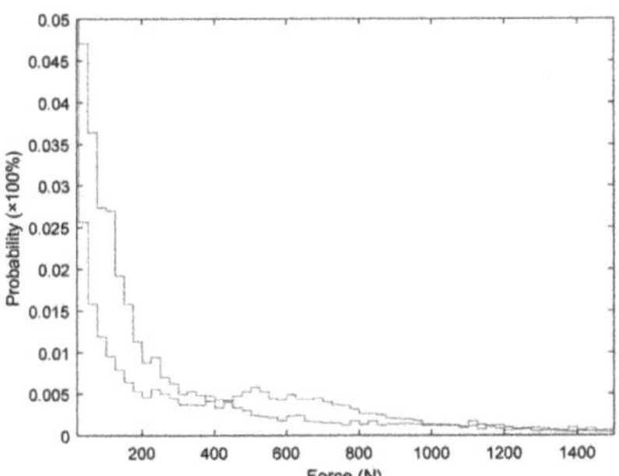

Figure 12. Histogram of force probability for the turbine (blue color) and pump (red color) modes; note forces below 20 N have been clipped, as they represent around 70% probability.

It is also notable that after the interaction with the runner, in the draft tube (in turbine mode) or the casing (in pump mode) the fish is subjected to swirl, often being sheared along the walls. This is on par with previous observations (see, e.g., [11]) stating the necessity to use smooth surfaces to avoid fish descaling and injury.

Potential design modifications could explore the reduction of stator number blades or smoothing thin edges so that the probability of collision, or at least the impact force, with a passing fish is reduced.

Apart from the collision statistics, other information that can be extracted is of the pressure rate of change, which can be an indication of barotrauma. Effectively, in all cases, the pressure change takes place mainly at the rotor part, within 0.2–0.3 s, even if the exact time interval is shifted forward/backward due to variations of the path taken. From the tracks observed (see Figure 13), no path was found to experience low pressures that could potentially justify internal cavitation.

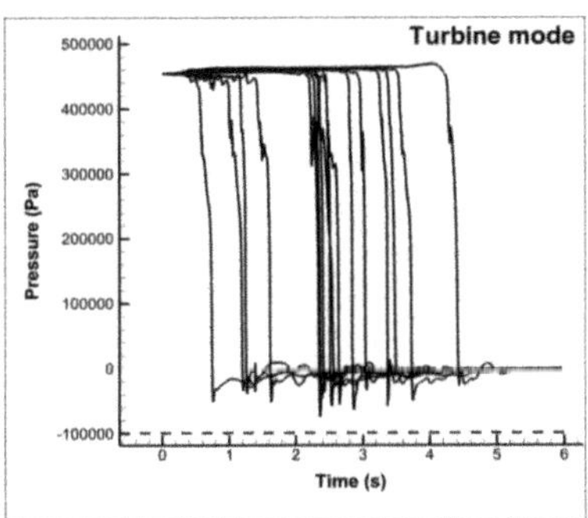
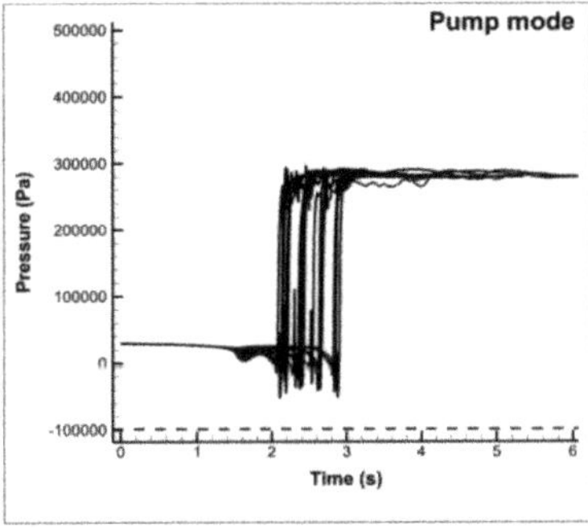

Figure 13. Gauge pressure history for different tracks at turbine mode (**left**) and pump mode (**right**). Cavitation threshold is at −98,985 Pa, shown with a dashed line.

Furthermore, aside from pressure and impact force, it is possible to evaluate localized shear stresses, at least from the flow. Shear stresses can be calculated as [33]:

$$\tau_{ij} = \mu_{eff} \cdot \left(\frac{\partial u_i}{\partial x_j} + \frac{\partial u_j}{\partial x_i} \right) \tag{19}$$

where μ_{eff} stands for the effective viscosity, i.e., both the laminar and turbulent contributions, whereas velocity gradients can be calculated over the tracked body surface from the interpolated flow field using the Gauss theorem:

$$\nabla\phi = \frac{1}{V}\sum_n \phi \cdot \mathbf{n} \cdot dS \tag{20}$$

where φ is a placeholder that can be any velocity component. It is highlighted here that this gradient is based solely on the flow field and not the interaction with the body, as the estimation using fast tracking cannot reproduce boundary layer effects. Nevertheless, it is an indication of the local flow conditions that affect fish passage when considering the maximum of the absolute value of all shear stress components.

An indicative comparison of the maximum shear stresses experienced by the body as it passes through the turbine/pump is shown in Figure 14. Shear stresses are notably smaller than pressure, with their maximum value being in the order of 5000 Pa at pump mode and 2500 Pa in turbine mode. However, shear damage thresholds are substantially lower; experiments determined a damage threshold at around 1600 Pa for minor injuries (descaling) [13]. Hence, 16% of the fish tracks exceed the minor damage threshold in pump mode, whereas 5% of the fish tracks exceed the minor damage threshold in turbine mode.

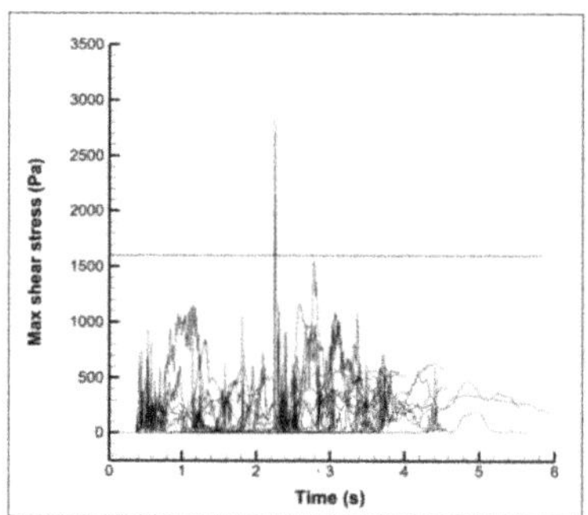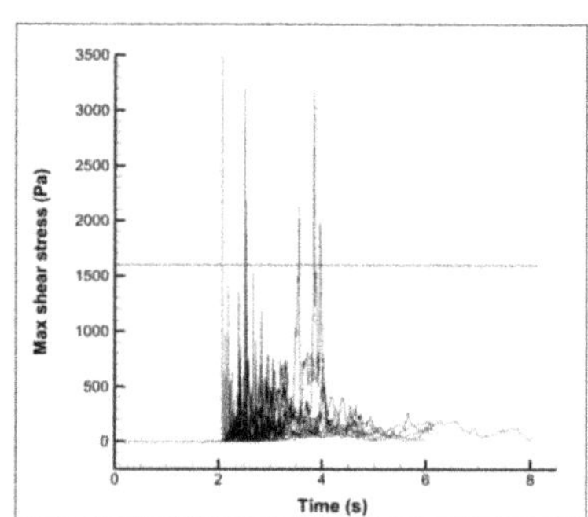

Figure 14. Maximum shear stress history for different tracks at turbine mode (**left**) and pump mode (**right**). The dotted horizontal line represents the minor injury threshold of 1600 Pa.

6. Conclusions

Given the trend towards increased utilization of unharnessed/unexploited hydraulic potential of water streams, it is expected that hydraulic turbine design should accommodate harmless interaction with aquatic life. Hence, the accurate estimation of the probability of fish passage in hydraulic machines of future hydropower will become another objective along with the efficiency requirements. Consequently, tools that can provide fast and accurate estimations for the design optimization of new fish-friendly pumps and turbines will be of great importance. The present work contributes towards this goal by proposing a method that is fast, flexible and considers the physical dimensions of the tracked body, instead of the simpler point-like perspective commonly used. At the same time it also incorporates collision detection that most commercial CFD software lacks, or such software relies on external coupling with other dedicated software. The proposed uncoupled 6-DoF method has a computational cost reduced by almost two orders of magnitude, while it also offers accuracy comparable to that of more detailed, fully-coupled methods, such as overset or Immersed Boundary 6-DoF. The fast execution of the proposed methodology enabled calculation of statistics, revealing problematic locations where fish collisions take place, while also making predictions of forces and potential damage or lethality. Hence, the present tool can be used alongside optimization methods for designing fish-friendly turbines.

Author Contributions: Conceptualization, P.K. and J.A.; methodology, P.K.; software, P.K.; validation, P.K.; formal analysis, P.K.; investigation, P.K. and J.A.; resources, J.A.; data curation, P.K. and J.A.; writing—original draft preparation, P.K. and J.A.; writing—review and editing, P.K. and J.A.; visualization, P.K.; supervision, J.A.; project administration, J.A.; funding acquisition, P.K. All authors have read and agreed to the published version of the manuscript.

Funding: This research has been co-financed by the European Union and Greek national funds through the Operational Program Competitiveness, Entrepreneurship & Innovation, under the call RESEARCH–CRE-ATE–INNOVATE (T1EDK-01334, 2018–21).

Data Availability Statement: Not applicable.

Conflicts of Interest: The authors declare no conflict of interest.

Nomenclature

m	Mass [kg]
$\mathbf{F}$	Force [N]
C_d	Drag coefficient [-]
ρ	Density [kg/m^3]
A	Surface area [m^2]
$\mathbf{u}$	Velocity [m/s]
V	Volume [m^3]
p	Pressure [Pa]
C_{Vm}	Virtual mass coefficient [-]
φ	Generic interpolated quantity (pressure or velocity here, [m/s] or [Pa])
$d, \mathbf{d}$	Distance [m]
w	Radial basis function/Linear regression weight (unit dependent on interpolated quantity)
$\mathbf{x}, \mathbf{r}$	Coordinate vector [m]
$\mathbf{n}$	Normal vector [-]
$\mathbf{T}$	Torque [N.m]
dS	Infinitesimal surface area [m^2]
$\ddot{x}$	Acceleration [m/s^2]
R	Transformation tensor [-]
φ, θ, ψ	Euler angles [rad]
ω	Angular velocity [rad/s]
$\dot{\omega}$	Angular acceleration [rad/s^2]
I	Moment of inertia tensor [kg.m^2]
G	Euler angular velocity transformation tensor [-]
k_{spring}	Spring constant [N/m]
τ_{ij}	Shear stress [Pa]
μ_{eff}	Effective dynamic viscosity [Pa.s]

References

1. BP, p.l.c.bp Statistical Review of World Energy. 2022. Available online: https://www.bp.com/content/dam/bp/business-sites/en/global/corporate/pdfs/energy-economics/statistical-review/bp-stats-review-2022-full-report.pdf (accessed on 24 November 2022).
2. IRENA. *Renewable Energy Statistics 2022*; International Renewable Energy Agency: Abu Dhabi, United Arab Emirates, 2022.
3. Jager, H.I.; De Silva, T.; Uria-Martinez, R.; Pracheil, B.M.; Macknick, J. Shifts in hydropower operation to balance wind and solar will modify effects on aquatic biota. *Water Biol. Secur.* **2022**, *1*, 100060. [CrossRef]
4. International Energy Agency. *How Rapidly Will the Global Electricity Storage Market Grow by 2026?* International Energy Agency: Paris, France, 2021.
5. Killingtveit, Å. *15—Hydroelectric Power*; Elsevier Ltd.: Amsterdam, The Netherlands, 2020; ISBN 9780081028865.
6. International Energy Agency. *Hydropower*; International Energy Agency: Paris, France, 2021.
7. Murdock, H.; Duncan, G.; Thomas, A. *Renewables Global Status Report*; International Atomic Energy Agency: Vienna, Austria, 2021.

8. Schmutz, S.; Sendzimir, J. *Riverine Ecosystem Management*; Schmutz, S., Sendzimir, J., Eds.; Springer International Publishing: Cham, Switzerland, 2018; ISBN 978-3-319-73249-7.
9. Gøtske, E.K.; Victoria, M. Future operation of hydropower in Europe under high renewable penetration and climate change. *Iscience* **2021**, *24*, 102999. [CrossRef] [PubMed]
10. Sale, M.J.; Cada, G.F.; Carlson, T.J.; Dauble, D.D.; Hunt, R.T.; Sommers, G.L.; Rinehart, B.N.; Flynn, J.V.; Brookshier, P.A. *DOE Hydropower Program Annual Report for FY 2001*; Idaho National Lab.: Idaho Falls, ID, USA, 2002.
11. Cada, G.F.; Coutant, C.C.; Whitney, R.R. *Development of Biological Criteria for the Design of Advanced Hydropower Turbines*; EERE Publication and Product Library: Washington, DC, USA, 1997.
12. Brown, R.S.; Pflugrath, B.D.; Colotelo, A.H.; Brauner, C.J.; Carlson, T.J.; Deng, Z.D.; Seaburg, A.G. Pathways of barotrauma in juvenile salmonids exposed to simulated hydroturbine passage: Boyle's law vs. Henry's law. *Fish. Res.* **2012**, *121–122*, 43–50. [CrossRef]
13. Garrison, L.A.; Fisher, J.R.K.; Sale, M.J.; Cada, G. Application of Biological Design Criteria and Computational Fluid Dynamics to Investigate Fish Survival in Kaplan Turbines. In Proceedings of the HydroVision 2002 Technical Papers, Portland, OR, USA, 29 July–2 August 2002; HCI Publications: Kansas City, MO, USA, 2002. [CrossRef]
14. Fu, T.; Deng, Z.D.; Duncan, J.P.; Zhou, D.; Carlson, T.J.; Johnson, G.E.; Hou, H. Assessing hydraulic conditions through Francis turbines using an autonomous sensor device. *Renew. Energy* **2016**, *99*, 1244–1252. [CrossRef]
15. Martinez, J.; Deng, Z.; Titzler, P.; Duncan, J.; Lu, J.; Mueller, R.; Tian, C.; Trumbo, B.; Ahmann, M.; Renholds, J. Hydraulic and biological characterization of a large Kaplan turbine. *Renew. Energy* **2018**, *131*, 240–249. [CrossRef]
16. Van Esch, B.; Spierts, I. Validation of a model to predict fish passage mortality in pumping stations. *Can. J. Fish. Aquat. Sci.* **2014**, *71*, 1910–1923. [CrossRef]
17. Romero-Gomez, P.; Lang, M.; Michelcic, J.; Weissenberger, S. Particle-based evaluations of fish-friendliness in Kaplan turbine operations. *IOP Conf. Series Earth Environ. Sci.* **2019**, *240*, 042016. [CrossRef]
18. Richmond, M.; Romero-Gómez, P. Fish passage through hydropower turbines: Simulating blade strike using the discrete element method. *IOP Conf. Series Earth Environ. Sci.* **2014**, *22*, 062010. [CrossRef]
19. Huang, Z.; Cheng, Y.; Wu, J.; Diao, W.; Huai, W. FSI simulation of dynamics of fish passing through a tubular turbine based on the immersed boundary-lattice Boltzmann coupling scheme. *J. Hydrodyn.* **2022**, *34*, 135–147. [CrossRef]
20. Kassanos, I.; Alexopoulos, V.; Anagnostopoulos, J. Numerical design methodology for reversible Deriaz turbine with high energy performance and reduced fish impacts. In Proceedings of the IAHR 2022, 31st Symposium on Hydraulic Machinery and Systems, Trondheim, Norway, 26 June–1 July 2022. [CrossRef]
21. ANSYS®. ANSYS Fluent v2021 User Manual, 12.2.1 Equations of Motion for Particles. ANSYS Inc. Available online: https://dl.cfdexperts.net/cfd_resources/Ansys_Documentation/Fluent/Ansys_Fluent_Theory_Guide.pdf (accessed on 24 November 2022).
22. Sharma, A.; Ananthan, S.; Sitaraman, J.; Thomas, S.; Sprague, M.A. Overset meshes for incompressible flows: On preserving accuracy of underlying discretizations. *J. Comput. Phys.* **2020**, *428*, 109987. [CrossRef]
23. Kim, W.; Choi, H. Immersed boundary methods for fluid-structure interaction: A review. *Int. J. Heat Fluid Flow* **2019**, *75*, 301–309. [CrossRef]
24. Friedman, J.H.; Bentley, J.L.; Finkel, R.A. An Algorithm for Finding Best Matches in Logarithmic Expected Time. *ACM Trans. Math. Softw.* **1977**, *3*, 209–226. [CrossRef]
25. Amidror, I. Scattered data interpolation methods for electronic imaging systems: A survey. *J. Electron. Imaging* **2002**, *11*, 157. [CrossRef]
26. Etkin, B. *Dynamics of Atmospheric Flight*; Dover Publications: Mineola, NY, USA, 2005; ISBN 978-0486445229.
27. Süli, E.; Mayers, D.F. *An Introduction to Numerical Analysis*; Cambridge University Press: Cambridge, UK, 2003; ISBN 9780521007948.
28. Ozdemir, R. Fish Surface Loft. Available online: https://grabcad.com/library/fish-surface-loft (accessed on 18 March 2022).
29. Aref, H.; Jones, S.W. Chaotic motion of a solid through ideal fluid. *Phys. Fluids A Fluid Dyn.* **1993**, *5*, 3026–3028. [CrossRef]
30. Borisov, A.V.; Mamaev, I.S. On the motion of a heavy rigid body in an ideal fluid with circulation. *Chaos Interdiscip. J. Nonlinear Sci.* **2006**, *16*, 13118. [CrossRef] [PubMed]
31. Aref, H. Chaotic advection of fluid particles. *Philos. Trans. R. Soc. London. Ser. A Phys. Eng. Sci.* **1990**, *333*, 273–288. [CrossRef]
32. Essmann, E.; Shui, P.; Popinet, S.; Zaleski, S.; Valluri, P.; Govindarajan, R. Chaotic orbits of tumbling ellipsoids. *J. Fluid Mech.* **2020**, *903*, A10. [CrossRef]
33. Versteeg, H.K.; Malalasekera, W. *An Introduction to Computational Fluid Dynamics: The Finite Volume Method*; Pearson Education Limited: London, UK, 2007; ISBN 978-0-13-127498-3.

*energies*

MDPI

Communication

Lubricant Oil Consumption and Opportunities for Oil-Free Turbines in the Hydropower Sector: A European Assessment

Emanuele Quaranta

European Commission, Joint Research Centre, 21027 Ispra, Italy; emanuele.quaranta@ec.europa.eu or quarantaemanuele@yahoo.it

Abstract: Lubricant oil is used in hydropower units to minimize friction, improving the turbine efficiency and reducing the wear. However, oil production is a pollutant process, while eventual spills may affect water quality and damage freshwater ecosystems. In this study, the lubricant oil consumption of the European hydropower fleet was estimated (considering its installed capacity of 254 GW). The energy required to extract and process the oil was also estimated based on available literature data. The oil consumption was estimated to be 22×10^3 tons/year, and the associated CO_2 emissions are 10^5 tons/year. The lubricant oil costs EUR 116 million per year. Although this is only 0.0022% of the oil consumed as a primary energy source in the European context, and less than 0.4% of the European industry consumption of lubricant oil, results show that new bearing types and oil-free turbines (e.g., self-lubricating or water-lubricated turbines) can improve the sustainability of the hydropower sector, minimizing the risks and impacts associated with incidental oil spills and leakages. The provided data can also be used for Life Cycle Assessment analyses.

Keywords: bearing; hydropower; lubricant; oil-free; turbine

Citation: Quaranta, E. Lubricant Oil Consumption and Opportunities for Oil-Free Turbines in the Hydropower Sector: A European Assessment. *Energies* **2023**, *16*, 834. https://doi.org/10.3390/en16020834

Academic Editors: Yongguang CHENG and Zhengwei Wang

Received: 1 December 2022
Revised: 5 January 2023
Accepted: 9 January 2023
Published: 11 January 2023

1. Introduction

Hydropower accounts for 1360 GW of installed power worldwide and 254 GW in Europe, and globally generates 4250 TWh/year [1]. Hydropower is a renewable energy technology and it provides flexibility to the electric grid and ancillary services. Multipurpose reservoirs provide several additional benefits, e.g., water storage and flood control. On the other hand, the alteration of aquatic ecosystems is perceived to be the main cause of hydropower-related impacts, with risks imposed on migrating fish and biodiversity. Therefore, hydropower needs to achieve a good balance between electricity generation, social benefits and impacts on the ecosystem and biodiversity, limiting the conflict with the environmental objectives of water policies. Within this context, novel emerging technologies are discussed in [2–4], where their benefits in a European context are estimated. The implementation of these technologies can mitigate the impacts of hydropower on the environment, while generating renewable energy and providing flexibility.

The impacts associated with turbine operation may be relevant in certain contexts. The turbine may affect water quality (e.g., oxygen deficit and oil spills) and may damage migrating fish which pass through it. With the aim of reducing these impacts, environmentally enhanced turbines (EETs), e.g., auto-venting turbines, self-aerated draft tubes, oil-free turbines and the so-called "fish-friendly" turbines, have gained attention in the past two decades [5]. However, although the benefits of these turbines (e.g., the reduced fish mortality of fish-friendly turbines), and the risks of traditional turbines, are clear and evident, the quantitative estimation of some of them is challenging on a large scale.

Among other things, the impacts of lubricant oil have not been quantitatively estimated yet on a large scale, e.g., in Europe. Oil lubricates the runner blade trunnion bearing and sliding parts of the operating mechanism in the hub [6], e.g., in the case of adjustable turbines (Kaplan and Deriaz turbines). Oil is also used in the gearbox, if present. The production of lubricant oil is highly pollutant and requires energy for extraction, production, maintenance and transport. González-Reyes et al. (2020) [7] discussed the different types of oil, distinguishing between mineral, biodegradable and synthetic. Biodegradable oil is a better option in terms of energy and CO_2 equivalent emissions than mineral or synthetic oils. However, biodegradable oil may pose potential risks in contact with water. Oil needs to be changed regularly, which entails a break in generation, causing economic losses. Furthermore, lubricant oil must be managed as a dangerous waste, with inherent risks of spillage during handling, that can generate negative impacts on the environment, and has some operational and maintenance problems [7–9]. In Francis turbines, the hydraulic oil system does not come into contact with the open water cycle, hence oil spills from the turbine directly to the open water are avoided. Oil spills may instead occur in the hydraulic cylinders of the intake gates. In Kaplan turbines, oil spills are more common, especially in the runner, because the blades are adjustable and the seals are not 100% tight. In older turbines, the whole runner is filled with hydraulic oil, whereas, in new runners, only parts of the runner are filled with oil [8].

Since most hydropower units use lubricant oil, the estimation of the annual amount of oil used is of high interest. This is especially valid for Europe, where the target of 100% renewables has been set by 2050 in the European Union (EU). Therefore, in this study, the lubricant oil consumption of the turbines installed in the European hydropower fleet is quantified, and the benefits that oil-free turbines can entail are quantified as well. Section 2 describes the implemented method to estimate oil consumption and impacts. Section 3 describes the results and Section 4 discusses the related implications. The European Union (EU) includes Austria, Belgium, Bulgaria, Croatia, Cyprus, Czechia, Denmark, Estonia, Finland, France, Germany, Greece, Hungary, Ireland, Italy, Latvia, Lithuania, Luxembourg, Malta, Netherlands, Poland, Portugal, Romania, Slovakia, Slovenia, Spain, Sweden. Europe also includes: Albania, Andorra, Belarus, Bosnia and Herzegovina, Faroe Islands, Gibraltar, Greenland, Iceland, Kosovo, Liechtenstein, Macedonia, Moldova, Monaco, Montenegro, Norway, San Marino, Serbia, Switzerland, Turkey, Ukraine, United Kingdom.

2. Materials and Methods

The open source JRC (Joint Research Centre) hydropower database, including 4030 European hydropower plants, 2429 of which are located in the European Union, was used in this study to carry out the calculations. The database mainly includes hydropower plants with installed power above 1 MW and from hereinafter called hydropower database. The hydropower database specifies, for each plant, the country, the type (run of river—ROR-, reservoir hydropower plant—RHP-, pumped hydropower storage—PHS), the installed power (P, kW), the gross head (H_{gross}, m) (in most cases, but not for all), the annual energy generation (E, kWh) and, for some of them, the reservoir volume (V, Mm^3). In this database, 194 GW are included, with respect to the total European installed power of 254 GW; therefore, although the database is not complete and was completed in 2019, it is a statistically representative sample of the hydropower fleet (the missing data are mainly related to small hydropower plants and to some countries where data are not available). This database was used in [3] to assess the potential of modernization strategies and in [10] to assess the energy potential associated with the recovery of the wasted energy (hydrokinetic energy at the tailrace, heat from the generators and degassing methane).

The number of units (to calculate the flow and power per unit) and the turbine type were estimated as described in [10], mostly considering indications and collected data from the hydropower company Voith Hydro (Figure 1 and Table 1). Therefore, we suggest referring to [10] for more information, where the methodology was validated and proved to adequately estimate these technical characteristics.

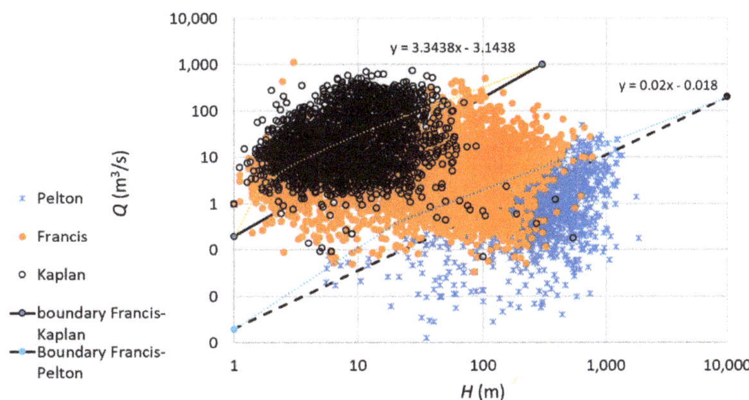

Figure 1. Range of Kaplan, Francis and Pelton turbines in the EU, based on flow rate Q and head H ([10], based on Voith Hydro), but without indication of the location.

Table 1. Average number of units (No.) per power class (K = Kaplan, F = Francis, P = Pelton) (Source: Voith Hydro).

Power Class P (MW) (Overall Installed Power per Plant)									
P	1–10	10–20	20–50	50–100	100–200	200–500	500–750	750–1000	1000–1500
No. of units	1 (K, P), 2 (F)	1.8	2.1	2.3	2.5	3.2	3 (K, P), 5 (F)	3.7	6.0

The required oil in each unit was estimated considering the power per unit, the number of units and the turbine type, according to Table 2 (RD 34.10.559 [11]) RD is a regulatory document that establishes rules and procedures for actions in a particular field of activity and is approved in accordance with the procedure established in the industry. This guidance document was developed by the production association for the adjustment, improvement of technology and operation of power plants and networks. RD 34.10.559 are designed to determine the annual oil consumption during the operation of hydraulic units for turbines of different types and capacities. The annual oil consumption is made up of its consumption for topping up during the operation of the hydraulic unit, for replacement and for compensation of losses during the overhaul.

Table 2. Oil consumption in hydropower units [11].

Turbine Type	Power, MW	Turbine Oil Consumption Rates (Tons per Year)				
		Topping Up	Replacement	Compensation of Losses	Total Annual	Collection Amount
Francis turbine	from 1 to 5 (incl.)	0.63	0.34	0.13	1.1	0.28
Propeller hydraulic	from 5 to 16 (incl.)	0.74	0.47	0.29	1.5	0.39
turbine	from 16 to 85 (incl.)	0.84	1.09	0.39	2.32	0.92
Pelton turbine	from 85 to 650 (incl.)	0.96	2.06	0.80	3.82	1.75
	from 1 to 5 (incl.)	0.7	0.23	0.3	1.23	0.19
	from 5 to 20 (incl.)	1.58	0.79	0.4	2.77	0.67
Kaplan turbine	from 20 to 40 (incl.)	1.99	1.42	0.5	3.91	1.20
(Vertical)	from 40 to 85 (incl.)	2.38	3.14	0.58	6.10	2.66
	from 85 to 220 (incl.)	2.67	4.92	0.9	8.49	4.18
Kaplan turbine (Horizontal)	from 19 to 48 (incl.)	1.4	1.42	0.5	3.32	1.20

Topping-up oil compensates for losses due to cleaning, evaporation, foam formation, leakage through leaks in the oil system and through the seals of the runners of Kaplan turbines, and when sampling for analysis. Oil consumption rates for topping up and compensation for irretrievable losses during the operation of hydraulic units are estimated by a statistical method. Oil consumption rates for replacement are estimated by an analytical method, taking into account the capacity of the oil system and the service life of the oil. Based on the results of a survey of hydroelectric power plants, the average oil service life for low-capacity oil systems is 8–10 years; for large ones it is 15 years. Oil losses during the overhaul of hydraulic units consist of losses in the form of an oily residue and emulsion at the bottom of the oil pans of the guide and thrust bearings of the turbine and generator when draining oil, as well as oil consumption for flushing the baths of the lubrication system and control units. Oil losses during overhauls are included in the oil consumption rates, taking into account the overhaul period of the hydraulic unit. When oil is replaced in the equipment, waste oil is collected. The collection of oil per unit during periodic replacement is determined by taking into account that the collected waste oil is about 0.85 of the volume of oil for replacement (from the normative RD 34.10.559 [11]).

The energy (MWh) used to manufacture and transport the oil was quantified according to [8]. The former was estimated as 8.26 kWh/L and the latter as 1.37 kWh/L, as average among mineral, synthetic and biodegradable oils. The associated emissions are 3.18 $kgCO_2$/L for the manufacturing, and 0.53 $kgCO_2$/L for the transport (the conversion factor from kWh to $kgCO_2$ is 0.385). The transport is very site-specific, and in this study, an average distance of 87 km was assumed. The average cost is estimated 4.3 EUR/L and a density of 0.8 kg/L was assumed. The emissions resulting from extraction and transport are insignificant compared with those related to the energy used in the manufacturing of the oil.

3. Results

Data of Table 2 were used to perform the assessment, implementing them to the European hydropower fleet data. Table 3 depicts the cumulative power per power class and the estimated turbine type (Figure 1) of the European hydropower fleet. Table 4 shows the oil consumption for each power class and turbine type. The composition of the European fleet has proven to be adequate for large-scale assessments, as already discussed in [3,10]. The Francis turbine is the most widely used in the European context, thanks to its wide operating range. It is the most suitable turbine for the typical European topographic/hydrologic context.

Table 3. Cumulative power P (MW) for each power range and turbine type.

	<5 MW	5 < P < 16	16 < P < 20	20 < P < 40	40 < P < 85	85 < P < 220	220 < P < 500
Pelton	1115	1550	194	2282	5244	9669	1269
Francis	3665	13,527	5466	26,820	37,436	48,219	11,176
Bulb	14	0	0	0	0	0	0
Kaplan	45	900	676	3361	9547	9969	2879

Table 4. Cumulative oil consumption (tons/year) for each power range and turbine type.

	<5 MW	5 < P < 16	16 < P < 20	20 < P < 40	40 < P < 85	85 < P < 220	220 < P < 500
Pelton	574	260	26	172	223	298	11
Francis	4028	2091	691	2069	1559	1413	168
Bulb	4	0	0	0	0	0	0
Kaplan	29	230	108	439	1043	662	93

The aggregated results at the European scale show that the amount of oil used in the European context is 16,200 tons/year with a required energy of 0.19 TWh/y and CO_2 emissions of 75,075 tons/year. The total cost is estimated to be EUR 87 million per year. These results refer to 195 GW, and can be extrapolated linearly to the whole 254 GW of European installed capacity. For the whole European context, the consumed oil is 21,665 tons per year and the required energy is 0.26 TWh/y, which represents 0.04% of the hydropower generation in Europe, roughly corresponding to one third of the maximum potential of hydrokinetic turbines in EU rivers estimated in [4]. The CO_2 emissions are 100,473 tons/year. The investment cost per year to satisfy the oil requirement becomes EUR 116 million per year. In the EU, the annual consumed oil is estimated to be 12,575 tons/year, corresponding to 0.0022% of the oil consumed as a primary energy source (582 Mtons per year).

Topping-up oil compensates for losses due to cleaning, evaporation, foam formation, leakage through leakages in the oil system and through the seals of the runners of Kaplan turbines, and when sampling for analysis. These losses and compensation for irretrievable losses represent 59% of the total amount of used oil. The collection amount is 85% of the replaced oil, which ranges between 15% and 49% (average, 34%) of the annual oil consumption of each unit.

4. Discussion

The impact of oil consumption in the European hydropower sector was assessed, identifying impacts in terms of CO_2 emissions, energy and costs. According to Mordor Intelligence (2022), in the most contributory European countries (France, Germany, Italy, Russia, Spain, United Kingdom), the lubricants market stood at 5.83 billion liters in 2021 and is projected to reach 6.53 billion liters in 2026. The automotive industry is the largest segment by end-user industry, accounting for around 49% of the total lubricant consumption in the region. Power generation is likely to be the fastest-growing end-user industry in the market, with a compound annual growth rate (CAGR) of 2.54% over the period 2021–2026, followed by automotive (2.49%). In Europe, the expanding power generation capacity, especially that of renewables, is likely to drive the consumption of lubricants in the power generation industry. The hydropower sector counts for less than 0.42% of the European lubricant oil consumption in the industrial sector [12].

The estimated values include the oil that is spilled into the environment (part of topping up item and losses, Table 1), that generally occurs as a consequence of incidents or in the form of oil droplets and oil mist leakage due to the centrifugal force and high temperatures [6]. However, the problem of oil mist leakage from bearings is complex, and it is difficult to assess exactly how much of this oil is spilled into the environment. Data collected in [7] show that this amount can reach 68 l/y/MW for a Kaplan turbine, with negative effects on the environment downstream, for example on fish, birds, mammals and plants. Spilled oil can contaminate the fish respiratory organs. Eggs and fish larvae can absorb toxic components from the oil. Spilled oil may stick to bird feathers; hence they may lose thermal insulation and waterproofing. The fur of mammals may become contaminated with oil, and it may also lose thermal insulation. In vegetation, a reduction in transpiration may occur, with a consequent reduction in photosynthesis. However, it must be noted that, according to [13], oil presence in aquatic ecosystems may not solely and mainly be attributed to the hydropower sector. Furthermore, other sectors, in addition to wind turbines, involve rotating components and desalination plants require oil [8,14].

The main solutions for minimizing the use of oil are (1) using oil free solutions, such as oil free bearings and oil free hubs, or water-lubricated turbines, and (2) implementing more environmental friendly solutions, such as an ester-based oil or other biodegradable oils, oil/water-based enhanced nano-lubricants, oil/greases and additives [15–17]. Sun et al. (2022) [6] discussed further conventional engineering measures to deal with oil mist leakage: reduction in the operating pad and oil temperature, optimization of the oil circulation loop in the oil tank, improvement of the sealing performance, and design of the oil mist emission device. Self or water-lubricated turbines can help in minimizing impacts, minimizing eventual environmental damages due to unexpected oil leakages and spills. Water-lubricated guide bearings also contribute to increasing the overall plant efficiency by reducing friction losses by about 50% and limiting maintenance in comparison to oil-lubricated ones [17]. To date, several Kaplan bulb and Francis turbines have been upgraded so as to make them work free from oil [5], but this technology can also be extended to Deriaz and Pelton turbines [5]. New materials and lubricants are being developed [18,19]. These technologies are developing more and more, especially in the European context. Some examples of their implementation are described in [20,21]. We suggest referring to the abovementioned references for further details.

However, some challenges exist. Given the low viscosity of water, bearings lubricated with water tend to operate in a boundary or mixed lubrication regime for relatively longer periods [5], especially when considering low sliding speeds and start/stop cycles. This can significantly increase friction while shortening the effective wear life of bearing components. Therefore, the use of water-based lubrication introduces several engineering challenges, foremost of which is the material choice for bearing surfaces. New materials are mainly at the development and R & D stage and have not fully met the operation and endurance targets set by the hydropower industry [19]. Oil-free rotor support systems have also enabled a broad range of high-speed machineries outside the hydropower sector, from air cycle machines, compressors and blowers to turbochargers, microturbines and small gas turbine engines [22].

Although, in this study, the impacts associated with oil consumption in hydropower plants was assessed, hydropower should not be regarded only in terms of impacts, especially when compared with the solar photovoltaic (PV) and wind sectors. For example, hydropower equipment does not contain critical materials such as lithium and cobalt (used in electric vehicles), or neodymium, praseodymium, and dysprosium (used in electric vehicles and wind power plants), or silver and silicon (used for photovoltaics). Hydropower is the best renewable energy for reducing pressure on mineral resources. The Extraction of Mineral Resources indicator is measured in kilograms of antimony equivalent (kgeq.Sb) per kilogram extracted to take into account existing reserves, the rate of extraction and the "depletion" of each mineral substance; the value for hydropower is 0.017, whereas it is 0.04 for coal, 0.3 for wind and 14 for solar PV [23]. Therefore, any energy technology should be comprehensively considered with regard to its impacts and benefits. Tables with sustainability indicators for each clean energy technology, benefits and impacts, with focus on the European Union, can be found in [24], and in [25] for the hydropower sector.

5. Conclusions

Hydropower benefits and impacts are at the center of major discussions. In order to mitigate the impacts, mitigation measures have to be developed and implemented.

In this study, a European assessment was carried out to estimate the oil consumption, and related impact, associated with hydropower operation. The aggregated results at the European scale show that the amount of oil used in the European context is 21,665 tons/year with a required energy of 0.26 TWh/year and CO_2 emissions of 100,473 tons/y. The total cost was estimated to be EUR 116 million per year. The required energy to satisfy the oil demand corresponds to 0.05% of the hydropower generation in Europe, roughly corresponding to one third of the maximum potential of hydrokinetic turbines in the EU rivers. In the EU, results reveal 12,575 tons/year of oil used, which is 0.0022% of oil consumption as a primary energy source and less than 0.42% of the lubricant oil production in the European industry.

This figure indicates that the pollution associated with the use of oil in the hydropower sector is marginal if compared with other sectors. Anyway, oil-free hydropower units could have positive impacts on the ecological status of rivers, supporting the clean and renewable energy targets set by the European Commission, and making hydropower further amendable, especially in light of the modernization needs of the European hydropower fleet.

Large-scale impact assessments should be also performed for the other sectors (e.g., wind and solar), whose impacts on the environment have been less studied, especially the impact of critical materials extraction and use.

Funding: The open access fee was paid by the European Commission Joint Research Centre, and the research was carried out within the Exploratory Activity SustHydro.

Data Availability Statement: Data are included in the article.

Conflicts of Interest: The author declares no conflict of interest.

References

1. International Hydropower Association (IHA). *Hydropower Status Report Sector Trends and Insights*; IHA Central Office: London, UK, 2022.
2. Kougias, I.; Aggidis, G.; Avellan, F.; Deniz, S.; Lundin, U.; Moro, A.; Theodossiou, N. Analysis of emerging technologies in the hydropower sector. *Renew. Sustain. Energy Rev.* **2019**, *113*, 109257. [CrossRef]
3. Quaranta, E.; Aggidis, G.; Boes, R.M.; Comoglio, C.; De Michele, C.; Patro, E.R.; Pistocchi, A. Assessing the energy potential of modernizing the European hydropower fleet. *Energy Convers. Manag.* **2021**, *246*, 114655. [CrossRef]
4. Quaranta, E.; Bódis, K.; Kasiulis, E.; McNabola, A.; Pistocchi, A. Is There a Residual and Hidden Potential for Small and Micro Hydropower in Europe? A Screening-Level Regional Assessment. *Water Resour. Manag.* **2022**, *36*, 1745–1762. [CrossRef]
5. Quaranta, E.; Pérez-Díaz, J.I.; Romero–Gomez, P.; Pistocchi, A. Environmentally Enhanced Turbines for Hydropower Plants: Current Technology and Future Perspective. *Front. Energy Res.* **2021**, *9*, 703106. [CrossRef]
6. Sun, J.; Zhang, Y.; Liu, B.; Ge, X.; Zheng, Y.; Fernandez-Rodriguez, E. Research on Oil Mist Leakage of Bearing in Hydropower Station: A Review. *Energies* **2022**, *15*, 2632. [CrossRef]
7. González-Reyes, G.A.; Bayo-Besteiro, S.; Vich Llobet, J.; Añel, J.A. Environmental and economic constraints on the use of lubricant oils for wind and hydropower generation: The case of NATURGY. *Sustainability* **2020**, *12*, 4242. [CrossRef]
8. Enzenhofer, K. Statkraft Hydro Power Plants–Oil Spills and Valuable Areas. Bachelor's Thesis, Mid Sweden University, Department of Engineering and Sustainable Development, Östersund, Sweden, 2022.
9. Inoue, K.; Deguchi, K.; Okude, K.; Fujimoto, R. Development of the water-lubricated thrust bearing of the hydraulic turbine generator. *IOP Conf. Ser. Earth Environ. Sci.* **2012**, *15*, 072022. [CrossRef]
10. Quaranta, E.; Muntean, S. Wasted and excess energy in the hydropower sector: A European assessment of tailrace hydrokinetic potential, degassing-methane capture and waste-heat recovery. *Appl. Energy* **2023**, *329*, 120213. [CrossRef]
11. RD 34.10.559 Individual Turbine Oil Consumption Rates for Repair and Maintenance Needs for Hydraulic Units. Available online: https://files.stroyinf.ru/Data2/1/4294816/4294816628.htm (accessed on 1 January 2023).
12. Mordor Intelligence. Europe Lubricants Market-Size, Share, COVID-19 Impact & Forecasts Up to 2026. 2022. Available online: https://www.mordorintelligence.com/industry-reports/europe-lubricants-market (accessed on 1 January 2023).
13. Masifwa, F.W.; Matuha, M.; Magezi, G.; Nabwire, R.; Amondito, B. Potential Impacts of Oil and Grease on Algae, Invertebrates and Fish in the Bujagali Hydropower Project Area. *Uganda J. Agric. Sci.* **2020**, *20*, 23–35. [CrossRef]
14. Elshorbagy, W.; Elhakeem, A.B. Risk assessment maps of oil spill for major desalination plants in the United Arab Emirates. *Desalination* **2008**, *228*, 200–216. [CrossRef]
15. Saini, V.; Bijwe, J. Polyaniline Nanoparticles: A Novel Additive for Augmenting Thermal Conductivity and Tribo-Properties of Mineral Oil and Commercial Engine Oil. *Lubricants* **2022**, *10*, 300. [CrossRef]
16. Saxena, A.; Kumar, D.; Tandon, N. Development of lubricious environmentally friendly greases using synergistic natural resources: A potential alternative to mineral oil-based greases. *J. Clean. Prod.* **2022**, *380*, 135047. [CrossRef]
17. Ingram, E.; Ray, R. Bearings & Seals: Examples of Innovations and Good Ideas. *Hydroworld* **2019**, *17*, 32–36.
18. Quaranta, E.; Davies, P. Emerging and Innovative Materials for Hydropower Engineering Applications: Turbines, Bearings, Sealing, Dams and Waterways, and Ocean Power. *Engineering* **2021**, *8*, 148–158. [CrossRef]
19. Somberg, J.; Saravanan, P.; Vadivel, H.S.; Berglund, K.; Shi, Y.; Ukonsaari, J.; Emami, N. Tribological characterisation of polymer composites for hydropower bearings: Experimentally developed versus commercial materials. *Tribol. Int.* **2021**, *162*, 107101. [CrossRef]
20. Hydro Review. Available online: https://www.hydroreview.com/world-regions/technologies-for-eliminating-oil-in-kaplan/#gref2022 (accessed on 1 January 2023).
21. Richard, C.; Groves, S. An Oil to Water Conversion of a Hydro Turbine Main Guide Bearing—Technical and Environmental Aspects. Available online: https://thordonbearings.com/docs/default-source/hydro-power/technical-papers/technical-paper---oil-to-water-conversion-of-a-hydro-turbine-main-guide-bearing.pdf?sfvrsn=9f11eb91_8 (accessed on 1 January 2023).
22. Della Corte, C. Oil-free shaft support system rotor dynamics: Past, present and future challenges and opportunities. *Mech. Syst. Signal Process.* **2012**, *29*, 67–76. [CrossRef]
23. Fry, J.J.; Schleiss, A.J.; Morris, M. Hydropower as a catalyst for the energy transition within the European Green Deal Part I: Urgency of the Green Deal and the role of Hydropower. *E3S Web Conf.* **2022**, *346*, 04015. [CrossRef]
24. Clean Energy Technology Observatory. 2022. Available online: https://setis.ec.europa.eu/publications/clean-energy-technology-observatory-ceto/ceto-reports-2022_en (accessed on 1 January 2023).
25. Quaranta, E.; Georgakaki, A.; Letout, S.; Kuokkanen, A.; Mountraki, A.; Ince, E.; Shtjefni, D.; Joanny, O.G.; Eulaerts, O.; Grabowska, M. *Clean Energy Technology Observatory: Hydropower and Pumped Hydropower Storage in the European Union—2022 Status Report on Technology Development, Trends, Value Chains and Markets*; Publications Office of the European Union: Luxembourg, 2022.

Article

IEC 62443 Standard for Hydro Power Plants

Jessica B. Heluany [1],* and Ricardo Galvão [2]

[1] Department of Information Security and Communication Technology, Norwegian University of Science and Technology, 2815 Gjøvik, Norway

[2] PECE—Industrial Automation, University of São Paulo, São Paulo 2373, Brazil

* Correspondence: jessica.b.heluany@ntnu.no or jessica.heluany@gmail.com

Abstract: This study approaches cyber security in industrial environments focusing on hydro power plants, since they are part of the critical infrastructure and are the main source of renewable energy in some countries. The theoretical study case follows the standard IEC 62443-2-1 to implement a cyber security management system (CSMS) in a hydro power plant with two generation units. The CSMS is composed of six steps: (1) initiate CSMS, (2) high level risk assessment, (3) detailed risk assessment, (4) establish policies, procedures, and awareness, (5) select and implement countermeasures, and (6) maintain the CSMS. To perform the high-level risk assessment, an overview of the most common activities and vulnerabilities in hydro power plants systems is presented. After defining the priorities, the detailed risk assessment is performed based on a HAZOP risk analysis methodology focusing on hackable digital assets (cyber-HAZOP). The analysis of the cyber-HAZOP assessment leads to mitigations of the cyber risks that are addressed proposing modifications in the automation architecture, and this also involves checking lists to be used by the stakeholders during the implementation of the solution, emphasizing security configurations in digital assets groups.

Keywords: HPPs cybersecurity; cyber-HAZOP; IEC 62443; CSMS; smart grid

Citation: Heluany, J.B.; Galvão, R. IEC 62443 Standard for Hydro Power Plants. *Energies* **2023**, *16*, 1452. https://doi.org/10.3390/en16031452

Academic Editors: Zhengwei Wang and Yongguang Cheng

Received: 25 November 2022
Revised: 18 January 2023
Accepted: 25 January 2023
Published: 1 February 2023

1. Introduction

The energy sector is a potential target for cyber-attacks given its role as critical infrastructure of a nation, which makes necessary the implementation of security strategies that hamper attacks such as ransomware, man-in-the-middle (MITM), denial-of-service (DoS), cross-site scripting (XXS), phishing, replay, false data insertion, jamming eavesdropping, spoofing, among others.

To facilitate the terminology understanding, in this paper, the common language established by the last release of NIST Framework and Roadmap of Smart Grid and Interoperability Standards (V4.0) [1] was adopted, where smart grids are conceptualized in a model with seven main domains: system transmission operators (TSOs); system distribution operators (DSOs); generation (including distributed energy resources (DERs); customer; markets; operations and service providers. Within each domain, actors and equipment have their respective roles and responsibilities in the electrical grid.

In countries where a considerable percentage of the produced energy comes from hydro power plants, it becomes very relevant to develop solutions and apply cybersecurity standards to this specific actor within the generation domain. The present paper is organized as follows: in Section 1, the motivation and the literature review are presented, justifying the purpose of this study and its contribution for the energy sector. In Section 2, some basic concepts are explained to give an overview of the steps that compose the study methodology, which involves the establishment of a cyber security management system (CSMS), which is detailed at Section 3. Then, a theoretical case study is developed in Section 4, and final considerations are discussed in Section 5.

1.1. Motivation

In recent discussions of cyber security in the energy world, considerable research attention has been directed toward smart grids. NIST conceptualization model makes clear that the term "smart grids" is broad and could be referring to any of the seven domains. On one hand, the high number of papers approaching cyber security in smart grids highlight the relevance of the topic. On the other hand, however, the broadness may hamper the applicability of such studies to specific actors within each of the seven domains. For this reason, attempting to collaborate by narrowing the topic in one of the domains, and this study focuses on generation, specifically hydro generation due to the background of the authors.

1.2. Literature Review

After identifying the energy market need for cyber security recommendations from practical experience, a brief review of the academic literature was conducted to better understand how the topic is being addressed and to acknowledge if there is a gap for the generation domain. Scopus was chosen as the database due to its broader range of publication sources. Filtering the field by "energy", format by "papers/articles", and running the query ((energy OR *grid) AND (cyber AND security)) in the title, 92 papers were collected for further screening.

Surprisingly, the first publications date from 2007, even before cyber-attacks rose in quantity and media attention. However, given that the aim of this study is to comprehend current approaches to smart grids, two exclusion criteria were applied to the set of papers: publication year was chosen from 2017 to 2022, and the content should be specific to any domain of smart grids, excluding end users of the "customers" domain, and other broad topics such as smart cities, smart homes, and electrical vehicles. This screening step resulted in 25 papers that were skimmed, resulting in 10 papers selected for detailed analysis.

In Table 1, the use case domain was identified and mapped against NIST domains together with the identified cyber security approach: risk vs. mitigation, attack vs. defence, network security, people management, and security management system. In some cases, the use case domain and the NIST domain coincide, while in others, the paper use case is referring to multiple NIST domains. It is interesting to note that, except for [2], the selected papers did not address an end actor within a domain, showing that there is a gap in the literature for narrower studies regarding cyber security in smart grid actors within each domain.

Table 1. Use case domain identified in analysed papers against NIST domains.

Paper title	Use Case Domain	NIST Domain	Cyber Security Approach
A microgrid ontology for the analysis of cyber-physical security	Microgrid	Generation; Distribution;	Attack vs. Defence
A novel actual time cyber security approach to smart grids	Substation	Distribution	Network security
Cyber security for multi-station integrated smart energy stations: Architecture and solutions	Smart Energy Stations (SESs)	Generation; Distribution;	Security management system
Cyber Security impact on energy systems	Virtual Power Plant (VPP)	Generation;	Risk vs. Mitigation; People management;
Cyber Security in the smart grid: challenges and solutions	Smart Grid	All	Risk vs. Mitigation
Cyber security in the energy world	Smart Grid	All	Network security
Grid cyber security strategy in an attacker–defender model	Power grid infrastructure	Distribution;	Attack vs. Défense
Improvement of cyber-security measures in national grid SA substation process control	Substation	Distribution;	Risk vs. Mitigation
Research on cyber security defence technology of power generation acquisition terminal in new energy plant	Power generation acquisition terminal	Generation;	Attack vs. defence
Smart grid cyber security enhancement: Challenges and solutions: a review	Smart Grid	All	Risk vs. Mitigation

Synthesizingpapers with a more generic smart grids security approach, refs. [3,4] report challenges and solutions using different perspectives. In [3], they conceptualize the smart grid and summarize the NIST security requirements regarding confidentiality, integrity, and availability to form the basis of the discussion about common risks and their corresponding mitigations according to the OSI layer. The following components are analysed: PMU (power metering unit), AMI (advanced metering infrastructure), smart meter, gateway, routing protocol, and control system. The risks mitigations recommended include input validation, DNSSEC, firewall, locking down ports, ICMP packet filtering, ARP inspection, disabling unused ports, and securing the physical infrastructure. Similarly, in [4], cyber attacks are mapped against confidentiality, integrity, and availability, but instead of suggesting techniques to overcome the security challenges according to the OSI layer, the countermeasures are classified according to an attack timeline divided into three steps: pre attack, under attack, and post attack. In addition to the mitigations listed in [3], this paper also explores more recent technologies, recommending IDS (intrusion detection system), blockchain, 5G, and AI (artificial intelligence). Both papers have an important role on demonstrating how vulnerable smart grids are and exemplifying some strategies to mitigate risks, but, at the same time, they emphasize that, besides the benefits brought by digital transformation, this also brings the need for a more robust security strategy.

In article [5], the authors mention diverse equipment that could be the target of cyber attacks, such as dispatch generators, electrical transformers, and circuit breakers. Moreover, one of the possible consequences cited, malfunction of a controller response, could lead to dangerous situations in terms of safety and security. Despite the use case is a microgrid, this equipment is also present in hydro power plants. The proposed solution, a combination of CIM (common interface model) and IEC 61850 for the development of an ontology, including AEGs (attack execution graphs), would allow the evaluation of cyber security issues through the generation of ADIVSE (ADVersary view security evaluation) models. Additionally, approaching network communication, ref. [6] examined standards and protocols most used in the power industry, highlighting IEC 60870-5 and DNP3 for SCADA systems and IEC 61850 for substations. In terms of substations security, in [7], the proposed solution is the implementation of homomorphic encryption in smart meters using MPI (message passing interface) with the Floyd-Warshall algorithm.

When it comes to the generation domain, ref. [8] recognized similar cyber attacks to the ones discussed so far, but also pointed that natural factors and human error/behaviour can lead to disruption in the energy supply. From a defence point of view, the authors claim that the responsibility of securing devices "is attached to the person" when, for example, applying patches or improving the company's personnel skills to manage incidents. For new power plants, ref. [9] applied an algorithm based on the Spark platform developed by AMP Laboratory of UCBerkeley to generate frequent item sets and extract association rules that were used to develop an anti-penetration strategy according to the results of the risk assessment. For lower impact systems, the strategy consisted of the packet filtering rule, while, for high-risk terminals with serious impact, the authors instantiated the anti-penetration strategy as address or port filtering rules.

Considering all these studies, but changing the perspective from risk vs. mitigation [2–4,8], attack-defence [5,9,10], or network security [6,7] to a security management view, only [11] proposed a holistic approach similar to the aim of this study. However, their focus is broader than generation, with the design a system architecture for smart energy stations composed of five entities: substation, photovoltaic station, energy storage station, electric vehicle charging station, and data centre station. Based on the data exchanges, they proposed cyber security protection solutions according to the principle of "security zoning, enhanced borders; dedicated network, multilayer protection; horizontal isolation, vertical authentication; classified storage, controlled sharing". Different zones are suggested, and the isolation devices range from industrial firewalls, forward/reverse isolators, and vertical encryption devices. Additionally, VLANs and service prioritization are addressed to guarantee real-time and no real-time communications.

1.3. Contribution

As observed in the literature review, there is limited research investigating cyber security in specific actors of smart grid domains. Nonetheless, given that knowledge building is complimentary, the previous studies analysed can be aggregated to transform approaches such as risk vs. mitigation or attack vs. defence into a holistic continuous security management approach, as suggested by IEC 62443-2-1 [12] with a cyber security management system (CSMS).

In the next topics, some basic concepts of safety, security, and risk analysis will be introduced to form the basis for the application of the IEC 62443-2-1 standard to a typical hydroelectric power plant, involving collaboration to cover the gap of domain/actor-focused studies. Systems and sub-systems of a two-generator hydro power plant will be evaluated considering the risk involved and the risk tolerance to make recommendations that are compliant with the IEC 62443-2-1 standard.

2. Conceptualization

2.1. Safety vs. Security

In the begging of the industrial revolution, plant safety was fundamental to strategic plans. The concern with injury or damage to the health of personnel, physical damage due to equipment incorrect operation according to input data, hardware failure, or error in operation were all taken into consideration to define the most appropriate technical solutions.

In industry 4.0, one of the main characteristics of the market solutions is the use of the internet to connect not only people, but systems themselves, being called IIoT (industrial internet of things). This technological evolution increases the cyber-attack surface. In this scenario besides safety, it is also important to focus on plant security, as well as investing in defence strategies of digital assets.

2.2. Risk Analysis

As defined at IEC 62443-1-1, risk is the "expectation of loss expressed as the probability that a particular threat will exploit a particular vulnerability with a particular consequence" [13]. Risk origins can be either physical or digital. Risk analysis methodologies can be focused on these origins, what enables the reutilization of a safety analysis to a security analysis through the classification of assets into hackable or non-hackable.

2.3. HAZOP vs. Cyber-HAZOP

The HAZOP (hazard and operability) analysis is a highly systematic methodology. The facility is divided into systems (nodes) for which many scenarios are evaluated through the selection of parameters and guidewords. The objective is to identify the qualitative potential of hazards and operating problems associated with a system or piece of equipment caused by deviations [14].

Following the method proposed by [15], it is possible to turn the HAZOP into a cyber-HAZOP risk analysis by selecting hackable scenarios. In this case, the step-by-step equivalent procedure can be observed below:

1. Define premises about the plant;
2. Define parameters to be considered;
3. Select a system to be evaluated;
4. Briefly explain the process;
4.1. Specify a parameter;
4.1.1. Specify a guideword;
4.1.1.1. Undertake a threat analysis;
4.1.1.2. For each threat, identify if the system is vulnerable to an incident (scenarios);
4.1.1.3. Identify the worst consequences associated to each scenario without safeguards;
4.1.1.4. Specify existing safeguards for each consequence;
4.1.1.5. Determine the probability and severity of each consequence;

4.1.1.6. Make recommendations to mitigate consequences if the probability or severity of occurrence are inacceptable according to the level of risk acceptance stablished by the company;

4.1.2. Repeat step 4.1.1 for each guideword;

4.2. Repeat step 4.1 for each parameter;

5. Repeat step 4 for all plant nodes.

2.4. Hydro Power Plants Systems and Its Vulnerabilities

Usually, the scope of supply for a hydroelectric power plant (HPP) involves more than one company. Table 2 contains a typical case for an automation project.

Table 2. Main activities in HPPs automation projects and its typical suppliers.

Activity	Supplier
Engineering, tests, and commissioning of field network, process network, and supervisory system	Suppliers of automation and control systems
Communication between process data and office network	Suppliers of automation and control systems or information technology with customer participation, possibly involving IT and OT teams to support the activity
Communication link for remote access	Suppliers of automation and control systems or information technology with customer participation, possibly involving IT and OT teams to support the activity
Communication link with regulation agent	Automation and control systems suppliers generally subcontract telecommunications companies for this scope. Customer participation can involve IT and OT teams to support the activity.

In order to ease the implementation of defence in depth strategies, each activity can be associated to one or more configuration group of devices. To exemplify the vulnerabilities, Table 3 contains six group categories and its main vulnerabilities, highlighting that some of them are common to more than one group.

Table 3. Configuration categories and its potential vulnerabilities. Based on [16].

Category	Potential Vulnerabilities
Supervisory System	Poor physical security; lack of system hardening; inadequate security awareness; poor password policies; poor account management; social engineering susceptibility; lack of patch management; lack of authentication; zero-day vulnerabilities; ineffective anti-virus/application whitelisting; insufficient access control; use of vulnerable ICS protocols; untested third-party applications; insecure embedded applications;
Communication link with operation centre or regulation agent	Poor physical security; lack of system hardening; inadequate port security; lack of authentication; unnecessary firewall rules; lack of patch/firmware management; poor configuration management; configuration errors;
Communication between process data and office network	Poor physical security; lack of system hardening; inadequate port security; lack of authentication; unnecessary firewall rules; poor configuration management; lack of patch/firmware management; configuration errors; uncontrolled file sharing;
Network equipment	Poor physical security; configuration errors; poor configuration management; inadequate port security; lack of firmware management; unnecessary firewall rules; lack of intrusion detection capabilities; use of vulnerable ICS protocols;
Electrical protection system	Poor physical security; lack of firmware management; configuration errors; poor configuration management; lack of authentication; use of vulnerable ICS protocols;
Controllers	Poor physical security; lack of system hardening; lack of firmware management; untested application integration; configuration errors; poor configuration management; lack of authentication; use of vulnerable ICS protocols;

This rearrangement of vulnerabilities is deeply related in the modus operandi of HPPs automation project once this separation proposal enables one to address risk mitigation for specific suppliers.

3. CSMS Application Guide for HPP's According to IEC 62443

The IEC 62443 standard is organized in four groups: General, policies and procedures, system, and component. Together, they provide a vast amount of information for the implementation of a cyber security management system (CSMS) in industrial control systems (ICS). In the following sessions, a series of questions are suggested as a starting point for the implementation of a CSMS in a hydro power plant.

3.1. Initiate CSMS Program

It is necessary to develop a business plan, determine the desired scope and obtain support from the organization. Suggested concerns:

- What is the expected result taking into consideration that a CSMS does not generate ROI (return of investment)?
- Are there financial resources in short term to enable the application of resulting countermeasures?
- Are there financial resources available in short, medium, and long term for the maintenance of the CSMS?
- Are there specialists available for each HPP system that can participate in the detailed risk assessment?
- Is it possible to initiate the CSMS only by reallocation of personnel activities or is it necessary to contract new personnel?
- Will the program be applied only by internal efforts, or will there be third-part suppliers?
- Will the CSMS be developed for a single HPP or for a group of them? Will reference automation architecture be developed?
- What is the timeline expectation? Is it feasible?

3.2. High Level Risk Assessment

The high-level risk assessment should be performed to indicate priorities and support the strategy definition regarding personnel and financial resources. Suggested concerns:

- Which HPP's systems can cause personnel injury in case of inadequate operation?
- Which HPP's systems can cause environmental damage in case of inadequate operation?
- Which HPP's systems can cause equipment damage in case of inadequate operation?
- What is the impact on company's reputation due to a well succeeded cyberattack?
- What is the financial and social impact if one or more generator unities stop working due to a successful cyber attack?
- What is the actual and pretended risk tolerance?
- How is the IT $\times$ OT integration? Which team acts in case of incidents?
- What does the governments agents require regarding cyber security in critical infrastructures?

3.3. Detailed Risk Assessment

The detailed risk assessment should be performed in accordance with the priorities and resources defined in the high-level risk assessment. It is important to determine a methodology and apply it for the specified systems. Suggested concerns:

- Is there an incident database to guide systems prioritization?
- Is there a safety risk analysis that can be adapted to a security risk analysis?
- How many defence layers will be applied for priority systems?
- What is the impact of the countermeasures on the systems?

3.4. Establish Policies, Procedures and Awareness

It is necessary to create policies, procedures, and awareness trainings that mitigate the risks evaluated. Suggested concerns:

- What kind of trainings will be created: theoretical, practical or both?
- Which personnel should be involved: IT, OT, or both? From which specialty?
- Will an external company be hired for specific trainings?
- Will there be mandatory trainings? What will be the frequency?
- How will be the means of divulgation of the new policies and procedures inside the HPP?
- How will the integration between IT and OT departments be approached in the policies, procedures, and awareness program?
- What will be the frequency of policies and procedures revision?
- Will there be mandatory policies and procedures?

3.5. Select and Implement Countermeasures

Once the risk tolerance, priorities, probabilities, and severities are known, they must be carefully evaluated for the selection of the most adequate countermeasures, and it is important to take into consideration the concept of defence in depth for the countermeasure's definition. Suggested concerns:

- Will it be given emphasis for internal or externa attacks for the HPP under consideration (intentional or unintentional)?
- Can the selected countermeasures be implemented in a single step or is it necessary to divide the implementation in more steps due to financial restrictions?
- Recent cyber attack cases in similar plants occurred due to what kind of lack of security?
- What are the existing safeguards for the worst-case consequences? Is it possible to improve by applying more defence layers?

3.6. Maintain the CSMS

As can be observed in Figure 1, the activity of maintaining the CSMS can go back to any other activity, highlighting the cyclic nature of security management. It is necessary to monitor the CSMS results and the real adherence to the policies and procedures, check changes in government regulations, look for best practices continuously, and revise the activities according to lessons learned during the implementation and maintenance of the program.

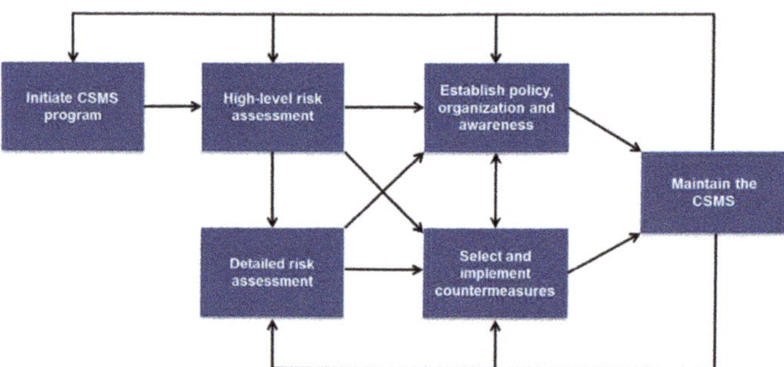

Figure 1. Top level activities for establishing a CSMS [12].

4. Study Case for a Typical HPP

4.1. Methodology and Contextualization

As detailed in the Section 2.3, ref. [15] proposed a generic method to turn a HAZOP into a cyber-HAZOP risk analysis by selecting hackable scenarios. In this paper, some basic concepts involving hydro power plants systems and CSMS were introduced to contextualize the theorical application of this method based on a real HPP with two Pelton generator unities that produce approximately 300 MW.

Some simple modifications were applied to the automation architecture shown in Figure 2 in order to represent a more typical solution of distributed automation and control system.

The devices are based on Siemens solutions, where the main controllers are from S7 400 family configured with hot stand-by redundancy. The field I/O devices are based on ET 200M and ET 200S solutions, and subsystems have S7 300 as the controller. The protection system is based on SIPROTEC equipment.

It is a client/server architecture where the operation network contains a historian, two operation stations, and tow engineering stations all communicating through proprietary protocol. The process network has double ring topology and optical fiber as physical mean for proprietary and non-proprietary communication protocols. The systems connected to the process network are: generator units, auxiliary systems (mechanical and electrical), and substation and electrical protection communicating through IEC 61850 protocol. Lastly, the field network is based on RS 485 as physical mean and Profibus DP as communication protocol. Moreover, there are two external connections: one with the office network and a gateway for the connection with the government regulation agent.

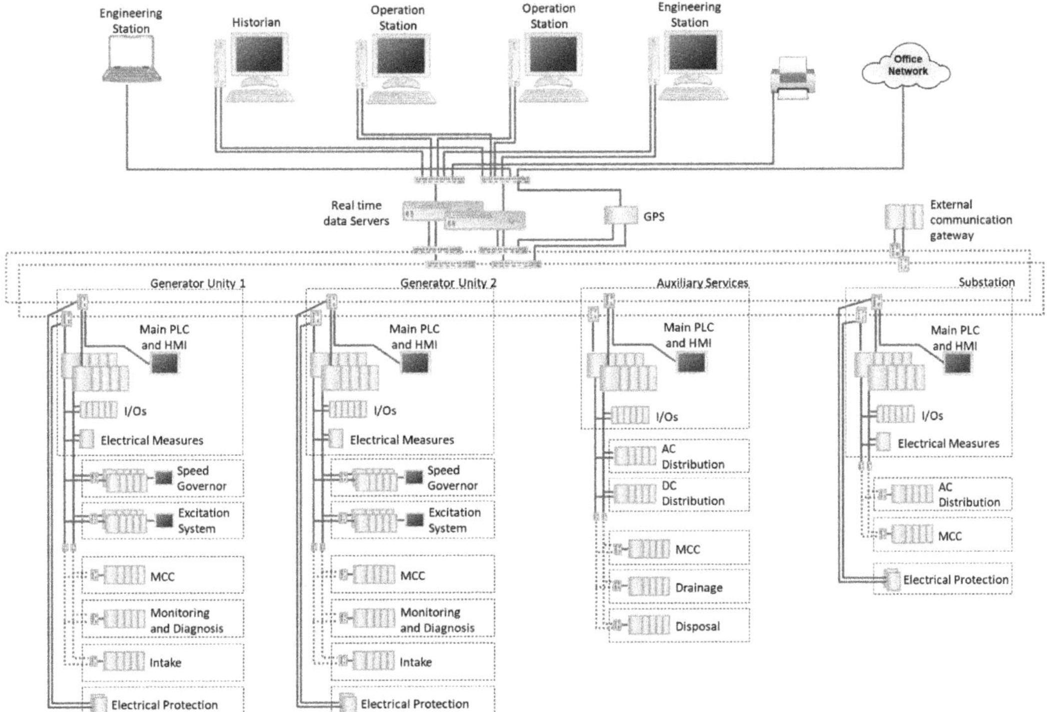

Figure 2. Study case automation architecture based on real HPP with two generator units.

4.2. Initiate CSMS Program

The company assumed that cyber security should be considered in all levels with a top-down strategy. Table 4 shows the topics selected and the decisions made regarding each one.

Table 4. Company decisions to initiate the CSMS.

Topic	Decision
Business plan	Enough investment for a dedicated team and for the implementation of a CSMS in this pilot plant.
CSMS scope	Complete automation architecture with selection of priority countermeasures according to the established risk tolerance and detailed risk analysis to be performed.
Stakeholders	Head office of the electrical company, regional unities, services, and product providers.
Organization support	Full support of all organization, given that cyber security has become a priority in the company's strategy.

Given that the business plan allows a dedicated team, the following departments are considered:

- Electrical maintenance and operation;
- Mechanical maintenance and operation;
- Electrical engineering;
- Mechanical engineering;
- Automation engineering;
- Information technology;

4.3. High Level Risk Assessment

Given the period of this study, the high frequency of ransomware attacks led to the conclusion that they have medium probability of occurrence in the supervisory system and should be considered in the program. Besides, there is also a medium probability of non-authorized remote access focusing on turning down the HPP more than aiming to destroy it, once it would require a deeper knowledge of the systems and existing solutions in the HPP.

Regarding regulations, although, in some countries, there is no mandatory government regulation on cyber security for critical infrastructures, the loss of revenue due to a non-operating unity generator, costs to restore the system, or possible taxes due to non-availability justify the investment in more defence layers. Given this context, the company also changed its risk tolerance from high to medium.

4.4. Detailed Risk Assessment

The following premises were adopted:

- Plant personnel would not attack or pass information to outside attackers intentionally because besides it is easier to detect, and, in the case of a dangerous incident, they would be exposed to possible injuries, so it is assumed that the leak of internal information is due to social engineering, reinforcing the need of trainings and awareness.
- Differently from process data that contains industrial secrets, it is assumed that data from a HPP is not the focus, reducing the need of defence layers related to data itself. Before applying solutions such as deep packet inspection (DPI), other defence layers would be priority.
- Taking the previous premises into consideration, it is presumed that the focus of a cyber attack on a HPP is to take control of the substation and/or generator unities to turn the system down, in other words, it is assumed that it would be an attack more similar to Blackenergy than to Stuxnet.

This plant already had a HAZOP analysis available, so it was decided to turn it into a cyber-HAZOP by filtering the hackable scenarios.

Table 5 exemplifies cases of the excitation system for two parameters: isolation and control system. If the isolation is inadequate and this event is not alarmed in the SCADA, the machine can be damaged. This is a hackable scenario because a cyber-attack can block alarms to the SCADA letting the excitation system under risk. The complete analysis can be found on [17].

Table 5. Company decisions to initiate the CSMS.

Subsystem	Guideword	Param.	Deviation	Possible Causes	Conseq.	P	S	R	Existing Safeguards
Excitation system	No/Low	Isolation	Low or inexistent isolation	Dust on the generator brushes. Damage in the internal equipment of the excitation system that are connected to the power electronics circuits.	Unity stops	0.1	5.0	0.5	SCADA alarm. Protection relay 64R
Excitation system	No	Control System	Control system out of work	CLP damage. Cabling failure. Failure on the 125Vcc powering system	Unity stops	0.1	3.0	0.3	Control system redundancy

4.5. Establish Policies, Procedures and Awareness

The IEC 62443 standard does not specify trainings frequency, so the suggestions are open to adaptations according to each plant needs and level of knowledge of personnel.

4.5.1. Awareness Trainings

Table 6 shows the trainings content that HPPs personnel must attend with 3 h of duration each 12 months.

Table 6. Training's content.

Topic	Content
Risks, threats and vulnerabilities	Conceptual definitions and approach of risks associated to the HPP's cyber assets.
Risk analysis	Conceptual definitions of the risk analysis utilized in the HPP (cyberHAZOP) to prepare the team for continuous documentation revision.
Standards	General view of the main international standards used as base for the HPP's policies: NERC CIP, IEC 62443 and NIST 800-82.
Social Engineering	Focus on behaviour issues exemplifying well succeeded social engineering attacks to promote awareness and improve the level of responsibility when dealing with internal data

4.5.2. Policies and Procedures

Once the HPP may have legacy systems, the policies and procedures are mandatory for all applicable systems. Table 7 shows the contents for each topic.

4.6. Select and Implement Countermeasures

Given that the cyber-HAZOP did not result in high-risk scenarios, the medium risk cases were filtered and commented with recommendations:

Discharge channel: The discharge channel level is measured through a sensor. If the level is very high, it can lower the turbine efficiency, damage it, and, in the worst case, inundate the powerhouse. Once this sensor can be physically targeted to send a wrong value to the system, it is recommended to be installed in a restricted area under access control.

Stator: if an overvoltage occurs and is not detected, the isolation system can be damaged. Besides project considerations that take this risk into consideration, some devices are also responsible to detect this situation: excitation digital control and protection relay, so it is recommended to control physical and configuration access to these assets.

Table 7. Policies and procedures content.

Topic	Content
Use of removable media	Policy describing allowed and forbidden attitudes and its respective procedures. If the utilization of removable media is absolutely necessary, before being connected to the system, it must be verified and sanitized
Backup	All applicable systems must have an associated document with the backup procedure, which must be conducted every three months or when some change is applied
Restore	The restore procedure must be tested every 12 months for all applicable systems. Examples: PLCs multi-project, servers' images, historian data, operation station images, devices configuration (e.g., firewall and protection relays), among others
Incidents	Incident report policy containing a procedure to be followed in case of compromised systems
Change	Change management policy with examples and templates to keep the documentation always in the last revision.
Inventory	Inventory policy containing a template document. The inventory procedure must be revised each 12 months or in case of change for both physical and cyber assets.
Patches	Patch management policy and related procedures focusing relevant systems to keep the plant safe. Only security patches should be applied for whatever system: Windows platforms, firmware, or automation software. The CSMS responsible must verify available patches every week and filter the applicable ones to include in the updates planning
Access control	Access control policy for both physical and cyber assets according to minimum privilege philosophy. Access levels must be documented with responsible names and the document must be revised every six months or in case of change
Logs	Policy describing which logs should be monitored for each system

Spherical valve: Inadequate pressure conditions during valve opening and closing can collapse the penstock. Theses sensors should be in a restricted area under access control with regards to the speed governor controller to avoid wrong configurations and commands.

Hydraulic governor: High pressure can cause pipes rupture. In addition to the existing SCADA alarm, the pressure sensor should be in a restrict area under access control.

Speed governor: If the PLC is unavailable or if there are failures for sending and receiving data, the unity must be stopped. A common safeguard is the system redundancy, but it is also recommended to control physical, configuration, and command access to this asset.

Electrical protection system: If there is a problem with input data, unknown devices status requires stopping the system, and the non-actuation of electrical protection can cause serious damages. It is recommended to apply many defence layer restriction for physical and configuration access, separate network, and constant security firmware updates.

Supervisory and control system: If this system is unavailable or if there are failures for sending and receiving data, the HPP must be stopped. It is recommended to control physical and configuration/operation access, follow the policies and procedures, and update Windows security patches and antivirus signatures.

Open cooling system: If the cooling system presents low flow or pressure, the circuit temperature can increase and/or the supply of services water can decrease. These sensors should be in a restricted area under access control.

Drainage: The HPPs drainage is performed through pumps whose commands are from a PLC or local operation. If this system does not work when needed, the powerhouse can inundate, so it is recommended to control physical, configuration, and command access to this system.

Services water: If there is no services water flow, many systems are affected, justifying restricting physical and operation access to this system.

Ventilation: If there is no ventilation flow, it is impossible to regulate the environment temperature and air quality, which can inhibit the permanency of the operators inside the powerhouse. It is recommended to control physical and command access to these assets.

As can be observed in the medium risk cases, most risks can be mitigated through physical access control, minimum privilege philosophy for configuration and operation, security patch updates, security firmware updates, antivirus, trainings, and personnel awareness.

Regarding the automation architecture, the following measures are suggested:

- Division into security cells protected through firewall to control activities that occur within and between then.
- Inclusion of a server with security functionalities: antivirus and Windows patches.
- Inclusion of a centralized log management system.
- Configuration of a DMZ (demilitarized zone) to limit data flow from process to office network containing the historian server, antivirus, Windows patch update server, and log management server.

Cell 1 contains the operation and engineering stations, while Cell 2 contains the real time data servers. The process and operation networks are physically distinct and, by separating them into different security cells, it is necessary to configure firewall rules to be compliant with IEC 62443 requirements. As a result, the modular switches must provide technical features that enable security configurations, or the PCs must have embedded security options such as firewall and VPN.

The communication gateway with the Operation Centre or Government Agent is extremely relevant, and it is an external connection. To be IEC 62443-compliant, another DMZ should be configured, but given that it involves a government agent and some countries still do not have a national mandatory standard, it is considered in Cell 3 without DMZ characteristics, but configured according to the government agent rules requirements.

As cited in the high-level risk assessment, it is admitted that external ransomware attacks and non-authorized remote access probability is higher than intentional internal attacks probability. For this reason, all systems connected to the process network (generator unities, auxiliary services, and substation) are considered in Cell 4.

In addition to the modifications suggested for the automation architecture, in order to keep a minimum security level, it is suggested to develop check lists for the six groups of devices configuration stated in Table 2:

- Supervisory system;
- Communication link with operation centre or regulation agent;
- Communication between process data and office network;
- Network equipment;
- Electrical protection system;
- Controllers.

The check lists should approach measures to mitigate identified vulnerabilities. Although Table 8 refers only to controllers, similar check lists should be developed for the other devices groups, taking into consideration Table 3 contents regarding common ICS vulnerabilities.

It can be said that the architecture suggest in Figure 3 is an intermediate solution. The architecture division into security cells can follow many criteria, and it impacts the amount and costs of security devices, and there must be at least one firewall for each cell. Figure 4 is an example of a lower level of security with only two security cells and a DMZ, while Figure 5 considers seven security cells and a DMZ.

Table 8. Controllers check list.

	Controllers			
Item	**Description**	**Yes**	**No**	**Comments**
1	Physical security: allocation of controllers inside cabinets with access control			
2	Hardening: disable non-utilized functionalities			
3	Hardening: disable logical and physical (USB and others) nonutilized ports.			
4	Firmware: guarantee that the device is running with the last homologated firmware version			
5	Configuration: keep the controllers configuration tool with physical access control and restrict users able to change configurations			

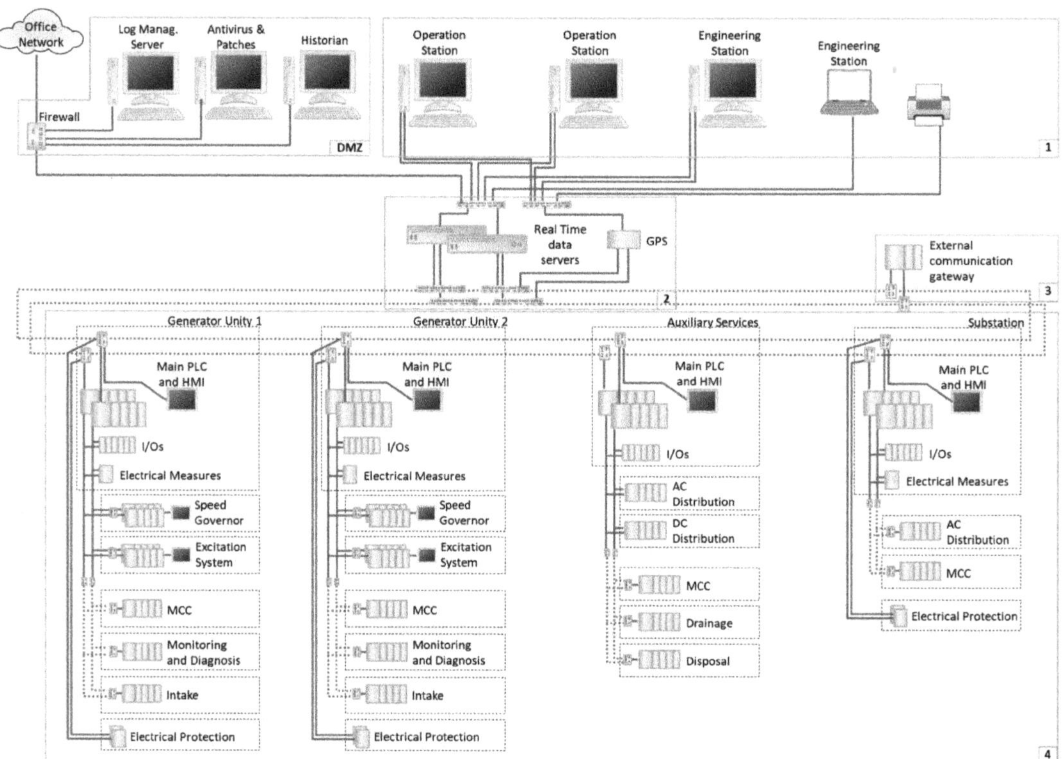

Figure 3. Architecture containing a DMZ and four security cells.

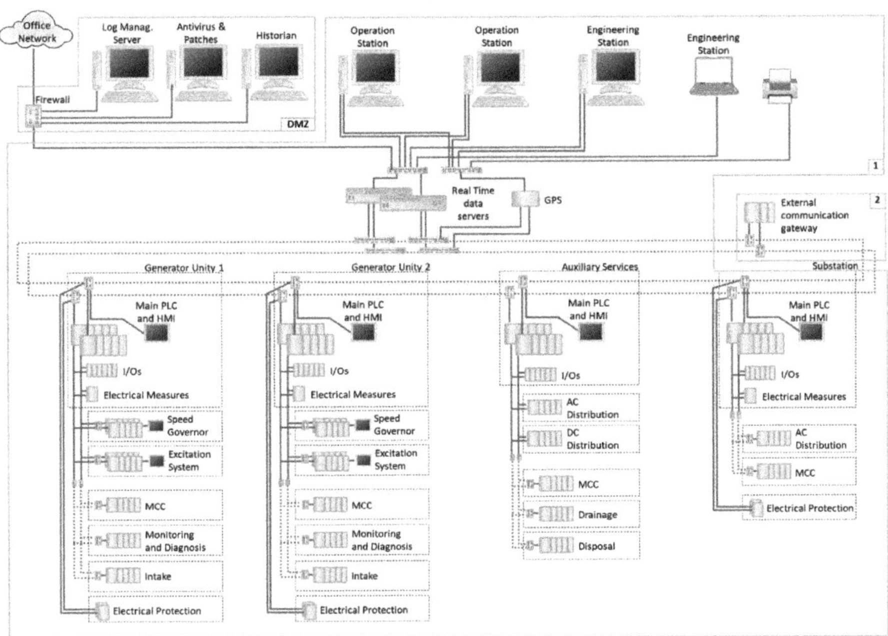

Figure 4. Architecture containing a DMZ and two security cells.

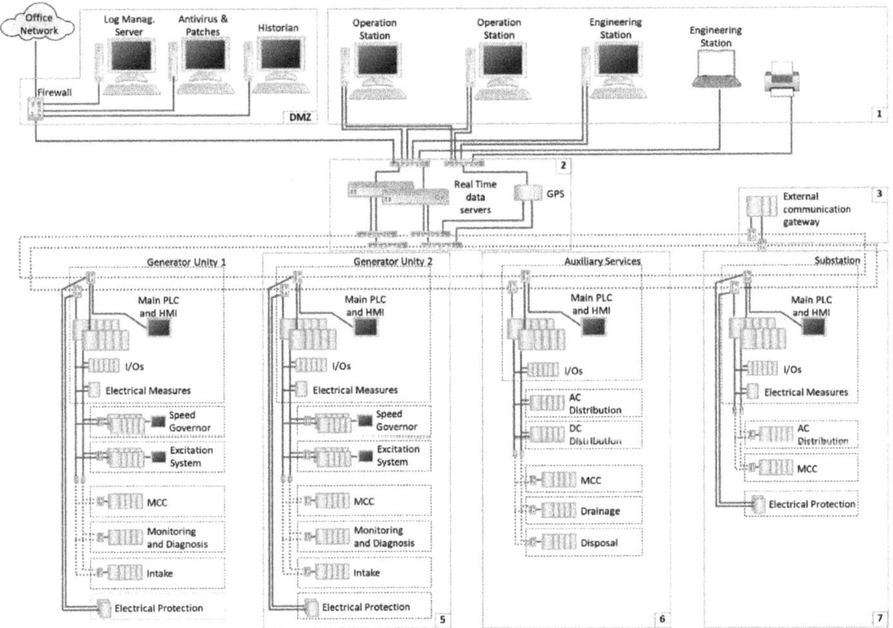

Figure 5. Architecture containing a DMZ and seven security cells.

5. Conclusions and Future Work

This study proposes the application of IEC 62443 concepts for hydro electrical power plants. As discussed in the literature review session, it adds value to the energy market given the gap for more domain/actor-based studies, and, in this case, the contribution is focused on hydropower owners and network operations. It is expected that the methodology developed in this paper can contribute to a better understanding of how to start a security management program, what are the implications of the chosen risk tolerance, and how it may change technical solutions and network equipment and configuration.

Before analysing the study case, common ICS vulnerabilities were considered in groups of devices according to a typical modus operandi of suppliers in an HPP automation project. To illustrate the application guide, a theoretical study case based on a real HPP was developed, and some premises were adopted to develop the CSMS. Once the HPP had a HAZOP risk analysis available, it was filtered for hackable cases exemplifying how to transform a safety risk analysis into a security risk analysis: cyber-HAZOP.

The medium risk cases were detailed, showing that most of them can be mitigated through physical access control, minimum privilege philosophy for configuration and operation access, security patch updates, security firmware updates, antivirus, trainings, and personnel awareness.

Regarding the automation architecture, it was suggested to implement a DMZ and divide the devices into four security cells. The DMZ was considered with the historian and servers for antivirus, Windows security patches, and log management. The external communication gateway was not considered in a DMZ because it would involve the government agent and, in many countries, there are no mandatory regulations regarding cyber security in critical infrastructures.

To exemplify other solutions, two more architectures were designed considering lower and higher risk tolerances. The lower the risk tolerance, more security cells should be implemented, thus increasing the solution costs and maintenance efforts.

Future studies on the energy generation domain could further investigate the cost—benefit relation resulting of the risk tolerance that each company is committed and how it impacts the cyber risks countermeasures implementation. For example, some mitigations from the literature review, such as IDS (intrusion detection system), blockchain, 5G, and AI (artificial intelligence) were not considered in this study due to the high implementation cost. Besides, limited budgets are common in smaller power plants, and more relevant plants may have a different scenario, being able to apply more security layers and invest in more expensive solutions. Furthermore, the literature review also shows little research on real case applications, where the results could be evaluated and improved according to this or other methodologies, deepening the analysis of different technical solutions and their efficacy to meet the challenge that the energy sector faces regarding cyber security.

Author Contributions: Writing and editing, J.B.H.; supervision, R.G. All authors have read and agreed to the published version of the manuscript.

Funding: This research received no external funding.

Data Availability Statement: The only external data source is the HAZOP safety assessment, which was obtained from 3rd party for study purposes and restrictions apply to these data, which is not publicly available.

Conflicts of Interest: The authors declare no conflict of interest.

References

1. Gopstein, A.; Nguyen, C.; O'Fallon, C.; Hastings, N.; Wollman, D.A. *NIST Framework and Roadmap for Smart Grid Interoperability Standards, Release 4.0, Special Publication (NIST SP)*; National Institute of Standards and Technology: Gaithersburg, MD, USA, 2021. Available online: chrome-extension://efaidnbmnnnibpcajpcglclefindmkaj/https://nvlpubs.nist.gov/nistpubs/SpecialPublications/NIST.SP.1900-206.pdf (accessed on 27 December 2022).
2. Jahil, A.A.A.; Giarratano, D. *Improvement of Cyber-Security Measures in National Grid SA Substation Process Control*; Institute of Electrical and Electronics Engineers Inc.: Piscataway, NJ, USA, 2017.
3. Faquir, D.; Chouliaras, N.; Sofia, V.; Olga, K.; Maglaras, L. *Cyber Security in Smart Grid: Challenges and Solutions*; Institute of Electrical and Electronics Engineers Inc.: Piscataway, NJ, USA, 2019; pp. 546–551.

4.	Alsuwian, T.; Shahid Butt, A.; Amin, A.A. Smart Grid Cyber Security Enhancement: Challenges and Solution—A Review. *Sustainability* **2022**, *14*, 14226. [CrossRef]
5.	Backes, M.; Keefe, K.; Valdes, A. *A Microgrid Ontology for the Analysis of Cyber-Physical Security*; Institute of Electrical and Electronics Engineers Inc.: Piscataway, NJ, USA, 2017.
6.	Ang, C.K.G.; Utomo, N.P. *Cyber Security in the Energy World*; Institute of Electrical and Electronics Engineers Inc.: Piscataway, NJ, USA, 2017; pp. 1–5.
7.	Buyuk, O.O.; Camurcu, A.Y. *A Novel Actual Time Cyber Security Approach to Smart Grids*; Institute of Electrical and Electronics Engineers Inc.: Piscataway, NJ, USA, 2018.
8.	Chobanov, V.; Doychev, I. Cyber Security impact on energy systems. In Proceedings of the 2022 International Congress on Human-Computer Interaction, Optimization and Robotic Applications (HORA), Ankara, Turkey, 9–11 June 2022; pp. 1–5.
9.	Liu, Y.; Qin, H.; Chen, Z.; Shi, C.; Zhang, R.; Chen, W. *Research on Cyber Security Defense Technology of Power Generation Acquisition Terminal in New Energy Plant*; Institute of Electrical and Electronics Engineers Inc.: Piscataway, NJ, USA, 2019; pp. 25–30.
10.	Chen, Y.C.; Mooney, V.; Grijalva, S. *Grid Cyber-Security Strategy in An Attacker-Defender Model*; Institute of Electrical and Electronics Engineers Inc.: Piscataway, NJ, USA, 2020.
11.	Chen, Y.; Li, J.; Lu, Q.; Lin, H.; Xia, Y.; Li, F. Cyber security for multi-station integrated smart energy stations: Architecture and solutions. *Energies* **2021**, *14*, 4287. [CrossRef]
12.	*IEC 62443-2-1*; Industrial Communication Networks–Network and System Security—Part 2-1: Establishing an Industrial Automation and Control System Security Program. IEC: Geneva, Switzerland, 2010.
13.	*IEC 62443-1-1*; Industrial Communication Networks–Network and System Security—Part 1-1: Terminology, Concepts and Models. IEC: Geneva, Switzerland, 2009.
14.	Nolan, D.P. *Safety and Security Review for the Process Industries–Application of HAZOP, PHA, What-If and SVA Reviews*; Elsevier: Amsterdam, The Netherlands, 2014.
15.	Marszal, E. Security process hazard analysis review. *ISA InTech Mag.* **2016**. Available online: https://www.isa.org/intech-home/2016/march-april/features/security-process-hazard-analysis-review (accessed on 24 November 2022).
16.	Knapp, E.D.; Langill, J.T. *Industrial Network Security–Securing Critical Infrastructure Networks for Smart Grid, SCADA, and Other Industrial Control Systems*; Syngress: Rockland, MA, USA, 2015.
17.	Heluany, J.B. Application of Cyber Security Standards in HPPs. Master's Thesis, University of São Paulo, São Paulo, Brazil, 2018.

Article

A Comparison Study of Hydro-Compact Generators with Horizontal Spiral Turbines (HSTs) and a Three-Blade Turbine Used in Irrigation Canals

Wiroon Monatrakul [1], Kritsadang Senawong [2,3], Piyawat Sritram [4] and Ratchaphon Suntivarakorn [1,*]

[1] Department of Mechanical Engineering, Faculty of Engineering, Khon Kaen University, Khon Kaen 40002, Thailand
[2] General Education Teaching Institute, Khon Kaen University, Khon Kaen 40002, Thailand
[3] Postharvest Technology Innovation Center, Science, Research and Innovation Promotion and Utilization Division, Office of the Ministry of Higher Education, Science, Research and Innovation, Bangkok 10400, Thailand
[4] Faculty of Agriculture and Technology, Rajamangala University of Technology Isan Surin Campus, Surin 32000, Thailand
* Correspondence: ratchaphon@kku.ac.th

Abstract: This study aimed to present the experimental results of two types of turbines and attachments used in a hydro-compact generator. Two Horizontal Spiral Turbines (HSTs) with blade angles of eighteen and twenty-one degrees, respectively, and a three-blade turbine were tested and experimented in a laboratory at five levels of water flow rate ranging from 1–2 m/s. After the efficiency and torque values of each turbine were identified, they were installed in two 200 W power generator systems: (1) with a "diffuser" attachment; and (2) with an "in-line+diffuser+nozzle chamber" attachment, and tested in a local irrigation canal with 1.2 m/s. The results from the laboratory indicated that the HST with a twenty-one degree blade angle had 38.10% efficiency at the water flow rate of 2 m/s. It could reach 120.0 rpm and produced 212 Nm of torque. The results from the field experiment revealed that the combination of the power generator with the twenty-one degree blade angle HST and the in-line + diffuser + nozzle chamber attachment was the most efficient, with 284 Nm of torque at 108 rpm and could generate 67.63 W of electrical power. When the water flow rate of the irrigation canal reached 1.5 m/s, it could reach 114 rpm and generate 129.2 W. This hydro-compact generator set is suitable for irrigation canals with a water flow rate ranging from 1–1.5 m/s.

Keywords: spiral turbine; electrical power; torque

Citation: Monatrakul, W.; Senawong, K.; Sritram, P.; Suntivarakorn, R. A Comparison Study of Hydro-Compact Generators with Horizontal Spiral Turbines (HSTs) and a Three-Blade Turbine Used in Irrigation Canals. *Energies* **2023**, *16*, 2267. https://doi.org/10.3390/en16052267

Academic Editors: Yongguang CHENG and Zhengwei Wang

Received: 24 December 2022
Revised: 12 February 2023
Accepted: 24 February 2023
Published: 27 February 2023

1. Introduction

Climate change causes incidents that challenge humankind to endure. The United Nation Office for Disaster Risk Reduction (UNDRR) states that some human activities have caused these drastic global disasters [1]. Global warming is one of the most serious problems faced by humankind, responsible for drastic weather changes caused by burning fossil fuels and population expansions. In the near future, there is expected to be a trend of power shortages that will lead to the increased usage of clean and renewable energy by every country in the world [2,3]. Clean and renewable energy usage is expected to have fewer effects on the environment than fossil fuels. However, clean and renewable energy from hydro-power plants can still harm the ecosystem, because their establishment requires immense areas of land [4,5].

In Thailand today, the capability of electrical power production from hydro-power plants throughout the country is 3107.51 megawatts (MW.), which accounts for 23.01% of Alternative Energy Consumption. Thailand aims to gather 3228 MW. from hydro-power plants by the year 2037 [6]. Although geographical limitations obstruct further establishment of large hydro-power plants, Thailand has other potential sources of hydro-kinetic power, such as local irrigation reservoirs, irrigation canals, or local administration sectors, that can provide aid for villages and communities to produce electrical power from water sources [7].

Many studies indicate that there are plenty of irrigation canals in Thailand. However, most of them have the low water flow speed rate of about 0.52–2.0 m/s [8]. Thus, extracting electrical power from these canals requires a special hydro-power generator that is designed to operate under low water flow speed (less than 2.0 m/s) or low head pressure (less than 0.3 m). For ultimate performance, the generator requires an effective small-size turbine within the system. This kind of turbine must be eco-friendly, and must be well-accepted by the majority of people [9–11]. However, leaves, waste, or debris in the canals must be considered because, as they flow along, they might clog up the generator's water inlets [12,13]. Some living creatures in the water may also be trapped inside and injured [14]. Despite these concerns, this Pico power generator system can benefit houses or small villages [15,16]. The output wattage can be predicted if there is sufficient water in the system [17]. Moreover, this kind of generator has a positive environmental impact [18,19].

River Current Energy Conversion Systems (RCECS) is a system that was proven to be able to convert the kinetic energy of water flow in rivers into other forms of useful power effectively. One key factor that makes RCECS successful is the cost of power production, including operation and maintenance costs. Moreover, the design, applications, capability, and practicality of RCECS have created accountability [20]. Thus, the adaptation of RCECS in this study will strengthen the proof of the system to be a powerful and effective alternative of gathering renewable energy.

The principle of water current power indicates that most of the water volume flowing horizontally in any natural source must have direction and speed. This can be a source of electrical power harvest [21,22] when a turbine hydro-power generator is installed [23]. The electricity is generated from a transformation of water flow power into kinetic power at the axle of the generator's turbine; still, there is a power loss in the system [24]. There are many types of water turbines in hydro-power generator systems, largely divided into two groups according to their axle configurations (vertical and horizontal). Horizontal-axle turbines are further divided into two types according to their applications; vertical and horizontal installations [25]. Turbines in whirlpool hydro-generator systems, as well as Induced Vibration (VIV) turbines [26], have also been developed. Therefore, to achieve the most practical and effective goal of extracting renewable energy from local irrigations requires choosing the right applications that suits the water flow characteristics and the size of the power generator. Betz's law indicates that the extracted peak power coefficient value from a turbine is 59 percent. However, in reality, the extracted power values are lower because of the loss from the turbine's characteristics. Therefore, the power loss value in the system must be included in the experiments [27]. A proper turbine type selection for this circumstance must be carefully selected for the highest potential and the lowest eco-impact, as well as for low production cost. Moreover, there are studies concerning different types of turbines used in hydro-generator systems in irrigation canals with water flow rates from 0.6–3 m/s, as shown in Table 1 [28].

Table 1. Turbine specifications [28].

Manufacturers	Device Name	Turbine Type	Min./Max. Speed	Power Output
Lucid Energy Pty., Ltd. (Dallas, TX, USA) [29]	Gorlov Helical turbine	Helical Darrieus cross-axis	(0.6 m/s)/no limit	Up to 20 kW, depends on size
Thropton Energy Services (Northumberland, UK) [30,31]	Water current turbine	Axis flow propeller	(0.6 m/s)/depends on diameter	Up to 2 kW at 240 V
Tidal EnergyPty., Ltd. (Canberra, Australia) [32]	Davidson–HillVenturi (DHV) Turbine	Cross-flow turbine	Min. 2 m/s	From 4.6 kW
Seabell Int.Co., Ltd. (Tokyo, Japan) [33]	Steam	Dual, cross-axis	(0.6 m/s)/no limit	0.5–10 kW models
New Energy Corporation Inc. (Calgary, AB, Canada) [34]	En Current Hydro Turbine	Cross-axis	Max. 3 m/s for maximum power	5 kW (and 10 kW)
Eclectic Energy Ltd. (Nottinghamshire, UK) [35]	DuoGen-3	Axial flow propeller	Min. (0.93 m/s)/(4.63 m/s) max.	8 amps at 3.09 m/s
Alternative Hydro Solutions Ltd. (Toronto, ON, Canada) [31]	Free-stream Darrieus water turbine	Cross-axis	(0.5 m/s)/depends on diameter	Up to 2–3 kW
Energy Alliance Ltd. (Ural region, Russia) [30]	Sub-merged hydro unit	Cross-axis	Min 3 m/s	1–5 kW (and 410 kW)

A properly designed Horizontal Spiral Turbine (HST) can achieve the highest efficiency. Its key feature is the ability to reduce the turbulence generated by the impact of high-pressure water flow with the surface of the blade. This is obviously a benefit to the generator's efficiency and performance. Another obvious unique feature is that HST can be applied to wind and water generators with low-speed flow. HST's advantages are the low cost of production, compactness, and an ability to operate effectively under low pressure head. A hydro-generator system with an HST application can be properly used in rivers, canals, and irrigation canals without doing serious harm to living creatures in the water. It can be harmlessly blended with ecosystems; moreover, the construction materials are not expensive [36]. HSTs have been designed based on Fibonacci functions [37] and the golden ratio [38]. They are an arithmetic series of natural phenomena according to Fibonacci sequence principles. One round of rotation (360 degree) of an HST axle is equal to a length. Thus, the angle of the blades alters the length. The last characteristic of efficient HST relates to the number of blades. If the number of blades is insufficient, the surface area for converting force will cause the HST to generate low torque. On the other hand, if the number of blades is too high, the HST will create a solid wall state. Too many blades also increase the HST mass and inertia that reduce the overall torque [39]. Ratchapol et al. (2016) discovered that, under low water velocity and with between two and six blades, an HST with three blades could achieve the maximum torque and most optimal performance [40].

Yasukuni Nishi Okubo and Norio Kikuchi designed a new hydro-compact generator system that had "a runner" and "a collection device" including "a diffuser section" to enhance the water flow through the turbine. This increased power extracting ability and flow rate by 2.76% in a hydroelectric power generator with a three-blade turbine. At 456 rpm and a water flow rate of 1.72 m/s, 156.4 watts of electric power could be generated [41,42]. Hidayat et al. (2020) proved that a hydro-spiral turbine could spin faster (90 rpm) than other types at the same water volume [43]. Ratchapol et al. (2016) used the golden ratio function to enhance the efficiency of spiral turbines by extending the blade's radius and adjusting the diameter/length (D/L) ratio to 2/3. They proved that a spiral turbine had proper efficiency when compared with the a three-blade turbine. A hydro-generator with a proper spiral turbine could effectively generate electrical power at water flow rate ranging from 0.5 to 2 m/s [40]. Wiroon and Ratchapol (2017) also proved that a horizontal spiral turbine in a hydro-generator system with a nozzle chamber inlet was more efficient than one with a free flow inlet. Moreover, the blade angle also significantly affected the system's performance and efficiency [44]. Uday Y. Bhenede (2015) had designed and developed a turbine in a hydro-generator system that could effectively operate under low water pressure. It was a small-size turbine in a generator system that could operate under variant head pressure [45].

Today, HST research and development is focused on identifying the best values of different blade angles, length/axis radius ratio, blade number, or torque. Therefore, in this study, two Horizontal Spiral Turbines (HSTs) with blade angles of eighteen and twenty-one degrees, respectively, and a three-blade turbine, were tested in a laboratory at five levels of water flow rate ranging from 1–2 m/s. After the efficiency and torque values of each turbine were identified, they were installed in two 200 W power generator systems: (1) with a "diffuser" attachment; and (2) with an "in-line+diffuser+nozzle chamber" attachment, and tested in a local irrigation canal with a water flow rate of 1.2 m/s.

2. Materials and Methods

Laboratory experiments and related equations

This study aimed to design 3-blade Horizontal Spiral Turbines with blade angles of 18 and 21 degrees used in a hydro-generator system [40]. The simulation was set as displayed in Figures 1 and 2 and was simulated in a laboratory. The generator was fitted with 6 inches of spiral turbine. It was attached with 6 inches of PVC pipes and its body was a clear acrylic for better performance observations. The simulator system could generate 5 levels of water flow speed ranging from 1–2 m/s, which represented the actual current in local irrigation canals with a flow speed rate of 0.5–2.0 m/s throughout the region [8]. The initial torque power and torque power was identified using Equation (1) [46–48]. Later, in the field experiments, the turbine size was increased to 15 inches.

$$P_{t,out} = \frac{2\pi\tau N}{60} \tag{1}$$

$$P_{t,out} = \text{Power Output (kW)}$$

$$\tau = \text{Torque (Nm)}$$

$$N = \text{Revolutions per minute (RPM)}$$

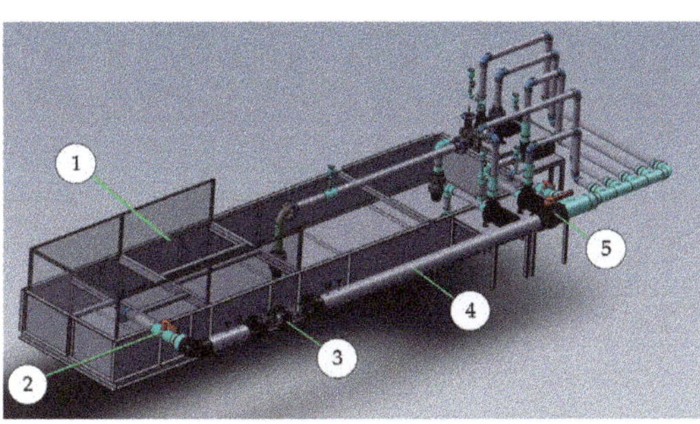

(a)

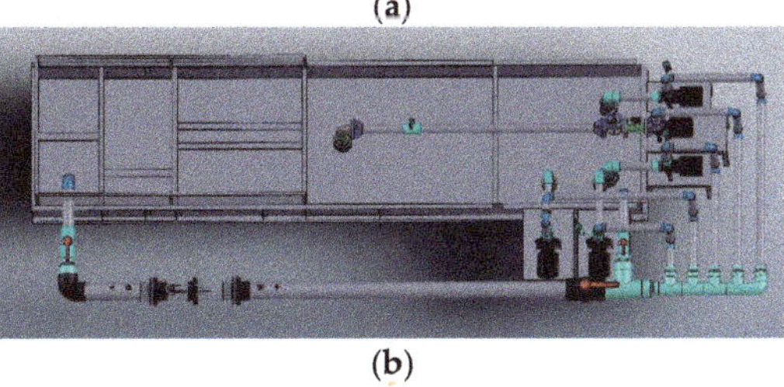

(b)

Figure 1. Sketch of the experimental setup. (**a**) Isometric view; (**b**) top view.

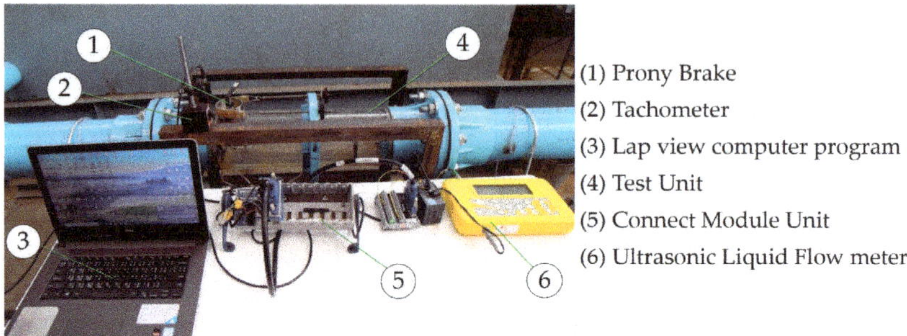

Figure 2. Simulation set.

Value measurements in the laboratory were operated by using: (1) an Ultrasonic Liquid Flow Meter; Micronics Portaflow PF300 for water flow rate (m/s), (2) a Tachometer Light Sensor Module for RPM, and (3) a Local Cell Module for torque at turbine axles. These values were processed by Lab View NATIONAL INSTRUMENTS NI cDAQ-9178 to identify the generated power and turbine efficiency. This laboratory simulation was carried out according to Gianluca Zitti et al. (2002) [49]. Torque values from each turbine model were studied and evaluated for appropriate uses in real contexts.

Power and efficiency of turbines at different water volumes and flow speeds were identified using Equations (2) and (3). The peak power of turbines was identified using Equation (4) [50,51].

$$P_{t,in} = \rho g Q H_n \tag{2}$$

$$\eta = \frac{P_{t,in}}{P_{t,out}} \tag{3}$$

$$P_{t,out} = \left[\frac{16}{27}\right]\left[\frac{1}{2}\right]\rho A v^3 \tag{4}$$

$$P_{t,in} = \text{Power input (kW)}$$

$$\rho = \text{Water Density}\left(\text{kg/m}^3\right)$$

$$g = \text{Gravitational of mass}\left(\text{m/s}^2\right)$$

$$Q = \text{Water Volume}\left(\text{m}^3/\text{s}\right)$$

$$H_n = \text{Head Pressure (m)}$$

$$\eta = \text{Efficiency}$$

$$\frac{16}{27} = \text{Max Power Coefficient (CPmax)}$$

$$A = \text{Turbine Surface}\left(\text{m}^2\right)$$

$$v = \text{Water Flow Speed (m/s)}$$

One rotation of the HST axle (360 degree) determined the width and the length of the turbine. The fixed pitch (L) and fixed diameter (D) are displayed in Table 2. HSTs with 18 and 21 degree blade angles, and the three-blade turbine, are displayed in Figures 3 and 4.

Table 2. Turbine specifications.

Turbine	Blade Area (mm²)	Pitch (mm)	Length (mm)	Width (mm)
18-degree	12,653.72	49	233	150
21-degree	10,587.91	49	197	150
3-blade	754.96	-	-	10

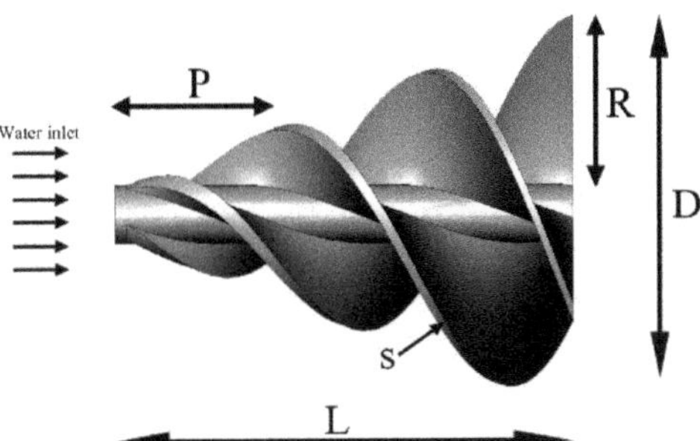

Figure 3. Horizontal spiral turbine design.

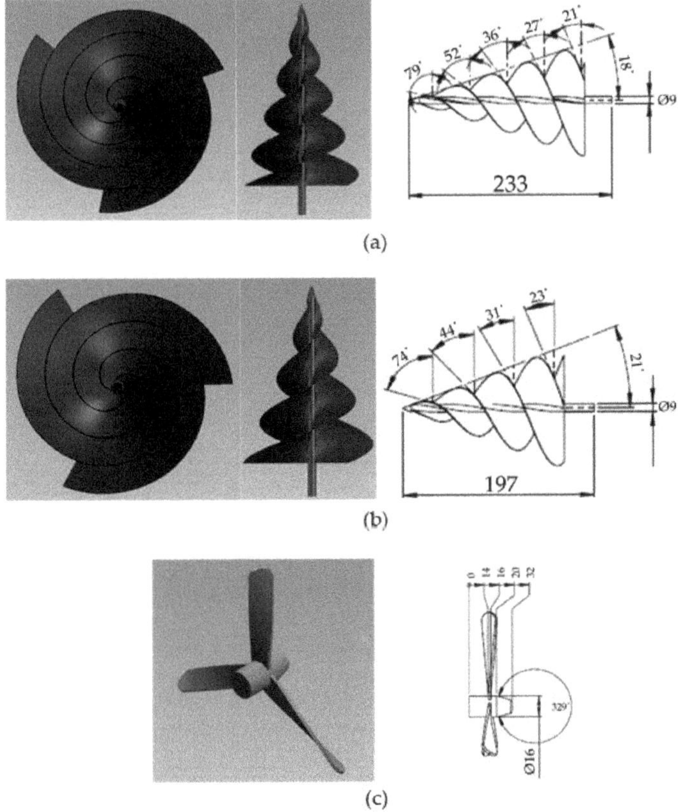

Figure 4. Turbines to be tested in the simulations. (**a**) 18-degree, (**b**) 21-degree, (**c**) 3-blade.

"L" represents the height at one round of rotation. "s" represents the peripheral length of the HST. "R" represents the HST radius. "N" represents the number of rounds of rotation. All the variables were used in Equations (5) and (6) for the design.

$$r'(t) = -\frac{2\pi NR}{L}\sin\left(\frac{2\pi Nt}{L}\right)i + \frac{2\pi NR}{L}\cos\left(\frac{2\pi Nt}{L}\right)j + k \tag{5}$$

$$s = \sqrt{4\pi^2 N^2 R^2 + L^2} \tag{6}$$

The collective chamber (see Figure 5) was designed according to Yasuyuki [41] to regulate the inlet fluid pressure of the system [52]. It was also attached with a rudder to enhance the water flow rate.

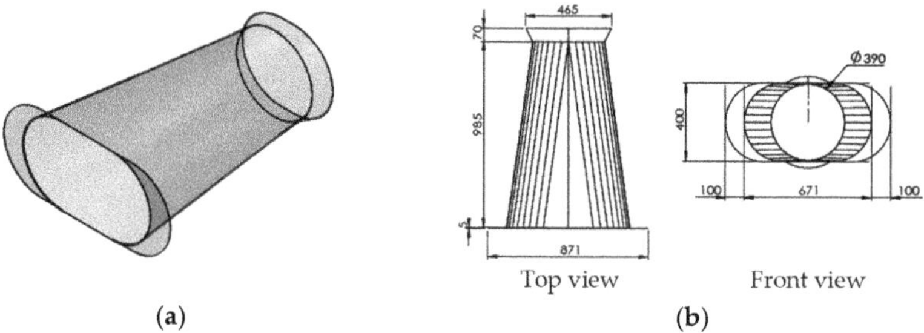

Figure 5. Diffuser/collective chamber. (a) Three-dimensional representation; (b) drawing view.

Power Output (η)

The power output was identified using Equations (7) and (8) [53].

$$\eta = \frac{P_e}{P_m} \tag{7}$$

$$P_e = \text{EI} \tag{8}$$

$$E = \text{Voltage (V)}$$

$$I = \text{Current (A)}$$

$$P_e = \text{Power Output (W)}$$

$$P_m = \text{Power Output from Water Power (W)}$$

3. The Field Experiment

A diffuser was attached to the turbine housing of a 200 W electrical power generator (Table 3) as displayed in Figure 6. A 15-inch-diameter HST with an 18-degree blade angle, a 15-inch-diameter HST with a 21-degree blade angle, and a three-blade turbine 15 inches in diameter were tested in this generator system. There were two generator systems used in the experiment: (1) a generator system with a diffuser as a trumpet-shaped collective chamber attachment (see Figure 7a), and (2) a generator system with an in-line pipe + diffuser + nozzle chamber attachment (see Figure 7b). They were tested in an irrigation canal in Baan Kota, Tambol Sila, Amphoe Muang, Khon Kaen Province (see Figure 8) with water velocity ranging from 0.8 to 1.2 m/s and a water volume of 2.10–3.15 m^3/s. They were planted using a rigid structure as a scaffold to set them in the middle and along the canal. The data of (1) water velocity, (2) the turbines' RPM, (3) the turbines' torque, and (4) the power output were analyzed using measuring devices, and this data reading process was carried out exactly as in the laboratory.

Table 3. Axial flux generator specification.

Rated Power	0.2 kW
Rated Rotation Speed	200 rpm
Rated Voltage	12VAC
Efficiency	90%
Start Torque	<0.05 Nm
Phase Type	3 Phase

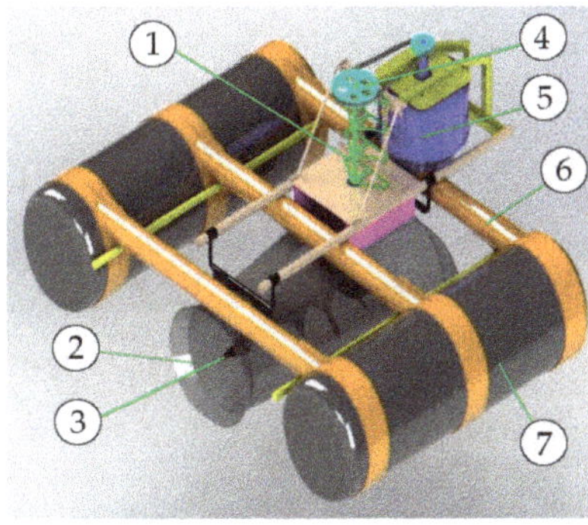

Figure 6. The prototype and its components.

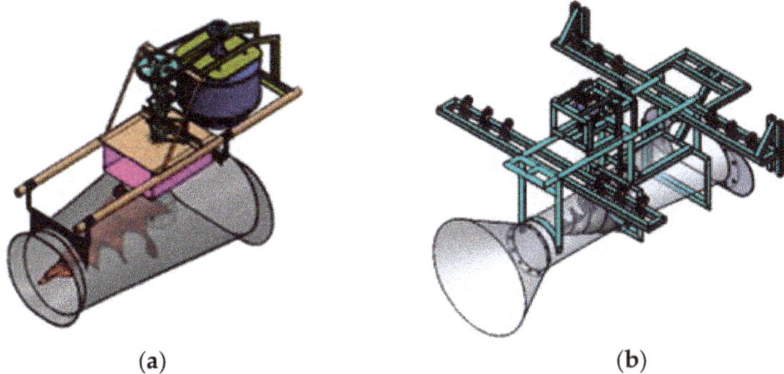

(a) (b)

Figure 7. Two generator systems. (**a**) The generator system with a diffuser as a trumpet-shaped collective chamber. (**b**) The generator system with an in-line pipe + diffuser + nozzle chamber.

(a) (b)

Figure 8. Field experiments. (**a**) The generator system with a diffuser as a trumpet-shaped collective chamber. (**b**) The generator system with an in-line pipe + diffuser + nozzle chamber.

4. Results and Discussions

Turbines' efficiency and torque from laboratory simulator

The comparison results of efficiency of the three turbine types with 6-inch diameters at different water velocities from the laboratory simulation are displayed in Figure 9. As shown below, the HST with the 21-degree blade angle reached the highest efficiency at all water flow speeds. The HST with the 18-degree blade angle's efficiency was lower than that of the 21-degree HST. However, the efficiency of the three-blade turbine could not be determined under water velocities ranging from 1 to 2 m/s. It needed higher water flow speed to be practical, so it had the lowest efficiency in this experiment. The HST with the 21-degree blade angle had 38.10% efficiency at 120.00 rpm of 2 m/s water velocity. It reached 39.05% at 156.63 rpm of 1.5 m/s water velocity. The results revealed that the operations of the HST with 21-degree blade angle at 1, 1.25, 1.50, 1.75, and 2.0 m/s indicated 1100, 1300, 1630, 1850, and 2100 L/s of water, respectively. Figure 10 displays the torque values generated from the HSTs and the three-blade turbine. All torque values were altered by water flow speed. The 21-degree HST generated the most torque at 212 Nm at 2 m/s water flow speed. This created data that can be used as a reference in enhancing systems for practical field applications.

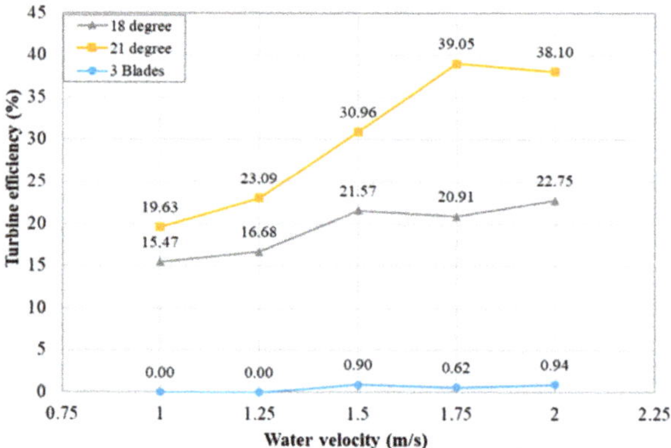

Figure 9. Comparison results of efficiency of the three turbine types in laboratory simulation at water flow speed of 1, 1.25, 1.50, 1.75, and 2.00 m/s.

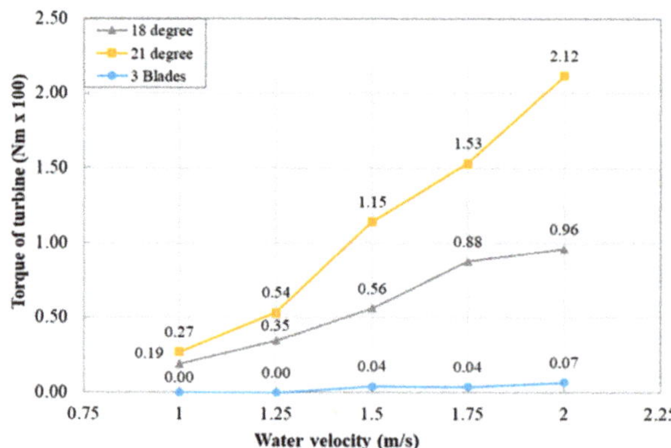

Figure 10. Comparison results of torque values of the three turbine types in laboratory simulation at water flow speed of 1, 1.25, 1.50, 1.75, and 2.00 m/s.

Field Experiments

The torque values of the three turbine types with 15-inch diameters are shown in Table 4. From the field experiments in an actual irrigation canal with the average water velocity of 1.2 m/s, the best turbine type was the HST with the 21-degree blade angle in a generator system with an in-line + collective chamber + inlet nozzle. It had a peak torque of 284 Nm. The HST with the 21-degree blade angle in a generator system with only the diffuser/collective chamber could only generate a torque value of 217.2 Nm.

Table 4. Torque results.

Turbine		Speed of Turbine (rpm)	Torque (Nm × 100)
Water velocity (m/s)		1.2	1.2
Spiral turbine + collective chamber	18-degree	28	13.65
	21-degree	41	21.72
	3-blade	26	4.56
Spiral turbine + inline + collective chamber + inlet nozzle 100 cm	18-degree	102	20.06
	21-degree	108	28.40
	3-blade	58	7.04

When torque values generated from the 6-inch-diameter turbines used in the laboratory simulation and the 15-inch-diameter turbines used in the field experiments from both generator systems were compared, it was obvious that the generator system with the additional modifications of collective chamber + inlet nozzle 100 cm could generate more torque. In the laboratory simulation, at a water flow speed of 1.2 m/s, the 21-degree HST with 6-inch diameter produced about 49 Nm of torque, but when the turbine diameter was increased to 15 inches, the torque values increased to 217.2 and 284.0 Nm in both generators. Obviously, the generator with higher torque could generate more power.

Power output

From the field experiments in an actual irrigation canal with the average water velocity of 1.2 m/s, the best turbine type among the rest that could generate the most power output (67.63 watts) was the HST with the 21-degree blade angle in a 200 W generator system with an in-line + collective chamber + inlet nozzle (as shown in Table 5). After the long-run operations at different water flow speeds, the results were as shown in Figures 11 and 12. These results were analyzed and compared with related studies as displayed in Table 6.

Table 5. Power output results.

Turbine		Electric Power (Watt)
Water velocity (m/s)		1.2
Spiral turbine + collective chamber	18-degree	21.90
	21-degree	28.43
	3-blade	6.92
Spiral turbine + inline + collective chamber + inlet nozzle 100 cm	18-degree	26.85
	21-degree	67.63
	3- blade	7.15

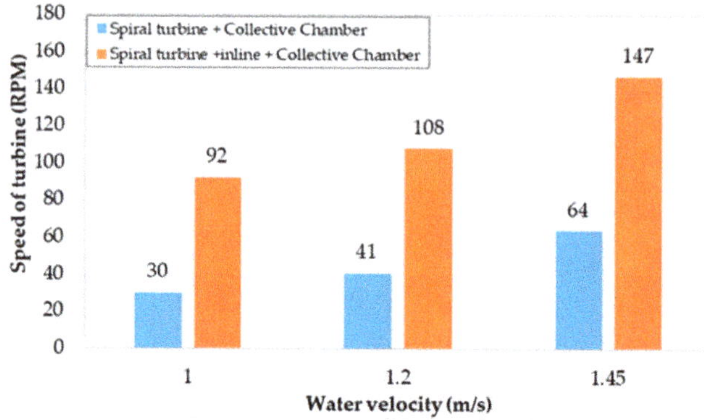

Figure 11. Comparison results of rpm generated by HST with 21-degree angle blade in the generator system with diffuser/collective chamber and the generator system with in-line + diffuser + nozzle chamber at water velocities of 1.0, 1.2, and 1.45 m/s.

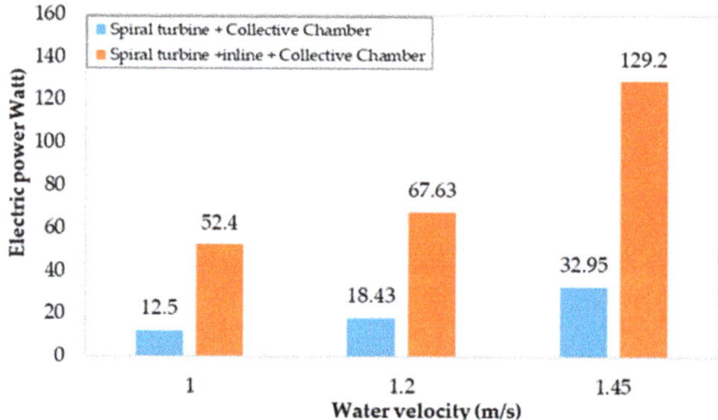

Figure 12. Comparison results of power output generated by HST with 21-degree angle blade in the generator system with diffuser/collective chamber and the generator system with in-line + diffuser + nozzle chamber at water velocities of 1.0, 1.2, and 1.45 m/s.

Table 6. Comparing results with related studies.

Ref.	Turbine	Water Velocity (V)	Section Inlet Area (A)	Work Output (Watt)
Wiroon	21 deg spiral Turbine + diffuser	1.45 m/s	0.16 m^2	67.63
Yasukuni Nishi et al. [41]	3-blades + diffuser	1.72 m/s	0.16 m^2	154.00
Erinofiardi [54]	Screw Turbine	0.077 m/s	0.0088 m^2	0.28
Tomomi Uchiyama [12]	Guide vane	0.159 m^3/s	0.005 m^2	222.00
Joel Titus [55]	Turbine blades	0.009 m^3/s	0.005 m^2	212.00
Budiarso [56]	Turgo turbine	n.d.	0.038 m^2	5.34
C.H. Achebe [57]	Crossflow turbine	0.0015 m^3/s	0.015 m^2	35.00
Gianluce Zitti et al. [49]	Screw turbine	1–2 m/s	0.003 m^2	500.00

5. Summary and Conclusions

The golden ratio function had the most important role in this study in order to verify the proper HST's blade angles. There were three turbines (an HST with an 18-degree blade angle, an HST with a 21-degree blade angle, and a three-blade turbine) designed to be tested in two generator systems in this study. Firstly, all turbines were tested in a simulation set in the laboratory. The simulator revealed that the HST with the 21-degree blade angle had the highest efficiency (38.10%) at all water flow speeds. When the two sets of generator systems (one with a diffuser/collective chamber attachment and another one with an in-line + diffuser + nozzle chamber) were tested in an actual irrigation canal at the average natural water flow speed of 1.2 m/s, the results revealed that the HST with the 21-degree blade angle in a generator system with an in-line + diffuser + nozzle chamber could generate the highest power input of 67.63 watts. In the long-run operation experiment, when the water velocity reached 1.5 m/s, it could generate electrical power up to 129.2 watts.

The turbines used the in laboratory and field experiments were different in size because the bigger turbine size employed in the actual irrigation canal had more inertia. Thus, the transmission set attached to the generator system had to be properly designed to fulfill the turbine's performance.

This study showed that a Horizontal Spiral Turbine (HST) in a hydro-compact generator system could effectively operate under low water flow speed. The HST blade angle could also generate high torque from low water flow speed. In conclusion, an HST in a generator system with an in-line + diffuser + nozzle chamber could significantly generate high torque and high power output in any actual irrigation canal with low water flow rate at the average of 1.5 m/s throughout the country.

6. Break-Even Point and Investments

From this study, the conditions that must be considered in order to meet the break-even point for investors are:

- 1.5 m/s average water flow speed;
- 24 h operation;
- 365 days production length;
- 0.8 (9.6 months) plant factor value;
- USD 0.13 per one power unit selling price.

The maximum cost of production of a 200-watt hydro-compact generator with an HST and attachments is USD 2434.83 and the break-even point is 22.86 years. Details are displayed in Table 7.

In conclusion, it is obvious that the main issue of applying this kind of technology is the correlation between the cost of production and the break-even point. However, the production cost of this compact hydro-generator system is still high, whereas the low-speed water flow rate of local irrigation canals provides "not much" renewable energy. Nicolas D et al. (2016) stated that the main concerns of using HSTs were lowering the production costs, enhancing the turbines' efficiency, environmental awareness, and public relations campaigns [58].

Table 7. A 200-watt hydro-compact generator with HST cost of production.

	Inventory	USD per Unit
1.	Turbine Cost of Production	429.31
2.	Materials Cost	572.41
3.	Generator Cost of Production	857.67
4.	Labor Cost	572.41

Remark: There will be variations in production cost since the cost has been converted from THB to USD (updated 15 December 2020).

Author Contributions: Conceptualization, W.M. and R.S.; methodology, W.M. and R.S.; software, W.M.; validation, R.S.; formal analysis, K.S.; investigation, W.M. and R.S.; resources, P.S.; data curation, P.S.; writing—original draft preparation, W.M. and K.S.; writing—review and editing, W.M. and R.S.; visualization, W.M.; supervision, R.S.; project administration, W.M. All authors have read and agreed to the published version of the manuscript.

Funding: This research received no external funding.

Institutional Review Board Statement: Not applicable.

Informed Consent Statement: Not applicable.

Data Availability Statement: Not applicable.

Acknowledgments: The researchers are grateful to the Farm Engineering and Automation Technology Research Group (FEAT), Department of Mechanical Engineering Khon Kaen University and Rajabhat Maha Sarakham University, Agricultural Machinery and Postharvest Technology Center Khon Kaen University Khon Kaen Province 40002 Thailand, Postharvest Technology Innovation Center Science Research and Innovation Promotion and Utilization Division Office of the Ministry of Higher Education Science Research and Innovation 10400 Thailand for their support with the tools and equipment used in this research.

Conflicts of Interest: The authors declare no conflict of interest.

References

1. UNDRR. Global Assessment Report on Disaster Risk Reduction. Available online: https://www.undrr.org/gar2022-our-world-risk (accessed on 10 December 2022).
2. Maji, I.K. Impact of Clean Energy and Inclusive Development on CO_2 Emissions in Sub-Saharan Africa. *J. Clean. Prod.* **2019**, *240*, 118186. [CrossRef]
3. Ghimire, A.; Dahal, D.; Kayastha, A.; Chitrakar, S.; Thapa, B.S.; Neopane, H.P. Design of Francis Turbine for Micro Hydropower Applications. *J. Phys. Conf. Ser.* **2020**, *1608*, 012019. [CrossRef]
4. Ezra, H. Unsustainable Development in the Mekong: The Price of Hydropower. *Consilience* **2014**, *12*, 63–76. [CrossRef]
5. Northrup, J.M.; Wittemyer, G. Characterising the Impacts of Emerging Energy Development on Wildlife with an Eye towards Mitigation. *Ecol. Lett.* **2013**, *16*, 112–125. [CrossRef]
6. Ministry of Energy. Percentage of Alternative Energy Consumption. Available online: https://www.dede.go.th/ewt_news.php?nid=48247 (accessed on 10 December 2022).
7. Ministry of Energy. Hydro Power. Available online: https://energy.go.th/th/interested/29228 (accessed on 10 December 2022).
8. Royal Irrigation Department. Water Situation Summary. Available online: http://hydro-3.rid.go.th/ (accessed on 23 January 2022).
9. EGAT. Environmental Report 2015. Available online: https://www.egat.co.th/home/wp-content/uploads/2021/08/environmental_report_2558.pdf (accessed on 10 December 2022).
10. Thailand Power Development. Thailand Power Development Plan 2015–2036 (PDP2015). Available online: https://www.eppo.go.th/images/POLICY/ENG/PDP2015_Eng.pdf (accessed on 10 December 2022).
11. Mishra, S.; Singal, S.; Khatod, D. Costing of a Small Hydropower Projects. *Int. J. Eng. Technol.* **2012**, *4*, 239–242. [CrossRef]
12. Tomomi, U.; Satoshi, H.; Tomoko, O.; Tomohiro, D. A Feasibility Study of Power Generation from Sewage Using a Hollowed Pico-Hydraulic Turbine. *Engineering* **2016**, *2*, 510–517. [CrossRef]
13. Tomomi, U.; Satoshi, H.; Tomohiro, D. Development of a Propeller-Type Hollow Micro-Hydraulic Turbine with Excellent Performance in Passing Foreign Matter. *Renew Energy* **2018**, *126*, 545–551. [CrossRef]
14. Čada, G.; Loar, J.; Garrison, L.; Fisher, R.; Neitzel, D. Efforts to Reduce Mortality to Hydroelectric Turbine-Passed Fish: Locating and Quantifying Damaging Shear Stresses. *Environ. Manag.* **2006**, *37*, 898–906. [CrossRef]
15. Zeleňáková, M.; Fijko, R.; Diaconu, D.C.; Remeňáková, I. Environmental Impact of Small Hydro Power Plant. A Case Study. *Environments* **2018**, *5*, 12. [CrossRef]
16. Cobb, B.R.; Sharp, K.V. Impulse (Turgo and Pelton) Turbine Performance Characteristics and Their Impact on Pico-Hydro Installations. *Renew Energy* **2013**, *50*, 959–964. [CrossRef]
17. Rakibuzzaman, M.; Suh, S.-H.; Kim, H.-H.; Ryu, Y.; Kim, K.Y. Development of a Hydropower Turbine Using Seawater from a Fish Farm. *Processes* **2021**, *9*, 266. [CrossRef]
18. International Energy Agency. *Key World Energy Statistics*; OECD: Paris, France, 2017; ISBN 9789264283213.
19. Tiago Filho, G.L.; dos Santos, I.F.S.; Barros, R.M. Cost Estimate of Small Hydroelectric Power Plants Based on the Aspect Factor. *Renew. Sustain. Energy Rev.* **2017**, *77*, 229–238. [CrossRef]
20. Khan, M.; Iqbal, M.; Quaicoe, J. River Current Energy Conversion Systems: Progress, Prospects and Challenges. *Renew. Sustain. Energy Rev.* **2008**, *12*, 2177–2193. [CrossRef]
21. Nyambuu, U.; Semmler, W. Climate Change and the Transition to a Low Carbon Economy-Carbon Targets and the Carbon Budget. *Econ. Model.* **2020**, *84*, 367–376. [CrossRef]
22. Lago, L.I.; Ponta, F.L.; Chen, L. Advance and Trends in Hydrokinetic Turbine System. *Energy Sustain. Dev.* **2010**, *14*, 287–296. [CrossRef]
23. Wanchat, S. *Design and Development of Water Free Vortex Hydro Power Plant*; Khon Kaen University: Khon Kaen, Thailand, 2014.
24. Lu, Q.; Li, Q.; Kim, Y.K. A Study on Design and Aerodynamic Characteristics of a Spiral-Type Wind Turbine Blade. *J. KSV* **2012**, *10*, 27–33. [CrossRef]
25. Yuce, M.I.; Muratoglu, A. Hydrokinetic Energy Conversion Systems: A Technology Status Review. *Renew. Sustain. Energy Rev.* **2015**, *43*, 72–82. [CrossRef]

26. Fernandes, A.C.; Rostami, A.B. Hydrokinetic Energy Harvesting by an Innovative Vertical Axis Current Turbine. *Renew Energy* **2015**, *81*, 694–706. [CrossRef]

27. Ragheb, M.; Ragheb, A.M. Wind Turbines Theory—The Betz Equation and Optimal Rotor Tip Speed Ratio. In *Fundamental and Advanced Topics in Wind Power*; Carriveau, R., Ed.; InTech: Rijeka, Croatia, 2011; ISBN 978-953-307-508-2.

28. Vermaak, H.J.; Kusakana, K.; Koko, S.P. Status of Micro-Hydrokinetic River Technology in Rural Applications: A Review of Literature. *Renew. Sustain. Energy Rev.* **2014**, *29*, 625–633. [CrossRef]

29. Johnson, J.B.; Pride, D.J. River, Tidal, and Ocean Current Hydrokinetic Energy Technologies: Status and Future Opportunities in Alaska. Available online: http://www.uaf.edu/files/acep/2010_11_1_State_of_the_Art_Hydrokinetic_Final.pdf (accessed on 11 December 2022).

30. Water current turbines pump drinking water. Caddet Renewable Energy. Available online: http://www.caddet-re.org/assets/no83.pdf (accessed on 11 December 2022).

31. Verdant Power. Technology Evaluation of Existing and Emerging Technologies—Water Current Turbines for River Applications. Available online: http://oreg.ca/web_documents/verdant_river_turbines_report.pdf (accessed on 12 December 2022).

32. Tidal Energy Pty Ltd. The Davidson-Hill Advantage. Available online: http://tidalenergy.net.au/ (accessed on 12 December 2022).

33. Seabell International Co., Ltd. Stream—Hydrokinetic Power Generation Systems. Available online: http://www.seabell-i.com/e/img/guide_pdf.pdf (accessed on 12 December 2022).

34. New Energy Corporation Inc. Encurrent Hydropower Turbines 5 k Wand 10 kW Specifications. Available online: http://www.newenergycorp.ca/Portals/0/documents/datasheets/ENC.005.010.DataSheet.pdf (accessed on 13 December 2022).

35. Eclectic Energy Ltd. Duogen-3: Combined Water and Wind Generator. Available online: http://www.duogen.co.uk/page14.html (accessed on 13 December 2022).

36. Safdari, A.; Kim, K.C. Aerodynamic and Structural Evaluation of Horizontal Archimedes Spiral Wind Turbine. *J. Clean Energy Technol.* **2015**, *3*, 15–20. [CrossRef]

37. Spira, M. On the Golden Ratio. In Proceedings of the ICME-12th International Congress on Mathematical Education COEX, Seoul, Republic of Korea, 8–15 July 2012.

38. Kanyanart, J.; Somjai, C. Some Evenly Divisible Properties of Fibonacci Numbers. *Thaksin Univ. J.* **2009**, *12*, 50–58.

39. Ragheb, M. Optimal Rotor Tip Speed Ratio. Available online: https://users.wpi.edu/~cfurlong/me3320/DProject/Ragheb_OptTipSpeedRatio2014.pdf (accessed on 12 December 2022).

40. Ratchaphon, S.; Sujate, W.; Wiroon, M. An Experimental Study of Electricity Generation Using a Horizontal Spiral Turbine Spiral M. On the Golden Ratio. In Proceedings of the CPESE 2016 3rd International Conference on Power and Energy Systems Engineering, Kitakyushu, Japan, 8–10 September 2016.

41. Yasuyuki, N.; Terumi, I.; Kaoru, O.; Norio, K. Study on an Axial Flow Hydraulic Turbine with Collection Device. *Hindawi Publ. Corp. Int. J. Rotating Mach.* **2014**, *2014*, 308058. [CrossRef]

42. Inoue, M.; Sakurai, A.; Ohya, Y. A Simple Theory of Wind Turbine with Brimmed Diffuser. *Turbomachinery* **2002**, *30*, 497–502. [CrossRef]

43. Hidayat, M.N.; Ronilaya, F.; Eryk, I.H.; Joelianto, G. Design and Analysis of a Portable Spiral Vortex Hydro Turbine for a Pico Hydro Power Plant. *IOP Conf. Ser. Mater. Sci. Eng.* **2020**, *732*, 12051. [CrossRef]

44. Monatrakul, W.; Suntivarakorn, R. Effect of Blade Angle on Turbine Efficiency of a Spiral Horizontal Axis Hydro. *Energy Procedia* **2017**, *138*, 811–816. [CrossRef]

45. Uday, Y.B. Case Study: Development of Low Head Turbine to Address the Micro Hydropower Market. In Proceedings of the International Conference on Hydropower for Sustainable Development, Dehradum, India, 5–7 February 2015.

46. Batchelor, G.K. *An Introduction to Fluid Dynamics*; Cambridge University Press: Cambridge, UK, 2000; ISBN 9780521663960.

47. Versteeg, H.K.; Malalasekera, W. An Introduction to Computational Fluid Dynamics. *Pearson Educ. Ltd.* **2007**, *1*, 1–49.

48. Sritram, P.; Suntivarakorn, R. Comparative Study of Small Hydropower Turbine Efficiency at Low Head Water. *Energy Procedia* **2017**, *138*, 646–650. [CrossRef]

49. Zitti, G.; Fattore, F.; Brunori, A.; Brunori, B.; Brocchini, M. Efficiency Evaluation of a Ductless Archimedes Turbine: Laboratory Experiments and Numerical Simulations. *Renew Energy* **2020**, *146*, 867–879. [CrossRef]

50. Schubel, P.J.; Crossley, R.J. Wind Turbine Blade Design. *Energies* **2012**, *5*, 3425–3429. [CrossRef]

51. Abolvafaei, M.; Ganjefar, S. Maximum Power Extraction from Fractional Order Doubly Fed Induction Generator Based Wind Turbines Using Homotopy Singular Perturbation Method. *Int. J. Electr. Power Energy Syst.* **2020**, *119*, 105889. [CrossRef]

52. Behrouzi, F.; Maimun, A.; Nakisa, M. Review of Various Designs and Development in Hydropower Turbines. World Academy of Science. *Eng. Technol. Int. J. Mechanical. Aerosp. Ind. Mechatron. Eng.* **2014**, *8*, 293–297.

53. Giorgio, R. *Principles and Applications of Electrical Engineering*, 5th ed.; McGraw-Hill Higher Education: New York, NY, USA, 2011.

54. Erinofiardi, E.; Syaiful, M.; Prayitno, A.H. Electric Power Generation from Low Head Simple Turbine for Remote Area Power Supply. *J. Teknol.* **2015**, *74*, 21–25. [CrossRef]

55. Joel, T.; Bakthavatsalam, A. Design and Fabrication of In-Line Turbine for Pico Hydro Energy Recovery in Treated Sewage Water Distribution Line. *Energy Procedia* **2019**, *156*, 133–138. [CrossRef]

56. Budiarso; Febriansyah, D.; Warjito; Adanta, D. The Effect of Wheel and Nozzle Diameter Ratio on the Performance of a Turgo Turbine with Pico Scale. *Energy Reports* **2020**, *6*, 601–605. [CrossRef]

57. Achebe, C.; Okafor, O.; Obika, E. Design and Implementation of a Crossflow Turbine for Pico Hydropower Electricity Generation. *Heliyon* **2020**, *6*, e04523. [CrossRef]
58. Laws, N.D.; Epps, B.P. Hydrokinetic Energy Conversion: Technology, Research, and Outlook. *Renew. Sustain. Energy Rev.* **2016**, *57*, 1245–1259. [CrossRef]

water

MDPI

Article

Study on Water Replacement Characteristics of Xinghai Lake Wetland Based on Landscape Water Quality Objectives

Mengdi Wu [1], Guobin Xu [1,*], Xiaoyu Niu [1], Zhen Fu [2] and Xianrong Liao [2]

[1] National Key Laboratory of Hydraulic Engineering Simulation and Safety, Tianjin University, Tianjin 300350, China
[2] Beifang Investigation, Design & Research Co., Ltd., Tianjin 300222, China
*   Correspondence: xuguob@tju.edu.cn; Tel.: +86-22-2740-1127

Abstract: Many issues with water quality and water ecology are caused by the Xinghai Lake's enormous catchment, significant evaporation rates, and one additional water supply. To quantitatively study Xinghai Lake's water displacement characteristics, a two-dimensional hydrodynamic-tracer coupling model based on MIKE21 was developed. The findings indicate that: (1) Xinghai Lake's water replacement cycle exhibits spatial heterogeneity, with a general characteristic of fast water renewal in the southern lake area and slow renewal in the northern lake area, and the gradient change of the water replacement cycle from south to north is influenced by a variety of factors, including the lake's flow field, flow, topography, and wind field. (2) The throughput flow has an impact on the majority of the waters in Xinghai Lake. When there is a high water flow, the lake region has a high flow velocity, rapid water transport, and a large capacity for water exchange; when there is a low water flow, the lake area has a slow flow velocity, poor water flow, and a lengthy water exchange period. (3) The flow field of Xinghai Lake is complicated, the flow velocity is low, and it is a lake system where quick water exchange and slow water exchange coexist. This flow field is influenced by the interplay of wind-generated flow and throughput flow. (4) To speed up the water body's rejuvenation, the Xinghai Lake wetland needs more inlets and exits to introduce new water sources.

Keywords: Xinghai Lake wetland; landscape water; MIKE21 two-dimensional hydrodynamic-tracer coupling; water displacement cycle; water demand

Citation: Wu, M.; Xu, G.; Niu, X.; Fu, Z.; Liao, X. Study on Water Replacement Characteristics of Xinghai Lake Wetland Based on Landscape Water Quality Objectives. *Water* **2023**, *15*, 1374. https://doi.org/10.3390/w15071374

Academic Editors: Yongguang Cheng and Zhengwei Wang

Received: 19 February 2023
Revised: 30 March 2023
Accepted: 1 April 2023
Published: 3 April 2023

1. Introduction

Landscape water bodies are typically found in the city's center, surrounded by other buildings such as homes and businesses, which have significant value for the city's ecological environment and economic growth. Urban landscape water bodies, on the other hand, are typically closed, slow-moving, static water bodies that are poorly mobile, easy to pollute, have little water environment capacities, and have poor self-cleaning [1,2]. The conflict between relatively poor environmental management and the growth of the urban population is becoming more and more obvious as urbanization progresses, and numerous urban lakes have experienced varying degrees of pollution. To avoid a decline in lake water quality, regulate pollution, and support the healthy growth of the urban water environment, it is important to clarify the flow field characteristics and water displacement capacity of artificial lakes. Ecological recharge, or the addition of comparatively clean water to polluted water bodies, is a technique for ecosystem restoration [3,4]. It can help the ecology of the water bodies by diluting the contaminants there. Research on hydrodynamics and water displacement characteristics can be used to create practical lake and wetland water environment regulatory policies, enhance lake water quality, and address other issues, including wetland degradation. The scientific basis for safeguarding and reestablishing the biological environment of lakes and wetlands can be found in a thorough understanding of the hydrodynamic and water body exchange features of lakes and the elements that influence them. To preserve the healthy and sustainable operation of the wetland water ecological environment, it is crucial to establish a hydrodynamic-tracer coupling model to realistically conduct recharge scheduling to satisfy the water quality and quantity of lakes and wetlands [5,6]. Jizhang Tang et al. [6] investigated water exchange capacity using the MIKE21 coupled hydrodynamic-staining agent model simulation and discovered that enhancing water exchange capacity is critical to maintaining the health of water ecosystems. Cucco et al. [7] created a two-dimensional hydrodynamic model of the Venice Lagoon to calculate the residence duration of the water column using the residual function of a passive tracer injected into the lagoon. Liping Xu et al. [5] employed computational methods to create hydrodynamic and convective diffusion models to investigate the renewal time and water exchange rate of manmade lakes. Delhezé et al. [8] created a two-dimensional estuary model and a one-dimensional river model to examine convective dispersion in the Skelt River estuary, obtaining the retention, transit, and influence time of estuarine water bodies. The majority of landscape water body research is focused on water quality prediction and water restoration approaches. Qingbin Meng et al. [9] used scenario analysis to simulate Daming Lake's flow rate and COD distribution under five different water diversion scenarios, in addition to the coupled water quality-quantity simulation method to simulate Daming Lake's current flow rate and COD distribution. These simulations provided a scientific basis for the reasonable water diversion scheduling of Daming Lake in Jinan City. Wenjie Xu et al. [10] developed a model to determine the optimal water diversion volume of Dongchang Lake after taking into account the water quality and quantity requirements for Dongchang Lake's main functions, such as landscape tourism and ecological environment. This model served as the foundation for the scientific diversion of water from the Yellow River.

Xinghai Lake is a significant lake in northern Ningxia, China. It has a huge surface area, high evaporation and water consumption in the lake region, and minimal water loss and waste. Since 2004, Xinghai Lake's primary source of water replenishment has been the Yellow River, due to the lake's limited capacity for self-purification and reliance on a single source of water recharge. The wetland ecology has a single structure and an unsound system, failing to create a healthy water ecosystem. The ecology is fragile, with low biodiversity and insufficient biological richness of microbes, plants, and fish. There are currently fewer studies on water demand computation and the water replacement cycle from the perspective of the water quality-volume balance of landscape water bodies than there are studies on Xinghai Lake's evaluation of water environment quality and analysis of pollution sources [11].

Many issues with the water environment and water ecology are caused by the poor mobility of lake water in its natural form, the low frequency of lake water replacement, the lengthy cycle of water replacement, and the declining quality of the lake water environment and water ecology. Welch [12] diverted water from the Columbia River into Moses Lake in 1992 and replaced 20% of the lake's water, greatly reducing the concentration of pollutants, such as total phosphorus, total nitrogen, and chlorophyll A, in the lake and increasing the lake's transparency, which made the water in the lake clearer and better. In various European nations, the water quality of polluted water bodies is frequently enhanced by adding clean water from the outside, diluting the pollutants already present in the waters, and then releasing the pollutants from the existing waters by scouring and water transportation. Water environment issues, including eutrophication in the ancient Danube Lake, were extensively managed and improved in 1999 by Donabaum et al. [13], using water diversion and replacement technologies.

The Xinghai Lake wetland is researched using the MIKE21 FM model (MIKE21 FM model, which was developed by the Hydraulic Research Institute of Denmark) created by the Danish Institute of Water Resources in this work. To guide the management of Xinghai Lake, numerical simulations are used to quantitatively examine the recharge cycle of the landscape water body of the Xinghai Lake wetland and to logically develop the water quality assurance and purification plan. This study aims to properly develop the water quality guarantee and purification plan, to research the recharging cycle of the Xinghai Lake wetland's landscape water body, and to serve as a guide for managing the lake's water quality and water ecology.

2. Data Source and Methods

2.1. Study Area

The geographic coordinates of Xinghai Lake (Dawukou flood control reservoir) are $105°58'–106°59'$ E and $38°22'–39°23'$ N. It has a total area of 43 km^2 and a lake area of around 23.38 km^2, of which 10.55 km^2 is the study area (Figure 1). It is situated in Dawukou District, Shizuishan City, Ningxia Hui Autonomous Region, China. It is also a crucial component of the "72 lakes" water system in the Yinchuan Plain, which primarily performs the tasks of flood control, drainage, improving the water environment, regulating microclimate, and maintaining the ecological balance of the region. It is also a significant protected wetland for the entire area. Xinghai Lake receives 200 to 400 mm of precipitation on average every year, with the amount rising from south to north. Water evaporates at a rate of, on average, between 800 and 1600 mm per year, which is four times more than it rains. As Xinghai Lake has grown and been used more recently, it has used an excessive amount of water; the water environment has become somewhat contaminated, and the water quality is declining.

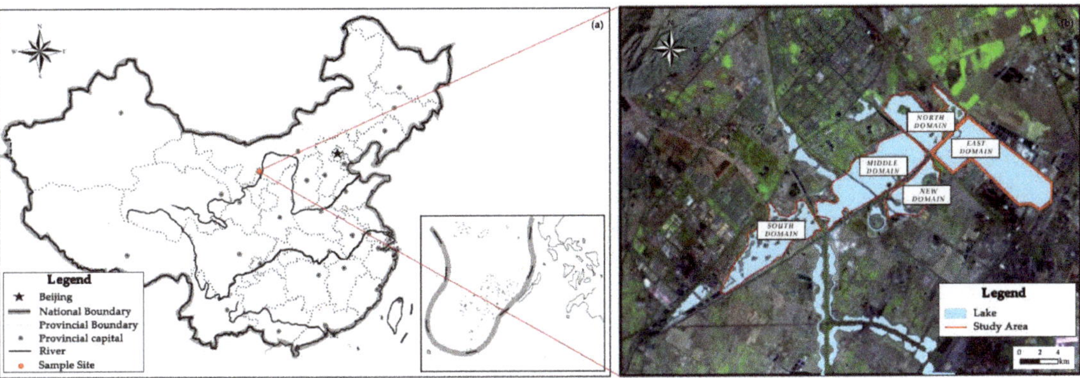

Figure 1. Study area. (**a**) Location of the project; (**b**) Scope of the study.

2.2. Data Source

The data for wind speed and wind direction is taken from the China Ground Cumulative Daily Value Dataset (1981–2010) of the National Meteorological Science Data Center; the data for water level and flow are all taken from the Feasibility Study Report on the Remediation of Ecological and Environmental Problems of Xinghai Lake; the data for precipitation are taken from the average measured data from December 1965 to December 2019 at Pingluo Rainfall Repr.

2.3. Methods

The Xinghai Lake wetland's landscape water needs should satisfy both the water requirements for wetland operation and the requirements for the tour's water quality. This is because the wetland's ecological and environmental functions must be protected as well as its public viewing, leisure entertainment, and aesthetic functions. Placement should take into account the following:

(1) The water requirements of the actual Xinghai Lake wetland, including those for the plants that grow there, for irrigation in the shoreline zone, and landscape water levels.

(2) The daily water replenishment of the Xinghai Lake wetland, including precipitation, seepage, and evaporation.

(3)   Starfish Lake wetland replacement water: maintains the landscape water body's water quality standards, preventing eutrophication and meets the needs of the water environment capacity, protecting the wetland's functions from being harmed by routine replacement water.

2.3.1. Water Demand Model of Xinghai Lake Landscape Water

In addition to surface water, groundwater and soil water are important sources of water for the landscape. The following factors should be given special attention when determining the water demand for the landscape water bodies in Xinghai Lake [14]:

(1)   Water requirement for evaporation W_W:

Since Xinghai Lake is situated in a dry and semi-arid region of northwest interior China, one of the main ways that water is lost from the lake is through substantial regional evaporation. The accounting average for evaporation during a multi-year period is 1107.3 mm, according to observation result data from a meteorological station's evaporation dish.

(2)   Water demand of wetland vegetation W_P:

To sustain their life and ability to reproduce, plants require a lot of water during their growth and development. Plant transpiration accounts for the majority of this water consumption, but soil evaporation also uses up a lot of water. The fundamental physiological water demand of plants in general is only a tiny component and is typically overlooked in the calculation procedure, whereas the ecological water demand of vegetation can be directly estimated by calculating the evapotranspiration of vegetation. The modified Penman formula technique and the Hargreaves algorithm [15] are typically employed in the computation of plant evapotranspiration. The equations can be expressed as:

$$W_P = A(t)ET \tag{1}$$

In the formula: W_P is the water requirements of wetland plants, $A(t)$ is the wetland area, E is plant cover, and T is evapotranspiration–emissions.

(3)   Leakage volume W_S:

When the water level of the wetland is higher than the groundwater level, infiltration will occur, and the seepage recharge of Xinghai Lake can be determined according to Darcy's law. The formula for calculating the amount of seepage is:

$$W_S = KAJ \tag{2}$$

In the formula: W_S is leakage: K is permeability coefficient, m/d; A is infiltration area, m²; J is hydraulic slope, dimensionless.

(4)   Precipitation W_R:

The findings of a lengthy series of meteorological observations show that Shizuishan City experiences relatively little precipitation, with an annual average of 179.8 mm.

2.3.2. A Two-Dimensional Hydrodynamic-Water Quality Model Based on the MIKE 21 Water System Connectivity

(1)   Hydrodynamic model

The model is based on the three-way incompressible Reynolds-valued homogeneous Navier–Stokes equations and obeys the Boussinesq assumption and the assumption of hydrostatic pressure. The two-dimensional non-constant shallow water equation set is [16]:

$$\frac{\partial h}{\partial t} + \frac{\partial h\overline{u}}{\partial x} + \frac{\partial h\overline{v}}{\partial y} = hS \tag{3}$$

$$\frac{\partial h\overline{u}}{\partial t} + \frac{\partial h\overline{u}^2}{\partial x} + \frac{\partial h\overline{uv}}{\partial y} = f\overline{v}h - gh\frac{\partial \eta}{\partial x} - \frac{h}{\rho_0}\frac{\partial p_a}{\partial x} -$$
$$\frac{gh^2}{2\rho_0}\frac{\partial \rho}{\partial x} + \frac{\tau_{xx}}{\rho_0} - \frac{\tau_{bx}}{\rho_0} - \frac{1}{\rho_0}\left(\frac{\partial s_{xx}}{\partial x} + \frac{\partial s_{xy}}{\partial y}\right) +$$
$$\frac{\partial}{\partial x}(hT_{xx}) + \frac{\partial}{\partial y}(hT_{xy}) + hu_s S \tag{4}$$

$$\frac{\partial h\overline{v}}{\partial t} + \frac{\partial h\overline{uv}}{\partial x} + \frac{\partial h\overline{v}^2}{\partial y} = -f\overline{u}h - gh\frac{\partial \eta}{\partial y} - \frac{h}{\rho_0}\frac{\partial p_a}{\partial y} -$$
$$\frac{gh^2}{2\rho_0}\frac{\partial p}{\partial y} + \frac{\tau_{sy}}{\rho_0} - \frac{\tau_{by}}{\rho_0} - \frac{1}{\rho_0}\left(\frac{\partial s_{yx}}{\partial x} + \frac{\partial s_{yy}}{\partial y}\right) +$$
$$\frac{\partial}{\partial x}(hT_{xy}) + \frac{\partial}{\partial y}(hT_{yy}) + hv_s S \tag{5}$$

In the formula: t is time; x, y are Cartesian coordinate system coordinates; η is the water level; d is the resting water depth; $h = \eta + d$ is the total water depth; u, v is the velocity components in x, y directions, respectively; f is the Gauche force coefficient, $f = 2\omega\sin\varphi$; ω is the angular velocity of the Earth's rotation; φ is the local latitude; g is the acceleration of gravity; ρ is the density of water; S_{ss}, S_{xy}, S_{yy}, are the radiation stress components; S is the source term; (u_s, v_s) is the source term current flow rate.

The letters with a crossbar are the average values. For example, is the average flow rate along the water depth, defined by:

$$h\overline{u} = \int_{-d}^{\eta} u\,dz, \, h\overline{v} = \int_{-d}^{\eta} v\,dz \tag{6}$$

T_{ij} is the horizontal viscous stress terms, including viscous forces, turbulent stresses, and horizontal convection, and these quantities are derived from the velocity gradient averaged along the water depth using the eddy viscosity equation [17].

$$T_{xx} = 2A\frac{\partial \overline{u}}{\partial x}, T_{xy} = A\left(\frac{\partial \overline{u}}{\partial y} + \frac{\partial \overline{v}}{\partial x}\right), T_{yy} = 2A\frac{\partial \overline{v}}{\partial y} \tag{7}$$

(2)   Water quality model

Convective diffusion is used in the water quality model, which has the following form in a Cartesian coordinate system [18]:

$$\frac{\partial C}{\partial t} + \frac{\partial uC}{\partial x} + \frac{\partial wC}{\partial y} + \frac{\partial wC}{\partial z} = F_c + \frac{\partial}{\partial z}\left(D_v\frac{\partial C}{\partial z}\right) - k_p C + C_s S \tag{8}$$

Among them:

$$F_c = \left[\frac{\partial}{\partial x}\left(D_h\frac{\partial}{\partial x}\right) + \frac{\partial}{\partial y}\left(D_h\frac{\partial}{\partial y}\right)\right]C \tag{9}$$

The integral form of the transport equation can be derived $\frac{\partial U}{\partial t} + \nabla\cdot F(U) = S(U)$ as follows:

$$U = h\overline{C}$$
$$F^1 = \left[h\overline{uC}, h\overline{vC}\right]$$
$$F^V = \left[hD_h\frac{\partial \overline{C}}{\partial x}, hD_h\frac{\partial \overline{C}}{\partial y}\right] \tag{10}$$
$$S = -hk_p\overline{C} + hC_s S$$

The discrete finite volume form of the transport equation is given by $\frac{\partial U_i}{\partial t} + \frac{1}{A_i}\sum_{j}^{NS} F\cdot n\Delta\Gamma_j = S_i$.

The numerical format is the same as for the hydrodynamic discrete part.

The MIKE21 FM model employs the finite volume method for the discrete solution of the study area, which is discretized into several irregular triangular meshes, and the mesh is encrypted in the local area to better reflect the complex boundary. The finite volume method ensures the conservation of water volume and momentum in the calculation area [19].

The MIKE21 hydrodynamic module is the foundation of the model simulation, primarily through the study area of the topographic grid processing, control of wet and dry grid conditions, selection of appropriate solution method, and consideration of the model input and output role to calculate the simulation of the study area water depth and flow field distribution law. The hydrodynamic module is built on top of the convective diffusion module, and the two are dynamically related. The convective diffusion module simulates the characteristics of diffusion phenomena occurring in the water body through convection and diffusion processes by setting different types of diffusion coefficients, which has the advantages of ignoring the complex physical, chemical, biological, and ecological processes that occur when pollutants are transported in the water and highlighting the role of the regional water body flow field. The hydrodynamic module and the convective diffusion module were chosen as research methods for this simulation based on the lake's water environment and the purpose of the study.

2.3.3. Water Displacement Cycle Coupled Hydrodynamic-Tracer Model

The convective diffusion equation in MIKE21 is used to compute the water replacement cycle, and an exponential decay function based on the change in concentration is used to calculate the tracer concentration decay rate (11).

$$C_t = C_0 e^{-t/Tt} \tag{11}$$

In the formula: C_0 is the initial tracer concentration value; C_t is the remaining tracer concentration value at the time t.

From equation (11), the concentration has decayed to e^{-1}, or 37% of the initial concentration, when $t = Tf(V/Q)$; therefore, the water change cycle is defined as the time required for the remaining concentration to decrease to 37% of the initial concentration [6].

2.4. Study on the Water Balance of the Water Bodies in Xinghai Lake Wetland

The Xinghai Lake Wetland's water requirement is the amount and quality of water necessary to maintain the wetland's ecological and environmental functions at a given stage of development and under various ecological and environmental preservation objectives. The following should be included in the water use positioning:

Wetland vegetation and landscape water level are the two main factors that contribute to the Xinghai Lake wetland's water requirement.

Wetland evaporation water demand and wetland seepage water demand are the two main components of wetland daily water replenishment.

Wetland water body replacement water: maintain the water quality standards of the landscape water body, to ensure that the water body does not experience eutrophication, and the capacity of the water environment to fulfill the periodic replacement water needs.

2.4.1. Water Demand of Xinghai Lake Wetland

Referring to the water balance principle proposed by Liu Jingling [20] and others, and combined with the current characteristics of Xinghai Lake, its ecological water demand can be expressed by the following equation:

$$W_L = W_W + W_P + W_S - W_R \tag{12}$$

In the formula: W_L is the ecological water demand of the lake wetland, W_W is the evaporation water demand of the water surface, W_p is the vegetation water demand of the wetland, W_S is the seepage water demand, and W_R is the precipitation.

(1) Evaporation water demand

Xinghai Lake is located in the arid and semi-arid region, and evaporation has emerged as one of the major means of water loss. According to the weather station data, the average annual evaporation of Xinghai Lake is 10,513,800 m^3.

(2)   Wetland vegetation water demand

As the reed community makes up the majority of the flora around Xinghai Lake, it is best to split the water demand level according to these fundamental reed features. Calculated from Equation (1), the water demand of Xinghai Lake vegetation is 5,968,800 m^3.

(3)   Seepage water demand

The foundation soil layer of the planned water body in this area is mainly vegetation fill, powder clay, sandy loam, fine sand, etc. The permeability coefficient of this study is 0.0001 m/d, and the hydraulic slope is 1. The annual seepage of Xinghai Lake is calculated by Equation (2) as 385,100 m^3.

(4)   Precipitation

One of the key sources of replenishment for the water of Xinghai Lake is natural precipitation. The Xinghai Lake region experiences an average of 179.8 mm of precipitation annually, as determined by data collected from meteorological stations in Shizuishan City between 1988 and 2019. According to calculations, the natural precipitation volume in Xinghai Lake's water is 1,896,900 m^3, based on the lake's 10.55 km^2 water surface area.

According to the above calculation results, the annual water demand of Xinghai Lake wetland is about 14,970,800 m^3.

2.4.2. The Water Requirement of Landscape Water Level in Xinghai Lake Wetland

There is not yet a standardized calculating technique or water need for landscape water level. This study primarily relies on the fundamental characteristics of Xinghai Lake as a flood control reservoir in the former plain of Helen Mountain and combines them with the ground elevation of the river bottom to establish the Xinghai Lake landscape water level. The five domains of Xinghai Lake are as follows: south, middle, north, east, and new domains. Three water levels are selected based on the wetland park's landscape purpose.

(1)   Minimum wetland water level

According to Table 1, the minimum water level in the south domain is 0.2 m, the minimum water level in the middle and north domains is 1.3 m, and the minimum water level in the east domain is 0.1 m. The one-time water demand for filling the reservoir = water surface area × water level = 982.29 million m^3.

Table 1. The domain and water level of Xinghai Lake.

Domain	Water Surface (km^2)	Minimum Water Level (m)	Standard Water Level (m)	Maximum Water Level (m)
South domain	1.54	0.2	0.3	0.4
Central domain	5.07	1.3	1.4	1.5
North domain	1.94	1.3	1.4	1.5
East domain	0.7	0.1	0.2	0.3
New domain	1	-	-	-

(2)   Standard wetland water level

Table 1 shows that the typical water level in the south domain is 0.3 m, the center and north domains are 1.4 m, the east domain is 0.2 m, and the total amount of water needed to fill the reservoir is 1081.79 million m^3.

(3)   Maximum wetland water level

Table 1 shows that the south domain's maximum water level is 0.4 m, the center and north domains' maximum water levels are 1.5 m, and the east domain's maximum water level is 0.3 m. A total of 1174.29 million m^3 of water will be needed to fill the reservoir all at once.

Calculating the water demand can be directly performed using water increment = water surface precipitation—water evaporation—water leakage = −9.002 million m^3. The investigation found that, after evaporation and seepage loss of the water, the quantity of natural precipitation entering Xinghai Lake's water could not satisfy the minimal demand for water needs. When the reservoir is filled all at once, 11,742,900 m^3 of water—considering the usual situation of water availability—is stored to sustain the wetland's functional landscape. The lake should be permitted to sink into other water sources when the water level reaches the minimal water level during operation.

2.5. Hydrodynamic Water Quality Modeling

2.5.1. Calculation Area and Grid

ArcGIS technology is used to vectorize the Xinghai Lake lake border, convert it to an XYZ file, create the lake boundary using MIKE ZERO, and then utilize a grid generator to create an unstructured grid representing the interior topography of the Xinghai Lake region, as shown in Figure 2a. The model uses a triangular mesh to adapt to the complex lake bottom topography and shoreline of Xinghai Lake, and the overall encryption of the mesh for part of the river is shown in Figure 2b. The topographic interpolation is carried out using the natural neighboring point interpolation method in the natural cell method, and the topographic data file of Xinghai Lake is obtained, and the elevation is shown in Figure 2c. A total of 41,341 grids and 24,447 nodes are generated for Xinghai Lake.

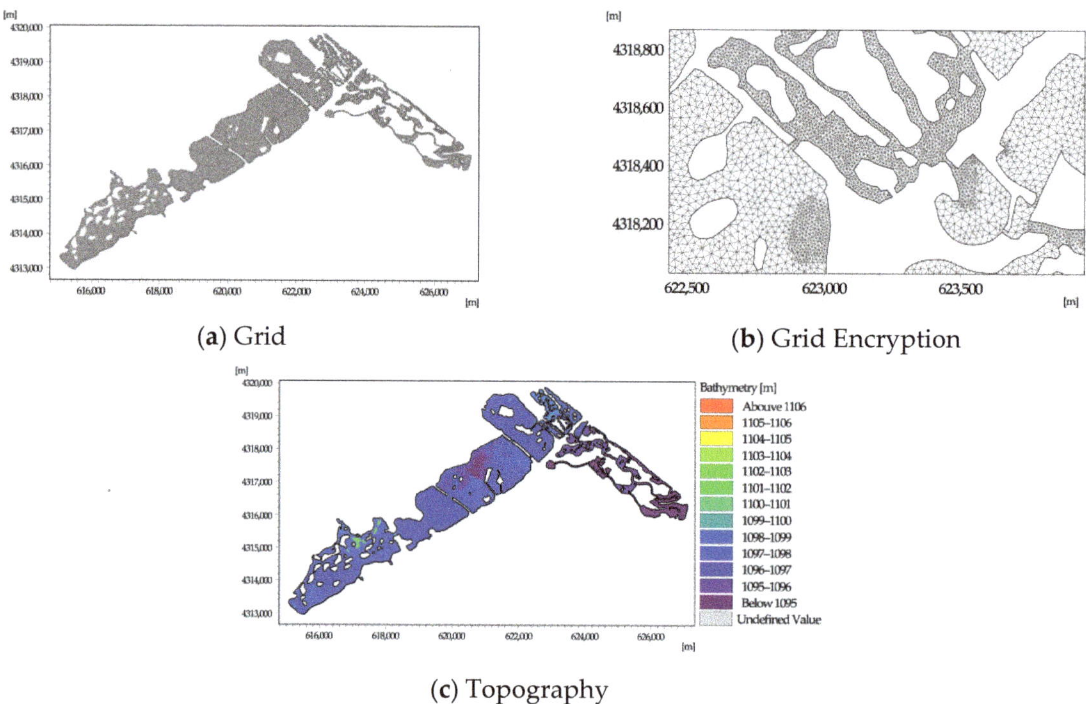

(a) Grid

(b) Grid Encryption

(c) Topography

Figure 2. Modeling the hydrodynamics and water quality of Xinghai Lake.

2.5.2. Boundary Conditions and Output

With the specified flow boundary conditions, the north domain C_1 serves as the inlet boundary, and the north domain C_2 serves as the outlet boundary. According to different water quality conditions (Figure 3), the model sets up eight source sinks from the east domain outflow S_6 to the south domain inlet S_1, from the east domain outflow S_8 to the middle domain inlet S_7, from the south domain outflow S_2 to the middle domain inlet

S_3, and from the north domain outflow S_4 to the east domain inlet. This is performed to account for the impact of wind on the local lake flow.

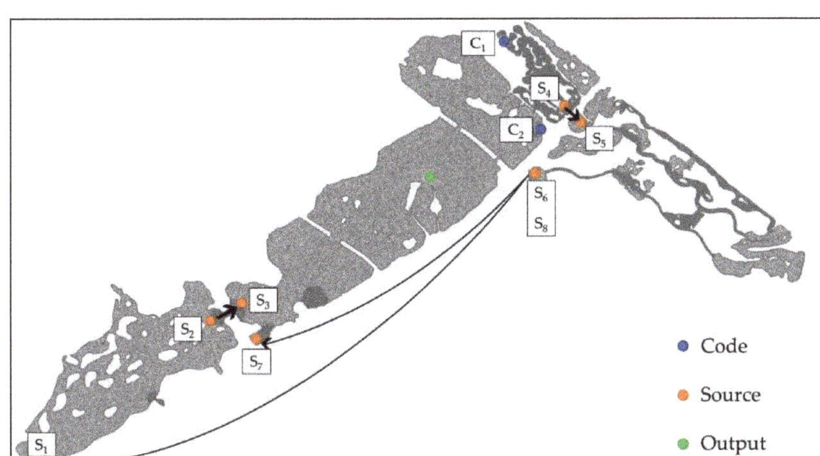

Figure 3. Boundary conditions and sources.

2.5.3. Simulation Scheme and Model

To simulate the water exchange cycle distribution of Xinghai Lake, the hydrodynamic module and convective diffusion module of MIKE21 FM are used and linked with:

(1) Initializing the concentration of the tracer material in Xinghai Lake's water column at 100 and its inflow concentration at 0.
(2) The simulation duration was chosen as 1st April 2020–1st November 2020, comprising 214 d, taking into account Xinghai Lake's chilly winter and the lengthy freezing period of the water body.
(3) For the simulation, six distinct flow conditions are chosen, as shown in Table 2, with the water quality in Xinghai Lake's middle domain serving as the study object.

Table 2. Diversion scheduling scheme of Xinghai Lake.

	Condition1/(m³/s)	Condition2/(m³/s)	Condition3/(m³/s)	Condition4/(m³/s)	Condition5/(m³/s)
Code	0.8	0.55	0.8	0.55	0.8
1	0.7	0.7	0.7	0.7	0.7
2	−0.7	−0.7	−0.35	−0.35	−0.7
3	0.7	0.7	0.35	0.35	0.7
4	−1.39	−1.39	−1.39	−1.39	−1.39
5	1.39	1.39	1.39	1.39	1.39
6	−0.7	−0.7	0.7	−0.7	−0.7
7	0.7	0.7	0.7	0.7	0
8	−0.7	−0.7	−0.7	−0.7	0

3. Results and Discussion

3.1. Flow Field Analysis

The MIKE21 Flow Model was used to execute the hydrodynamic simulation of the Xinghai Lake diversion scheduling scheme, and the results of the simulation of the Xinghai Lake water's flow field were obtained. Table 3 and Figure 4 shows that the wind field, throughput flow, and lake shape all influence the Xinghai Lake flow field's structure, which, in turn, has a significant bearing on the pollution transport process and the vast shallow lake's capacity for water regeneration [21].

Table 3. Hydrodynamic simulation for each working condition.

Condition	Average Water Level/m				Average Flow Rate (m/s)	Maximum Flow Rate (m/s)
	South Domain	Central Domain	North Domain	East Domain		
1	1096.97	1098.66	1098.71	1095.71	0.0126	0.69
2	1096.97	1098.3	1098.41	1095.71	0.0130	0.69
3	1098.28	1098.24	1098.34	1095.71	0.0112	0.69
4	1098.65	1098.24	1098.33	1095.71	0.0112	0.65
5	1096.98	1098.26	1098.38	1100.13	0.0059	0.65

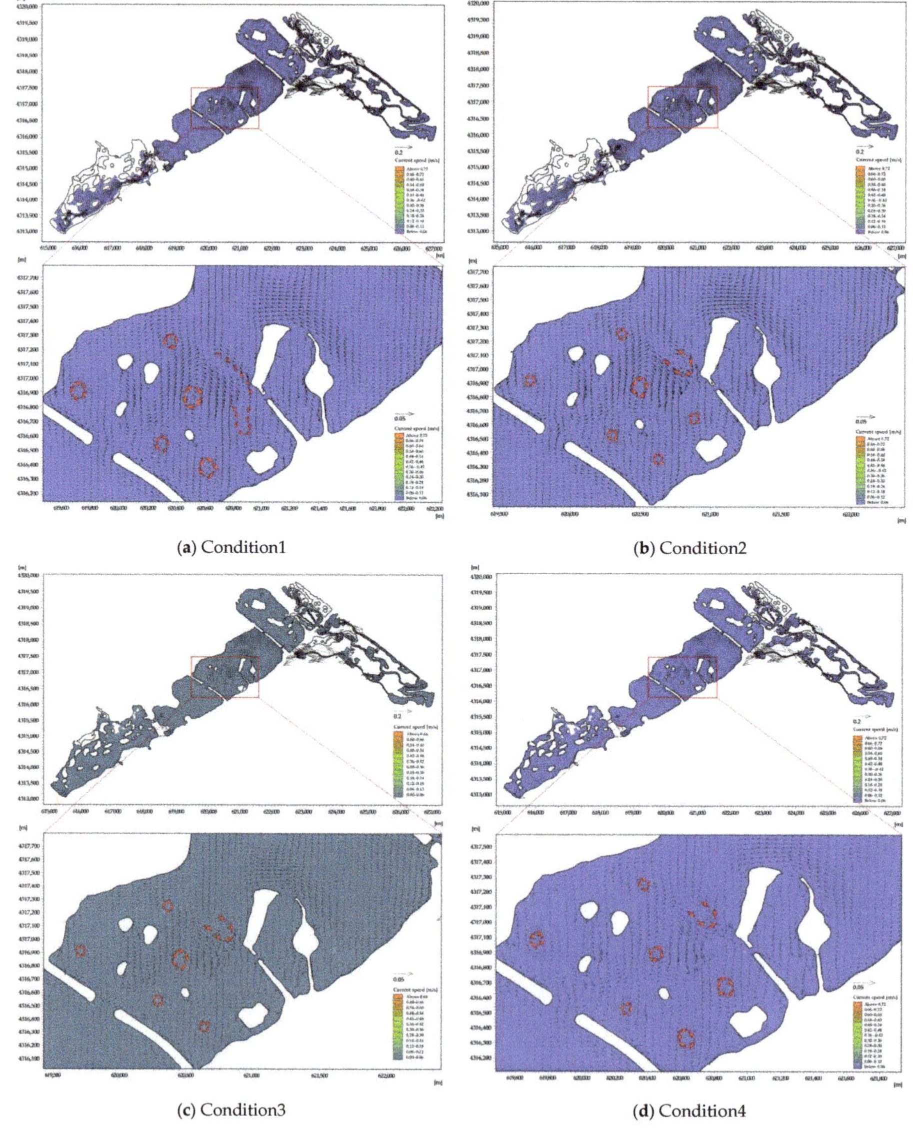

(**a**) Condition1　　　　(**b**) Condition2

(**c**) Condition3　　　　(**d**) Condition4

Figure 4. *Cont.*

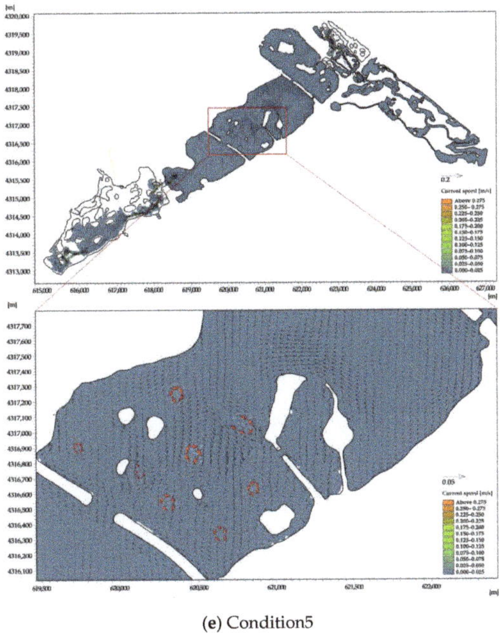

(**e**) Condition5

Figure 4. Flow field distribution.

The average water level in the south domain is lower when the S2 outflow flow is 0.7 m/s, the average value is 1096.97m, and the flow velocity is higher in the area with lower topography, which is 1.6 times higher than that when the outflow flow is 0.35 m/s, essentially forming a "river form." The flow field of Xinghai Lake is primarily influenced by the inlet flow, as seen in Figure 4. The water area increased 1.2 times and 1.3 times in comparison to the outflow of S2 at 0.7 m/s, resulting in a water level in the south region of 1098.28 m and an area covered by water accounting for 83.18%. This suggests that when the water level is lower, the water flow is quicker, increasing the dilution of pollutants, lowering the danger of local eutrophication, and enhancing the water's ability to interchange water.

The water flow may have been introduced into the middle domain at S3 and S7, and the collision between the artificial islands and the water flow in the path of water movement may have caused the water flow to be disturbed and change direction to form the circulation, as can be seen in Figure 4. Several obvious circulation areas are formed close to the artificial islands in the middle domain. The near-shore lake region is affected by the wind field under the wind velocity of 1.8 m/s, which causes the overall circulation to be strengthened and a smaller circulation to emerge. This causes positive circulation in the northeastern section of the middle domain with a flow velocity of 0.002~0.3 m/s. Because of the tiny diversion flow, sluggish flow velocity, and limited effect of the throughput flow, which is mostly driven by the wind field, the area of the stagnant region in the middle domain in Condition 3 grows by 4.7% compared to Condition 1, and by 35.7% compared to Condition 2. In Condition 4, the diversion volume in the middle domain S3 is less, the lake's water flow condition is more steady than it was in Condition 1, and less gyratory area is generated. A positive gyratory zone with a flow velocity of around 0.003–0.02 m/s forms in the northeastern nearshore area of the lake in the northern portion of the middle domain as a result of the wind field's influence. The hydrodynamic force of the water at the entrance and outflow, as well as at the artificial island, is stronger, the water area in the central domain of Xinghai Lake is bigger, the flow velocity in the central region of the water body is lower, and the flow of the water body is more visible.

When the water is abundant in Condition 1 and scarce in Condition 2, the average flow velocity in the north of Xinghai Lake is 0.0043 m/s, which is virtually stagnant, and 56% less than the average flow velocity of 0.0098 m/s in the outgoing lake region. Several clockwise circulation vortices are seen along the beach underneath the wind field, indicating that a portion of the north domain is primarily driven by the wind-generated flow. The north domain's stagnant area is higher than that of the other domains due to the slower flow velocity (Table 4), which suggests that the influence of external input pollution load on changes in the north domain's water quality is less pronounced. At the same time, under the hydrodynamic conditions of slow flow velocity, the pollutants in the water body are easily accumulated, increasing the risk of water wars in the water body.

Table 4. Area of the water retention zone.

Condition	Stagnant Water Zone Area Ratio (%)			
	South Domain	**Central Domain**	**North Domain**	**East Domain**
1	9.40	16.63	19.57	6.37
2	9.30	12.83	19.80	6.04
3	20.52	17.41	21.95	7.04
4	18.32	12.20	23.99	6.04
5	9.38	13.46	21.51	73.47

According to the findings of the previous calculations, the east domain's highest water level is 0.3 m. Table 3 displays the simulated water level in the east domain for each operational scenario. According to Table 4, when the outflow water volume of condition 5 is cut in half compared to other conditions, and the average flow velocity is 0.003 m/s, which is roughly 88% lower than that of condition 1's 0.025 m/s, the stagnant area in the east domain exceeds 10 times the area of other conditions. It shows that when the east domain's water volume is smaller and the water level is lower, throughput flow has a greater influence and the flow velocity is faster, but when the east domain's water volume is larger and the water level is higher, the wind-generated flow has a greater influence and the flow velocity is slower. The nitrogen and phosphorus nutrients in the water body are primarily present in a dissolved state and under conditions of high concentration and slow water flow.

The average flow velocity is greater than 0.01m/s in all conditions (except Condition 5), and the flow field distribution is similar, but the flow velocity is slower in the near-shore and lake island areas of the central domain, and small eddies appear. This indicates that, in large landscape lakes, the flow velocity of the water body in some areas does not change much with increasing diversion flow, and the main driving force for pollutant transport in these areas is wind-generated flow, which is consistent with Tang et al. [22] conclusion's that wind is the main force affecting water flow when there is no water input. As a result, when determining the diversion flow rate, the average flow rate cannot be considered, as in the case of a flat landscape channel, but rather the flow rate distribution of the entire water body. Increasing the diversion flow rate will only result in a waste of energy in local stagnant areas. Managers can consider increasing the number of diversion points to more effectively promote the flow of water bodies, or they can implement ecological water quality treatment methods such as ecological berms and the addition of ecological floating islands to reduce the risk of eutrophication in the stagnant water area. The flow field analysis model can be combined with other physical, chemical, and ecological methods in future studies to provide a reference for the application of other water quality treatment measures.

3.2. Water Replacement Cycle Analysis

The number of days of water replacement in each domain of Xinghai Lake under six conditions was calculated by simulating 214 d of the water replacement cycle. The water exchange cycle of Xinghai Lake is spatially heterogeneous, as can be observed in

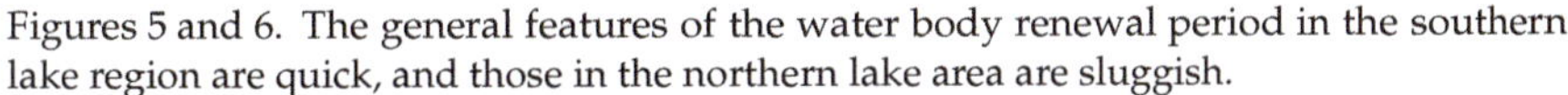

Figures 5 and 6. The general features of the water body renewal period in the southern lake region are quick, and those in the northern lake area are sluggish.

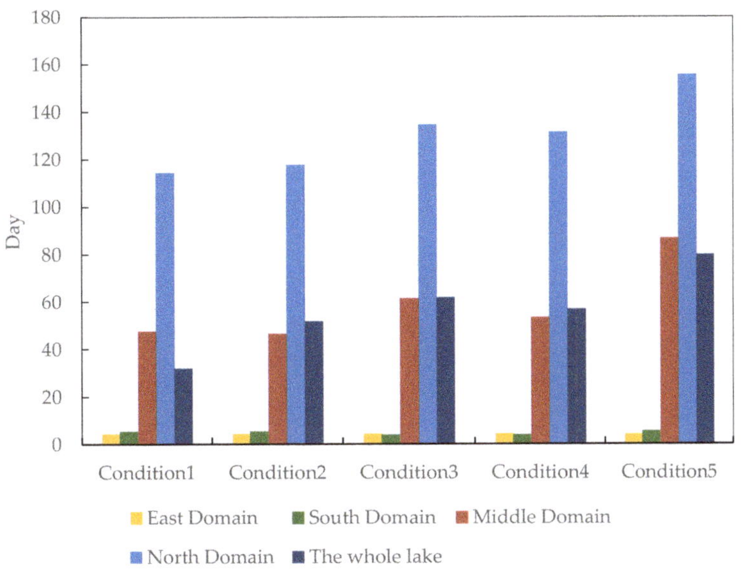

Figure 5. Average water change interval of each domain of Xinghai Lake.

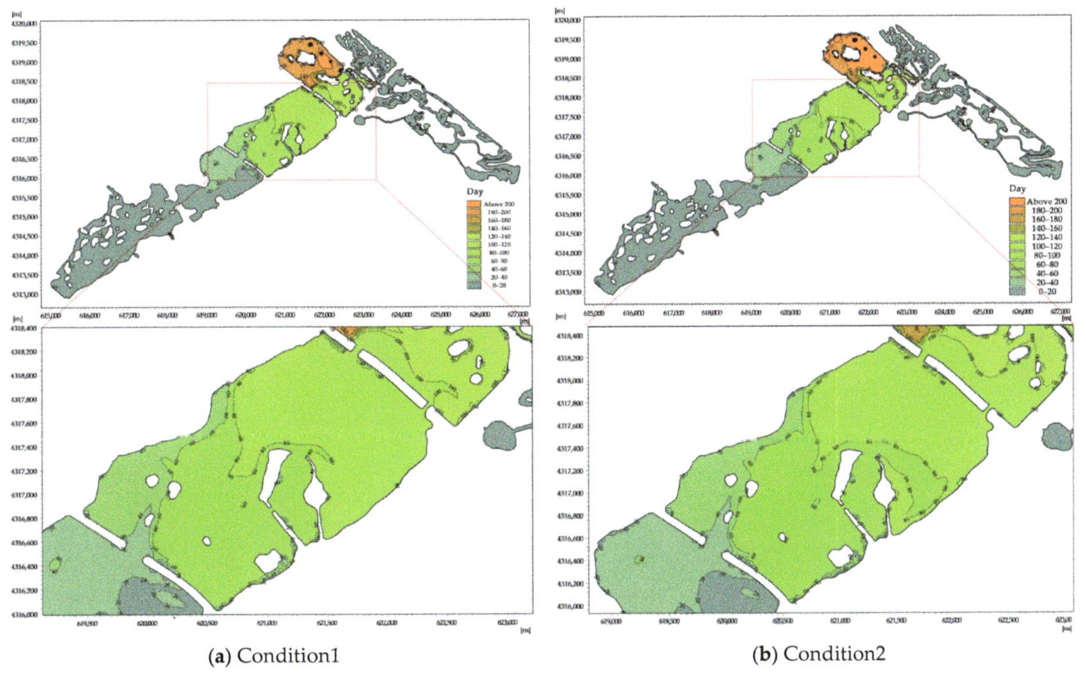

(a) Condition1

(b) Condition2

Figure 6. *Cont.*

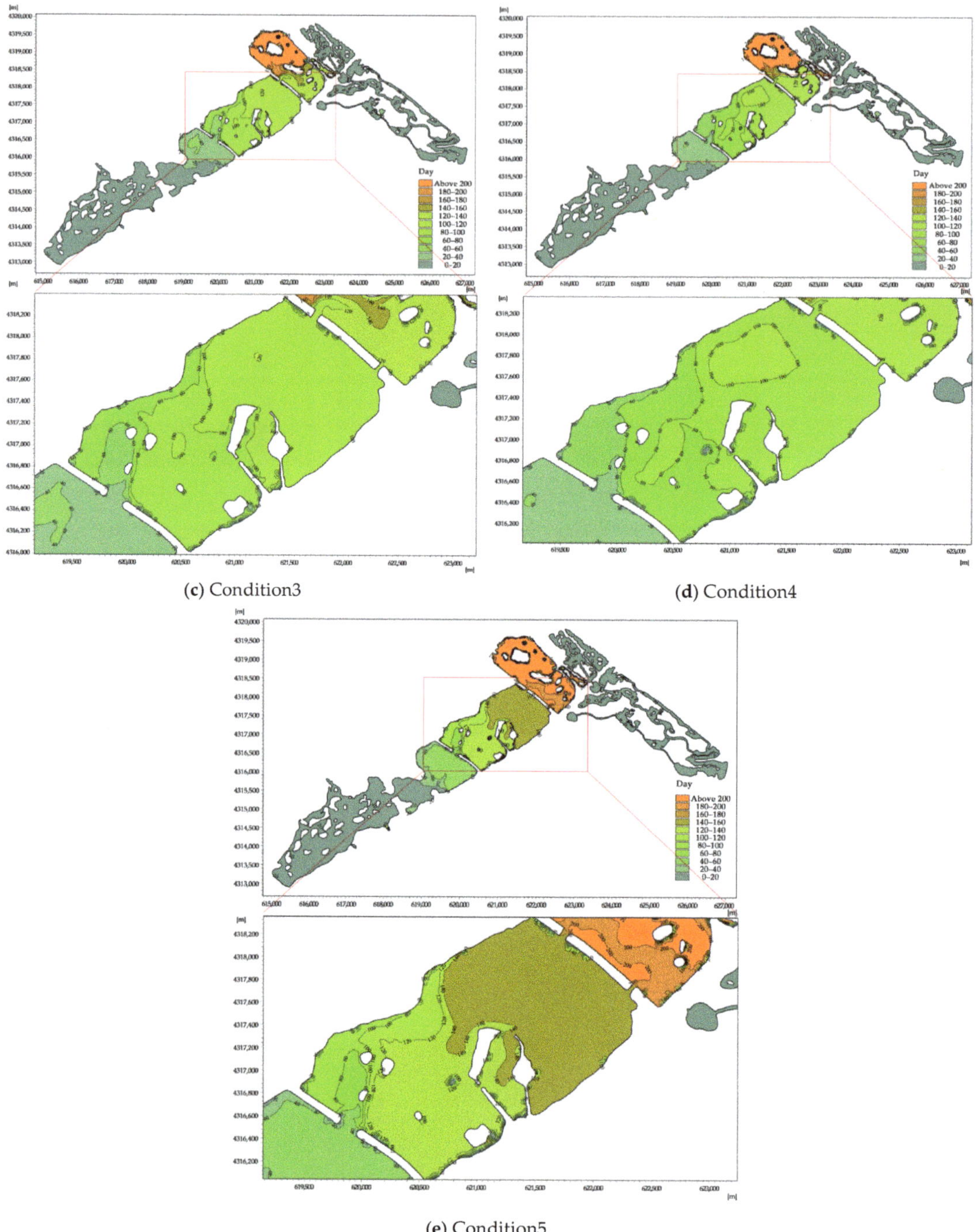

(**c**) Condition3   (**d**) Condition4

(**e**) Condition5

Figure 6. The water transfer process of Xinghai Lake.

The water exchange capacity of Xinghai Lake benefits from the high flow rate, and the throughput flow might encourage the transit and dispersion of contaminants. Generally, the mean value of the lake's water exchange cycle decreases as the flow increases, as shown in Figure 5. In Condition 1 and 2, the lake's whole water exchange period's geographical distribution is comparable. The middle domain exhibits increasing characteristics from the southern inlet to the northern outlet area, with average water exchange periods of 14 d and 74 d, respectively, and there is no discernible difference between the observation points in the middle domain. The south and east domains have a stronger water exchange capacity, with water exchange periods of 6 d and 4 d, respectively, as shown in Figure 5. The northern region has the longest average water exchange times, between 114 and 118 days. While the water flow in other regions is slow and the water exchange capacity is poor, with an average value of 200 d, the local water exchange near the lake's mouth is quicker, within 100 d. The average water exchange periods of the whole lake in Condition 2 and Condition 1 are 52 d and 32 d, respectively, and the diversion flow in Condition 2 is 46% smaller than that in Condition 1, which increases the water exchange period by 38%.

The diversion flow at S3 in the middle domain in Condition 3 is reduced by 50% compared with Condition 1, resulting in a weakened renewal capacity of the water body in the whole lake, and the mean value of the water exchange cycle is 62 d, an increase of 94% compared with Condition 1. Among them, the average value of the water exchange cycle in the northern lake area of the middle domain is 97 d, an increase of 31% compared with Condition 1, indicating that the flow accelerates the water flow in the middle domain of Xinghai Lake and enhances the dilution capacity of pollutants in this area, which is an important driving force affecting the transport of pollutants in Xinghai Lake. The average water exchange period in the outgoing area of the northern domain is 155 d, which is 55% higher than that of working condition 1, and the water exchange period in other areas is 210 d, which is 5% higher than that of working condition 1. The water exchange capacity of the outgoing lake area is significantly weakened, indicating that the throughput flow is the most important factor affecting the water renewal in this area.

The average water exchange duration increased by 16% while the average water diversion in condition 4 reduced by 46% compared to condition 3, showing a positive correlation between the two variables. The average water exchange duration in the middle domain is 43 days, which is 31% and 20% greater than that of Condition 3. The average value in the southern lake region is 17 days, while the average value in the northern lake area is 83 days. As seen in Figure 6d, the lake area of the active line of water flow in the northern part of the middle domain has a higher water exchange capacity than the near-shore circulation area, and the average water exchange period is about 21% longer. The lake area out of the northern domain also has a significantly longer water exchange period than the non-outflow area, and the average value is about 99 d less.

The average water exchange period in Condition 5 is 80 d, which is 1.5 times higher than that in Condition 1. Among them, the water exchange period in the middle domain is 86 d, and that in the north domain is 156 d, which are 79% and 37% higher than that in Condition 1, respectively. It indicates that the reduction of the inlet flow will significantly affect the regional water exchange capacity and cause the water exchange cycle to be prolonged.

Under the aforementioned circumstances, Xinghai Lake's water renewal period is simulated. Figure 7 shows the concentration–time curve for the middle domain observation point. According to the findings, the residual mass of preserved material in the lake is a time-dependent function. The residual conservative material in the lake steadily diminishes with time and goes to zero, but its value is never zero, indicating that residual un-excluded material exists in the lake at all times. Because the observation point is closest to the active water flow line and has the shortest water flow path among them, the water flow diffusion to this location collides with the artificial island and changes direction, impeding the diffusion to the north and resulting in the oscillation of water body exchange. The water renewal time of the observation points for Condition 3 and 5 slows down significantly when the diversion flow of S3 decreases. This is because the lake area in which the observation point in the middle domain is located is affected by the throughput flow, and when the incoming water flow is low, there is not a significant difference in the water renewal period between the observation points for Condition 1 and 2, as can be seen from Figure 7.

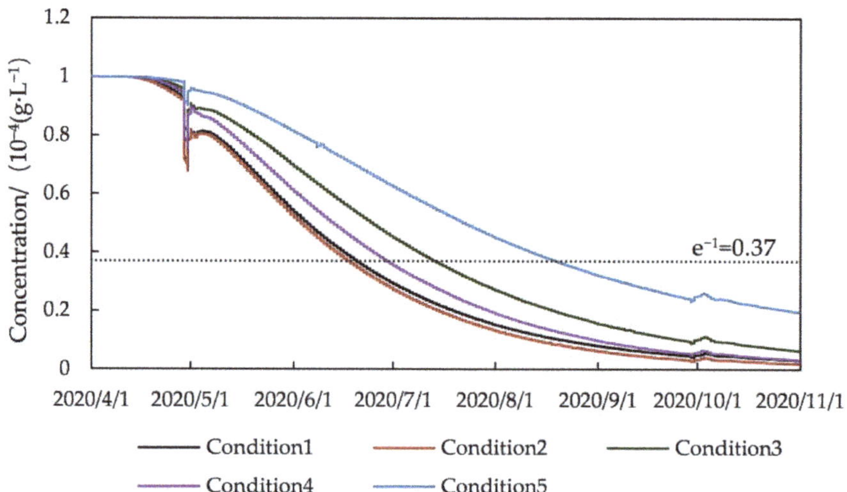

Figure 7. Concentration curves with time at the observation sites in the central domain of Xinghai Lake.

There is a clear regional variation in the Xinghai Lake's water renewal time. The water exchange cycle is shorter the closer the water is to the inlet and outlet because the water flow is higher and the flow path is the shortest; next comes the artificial island close to the main flow path because the water flow collides with the artificial island to create a larger circulation; and, finally, the near-shore area is the slowest, because the water flow is less influenced by the throughput flow, the hydrodynamic force is insufficient, and the water body is closer to the shore. According to the simulation study of the water body renewal cycle, the main driving force of Xinghai Lake is the clean water source introduced from outside, while the driving force of surrounding rainfall and wind is very limited. If the external water source carries pollutants into the lake, it will pose a serious threat to the water quality of Xinghai Lake. As a result, other engineering measures, such as increasing the purification capacity of the east domain wetland, introducing the water body from the north domain outlet to the east domain wetland via a pumping station, and then introducing the purified water body via the pumping station after the wetland has dissipated pollutants and improved water quality must be considered to order to provide a stable water body for Xinghai Lake.

4. Conclusions

(1) Xinghai Lake's water exchange cycle is spatially heterogeneous, exhibiting general characteristics of rapid water renewal in the southern lake area and slow renewal in the northern lake area. The gradient of the water exchange cycle from south to north is closely related to Xinghai Lake's hydrological situation.

(2) The flow rate, terrain, and wind field all have an impact on the Xinghai Lake flow field. The topography and flow velocity have an impact on the development of many complicated circulations close to the artificial island. The wind field affects the near-shore region, which creates a tiny circulation with a modest flow velocity.

(3) Xinghai Lake is influenced by throughput flow; under high flow conditions, the flow velocity in the lake area is larger, the water body transport speed is faster, and the water exchange capacity is strong; under low flow conditions, the flow velocity in the

(4) Without new water inlets and outlets, the center and northern domain of Xinghai Lake will create a significant area of stagnant water, which does not fulfill the requirements for the water quality of the landscape water body. To optimize the renewal time of the water body, it is recommended to enhance the inlet and outlet and incorporate additional alternative water sources.

(5) The numerical model can be used to determine the flow velocity distribution and hydraulic residence time in the water body, which can be used as an important tool for future water environment management in urban landscape lakes.

Author Contributions: Conceptualization, G.X. and M.W.; methodology, M.W.; software, M.W. and X.N.; validation, M.W.; formal analysis, M.W.; investigation, M.W. and X.N.; resources, Z.F. and X.L.; data curation, M.W. and X.N.; writing—original draft preparation, M.W.; writing—review and editing, M.W.; visualization, M.W.; supervision, G.X.; project administration, Z.F.; funding acquisition, Z.F. All authors have read and agreed to the published version of the manuscript.

Funding: This research received no external funding.

Data Availability Statement: All data generated or analyzed during this study are included in this published article.

Conflicts of Interest: The authors declare no conflict of interest.

References

1. Whipple, W., Jr.; Duflois, D.; Grigg, N.; Herricks, E.; Holme, H.; Jones, J.; Keyes, C., Jr.; Ports, M.; Rogers, J.; Strecker, E.; et al. A proposed approach to coordination of water resource development and environmental regulations. *JAWRA J. Am. Water Resour. Assoc.* **1999**, *35*, 713–716. [CrossRef]
2. Tisdell, J.G. The environmental impact of water markets: An Australian case study. *Jocase Study Environ. Manag.* **2001**, *62*, 113–120. [CrossRef] [PubMed]
3. McCallum, B.E. Areal extent of fresh water from an experimental release of Mississippi RiMississippinto Lake Pontchartrain. In Proceedings of the 9th 1995 Conference on Coastal Zone, Tampa, FL, USA, 16–21 July 1995; pp. 363–364.
4. Cox, W.E. Interbasin water transfer in the United States: Overview of the institutional framework. In Proceedings of the IWRA Seminar on Interbasin Water Transfer, Beijing, China, 15–19 June 1986.
5. Xu, L.; Gao, X.; Zhang, C.; Wang, C. Numerical simulation-based study on exchange of water in urban artificial lake. *Water Resour. Hydropower Eng.* **2018**, *49*, 94–100.
6. Tang, J.Z.; Zhou, W.B.; An, B.J.; Yang, H.; Li, Y. Research on characteristics of water change capability of Kunming Lake (test section) in Xi'an based on the MIKE 21 Model. *J. Water Resour. Water Eng.* **2020**, *31*, 58–63.
7. Cucco, A.; Umgiesser, G. Modeling the Venice Lagoon residence time. *Ecol. Model.* **2006**, *193*, 34–51. [CrossRef]
8. Delhez, É.J.; de Brye, B.; de Brauwere, A.; Deleersnijder, E. Residence time vs influence time. *J. Mar. Syst.* **2014**, *132*, 185–195. [CrossRef]
9. Meng, Q.B.; Liu, Y.Z.; Tan, Y.M.; Guo, L.; Zhou, Q.J. Study on Water Quantity and Quality Coupled Simulation of Daming Lake and Delivery Scheme. *Water Resour. Power* **2008**, *26*, 24–26.
10. Xu, W.J. Comprehensive Water Requirement Analysis and Ecosystem Health Assessment of Urban Lakes. Ph.D. Thesis, Dalian University of Technology, Dalian, China, 2009.
11. Wang, S.Q.; Shi, W.; Ouyang, H.; Qiu, X.C.; Zhao, Z.F.; Li, Y.L. Study on the ecological water supply of Xinghai Lake based on water quantity and quality model. *J. Gansu Agric. Univ.* **2020**, *55*, 180–186.
12. Welch, E.B.; Barbiero, R.P.; Bouchard, D.; Jones, C.A. Lake trophic state change and constant algal composition following dilution and diversion. *Ecol. Eng.* **1992**, *1*, 173–197. [CrossRef]
13. Donabaum, K.; Schagerl, M.; Dokulil, M.T. Integrated management to restore macrophyte domination. *Hydrobiologia* **1999**, *395–396*, 87–97. [CrossRef]
14. Yang, W.; Sun, Z.F.; Sun, T. A Review of Requirement Quantity and Allocation of Ecological Water for Wetland. *Wetl. Sci.* **2008**, *6*, 531–535.
15. Hu, G.L.; Zhao, W.Z. Reviews on calculating methods of vegetation ecological water requirement in arid and semiarid regions. *Acta Ecol. Sin.* **2008**, *28*, 6282–6291.
16. Xu, M.J.; Yu, L.; Zhao, Y.W.; Li, M. The Simulation of Shallow Reservoir Eutrophication Based on MIKE21: A Case Study of Douhe Reservoir in North China. *Procedia Environ. Sci.* **2012**, *13*, 1975–1988. [CrossRef]

17. Li, X.; Huang, M.; Wang, R. Numerical Simulation of Donghu Lake Hydrodynamics and Water Quality Based on Remote Sensing and MIKE 21. *ISPRS Int. J. Geo-Inf.* **2020**, *9*, 94. [CrossRef]

18. Zhu, C.; Liang, Q.; Yan, F.; Hao, W. Reduction of Waste Water in Erhai Lake Based on MIKE21 Hydrodynamic and Water Quality Model. *Sci. World J.* **2013**, *2013*, 958506. [CrossRef] [PubMed]

19. Wang, Q.; Peng, W.; Dong, F.; Liu, X.; Ou, N. Simulating Flow of An Urban River Course with Complex Cross Sections Based on the MIKE21 FM Model. *Water* **2020**, *12*, 761. [CrossRef]

20. Liu, J.L.; Yang, Z.F. A study on the calculation methods of the minimum eco-environmental water demand for lakes. *J. Nat. Resour.* **2002**, *17*, 604–609.

21. Schoen, J.H.; Stretch, D.D.; Tirok, K. Wind-driven circulation patterns in a shallow estuarine lake: St Lucia, South Africa. *Estuar. Coast. Shelf Sci.* **2014**, *146*, 49–59. [CrossRef]

22. Tang, C.; Yi, Y.; Yang, Z.; Zhang, S.; Liu, H. Effects of ecological flow release patterns on water quality and ecological restoration of a large shallow lake. *J. Clean. Prod.* **2018**, *174*, 577–590. [CrossRef]

Article

Advancing Medium-Range Streamflow Forecasting for Large Hydropower Reservoirs in Brazil by Means of Continental-Scale Hydrological Modeling

Arthur Kolling Neto [1,*], Vinícius Alencar Siqueira [1], Cléber Henrique de Araújo Gama [1], Rodrigo Cauduro Dias de Paiva [1], Fernando Mainardi Fan [1], Walter Collischonn [1], Reinaldo Silveira [2], Cássia Silmara Aver Paranhos [3] and Camila Freitas [3]

[1] Instituto de Pesquisas Hidráulicas, Universidade Federal do Rio Grande do Sul, Porto Alegre 90010-150, RS, Brazil
[2] Sistema de Tecnologia e Monitoramento Ambiental do Paraná—SIMEPAR, Curitiba 81531-980, PR, Brazil
[3] Companhia Paranaense de Energia—COPEL, Curitiba 80420-170, PR, Brazil
* Correspondence: arthur.kolling@ufrgs.br

Abstract: Streamflow forecasts from continental to global scale hydrological models have gained attention, but their performance against operational forecasts at local to regional scales must be evaluated. This study assesses the skill of medium-range, weekly streamflow forecasts for 147 large Brazilian hydropower plants (HPPs) and compares their performance with forecasts issued operationally by the National Electric System Operator (ONS). A continental-scale hydrological model was forced with ECMWF medium-range forecasts, and outputs were corrected using quantile mapping (QM) and autoregressive model approaches. By using both corrections, the percentage of HPPs with skillful forecasts against climatology and persistence for 1–7 days ahead increased substantially for low to moderate (9% to 56%) and high (72% to 94%) flows, while using only the QM correction allowed positive skill mainly for low to moderate flows and for 8–15 days ahead (29% to 64%). Compared with the ONS, the corrected continental-scale forecasts issued for the first week exhibited equal or better performance in 60% of the HPPs, especially for the North and Southeast subsystems, the DJF and MAM months, and for HPPs with less installed capacity. The findings suggest that using simple corrections on streamflow forecasts issued by continental-scale models can result in competitive forecasts even for regional-scale applications.

Keywords: ensemble forecasting; post-processing; bias correction; South America

Citation: Kolling Neto, A.; Siqueira, V.A.; Gama, C.H.d.A.; Paiva, R.C.D.d.; Fan, F.M.; Collischonn, W.; Silveira, R.; Paranhos, C.S.A.; Freitas, C. Advancing Medium-Range Streamflow Forecasting for Large Hydropower Reservoirs in Brazil by Means of Continental-Scale Hydrological Modeling. *Water* **2023**, *15*, 1693. https://doi.org/10.3390/w15091693

Academic Editors: Yongguang CHENG and Zhengwei Wang

Received: 11 March 2023
Revised: 17 April 2023
Accepted: 18 April 2023
Published: 27 April 2023

1. Introduction

Hydropower is an important source of renewable and clean energy. Brazil is the country with the second-largest installed capacity of hydropower globally [1], hosting six out of the twenty largest hydropower plants in the world [2]. While hydropower contributes approximately 60% of the total power capacity in the country [1], the largest percentage of the energy produced by this source comes from the large hydropower plants (HPPs), which are part of a very extensive and complex hydrothermal system called the Brazilian National Interconnected System (SIN). The SIN is optimized by a chain of models addressing long-range (5 years), seasonal (12 months), and monthly operational planning, as well as short-range for making operational decisions in the coming weeks [3]. The planning and coordination of the SIN are conducted by the Brazilian National Electric System Operator (ONS), which routinely issues natural inflow forecasts for the SIN reservoirs required as input to the optimization models [4].

Natural inflow forecasts play a crucial role in planning the operation of the SIN, which aims to meet energy demand and maximize overall efficiency by minimizing spillover losses and reducing additional fuel costs [5–7]. Methodologies to produce streamflow forecasts for the few weeks ahead have long been based on statistical methods based on observed discharge [8]. In recent years, operational forecasting methods have been gradually changing by incorporating precipitation forecasts up to 14 days ahead into rainfall-runoff models [9], although natural inflow forecasts used in the optimization models of the SIN are still deterministic.

On the other hand, progress in the field of catchment-scale streamflow forecasting has been toward the use of multiple future streamflow scenarios in the form of ensembles (e.g., [10–18]). Ensemble methods can account for uncertainties in the forecast chain that arise from multiple sources, such as errors in meteorological forcing, the inability of models to adequately represent hydrological processes, and deficiencies in parameter estimation [19–24]. While ensemble hydrological forecasts have shown advantages over single-value ones for hydropower purposes, for instance, by improving operational decisions and leveraging economic benefits (e.g., [17,25–29]), the development of studies on this topic is still slow in South America compared to other regions of the world, especially in the northern hemisphere [30–32]. In parallel, the scientific community has dedicated substantial efforts to developing ensemble streamflow forecasting methods also at continental and global scales [33–41], and analyses of streamflow forecasts produced with such techniques have been possible for Brazil as a whole [42].

Less attention has been given to how competitive continental (or global) scale forecasts are compared with those made operationally at the regional scale. It is intuitive that streamflow forecasts generated by large-scale hydrological models are less accurate than locally calibrated ones [43], which reflect limitations in the forcing data used, parameterization, and level of detail. However, there are currently several techniques that can improve the accuracy of flow prediction from local information, ranging from simple bias correction methods (e.g., [44]) to more complex methods such as ensemble calibration, statistical postprocessing (e.g., [14,18,45]), and data assimilation (e.g., [13,28,46]), although simple methods are generally more attractive because of their efficiency and ease of operational application [47]. In this sense, Lozano et al. [48] showed that a simple bias correction on the outputs of a global-scale system was able to effectively transform historical runoff simulations and forecasts for local-scale use in Brazil, while Wang et al. [49] found that a global forecast system (GloFAS) outperformed a regional system in predicting high runoff and even performed reasonably well in predicting low to moderate runoff after bias correction on forecast runoff. Combining simple bias correction with autoregressive models that can use newly available local information [50,51] also proved suitable for hydrological forecasting. It is therefore of interest to know the extent to which it is possible to produce accurate natural inflow forecasts for the SIN reservoirs using continental-scale modeling techniques and whether these forecasts can be used as additional information to the forecasts currently produced on an operational basis.

The objective of this study is twofold: (i) To assess the skill of medium-range, weekly streamflow forecasts issued for 147 SIN HPPs by a continental-scale hydrological model and (ii) to compare these forecasts with those being generated by ONS to support operation at the HPP sites. The next sections are organized in the following order: Section 2 presents the study area of the SIN, while Section 3 presents the methods used to generate and correct the streamflow forecasts, as well as the ONS forecasts used for comparison and the metrics adopted for the analysis. The results, discussion, and final conclusions are presented in Sections 4–6, respectively.

2. Materials and Methods

The overview of the main methodological steps is shown in Figure 1. A continental-scale hydrologic model was forced by ECMWF precipitation forecasts to produce streamflow forecasts at the sites of the SIN HPPs. The raw forecasting performance was initially evaluated against observed (naturalized) flows at these locations, and the performance gain obtained by applying bias correction and autoregressive model output correction was also evaluated. Next, the corrected streamflow forecasts were compared with the operational forecasts issued by ONS.

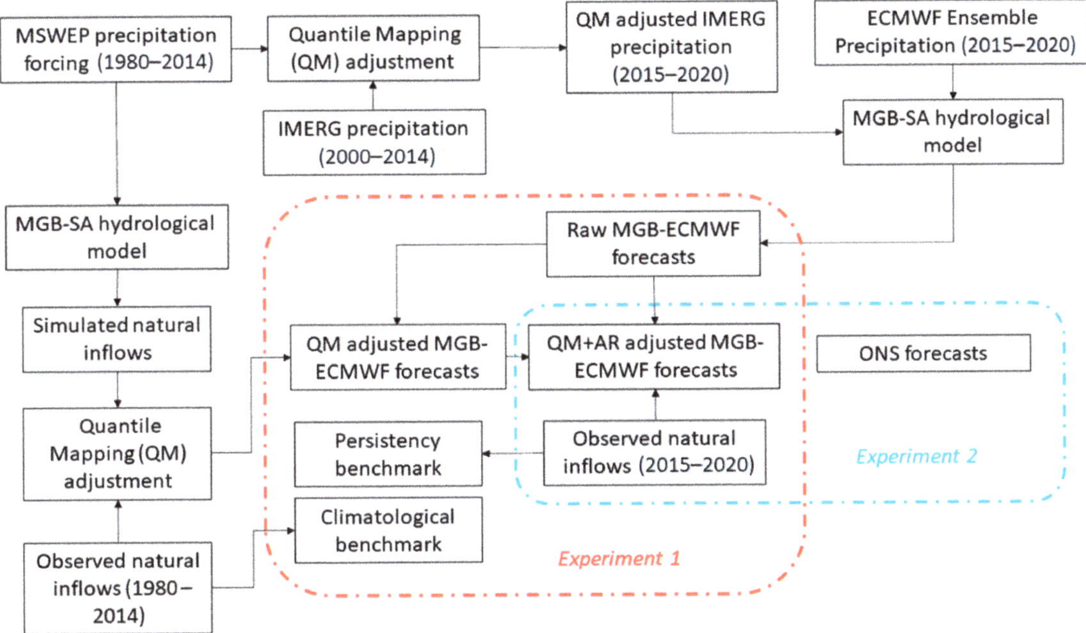

Figure 1. Flowchart of the methodology used in this study, highlighting the two experiments, which are demarcated by the dashed lines in red and light blue.

3. Study Area

The study encompasses 147 HPPs of the Brazilian National Interconnected System (SIN) in the domain between latitudes 5° N–30° S and longitudes 65° W–35° W. These HPPs were selected based on the availability of natural streamflow data, as some HPPs receive most of the inflows from other rivers through artificial channels and pumping stations. Currently, the SIN has an installed capacity of more than 179,366 MW, and hydroelectric plants account for 109,190 MW (60.9%) and are located in river basins with different hydrological characteristics and climate variability [52].

The SIN is a large hydro-thermoelectric system for the production and transmission of electricity, composed of four subsystems: the South, the Southeast/Central West, the Northeast, and most of the North. The installed generation capacity of the SIN is composed mainly of hydroelectric plants distributed in sixteen hydrographic basins in the different regions of the country [53]. Figure 2 shows a map of the locations of the SIN HPPs and their respective subsystems.

Until 2006, ONS flow forecasts were produced only by stochastic approaches such as the Periodic Auto-Regressive or Auto-Regressive Moving Average models [8]. From 2006 onwards, other forecasting methods in addition to stochastic models have been recommended, including lumpedand distributed conceptual models, artificial intelligence techniques, as well as the incorporation of future rainfall from Numerical Weather Prediction [6]. During 2018, the ONS formally adopted the Soil Moisture Accounting Procedure model as the only rainfall-runoff model for the first week of the forecast horizon at several SIN sites, and the use of stochastic models has declined in recent years [54,55].

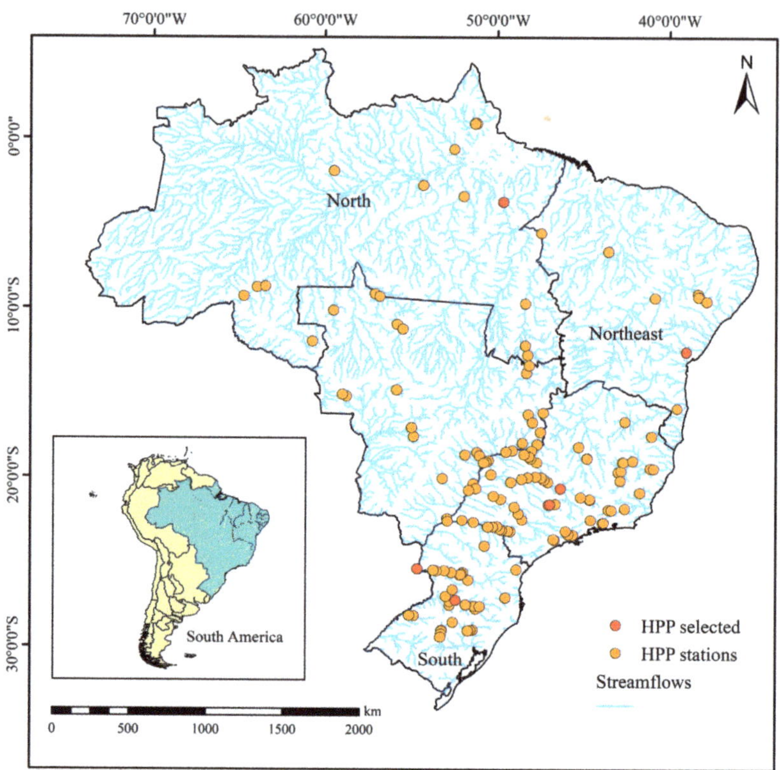

Figure 2. Locations of the 147 Hydroelectric power plants of the SIN for which forecasts were evaluated.

3.1. Observed Streamflow Data

Daily time series of naturalized streamflow spanning from January 1980 to December 2020 were obtained for the selected 147 SIN HPPs through the SINtegre portal (https://sintegre.ons.org.br, accessed on 1 August 2022). Naturalized streamflow at dam locations is computed by routing downstream natural incremental reservoir inflows, which are reconstructed through water balance using evapotranspiration estimates and operation data from SIN reservoirs such as water levels, volumes, and outflows, as well as water withdrawals across the basin [56]. Natural flows can be used to evaluate the effects of human interventions in rivers, such as the implementation of reservoirs, but for the ONS, they are relevant information for the planning and operation of the SIN.

3.2. Streamflow Forecasts

3.2.1. Forecast Input Data and Hydrological Model

Daily ensemble precipitation forecasts from the European Center for Medium-Range Weather Forecasting (ECMWF) were achieved for the period between May 2015 and December 2020 (initialization of 00 UTC) through the Thorpex Interactive Grand Global Ensemble (TIGGE) platform (http://tigge.ecmwf.int/, accessed on 12 March 2021). These ensemble forecasts consist of 50 perturbed members with a forecast time horizon of 15 days, and their spatial resolution depends on the ECMWF model cycle, for instance, it was changed from 36 km (in 2015) to 18 km (from Mar 2016 onwards).

To obtain streamflow forecasts at SIN reservoir locations, the ECMWF predicted precipitation data were used as inputs to the continental-scale hydrologic-hydrodynamic MGB model for South America (MGB-SA) [57]. The MGB-SA is a conceptual, semi-distributed model that discretizes the domain into unit-catchments, each containing a ~15 km-long river segment, and further into Hydrological Response Units (HRU), which are subdivisions according to combinations of land cover and soil type. Evapotranspiration (based on Penman-Monteith) and runoff generation (based on the ARNO model) are computed at a daily time step at the HRU level. Surface, subsurface, and groundwater runoff are routed to the main channel through linear reservoirs, and the propagation in river channels is computed by using an explicit 1D inertial approximation of the Saint-Venant equations. The MGB-SA has been calibrated with more than 600 in situ gauges and validated with remote sensing-based datasets [57]. Details on MGB-SA model performance at each HPP location can be found in the Supplementary Material (Table S1). For further information on general MGB equations regarding water balance and river routing, the reader is referred to [58,59].

For model initialization along the forecast period (2015–2020), we used daily precipitation data from the Integrated Multi-satellite Retrievals for GPM (IMERG) v06 final run [60]. Herein, IMERG data is further bias-corrected to match the climatological distribution of the Multi-Source Weighted Ensemble Precipitation (MSWEP) v1.1 [61], as the MSWEP was used to calibrate the MGB-SA model [57]. Other climate variables used to compute evapotranspiration in the forecast period are assumed to be equal to their long-term monthly means computed with CRU v.2 data, according to previous applications (e.g., [42]). For historical MGB-SA simulations, which are required for streamflow correction approaches (i.e., before 2015; see Section 3.2), the model was forced to use MSWEP v1.1 data to maintain coherence with its original configuration.

3.2.2. Quantile Mapping Applied to Streamflow Forecasts

To correct biases in the streamflow forecasts, a quantile mapping (QM) procedure was applied in a similar way to Wood and Schaake [62] and Hashino et al. [44]. This is a simple correction method that matches both the mean and variance (including higher moments) of the hydrological model outputs to those of the observed climatology. Thus, cumulative distribution functions (CDF) are obtained from the observed and simulated discharges, and for each forecast ensemble trace, the QM replaces the predicted discharge with the observed value that has the same no exceedance probability, according to:

$$\hat{Z}_i = F_o^{-1}[F_S(Z_i)] \tag{1}$$

where $\hat{Z}_i$ is the bias-corrected forecast ensemble trace I, F_o is the inverse of the CDF of the observed discharge, F_s is the CDF of the simulated discharge, and Z_i is the raw forecast ensemble trace.

Before applying QM to the forecasts, we first analyzed the performance of empirical and gamma distributions to construct/fit the CDF curves over a historical, independent period (1980 to 2014). For that, we applied a leave-one-year-out cross-validation by constructing/fitting CDF curves of observed and simulated flows using data from the entire historical period except the target year and then applying QM to correct the simulated flows for this same year. For each HPP, the Nash-Sutcliffe (NSE) and logarithm of NSE were computed for the cross-validated, bias-corrected simulated flows, and the distribution that resulted in the best performance was then selected to correct the ensemble traces in the forecast verification period (2015–2020).

In our initial assessment (see the Supplementary Material), we noted that both gamma and empirical distributions frequently result in improved simulated discharges with only minor variations in performance. As such, we prioritized the gamma distribution (parametric), as it allows for extrapolation to values outside the range of historical observations. For certain SIN gauges, we either applied empirical distribution-based corrections to discharges or no correction was made. The final CDFs employed for QM correction in the forecast verification period were derived from data spanning the entire historical period (i.e., without cross-validation).

3.2.3. Autoregressive Model (AR)

After applying QM to the streamflow forecasts, these were further corrected by using a simple AR model. The autoregressive model uses error updating to anticipate the errors in a forecast period as a linear function of the known errors in previous steps [26,50]. In this way, according to Liu et al. [63], error updating is based on using data to generate predictions of future differences between the model prediction and future observations, so it is not restricted to the goal of producing improved predictions in the hydrologic model.

In this method, the current value of the time series (Q_t) is defined as a combination of past values of the time series itself plus a random noise (ε_t), where t is the time index. Thus, in the $AR_{(p)}$ model, where p is the order of the model, one has as input the past values $Q_{t-1}, Q_{t-2}, \ldots, Q_{t-p}$, multiplied by optimized parameters α to predict the next value Q_t. In Equation (2), an example is given of what an *AR*-only model would look like:

$$Q_{\text{prev}}(t)' = Q_{\text{prev}}(t) - \alpha_1^t * \left(Q_{\text{prev}}(t_0) - Q_{\text{Obs}}(t_0) \right) \tag{2}$$

where $Q_{prev}(t)'$ is the forecast streamflow corrected with a lag-1 autoregressive model; $Q_{Obs}(t_0)$ and $Q_{prev}(t_0)$ are observed streamflow and model simulated streamflow, respectively, at initial time t_0. For example, εt is an identically and independently distributed Gaussian deviation with a mean of zero and a constant standard deviation.

In the autoregressive model, one can consider the autocorrelation function of the process, $(p_t = \alpha_1^t * t)$, where t is the number of observations to be included in the correction; in this case, t is equal to 1 [64]. The parameter α_1^t can be determined by calculating the autocorrelation function between the lead times of the forecast data. The value of α_1^t for each HPP was assumed as the lag-1 autocorrelation of the time series of observed natural flows.

3.3. Operational Forecasts from the Brazilian National Electric Service Operator (ONS)

ONS routinely produces streamflow forecasts to support the Monthly Operation Programs (PMO—Programação Mensal de Operação in Portuguese). The PMO allows the establishment of energy production policies and regional exchanges between SIN subsystems, providing directives on which hydropower plants will be dispatched as well as information on energy pricing in the short-term market. The PMO is conducted monthly and is revised every week [55].

Operational streamflow forecasts produced for PMOs along May/2015–December/2020 were obtained from the SINtegre data portal (https://sintegre.ons.org.br/sites/9/13/79/Produtos/245, accessed on 5 September 2022). These forecasts are given in weekly average discharges with a maximum lead time of 6 weeks (also called "operation weeks"), each one starting at 00:00 on Saturday and ending at 24:00 on the following Friday. The first operation week of a given month is the one that includes the first day of that month. Streamflow forecasts are issued on a weekly basis, always on the day immediately preceding the PMO or one of its revisions [65]. As a rule, forecasts are officially issued on Thursdays; in the event of a holiday occurring on the forecasting day (Thursday) or on the PMO/revision day (Friday), forecasts are issued earlier on Tuesday and Wednesday, respectively. Figure 3 shows an example of the forecast generation schedule.

A preliminary review of the forecast dates was conducted to identify potential changes due to holidays. Since forecast calendars were not available for all years analyzed in the official reports, we chose to use the file creation date as an indication of the forecast issue date, both for the PMO and its weekly revisions, after consulting with ONS technicians. Additionally, as forecasts must be produced one day before the PMO (which takes place on Friday), any forecast issue date listed in the files as Friday was adjusted to the previous day (Thursday). A total of 268, 18, and 6 weekly forecasts were issued on Thursday, Wednesday, and Tuesday, respectively.

Forecast for / Day of week	Sat	Sun	Mon	Tue	Wed	Thu	Fri	Operation week number
PMO (month m)	-	-	-	-	-	27	28	-
Revision 1 (m)	29	30	31	1	2	3	4	Op. week 1 (m)
Revision 2 (m)	5	6	7	8	9	10	11	Op. week 2 (m)
Revision 3 (m)	12	13	14	15	16	17	18	Op. week 3 (m)
PMO (month m+1)	19	20	21	22	23	24	25	Op. week 4 (m)
Revision 1 (m+1)	26	27	28	29	30	1	2	Op. week 5 (m) / Op. week 1 (m+1)
Revision 2 (m+1)	3	4	5	6	7	8	9	Op. week 6 (m) / Op. week 2 (m+1)
Revision 3 (m+1)	10	11	12	13	14	15	16	Op. week 3 (m+1)
Revision 4 (m+1)	17	18	19	20	21	22	23	Op. week 4 (m+1)
PMO (month m+2)	24	25	26	27	28	29	30	Op. week 5 (m+1)
Revision 1 (m+2)	1	2	3	4	5	6	7	Op. week 6 (m+1) / Op. week 1 (m+2)

Figure 3. Typical schedule of forecast generation for Monthly Operation Programs (PMO) and their weekly revisions. Forecast issue dates for a given PMO/revision are indicated by the corresponding-colored boxes. Source: adapted from [55].

3.4. Forecast Assessment

Firstly, the impact of the correction schemes on raw streamflow forecasts produced by MGB-SA using ECMWF data as input (hereafter referred to as MGB-ECMWF) was analyzed. The assessment was performed considering weekly average discharges for 1–7 and 8–15 days in advance, and raw streamflow forecasts were compared to those corrected by using the QM individually and both corrections (QM+AR). In addition, the potential performance gains by applying QM and QM+AR on streamflow forecasts at SIN locations were further investigated in terms of high (>Q_{75} of non-exceedance flows) and moderate to low discharges (<Q_{50}), as well as in terms of characteristics such as streamflow seasonality and flashiness (Appendix A).

To assess the raw, QM, and QM+AR configuration strategies for MGB-ECMWF streamflow forecasts, we used the percent bias (PBIAS) and the Continuous Ranked Probability Score (CRPS) [66]. PBIAS measures the tendency of forecast values to overestimate or underestimate the observed ones, and it is computed for the ensemble mean:

$$\text{PBIAS} = 100 \frac{\sum_{i=1}^{N}(Fcst_i - Obs_i)}{\sum_{i=1}^{N}(Obs_i)} \tag{3}$$

where Obs_i and $Fcst_i$ are the observed and predicted discharges, respectively, and i and N are the current and total number of forecasts.

The CRPS summarizes the overall performance of a probabilistic forecast. It is defined by the quadratic difference between the cumulative distribution function (CDF) of the forecast and the empirical CDF of the observation and is typically averaged over a set of forecasts:

$$\text{CRPS} = \frac{1}{N}\sum_{i=1}^{N}\int_{-\infty}^{\infty}[F_i(x) - 1(x \geq y_i)]^2 dx \tag{4}$$

where $F_i(x)$ is the CDF of the forecast ensemble x and forecast day i, $1(x \geq y_i)$ is a Heaviside step function that equals one when forecast values are greater than the observed value y_i and zero otherwise, and N is the total number of forecasts.

Following Siqueira et al. [42], CRPS was transformed into an overall skill score (CRPSS = $1 - \text{CRPS}_{\text{fcst}}/\text{CRPS}_{\text{benchmark}}$) using both daily streamflow climatology and persistence as benchmarks. Streamflow climatology was computed for each calendar day by sampling 50 equally distanced quantiles ($1/51, 2/51, 3/51, \ldots, 50/51$) from the empirically observed CDF, that is, the same number of ensemble members as the MGB-ECMWF, since the number of ensemble members is known to affect the CRPS value [67]. In turn, for the persistence, it is assumed that all forecast lead times have the same predicted value equal to the last observed discharge (i.e., a deterministic forecast), so the CRPS reduces to the mean absolute error [66]. Maximum skill is achieved when CRPSS = 1, and values below 0 indicate no skill.

In a second experiment, the corrected streamflow forecasts (QM+AR) produced by the Continental-Scale Hydrological model were compared with the Operational model issued by the ONS. The forecast assessment was carried out only for the lead time of 1 week ahead, and it was performed in such a way to keep coherence with the evaluations presented in the official ONS reports [65]. By taking the calendar in Figure 3 as an example, the 1-week lead time forecast issued on day 3 for revision 1 of month m corresponds to the average of discharges predicted for days 5–11. Similarly, the 1-week lead time forecast issued on day 29 for the PMO of month $m+2$ corresponds to the average of discharges predicted for days 1–7. No distinctions are made for streamflow forecasts issued for PMOs or revisions, so that both are equally treated in a single verification set (May 2015–December 2020). Note that there is always a gap of at least 2 days between the forecast issue date and the start of the operation week, which can be larger due to the possible occurrence of holidays as mentioned in Section 3.3.

For this assessment, we adopted standard metrics that are routinely used for forecast evaluation by ONS, namely the Mean Absolute Percent Error (MAPE) and the Nash-Sutcliffe Efficiency (NSE) [68]:

$$\text{MAPE} = \frac{1}{N} \sum_{i=1}^{N} \left| \frac{Fcst_i - Obs_i}{Obs_i} \right| \tag{5}$$

$$\text{NSE} = 1 - \frac{\sum_{i=1}^{N} (Fcst_i - Obs_i)^2}{\sum_{i=1}^{N} (Obs_i - \overline{Obs_i})^2} \tag{6}$$

where Obs_i and $Fcst_i$ are the observed and predicted discharges, respectively; i and N are the current and total number of forecasts, and $\overline{Obs_i}$ is the mean of observed values.

In addition, ONS [69] developed an overall performance index called Multicriteria Distance (MD), which uses the above metrics as an ordered pair (1—NSE, MAPE) and calculates its Euclidean distance to the origin of a cartesian coordinate system:

$$\text{MD} = \sqrt{(1 - \text{NSE})^2 + (\text{MAPE})^2} \tag{7}$$

The MD ranges from 0 to ∞ and values close to zero indicate better performance.

The implementation of prediction correction methods and statistical metrics analysis were carried out using Matlab software.

4. Results

4.1. Skill Assessment of Raw and Corrected Continental-Scale Streamflow Forecasts

Examples of hydrographs with raw and corrected weekly averaged streamflow forecasts from the continental-scale hydrological model are presented in Figure 4. Results are shown considering a fixed lead time of 1–7 days, and the HPPs used as examples were chosen based on different characteristics of seasonality and day-to-day discharge variations. Figure 4a shows the hydrographs for Pedra do Cavalo HPP, which has high seasonality and flashiness, and Figure 4b shows the hydrographs for Itaipu HPP, where both indexes are low. Figure 4c,d display hydrographs for Itá and Tucuruí HPPs, respectively, where the former is characterized by high daily variation of discharges and low seasonality, and the latter by low flashiness and a high seasonal index. Finally, in Figure 4e,f, the Euclides da Cunha and Furnas HPP exhibit average values for these streamflow indexes.

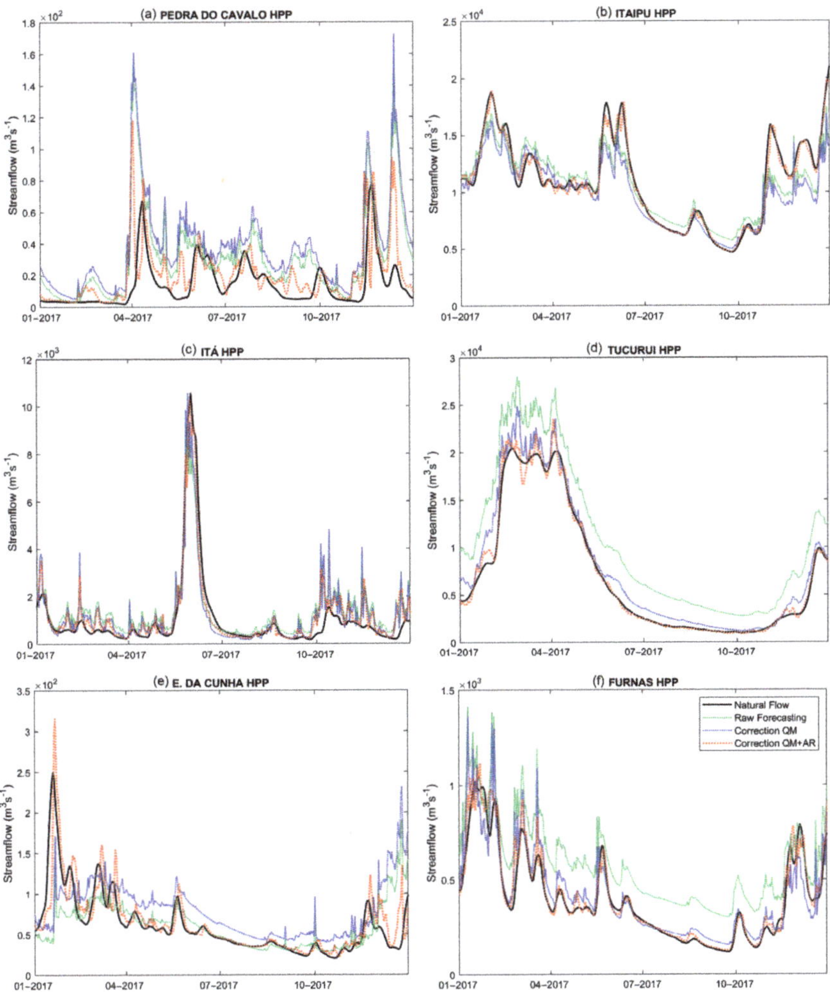

Figure 4. Forecasts of average weekly flow (ensemble mean) for the lead time of 1–7 days at (**a**) Pedra do Cavalo, (**b**) Itaipu, (**c**) Itá, (**d**) Tucuruí, (**e**) Euclides da Cunha, and (**f**) Furnas hydropower plant.

The corrections with the QM method show variable performances. For instance, there is a notably positive effect on the rise and recession of hydrographs at Tucuruí and Furnas HPPs (and to a minor extent on the recession at Itaipu), while at Pedra do Cavalo and Itá HPPs there is apparently no benefit from this correction. In some cases, as in Euclides da Cunha HPP, the application of QM apparently causes a reduction in the accuracy of the forecasts, which can be explained by the very low performance of the hydrological model at this location (see Table S1 in Supplementary Materials). On the other hand, the application of QM+AR resulted in substantial performance gains in all the analyzed cases.

The performance of MGB-ECMWF forecasts was verified from May 2015 to December 2020, considering average weekly predicted discharges and lead times of 1–7 and 8–15 days. The biases of raw and corrected (QM and QM+AR) forecasts were categorized into bins, and the relative frequency of HPPs with forecasts falling in each category is shown in Figure 5. In general, the MGB-ECMWF raw forecasts have a predominantly positive bias. The QM contributes to reducing the percentage of HPPs for which predicted flows are overestimated by 20–40% and >40%, although it increases the frequency of underestimation in the bias range of −20 to −10%. For the QM+AR configuration and lead time of 1–7 days, bias is mostly concentrated between -10 and 10% (~70% of the HPPs), while for 8–15 days in advance, values tend to be closer to those observed for QM but still exhibit lower biases than the latter. Moreover, for longer lead times, a larger number of HPPs show (high) positive PBIAS values when bias correction is applied, probably due to the miscorrection of runoff response biases from predicted precipitation.

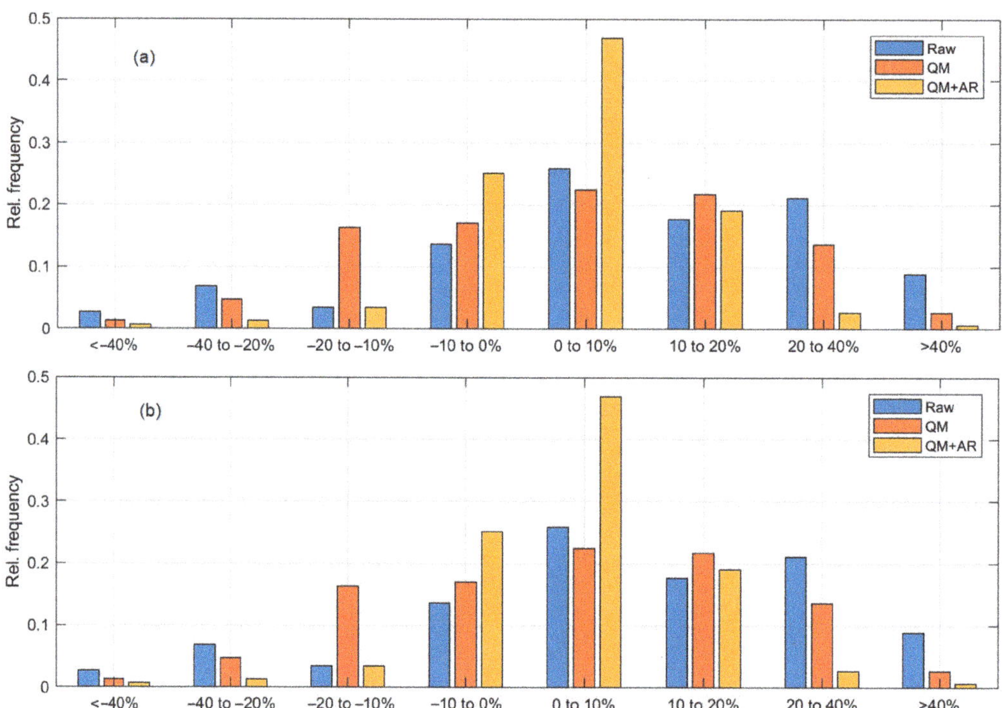

Figure 5. Percent bias of MGB-ECMWF streamflow forecasts for raw, bias correction (QM), and bias correction + updating (QM+AR) configurations. The relative frequency refers to the number of SIN hydropower plants falling into each category. The graphs show results for the lead times of (**a**) 1–7 and (**b**) 8–15 days.

The assessment of MGB-ECMWF forecast skill at each HPP was carried out using climatology ($CRPSS_{clim}$) and persistence ($CRPSS_{pers}$) as benchmarks, and the results were conditioned on high (>Q_{75} of non-exceedance) (Figure 6) and low to moderate (<Q_{50}) flows (Figure 7). For high flows and a lead time of 1–7 days, both the raw forecasts and the QM exhibit positive skill relative to persistence in 72% of the HPPs, which increases to 94% after applying the QM+AR corrections. For the 8–15-day lead time, raw MGB-ECMWF forecasts already show significant positive $CRPSS_{pers}$ for the majority of SIN locations (>90%), and the overall skill improvement by using QM or QM+AR correction is relatively small. When compared to climatology (8–15 days ahead), the raw forecasts exhibit positive skill in 70% of the HPPs, and performance increases slightly for the QM (77%) and QM+AR (87%) configurations, while for the lead time of 1–7 days, the patterns of skill are similar to those observed against persistence.

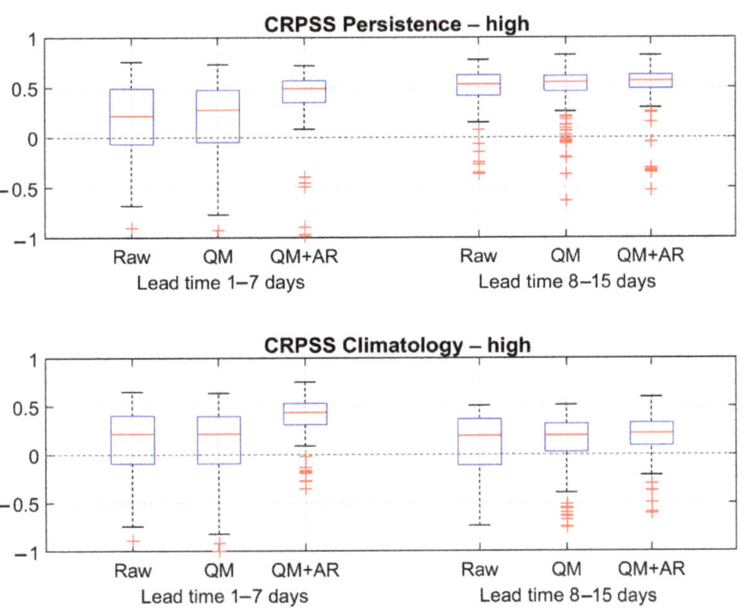

Figure 6. CRPS skill of MGB-ECMWF forecasts for the raw, bias correction (QM), and bias correction + updating (QM+AR) configurations, considering only high flows ($>Q_{75}$ of non-exceedance flows). The gray dashed horizontal line denotes skill = 0.

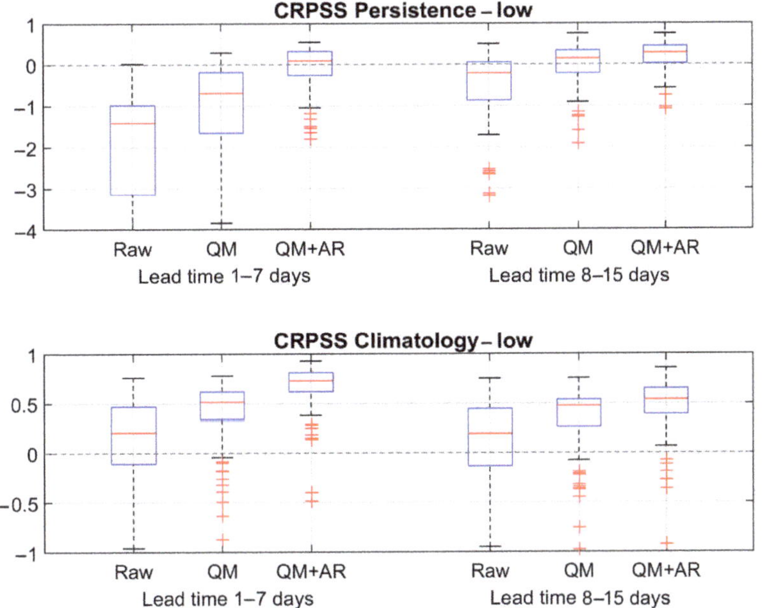

Figure 7. CRPS skill of MGB-ECMWF forecasts for the raw, bias correction (QM), and bias correction + updating (QM+AR) configurations, considering only low to moderate flows ($<Q_{50}$, non-exceedance flows). The gray dashed horizontal line denotes skill = 0.

In general, the MGB-ECMWF forecast skill for low to moderate flows (Figure 7) tends to be lower than that observed for the higher flows. For the lead time of 1–7 days, the raw MGB-ECMWF forecasts exhibit virtually no positive skill, and a few HPPs (9%) show $CRPSS_{pers} > 0$ after applying the QM method, despite the substantial performance gain over the no correction configuration. Even when both correction approaches (QM+AR) are used, MGB-ECMWF forecasts exhibit positive skill in only 56% of the HPPs, which indicates difficulty in overcoming a naive forecast. For 8–15 days in advance, the percentage of HPPs where forecasts exhibit positive skill relative to persistence (climatology) improves from 29% (69%) to 64% (91%) and 76% (96%) for the QM and QM+AR configurations, respectively.

Figure 8 presents the relative skill improvement (ΔCRPSS) between the raw MGB-ECMWF forecasts and those corrected with the QM+AR methods, plotted against seasonality and flashiness indices that were calculated from the observed naturalized discharges at each SIN HPP. Results are also separated by low to moderate ($<Q_{50}$) and high ($>Q_{75}$, no exceedance) flows. In general, the results relative to climatology and persistence are similar. For higher flows, skill improvements are larger (ΔCRPSS > 1) for flashiness usually lower than 0.1 (i.e., rapid day-to-day discharge variations) regardless of seasonality, but in some HPPs, larger skill gains can be observed for flashiness values closer to 0.2 and a seasonality index around 5 (moderate seasonality). For flashiness larger than 0.2, smaller performance gains are obtained (ΔCRPSS < 0.3). For low to moderate flows, ΔCRPSS > 1 is observed even in locations where rapid flow variations may occur (flashiness ~0.5), but with some degree of seasonality.

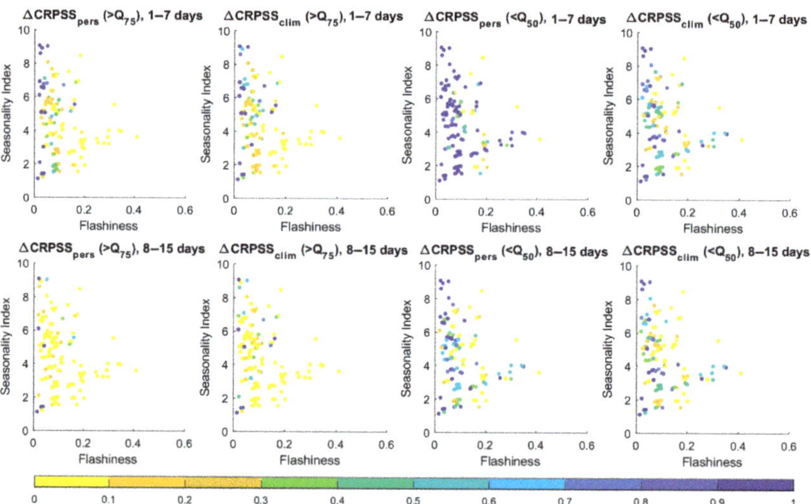

Figure 8. Difference in CRPS skill between raw and corrected forecasts with bias correction + updating (QM+AR) MGB-ECMWF forecasts according to streamflow seasonality and flashiness. Results are shown for low to moderate ($<Q_{50}$) and high ($>Q_{75}$ of non-exceedance) flows and lead times of 1–7 and 8–15 days.

4.2. Comparison between Continental-Scale and ONS Operational Streamflow Forecasts

A comparison between the MGB-ECMWF and ONS forecasts according to season is shown in Figure 9. The box plots include the performance of the 147 HPPs. Overall, the largest differences in global accuracy (as measured by MD) are found mainly in DJF and MAM, where there is a positive performance for the MGB-ECMWF forecasts. Percentual errors tend to be larger during the austral spring (SON) and smaller during the austral winter (JJA), and in both seasons MGB-ECMWF forecasts have lower performance compared with ONS forecasts.

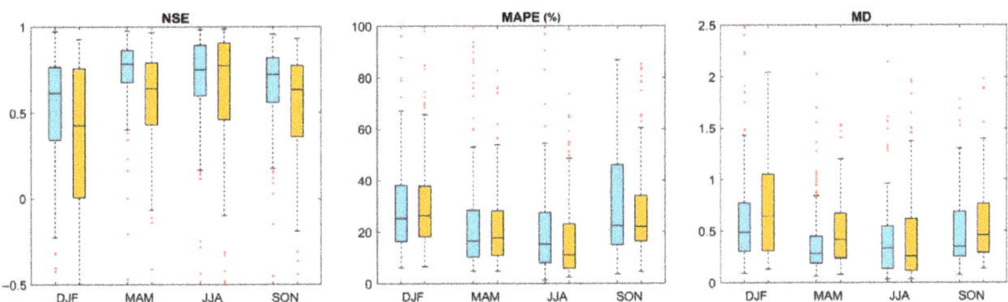

Figure 9. Comparison of forecast performance (MGB-ECMWF × ONS) for 1 week in advance, according to season.

Figure 10 shows the differences in performance (ΔMD) between ONS and MGB-ECMWF forecasts according to the installed capacity of the SIN HPPs. The installed capacity ranges were assigned in such a way to encompass a similar number of HPPs in each class. Better performances of the MGB-ECMWF forecasts are observed for HPPs with smaller installed capacity, where the median values of ΔMD are close to 0.1 and the 75th percentile reaches 0.15, 0.3, and 0.37 for the <85, 85–150, and 150–350 MW ranges, respectively. For the larger HPPs, with installed capacity larger than 350 MW, the differences are smaller (median close to zero), and there is a generally better performance of the ONS forecasts (a larger spread of ΔMD for negative values).

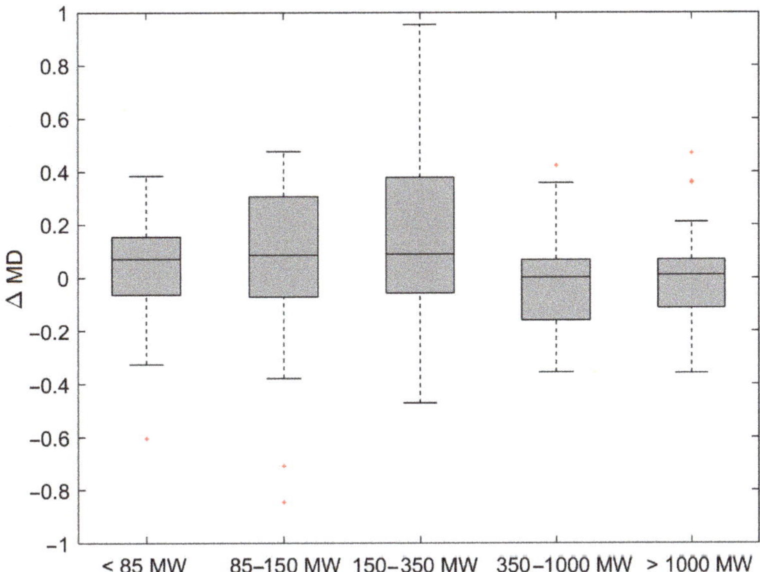

Figure 10. Multicriteria Distance (MD) differences between ONS and MGB-ECMWF forecasts (May 2015–December 2020) for 1 week in advance, according to the installed capacity of the SIN hydropower plants. Positive differences represent better overall performance of the continental-scale forecasts.

To spatially evaluate the relative performance between MGB-ECMWF and ONS streamflow forecasts, Figure 11 shows the ΔMD values geographically distributed over the SIN. Forecast performances are quite variable in space, with hotspots of ΔMD usually alternating (between negative and positive values) over southern and southeastern Brazil, although the corrected (QM+AR) continental-scale forecasts exhibit higher relative accuracy (ΔMD > 0) in most of the analyzed locations (~60% of all HPPs). MGB-ECMWF forecasts mostly outperform those of ONS in the North and Midwest/Southeast subsystems, where positive

ΔMD values are observed in 100% and 69% of the HPPs, respectively. On the other hand, MGB-ECMWF forecasts exhibit lower accuracy in the Northeast and South subsystems, with positive ΔMD in only 28% and 38% of the HPPs, respectively.

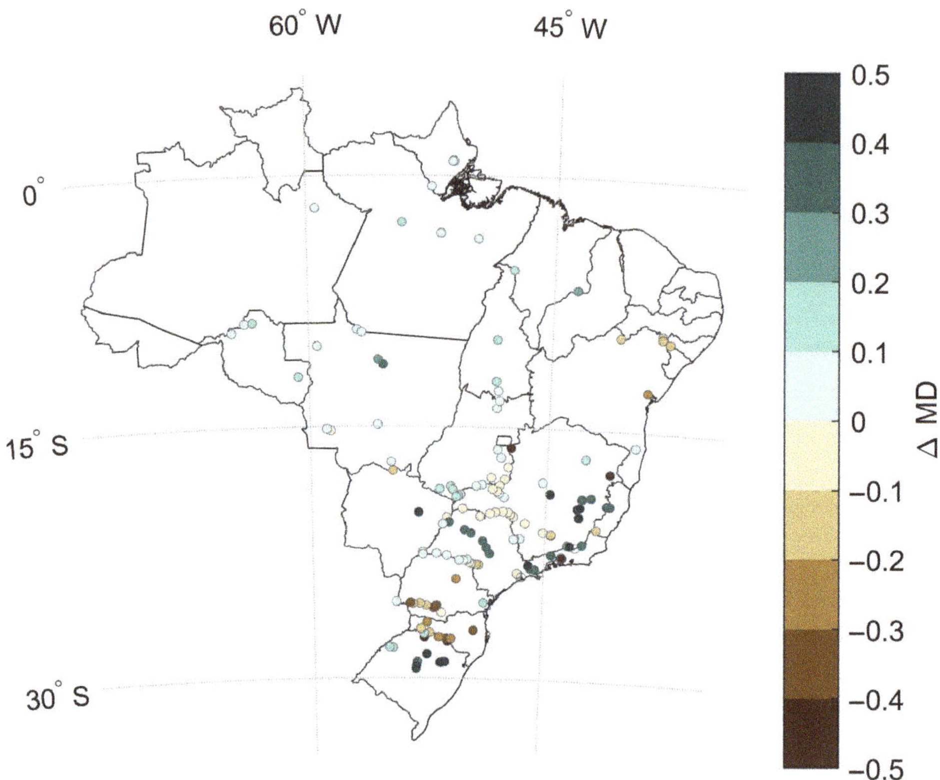

Figure 11. Spatial patterns of Multicriteria Distance differences (ΔMD between ONS and MGB-ECMWF forecasts (May 2015–December 2020) over SIN locations. Positive differences represent better overall performance of the continental-scale forecasts.

The performance metrics were also calculated for each year individually between 2015 and 2020 (Figure 12). The median accuracy (NSE) of the MGB-ECMWF forecasts is generally higher than that of the ONS, and the 25–75% range of the former is considerably higher for the first years of analysis. On the other hand, in later years (2019 and 2020), the accuracy of the MGB-ECMWF and ONS tends to become closer. The median MAPE values between the two forecasting approaches are quite similar (~20%), although the interquartile range for the MGB-ECMWF encompasses higher percentual errors (35–40%) compared to that of the ONS (25–30%), especially from the year 2019 onward. In terms of MD, a very similar pattern to NSE is observed, where the overall performance is higher for MGB-ECMWF before 2019 and similar for later years.

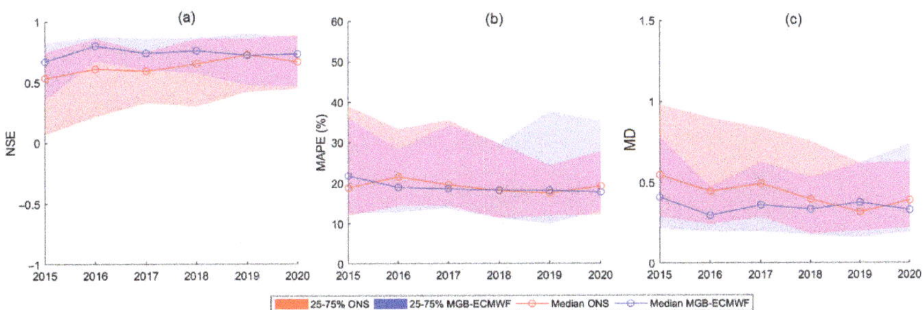

Figure 12. Year-to-year performance comparison between MGB-ECMWF and ONS streamflow forecasts for 1 week ahead over the verification period. The graphs show the median and the 25–75% range of performance considering the 147 SIN hydropower plants and are presented for (**a**) NSE, (**b**) MAPE, and (**c**) Multicriteria Distance (MD) metrics.

5. Discussion

The analysis showed that there are important challenges regarding the use of continental-scale streamflow forecasts for operational purposes in large Brazilian HPPs. Applying QM on the predicted flows allowed improvements generally for low to moderate discharges, although forecasts often exhibited lower skill when compared to simple benchmarks (e.g., persistence) for such conditions. Nevertheless, the application of QM on predicted discharges can lead to reductions in accuracy relative to that of raw forecasts if the performance of the hydrologic model is too low, as noted in Euclides da Cunha HPP. On the other hand, correcting the predicted discharges with both QM and AR updating using the observed natural flows led to substantial improvement in forecast skill, and this correction was more effective for locations with less day-to-day discharge variability. Seasonality affected the QM+AR correction performance only to a minor extent; we observed improvements in forecast skill in a few HPPs with some degree of seasonality but only for low to moderate flow conditions. This means that the ability to correct predicted discharges produced by continental-scale hydrological models will depend more on the time lag between rainfall and runoff than the existence (or absence) of a well-defined, wet-dry hydrological behavior.

Regarding the comparison between forecast performance and installed capacity, it is possible that the ONS models are more focused on HPPs with higher installed capacity, which play a more important role in setting energy prices and meeting demand. This could be one of the reasons that explain the higher accuracy of ONS predictions relative to continental-scale forecasts for such HPPs. Continental-scale predictions also showed relatively lower performance during the JJA months, which is the dry season in much of the SIN. During the dry season, the benefit of one-week ahead flow predictions is typically less than that of forecasts issued during the wet period, as predictability is higher due to the long hydrograph recession and consequent high persistence of flows. On the other hand, there seems to be room for improvement in flow predictions for HPPs with lower installed capacity, especially for the wet season (DJF).

The regional performance differences between ONS and MGB-ECMWF forecasts are possibly explained by the type of model used by ONS for a given location or even the quality of the model adjustment. Although the MGB-ECMWF forecasts exhibit higher overall accuracy over the entire verification period (2015–2020) at several SIN locations (Figure 12), it is worth mentioning that the performance of ONS forecasts has increased over those years. Possible reasons for this improvement may include the gradual replacement of ONS forecast models (including a stochastic one with a lumped rainfall-runoff hydrological model that uses precipitation forecasts from numerical weather prediction) [54,70] and the potential calibration of the hydrological models using data from more recent years, which were characterized by a drier than normal period in many regions of Brazil [71].

It should be noted that the naturalized flow estimated at the time that the ONS forecasts are updated, that is, in (near) real time, may differ slightly from that used in the AR model to update the MGB-SA streamflow forecasts. Near real-time naturalized flows are further quality checked before being made available on the SINtegre portal, so the corrected MGB-SA forecasts may have benefited positively since the consistent naturalized streamflow data (used in the AR model) are the same as those used here for forecast performance evaluation. Therefore, access to the naturalized streamflow estimated at the time of the forecast would be necessary to approximate the performance of the AR correction to that which could be obtained in (near) real time.

The bias correction on the streamflow forecasts was based on the assumption of stationarity, which is a limiting factor since the analyzed period was a critical one in most of the SIN HPPs. Moreover, some studies have already pointed out the non-stationarity of the streamflow time series at several SIN locations [72]. Nevertheless, the use of QM as the only bias correction method is justified by the focus on the ensemble mean, especially in the comparison with ONS forecasts. On the other hand, studies have shown that the QM approach alone is not able to improve the reliability of the forecasts when the focus is on the predicted probabilities [62,73], and in this case the inclusion of post-processing techniques designed for ensemble calibration would be more appropriate (e.g., [18,42,45]).

No bias correction was performed on the predicted precipitation. It is worth mentioning that freely available ECMWF reforecasts, i.e., forecasts for multiple years in the past produced with the same version of the operational numeric weather model—which could be used for bias correction—are archived at a coarser resolution (1.5°) than that of real-time forecasts, which makes it difficult to preprocess the predicted precipitation before it is propagated through the hydrological model.

6. Conclusions

Streamflow forecasts produced by continental and global scale hydrological models have gained increasing attention in the scientific community, and there is a need to evaluate the quality of these forecasts and the (potential) performance gains through the use of correction approaches, as well as to assess how such forecasts compare with those issued by institutions operating at the local to regional scales. These analyses are particularly relevant for the Brazilian context, where streamflow forecasting plays a key role due to the high dependence on water resources for energy production. In this study, we evaluated the performance of medium-range, weekly average streamflow forecasts (up to 15 days ahead) for 147 hydropower plants (HPPs) of the Brazilian National Interconnected System (SIN). The streamflow forecasts were generated by a continental-scale hydrologic-hydrodynamic model (MGB-SA) forced with ECMWF precipitation forecasts (referred to as MGB-ECMWF), while bias correction and updating procedures were applied to the model outputs by using quantile mapping (QM) and autoregressive models (AR), respectively.

The results showed that the MGB-ECMWF streamflow forecasts issued for the SIN HPPs are mostly affected by positive bias, and their skill is generally low. However, with the introduction of both output correction methods (QM+AR), the percentage of HPPs exhibiting skillful forecasts for the lead time of 1–7 days increased substantially for both low to moderate flows and high flows, whereas using only the QM correction allowed positive skill mainly for low to moderate flows and for 8–15 days ahead. Although differences in forecast skill (between correction and no correction) were less dependent on streamflow seasonality, for high discharges the skill improvements were larger at locations with slow day-to-day variations in river discharge (i.e., lower flashiness), while for low to moderate flows the improvements in skill were obtained even for locations characterized by relatively high daily discharge variability.

The forecasts generated by the continental scale model were subjected to a comparative analysis with the operational forecasts released by the Brazilian National Electric Service Operator (ONS). The evaluation encompassed the first week in advance, with due consideration given to the ONS-provided data on the forecast production dates and the operating weeks. Considering the verification period from 2015 to 2020, we observed that the relative performance between ONS and MGB-ECMWF was quite variable (exhibiting positive and negative values) over the geographical extent of the SIN, but in most locations the MGB-ECMWF forecasts performed equal to or even better than those issued by ONS, especially in HPPs with lower installed capacity (typically < 350 MW) and during the months of DJF and MAM. In addition, better performances of the continental-scale forecasts were observed for the North and Midwest/Southeast SIN subsystems. On the other hand, the results indicated that the overall performance of ONS forecasts produced for SIN locations has improved over time, and in the final years of the assessment, the overall performance of ONS forecasts was similar to that observed for MGB-ECMWF.

In future studies, especially regarding comparisons with ONS forecasts, we recommend evaluating longer lead times (e.g., subseasonal forecasts up to week 6) and exploring the probabilistic information of the ensemble. With respect to the continental hydrological-hydrodynamic model, some improvements in the accuracy of the streamflow forecasts could be achieved through better calibration as well as by ensuring consistency of rainfall data between the verification period and the historical one used as the baseline for the QM and AR corrections. Future model calibration could take advantage of datasets with precipitation data available from decades in the past to the present, for example, MERGE [74] or MSWX [75].

Finally, the findings suggest that the use of local data to correct outputs from continental-scale models can result in forecasts with competitive accuracy for regional-scale applications. In addition, there are opportunities for improvement in the performance of operational streamflow forecasts issued for the SIN HPP locations, even for forecasts produced with the current ONS models. This would be possible as the methods used in this study can be directly applied to other rainfall-runoff models (e.g., lumped or semi-distributed) developed to operate at smaller spatial scales (local or regional).

Supplementary Materials: The following supporting information can be downloaded at: https://www.mdpi.com/article/10.3390/w15091693/s1, Table S1: Performance statistics for the simulated flows at the SIN gauges. Table S2: SIN gauges for which the predicted flows were corrected based on QM + empirical distribution. Table S3: SIN gauges for which no bias correction was performed on the predicted flows.

Author Contributions: A.K.N.: literature review, data curation, analysis, and writing of the manuscript. V.A.S.: conceptualization, data curation, software, analysis, discussion, writing and review of the manuscript. C.H.d.A.G.: conceptualization, data curation, software, analysis, discussion, and manuscript review. R.C.D.d.P.: conceptualization and manuscript review. F.M.F.: supervision, funding acquisition, and manuscript review, W.C.: supervision and funding acquisition. R.S., C.S.A.P. and C.F.: discussion and manuscript review. All authors have read and agreed to the published version of the manuscript.

Funding: This work presents part of the results obtained during the project granted by the Brazilian Agency of Electrical Energy (ANEEL) under its Research and Development program, Project PD 6491-0503/2018– "Previsão Hidroclimática com Abrangência no Sistema Interligado Nacional de Energia Elétrica", developed by the Paraná State Electric company (COPEL GeT), the Meteorological System of Paraná (SIMEPAR), and the RHAMA Consulting company. The Hydraulic Research Institute (IPH) from the Federal University of Rio Grande do Sul (UFRGS) contributes to part of the project through an agreement with the RHAMA company (IAP-001313).

Data Availability Statement: Natural flows from ONS and forecast data from ECMWF can be obtained from the websites indicated in the methods section. The source code of the MGB model can be downloaded from www.ufrgs.br/hge.

Acknowledgments: Authors thank Copel for funding the project granted by the Brazilian Agency of Electrical Energy (ANEEL). The first author also thanks the Brazilian Coordination for the Improvement of Higher Education Personnel (CAPES) for partially funding the project with a scholarship. The authors are also grateful to the Copernicus Climate Change and Atmosphere Monitoring Services for providing the medium-range forecasts generated by the ECMWF, and to the three anonymous reviewers for their comments and suggestions that improved the quality of this manuscript.

Conflicts of Interest: The authors declare no conflict of interest.

Appendix A

Appendix A.1. Seasonality Index (SI)

In order to verify the influence of seasonality on the performance of the flow forecast corrections, the Seasonality Index has been used [76]. This index can be defined as the sum of the absolute deviations of monthly average flows from the overall monthly average, divided by the annual average of flows:

$$\text{SI} = \sum_{i=12}^{1} \frac{|Q - Q_{hist}|}{Q_{hist}} \tag{A1}$$

where, Q is the mean monthly streamflow and Q_{hist} is the long-term mean streamflow.

Appendix A.2. Richard-Baker Flashiness Index (RBI Index)

The Richard-Baker Flashiness Index (R-BI Index) was used to measure the degree of variability of flow (day-to-day) relative to total flow, i.e., it reports changes in short-term daily flows relative to average yearly flows. The resulting index is dimensionless, and its value is independent of the units chosen to represent flow [77].

$$\text{R} - \text{BI Index} = \frac{\sum_{i=1}^{n} |q_i - q_{i-1}|}{\sum_{i=1}^{n} q_i} \tag{A2}$$

References

1. Empresa de Pesquisa Energética (EPE). Anuário Estatístico de Energia Elétrica. 2022. Available online: http://shinyepe.brazilsouth.cloudapp.azure.com:3838/anuario-livro/#23_Gera%C3%A7%C3%A3o_el%C3%A9trica_por_fonte_(GWh) (accessed on 28 January 2023).
2. Cazzaniga, R.; Rosa-Clot, M.; Rosa-Clot, P.; Tina, G.M. Integration of PV Floating with Hydroelectric Power Plants. *Heliyon* **2019**, *5*, e01918. [CrossRef] [PubMed]
3. Collischonn, W.; Morelli Tucci, C.E.; Clarke, R.T.; Chou, S.C.; Guilhon, L.G.; Cataldi, M.; Allasia, D. Medium-Range Reservoir Inflow Predictions Based on Quantitative Precipitation Forecasts. *J. Hydrol.* **2007**, *344*, 112–122. [CrossRef]
4. Maciel, G.M.; Cabral, V.A.; Marcato, A.L.M.; Júnior, I.C.S.; Honório, L.D.M. Daily Water Flow Forecasting via Coupling between SMAP and Deep Learning. *IEEE Access* **2020**, *8*, 204660–204675. [CrossRef]
5. Costa, F.; Damazio, J.; Maceira, M. Modelos de Previsão Hidrológica Aplicados Ao Planejamento Da Operação Do Sistema Elétrico Brasileiro. *Rev. Bras. Recur. Hídr.* **2007**, *12*, 21–30. [CrossRef]
6. Guilhon, L.G.F.; Rocha, V.F.; Moreira, J.C. Comparação de Métodos de Previsão de Vazões Naturais Afluentes a Aproveitamentos Hidroelétricos. *Rev. Bras. Recur. Hídr.* **2007**, *13*, 13–20.
7. de Oliveira, V.G.; Lima, C.H.R. Previsões Multiescala de Vazões Para o Sistema Hidrelétrico Brasileiro Utilizando Ponderação Bayesiana de Modelos (BMA). *RBRH* **2016**, *21*, 618–635. [CrossRef]
8. Maceira, M.E.P.; Damazio, J. Periodic Auto-Regressive Streamflow Models Applied to Operation Planning for the Brazilian Hydroelectric System. In *Regional Hydrological Impacts of Climatic Change—Impact Assessment and Decision Making*; International Assn of Hydrological Sciences: Wallingford, UK, 2005.
9. Tucci, C.E.M.; Collischonn, W.; Fan, F.M.; Schwanenberg, D. Hydropower Forecasting in Brazil. In *Handbook of Hydrometeorological Ensemble Forecasting*; Springer: Berlin/Heidelberg, Germany, 2019; pp. 1307–1328. [CrossRef]
10. Valdés-Pineda, R.; Valdés, J.B.; Wi, S.; Serrat-Capdevila, A.; Roy, T. Improving Operational Short- to Medium-Range (SR2MR) Streamflow Forecasts in the Upper Zambezi Basin and Its Sub-Basins Using Variational Ensemble Forecasting. *Hydrology* **2021**, *8*, 188. [CrossRef]
11. Pagano, T.C.; Shrestha, D.L.; Wang, Q.J.; Robertson, D.; Hapuarachchi, P. Ensemble Dressing for Hydrological Applications. *Hydrol. Process.* **2013**, *27*, 106–116. [CrossRef]
12. Sharma, V.C.; Regonda, S.K. Multi-Spatial Resolution Rainfall-Runoff Modelling—A Case Study of Sabari River Basin, India. *Water* **2021**, *13*, 1224. [CrossRef]
13. Piazzi, G.; Thirel, G.; Perrin, C.; Delaigue, O. Sequential Data Assimilation for Streamflow Forecasting: Assessing the Sensitivity to Uncertainties and Updated Variables of a Conceptual Hydrological Model at Basin Scale. *Water Resour. Res.* **2021**, *57*, e2020WR028390. [CrossRef]
14. Reggiani, P.; Renner, M.; Weerts, A.H.; Van Gelder, P.A.H.J.M. Uncertainty Assessment via Bayesian Revision of Ensemble Streamflow Predictions in the Operational River Rhine Forecasting System. *Water Resour. Res.* **2009**, *45*, W02428. [CrossRef]

15. Cloke, H.L.; Pappenberger, F. Ensemble Flood Forecasting: A Review. *J. Hydrol.* **2009**, *375*, 613–626. [CrossRef]
16. Casagrande, L.; Tomasella, J.; dos Santos Alvalá, R.C.; Bottino, M.J.; de Oliveira Caram, R. Early Flood Warning in the Itajaí-Açu River Basin Using Numerical Weather Forecasting and Hydrological Modeling. *Nat. Hazards* **2017**, *88*, 741–757. [CrossRef]
17. Fan, F.M.; Schwanenberg, D.; Collischonn, W.; Weerts, A. Verification of Inflow into Hydropower Reservoirs Using Ensemble Forecasts of the TIGGE Database for Large Scale Basins in Brazil. *J. Hydrol. Reg. Stud.* **2015**, *4*, 196–227. [CrossRef]
18. Verkade, J.S.; Brown, J.D.; Davids, F.; Reggiani, P.; Weerts, A.H. Estimating Predictive Hydrological Uncertainty by Dressing Deterministic and Ensemble Forecasts; a Comparison, with Application to Meuse and Rhine. *J. Hydrol.* **2017**, *555*, 257–277. [CrossRef]
19. Thirel, G.; Regimbeau, F.; Martin, E.; Noilhan, J.; Habets, F. Short-and Medium-Range Hydrological Ensemble Forecasts over France. *Atmos. Sci. Lett.* **2010**, *11*, 72–77. [CrossRef]
20. Renard, B.; Kavetski, D.; Kuczera, G.; Thyer, M.; Franks, S.W. Understanding Predictive Uncertainty in Hydrologic Modeling: The Challenge of Identifying Input and Structural Errors. *Water Resour. Res.* **2010**, *46*, W05521. [CrossRef]
21. Cuo, L.; Pagano, T.C.; Wang, Q.J. A Review of Quantitative Precipitation Forecasts and Their Use in Short- to Medium-Range Streamflow Forecasting. *J. Hydrometeorol.* **2011**, *12*, 713–728. [CrossRef]
22. Pappenberger, F.; Beven, K.J.; Hunter, N.M.; Bates, P.D.; Gouweleeuw, B.T.; Thielen, J.; de Roo, A.P.J. Cascading Model Uncertainty from Medium Range Weather Forecasts (10 Days) through a Rainfall-Runoff Model to Flood Inundation Predictions within the European Flood Forecasting System (EFFS). *Hydrol. Earth Syst. Sci.* **2005**, *9*, 381–393. [CrossRef]
23. Pappenberger, F.; Bartholmes, J.; Thielen, J.; Cloke, H.L.; Buizza, R.; de Roo, A. New Dimensions in Early Flood Warning across the Globe Using Grand-Ensemble Weather Predictions. *Geophys. Res. Lett.* **2008**, *35*, L10404. [CrossRef]
24. Velázquez, J.A.; Anctil, F.; Ramos, M.H.; Perrin, C. Can a Multi-Model Approach Improve Hydrological Ensemble Forecasting? A Study on 29 French Catchments Using 16 Hydrological Model Structures. *Adv. Geosci.* **2011**, *29*, 33–42. [CrossRef]
25. Ficchì, A.; Raso, L.; Dorchies, D.; Pianosi, F.; Malaterre, P.-O.; Van Overloop, P.-J.; Jay-Allemand, M. Optimal Operation of the Multireservoir System in the Seine River Basin Using Deterministic and Ensemble Forecasts. *J. Water Resour. Plan. Manag.* **2015**, *142*, 05015005. [CrossRef]
26. Liu, J.; Yuan, X.; Zeng, J.; Jiao, Y.; Li, Y.; Zhong, L.; Yao, L. Ensemble Streamflow Forecasting over a Cascade Reservoir Catchment with Integrated Hydrometeorological Modeling and Machine Learning. *Hydrol. Earth Syst. Sci.* **2022**, *26*, 265–278. [CrossRef]
27. Arsenault, R.; Côté, P. Analysis of the Effects of Biases in Ensemble Streamflow Prediction (ESP) Forecasts on Electricity Production in Hydropower Reservoir Management. *Hydrol. Earth Syst. Sci.* **2019**, *23*, 2735–2750. [CrossRef]
28. Boucher, M.A.; Quilty, J.; Adamowski, J. Data Assimilation for Streamflow Forecasting Using Extreme Learning Machines and Multilayer Perceptrons. *Water Resour. Res.* **2020**, *56*, e2019WR026226. [CrossRef]
29. Cassagnole, M.; Ramos, M.H.; Zalachori, I.; Thirel, G.; Garçon, R.; Gailhard, J.; Ouillon, T. Impact of the Quality of Hydrological Forecasts on the Management and Revenue of Hydroelectric Reservoirs-a Conceptual Approach. *Hydrol. Earth Syst. Sci.* **2021**, *25*, 1033–1052. [CrossRef]
30. Schwanenberg, D.; Fan, F.M.; Naumann, S.; Kuwajima, J.I.; Montero, R.A.; Assis dos Reis, A. Short-Term Reservoir Optimization for Flood Mitigation under Meteorological and Hydrological Forecast Uncertainty: Application to the Três Marias Reservoir in Brazil. *Water Resour. Manag.* **2015**, *29*, 1635–1651. [CrossRef]
31. Boucher, M.A.; Tremblay, D.; Delorme, L.; Perreault, L.; Anctil, F. Hydro-Economic Assessment of Hydrological Forecasting Systems. *J. Hydrol.* **2012**, *416–417*, 133–144. [CrossRef]
32. Avesani, D.; Zanfei, A.; Di Marco, N.; Galletti, A.; Ravazzolo, F.; Righetti, M.; Majone, B. Short-Term Hydropower Optimization Driven by Innovative Time-Adapting Econometric Model. *Appl. Energy* **2022**, *310*, 118510. [CrossRef]
33. Wu, W.; Emerton, R.; Duan, Q.; Wood, A.W.; Wetterhall, F.; Robertson, D.E. Ensemble Flood Forecasting: Current Status and Future Opportunities. *Wiley Interdiscip. Rev. Water* **2020**, *7*, e1432. [CrossRef]
34. Troin, M.; Arsenault, R.; Wood, A.W.; Brissette, F.; Martel, J.L. Generating Ensemble Streamflow Forecasts: A Review of Methods and Approaches Over the Past 40 Years. *Water Resour. Res.* **2021**, *57*, e2020WR028392. [CrossRef]
35. Das, J.; Manikanta, V.; Nikhil Teja, K.; Umamahesh, N.V. Two Decades of Ensemble Flood Forecasting: A State-of-the-Art on Past Developments, Present Applications and Future Opportunities. *Hydrol. Sci. J.* **2022**, *67*, 477–493. [CrossRef]
36. Ayzel, G. OpenForecast v2: Development and Benchmarking of the First National-Scale Operational Runoff Forecasting System in Russia. *Hydrology* **2021**, *8*, 3. [CrossRef]
37. Hapuarachchi, H.A.P.; Bari, M.A.; Kabir, A.; Hasan, M.M.; Woldemeskel, F.M.; Gamage, N.; Sunter, P.D.; Zhang, X.S.; Robertson, D.E.; Bennett, J.C.; et al. Development of a National 7-Day Ensemble Streamflow Forecasting Service for Australia. *Hydrol. Earth Syst. Sci.* **2022**, *26*, 4801–4821. [CrossRef]
38. Snow, A.D.; Christensen, S.D.; Swain, N.R.; Nelson, E.J.; Ames, D.P.; Jones, N.L.; Ding, D.; Noman, N.S.; David, C.H.; Pappenberger, F.; et al. A High-Resolution National-Scale Hydrologic Forecast System from a Global Ensemble Land Surface Model. *JAWRA J. Am. Water Resour. Assoc.* **2016**, *52*, 950–964. [CrossRef]
39. Thielen, J.; Bartholmes, J.; Ramos, M.-H.; De Roo, A. Hydrology and Earth System Sciences the European Flood Alert System-Part 1: Concept and Development. *Hydrol. Earth Syst. Sci.* **2009**, *13*, 125–140. [CrossRef]
40. Alfieri, L.; Burek, P.; Dutra, E.; Krzeminski, B.; Muraro, D.; Thielen, J.; Pappenberger, F. GloFAS-Global Ensemble Streamflow Forecasting and Flood Early Warning. *Hydrol. Earth Syst. Sci.* **2013**, *17*, 1161–1175. [CrossRef]

41. Emerton, R.; Zsoter, E.; Arnal, L.; Cloke, H.L.; Muraro, D.; Prudhomme, C.; Stephens, E.M.; Salamon, P.; Pappenberger, F. Developing a Global Operational Seasonal Hydro-Meteorological Forecasting System: GloFAS-Seasonal v1.0. *Geosci. Model Dev.* **2018**, *11*, 3327–3346. [CrossRef]

42. Siqueira, V.A.; Weerts, A.; Klein, B.; Fan, F.M.; de Paiva, R.C.D.; Collischonn, W. Postprocessing Continental-Scale, Medium-Range Ensemble Streamflow Forecasts in South America Using Ensemble Model Output Statistics and Ensemble Copula Coupling. *J. Hydrol.* **2021**, *600*, 126520. [CrossRef]

43. Crochemore, L.; Ramos, M.H.; Pechlivanidis, I.G. Can Continental Models Convey Useful Seasonal Hydrologic Information at the Catchment Scale? *Water Resour. Res.* **2020**, *56*, e2019WR025700. [CrossRef]

44. Hashino, T.; Bradley, A.A.; Schwartz, S.S. Evaluation of Bias-Correction Methods for Ensemble Streamflow Volume Forecasts. *Hydrol. Earth Syst. Sci.* **2007**, *11*, 939–950. [CrossRef]

45. Hemri, S.; Lisniak, D.; Klein, B. Multivariate Postprocessing Techniques for Probabilistic Hydrological Forecasting. *Water Resour. Res.* **2015**, *51*, 7436–7451. [CrossRef]

46. Vrugt, J.A.; Gupta, H.V.; Nualláin, B.Ó.; Bouten, W. Real-Time Data Assimilation for Operational Ensemble Streamflow Forecasting on JSTOR. *J. Hydrometeorol.* **2006**, *7*, 548–565. [CrossRef]

47. Boucher, M.A.; Perreault, L.; Anctil, F.; Favre, A.C. Exploratory Analysis of Statistical Post-Processing Methods for Hydrological Ensemble Forecasts. *Hydrol. Process.* **2015**, *29*, 1141–1155. [CrossRef]

48. Lozano, J.S.; Bustamante, G.R.; Hales, R.C.; Nelson, E.J.; Williams, G.P.; Ames, D.P.; Jones, N.L. A Streamflow Bias Correction and Performance Evaluation Web Application for GEOGloWS ECMWF Streamflow Services. *Hydrology* **2021**, *8*, 71. [CrossRef]

49. Wang, H.; Zhong, P.; Zsoter, E.; Prudhomme, C.; Pappenberger, F.; Xu, B. Regional Adaptability of Global and Regional Hydrological Forecast System. *Water* **2023**, *15*, 347. [CrossRef]

50. Li, M.; Wang, Q.J.; Bennett, J.C.; Robertson, D.E. A Strategy to Overcome Adverse Effects of Autoregressive Updating of Streamflow Forecasts. *Hydrol. Earth Syst. Sci.* **2015**, *19*, 1–15. [CrossRef]

51. Sharma, S.; Siddique, R.; Reed, S.; Ahnert, P.; Mendoza, P.; Mejia, A. Relative Effects of Statistical Preprocessing and Postprocessing on a Regional Hydrological Ensemble Prediction System. *Hydrol. Earth Syst. Sci.* **2018**, *22*, 1831–1849. [CrossRef]

52. ONS—Operador Nacional do Sistema Elétrico. O Sistema em Números. Available online: https://www.ons.org.br/paginas/sobre-o-sin/o-sistema-em-numeros (accessed on 23 January 2023).

53. ONS—Operador Nacional do Sistema Elétrico. O Que é o Sin. Available online: https://www.ons.org.br/paginas/sobre-o-sin/o-que-e-o-sin (accessed on 23 January 2023).

54. ONS—Operador Nacional do Sistema Elétrico. *Relatório Anual de Avaliação Das Previsões de Vazões e Energias Naturais Afluentes de 2017*; ONS: Rio de Janeiro, Brazil, 2018; pp. 1–100.

55. ONS—Operador Nacional do Sistema Elétrico. *Relatório Anual de Avaliação Das Previsões de Vazões e Energias Naturais Afluentes de 2019*; ONS: Rio de Janeiro, Brazil, 2020; pp. 1–105.

56. ONS—Operador Nacional do Sistema Elétrico. *Procedimentos de Rede Assunto Submódulo Revisão Data de Vigência*; ONS: Rio de Janeiro, Brazil, 2020.

57. Siqueira, V.A.; Paiva, R.C.D.; Fleischmann, A.S.; Fan, F.M.; Ruhoff, A.L.; Pontes, P.R.M.; Paris, A.; Calmant, S.; Collischonn, W. Toward Continental Hydrologic-Hydrodynamic Modeling in South America. *Hydrol. Earth Syst. Sci.* **2018**, *22*, 4815–4842. [CrossRef]

58. Collischonn, W.; Allasia, D.; da Silva, B.C.; Tucci, C.E.M. The MGB-IPH Model for Large-Scale Rainfall-Runoff Modelling. *Hydrol. Sci. J.* **2007**, *52*, 878–895. [CrossRef]

59. Pontes, P.R.M.; Fan, F.M.; Fleischmann, A.S.; de Paiva, R.C.D.; Buarque, D.C.; Siqueira, V.A.; Jardim, P.F.; Sorribas, M.V.; Collischonn, W. MGB-IPH Model for Hydrological and Hydraulic Simulation of Large Floodplain River Systems Coupled with Open Source GIS. *Environ. Model. Softw.* **2017**, *94*, 1–20. [CrossRef]

60. Skofronick-Jackson, G.; Petersen, W.A.; Berg, W.; Kidd, C.; Stocker, E.F.; Kirschbaum, D.B.; Kakar, R.; Braun, S.A.; Huffman, G.J.; Iguchi, T.; et al. The Global Precipitation Measurement (GPM) Mission for Science and Society. *Bull. Am. Meteorol. Soc.* **2017**, *98*, 1679–1695. [CrossRef] [PubMed]

61. Beck, H.; Van Dijk, A.; Levizzani, V.; Schellekens, J.; Miralles, D.; Martens, B.; de Roo, A.; Pappenberger, F.; Huffman, G.; Wood, E. MSWEP: 3-Hourly 0.1° Fully Global Precipitation (1979–Present) by Merging Gauge, Satellite, and Weather Model Data. *EGUGA* **2017**, *19*, 18289.

62. Wood, A.W.; Schaake, J.C. Correcting Errors in Streamflow Forecast Ensemble Mean and Spread. *J. Hydrometeorol.* **2008**, *9*, 132–148. [CrossRef]

63. Liu, Y.; Weerts, A.H.; Clark, M.; Hendricks Franssen, H.J.; Kumar, S.; Moradkhani, H.; Seo, D.J.; Schwanenberg, D.; Smith, P.; Van Dijk, A.I.J.M.; et al. Advancing Data Assimilation in Operational Hydrologic Forecasting: Progresses, Challenges, and Emerging Opportunities. *Hydrol. Earth Syst. Sci.* **2012**, *16*, 3863–3887. [CrossRef]

64. Maidment, D.R. *Handbook of Hydrology*; Mc-Graw-Hill, Inc.: New York, NY, USA, 1993; ISBN 0070397325.

65. ONS—Operador Nacional do Sistema Elétrico. *Relatório Anual de Avaliação Das Previsões de Vazões e Energias Naturais Afluentes de 2020*; ONS: Rio de Janeiro, Brazil, 2021.

66. Hersbach, H. Decomposition of the Continuous Ranked Probability Score for Ensemble Prediction Systems. *Weather Forecast.* **2000**, *15*, 559–570. [CrossRef]

67. Ferro, C.A.T.; Richardson, D.S.; Weigel, A.P. On the Effect of Ensemble Size on the Discrete and Continuous Ranked Probability Scores. *Meteorol. Appl.* **2008**, *15*, 19–24. [CrossRef]
68. Nash, J.E.; Sutcliffe, J.V. River Flow Forecasting through Conceptual Models Part I—A Discussion of Principles. *J. Hydrol.* **1970**, *10*, 282–290. [CrossRef]
69. ONS—Operador Nacional do Sistema Elétrico. *Operação Do Sistema Interligado Nacional: Relatório Anual de Avaliação Das Previsões de Vazões—2010*; ONS: Rio de Janeiro, Brazil, 2010; pp. 1–230.
70. ONS—Operador Nacional do Sistema Elétrico. *Submódulo 9.2—Acompanhamento, Análise e Tratamento Dos Dados Hidroenergéticos Do Sistema Interligado Nacional*; ONS: Rio de Janeiro, Brazil, 2020; pp. 1–10.
71. Cuartas, L.A.; Cunha, A.P.M.D.A.; Alves, J.A.; Parra, L.M.P.; Deusdará-Leal, K.; Costa, L.C.O.; Molina, R.D.; Amore, D.; Broedel, E.; Seluchi, M.E.; et al. Recent Hydrological Droughts in Brazil and Their Impact on Hydropower Generation. *Water* **2022**, *14*, 601. [CrossRef]
72. Detzel, D.; Bessa, M.; Vallejos, C.; Santos, A.; Thomsen, L.; Mine, M.; Bloot, M.; Estrocio, J. Estacionariedade Das Afluências Às Usinas Hidrelétricas Brasileiras. *Rev. Bras. Recur. Hídr.* **2011**, *16*, 95–111. [CrossRef]
73. Madadgar, S.; Moradkhani, H.; Garen, D. Towards Improved Post-Processing of Hydrologic Forecast Ensembles. *Hydrol. Process.* **2014**, *28*, 104–122. [CrossRef]
74. Rozante, J.R.; Moreira, D.S.; de Goncalves, L.G.G.; Vila, D.A. Combining TRMM and Surface Observations of Precipitation: Technique and Validation over South America. *Weather Forecast.* **2010**, *25*, 885–894. [CrossRef]
75. Beck, H.E.; Van Dijk, A.I.J.M.; Larraondo, P.R.; McVicar, T.R.; Pan, M.; Dutra, E.; Miralles, D.G. MSWX: Global 3-Hourly 0.1° Bias-Corrected Meteorological Data Including Near-Real-Time Updates and Forecast Ensembles. *Bull. Am. Meteorol. Soc.* **2022**, *103*, E710–E732. [CrossRef]
76. Walsh, R.P.D.; Lawler, D.M. Rainfall Seasonality: Description, Spatial Patterns and Change through Time. *Weather* **1981**, *36*, 201–208. [CrossRef]
77. Baker, D.B.; Richards, R.P.; Loftus, T.T.; Kramer, J.W. A New Flashiness Index: Characteristics and Applications to Midwestern Rivers and Streams. *JAWRA J. Am. Water Resour. Assoc.* **2004**, *40*, 503–522. [CrossRef]

Article

Discussion on Operational Stability of Governor Turbine Hydraulic System Considering Effect of Power System

Jianxu Zhou *, Chaoqun Li and Yutong Mao

College of Water Conservancy and Hydropower Engineering, Hohai University, 1 Xikang Road, Nanjing 210098, China; 200202030004@hhu.edu.cn (C.L.); 190802030003@hhu.edu.cn (Y.M.)
* Correspondence: jxzhouhhu@163.com

Abstract: Hydropower has grown to play an important role in power systems including increasing clean and low-carbon energies, and the effect of electric loads should be basically evaluated for the reliable operation of these systems. For the hydraulic–mechanical–electrical system of the hydropower station, the state equation model for stability evaluation was derived with typical electric load models and elastic models for pipe flow, and after experimental confirmation with a built single-unit setup for a system, the effects of different electrical loads and pipe flow models on typical hydropower systems stability were investigated in detail. The results indicate that for the built single-unit system with different load characteristics, the numerical results were basically consistent with experimental research, and the unit's regulation performance for the dynamic load was superior to that of the static load. Evident differences existed in the effects of different electric loads on the operational stability, mainly depending on the pipe length and the corresponding models, and an optimum-order elastic model of pipe flow was preferred to reveal the dynamic interactions between different systems. Furthermore, for a typical two-unit system, the potential coupling resonance hydraulic–mechanical–electrical system is pointed out with the preferred-order elastic model of pipe flow.

Keywords: hydropower; stability; elastic model; experimental research

Citation: Zhou, J.; Li, C.; Mao, Y. Discussion on Operational Stability of Governor Turbine Hydraulic System Considering Effect of Power System. *Energies* **2023**, *16*, 4459. https://doi.org/10.3390/en16114459

Academic Editors: Zhengwei Wang and Yongguang Cheng

Received: 17 April 2023
Revised: 25 May 2023
Accepted: 29 May 2023
Published: 31 May 2023

1. Introduction

With the gradually growing energy demand, renewable and low-carbon energy sources have become a considerable part of the global energy mix [1]. Among them, for the utilization of water energy, many hydropower stations were successively constructed and brought into service, including the power transmission projects from west to east in China [2,3]. These hydropower stations are well suited to provide scheduling flexibility and play an important role in the normal operation of the power system after the grid connection [4–8]. Then, the stable operation of hydropower systems not only is beneficial to improve its power supply quality but also ensures reliable service of the parallel power system [9,10].

For the hydropower system, its normal operation and regulation is a dynamic process in which the governor turbine hydraulic system and the parallel power system inevitably have a dynamic interaction. During the generation condition of the hydropower system, as a consumer for water energy utilization, because the power system includes different electric loads, the effect of the power system on the operational stability of the governor turbine hydraulic system in the hydropower stations basically depends on the electric loads with different characteristics. Therefore, the assessment of different electric loads is indispensable to analyze the operational stability of the hydropower system [11]. Typically, two typical electric loads in power systems are often concerned, namely, the static load with basically unchanged impedance [12] and the dynamic load with quick response to frequency changes in dynamic studies [13]. Recently, the increasing complexity of new generation units and water systems has caused the load instability to become a new problem in many countries [14–17], mainly because of the complex interactions of hydro-mechanical power subsystems in the hydropower stations.

Hence, proper models of the electrical loads are the premise to understand their dynamic behaviors in different systems, and their importance in power system stability studies has been emphasized over the long term [18–20]. Relevant achievements have been obtained by numerical analysis in recent years. It has enabled researchers to further present the effect of load characteristic on the damping for low-frequency oscillation in the power system [21]. Similarly, the influence of different loads on the damping analysis of the power system was studied by numerical analysis [22]. In fact, the main motivation for the implementation of these load models is that static load models cannot predict instability in different subsystems. For instance, Nomikos analyzed the impact of a dynamic load, that is, an induction motor, on a mechanical–electrical oscillation in the power system and emphasized that the dynamic load had an evident impact on inter-area modes of power systems [23]. Hiskens and Milanović also studied the impact of these load models on the stability and analytical damping of the power system [24,25]. Their research shows that the recovery time constant of the dynamic load has a major effect on the eigenvalue analysis of the system.

Furthermore, alternative approaches are adopted based on the experimental setup and laboratory measurements instead of numerical simulation of the electrical loads. To investigate the dynamic stability of the governor turbine hydraulic system at a large hydropower station, Yang et al. carried out experimental research with simulated units [26]. Zeng et al. performed a full-load rejection test to confirm the effect of the S-shaped property on the water hammer calculation, which provided important experimental data to ensure the stable operation of the pumped storage station [27]. In addition, based on the physical model experiment, an analysis method considering the theoretical model was provided by Yang et al. [28], and the sustained ultra-low-frequency oscillations and frequency instability of hydropower units were revealed.

Worth noting is that the integration of pipe flow models and various electrical loads in hydropower systems remains a relatively unexplored research field. The effect analysis of load characteristics on operational stability has been preliminarily discussed in detail before [29], but it is still insufficient to meet the present demand or provide a comprehensive analysis of the influence mechanisms of different-order elastic models of pipe flow together with various electrical loads on typical hydropower systems. Most importantly, the load-influencing mechanism on the power system stability should be clarified in detail, in the case of complex load characteristics after the integration of the hydraulic–mechanical system with the power system. Therefore, further analysis with different load characteristics is urgent to ensure the system's stable operation in different operating boundaries. After the mathematical model of the hydraulic–mechanical system is derived and organized, with further introduction of load models with different characteristics, the simulation model presented by state equations is derived to reveal the effects of different load characteristics on the system's stability. Then, a built experiment setup of the hydraulic–mechanical–electrical system for a hydropower station [30] is also introduced for further confirmation analysis with numerical analysis. Finally, the effects of elastic models with different orders of pipe flow with different load characteristics on the system's stability are investigated in detail based on two case analyses.

The rest of this paper is organized as follows. In Section 2, the mathematical models of static and dynamic loads are established in the form of state equations, together with the derived mathematical models of the governor turbine hydraulic system. In Section 3, the experimental-based stability study and confirmation of the numerical model are carried out through the experimental setup of hydro-mechanical power systems. In Section 4, based on a single-unit system and a two-unit system, the sensitivity analysis of the model order of pipe flow together with different load characteristics was conducted, followed by further numerical analysis. Finally, the conclusions are presented in Section 5.

2. Mathematical Models

For the hydropower station paralleled in the power system, its dynamic characteristics were mainly concerned with the hydraulic pipelines, the turbine generator, the governor, the exciter, and the electric loads. Therefore, in order to clarify the inherent dynamic interaction of the governor turbine hydraulic system and the power system and accurately reveal the effect of different pipe flow models and different electrical loads on the system's operational stability, their mathematical models should be accurately established with the inter-variables.

2.1. Hydraulic System

For a simple hydropower system with a water diversion pipe and a hydraulic unit at the downstream end, different linear models, including rigid models and elastic models, can be deduced for the pipe flow provided in [31]. Then, as a single-unit system is considered, the state equations for state variables at the pipe's outlet section are

$$
\begin{cases}
\frac{dT_i}{dt} = q_i \\
\frac{dq_i}{dt} = -\frac{fQ_0}{DA}q_i - \frac{(i\pi a)^2}{l^2}T_i - \frac{2gAH_0(-1)^i}{l}\varsigma \\
\frac{dq}{dt} = -\frac{fQ_0}{DA}q + \frac{1}{Q_0}\sum_{i=1}^{n}\frac{(i\pi a)^2(-1)^i}{l^2}T_i - \frac{H_0}{Q_0}\frac{(2n+1)gA}{l}\varsigma
\end{cases}
\tag{1}
$$

where q_i and T_i are the ith-order oscillatory flow rate and introduced conversion variable; l is the pipe length; f, D, A, and a are the pipe's friction coefficient, diameter, sectional area, and wave speed, respectively; H_0 and Q_0 are the initial pressure and flow rate at the pipe's end section of pressurized pipe; q and ς are unit's relative variation of flow rate and operating head; and n is the model order.

Here, based on the linearization of the turbine's flow equations at a given operation condition [32], the unit's relative variation of operating head in Equation (1) can be described by

$$
\varsigma = C_1 q + C_2 \varphi + C_3 \mu,
\tag{2}
$$

where φ and μ are unit's relative variation of rotational speed and wicket opening, and C_i ($i = 1, 2, 3$) are parameters obtained from the model hill charts of the hydraulic turbine according to a given operating condition.

By the introduction of Equation (2), Equation (1) can be rewritten as follows:

$$
\begin{cases}
\frac{dT_i}{dt} = q_i \\
\frac{dq_i}{dt} = -\frac{fQ_0}{DA}q_i - \frac{(i\pi a)^2}{l^2}T_i - \frac{2gAH_0(-1)^iC_1}{l}q - \frac{2gAH_0(-1)^iC_2}{l}\varphi - \frac{2gAH_0(-1)^iC_3}{l}\mu \\
\frac{dq}{dt} = \frac{1}{Q_0}\sum_{i=1}^{n}\frac{(i\pi a)^2(-1)^i}{l^2}T_i - \left(\frac{fQ_0}{DA} + \frac{H_0}{Q_0}\frac{(2n+1)gAC_1}{l}\right)q - \frac{H_0}{Q_0}\frac{(2n+1)gAC_2}{l}\varphi - \frac{H_0}{Q_0}\frac{(2n+1)gAC_3}{l}\mu
\end{cases}
\tag{3}
$$

Equation (3) is a typical state equation group mainly involving a different-order elastic model of pipe flow, which provides the dynamic interaction between the hydraulic system and the mechanical system with some inter-variables.

Similarly, for different hydropower systems, the elastic model can also be derived with state variables at the outlet section for the long water diversion pipelines connected to an upstream surge tank or bifurcation or with state variables at the inlet section for the tail pipelines introduced from a downstream surge tank or bifurcation.

2.2. Turbine Generator

For the turbine generator unit, as a typical salient pole generator, the commonly used stability computation model is the five order equations recommended by IEEE [33], which can exactly simulate its transient characteristics and dynamic responses and conveniently introduce the interaction of hydraulic–mechanical system. Its basic equations include

$$\frac{dE'_q}{dt} = \frac{1}{T'_{d0}}\left[E_{fd} - E'_q - (x_d - x'_d)i_d\right],\tag{4}$$

$$\frac{dE''_q}{dt} = \left(\frac{1}{T''_{d0}} - \frac{1}{T'_{d0}}\right)E'_q - \frac{1}{T''_{d0}}E''_q - \left(\frac{x_d - x'_d}{T'_{d0}} + \frac{x'_d - x''_d}{T''_{d0}}\right)i_d + \frac{1}{T'_{d0}}E_{fd},\tag{5}$$

$$\frac{dE''_d}{dt} = \frac{1}{T''_{q0}}\left[-E''_d + (x_q - x''_q)i_q\right],\tag{6}$$

$$\frac{d\omega}{dt} = \frac{1}{M}[P_m - P_e],\tag{7}$$

$$\frac{d\delta}{dt} = \omega\omega_0,\tag{8}$$

where T_{d0}', T_{d0}'', and T_{q0}'' are all time constants corresponding to the d-axis and the q-axis; M is the unit's inertia time constant; ω and δ are the rotor's angular speed and power angle; E_q', E_q'', and E_{fd} are the q-axis transient potential and sub-transient potential and excitation potential respectively; E_d'' is the d-axis sub-transient potential; x_d, x_d', x_d'', and x_q, x_q'' are all known parameters corresponding to the d-axis and q-axis; i_d and i_q are the d-axis and q-axis current; V_d and V_q are d-axis and q-axis voltage, $V_q = E_q'' - x_d''i_d$, $V_d = E_d'' + x_q''i_q$; V_t is the terminal voltage, $V_t^2 = V_q^2 + V_d^2$; and P_m and P_e are the mechanical and electromagnetic power.

For the hydraulic turbine at a certain operating condition, based on the linearized equation of the unit's output [32] and further introduction of Equation (2), the relative power variation corresponding to the mechanical power P_m in Equation (7) can be obtained as

$$p_m = C_4 q + C_5 \varphi + C_6 \mu,\tag{9}$$

where P_m is the relative power variation, and C_i ($i = 4, 5, 6$) are parameters obtained from the model hill charts of the hydraulic turbine according to a given operating condition. Similarly, the relative power variation corresponding to the electromagnetic power P_e in Equation (7) can be derived according to electric loads with different characteristics.

2.3. Electric Load in Power System

The electric loads mainly include the static load, dynamic load, and integrated load. Because the integrated load has the dynamic characteristics combined with the static load and dynamic load, and in most cases, it is difficult to introduce an exact simulation model with defined transient parameters, for their effect analysis on the system's operational stability, the static load and dynamic load model is typically emphasized. Further, the induction motor is the commonly used dynamic load in a power system, and its dynamic characteristics are basically represented by the transient of the induction motor.

Static Load. The description form for the static load model can be rewritten by simplification.

$$P_l = P_{l0}\left[a_p\left(\frac{E''_q}{E''_{q0}}\right)^2 + b_p\left(\frac{E''_q}{E''_{q0}}\right) + c_p\right],\tag{10}$$

where coefficients a_p, b_p, and c_p meet $a_p + b_p + c_p = 1$; P_l and P_{l0} are the transient and rated active power, respectively; and E_{q0}'' is the q-axis initial sub-transient potential. The specified characteristics of the static load can be described by different combinations of a_p, b_p, and c_p.

Then, with the introduction of the static load model and by the Newton–Raphson linearization, the relative power variation corresponding to electromagnetic power P_e in Equation (7) can also be derived and reorganized as below [29]:

$$p_e = k_e e_q'' + k_s \theta, \tag{11}$$

where e_q'' is the q-axis relative sub-transient potential an θ is the generator's rotor angle. The derived coefficients are defined as below:

$$k_e = \frac{V_t}{x_d''} \sin \theta_0 + \frac{P_{l0}}{E_{q0}''}(2a_p + b_p), \quad k_s = \frac{E_{q0}'' V_t}{x_d''} \cos \theta_0 + \frac{V_t^2 (x_d'' - x_q'')}{x_d'' x_q''} \cos(2\theta_0).$$

With the substitution of the mechanical power P_m, Equation (9), and the electromagnetic power P_e, Equation (11), and further linearization, the unit's motion Equation (7) is rewritten into

$$\frac{d\varphi}{dt} = \frac{1}{T_a}(C_4 q + C_5 \varphi + C_6 \mu - k_e e_q'' - k_s \theta), \tag{12}$$

where T_a is inertia time constant of the turbine generator.

Dynamic Load. In order to clearly reveal the effect of the dynamic load on the system's operational stability with a simplified simulation model, considering an induction motor with absolutely symmetrical d- and q-axes, after ignoring its transient process and defining positive inflow of current, the obtained state equations for the dynamic load are as follows:

$$\frac{dE_{qe}'}{dt} = -\frac{1}{T'}E_{qe}' + \frac{1}{T'}(x - x')i_{de} \tag{13}$$

$$\frac{dE_{de}'}{dt} = -\frac{1}{T'}E_{de}' - \frac{1}{T'}(x - x')i_{qe} \tag{14}$$

$$\frac{ds}{dt} = \frac{1}{2M'}(T_m - T_e), \tag{15}$$

where T' is the time constant; E_{qe}' and E_{de}' are the transient potentials of the q- and d-axes, respectively; x and x' are the reactance and transient reactance; i_{de} and i_{qe} are the q- and d-axis currents; s is the slip of the motor; M' is the inertia time constant; T_e is the electromagnetic torque; and T_m is the mechanical torque. Then, based on the voltage equations, the electromagnetic power formula, and the interface models for the generator and the electric load, the relative power variation corresponding to electromagnetic power P_e can also be derived for the dynamic load [12].

2.4. Stability Analysis Based on State Equations

Therefore, for a simple hydropower system with a water diversion pipe and a hydraulic unit at the downstream end, by further introduction of the mathematical models for the exciter and speed governor provided in [33,34], the whole state equations group can be organized to directly describe the dynamic characteristic of the governor turbine hydraulic system and power system, and the corresponding matrix form is

$$\frac{dX}{dt} = AX + BU, \tag{16}$$

where A is the coefficient matrix, X is the state vector, B is the input matrix, and U is the disturbance vector.

For this single-unit system, the derived state vector X with the static load in Equation (16) is

$$X = [T_1, q_1, \cdots, T_n, q_n, q, \varphi, y, \mu, x_I, x_D, u_a, u_f, e_{fd}, e_q', e_q'', e_d'', \theta]^T, \tag{17}$$

where y, x_I, and x_D are the governor's state variables; u_a and u_f are the output voltage and exciting feedback voltage of the voltage regulator; e_{fd} is relative the exciting potential for the exciter; e_q' is the q-axis relative transient potential; and e_d'' is the d-axis relative subtransient potential.

Similarly, for this single-unit system, the derived state vector with the dynamic load in Equation (16) is

$$X = [T_1, q_1, \cdots, T_n, q_n, q, \varphi, y, \mu, x_I, x_D, u_a, u_f, e_{fd}, e_q', e_q'', e_d'', e_{qe}', e_{de}', s, \theta_{12}, \theta_1]^T, \qquad (18)$$

where e_{qe}' and e_{de}' are the q- and d-axis relative transient potentials; θ_{12} meets θ_1-θ_2; and θ_1 and θ_2 are the generator's and the induction motor's rotor angles respectively.

In particular, if the dynamic characteristics of the generator, exciter, and electrical load are not considered, the simplified state vector in Equation (16), which is commonly used in stability analysis on the governor turbine hydraulic system, is

$$X = [T_1, q_1, \cdots, T_n, q_n, q, \varphi, y, \mu, x_I, x_D]^T, \qquad (19)$$

With the obtained state vectors for different considerations, the corresponding linear state equations can be built to describe the dynamic characteristics of the hydropower system, and then the system's operational stability with dynamic responses will be investigated under small disturbance. If different electric loads are involved with the state vectors presented in Equations (17) and (18), with further eigenvalue and sensitivity analysis, the influence of different-order elastic models of pipe flow and different load characteristics on the system's operational stability can be revealed.

3. Experimental Research on System Stability

3.1. Experimental Setup

Considering the detailed layout of a given simple hydropower system, a small-scale experimental setup focusing on operating stability of the hydro-mechanical–electrical system was designed and established [30]. This equipment mainly consisted of an upstream reservoir, a pressure conduit, turbine generator, a tail tube, a governor, an exciter, a simplified electrical system, and so on, shown in Figure 1, and a short open channel was used to induce water flow from unit's draft tube to the tail water. The total length of the water diversion conduit was 18.0 m with a 22.5 cm diameter, and the other experimental equipment parameters are shown in Appendix A. An electromagnetic flowmeter was set close to the inlet of the unit's spiral case for the measurement of the unit's operating flow rate.

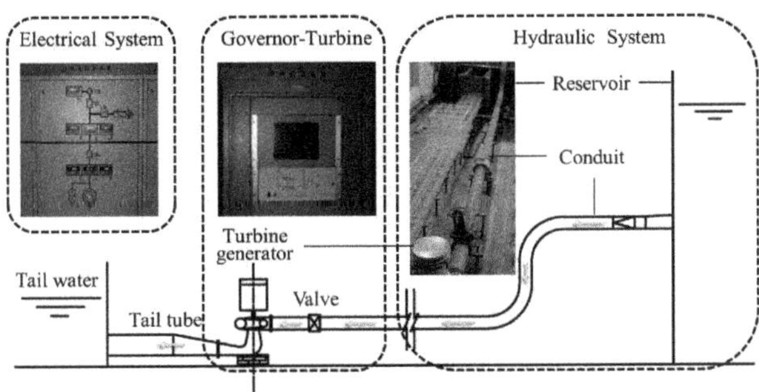

Figure 1. Longitudinal section of the simulated hydro-mechanical system.

The simulated hydropower system can be completely monitored under stable operation, normal starting and stopping, emergency load rejection, and small disturbances. The pressure sensors are preset at the corresponding sections to collect the real-time data of the simulated hydraulic system. With the equipped speed governor, exciter, and synchronization system, the simulated hydropower system can be synchronized and paralleled into the external power system or be put into isolated operation with different electric loads.

3.2. Electric Load Design

Considering to the research achievements on electric loads' models in recent years [25], by comparing with the static load and dynamic load, there is still a bottleneck on how to ascertain the integrated loads and their parameters in the stability analysis.

Therefore, in order to clearly reveal the influence of the electric loads with different characteristics on the system's stability, the static load and dynamic load were preferred instead of the integrated loads to realize the system's stability evaluation. As an additional and indispensable part for the simulated hydropower system, the physical models were designed and established for the static load (electric resistance) and the dynamic load, which is a water pumping system with a three-phase induction motor [30] shown in Figure 2.

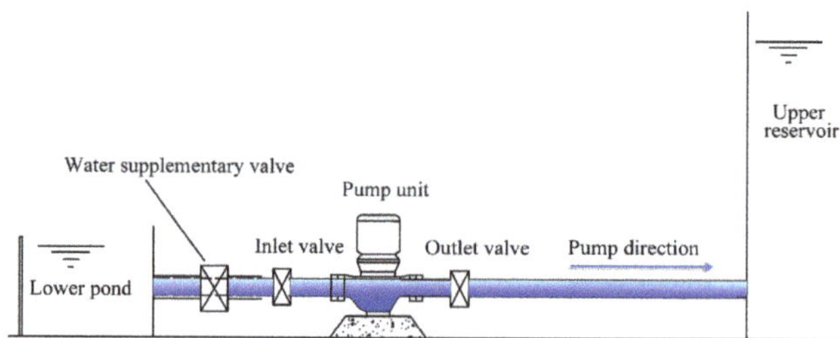

Figure 2. Longitudinal plan of the pumping system.

3.3. Computational Verification Based on the Simulated System

Based on the simulated hydropower system in Figure 1 running with rated head $H_r = 3.2$ m, rated flow rate $Q_r = 0.045$ m^3/s, and rated rotational speed $n_r = 600$ rpm and then serving for different loads in Figure 2 paralleled into the external power system, considering electrical loads with different characteristics under grid-connected operation or isolated operation, the effect of power system and different load characteristics on system's operational stability was experimentally studied. Then, for the simulated hydropower system, further numerical simulation was investigated on the system's operational stability, and verification analysis for the numerical simulation was also conducted by comparative analysis with the experimental results [29].

For the simulated hydropower system in Figures 1 and 2, under grid-connected operation or isolated operation with the static load or dynamic load, according to its relatively short pipe length, the rigid model was preferred to simulate the dynamic characteristics of pipe flow [31] instead of the elastic model, derived from Equation (3). Based on the derived state equations for the hydropower system with a state vector (17) for the static load and a state vector (18) for the dynamic load, respectively, with eigenvalue analysis and load disturbance computation, the effects of different load characteristics on the system's stability and regulation performance could be numerically analyzed, including a comparative analysis with experimental results. For the simulated hydropower system, in order to be basically consistent with the experimental boundaries, for the static load, only resistance with $a_p = 1$ and $b_p = 0$ was concerned for the numerical computation, so based on the eigenvalue computation and further correlation analysis, the computed oscillation mode strongly related with the low-frequency oscillation was $-0.0761 \pm j2.041$ for the static load (resistance), and it was $-0.2492 \pm j0.724$ for the dynamic load. By a comparative analysis on the attenuation factor (real part of low-frequency oscillation mode) for different loads, it was revealed that the unit serving for a water pumping system (dynamic load) was more stable.

Considering the unit under grid interconnection operation and isolated operation and then with -10% step load disturbance, time histories of the unit's relative speed variation considering the dynamic load for further comparative analysis with the experimental results are provided in Figure 3.

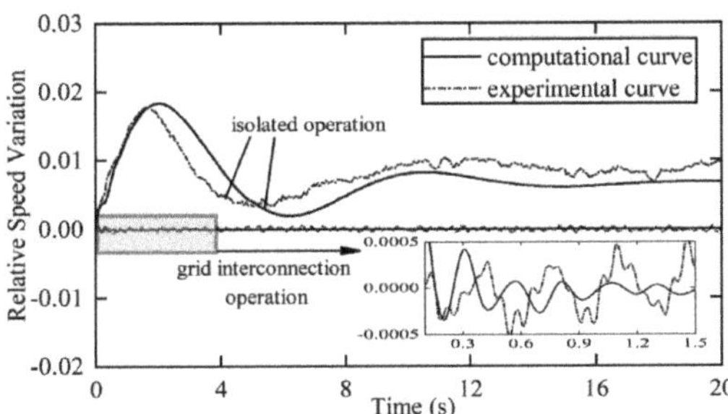

Figure 3. Time histories of the unit's relative speed variation considering dynamic load.

From the dynamic characteristics of the unit's rotational frequency in Figure 3 under -10% step load disturbance considering grid-connected operation, it was obviously confirmed that the power system was beneficial to the unit's stable operation. Because the simulated unit's capacity had a very small proportion of the external power system, the electric load with different characteristics had a negligible influence on the system's stability. As far as isolated operation was concerned, Figure 3 shows the similar oscillation trend and approximately equal peak values for numerical simulation and experimental research, and there was also some difference in the oscillation period and the attenuation trend, which was basically originated from the calibration error of the dynamic load's technical parameters, such as the time constant, besides the experimental error.

Similarly, under a -10% step load disturbance, considering the unit under isolated operation, time histories of the unit's relative speed variation considering the static load for further comparative analysis with the experimental results are shown in Figure 4.

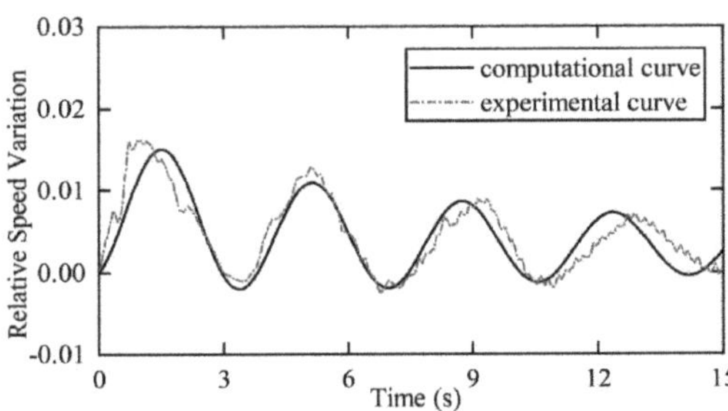

Figure 4. Time histories of unit's relative speed variation considering static load.

From Figures 3 and 4 and further unit regulation performance analysis, it can be drawn that, with setting parameters, under load disturbance, the unit serving for different loads under isolated operation was stable. For the unit serving for a water pumping system (dynamic load), the system's stability and regulation performance were better than those of resistance (static load), including an obviously shorter regulating time, less oscillation, and a larger attenuation degree, which was 84.8% greater than that of the static load, 24.8%, well in agreement with the above experimental results.

Furthermore, Figures 3 and 4 also clarify that the numerical computed curves were basically identical to the experimental results, which confirms that the established mathematical models can accurately reveal the system's dynamic characteristics and the effects of different loads on the hydropower system stability performance.

4. Numerical Analysis

4.1. Case Description for a Simple Hydropower System

Considering that the unit's speed was an important inter-variable for the governor turbine hydraulic system and the power system, for further numerical simulation analysis, it was preferable to investigate the effect of different electric loads on the dynamic characteristics of the unit's speed in detail.

Figure 5 shows a simple hydropower system with a pressurized pipe 250 m in length and 5.0 m in diameter, and the unit's rotational inertia is 11,000 t·m^2 with rated speed of 150.0 rpm. It was running with an operating head of 73.0 m and a flow rate of 116.0 m^3/s.

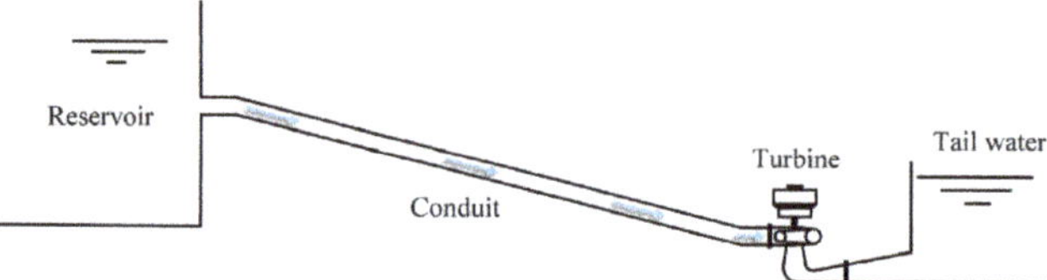

Figure 5. Sketch of a single-pipe and single-unit system.

4.2. Further Numerical Analysis with Different Loads

For the hydropower system shown in Figure 5, the first-order elastic model derived from Equation (3) was introduced to simulate the water flow in the diversion pipe according to its relatively short length. In order to reveal the effects of the dynamic characteristics of the generator, the exciter, and the electrical load on system's stability, first, based on the state vector, Equation (19), without the consideration of their effects, the stable region is shown in Figure 6a with the governor parameters, including the dashpot time constant T_d and the temporary speed drop b_t. In Figure 6a, a critical combination of the governor parameters (T_d, b_t) is also provided for further comparative analysis, and then its corresponding computational curve of the relative speed variation under a -10% step load disturbance is yielded in Figure 6b.

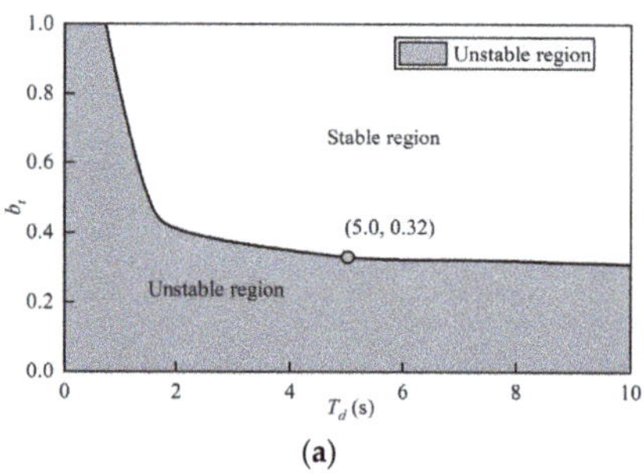

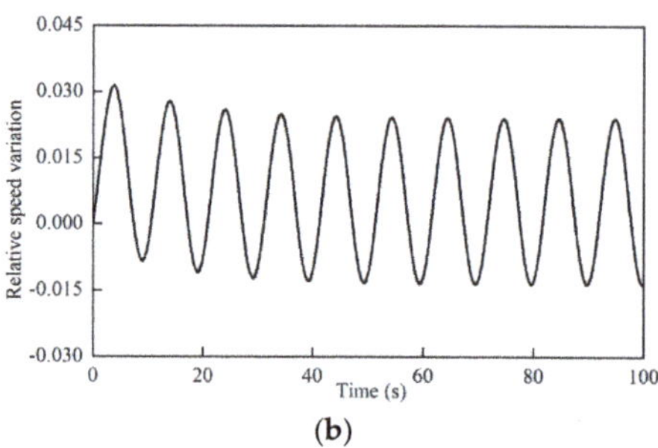

Figure 6. Stability analysis without the effect of electrical system: (**a**) Stable region. (**b**) Relative speed variation.

Figure 6 illustrates that if the effect of the dynamic characteristic of the generator, excitor, and electrical loads on the system's stability was not considered, the T_d-b_t plane was divided into a stable region and an unstable region with a critical curve; if the combination of the governor's parameters (T_d, b_t) was just located along the critical curve, the system's dynamic variables, mainly including relative speed variation, tended to equal-amplitude oscillation.

Furthermore, the effect analysis of the load characteristics on low-frequency oscillation was conducted based on the dynamic characteristics of the unit's speed considering the static load or the dynamic load. Table 1 provides various groups of a_p and b_p values to describe different computational cases considering the static load in addition to the dynamic load and the computed corresponding low-frequency oscillation modes. In Table 1, based on the sensitivity analysis of the governor's parameters, the effects of governor parameters on the low-frequency oscillation mode is also analyzed, including the obtained critical combination of the governor parameters (T_d, b_t) in Figure 6.

Table 1. Low-frequency oscillation mode of different load characteristics.

Load Model		a_p	b_p	$2\,a_p + b_p$	Low-Frequency Oscillation Mode	
					$T_d = 5.0$ s, $b_t = 0.50$	$T_d = 5.0$ s, $b_t = 0.32$
Static load	Constant resistance	1.0	0.0	2.0	$-0.4316 \pm j12.94$	$-0.4017 \pm j12.94$
	Constant current	0.0	1.0	1.0	$-0.4005 \pm j13.02$	$-0.3723 \pm j13.01$
	Constant output	0.0	0.0	0.0	$-0.3697 \pm j13.09$	$-0.3431 \pm j13.08$
		0.6	0.2	1.4	$-0.4129 \pm j12.99$	$-0.3840 \pm j12.98$
Dynamic load					$-0.5599 \pm j12.50$	$-0.5198 \pm j12.54$

It can be drawn from Table 1 that if the first-order elastic model of the pipe flow was introduced, the absolute value of the attenuation factor considering the dynamic load was obviously greater than that considering the static load. The absolute value of the attenuation factor increased, and the system's stability became better as the value of $2\,a_p + b_p$ for the static load was larger; considering the decreasing of b_t with the same T_d, the absolute value of the attenuation factor was slightly reduced, and its stability became a bit worse.

As the hydropower station was running and serving for the dynamic load or the static load (constant resistance), respectively, then by introducing a -10% step disturbance from the governor side, the time histories of the relative speed variations under isolated operation are shown in Figure 7 with $T_d = 5.0$ s and $b_t = 0.50$. Meanwhile, Figure 8 shows the time histories of the relative speed variation under a -10% step load disturbance with $T_d = 5.0$ s and $b_t = 0.32$ for further comparative analysis.

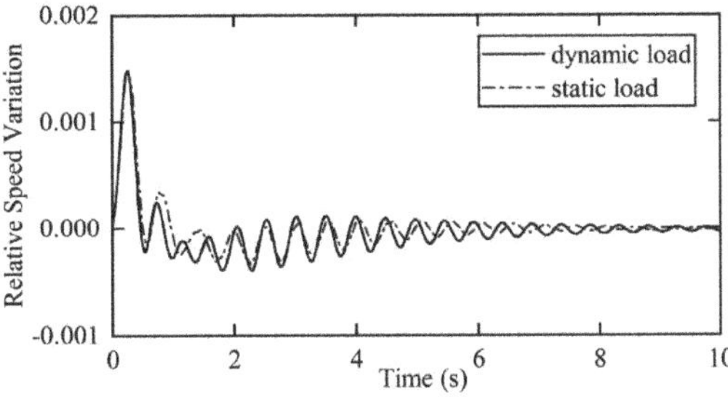

Figure 7. Computational curves of unit's relative speed variation under isolated operation.

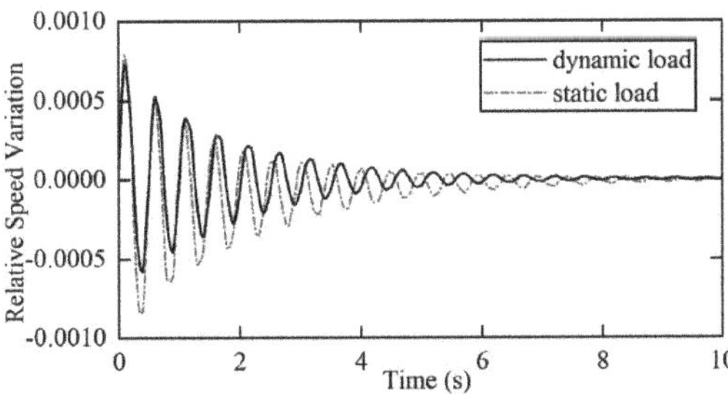

Figure 8. Computational curves of relative speed variation under -10% step load disturbance.

From Figures 7 and 8, the dynamic process of the hydraulic–mechanical system was significantly different considering the influence of the static load or the dynamic load. At the same time, considering the dynamic load, the system's operational stability was better with the faster speed fluctuation attenuation and relatively short regulation time as compared with the static load, which was consistent with the results in Table 1. From further comprehensive analysis with Figures 7 and 8, there were obviously different behaviors for different disturbances from the governor side or the load disturbance, including different oscillation processed and attenuation rates. By comparative analysis with Figures 7 and 8, with the same combination of the governor's parameters (T_d, b_t), that is, $T_d = 5.0$ s and $b_t = 0.32$, the system was stable with better regulation performance considering the effects of the dynamic characteristics of the generator, exciter, and electrical loads, while it was in a critical state with equal-amplitude oscillation without their effects; therefore, it was concluded that the dynamic characteristics of the generator, exciter, and electrical loads could improve the system's stability and regulation performance.

4.3. Effect of Pipe Flow Models on Low-Frequency Oscillation with Different Loads

In the above numerical analysis, the first-order elastic model for pipe flow was introduced to evaluate the effect of load characteristics on low-frequency oscillation and further the system's stability. Furthermore, for the simple unit system shown in Figure 5, considering the detailed hydraulic characteristics of the pipe flow illustrated by different-order elastic models, that is, the stiff model, the first-order model, the second-order model, and the third-order model, which can be derived from Equation (1) with $n = 0, 1, 2$, and 3, respectively, the eigenvalue computation and the analysis of further dynamic characteristics was carried out to reveal the effects of different models of pipe flow on low-frequency oscillation and the system's stability, together with the effect analysis of different load characteristics. The corresponding modes for low-frequency oscillation with different-order elastic models for pipe flow and different load characteristics are provided in Table 2, and considering the static load, under a −10% step load disturbance, the dynamic curves of the relative speed variation are shown in Figure 9.

Table 2. Low-frequency oscillation modes of different-order elastic models for pipe flow.

Flow Model	Static Load	Dynamic Load
Stiff model (zeroth-order)	$-0.5411 \pm j12.80$	$-0.5036 \pm j12.31$
First-order elastic model	$-0.4316 \pm j12.94$	$-0.5599 \pm j12.50$
Second-order elastic model	$-0.4278 \pm j12.95$	$-0.5695 \pm j12.50$
Third-order elastic model	$-0.4264 \pm j12.95$	$-0.5735 \pm j12.49$

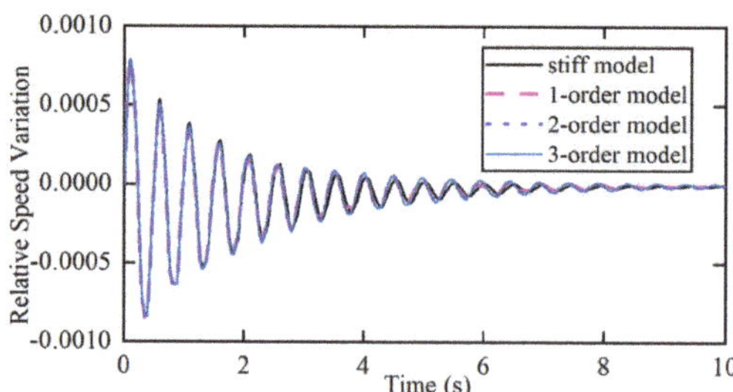

Figure 9. Dynamic curves of unit's relative speed variation of different elastic models for pipe flow with static load.

From Table 2 and Figure 9, it can be seen that for the static load, with the increasing of the model order for the pipe flow, the absolute value of the attenuation factor of the low-frequency oscillation mode gradually reduced and then tended to be a constant, which was illustrated by the different attenuation ratios for the relative speed variation. Compared with the static load (constant resistance), for dynamic loads, with the increasing of the model order for the pipe flow, the absolute value of the attenuation factor of the low-frequency oscillation mode gradually increased and then tended to be a constant. Therefore, for both the static load and the dynamic load, instead of a higher-order elastic model, an optimum-order elastic model, a three-order elastic model, of the pipe flow was preferred to exactly evaluate the effects of different load characteristics on the low-frequency oscillation mode and the system's regulation performance.

4.4. Sensitivity Analysis of Pipe Length on Low-Frequency Oscillation with Different Loads

Based on the effect analysis of the pipe flow models on low-frequency oscillation with different loads, it was also necessary to conduct further sensitivity analyses of the pipe length on low-frequency oscillation. For the simple unit system in Figure 5, considering the reasonable change of the pipe length from 150 to 400 m, further sensitivity analysis was implemented focusing on the effect of the pipe length on low-frequency oscillation and further the system's stability, along with different models of the pipe flow and different load characteristics. Based on detailed eigenvalue computation with different pipe lengths, the corresponding modes for low-frequency oscillation with different-order elastic models for the pipe flow and the static load (constant resistance) are provided in Table 3, and the changing trends for the real part of the low-frequency oscillation mode of different pipe lengths with the dynamic load are shown in Figure 10.

Table 3. Low-frequency oscillation mode of different pipe length with static load.

Flow Model	150 m Pipe	250 m Pipe	400 m Pipe
Stiff model (zeroth-order)	$-0.5347 \pm j12.80$	$-0.5411 \pm j12.80$	$-0.5447 \pm j12.80$
First-order elastic model	$-0.5537 \pm j12.79$	$-0.4316 \pm j12.94$	$-0.5258 \pm j12.80$
Second-order elastic model	$-0.5571 \pm j12.80$	$-0.4278 \pm j12.95$	$-0.5483 \pm j12.79$
Third-order elastic model	$-0.5585 \pm j12.80$	$-0.4264 \pm j12.95$	$-0.5539 \pm j12.79$

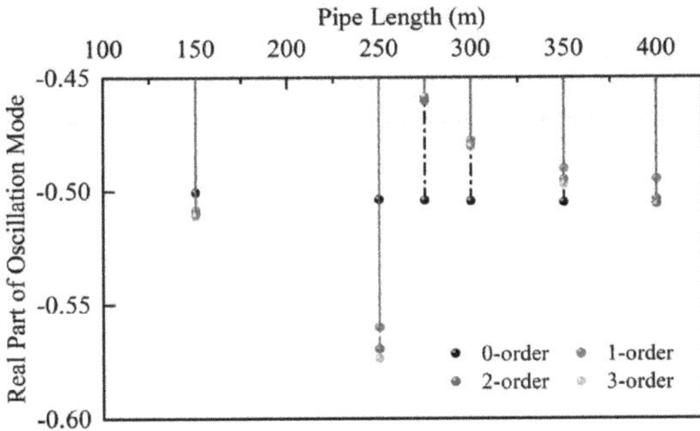

Figure 10. Changing trends for real part of low-frequency oscillation mode of different pipe lengths with dynamic load.

It can be analyzed from Table 3 and Figure 10 that for both the static load and the dynamic load, if the stiff model of the pipe flow was introduced, the absolute value of the real part of the low-frequency oscillation mode slightly increased with the increasing pipe length. With the increasing model order for the pipe flow, considering different pipe lengths, the absolute value of the real part of the low-frequency oscillation mode followed a different variation tendency, and then all tended to be a constant. Therefore, in order to completely evaluate the effects of different load characteristics on the low-frequency oscillation mode, an optimum-order elastic model of the pipe flow was preferred according to different pipe lengths, and in this case, the first- to second-order elastic model would exactly reveal the effects of the pipe flow characteristics on the system's regulation performance.

4.5. Discussion on Stability for Two-Unit System with Different Static Loads

In recent years, many hydropower stations have been successively constructed, paralleled into different power systems, and served for local power supply, which has strongly advanced the clean, low-carbon energy utilization and development in the world. Besides the simple unit hydropower systems, some typical and relatively complex hydropower systems play an important role in these hydropower systems, including the two-unit system sharing a common water diversion pipeline or tail tunnel with the necessary surge chambers. Therefore, Figure 11 shows a two-unit system sharing a downstream surge chamber and tail tunnel, with a water level in the reservoir of 150 m, a tail water level of 29.0 m, a sectional area of downstream surge tank of 233.4 m^2, a unit output of 122.6 MW, a unit rotational inertia of 8900 t·m^2, and a rated speed of 230.8 rpm. For the pipe parameters, the water diversion pipes were 466.3 m and 450.3 m in length with 5.7 m in diameter, the tail branches were 101.1 m and 111.1 m in length with 6.2 m in diameter, and the tail tunnel was 945.9 m in length and 8.8 m in diameter.

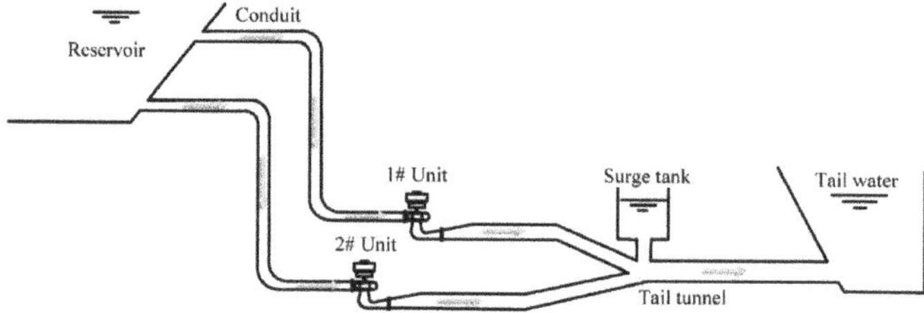

Figure 11. Sketch of a typical two-unit system.

Considering two units in Figure 11 serving for the static load and by introducing different-order models for the tail tunnel flow and stiff models for the branch pipes, a detailed eigenvalue computation and analysis of further dynamic characteristics was carried out to discuss the effects of different models of pipe flow on low-frequency oscillation and the system's stability, together with the effect analysis of different load characteristics. The corresponding modes for low-frequency oscillation and hydraulic oscillation with third-order elastic models for tail tunnel flow and different load characteristics are provided in Table 4. Based on constant resistance, considering a -10% step load disturbance with a 1# unit, the dynamic curves of the relative state variables are shown in Figure 12, including the unit's relative speed variation, the relative water level variation in the surge tank, and the oscillating flow rates in the tail tunnel.

Table 4. Oscillation mode analysis with third-order elastic models for tail tunnel flow.

Oscillation Modes	Static Load (Constant Output)	Static Load (Constant Resistance)
Low-frequency oscillation	$-0.9480 \pm j6.834$ $-0.8839 \pm j6.693$	$-0.9569 \pm j6.838$ $-1.1060 \pm j5.471$
Hydraulic oscillation	$-0.005636 \pm j3.322$ $-0.005635 \pm j6.643$ $-0.005627 \pm j9.964$	$-0.005365 \pm j3.322$ $-0.005632 \pm j6.643$ $-0.005637 \pm j9.964$
Water level oscillation	$-0.008115 \pm j0.5201$	$-0.008112 \pm j0.5202$

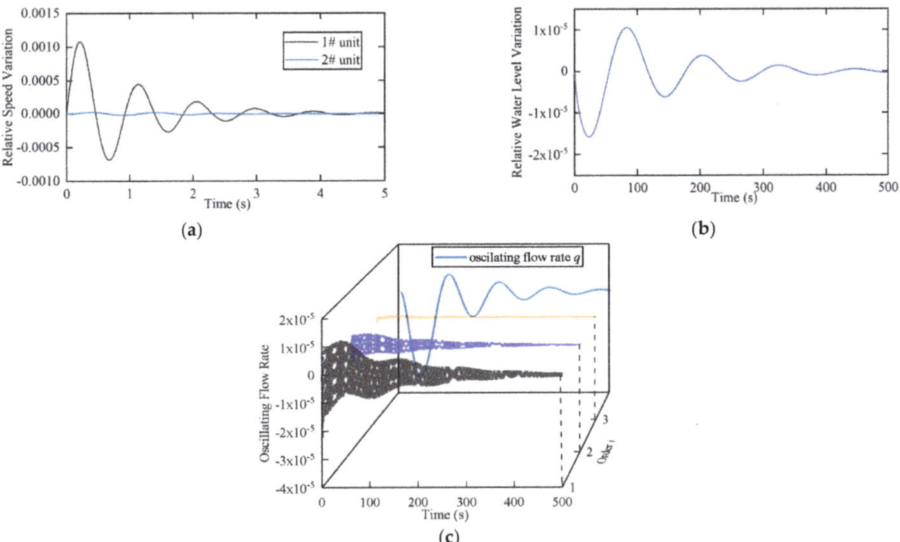

Figure 12. Dynamic curves of state variables of third-order elastic models for tail tunnel flow: (**a**) Relative speed variation. (**b**) Water level in downstream surge tank. (**c**) Different-order oscillating flow rate in tail tunnel.

From Table 4 and Figure 12, it can be concluded that for the two-unit system sharing a downstream surge tank and tail tunnel, if the third-order elastic model for the tail tunnel flow was introduced, the system's oscillation modes mainly included two low-frequency oscillations modes strongly related with the two units, three hydraulic oscillation modes illustrating the dynamic characteristics of three oscillating flow rates, and one water level oscillation mode directly representing the downstream surge tank's surge characteristics. The two low-frequency oscillation modes were not changed with the different model orders for the tail tunnel flow, mainly because of the reflection of the downstream surge tank, which was obviously different from that of the simple unit system. By comparative analysis, one of the low-frequency oscillation modes was basically consistent with one of the hydraulic oscillation modes, which may have resulted in coupling resonance in this hydropower system, which should be exactly evaluated with the premise of considering a reasonable model order for the tail tunnel flow.

Under a -10% load disturbance on a 1# unit, the detailed computation shows that the relative speed variation of the 1# unit decayed accompanied by obvious oscillation, while that of the 2# unit had less oscillation. The water level in the downstream surge tank also tended to a steady state gradually. Especially in Figure 12c, three oscillating flow rates q_i (i = 1, 2, and 3) showed an attenuated oscillation close to the water level oscillation period in the surge tank, combined with a high-frequency oscillation with an oscillation period decided by the ith-order hydraulic oscillation.

Further, considering static loads with different characteristics, from constant output to constant resistance, the absolute value of the real part of the low-frequency oscillation mode increased, and the relative speed variation decayed faster, while there was less effect on the hydraulic oscillation mode and the water level oscillation.

Therefore, similar to the single-unit system, for the typical two-unit system, the different characteristics of the electric loads also had obvious effects on the system's operational stability, especially low-frequency oscillation, while the higher-order elastic model of the tail tunnel flow had no effect on low-frequency oscillation but revealed the potential coupling resonance between the hydraulic system and the power system.

5. Conclusions

After introducing the elastic model with a reasonable order for the pipe flow and the linear models of the hydro-mechanical–electrical system of the hydropower stations, the analytical model considering the electric loads with different characteristics was comprehensively derived with state equations. Further, based on the numerical simulation model confirmed by an experimental setup of a single-unit hydropower system, the effects of load characteristics and pipe flow models on the operational stability were investigated in detail, and the following can be concluded.

For a given single-unit hydropower system running with different electrical loads, if a first-order elastic model of the pipe flow is introduced, the system's stability and the regulation performance of the unit running with the dynamic load under isolated operation was superior to that of the unit running with different static loads. Based on the stable region and sensitivity analysis of the governor's parameters, it was confirmed that the dynamic characteristics of the generator, excitor, and electrical loads could improve the system's stability and regulation performance.

Based on the influence analysis of different load characteristics on the single-unit hydropower system, it was pointed out that a relatively higher-order elastic model could exactly represent the dynamic characteristics of the pipe flow, while an optimum-order elastic model of the pipe flow according to different pipe lengths was preferred to evaluate the effects of different load characteristics on the low-frequency oscillation mode and the system's regulation performance.

For a typical two-unit system, the load characteristics also had obvious effects on low-frequency oscillation, the low-frequency oscillation modes were not changed with the different model orders for the tail tunnel flow mainly because of the reflection of the downstream surge tank compared with the single-unit system, and considering the preferred-order model of the pipe flow, the possible coupling resonance between the hydraulic system and the power system could be exactly evaluated.

Author Contributions: Conceptualization, J.Z. and C.L.; methodology, J.Z.; validation, J.Z. and C.L.; writing—original draft preparation, J.Z.; writing—review and editing, J.Z., C.L. and Y.M. All authors have read and agreed to the published version of the manuscript.

Funding: This research was funded by the Postgraduate Research & Practice Innovation Program of Jiangsu Province, grant number No. KYCX22_0655.

Data Availability Statement: Relevant data supporting this study can be found in the article chart and Appendix A.

Acknowledgments: The authors gratefully acknowledge the comments of all the reviewers, which led to significant improvement of the paper.

Conflicts of Interest: The authors declare no conflict of interest.

Appendix A Data of Experimental Setup

Table A1. Main parameters of the generator and exciter in Figure 1.

Rated Speed n_r/(r/min)	Rated Output P_r/KW	Rated Voltage U_r/V	Rated Current i_r/A	Rated Frequency f_r/Hz	Power Factor $\cos\varphi$	Rotary Inertia GD^2/(t·m^2)
600	1.0	400	1.8	50	0.8	0.0056
Exciting Voltage u_f/V	**Reactance** x_d (**d-Axis**)	**Transient Reactance** x_d'	**Sub-Transient Reactance** x_d''	**Reactance** x_q (**q-Axis**)	**Transient Reactance** x_q'	**Sub-Transient Reactance** x_q''
21	1.0222	0.2505	0.1336	0.6074	0.6074	0.1306

Transient Open-Circuit Time Constant T_{d0}'/s (*d*-Axis)		Sub-Transient Open-Circuit Time Constant T_{d0}''/s (*d*-Axis)		Sub-Transient Open-Circuit Time Constant T_{q0}''/s (*q*-Axis)	
1.5~9.0		0.01~0.05		0.01~0.09	

Table A2. Basic parameters of the turbine in Figure 1.

Rated Head H_r/m	Rated Flow Q_r/(m^3/ s)	Rated Speed n_r/(r/min)	Diameter D_1/cm	Designed Output P_{sr}/KW	Runaway Speed n_f/(r/min)	Design Efficiency η/%
3.2	0.045	600	20 cm	1.15	1120	78.2

References

1. Zhang, D.; Wang, J.; Lin, Y.; Si, Y.; Huang, C.; Yang, J.; Huang, B.; Li, W. Present situation and future prospect of renewable energy in China. *Renew. Sustain. Energy Rev.* **2017**, *76*, 865–871. [CrossRef]
2. Chang, X.; Liu, X.; Zhou, W. Hydropower in China at present and its further development. *Energy* **2010**, *35*, 4400–4406. [CrossRef]
3. Zhang, L.; Wu, Q.; Ma, Z.; Wang, X. Transient vibration analysis of unit-plant structure for hydropower station in sudden load increasing process. *Mech. Syst. Sig. Process.* **2019**, *120*, 486–504. [CrossRef]
4. Hamann, A.; Hug, G.; Rosinski, S. Real-Time Optimization of the Mid-Columbia Hydropower System. *IEEE Trans. Power Syst.* **2017**, *32*, 157–165. [CrossRef]
5. Xu, B.; Zhang, J.; Egusquiza, M.; Chen, D.; Li, F.; Behrens, P.; Egusquiza, E. A review of dynamic models and stability analysis for a hydro-turbine governing system. *Renew. Sustain. Energy Rev.* **2021**, *144*, 110880. [CrossRef]
6. Xu, J.; Ni, T.; Zheng, B. Hydropower development trends from a technological paradigm perspective. *Energy Convers. Manag.* **2015**, *90*, 195–206. [CrossRef]
7. Yu, X.; Zhang, J.; Fan, C.; Chen, S. Stability analysis of governor-turbine-hydraulic system by state space method and graph theory. *Energy* **2016**, *114*, 613–622. [CrossRef]
8. Yu, X.D.; Zhou, Q.; Zhang, L.; Zhang, J. Hydraulic Disturbance in Multiturbine Hydraulically Coupled Systems of Hydropower Plants Caused by Load Variation. *J. Hydraul. Eng.* **2019**, *145*, 04018078. [CrossRef]
9. Cassano, S.; Sossan, F. Model predictive control for a medium-head hydropower plant hybridized with battery energy storage to reduce penstock fatigue. *Electr. Power Syst. Res.* **2022**, *213*, 108545. [CrossRef]
10. Reigstad, T.I.; Uhlen, K. Optimized Control of Variable Speed Hydropower for Provision of Fast Frequency Reserves. *Electr. Power Syst. Res.* **2020**, *189*, 106668. [CrossRef]
11. Makarov, Y.V.; Maslennikov, V.A.; Hill, D.J. Revealing loads having the biggest influence on power system small disturbance stability. *IEEE Trans. Power Syst.* **1996**, *11*, 2018–2023. [CrossRef]
12. Kundur, P. *Power System Stability and Control*; McGraw-Hill Professional: New York, NY, USA, 1994; pp. 377–417.
13. Garmroodi, M.; Hill, D.J.; Verbic, G.; Ma, J. Impact of Load Dynamics on Electromechanical Oscillations of Power Systems. *IEEE Trans. Power Syst.* **2018**, *33*, 6611–6620. [CrossRef]
14. Jiang, C.X.; Zhou, J.H.; Shi, P.; Huang, W.; Gan, D.Q. Ultra-low frequency oscillation analysis and robust fixed order control design. *Int. J. Electr. Power Energy Syst.* **2019**, *104*, 269–278. [CrossRef]
15. Mo, W.; Chen, Y.; Chen, H.; Liu, Y.; Zhang, Y.; Hou, J.; Gao, Q.; Li, C. Analysis and Measures of Ultra-Low-Frequency Oscillations in a Large-Scale Hydropower Transmission System. *IEEE J. Emerg. Sel. Top. Power Electron.* **2017**, *6*, 1077–1085. [CrossRef]
16. Villegas Pico, H.; Mccalley, J.D.; Angel, A.; Leon, R.; Castrillon, N.J. Analysis of Very Low Frequency Oscillations in Hydro-Dominant Power Systems Using Multi-Unit Modeling. *IEEE Trans. Power Syst.* **2012**, *27*, 1906–1915. [CrossRef]
17. Wang, G.; Zheng, X.; Guo, X.; Liang, X. Mechanism analysis and suppression method of ultra-low-frequency oscillations caused by hydropower units. *Int. J. Electr. Power Energy Syst.* **2018**, *103*, 102–114. [CrossRef]
18. Li, Y.; Chiang, H.D.; Choi, B.K.; Chen, Y.T.; Huang, D.H.; Lauby, M.G. Load models for modeling dynamic behaviors of reactive loads: Evaluation and comparison. *Int. J. Electr. Power Energy Syst.* **2008**, *30*, 497–503. [CrossRef]
19. Maslennikov, V.A.; Milanovic, J.V.; Ustinov, S.M. Robust ranking of loads by using sensitivity factors and limited number of points from a hyperspace of uncertain parameters. *IEEE Trans. Power Syst.* **2002**, *22*, 565–570. [CrossRef]
20. Price, A.W.W.; Taylor, C.W.; Rogers, G.J. Bibliography on load models for power flow and dynamic performance simulation. *IEEE Trans. Power Syst.* **1995**, *10*, 523–538. [CrossRef]
21. Yang, Y.; Zhao, S.Q. Load characteristics' effect on low-frequency oscillation damping of power systems. *Electr. Power Autom. Equip.* **2003**, *23*, 13–16. [CrossRef]
22. Kao, W.S. The effect of load models on unstable low-frequency oscillation damping in taipower system experience w/wo power system stabilizers. *IEEE Trans. Power Syst.* **2001**, *16*, 463–472. [CrossRef]
23. Nomikos, B.M.; Vournas, C.D. Evaluation of motor effects on the electromechanical oscillations of multimachine systems. In Proceedings of the 2003 IEEE Bologna Power Tech Conference, Bologna, Italy, 23–26 June 2003. [CrossRef]
24. Hiskens, I.A.; Milanovic, J.V. Load modelling in studies of power system damping. *IEEE Trans. Power Syst.* **1995**, *10*, 1781–1788. [CrossRef]
25. Milanovic, J.V.; Hiskens, I.A. Effects of load dynamics on power system damping. *IEEE Trans. Power Syst.* **1995**, *10*, 1022–1028. [CrossRef]
26. Yang, J.; Li, J.; Wang, D.; Chen, J.; Wu, R. Study on the physical simulation of transient process in conduits of hydropower stations. *J. Hydroelectr. Eng.* **2004**, *23*, 57–63.

27. Zeng, W.; Yang, J.; Hu, J. Pumped storage system model and experimental investigations on S-induced issues during transients. *Mech. Syst. Sig. Process.* **2017**, *90*, 350–364. [CrossRef]
28. Yang, W.; Yang, J.; Zeng, W.; Tang, R.; Hou, L.; Ma, A.; Zhao, Z.; Peng, Y. Experimental investigation of theoretical stability regions for ultra-low frequency oscillations of hydropower generating systems. *Energy* **2019**, *186*, 115816. [CrossRef]
29. Zhou, J.; Chen, G. Effect Analysis of Load Characteristic on Operation Stability of Hydropower Stations. In Proceedings of the 2012 International Conference on Future Electrical Power and Energy Systems, Sanya, China, 21–22 February 2012. [CrossRef]
30. Zhou, J.; Hu, M.; Cai, F.; Hu, R. Experimental Research on Stability of Hydro-Mechanical-Electrical System in Hydropower Station. In Proceedings of the 2009 Asia-Pacific Power and Energy Engineering Conference, New York, NY, USA, 28–30 March 2009. [CrossRef]
31. Zhou, J.; Cai, F.; Wang, Y. New Elastic Model of Pipe Flow for Stability Analysis of the Governor-Turbine-Hydraulic System. *J. Hydraul. Eng.* **2011**, *137*, 1238–1247. [CrossRef]
32. Zhou, J.X.; Mao, Y.T.; Shen, A.L.; Zhang, J. Modeling and stability investigation on the governor-turbine-hydraulic system with a ceiling-sloping tail tunnel. *Renew. Energy* **2023**, *204*, 812–822. [CrossRef]
33. Jaeger, E.D.; Janssens, N. Hydro turbine model for system dynamic studies. *IEEE Trans. Power Syst.* **1994**, *9*, 1709–1715. [CrossRef]
34. Hannett, L.N.; Feltes, J.W.; Fardanesh, B. Field tests to validate hydro turbine-governor model structure and parameters. *IEEE Trans. Power Syst.* **1994**, *9*, 1744–1751. [CrossRef]

Article

Optimal Scheduling of Cascade Reservoirs Based on an Integrated Multistrategy Particle Swarm Algorithm

Yixuan Liu [1,2,3], **Li Mo** [2,3,*], **Yuqi Yang** [1] and **Yitao Tao** [2,3]

1 Hubei Key Laboratory of Intelligent Yangtze and Hydroelectric Science, China Yangtze Power Co., Ltd., Yichang 443000, China
2 School of Civil and Hydraulic Engineering, Huazhong University of Science and Technology, Wuhan 430074, China
3 Hubei Key Laboratory of Digital River Basin Science and Technology, Huazhong University of Science and Technology, Wuhan 430074, China
* Correspondence: moli@hust.edu.cn

Abstract: The optimal scheduling of cascade reservoirs is an important water resource management and regulation method. In the actual operation process, its nonlinear, high-dimensional, and coupled characteristics become increasingly apparent under the influence of multiple constraints. In this study, an integrated multistrategy particle swarm optimization (IMPSO) algorithm is proposed to realize the optimal operation of mid- and long-term power generation in cascade reservoirs according to the solution problem in the scheduling process of cascade reservoirs. In IMPSO, a variety of effective improvement strategies are used, which are combined with the standard PSO algorithm in different steps, among which beta distribution initialization improves population diversity, parameter adaptive adjustment accelerates convergence speed, and the Lévy flight mechanism and adaptive variable spiral search strategy balance the global and local search capabilities of the algorithm. To handle complex constraints effectively, an explicit–implicit coupled constraint handling technique based on constraint normalization is designed to guide the update process into the feasible domain of the search space. The feasibility of the proposed method is verified in the mid- and long-term power generation optimization scheduling of the lower reaches of the Jinsha River–Three Gorges cascade hydropower reservoirs. The results show that the proposed method outperforms the other methods in terms of search accuracy and has the potential to improve hydropower resource utilization and power generation efficiency significantly.

Keywords: optimal scheduling; cascade reservoirs; particle swarm algorithm; constraint handling technique

Citation: Liu, Y.; Mo, L.; Yang, Y.; Tao, Y. Optimal Scheduling of Cascade Reservoirs Based on an Integrated Multistrategy Particle Swarm Algorithm. *Water* **2023**, *15*, 2593. https://doi.org/10.3390/w15142593

Academic Editor: Gwo-Fong Lin

Received: 12 June 2023
Revised: 9 July 2023
Accepted: 13 July 2023
Published: 16 July 2023

1. Introduction

Hydropower energy, which is a source of renewable energy, exhibits the qualities of being environmentally friendly, cost-effective, flexible, and reliable. These properties effectively contribute towards reducing carbon emissions while simultaneously promoting the harmonious development of economic and environmental benefits [1,2]. With the reorganization of the global energy system and the restructuring of energy sources, the advantages of hydraulic power have become increasingly apparent. Therefore, to facilitate the efficient allocation of resources, rational utilization and the development of hydropower energy are crucial. In this context, the optimal operation of cascade reservoirs has emerged as a major research area in recent years [3], with research efforts mainly focused on deriving reservoir scheduling regulations [4–6], identifying targets for reservoir scheduling models [7–10], and revising and utilizing optimization techniques for reservoir scheduling [11–13].

The connections of hydraulic and electrical systems among cascade reservoirs are closely related, and the topological relationships are extremely complicated. The optimal operation of cascade reservoirs is a type of nonlinear, multidimensional, multistage, multivariable, and multiconstrained complex optimization decision problem. To solve these problems effectively, scientists have conducted extensive research using traditional mathematical methods such as linear programming [14], nonlinear programming [15,16], dynamic programming [17,18], and other methods to solve the overall nonlinear problematic release of reservoirs. They also designed a set of metaheuristic algorithms [19–21] that combine physical mechanisms, natural inspiration, social laws, and evolutionary technology in order to improve the poor accuracy and vulnerability of traditional mathematical methods relative to the dimensionality problem and obtained a better solution. In addition, the optimal operation of cascade reservoirs also requires complex constraint treatment [22], and the relevant constraint processing methods are mainly the penalty function method [23], the ranking method [24], the variable repair method [25], and the boundary limitation method [26], which effectively improves the solution's quality in practical engineering problems. However, using these methods results in greater computation or reduced population diversity, and the global optimal solution of the problem cannot be reached in a given period of time, falling to balance algorithmic robustness [27] with effectiveness in complex scenarios.

The particle swarm optimization (PSO) algorithm is a type of metaheuristic algorithm. Because of its simple structure, easy implementation, and fast convergence speed, it is widely used in practical engineering. However, due to its evolution mechanism and implementation style, this algorithm is prone to early maturation and delayed convergence in later stages, and easily falls into local optimal solutions. Numerous studies have implemented improvements relative to the PSO, such as the introduction of linear or nonlinear inertia weights and learning factors [28,29], the integration of mathematical probability distribution functions [30], and the integration with other optimization approaches [31]. These interventions have obviously increased the precision of the algorithm and enhanced its global exploration capabilities. Although the PSO algorithm has achieved some success in many aspects, the overall optimization effect of the algorithm is poor and lacks stability because of the stochasticity of the initial process and the flaw in the exploration ability. Therefore, it is necessary to study the defects of the PSO algorithm in order to improve its performance and better apply it to practical engineering cases. Based on this, a series of enhancements to the PSO method have been created to overcome the algorithm's inadequacies in addressing high-dimensional difficult optimization issues. In particular, a beta distribution strategy is adopted to improve the monotonicity of the stochastic initialization of the PSO algorithm, a parametric nonlinear adaptive strategy is used to cope with the attraction of locally optimal neighbors in the iterative process, and a combination of the Lévy flight mechanism and a variable spiral search strategy is used to improve the algorithm exploration and exploitation performance. Meanwhile, an implicit constraint processing method is introduced to deal with the reservoir operation problem. It is combined with the explicit constraint processing technique of the static penalty function by restricting the feasible space of variables to solve the multidimensional constraint problem of cascade reservoir operation. Simulation results in practice show that the proposed method is robust and effective.

In IMPSO, the populations are first selected by beta distribution. The evolutionary process is then split into two updates. The first update employs a parameter adaptive change approach to update population velocity and location; the second employs a random number and either a Lévy flying mechanism or a variable helix search strategy to optimize population position based on the size of the random number. Finally, the answers to the two evolutions are compared, and the better option is chosen to maintain the population's ability to explore and diversity.

The proposed scheme is experimentally evaluated by combining IMPSO with the explicit–implicit coupled constraint treatment technique and applying it to the problem of graded reservoir generation and scheduling under different runoff conditions based on various experiments. First, statistical results such as box plots, Friedman test, convergence analysis, and population diversity evaluation are used to verify the validity of the proposed scheme; second, the applicability of the proposed scheme is evaluated by analyzing the results of water abandonment, reservoir level fluctuation and output, and power generation of the step system generated during the practical application. The acquired results are compared with many other regularly used algorithms to demonstrate the suggested scheme's superior performance in the field of reservoir scheduling.

In summary, the main contributions of this research are outlined as follows:

1. An IMPSO method is proposed that uses initialization, adaptive parameters, Lévy flight, and variable spiral search strategies to efficiently maximize performance.
2. In conjunction with the proposed IMPSO algorithm, an explicit–implicit coupled constraint handling technique is introduced to solve the generation scheduling model of cascaded hydropower systems.
3. The proposed scheme provides an effective tool for solving the complex problem of reservoir operation. Compared with several other schemes, it has the advantages of strong searching ability and robustness, which also can better utilize the collaborative generation effect of cascade hydropower units.

The remainder of this paper is organized as follows: Section 2 presents the optimization model for the operation of cascade reservoirs. Section 3 describes the IMPSO method. Section 4 briefly introduces the techniques used to process constraints. Section 5 testifies the proposed scheme's performances in a real cascade hydropower system. Finally, Section 6 summarizes the study and outlines future research directions.

2. Operation Model of Cascade Reservoirs

2.1. Objective Function

Taking the maximum annual power generation of cascade reservoirs as the objective function, the objective function can be expressed mathematically as follows:

$$E = \max \sum_{i=1}^{I} \sum_{t=1}^{T} N_{i,t} \Delta t \tag{1}$$

$$N_{i,t} = A_i Q_{i,t} H_{i,t} \tag{2}$$

where E is all the power generation of the cascade reservoirs. The cascade reservoirs are numbered from upstream to downstream in a sequence like $1, 2, 3 \cdots$, and I is the number of reservoirs. t is the subperiod of the scheduling period, and T is the total number of stages over all scheduling periods. $N_{i,t}$ is the output at the tth stage of the ith hydropower reservoir. A_i is the comprehensive output coefficient of the ith hydropower station. $Q_{i,t}$ and $H_{i,t}$ are the generation flow and the corresponding water head at the tth stage of the ith hydropower reservoir, respectively. In the reservoir scheduling field, power generation scheduling takes the maximum power generation as the objective function, and the final power generation result is obtained by calculating the output of each subdispatch period of the reservoir and multiplying it with the length of the subdispatch period. Therefore, Δt is the length of each subperiod of the operation, and it is a constant value, taking different values according to different scheduling periods. In this paper, its duration is one decade.

2.2. Constraints

(1) Water balance constraints

$$V_{i,t+1} = V_{i,t} + (I_{i,t} - R_{i,t}) \cdot \Delta t \tag{3}$$

$$R_{i,t} = Q_{i,t} + S_{i,t} \tag{4}$$

where $V_{i,t}$, $I_{i,t}$, $R_{i,t}$, $Q_{i,t}$, and $S_{i,t}$ denote the storage capacity, inflow, outflow, generation flow, and abandoned water flow of the ith hydropower reservoir in the tth stage, respectively.

(2) Storage capacity constraint

$$V_{i,t}^{\min} \leq V_{i,t} \leq V_{i,t}^{\max} \tag{5}$$

where $V_{i,t}^{\min}$ and $V_{i,t}^{\max}$ denote the maximum and minimum storage capacity of the ith hydropower reservoir in the tth stage, respectively.

(3) Hydraulic constraints

$$I_{i,t} = R_{i-1,t-\tau} + q_{i,t} \tag{6}$$

where $q_{i,t}$ denotes the interval inflow of the ith hydropower reservoir in the tth stage. τ denotes the lagging periods of water flow.

(4) Outflow constraints

$$R_{i,t}^{\min} \leq R_{i,t} \leq R_{i,t}^{\max} \tag{7}$$

where $R_{i,t}^{\max}$ and $R_{i,t}^{\min}$ denote the maximum and minimum outflow of the ith hydropower reservoir in the tth stage.

(5) Generation flow constraints

$$Q_{i,t}^{\min} \leq Q_{i,t} \leq Q_{i,t}^{\max} \tag{8}$$

where $Q_{i,t}^{\max}$ and $Q_{i,t}^{\min}$ denote the upper and lower limits of the generation flow of the ith hydropower reservoir in the tth stage, respectively.

(6) Water level constraints

$$Z_{i,t}^{\min} \leq Z_{i,t} \leq Z_{i,t}^{\max} \tag{9}$$

$$\Delta Z_i^{down} \leq Z_{i,t+1} - Z_{i,t} \leq \Delta Z_i^{up} \tag{10}$$

$$Z_{i,t+1} = f^{ZV}(I_{i,t} - R_{i,t}, Z_{i,t}) \tag{11}$$

where $Z_{i,t}$ denotes the forebay water level of the ith hydropower reservoir in the tth stage. $Z_{i,t}^{\max}$ and $Z_{i,t}^{\min}$ are the upper and lower limits of forebay water levels of the ith hydropower reservoir in the tth stage, respectively. ΔZ_i^{up} and ΔZ_i^{down} are the upper and lower limits of the forebay water level variation of the ith hydropower reservoir in the tth stage, respectively. $f^{ZV}(\cdot)$ denotes the characteristic curve of the reservoir forebay water level and storage.

(7) Water level head constraints

$$H_{i,t} = \frac{Z_{i,t} + Z_{i,t+1}}{2} - Z_{i,t}^{down} \tag{12}$$

$$Z_{i,t}^{down} = f^{ZR}(R_{i,t}) \tag{13}$$

where $H_{i,t}$ and $Z_{i,t}^{down}$ denote the water level head and downstream water level of the ith hydropower reservoir in the tth stage. $f^{ZR}(\cdot)$ denotes the function of downstream water level and outflow.

(8) Output constraints

$$N_{i,t}^{\min} \leq N_{i,t} \leq N_{i,t}^{\max} \tag{14}$$

where $N_{i,t}^{min}$ and $N_{i,t}^{max}$ denote the upper and lower limits of the output of the *i*th hydropower reservoir in the *t*th stage, respectively.

(9)   Initial and final water level control constraints

$$Z_{i,0} = Z_{i,start} \tag{15}$$

$$Z_{i,T} = Z_{i,end} \tag{16}$$

where $Z_{i,start}$ and $Z_{i,end}$ denote the water levels at the beginning and end of the operation period required by the corresponding regulation of the *i*th hydropower reservoir, respectively.

3. Integrated Multistrategy Particle Swarm Optimization Algorithm (IMPSO)

3.1. Introduction of the Particle Swarm Optimization Algorithm (PSO)

The PSO algorithm [32] is a metaheuristic algorithm based on population evolution that simulates the natural foraging behavior of birds to try to achieve global optimization [33] in the search zone. Therefore, by analogy with the foraging behavior of a flock of birds, in the PSO algorithm, the particles usually represent the birds, and the solution in space, i.e., the location of the particles, is the location where each bird is located; moreover, the globally optimal solution is the location with the most amount of food. The PSO algorithm consists of three phases: random population initialization, particle velocity update, and particle position update. Figure 1 shows the basic structure of the PSO algorithm.

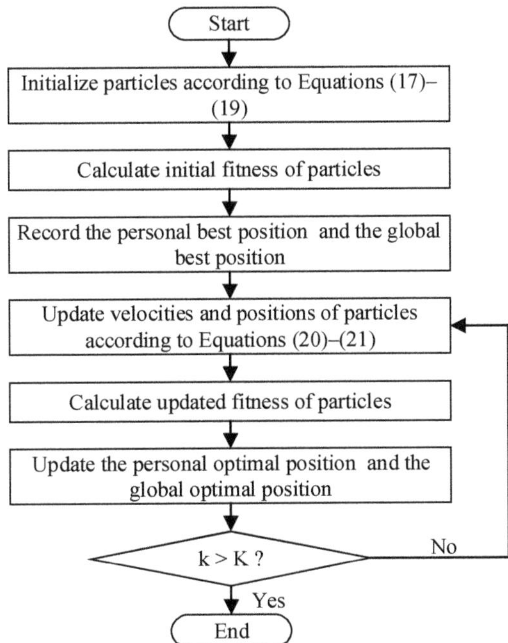

Figure 1. Flowchart of the PSO algorithm.

(1)   Random population initialization

First, the initial population is generated at random by specifying the number of particles, the range of velocities, and positions and restricting the search space to the upper and lower bounds of velocity and position constraints. The initial particle velocity and position of the PSO algorithm are as follows:

$$v_i^0 = v_{i,\min} + rand(v_{i,\max} - v_{i,\min}) \tag{17}$$

$$x_i^0 = x_{i,\min} + rand(x_{i,\max} - x_{i,\min}) \tag{18}$$

The initial population of the PSO algorithm can be represented as follows:

$$X^0 = \left[x_1^0, x_2^0, \ldots, x_N^0 \right] \tag{19}$$

where v_i^0 and x_i^0 denote the initial velocity and position of the particle, respectively. $v_{i,\min}$ and $v_{i,\max}$ denote the lower and upper limits of the velocity of the particle, respectively. $x_{i,\min}$ and $x_{i,\max}$ are the bottom and upper boundaries of the position, respectively. *rand* is the random number of the interval of $[0, 1]$, and N represents the number of particles in the population.

(2)   Particle velocity update

After the population is initialized, the global and personal optimal positions are recorded, and the velocities are updated. The particle velocity update process of the PSO algorithm can be formally stated as follows:

$$v_i^{k+1} = \omega v_i^k + c_1 r_1 (x_i^p - x_i^k) + c_2 r_2 (x^g - x_i^k) \tag{20}$$

where v_i^k denotes the velocity of the kth iteration of the ith particle. ω stands for the inertia weight for the population iteration, and its general values fall between 0.4 and 0.9. c_1 and c_2 represent the self-learning factor and social learning factor of particles, respectively; they commonly take a value of 2. r_1 and r_2 are two randomly selected numbers within the range of 0 and 1. x_i^k denotes the position in the kth iteration of the ith particle. x_i^p and x^g represent the personal optimal position of the ith particle and the global optimal position, respectively.

(3)   Particle location update

The particle position is updated according to the updated particle velocity. The formula for updating the PSO particle position is as follows:

$$x_i^{k+1} = v_i^{k+1} + x_i^k \tag{21}$$

where x_i^{k+1} represents the position in $k + 1$th iteration of the ith particle.

3.2. Integrated Multistrategy Particle Swarm Optimization Algorithm (IMPSO)

The PSO algorithm moves fast in the early phase but moves slowly in the late phase, which facilitates convergence to the local optimal solution in the early phase and cannot precisely converge to the global optimal solution. To solve this problem, an integrated multistrategy particle swarm optimization algorithm is developed that integrates the beta distribution, nonlinear adaptive parameter fitting strategy, Lévy flight mechanism, and adaptive spiral search strategy in the optimization process of PSO.

3.2.1. Population Initialization Based on the Beta Distribution

The initial population of PSO is randomly generated as Equation (18) in the search space, and the initial population generated by this method does not have small differences, so it cannot effectively cover the entire search space, the diversity of the population optimization is reduced, and the algorithm easily falls into the local optimal solution. The beta distribution is a set of continuous probability distributions defined on intervals, and the distribution [34] function is described as follows:

$$P(x; \alpha, \beta) = \frac{\Gamma(\alpha + \beta)}{\Gamma(\alpha) + \Gamma(\beta)} x^{\alpha-1} (1 - x)^{\beta-1} \tag{22}$$

where $\Gamma(\cdot)$ is the standard gamma function. The two important parameters—α, $\beta(\alpha > 0, \beta > 0)$—of the beta distribution are parameters that control the shape of the distribution. In general, the beta distribution function's representation can also be described as $X \sim Be(\alpha, \beta)$.

Figure 2 shows the structure of various beta distributions. The shape of the beta distribution produced by different parameters may be observed in the image; when $\alpha = \beta = 1$, its shape is comparable to a uniform distribution; if $\alpha > \beta$, its distribution is left skewed; and if $\alpha < \beta$, its distribution is right skewed. Many studies [34–36] have shown that the beta distribution is more competitive than the random sequence and uniform distribution. Therefore, we used the beta distribution to initialize the population. The initial particle distribution of the population is more even. The initial population sequence of the beta distribution is presented as follows:

$$x_i^0 = x_{i,\min} + Be(\alpha, \beta)(x_{i,\max} - x_{i,\min}) \tag{23}$$

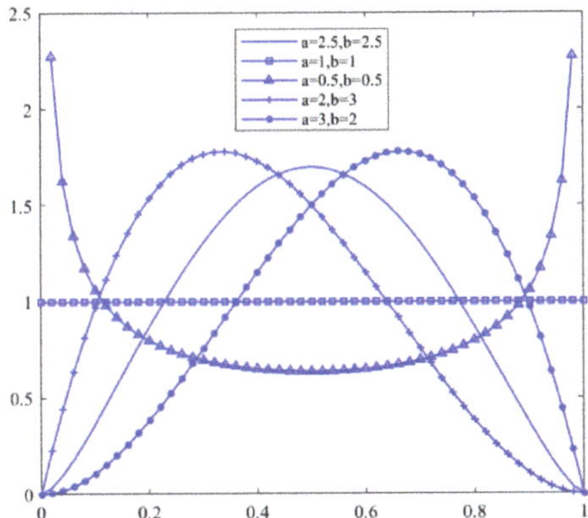

Figure 2. The structure of the different beta distributions.

3.2.2. Nonlinear Adaptive Parameter Fitting Strategy

To optimize the performance of PSO, the parameters of inertia weight ω, self-learning factor c_1, and social learning factor c_2 are adjusted adaptively. Inertia weight ω keeps particles moving with inertia and tends to expand the search space. When the ω value is large, the algorithm's global optimization ability is greater, but its local optimization ability is weak, which means that the convergence speed is high but the search accuracy is low. Although a more accurate global optimal solution can be found while the ω value is low, the convergence ability of the algorithm is weak. The self-learning factor and the social learning factor guide the particles in the iteration through the optimal position of the particles themselves and the optimal position of the population. In this study, we nonlinearly and dynamically adjust the inertia weight, self-learning factor, and social learning factor according to the number of iterations. This measure can prevent particles from falling into the local optimal solution, realize the fast convergence of particles to the global optimal solution, and effectively improve the convergence ability and optimization ability of the algorithm. The adjustment process is expressed using the following formulas:

$$\omega_k = \omega_{\max} + k(\omega_{\max} - \omega_{\min})\frac{(k - 2K)}{K^2} \tag{24}$$

$$c_1^k = c_1^{\max} + k(c_1^{\max} - c_1^{\min})\frac{(k - 2K)}{K^2} \tag{25}$$

$$c_2^k = c_2^{\min} - k(c_2^{\max} - c_2^{\min})\frac{(k - 2K)}{K^2} \tag{26}$$

where ω_k, c_1^k, and c_2^k denote the inertia weight, self-learning factor, and social learning factor of the kth iteration, respectively. $\omega_{\max}$ and $\omega_{\min}$ are the upper and lower limits of inertia weight, respectively. $c_1^{\max}$ and $c_1^{\min}$ are the maximum and minimum values of the self-learning factor, respectively. $c_2^{\max}$ and $c_2^{\min}$ are the maximum and minimum values of the social learning factor, respectively. K stands for the maximum value of the iteration.

Figure 3 depicts the nonlinear variation patterns of the three PSO algorithm parameters after improvement. The figure shows that the change patterns for all three parameters are steep at first and thereafter flat, with ω and c_1 nonlinearly falling and c_2 nonlinearly increasing. This dynamic tendency may cause the algorithm to have good exploitation ability early in the optimization search and good global exploration ability later in the search to reach a balance of individual exploitation and global exploration.

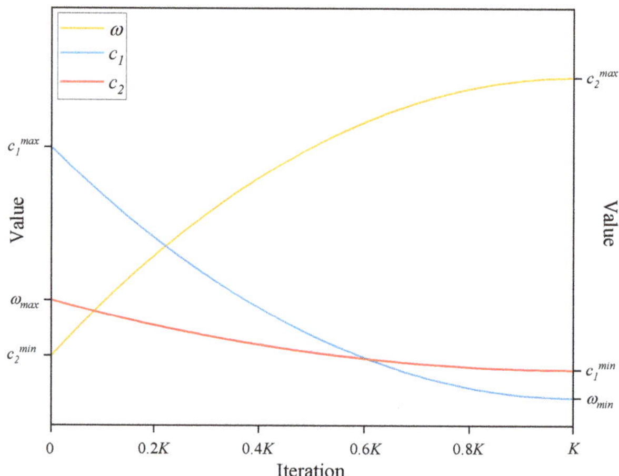

Figure 3. The nonlinear trend of parameters.

The particle's velocity and position are updated using the updated parameters. The position formula is shown in Equation (21), and the velocity formula is updated as follows:

$$v_i^{k+1} = \omega_k v_i^k + c_1^k r_1(x_i^p - x_i^k) + c_2^k r_2(x^g - x_i^k) \tag{27}$$

3.2.3. Lévy Flight Mechanism

As demonstrated in Figure 4a, Lévy flight [37] is a random wandering mechanism with a heavy-tailed distribution of step sizes. It has two characteristics: (1) small step length tracking, which can assist the algorithm in performing local neighborhood search, and (2) long distance spanning, which can jump at local locations with high probability and impose perturbation on the algorithm, allowing it to escape the local optimal solution. Therefore, we introduce the Lévy flight strategy to accelerate the particles to the global optimal position, shorten the iteration time, improve the algorithm's convergence speed, and prevent the algorithm from stagnating in the local optimal solution. The improved position update formula is as follows:

$$x_{i,\text{new}}^{k+1} = x_i^k + (x^g - x_i^k) \oplus Levy(\beta) \tag{28}$$

where $x_{i,new}^{k+1}$ denotes the position of the ith particle in the kth iterative Lévy flight. $\oplus$ denotes point-to-point multiplication. $Levy(\beta)$ is a random search path for particles that follow the Lévy distribution with parameter β. The formulas of the Lévy distribution [37] are satisfied as follows:

$$Levy(\beta) \sim \frac{\mu}{|v|^{\frac{1}{\beta}}} \tag{29}$$

$$\mu \sim N\left(0, \sigma_\mu^2\right) \tag{30}$$

$$v \sim N\left(0, \sigma_v^2\right) \tag{31}$$

$$\sigma_\mu = \left\{ \frac{\Gamma(1+\beta) + \sin\left(\pi \times \frac{\beta}{2}\right)}{\Gamma\left(1+\frac{\beta}{2}\right) \times \beta \times 2^{\frac{\beta-1}{2}}} \right\} \tag{32}$$

$$\sigma_v = 1 \tag{33}$$

where μ and v represent the standard normal distribution. β is usually valued at a constant of 1.5.

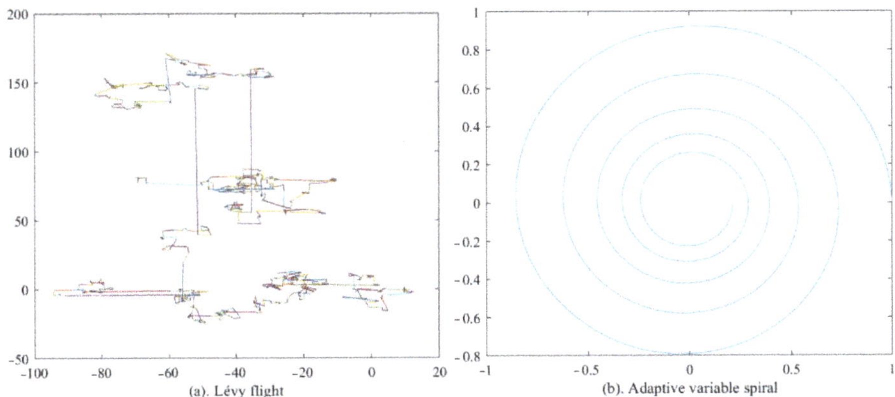

Figure 4. The Lévy flight mechanism and adaptive variable spiral policy of the IMPSO algorithm illustrated. Lévy flight trajectory is shown, where the different colored lines represent the direction and distance of each flight movement.

Because the Lévy flight mechanism is widely employed in particle swarm optimization, the algorithm's temporal complexity increases, causing the particle search strategy to be shifted by the size of the random number r of the interval of (0, 1]. For r > 0.5, we update the position using the Lévy flight mechanism; otherwise, the following search strategy is added.

3.2.4. Adaptive Variable Spiral Policy

To develop more diverse search paths for particle position updates, variable spiral search [38] is recommended, as shown in Figure 4b. It can be seen that this policy can explore and develop globally in the form of a large spiral shape in the early stage and focus on the optimal solution neighborhoods in the form of a small spiral in the later stage by adjusting the adaptive spiral coefficient. It also improves the ability of particles to explore unknown search domains. The variable spiral search position update formula designed in this study is as follows:

$$\begin{cases} x_{i,\text{new}}^{k+1} = x_i^k + e^{zl}\left(x^g - x_i^k\right)\cos(2\pi l) \\ z = e^{s\cos(\pi(1-\frac{k}{K}))} \end{cases} \tag{34}$$

where z is a variable that changes the shape of the spiral from large to small as the number of iterations increases. l is a uniformly distributed random number between $[-1,1]$. s is the coefficient of variation, which takes a constant number of 5. In the adaptive variable spiral policy, the scope of the spiral decreases with an increase in the number of iterations. The particle swarm can search for food using a large-scale spiral pattern early in the search, which effectively increases its diversity; later in the search, the bounding circle is gradually reduced to approximate the optimal solution, which can improve the program's optimization performance.

3.3. Framework of IMPSO

The technological process of the integrated multistrategy particle swarm algorithm designed in Section 3.2 is shown in Figure 5, which represents the overall framework of the algorithm. The execution steps of IMPSO for the optimization problems are listed below:

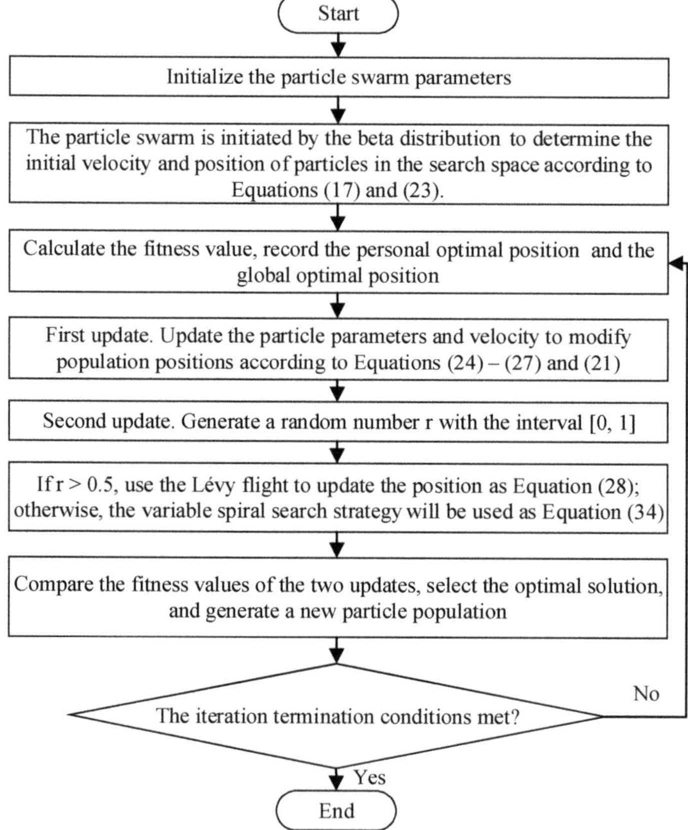

Figure 5. Flowchart of the IMPSO algorithm.

Step 1: The initial parameters of the population are set, and the population's velocities are initialized.

Step 2: The beta distribution in Section 3.2.1 is used to generate the initial positions of the population, as shown in Equation (23).

Step 3: The fitness value of each individual in the initial population is calculated and recorded, and the global optimal value for storage is found.

Step 4: The first update is performed. The inertial weights, self-learning factors, and social learning factors are updated according to Equations (24)–(26) in Section 3.2.2, and individual velocities and positions are updated according to Equations (21) and (27).

Step 5: The second update is performed. A random number, r, is selected with interval [0, 1]; if $r > 0.5$, the original particle position should be updated using Equation (28) in Section 3.2.3. If $r \leq 0.5$, the original particle position should be updated using Equation (34) in Section 3.2.4.

Step 6: Each individual fit value derived from the two updates is calculated separately, and the better values are chosen for storage. Then, the individual optimal values and global optimal values are recorded for each iteration.

Step 7: Whether the iteration termination condition is met is determined. If the condition is met, the procedure ends and the optimization result is exported. Otherwise, return to Step 4 to proceed with the update.

4. Constraint Handling Methods

4.1. Review of Constraint Handling Methods

The optimal operation of cascade reservoirs involves complex nonlinear constraints due to the complex system and numerous coupling connections among reservoirs. The general solution to these constraints is to enforce control over the search space, explicitly discard the solution that violates the constraint, and select only the workable solution. In the case of the metaheuristic algorithm, the abandonment of the infeasible solution limits the exploration and exploitation of the search domain, resulting in a loss of population diversity. Additionally, it is simple to settle for the local optimal value, and the global optimal solution cannot be found. As a result, solutions that exceed constraints must be properly processed to increase the precision of the solution.

Different methods of constraint processing have different effects. There are numerous techniques [39,40] that have been developed to solve constraint optimization problems, including the penalty function, feasibility rule, stochastic ranking, repair factors, and ε-constraint methods. In order to approach the optimal solution, the penalty function method converts the constrained optimization problem into an unconstrained one, and the variable violating the constraint is penalized by introducing a penalty factor. The feasibility rule is used in the evolution process to select the better solution between feasible and nonfeasible options. Due to its fundamental and flexible properties, it is suitable for use with any optimization algorithm. The stochastic ranking method sorts all solutions in the population in a manner similar to bubble sort in order to balance the target value and the penalty value for the constraint violation. The ε-constraint method changes the search bias dynamically, allowing the system to explore unfeasible domains. To solve the complex constrained optimization problem of cascade reservoirs, the static penalty function is combined with a new implicit constraint processing method [41] to update the constraint variable's boundary and reduce the probability of the unfeasible solution.

4.2. Explicit–Implicit Coupled Constraint Handling Technique

In general, the constrained optimization problem can be mathematically described as a minimization or maximization problem as follows:

$$Min\ f(X)\ or\ Max\ f(X),\ X = (x_1, x_2, \ldots, x_n) \tag{35}$$

$$\begin{cases} g_l(X) \leq 0, \; l = 1, 2, \ldots, L \\ h_k(X) = 0, \; k = 1, 2, \ldots, K \\ lb_i \leq x_i \leq ub_i \end{cases} \tag{36}$$

where X is the set of decision variables. $g_l(X)$ and $h_k(X)$ are the inequality constraints and equations, respectively. ub_i and lb_i are the upper and lower bounds of the decision variables x_i, respectively.

The implicit constraint handling method presented in this study further limits the search space by turning each inequality constraint into an upper and lower bound with respect to decision variables and restricts the search space via the intersection as follows:

$$\begin{cases} lb_i^u = \min(\max[l_{i,1}(x_{\neq i}), \ldots, l_{i,k_i}(x_{\neq i}), lb_i], ub_i) \\ ub_i^u = \max(\min[u_{i,1}(x_{\neq i}), \ldots, u_{i,k_i}(x_{\neq i}), ub_i], lb_i) \end{cases} \tag{37}$$

where u_{i,k_i} and l_{i,k_i} are the upper and lower bounds of the decision variables x_i under the inequality constraint k_i, respectively, and $k_i \leq L$. ub_i^u and lb_i^u are the upper and lower bounds of the update process of the decision variables x_i, respectively. ub_i and lb_i are the original upper and lower bounds of the decision variables x_i, respectively.

The boundary update formula is modified as follows in practical applications to prevent the upper bounds of the decision variable update process from being smaller than the lower bounds.

$$\begin{cases} lb_i^u = \min(\max[l_{i,1}(x_{\neq i}), \ldots, l_{i,k_i}(x_{\neq i}), lb_i], ub_i) \\ ub_i^u = \max(\min[u_{i,1}(x_{\neq i}), \ldots, u_{i,k_i}(x_{\neq i}), ub_i], lb_i^u) \end{cases} \tag{38}$$

Following the boundary update on decision variables, they are assigned to the search domain to ensure continuity across iterations, as shown below:

$$x_i = lb_i^u + p_i(ub_i^u - lb_i^u) \tag{39}$$

where p_i denotes a random number between $[0, 1]$.

In this study, a static penalty function is presented that can be combined with the above implicit constraint processing approach to penalize the solution that exceeds the bounds of the decision variables during operations in order to achieve a more suitable result. The static penalty function can be expressed as follows:

$$F(X) = f(X) + \theta \times \sum_{i=1}^{N} violation_i \tag{40}$$

where $F(X)$ and $f(X)$ are the objective values after increasing the penalty function and the original function. θ is the static penalty factor, according to the specific choice; $\sum_{i=1}^{N} violation_i$ denotes the total constraint violation.

5. Case Study

5.1. Study Area

The lower reaches of the Jinsha River–Three Gorges cascade hydropower system is selected as the study area in this section. With its abundant water resources and favorable development environment, the lower reaches of the Jinsha River have become one of the country's largest hydropower bases and the hub of China's "West-to-East power transmission", providing an important contribution to local economic and social sustainability. The lower reaches of the Jinsha River–Three Gorges cascade reservoir group perform a series of social functions, including power generation, flood control, navigation, water supply, ecology, etc., providing substantial social and economic benefits for the Yangtze River basin. However, in order to maximize the economic benefits of these cascade reservoirs, establishing a joint dispatch management model of the lower reaches of the Jinsha River–Three Gorges cascade reservoirs is necessary. As a result, in order to maximize the benefits of power generation in the lower reaches of the Jinsha River–Three Gorges cascade reservoirs, this section examines the joint optimal operation of the cascade hydropower system, which provides technical support and theoretical foundations for the cascade economic operation of the system. In this paper, the Xiluodu, Xiangjiaba, Three Gorges, and Gezhouba reservoirs are mainly tested as the objects of investigation. These reservoirs are located in the lower reaches of the Jinsha River and have a total installed capacity of 44,410 MW. They are the largest reservoirs in the Yangtze River basin, and hold the top position in the hydroelectric power industry in terms of annual electricity generation and installed capacity. The topological structure of the four reservoirs in the study area is shown in Figure 6, and the main characteristic parameters are shown in Table 1.

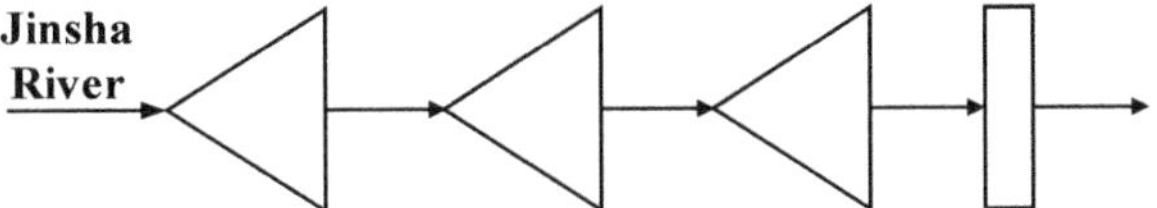

Figure 6. Topology map of the reservoir group in the lower reaches of the Jinsha River–Three Gorges basin.

Table 1. Main characteristic parameters of the lower reaches of Jinsha River–Three Gorges cascade reservoirs.

Reservoir Feature	Xiluodu	Xiangjiaba	Three Gorges	Gezhouba
Regulation performance	Annually	Seasonally	Annually	Daily
Generation coefficient	8.5	8.35	8.8	8.5
Normal water level (m)	600	380	175	66
Dead water level (m)	540	370	145	63
Guaranteed capacity (MW)	3795	2009	4990	1040
Installed capacity (MW)	12,600	6400	22,500	2910
Minimum outflow (m^3/s)	1700	1700	6000	5500

5.2. Detailed Technical Procedures

5.2.1. Solution Structure and Population Initialization

It was mentioned in Section 3 that particles in the PSO algorithm usually refer to birds; however, the meaning of the referent can vary in specific problems. For the cascaded hydropower system of a river basin, taking the water level as the particle, i.e., the decision variable X of the optimization problem, the variable matrix is defined as follows:

$$X = \begin{bmatrix} Z_{1,1}, & Z_{1,2}, & \cdots, & Z_{1,T} \\ Z_{2,1}, & Z_{2,2}, & \cdots, & Z_{2,T} \\ \vdots & \vdots & \ddots & \vdots \\ Z_{N,1}, & Z_{N,2}, & \cdots, & Z_{N,T} \end{bmatrix}_{N \times T} \tag{41}$$

where $Z_{i,t}(i = 1, 2, \ldots, N; t = 1, 2, \ldots, T)$ denotes the water level of the ith particle in the tth dim and the variables are continuous.

For the initial population of the optimization problem, Equation (23) of Section 3.2.1 is used to initialize the water level. The formula is expressed as follows:

$$Z_{i,0} = Z_{i,0}^{\min} + Be(\alpha, \beta)(Z_{i,0}^{\max} - Z_{i,0}^{\min}) \tag{42}$$

where the parameters α and β of the beta distribution function are both 2.5 [34], which is obtained from some probability distribution initialization experiments.

5.2.2. Explicit–Implicit Coupled Constraint Handling Method Based on Constraint Normalization

In this article, the output-related constraints and flow boundary constraints are transformed into water level limit constraints using the relationship of the water level–capacity curve, ensuring the upper and lower limits of the output, installed capacity, and water level fluctuation. The feedback constraint processing method is referred to as the applied mathematical model in Section 2.1:

$$
\begin{cases}
lb_{i,t}^{u} = \min(\max[Z_{i,t,\min}^{f^{Z\sim R}}, Z_{i,t,\min}^{f^{Z\sim N}}, Z_{i,t,\min}^{f^{\Delta Z\sim Z}}, Z_{i,t,\min}^{f^{Z\sim Q}}, Z_{i,t,\min}^{f^{Z\sim V}}, lb_{i,t}], ub_{i,t}) \\
ub_{i,t}^{u} = \max(\min[Z_{i,t,\max}^{f^{Z\sim R}}, Z_{i,t,\max}^{f^{Z\sim N}}, Z_{i,t,\max}^{f^{\Delta Z\sim Z}}, Z_{i,t,\max}^{f^{Z\sim Q}}, Z_{i,t,\max}^{f^{Z\sim V}}, ub_{i,t}], lb_{i,t}^{u})
\end{cases}
\tag{43}
$$

$$
E = \max \sum_{i=1}^{I} \sum_{t=1}^{T} A_i Q_{i,t} H_{i,t} \Delta t + \theta \times \sum_{i=1}^{N} violation_i
\tag{44}
$$

$$
violation_i = \begin{cases}
0, & othervise \\
Z_{i,t} - lb_{i,t}^{u}, & if\ Z_{i,t} < lb_{i,t}^{u} \\
ub_{i,t}^{u} - Z_{i,t}, & if\ Z_{i,t} > ub_{i,t}^{u}
\end{cases}
\tag{45}
$$

where $Z_{i,t,\min}^{f^{Z\sim R}}$, $Z_{i,t,\min}^{f^{Z\sim N}}$, $Z_{i,t,\min}^{f^{\Delta Z\sim Z}}$, $Z_{i,t,\min}^{f^{Z\sim Q}}$, $Z_{i,t,\min}^{f^{Z\sim V}}$ denote the lower boundary of the water level after the conversion of the level–outflow function, level–output function, level–range of the water level function, level–generation flow function, and level–capacity function of hydropower station i in the tth stage, respectively. $Z_{i,t,\max}^{f^{Z\sim R}}$, $Z_{i,t,\max}^{f^{Z\sim N}}$, $Z_{i,t,\max}^{f^{\Delta Z\sim Z}}$, $Z_{i,t,\max}^{f^{Z\sim Q}}$, $Z_{i,t,\max}^{f^{Z\sim V}}$ denote the upper boundary of the water level after the conversion of the level–outflow function, level–output function, level–range of the water level function, level–generation flow function, and level–capacity function of hydropower station i in the tth stage, respectively. The penalty coefficient θ in this study is taken as a constant of 0.01, and it is determined by repeated trial calculation.

Figure 7 depicts a graphical depiction of the decision variable for this research issue, namely, water level. As can be observed, the feasible range of water level is $[lb_{i,t}, ub_{i,t}]$, a vast range of the search for the best is not favorable to the solution in the actual engineering problem, and the probability of an infeasible solution is higher. Following the improvement of the constraints, the feasible range of water level is gradually lowered under the management of different constraints, and eventually, the intersection of the constraints is taken to achieve the final viable range of water level, reducing the likelihood of infeasible solutions.

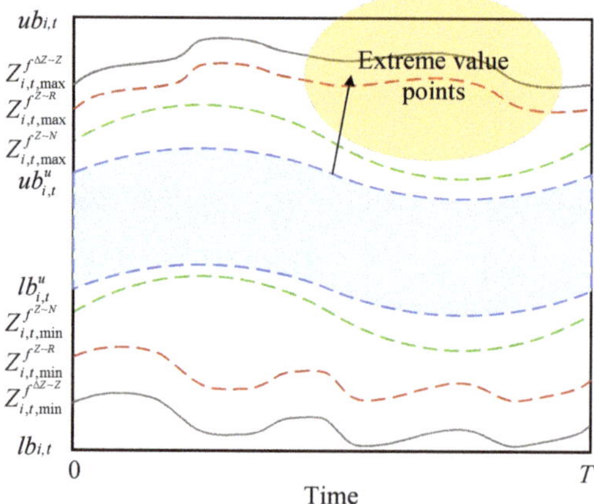

Figure 7. Graphical representation of decision variables (water level). Each color line is an upper and lower limit of a constraint corresponding to the water level, respectively. The solid gray line

represents the water level range under the water level variation constraint; the red dashed line represents the water level range under the outflow constraint; the green dashed line represents the water level range under the outflow constraint; and the blue shaded area is the final water level feasible range. In addition, the yellow shaded area is the feasible solution that exists outside the feasible range, and the arrows represent the search agent jumping out of the feasible range to find the feasible solution.

5.3. Overall Implementation Framework

The ideal operating method and process for maximizing the power generation benefit of the river basin cascade hydropower system is summarized as follows:

Step 1: The basic parameters of the optimization problem are defined, and the initialization method in Section 5.2.1 is used to generate initial groups in the problem space, i.e., the process of generating the initial water level of the cascade reservoir group.

Step 2: The fitness of all solutions in the current group is evaluated using the explicit–implicit coupled constraint handling method in Section 5.2.2.

Step 3: The individual best position and the global best position for each solution are updated. Then, the location of all solutions using the IMPSO approach in Section 3.2 is updated.

Step 4: If the terminal condition of the algorithm is not satisfied, return to Step 2 for the next cycle; otherwise, the iteration is stopped. Then, the optimal solution is taken as the final solution for the optimal scheduling of cascade reservoirs.

The procedure for solving the problem of the optimal operation strategy using the IMPSO algorithm is shown in Figure 8.

5.4. Results and Analysis

In order to verify the technical feasibility and efficiency of the proposed method in cascade reservoir power generation dispatching, three different typical runoff scenarios (wet, normal, and dry) were studied. In this paper, six methods were selected for the calculation to compare and analyze the application effect of the proposed method in cascade reservoir power generation dispatching, namely, IMPSO and the explicit–implicit coupled constraint handling technique (IMPSO+CHT), such as IMPSO, PSO, SAPSO [42], LFPSO [43], and DE [44].

5.4.1. Statistical analysis

For all parameter settings of the algorithms, the maximum number of iterations for each algorithm is $K = 500$, and the population size is $N = 50$. In addition, we set the mutation probability of DE to be 0.5 and the crossover probability to be 0.8; the inertia weight ω of the PSO algorithm decreases linearly from 0.9 to 0.4, and the self-learning factor c_1 and social learning factor c_2 are both 2; the SAPSO and LFPSO parameters were established by referring to Harrison H R et al. [43] and Haklı H et al. [44]; the inertia weight ω in IMPSO declined nonlinearly from 0.9 to 0.4, and the self-learning factor c_1 decreased nonlinearly from 2 to 0.2, while the social learning factor c_2 increased nonlinearly from 0.5 to 2.5. Considering the variation in the results of the intelligent optimization algorithm, each algorithm runs 10 times independently using different input data, and the average, median, best value, worst value, and standard deviation are counted. The detailed results are shown in Table 2.

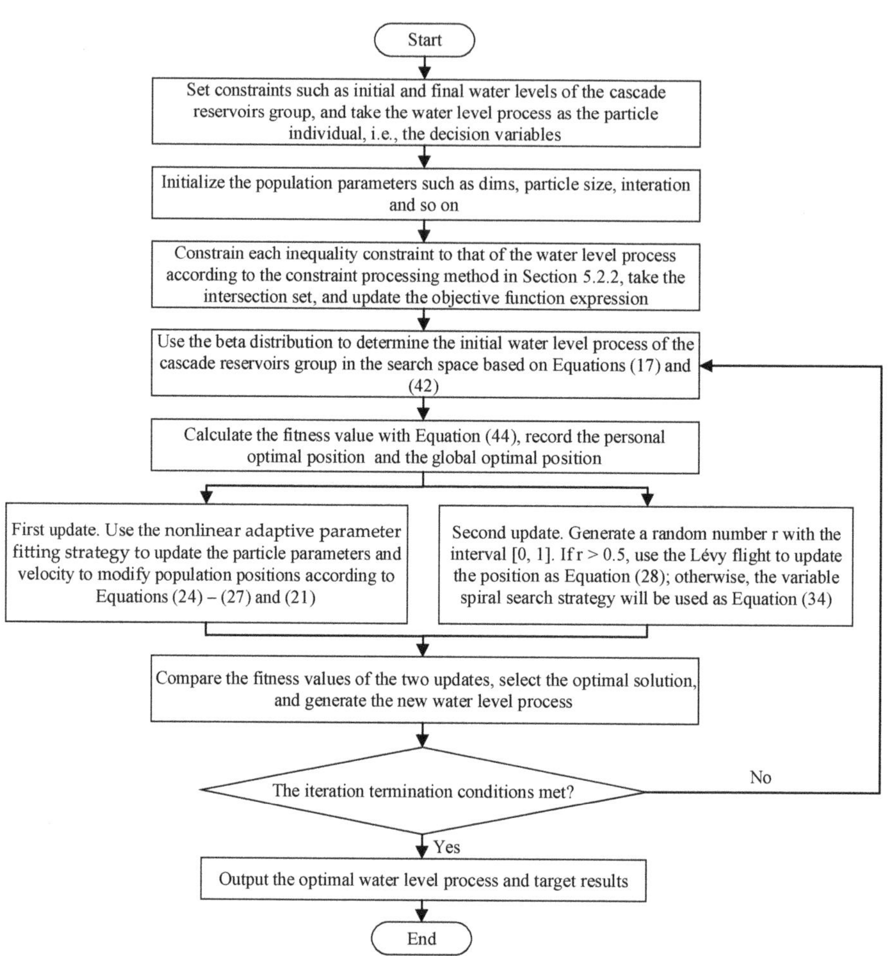

Figure 8. Flowchart of the procedure for solving the problem of the optimal operation strategy using the IMPSO algorithm.

Table 2. Statistical results of 10 independent operations for each year.

Runoff Scenarios	Methods	Objective Value (10^8 kW·h)				
		Average	Median	Best	Worst	Standard Deviation
Wet year	IMPSO+CHT	**2244.86**	**2244.41**	**2267.39**	**2224.81**	**15.95**
	IMPSO	2229.94	2240.86	2254.41	2173.14	25.12
	SAPSO	1927.66	1941.31	2003.24	1839.86	54.99
	LFPSO	1944.84	1933.59	2042.93	1867.86	56.08
	PSO	1942.49	1963.76	2028.84	1845.59	55.88
	DE	1942.35	1944.82	2001.02	1894.85	34.07

Table 2. *Cont.*

Runoff Scenarios	Methods	Objective Value (10^8 kW·h)				
		Average	Median	Best	Worst	Standard Deviation
Normal year	IMPSO+CHT	**2091.76**	**2091.57**	**2105.61**	**2075.34**	**8.59**
	IMPSO	2086.25	2089.53	2101.46	2045.12	16.08
	SAPSO	1847.68	1820.32	1987.33	1697.08	103.22
	LFPSO	1899.35	1899.90	2015.22	1745.75	90.79
	PSO	1830.97	1801.22	2023.52	1699.34	102.77
	DE	1842.75	1846.89	1916.32	1781.71	41.10
Dry year	IMPSO+CHT	**1859.84**	**1860.77**	**1864.86**	**1853.50**	**3.14**
	IMPSO	1856.33	1859.99	1864.10	1840.60	7.91
	SAPSO	1706.79	1690.30	1860.61	1556.13	83.64
	LFPSO	1766.92	1766.00	1875.46	1652.74	64.51
	PSO	1717.39	1712.80	1859.58	1572.45	100.16
	DE	1707.64	1722.54	1771.95	1626.00	49.92

It can be observed from Table 2 that under three runoff scenarios, the statistical data of the proposed IMPSO algorithm combined with the constraint processing method are mostly better than other algorithms. Taking the average power generation of 10 independent runs as an example, compared with IMPSO, SAPSO, LFPSO, PSO, and DE algorithms, the proposed scheme increased by 1.49, 31.72, 30.00, 30.24, and 30.25 billion, respectively, in the year of abundant water. Compared with IMPSO, SAPSO, LFPSO, PSO, and DE algorithms, the proposed program has increased by 0.55, 24.41, 19.24, 26.07, and 24.90 billion, respectively. Compared with IMPSO, SAPSO, LFPSO, PSO, and DE algorithms, the proposed scheme adds 0.35, 15.31, 9.29, 14.25, and 15.22 billion, respectively, in the dry year. The data shown above demonstrate that the proposed technique has excellent iterative performance and optimization capacity, and that it can significantly increase the economic benefits of cascade reservoirs aimed at generating electricity.

Figure 9 shows the distribution of the annual power generation results for each of the six methods under three typical runoff scenarios. First, it can be observed in the graph that the outcomes of the IMPSO+CHT and IMPSO algorithms are very stable compared with the other four algorithms in three runoff scenarios. The IMPSO+CHT and IMPSO algorithms are shown to be superior to the other four methods. Second, compared with IMPSO alone, the volatility of IMPSO+CHT was lower in the three runoff scenarios and the upper and lower quartile shadow area was smaller, and no outliers appeared. This indicates that IMPSO, when used in conjunction with the proposed constraint processing mechanism, is more centralized and robust in the three runoff situations. The SAPSO, LFPSO, PSO, and DE schemes all have average performance in the three runoff scenarios, with the LFPSO scheme having more outliers in the dry year scenario, and the maximum values of SAPSO and PSO are close to the two IMPSO-related schemes, but the minimum value is also lower than the other schemes, indicating that the SAPSO and PSO algorithms are less stable. The box plots along with the statistical information in Table 2 show that the IMPSO+CHT scheme is more robust and optimal compared with the IMPSO scheme.

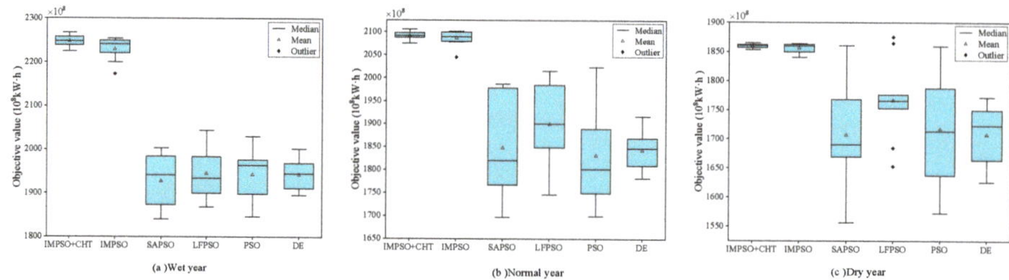

Figure 9. Box diagram of power generation operation in each typical runoff scenario.

Table 3 displays the Friedman test results for several scheduling methods under various runoff scenarios. The Friedman test ranking findings compare the overall performance of the six methods in the reservoir power generation scheduling problem. In dry and normal year scenarios, the IMPSO+CHT scheme and the IMPSO scheme rank significantly better than the rest of the schemes, and the IMPSO+CHT scheme is better than the IMPSO scheme. In the wet year, the IMPSO+CHT scheme and the IMPSO scheme have the same ranking results. Furthermore, by integrating the final ranking results, we can see that LFPSO outperforms SAPSO, LFPSO, PSO, and DE schemes, whereas PSO and DE optimization results are low overall.

Table 3. Mean ranking results in different runoff scenarios using the Friedman test.

Runoff Scenarios	Methods					
	IMPSO+CHT	**IMPSO**	**SAPSO**	**LFPSO**	**PSO**	**DE**
Wet year	**1.5**	**1.5**	4.4	4.5	4.4	4.8
Normal year	**1.4**	1.6	4.4	3.8	5.3	4.5
Dry year	**1.7**	1.8	4.6	3.7	4.5	4.7
Mean rank	**1.53(1)**	1.63(2)	4.67(4)	4(3)	4.73(5)	4.67(4)

Figure 10 depicts the convergence process of the six schemes when the power generation scheduling model is solved using inflow data from various runoff scenarios as model inputs. It is clear that the IMPSO+CHT scheme and the IMPSO scheme not only have high initial values but also have the highest growth rate at the start of the iterations, allowing the objective function values to converge to near-optimal values within 200 iterations. SAPSO, LFPSO, and DE schemes, on the other hand, have slower growth rates and poorer optimization results, while the PSO algorithm has the slowest growth rate and readily falls into a local optimum solution. The numbers (1), (2), etc. are a general ranking according to the mean rank.

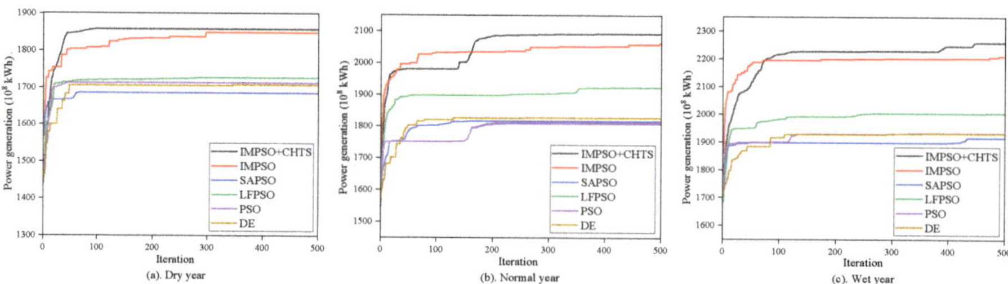

Figure 10. Convergence process of power generation operation in each typical runoff scenario.

In addition, population diversity analysis is an important component of the algorithm's optimization search process. In general, a higher population diversity number reflects a more dispersed population, indicating that the algorithm is capable of global exploration and less likely to fall into the trap of the local optimal solutions [45]. Equation (46) can be used to express population diversity:

$$Diversity(t) = \sqrt{\sum_{i=1}^{N}\sum_{d=1}^{D}(x_{id}(t) - \overline{x}_d(t))^2} \tag{46}$$

where $Diversity(t)$ is the value of population diversity at the tth iteration, the parameter N is the number of population size, D is the dimension of the problem, $x_{id}(t)$ is the position of the ith agent in the dth dimension at the tth iteration, and $\overline{x}_d(t)$ implies the average position of the population in the dth dimension at the tth iteration.

Figure 11 depicts the population diversity study results for all scenarios under various typical runoff conditions. The figure shows that the overall results of the IMPSO+CHT scheme and the IMPSO scheme are better than the other algorithms in the three scenarios, with the IMPSO scheme outperforming the IMPSO+CHT scheme due to the use of the constraint treatment method, which weakens the algorithm performance to some extent. The outcomes displayed in different circumstances for the other four systems are not exactly the same. In the dry year, the DE scheme is the worst, the LFPSO scheme is better; in the wet year, the DE, PSO, and LFPSO schemes have similar results, and the SAPSO scheme is the best; in the wet year, the LFPSO and PSO schemes are better, then the SAPSO, and finally the DE scheme. Overall, the IMPSO scheme produced the best population diversity analysis results, followed by the IMPSO+CHT scheme, and the DE scheme produced the worst.

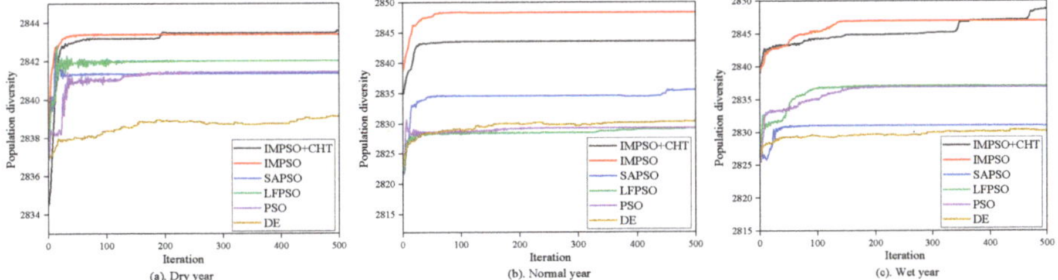

Figure 11. Population diversity analysis of six schemes in each typical runoff scenario.

Based on the statistical results discussed above, it is shown that the four improvement strategies comprising beta distribution initialization, nonlinear parameter change, Lévy flight, and adaptive spiral update are feasible, and that the algorithm can effectively improve the ability of local disturbance optimization, avoid individual extreme values falling into local optimization too early, and extend the global search range scope of the algorithm. This shows that the IMPSO algorithm has broad engineering applicability in cascade reservoir optimal scheduling, which can increase the efficiency of cascade reservoir power generation and attain optimal hydropower resource assignment. At the same time, the constraint handling technique combining a variable boundary update and a penalty function is introduced, which can effectively reduce the out-of-bounds behavior of particles in the optimization process and improve the algorithm's solution accuracy, thereby achieving the unity of robustness and exploring performance in the optimization mechanism.

5.4.2. Analysis of Optimization Scheduling Results

In this section, the disparities in water abandonment of cascade reservoirs under three typical runoff situations, the optimization results of the IMPSO+CHT scenario, and the generation rules are carefully studied based on the optimization outcomes of the six schemes.

Table 4 compares the average volume of abandoned water generated by all methods for each typical runoff scenario in the results of 10 independent operations. It can be seen that under various runoff scenarios, the IMPSO+CHT scheme and the IMPSO scheme both generate significantly less abandoned water than the other schemes, with the IMPSO+CHT plan producing the least amount of waste water. Furthermore, the LFPSO scheme performs rather well, whereas the PSO and DE schemes dispose of the most water and have the lowest usage rate for water resources. Figures 12–14 show the hydrological processes of the four reservoirs of the lower reaches of the Jinsha River–Three Gorges cascade hydropower system in typical runoff scenarios. It can be observed that in the flood scenario, due to the abundant flow of water, the peak value of the reservoir is large; except for the Three Gorges reservoir, the other three reservoirs have abandoned water. Due to the limited installed capacity and storage capacity, there are more water discharges in Xiangjiaba and Gezhouba, especially Gezhouba, which is a daily regulated reservoir. Under the scenario of the normal year, there is no abandoned water in Xiluodu and Three Gorges reservoirs, the water stock of Xiangjiaba consists of a small amount of abandoned water, and Gezhouba has more abandoned water, especially when the water rises sharply during the flood season, and the unit overflow of the has reached the limit, which is consistent with the trend of the inflow process. In the dry year, only Gezhouba has abandoned water in the cascade reservoirs, because the selected year has a large amount of water in the flood season, which is similar to the normal year, and the Gezhouba units, which have a low flow capacity, cannot bear a large amount of water during the flood season. Given their high regulating capacity, the Xiluodu, Xiangjiaba, and Three Gorges reservoirs can release the equivalent amount of water through reasonable operation throughout the nonflood season, lowering the quantity of water released during the regulation period. According to the findings of the preceding research, cascade reservoirs may effectively minimize water abandonment, improve hydropower utilization, and promote energy upgrading and transformation after IMPSO+CHT optimization.

Table 4. Average abandoned water flow from 10 operations carried out each year (m^3/s).

Methods	Runoff Scenarios		
	Wet Year	Normal Year	Dry Year
IMPSO+CHT	**102,496.94**	**58,496.09**	**46,525.09**
IMPSO	10,9926.80	58,995.86	53,199.21
SAPSO	150,642.80	98,511.19	69,218.02
LFPSO	111,575.85	94,364.46	69,667.44
PSO	155,965.26	139,110.53	69,218.02
DE	159,466.37	134,618.88	71,089.28

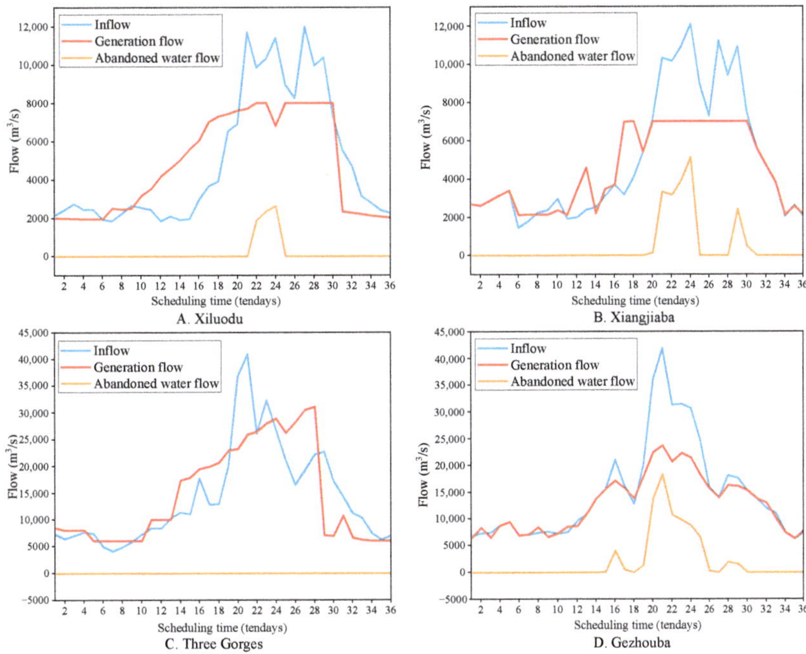

Figure 12. The hydrological process of reservoirs in the wet year.

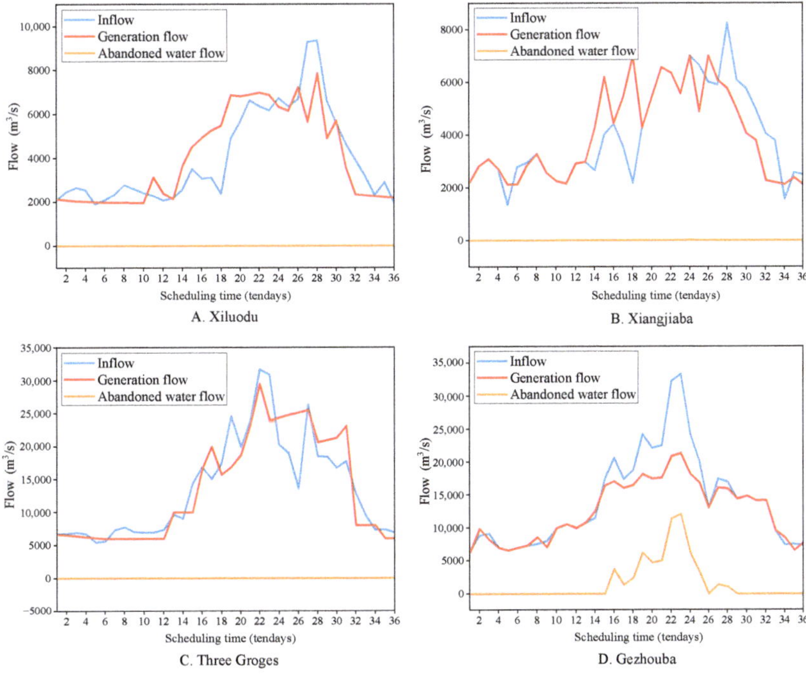

Figure 13. The hydrological process of reservoirs in the normal year.

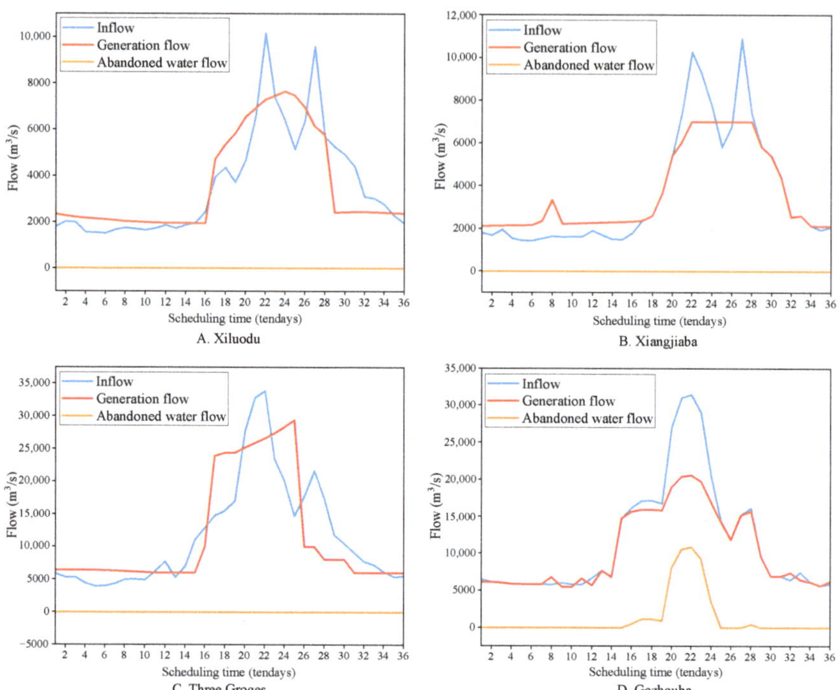

Figure 14. The hydrological process of reservoirs in the dry year.

Figures 15–17 depict the water level–output process of four reservoirs in the lower reaches of the Jinsha River–Three Gorges cascade hydropower system in each typical runoff scenario. It can be observed that in the case of a rainy year, the overall production degree of cascade reservoirs is rather high due to there being sufficient water resources. In the scenario of a dry year, the peak output of each hydropower station is from mid- and late June to the end of September, and the remainder of the period essentially ensures operation with the guaranteed output, with significant differences in performance between the two flood and drought years. In conjunction with the three runoff scenarios, it can be observed that the hydropower station produces less electricity in the early phase of dispatching, and the water level drops to its lowest point in mid- to late June. The production in the later period also increases, and the water level necessarily increases to almost normal water storage level from the end of October to the beginning of November. Figure 18 shows the power generation process of the lower reaches of the Jinsha River–Three Gorges cascade hydropower system under different runoff scenarios. It can be observed that the power generation process of four hydropower stations is consistent with the output process, with increased power generation during the flood period and decreased power generation during the nonflood period, and the power generation capacity of the reservoir in the nonflood period is also higher than that of other runoff scenarios under the influence of inflow in the wet year. In addition, the power generation trend of Xiluodu and Xiangjiaba is relatively close, and the power generation trend of Three Gorges and Gezhouba is relatively similar, which is caused by the different inflow situations in different regions. The modification process of the above study indicates that the output and power generation level of different reservoirs are limited by their own capacities, and the output and power generation level also change under different runoff conditions, but the overall trend and water level change are consistent with the scheduled regulations of the hydropower system in the river basin.

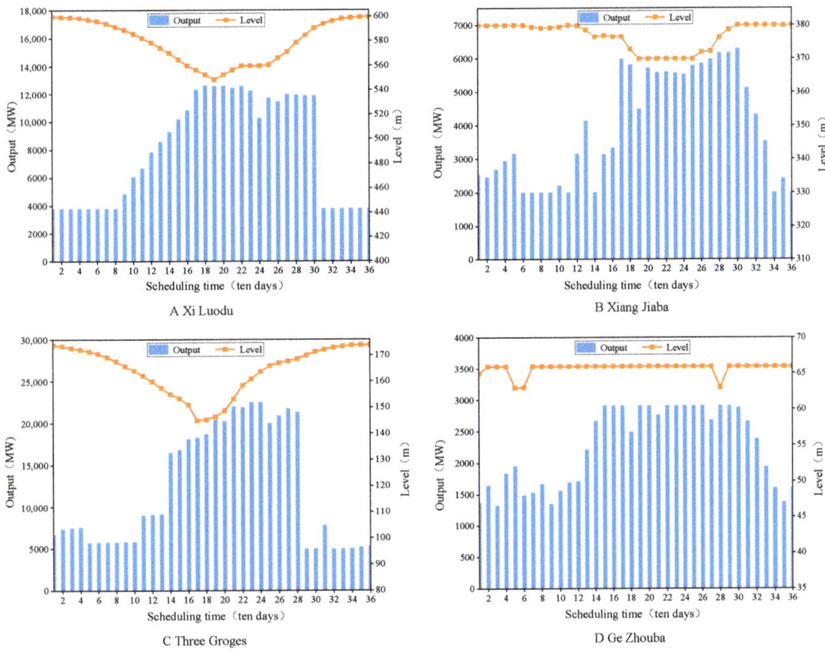

Figure 15. Water level–output process of reservoirs in the wet year.

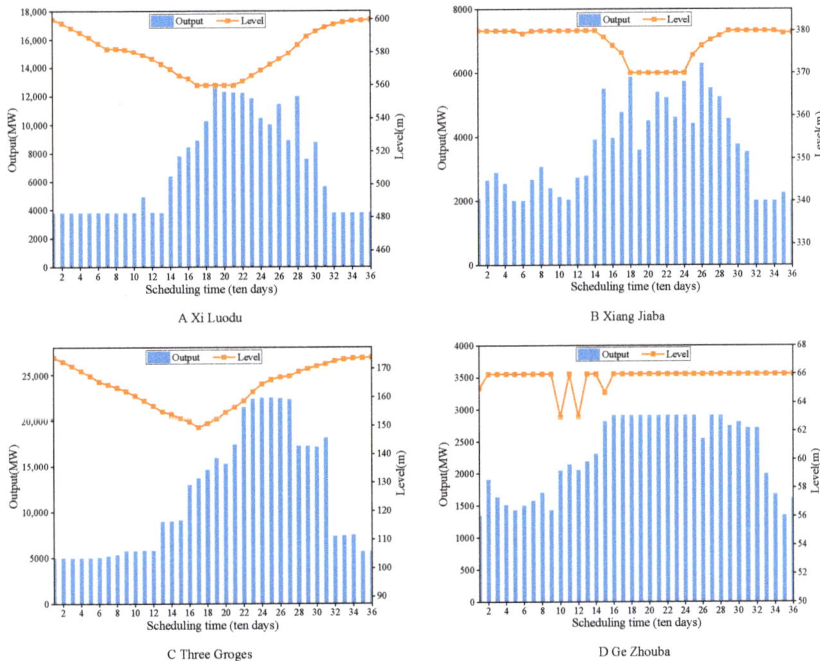

Figure 16. Water level–output process of reservoirs in the normal year.

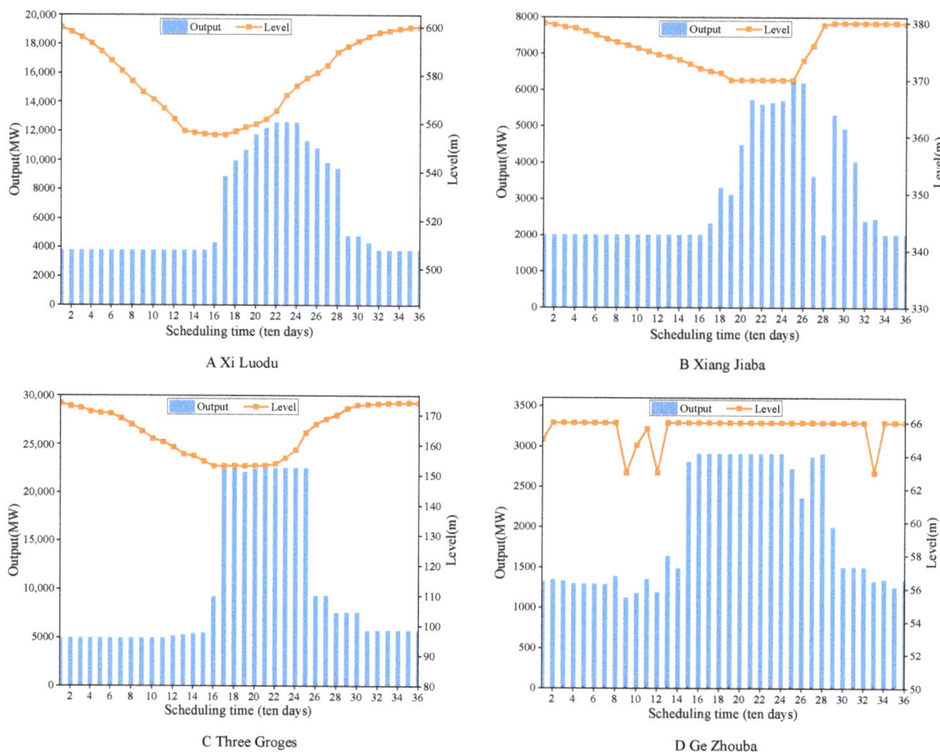

Figure 17. Water level–output process of reservoirs in the dry year.

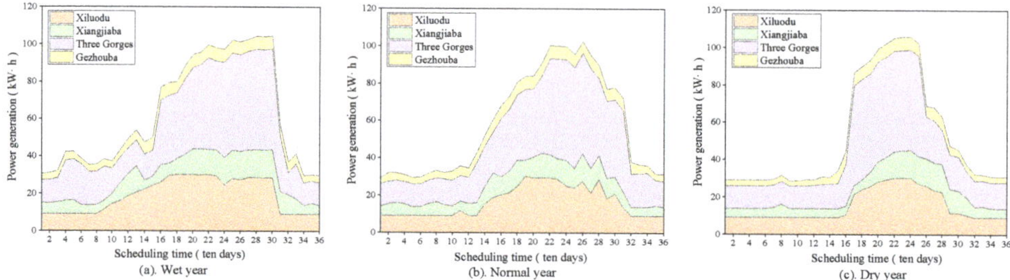

Figure 18. Power generation process of reservoirs in different runoff scenarios.

6. Conclusions

In this study, an integrated multistrategy particle swarm algorithm (IMPSO) is developed, and a joint explicit–implicit coupled constraint handling technique is applied to build a mathematical model of joint power generation scheduling to maximize the total power generation of a cascade hydropower station, and the operating water level, output capacity, outgoing flow, and generation flow are used as constraints. Using the example of four reservoirs in the lower reaches of the Jinsha River—Three Gorges cascade hydropower system, the model was solved by applying IMPSO+CHT, IMPSO, SAPSO, LFPSO, PSO, and DE methods according to three runoff scenarios. By carrying out the statistical analysis of the optimization results, the following conclusions are drawn.

(1)   Using beta distribution initialization to generate candidate solutions and adaptive nonlinear variation of the speed update-related parameters improves the proposed algorithm's population diversity, resulting in faster convergence to the global optimal solution than the comparison schemes. Meanwhile, introducing the Lévy flight mechanism and variable helix search strategy, the optimization search is performed in a diverse exploration manner, which effectively enhances the local exploration capability and global exploitation capability of the algorithm.

(2)   The combined statistic results in Tables 2 and 3 and Figures 9–11 show that using the IMPSO algorithm in reservoir power generation scheduling engineering can achieve more robust and accurate results, increased population diversity, and algorithmic exploration than the comparison algorithms. Meanwhile, the IMPSO method combined with the explicit–implicit coupled constraint processing technique achieves superior results but falls short in terms of population variety when compared with the IMPSO scheme.

(3)   The combined results in Table 4 and Figures 12–14 show that both IMPSO scenarios outperform the comparison scenario under various runoff situations, allowing for a more efficient use of and increased water energy utilization.

(4)   The water level and treatment results from Figures 15–17 obtained using the IMPSO+CHT method can not only meet the boundary conditions of cascade hydropower reservoirs, but also make better use of the abundant water resources in the catchment area and take full advantage of the regulatory and balancing functions of cascade reservoirs; moreover, these results produce a more efficient and reasonable regulation process for the operations.

In three typical runoff scenarios, IMPSO+CHT and IMPSO are superior to the other algorithms and have absolute benefits. The IMPSO algorithm can lead to more robust benefits in power generation when a new constraint processing approach is introduced, demonstrating the advantages of the suggested method in terms of solution capacity and engineering application. Despite the fact that the optimization approach proposed in this study enhances the utilization share of hydropower resources, there are certain issues. It should be noted that the proposed method has advantages in terms of optimization capability and performance, and it can meet the operational demands of hydropower systems. However, its structure is relatively complex, and the consideration of ecological river flow restriction is required in order to satisfy the coordinated operation of basin ecology and power generation demand. Therefore, future research can concentrate on the following aspects: (1) the performance of the algorithm—developing more efficient strategies can reduce the structural complexity and improve the robustness [27] and experimental accuracy of the algorithm while optimizing its goal and taking into account learning and comparison with more advanced algorithms [45,46]; (2) model construction—carrying out model construction according to actual demands and considering the multiple constraints in order to achieve the further integration of theoretical research and practical production.

Author Contributions: Conceptualization, Y.L.; methodology, Y.L.; software, Y.L.; validation, L.M. and Y.Y.; data curation, L.M.; writing—original draft preparation, Y.L.; writing—review and editing, Y.T.; visualization, Y.T.; supervision, L.M. and Y.Y; funding acquisition, L.M. and Y.Y. All authors have read and agreed to the published version of the manuscript.

Funding: This research was funded by the National Natural Science Foundation of China (Grant No. 51979114) and the Open Research Fund of Hubei Key Laboratory of Intelligent Yangtze and Hydroelectric Science (Grant No. 242202000913).

Data Availability Statement: Not applicable.

Acknowledgments: Special thanks are given to the anonymous reviewers and editors for their constructive comments.

Conflicts of Interest: The authors declare no conflict of interest.

References

1. Sun, L.; Niu, D.; Wang, K.; Xu, X. Sustainable development pathways of hydropower in China: Interdisciplinary qualitative analysis and scenario-based system dynamics quantitative modeling. *J. Clean. Prod.* **2020**, *287*, 125528. [CrossRef]
2. Li, X.-Z.; Chen, Z.-J.; Fan, X.-C.; Cheng, Z.-J. Hydropower development situation and prospects in China. *Renew. Sustain. Energy Rev.* **2018**, *82*, 232–239. [CrossRef]
3. Xie, Y.; Liu, S.; Fang, H.; Ding, M.; Wang, J. Three-Parameter Regulation Rules for the Long-Term Optimal Scheduling of Multiyear Regulating Storage Reservoirs. *Water* **2021**, *13*, 3593. [CrossRef]
4. Jiang, Z.; Qin, H.; Wu, W.; Qiao, Y. Studying Operation Rules of Cascade Reservoirs Based on Multi-Dimensional Dynamics Programming. *Water* **2018**, *10*, 20. [CrossRef]
5. Chang, J.; Wang, X.; Li, Y.; Wang, Y.; Zhang, H. Hydropower plant operation rules optimization response to climate change. *Energy* **2018**, *160*, 886–897. [CrossRef]
6. Suwal, N.; Huang, X.; Kuriqi, A.; Chen, Y.; Pandey, K.P.; Bhattarai, K.P. Optimisation of cascade reservoir operation considering environmental flows for different environmental management classes. *Renew. Energy* **2020**, *158*, 453–464. [CrossRef]
7. Guo, Y.; Ming, B.; Huang, Q.; Wang, Y.; Zheng, X.; Zhang, W. Risk-averse day-ahead generation scheduling of hydro–wind–photovoltaic complementary systems considering the steady requirement of power delivery. *Appl. Energy* **2022**, *309*, 118467. [CrossRef]
8. Qiu, H.; Chen, L.; Zhou, J.; He, Z.; Zhang, H. Risk analysis of water supply-hydropower generation-environment nexus in the cascade reservoir operation. *J. Clean. Prod.* **2021**, *283*, 124239. [CrossRef]
9. Liu, B.; Yao, K.; Wang, F.; Chi, X.; Gong, Y. Benefit Sharing in Hydropower Development: A Model Using Game Theory and Cost–Benefit Analysis. *Water* **2022**, *14*, 1208. [CrossRef]
10. Ren, Y.; Yao, X.; Liu, D.; Qiao, R.; Zhang, L.; Zhang, K.; Jin, K.; Li, H.; Ran, Y.; Li, F. Optimal design of hydro-wind-PV multi-energy complementary systems considering smooth power output. *Sustain. Energy Technol. Assess.* **2022**, *50*, 101832. [CrossRef]
11. Nazari-Heris, M.; Mohammadi-Ivatloo, B.; Gharehpetian, G.B. Short-term scheduling of hydro-based power plants considering application of heuristic algorithms: A comprehensive review. *Renew. Sustain. Energy Rev.* **2017**, *74*, 116–129. [CrossRef]
12. He, S.; Guo, S.; Chen, K.; Deng, L.; Liao, Z.; Xiong, F.; Yin, J. Optimal impoundment operation for cascade reservoirs coupling parallel dynamic programming with importance sampling and successive approximation. *Adv. Water Resour.* **2019**, *131*, 103375. [CrossRef]
13. Zhou, B.; Feng, S.; Xu, Z.; Jiang, Y.; Wang, Y.; Chen, K.; Wang, J. A Monthly Hydropower Scheduling Model of Cascaded Reservoirs with the Zoutendijk Method. *Water* **2022**, *14*, 3978. [CrossRef]
14. Needham, J.T.; Watkins, D.W.; Lund, J.R.; Nanda, S.K. Linear Programming for Flood Control in the Iowa and Des Moines Rivers. *J. Water Resour. Plan. Manag. Am. Soc. Civ. Eng.* **2000**, *126*, 118–127. [CrossRef]
15. Barros, M.T.L.; Tsai, F.T.-C.; Yang, S.-L.; Lopes, J.E.G.; Yeh, W.W.-G. Optimization of Large-Scale Hydropower System Operations. *J. Water Resour. Plan. Manag. Am. Soc. Civ. Eng.* **2003**, *129*, 178–188. [CrossRef]
16. Dogan, M.S.; Lund, J.R.; Medellin-Azuara, J. Hybrid Linear and Nonlinear Programming Model for Hydropower Reservoir Optimization. *J. Water Resour. Plan. Manag.* **2021**, *147*, 06021001. [CrossRef]
17. Van Dau, Q.; Kangrang, A.; Kuntiyawichai, K. Probability-Based Rule Curves for Multi-Purpose Reservoir System in the Seine River Basin, France. *Water* **2023**, *15*, 1732. [CrossRef]
18. Hjelmeland, M.N.; Zou, J.; Helseth, A.; Ahmed, S. Nonconvex Medium-Term Hydropower Scheduling by Stochastic Dual Dynamic Integer Programming. *IEEE Trans. Sustain. Energy* **2018**, *10*, 481–490. [CrossRef]
19. Agushaka, J.O.; Ezugwu, A.E. Initialisation Approaches for Population-Based Metaheuristic Algorithms: A Comprehensive Review. *Appl. Sci.* **2022**, *12*, 896. [CrossRef]
20. Lai, V.; Huang, Y.F.; Koo, C.H.; Ahmed, A.N.; El-Shafie, A. A Review of Reservoir Operation Optimisations: From Traditional Models to Metaheuristic Algorithms. *Arch. Comput. Methods Eng.* **2022**, *29*, 3435–3457. [CrossRef]
21. Azad, A.S.; Rahaman, S.A.; Watada, J.; Vasant, P.; Vintaned, J.A.G. Optimization of the hydropower energy generation using Meta-Heuristic approaches: A review. *Energy Rep.* **2020**, *6*, 2230–2248. [CrossRef]
22. Thaeer Hammid, A.; Awad, O.I.; Sulaiman, M.H.; Gunasekaran, S.S.; Mostafa, S.A.; Manoj Kumar, N.; Khalaf, B.A.; Al-Jawhar, Y.A.; Abdulhasan, R.A. A Review of Optimization Algorithms in Solving Hydro Generation Scheduling Problems. *Energies* **2020**, *13*, 2787. [CrossRef]
23. He, Z.; Zhou, J.; Xie, M.; Jia, B.; Bao, Z.; Qin, H.; Zhang, H. Study on guaranteed output constraints in the long term joint optimal scheduling for the hydropower station group. *Energy* **2019**, *185*, 1210–1224. [CrossRef]
24. Mao, J.-Q.; Tian, M.-M.; Hu, T.-F.; Ji, K.; Dai, L.-Q.; Dai, H.-C. Shuffled complex evolution coupled with stochastic ranking for reservoir scheduling problems. *Water Sci. Eng.* **2019**, *12*, 307–318. [CrossRef]
25. Hu, H.; Yang, K.; Su, L.; Yang, Z. A Novel Adaptive Multi-Objective Particle Swarm Optimization Based on Decomposition and Dominance for Long-term Generation Scheduling of Cascade Hydropower System. *Water Resour. Manag.* **2019**, *33*, 4007–4026. [CrossRef]
26. Wang, C.; Zhou, J.; Lu, P.; Yuan, L. Long-term scheduling of large cascade hydropower stations in Jinsha River, China. *Energy Convers. Manag.* **2015**, *90*, 476–487. [CrossRef]
27. Baldo, A.; Boffa, M.; Cascioli, L.; Fadda, E.; Lanza, C.; Ravera, A. The polynomial robust knapsack problem. *Eur. J. Oper. Res.* **2022**, *305*, 1424–1434. [CrossRef]

28. Lu, Y.; Liang, M.; Ye, Z.; Cao, L. Improved particle swarm optimization algorithm and its application in text feature selection. *Appl. Soft Comput.* **2015**, *35*, 629–636. [CrossRef]
29. Liu, D.; Xiao, Z.; Li, H.; Hu, X.; Malik, O. Accurate Parameter Estimation of a Hydro-Turbine Regulation System Using Adaptive Fuzzy Particle Swarm Optimization. *Energies* **2019**, *12*, 3903. [CrossRef]
30. Cao, Y.; Pan, H. Energy-Efficient Cooperative Spectrum Sensing Strategy for Cognitive Wireless Sensor Networks Based on Particle Swarm Optimization. *IEEE Access* **2020**, *8*, 214707–214715. [CrossRef]
31. Dahmani, S.; Yebdri, D. Hybrid Algorithm of Particle Swarm Optimization and Grey Wolf Optimizer for Reservoir Operation Management. *Water Resour. Manag.* **2020**, *34*, 4545–4560. [CrossRef]
32. Shi, Y.H.; Eberhart, R. A modified particle swarm optimizer. In Proceedings of the IEEE World Congress on Computational Intelligence, 1998 IEEE International Conference on Evolutionary Computation Proceedings, Anchorage, AK, USA, 4–9 May 1998; pp. 69–73. [CrossRef]
33. Yang, X.-S.; Deb, S.; Zhao, Y.-X.; Fong, S.; He, X. Swarm intelligence: Past, present and future. *Soft Comput.* **2018**, *22*, 5923–5933. [CrossRef]
34. Agushaka, J.O.; Ezugwu, A.E. Evaluation of several initialization methods on arithmetic optimization algorithm performance. *J. Intell. Syst.* **2022**, *31*, 70–94. [CrossRef]
35. Li, Q.; Liu, S.-Y.; Yang, X.-S. Influence of initialization on the performance of metaheuristic optimizers. *Appl. Soft Comput.* **2020**, *91*, 106193. [CrossRef]
36. Rauf, H.T.; Shoaib, U.; Lali, M.I.; Alhaisoni, M.; Irfan, M.N.; Khan, M.A. Particle Swarm Optimization with Probability Sequence for Global Optimization. *IEEE Access* **2020**, *8*, 110535–110549. [CrossRef]
37. Li, J.; An, Q.; Lei, H.; Deng, Q.; Wang, G.-G. Survey of Lévy Flight-Based Metaheuristics for Optimization. *Mathematics* **2022**, *10*, 2785. [CrossRef]
38. Liu, L.; Bai, K.; Dan, Z.; Zhang, S.; Liu, Z. Whale Optimization Algorithm with Global Search Strategy. *J. Chin. Comput. Syst.* **2020**, *41*, 1820–1825. Available online: https://scholar.google.com/scholar_lookup?title=Whale+Optimization+Algorithm+with+Global+Search+Strategy&author=Liu,+L.&author=Bai,+K.&author=Dan,+Z.&author=Zhang,+S.&author=Liu,+Z.&publication_year=2020&journal=J.+Chin.+Comput.+Syst.&volume=41&pages=1820%E2%80%931825 (accessed on 4 September 2020).
39. Zhang, C.; Lin, Q.; Gao, L.; Li, X. Backtracking Search Algorithm with three constraint handling methods for constrained optimization problems. *Expert Syst. Appl.* **2015**, *42*, 7831–7845. [CrossRef]
40. Li, Z.; Chen, S.; Zhang, S.; Jiang, S.; Gu, Y.; Nouioua, M. FSB-EA: Fuzzy search bias guided constraint handling technique for evolutionary algorithm. *Expert Syst. Appl.* **2019**, *119*, 20–35. [CrossRef]
41. Gandomi, A.H.; Deb, K. Implicit constraints handling for efficient search of feasible solutions. *Comput. Methods Appl. Mech. Eng.* **2020**, *363*, 112917. [CrossRef]
42. Harrison, K.R.; Engelbrecht, A.P.; Ombuki-Berman, B.M. Self-adaptive particle swarm optimization: A review and analysis of convergence. *Swarm Intell.* **2018**, *12*, 187–226. [CrossRef]
43. Haklı, H.; Uğuz, H. A novel particle swarm optimization algorithm with Levy flight. *Appl. Soft Comput.* **2014**, *23*, 333–345. [CrossRef]
44. Yan, Z.; Zhaobin, L.; Zhang, B. A Novel Differential Evolution Algorithm for Constrained Optimization. In Proceedings of the 2017 IEEE International Conference on Computational Science and Engineering (CSE) and IEEE International Conference on Embedded and Ubiquitous Computing (EUC), Guangzhou, China, 21–24 July 2017. [CrossRef]
45. Zamani, H.; Nadimi-Shahraki, M.H.; Gandomi, A.H. QANA: Quantum-based avian navigation optimizer algorithm. *Eng. Appl. Artif. Intell.* **2021**, *104*, 104314. [CrossRef]
46. Zamani, H.; Nadimi-Shahraki, M.H.; Gandomi, A.H. Starling murmuration optimizer: A novel bio-inspired algorithm for global and engineering optimization. *Comput. Methods Appl. Mech. Eng.* **2022**, *392*, 114616. [CrossRef]

Article

Hydropower and Pumped Storage Hydropower Resource Review and Assessment for Alaska's Railbelt Transmission System

Leif Bredeson [1] and Phylicia Cicilio [2,]*

[1] Department of Electrical Engineering and Computer Science, South Dakota State University, Brookings, SD 57007, USA; leif.bredeson@jacks.sdstate.edu

[2] Alaska Center for Energy and Power, University of Alaska, Fairbanks, AK 99775, USA

* Correspondence: pcicilio@alaska.edu

Abstract: The Alaska Railbelt transmission system runs from Fairbanks to Anchorage to Homer and supplies 75% of the state's population with power. In the near future, this system will experience significant increases in load due to electrification of the transportation and heating sectors. To account for this, several state organizations are working towards the creation of an integrated resource plan and reliability standards.This work encompasses the efforts of researching the operations, cost, and locations of desirable hydropower and pumped storage hydropower (PSH) resources in the areas surrounding the Railbelt transmission system. The aspects of conventional hydropower and PSH as well as adjustable-speed and ternary PSH were analyzed. With Alaska's diverse and rugged landscape, QGIS was utilized to delineate the positions of energy resources within a reasonable distance from the Railbelt. By incorporating Digital Terrain Models and the QGIS processing toolbox, a least cost path analysis was completed to filter out resources within the designated distance of the Railbelt. Applying existing cost models to the data in this work helped to decide the energy resources that would be studied further. The future of this project includes modeling the selected energy sources in the PSS/e Railbelt model to examine their effects on the reliability and stability.

Keywords: hydropower; pumped storage hydro; resource assessment; power system planning

Citation: Bredeson, L.; Cicilio, P. Hydropower and Pumped Storage Hydropower Resource Review and Assessment for Alaska's Railbelt Transmission System. *Energies* **2023**, *16*, 5494. https://doi.org/10.3390/en16145494

Academic Editors: Yongguang Cheng and Zhengwei Wang

Received: 17 April 2023
Revised: 23 June 2023
Accepted: 17 July 2023
Published: 20 July 2023

1. Introduction

In the United States, the federal government has set a net zero carbon emissions goal for 2050 [1]. Hydropower currently provides 38% of the renewable generation in the United States, and pumped storage hydro (PSH) currently provides 93% of grid storage [2]. Hydropower and PSH are currently key components of the net zero carbon emission goal for the United States. The state of Alaska does not currently have any state carbon emission goals; however, individual electric utilities do have carbon reduction goals, and hydropower plays in an important role in energy generation in the state.

Conventional hydropower first began its United States utility-scale journey in the early 1880s [3]. In 2019, there was a hydropower installed capacity of 80.25 gigawatts in the United States, which accounted for 6.7% of the installed generation capacity and 6.6% of all electricity generated in the United States [4]. In Alaska, hydroelectricity has been a cornerstone of powering communities and advancements since the early 1900s. Alaska's energy portfolio is composed of about 25% hydropower with a capacity of about 483 MW across 50 power plants [5], and it accounts for 90% of the renewable energy generated in Alaska [6].

There is no history of pumped storage hydro methods, such as conventional hydropower, in Alaska; however, it has played a larger role in the contiguous United States since 1930. PSH accounts for 93% of utility-scale storage across 43 PSH plants. This amounts to a generation capacity of about 23 GW [7]. PSH was originally developed in the 1960s to assist with new nuclear generation with limited ramping capabilities being built at the time [8]. The US has not brought any new plants online since 2000. Most plants were brought online in the 1970s [9]. PSH is traditionally used to shift energy between day and night electricity rates; however, in scenarios with high renewables, particularly those with high wind generation, PSH could provide cost savings and reduce wind curtailment [10–12].

This paper explores the feasibility of an increased renewable energy capacity in Alaska, specifically in the regional transmission system called the Railbelt outlined in Section 2. Hydropower and PSH were researched in particular because of the large resources available in Alaska. Hydropower has also been a staple for power on the Railbelt and in more remote areas. The hydropower and pumped storage hydro (PSH) resource availability and resource costs were assessed for this region in this paper. Impoundment hydro facilities and three types of PSH are included: conventional, adjustable-speed, and ternary. Multiple characteristics of the resources were collected and calculated, including the operation behavior, location, capacity, and cost.

The purpose of this work is to gather information on these resources that can be used to help make informed decisions on these technologies' usefulness in Alaska. The contribution of this paper is (1) to provide a review of hydropower and PSH technologies, including how they operate and resource locations in Alaska's Railbelt, and (2) to provide updated cost estimates for identified PSH sites in Alaska's Railbelt and outline these technologies' advantages and disadvantages for the purpose of power system planning and decision making on Alaska's Railbelt transmission system.

This paper is organized into the following sections: Section 2 details the study area used in this paper, Alaska's Railbelt transmission system. Section 3 provides a review of the technical operation of hydropower and PSH. Section 4 details the methodologies used to perform the resource assessment and availability and cost assessments in the region. Section 5 provides the resulting hydropower and PSH site locations and costs. Section 6 provides a discussion of other considerations for the development of hydropower and PSH. Section 7 concludes the paper and details future work.

2. Study Area

Alaska's Railbelt Transmission system, referred to as the Railbelt, is the largest electrical grid in Alaska, serving about 900 MW of the peak winter load from Homer, AK to Fairbanks, AK. The Alaska Railbelt transmission system supplies 75% of the state's population with power. Figure 1 illustrates the approximate location and layout of the Railbelt in Alaska.

The Railbelt has four electric cooperatives: the Golden Valley Electric Association, the Matanuska Electric Association, the Chugach Electric Association, and the Homer Electric Association. These utilities are grouped into three main load regions: northern, central, and southern.

Three hydropower plants on the Railbelt account for 18% of the total generation [13]. These three power plants are Bradley Lake, Cooper Lake, and Eklutna Lake. Bradley Lake began its operation in 1991 and currently supplies the Railbelt with a generation capacity of 120 MW. Cooper Lake went online in 1960 and generates 16.7 MW. Eklutna Lake came online in 1955 and currently generates 30 MW [14].

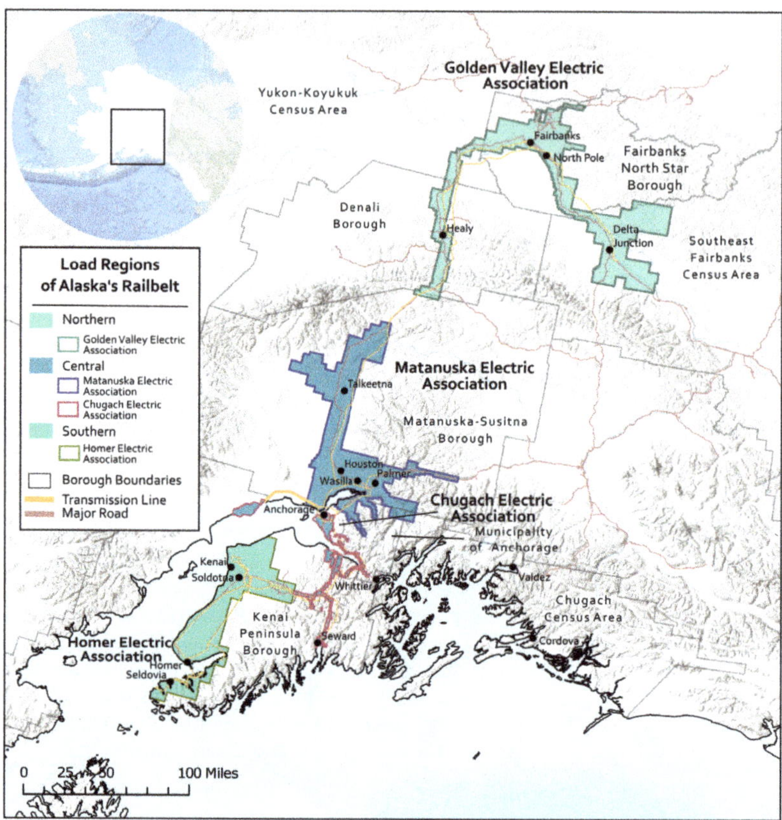

Figure 1. Alaska's Railbelt transmission system spanning Fairbanks to Homer, AK.

3. Technology Review

Conventional hydropower generates power by converting the gravitational potential energy of raised water to kinetic energy, which flows through a turbine connected to a generator. Hydropower uses one "reservoir" that is usually situated above the powerhouse. The reservoir can change based on the type of hydropower plant. Three types of hydropower exist: impoundment, diversion, and run-of-river [15]. Impoundment is the most common of the three. A dam is created to impound the water and create a reservoir with a much higher head pressure. Diversion does not create a reservoir; it diverts a portion of the river or stream into a penstock or another channel that flows into the powerhouse. Run-of-river hydro creates a low-profile solution that does not require a large impoundment of water, as opposed to full impoundment dams.

Conventional hydropower uses several types of turbines as well as accessory systems to harness the kinetic energy of water and efficiently control it. Hydroelectric turbines are separated into two categories, impulse and reaction. These categories separate the turbines by characterizing the method of water delivery. Each one also has a unique combination of flow and head. Impulse turbines utilize the velocity of the water to discharge into a bucket to transfer the energy to the turbine. Pelton and cross-flow turbines are the most common types of impulse turbines.

Reaction turbines use the pressure generated by the water to turn the turbine. This water flows through a curved penstock called a scroll case. This case directs the water into the turbine. This water is discharged perpendicular to its angle of entry. Two common reaction turbines are Francis and Kaplan, Francis being one of the most common models [15].

Each type of turbine has its own method for controlling the flow of water. For the Pelton turbine, an adjustable "spear" controls the amount of water moving through the nozzle; there is also a deflector that will move in front of the nozzle to control the flow. For the Francis turbine, wicket gates are aligned around the edge of the turbine. The gates are rotated to control the flow, ranging from fully closed to fully open, and are located parallel to the water flow. These controls allow the hydropower plant to regulate its power output and respond to the load demand. Hydropower has the ability to ramp to full generation on a scale of 12–50 MW/s [16].

PSH can be described as a gravity-driven water battery. PSH installations involve two or more reservoirs of water with a substantial amount of elevation difference. Often, this can be conducted by utilizing lakes, existing reservoirs, or even oceans and seas. Another method consists of creating reservoirs using geographic features, such as large "bowl-shaped" features separated by elevation. Depending on the reservoir type, a PSH plant can be characterized as closed- or open-loop [7]. Closed-loop systems are systems that are not connected to any natural water source i.e., lakes, rivers, and oceans. Open-loop systems are systems that utilize a natural reservoir, for example, pumping between two lakes [17].

PSH moves volumes of water between the two reservoirs to follow the demand. When the power demand is high, water flows from the upper reservoir to the lower reservoir to generate power. When the demand is low, water is pumped from the lower reservoir to the upper reservoir using cheap or excess power. The water is transported, via penstocks, to and from the powerhouse at the lower reservoir using pump–turbines.

The configuration of the pump/turbine depends on the type of turbine used. The setups researched for this study included conventional (C-PSH), adjustable-speed (AS-PSH), ternary (T-PSH) [18], and quaternary (Q-PSH) [19]. Conventional and adjustable-speed units utilize a reversible pump–turbine, sometimes referred to as pump-as-turbines. Due to the similarity in the design of pumps and turbines, slight modifications can be made to enable the turbine to reverse its direction and pump the water, instead of allowing it to flow through. This also means that the generator, when reversed, acts as a motor. C-PSH most often uses a synchronous machine for its operation. Configurations using synchronous machines are restricted to operating at one rated speed for maximum efficiency, limiting their capability to "Full Off" or "Full On". AS-PSH can be configured with a Doubly Fed Induction Machine, an asynchronous machine, or a synchronous machine, with the only item changing being the location and size of its power converter. The asynchronous machine is similar to a Type 3 Wind Turbine, while the synchronous machine is similar to a Type 4.

The asynchronous machine has a wider operating range with a high efficiency. Being able to change the speed at which the reversible pump-turbine generates or pumps enables a stronger ability to regulate the frequency or respond to the load. T-PSH can also respond effectively due to its unique setup. T-PSH utilizes a pump and turbine separately on a single shaft connected to a synchronous machine. The separate machines allow the system to pump and generate without reversing the rotation of the shaft. The pump and turbine are fed by the same penstock split in two. C-PSH and AS-PSH are limited to two modes, pump and turbine, allowing only one direction for water flow in the system. Ternary has three modes, pump, turbine, and hydraulic short circuit. This third mode is enabled due to the split penstock and single direction rotation of the shaft. A hydraulic short circuit operates the pump and the turbine at the same time while also supplying the pump with energy from the grid. Volumes of water being moved are segmented into blocks of power, as shown in the Figure 2. This short-circuit mode enables T-PSH to have a very competitive advantage over C-PSH and AS-PSH due to decreased mode switching times and increased frequency regulation capabilities.

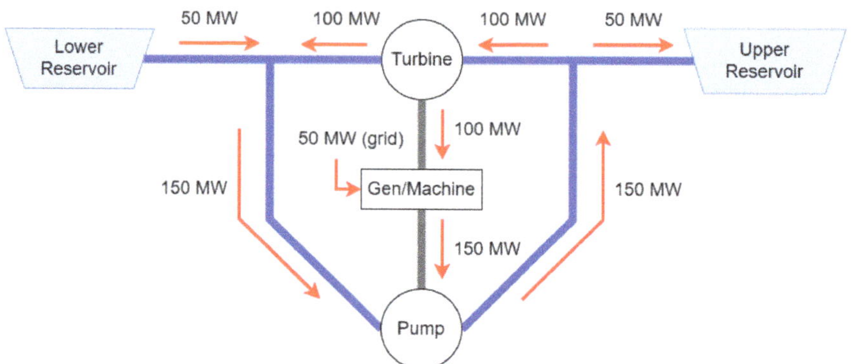

Figure 2. Control diagram of the ternary PSH.

Q-PSH has two shafts compared to the one shaft that is available in the other three types of PSH. The two shafts allow for a pump–motor combination and a turbine–generation combination; with a full sized converter between the pump and motor and a direct grid connection with the turbine–generator combination. This two-shaft structure, with the combinations of the pump–motor and turbine–generator, combines the benefits of the AS- and T-PSH and allows faster mode change times. Therefore, it provides the best frequency regulation of the types of PSH available, particularly with high penetration of renewable energy resources [19].

When selecting a PSH configuration, system demand and cost are the primary factors. Depending on the demand, the ability to transition between modes can also quickly become an important factor in the decision-making process. C-PSH is the most basic of the technologies. It also has the slowest switching time with averages in the range of minutes. AS-PSH is slightly faster, ranging from 1 to 4 min for a transition between pumping and generating. T-PSH has a transition time of 30 s to 1.5 min [20]. Q-PSH has the fastest switching times and provides the greatest frequency support [19].

4. Data and Research Methodology

4.1. Resource Assessment and Availability

For conventional hydropower, a database from the Alaska Energy Authority distributed through the State of Alaska Open Data Geoportal [21] was used to map proposed hydroelectric sites in Alaska. For PSH, datasets from the National Renewable Energy Laboratory (NREL) were used [22]. This dataset was created using a GIS-based analysis of potential closed-loop systems in the United States. The PSH sites were filtered in an assessment performed by NREL [22,23], and the number of sites was narrowed down to 1819 sites. The goal of the generated resource maps was to estimate the possible size and number of PSH and hydro resources potentially usable for the Railbelt. The following criteria were used to select sites:

- Cost ($/kW);
- Distance from Railbelt (miles);
- Capacity (MW);
- Land ownership;
- Previously researched potential sites.

To apply these criteria, publicly available infrastructure data for Alaska's high-voltage transmission lines were imported from the Alaska Energy Authority's database hosted on the State of Alaska Geoportal. This public transmission data were selected to exclusively show the Railbelt transmission lines. The potential resources were filtered based on their straight-line proximity to the transmission line; for this study, 20 miles was selected as the maximum distance from a Railbelt transmission line. However, due to Alaska's topography, if a spur line cost was to be estimated, the straight-line distance would not be correct.

To estimate the cost of a spur line, a possible path for this line must be created. To avoid mountains and rapid elevation changes, it was assumed that the spur line would follow a path with the least amount of slope. To simulate this path in QGIS, a least cost path analysis was completed with the cost being the slope. A Digital Terrain Model of the area of interest surrounding the Railbelt was imported from the State of Alaska's Geoportal. The Digital Terrain Model held elevation data about the area of interest's terrain. This excluded trees and buildings that would be included in a Digital Surface Model. Using the slope tool in QGIS, a slope map was created. To keep the paths reasonable, all bodies of water were excluded from the slope map using the intrinsic slope of water, 0, and converting all 0 values to NULL values, leading them to be ignored by the least cost path tool. The least cost path tool assessed the values held in each pixel of the slope map. This assessment was used to draw a path from the resource site to the nearest point on the Railbelt whilst collecting the least amount of cost (slope).

The Railbelt-wide hydro and PSH sites with the least cost paths are shown in Figure 3. A close-up view of Figure 3 is shown in Figure 4 to illustrate this method.

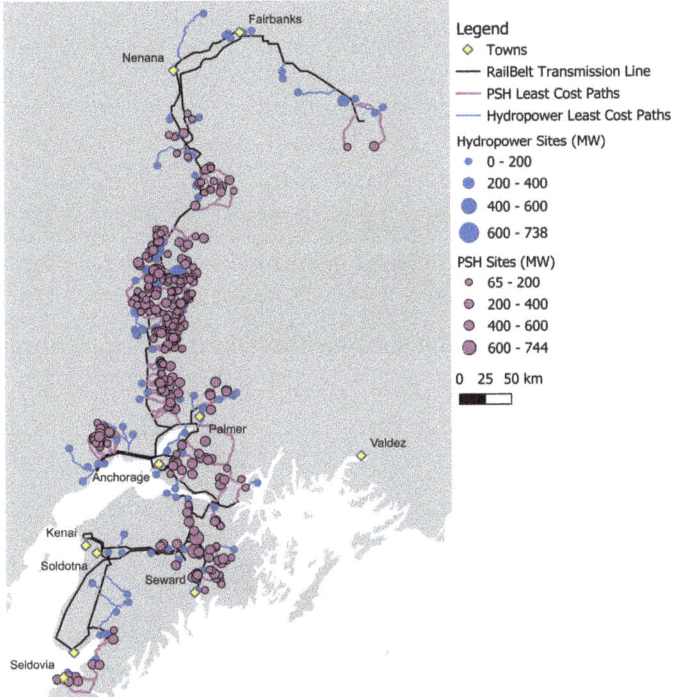

Figure 3. Least cost path of the hydro and PSH sites to the Railbelt transmission system.

Alaska's land ownership was the next factor considered for resource location. Alaska has native, federal, state, and borough lands, protected wildlife, national parks, military bases, etc. To visualize the land ownership, a map of Administered Lands, courtesy of the Alaska division of the US Bureau of Land Management, was used [24]. This map specifies the owner of each parcel of land in Alaska, allowing potential resource sites to be removed if they are on non-buildable land, such as National Parks, military installations, and other protected areas. Locations that exist on public lands were selected as these are more likely to be used for potential development. The map of Administered Lands was also used for potential resource sites from the Alaska Energy Authority.

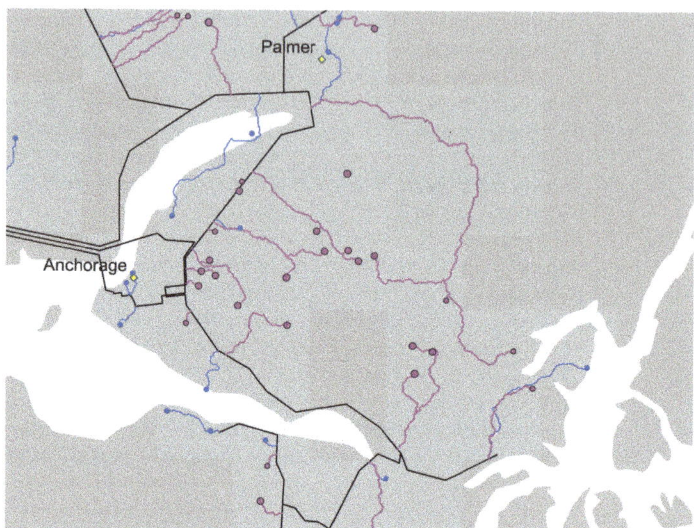

Figure 4. Close-up view of the least cost path of the hydro and PSH sites to the Railbelt transmission system. The purple lines and dots represent PSH least cost paths and location, and the blue lines and dots represent hydropower least cost paths and locations.

4.2. Resource Cost

Water-based energy resources are commonly expensive to construct due to their low energy density (dams, reservoirs, etc.) and environmental impact. The conventional hydropower initial capital cost (ICC) is dependent on several factors. Hydropower projects are often filtered based on their generation capacity, which contributes to other cost factors, such as the head and reservoir size. Capital costs of hydropower include electromechanical systems, C&I, environmental impact studies, permitting, and environmental impact mitigation procedures. A cost analysis of hydropower by the International Renewable Energy Agency identified the largest contributors to capital costs for large- and small-scale hydropower projects [25]. Large-scale hydropower projects have capital costs consisting mainly of C&I, such as the impoundment, tunneling, and powerhouse costs associated with moving large volumes of water. Capital costs of small-scale hydropower projects are dominated by electromechanical systems, such as generators, turbines, control systems, etc. Multiple models and studies have shown a general relationship between the generation capacity and cost. Head also has an effect on this relationship [25]. For most projects, the ICC ($/kW) will increase as the capacity decreases. For small hydro projects, 5–100 MW, the cost can range from $1300 to $8000/kW ($2010 USD). Large hydro projects, greater than 100 MW, have a slightly lower range of $1050–7650/kW [25]. The Oak Ridge National Laboratory (ORNL) completed their own analysis of hydropower costs and created a model to estimate new project costs [26]. The ORNL analyzed 84 New-Stream Development projects, meaning situations where no dam existed prior to the project. The ORNL assessed this database using project characteristics and a confidence score system to develop an ICC model equation, Equation (1), where P is the capacity in MW, and H is the head in feet. The ICC is measured in USD for 2012.

$$ICC = 8717830 \, P^{0.975} H^{-0.12} \tag{1}$$

Using this model, the ORNL estimated that, on average, new-stream development projects cost 3882 $/kW. This cost was verified against 17 constructed projects with an average cost of 3885 $/kW [26].

The pumped storage hydropower capital cost is dependent on multiple factors. Those with the greatest effect include the head height, flow rate, and storage/generation capacity. The NREL modified the simplified PSH cost model from the Australian National University [27], which was adapted from a detailed model created in a report from Entura [28]. Equation (2) is the total cost model, and seven equations from [22] go into this equation.

$$C_T = (C_P + C_t + C_u + C_l)\,\frac{1.33}{1.2} + C_s \tag{2}$$

where C_T is the total cost in 2018\$, C_P is the cost of the powerhouse in 2018\$, C_t is the cost of the tunnel in 2018\$, C_u is the cost of the upper reservoir in 2018\$, C_l is the cost of the lower reservoir in 2018\$, the ratio 1.33/1.2 is the contingency cost adjustment factor used to adjust the contingency cost of 20% used in the ANU model to the contingency cost of 33% used for other NREL technologies, and C_s is the cost of the spur line in 2018\$. The NREL made two adjustments in Equation (2): C_s was added, as the ANU's model did not include spur line costs, and a cost correction factor of 1.51 was added to this simplified model for plants with a capacity of 900–1100 MW. This cost correction factor was arrived at by comparing the median cost of 900–1100 MW plants, created by the ANU's cost model, to the average cost of a 1000 MW plant modeled as a result of the Energy Storage Grand Challenge. The final range of costs for PSH plants from the NREL [28] was 1407–2137\$/kW. The Energy Storage Grand Challenge created a comprehensive table outlining the costs and lifetime performance for a 100 MW, 1000 MWh system [16]. This table includes the powerhouse C&I, reservoir, contingency, O&M, electromechanical powertrain, and performance metrics based on the lifespan, efficiency, and ramp rates. It does not include O&M, which was modeled using a function of the plant capacity (P) in MW and the annual energy throughput (AE) in MWh, as shown in Equation (3).

$$C_{\text{O\&M}} = 34730\,P^{0.32}AE^{0.33} \tag{3}$$

Alaska's construction costs were calculated based on the cost assessments from the NREL [28]. Alaska's construction costs included an adjustment for inflation from 2018 to 2023. An estimated increased cost of construction in Alaska was determined by comparing average new-stream development costs from the United States, 4800 \$/kW (2015\$) [26] adjusted to 6144 \$/kW (2023\$), to the most recently proposed hydro development in Alaska. The most recently proposed Alaskan hydro project is Grant Lake, which has a capacity of 5 MW. Grant Lake's estimated construction costs total \$53,878,050 (2018\$), adjusted to \$65,192,441 (2023\$) [29], which equates to 13,038 \$/kW. Therefore, the an Alaskan construction cost factor of 2.12 was used to calculate the increased cost of construction in Alaska.

5. Results

5.1. Pumped Storage Hydro

Resource locations for each type of technology were narrowed down from the locations identified by the Alaska Energy Authority for hydro and the NREL locations for PSH as well as the Eklutna and Right Mountain locations identified by the Alaska Institute for Climate and Energy, which are located on the Railbelt [30]. Cost was the ultimate deciding factor for the sites selected. The PSH site locations were narrowed down to the top ten lowest-cost sites by \$/kW using the Alaska construction costs. The capital costs of the Eklutna and Right Mountain sites reported in the Alaska Institute for Climate and Energy report were increased to adjust to 2023\$, and no additional Alaska construction cost multiplier was added; however, it should be noted that these costs were calculated differently and may not be suited for direct comparison to the other site costs. The details of each of the top ten lowest-cost sites within 3 miles of the Railbelt are listed in Table 1, where the first site is the Eklutna PSH site.

Table 1. Top ten lowest-cost ($/kW) PSH sites within 3 miles of the Railbelt.

Site	Nearest Water Body	Power Capacity (MW)	Energy Storage (GWh)	Capital Cost Estimate (2023 $/kW)
1	Eklutna Lake	426	507	2328
2	Sustina River	744	7.44	2979
3	Lewis River	707	7.07	3400
4	Nellie Juan River	461	4.61	3540
5	Ship Creek	542	5.42	3645
6	Jakolof Creek	432	4.32	3753
7	Yanert Fork	502	5.02	3963
8	Pierce Creek	407	4.08	3997
9	Willow Creek	296	2.96	4146
10	Little Willow Creek	445	4.45	4176

5.2. Hydropower

Hydropower locations were identified based solely on previously studied sites as well as some upgrades to current hydropower sites. The largest site identified was the Susitna–Watana Hydro Project. This project has an annual energy budget of 2.8 million MWh with an installed generation capacity of 459 MW [14]. The Dixon Diversion was also included. This is a project to divert glacial meltwater from the Dixon Glacier, increasing the annual energy budget of Bradley hydro by 150 GWh. This additional capacity would require the Bradley Lake dam to be raised; however, the total nameplate capacity of the Bradley Lake power plant would stay the same. The Grant Lake hydro project on the Kenai Peninsula was also identified. This project is scheduled to begin construction in 2023 and has a 5 MW generation capacity and an annual energy production of 18,600 MWh. The selected hydro sites along with the top ten lowest-cost PSH sites are shown in Figure 5.

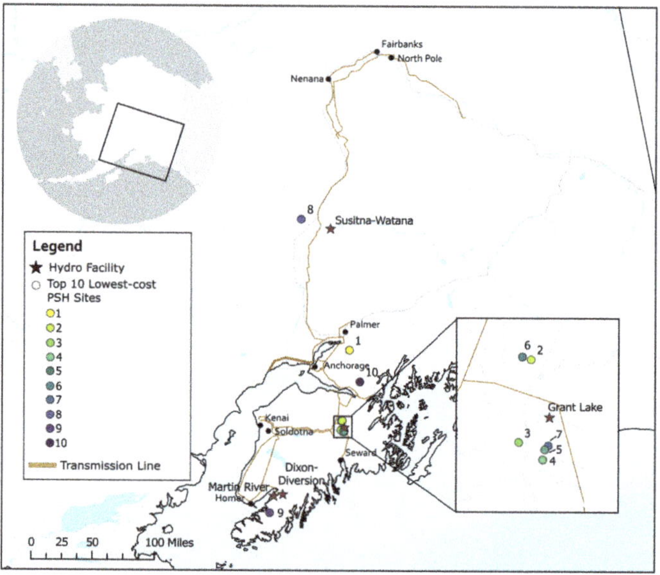

Figure 5. Top ten lowest-cost PSH sites within 3 miles of the Railbelt transmission system and selected hydro sites.

6. Other Considerations

Hydropower can also provide additional services to electric grids beyond energy. There are also negative consequences associated with both hydropower and PSH. These other considerations are further outlined here and classified as either a strength or weakness of hydropower or PSH.

6.1. Strengths

Hydropower generators are capable of providing frequency regulation and reserves, and their hourly ramping flexibility is the most extensively used out of any other type of resource in the United States [2]. This flexibility and frequency regulation and reserve capability are key grid services, especially when integrating variable renewable energy resources. Hydropower generators can also be black-start capable, meaning that they can be used to restart the electrical grid after a system-wide blackout [2]. Black-start capability is not found with all types of resources and is a critical component of grid resilience.

Hydropower also has non-energy-related benefits including food security through providing a source of water for irrigation, water quality and quantity through the control of water release, high-paying jobs, flood protection through capturing floodwaters, and recreation and tourism [31].

6.2. Weaknesses

Energy resources that use water as the generation medium can have ecological and environmental (flora, fauna, landscape, and historical remains [32]) impacts. Water must be diverted or stored to create power. This often creates reservoirs that can span hundreds of miles. In Alaska, there are several factors that come into consideration when deciding where to build a PSH or hydropower plant. For hydropower, much of Alaska's most energy dense river systems are considered anadromous streams [33]. These streams are used by salmon for spawning, meaning that interference would impact salmon runs, which is a subsistence resource and a cornerstone of Alaska's economy. Other issues include the obstruction of natural views, which is also important for tourism.

PSH can have a similar effect to conventional hydropower, depending on how it is configured. Closed-loop systems have the potential to impact habitats and species, depending on where the reservoirs are placed. This, for example, could be a concern for caribou habitats in much of Alaska. Open-loop systems pose the same threat as hydropower due to constantly changing water levels disabling the ability for the reservoirs to be used by wildlife.

Studies have identified that the water bodies of dams of hydropower facilities release CO_2 and CH_4 emissions due to both human-made and natural processes [34,35]. Therefore, hydropower generation is not a completely carbon-free resource. However, it has been found that hydropower generation facilities do have lower emissions in comparison to thermo-based generation methods, such as coal, oil, diesel, and natural gas generation facilities [35].

7. Conclusions

This paper covered how PSH and hydropower operate on a technical basis as well as the cost, and potential locations of these resources. PSH and hydropower generate and store power using the potential energy of elevated water sources. Hydropower uses conventional turbine generators for operation and can ramp up to full generation on a scale of 12–50 MW/s [16]. PSH is similar to hydropower; however, it utilizes two reservoirs of water separated by an elevation difference to create a large water battery. The water is pumped to the upper reservoir when the demand is low and generated down to the lower reservoir when the demand is high. PSH powerhouses use reversible pump–turbines to pump and generate between the two reservoirs. The transition time between pumping and generating changes depending on the PSH configuration: conventional, adjustable-speed, ternary, and quaternary. Quaternary has the fastest switching times and greatest frequency regulation capabilities.

PSH and hydropower have comparable ICC categories: C&I, electromechanical systems, environmental protection procedures, and permitting. Cost models for hydropower plants created by ORNL showed that new-stream development hydropower has an average ICC of around 3882 $/kW. PSH cost models from the NREL HydroWIRES report resulted in a ICC range of 1407–2137 $/kW. The construction costs accounting for inflation and Alaskan construction costs resulted in a range of 2090–2931 $/KW.

The potential locations of hydropower and PSH were analyzed from several sets of data. Ultimately, two hydropower sites were identified: the Susitna–Watana Hydro and Grant Lake. Ten potential PSH sites were also identified based on cost.

Author Contributions: Conceptualization, L.B. and P.C.; Methodology, L.B.; Formal analysis, L.B.; Investigation, L.B.; Writing—original draft, L.B.; Writing—review & editing, L.B. and P.C.; Visualization, L.B.; Supervision, P.C.; Project administration, P.C.; Funding acquisition, P.C. All authors have read and agreed to the published version of the manuscript.

Funding: This project is part of the Arctic Regional Collaboration for Technology Innovation and Commercialization (ARCTIC) 2 Program-Innovation Network, an initiative supported by the Office of Naval Research (ONR) Award #N00014-22-1-2049. This project is funded by the state of Alaska FY23 economic development capital funding.

Acknowledgments: The authors would like to thank Noelle K. Helder for the creation of the maps used in this paper and Erin Trochim for technical guidance.

Conflicts of Interest: The funders had no role in the design of the study; in the collection, analyses, or interpretation of data; in the writing of the manuscript; or in the decision to publish the results.

Abbreviations

The following abbreviations are used in this manuscript:

Pumped storage hydro	PSH
Conventional pumped storage hydro	C-PSH
Adjustable-speed pumped storage hydro	AS-PSH
Ternary pumped storage hydro	T-PSH
Quaternary pumped storage hydro	Q-PSH
Initial capital cost	ICC
Oak Ridge National Laboratory	ORNL
National Renewable Energy Laboratory	NREL

References

1. The White House. FACT SHEET: President Biden to Catalyze Global Climate Action through the Major Economies Forum on Energy and Climate. 2023. Available online: https://www.whitehouse.gov/briefing-room/statements-releases/2023/04/20/fact-sheet-president-biden-to-catalyze-global-climate-action-through-the-major-economies-forum-on-energy-and-climate/ (accessed on 2 February 2022).
2. Uría-Martínez, R.; Johnson, M.M.; Shan, R.; Samu, N.M.; Oladosu, G.; Werble, J.M.; Battey, H. U.S. Hydropower Market Report. 2021. Available online: https://www.energy.gov/sites/default/files/2021/01/f82/us-hydropower-market-report-full-2021.pdf (accessed on 2 February 2022).
3. U.S. Energy Information Administration. History of Hydropower. 2022. Available online: https://www.eia.gov/energyexplained/hydropower/ (accessed on 2 February 2022).
4. Discover Hydropower. Available online: https://www.hydro.org/waterpower/hydropower/ (accessed on 2 February 2022).
5. Renewable Energy Alaska Project. Hydroelectric. 2022. Available online: https://alaskarenewableenergy.org/technologies/hydroelectric/ (accessed on 2 February 2022).
6. Existing Hydropower. Available online: https://www.hydro.org/map/hydro/existing-hydropower/ (accessed on 14 June 2023).
7. Department of Energy, Office of Energy Efficiency and Renewable Energy. Pumped Storage Hydropower. 2022. Available online: https://www.energy.gov/eere/water/pumped-storage-hydropower (accessed on 2 February 2022).
8. Yang, C.J.; Jackson, R.B. Opportunities and barriers to pumped-hydro energy storage in the United States. *Renew. Sustain. Energy Rev.* **2011**, *15*, 839–844. [CrossRef]

9. U.S. Energy Information Administration. Most Pumped Storage Electricity Generators in the U.S. Were Built in the 1970s. 2019. Available online: https://www.eia.gov/todayinenergy/detail.php?id=41833 (accessed on 2 February 2022).
10. Tuohy, A.; O'Malley, M. Pumped storage in systems with very high wind penetration. *Energy Policy* **2011**, *39*, 1965–1974. [CrossRef]
11. Shao, M.; Guo, X.; Bisceglia, C.; de Mijolla, G.; Rao, S.; Pajic, S.; Ibanez, E.; Bringolf, M.; Havard, D. *Value and Role of Pumped Storage Hydro under High Variable Renewables*; U.S. Department of Energy: Washington, DC, USA, 2021. [CrossRef]
12. Golshani, A.; Sun, W.; Zhou, Q.; Zheng, Q.P.; Wang, J.; Qiu, F. Coordination of Wind Farm and Pumped-Storage Hydro for a Self-Healing Power Grid. *IEEE Trans. Sustain. Energy* **2018**, *9*, 1910–1920. [CrossRef]
13. Denholm, P.; Schwarz, M.; DeGeorge, E.; Stout, S.; Wiltse, N. *Renewable Portfolio Standard Assessment for Alaska's Railbelt*; National Renewable Energy Laboratory: Golden, CO, USA, 2022. [CrossRef]
14. Susitna-Watana Hydro. Hydropower in Alaska. 2022. Available online: https://www.susitna-watanahydro.org//index_769.html (accessed on 2 February 2022).
15. U.S. Department of Energy. Types of Hydropower Turbines. Available online: https://www.energy.gov/eere/water/types-hydropower-turbines (accessed on 2 February 2022).
16. Mongird, K.; Viswanathan, V.; Alam, J.; Vartanian, C.; Sprenkle, V. 2020 Grid Energy Storage Technology Cost and Performance Assessment. 2020. Available online: https://www.pnnl.gov/sites/default/files/media/file/Final%20-%20ESGC%20Cost%20Performance%20Report%2012-11-2020.pdf (accessed on 2 February 2022).
17. Saulsbury, J.W. *A Comparison of the Environmental Effects of Open-Loop and Closed-Loop Pumped Storage Hydropower*; U.S. Department of Energy: Washington, DC, USA, 2020. [CrossRef]
18. Feltes, J.W.; Koritarov, V.; Kazachkov, Y.; Gong, B.; Donalek, P.J.; Gevorgian, V.; Pti, S.; Americas, M. *Testing Dynamic Simulation Models for Different Types of Advanced Pumped Storage Hydro Units*; Number No. ANL/DIS-13/08; UNT Libraries Government Documents Department: Denton, TX, USA, 2013.
19. Nag, S.; Dong, Z.; Tan, J.; Kim, J.; Muljadi, E.; Lee, K.Y.; Jacobson, M. *Impact of Quaternary-Pumped Storage Hydropower on Frequency Response of U.S. Western Interconnection with High Renewable Penetrations: Preprint*; National Renewable Energy Laboratory (NREL): Golden, CO, USA, 2022.
20. Fisher, R.; Koutnik, J.; Meier, L.; Loose, V.; Engels, K.; Beyer, T. A Comparison of Advanced Pumped Storage Equipment Drivers in the US and Europe. 2012. Available online: https://www.researchgate.net/publication/265907090_A_Comparison_of_Advanced_Pumped_Storage_Equipment_Drivers_in_the_US_and_Europe?channel=doi&linkId=5420834e0cf203f155c5e407&showFulltext=true (accessed on 2 February 2022).
21. The State of Alaska. State of Alaska Open Data Geoportal. 2020. Available online: https://gis.data.alaska.gov/ (accessed on 2 February 2022).
22. Rosenlieb, E.; Heimiller, D.; Cohen, S. *Closed-Loop Pumped Storage Hydropower Resource Assessment for the United States. Final Report on HydroWIRES Project D1: Improving Hydropower and PSH Representations in Capacity Expansion Models*; National Renewable Energy Laboratory (NREL): Golden, CO, USA, 2022. [CrossRef]
23. Ingram, M.R. *Large Scale Pumped Storage to Support Renewable Deployment; PSH Resource Assessment and Complementary Analysis for Alaska*; National Renewable Energy Laboratory (NREL): Golden, CO, USA, 2022.
24. U.S. Department of Interior, Bureau of Land Management. BLM Alaska. 2022. Available online: https://www.blm.gov/alaska (accessed on 2 February 2022).
25. International Renewable Energy Agency. Renewable Energy Technologies: Cost Analysis Series, Hydropower. 2012. Available online: https://www.irena.org/publications/2012/Jun/Renewable-Energy-Cost-Analysis---Hydropower (accessed on 2 February 2022).
26. O'Connor, P.W. *Hydropower Baseline Cost Modeling, Version 2*; Oak Ridge National Laboratory (ORNL): Oak Ridge, TN, USA, 2015. [CrossRef]
27. Lu, B.; Stocks, M.; Blakers, A.; Anderson, K. Geographic information system algorithms to locate prospective sites for pumped hydro energy storage. *Appl. Energy* **2018**, *222*, 300–312. [CrossRef]
28. Entura, Hydro-Electric Corporation. PUmped Hydro Cost Modelling. 2018. Available online: https://www.aemo.com.au/-/media/Files/Electricity/NEM/Planning_and_Forecasting/Inputs-Assumptions-Methodologies/2019/Report Pumped Hydro-Cost-Modelling.pdf (accessed on 2 February 2022).
29. *Final Environmental Impact Statement for Hydropower Licenses, Grant Lake Hydroelectric Project—FERC Project No. 13212-005*; Technical Report FERC/FEIS-0283F; Federal Energy Regulatory Commission: Washington, DC, USA, 2019.
30. Williams, K.; Smith, C.; Higman, B. Pumped Energy Storage for Alaska. 2020. Available online: https://www.alaskapublic.org/wp-content/uploads/2020/09/5e38b171d8ca1f77fdf7c3b3_Pumped-Energy-Storage-for-Alaska-3-Feb-2020.pdf (accessed on 2 February 2022).
31. U.S. Department of Energy. Six Non-Power Benefits of Hydropower. 2020. Available online: https://www.energy.gov/eere/articles/six-non-power-benefits-hydropower (accessed on 2 February 2022).
32. Botelho, A.; Ferreira, P.; Lima, F.; Pinto, L.M.C.; Sousa, S. Assessment of the environmental impacts associated with hydropower. *Renew. Sustain. Energy Rev.* **2017**, *70*, 896–904. [CrossRef]
33. The State of Alaska. Overview-Anadromous Waters Catalog-Sport Fish. Available online: https://www.adfg.alaska.gov/sf/sarr/awc/ (accessed on 2 February 2022).

34. Guérin, F.; Abril, G.; Richard, S.; Burban, B.; Reynouard, C.; Seyler, P.; Delmas, R. Methane and carbon dioxide emissions from tropical reservoirs: Significance of downstream rivers. *Geophys. Res. Lett.* **2006**, *6*.
35. Santos, M.A.d.; Rosa, L.P.; Sikar, B.; Sikar, E.; Santos, E.O.d. Gross greenhouse gas fluxes from hydro-power reservoir compared to thermo-power plants. *Energy Policy* **2006**, *34*, 481–488. [CrossRef]

Article

A Method for Rotor Speed Measurement and Operating State Identification of Hydro-Generator Units Based on YOLOv5

Jiajun Liu, Lei Xiong, Ji Sun *, Yue Liu, Rui Zhang and Haokun Lin

School of Electrical Engineering, Xi'an University of Technology, Xi'an 710000, China
* Correspondence: 1211911003@stu.xaut.edu.cn; Tel.: +86-185-2174-4707

Abstract: With the rapid development of artificial intelligence, machine vision and other information technologies in the construction of smart power plants, the requirements of power plants for the state monitoring of hydro-generator units (HGU) are becoming higher and higher. Based on this, this paper applies YOLOv5 to the state monitoring scenario of HGU, and proposes a method for rotor speed measurement (RSM) and operating state identification (OSI) of HGUs based on the YOLOv5. The proposed method is applied to the actual RSM and OSI of HGUs. The experimental results show that the Precision and Recall of the proposed method for rotor image are 99.5% and 100%, respectively. Compared with the traditional methods, the online image monitoring based on machine vision not only realizes high-precision RSM and the real-time and accurate determination of operating states, but also realizes video image monitoring of the rotor, the operation trend prediction of the rotor and the early warning of abnormal operating states, so that staff can find the hidden dangers in time and ensure the safe operation of the HGU.

Keywords: artificial intelligence; hydro-generator unit; YOLOv5; rotor speed measurement; online monitoring

Citation: Liu, J.; Xiong, L.; Sun, J.; Liu, Y.; Zhang, R.; Lin, H. A Method for Rotor Speed Measurement and Operating State Identification of Hydro-Generator Units Based on YOLOv5. *Machines* **2023**, *11*, 758. https://doi.org/10.3390/machines11070758

Academic Editors: Zhengwei Wang and Yongguang Cheng

Received: 30 May 2023
Revised: 6 July 2023
Accepted: 14 July 2023
Published: 20 July 2023

1. Introduction

An HGU is a complex coupling system of hydraulic, mechanical, and electrical systems. As the service life of the units increases, the issues of structural fatigue and deterioration become increasingly prominent. At present, large and medium-sized hydropower plants are developing an unmanned management mode and one with few personnel on duty, and the equipment maintenance method is gradually transitioning from regular preventive maintenance based on time to predictive maintenance based on state monitoring [1–3]. How to accurately monitor the unit, judge its operation state, and detect unit operation problems on time is an important issue in the state maintenance of HGU. The monitoring parameters (nonelectrical quantity) of HGUs can generally be vibration, noise signals, temperature, and rotor speed [4–6]. The rotor speed of the HGU can reflect both the state and frequency of the HGU, which is a very important detection quantity in HGU monitoring. Therefore, accurately calculating the rotor speed of the HGU is of great significance for monitoring and judging the state of HGU [7,8].

At home and abroad, the methods of RSM for HGUs include direct method and indirect method. The indirect method mainly converts mechanical rotation into other physical quantities, and converts physical quantities into velocity quantities according to the corresponding calculation formula. The mainstream indirect method is PT residual pressure velocimetry. As the name implies, the direct method measures the mechanical rotation of the object directly through the corresponding sensor. The mainstream method installs a toothed disc on the main shaft of the HGU, so that the toothed disc is connected with the main shaft of the HGU. The rotation of the main shaft of the HGU will drive the toothed disc to rotate synchronously. The toothed disc sensor will collect the corresponding pulse signal, and the current speed value of the HGU will be calculated through the processing of the signal by the single-chip microcomputer in the later stage [9]. The above methods all have certain drawbacks. On the one hand, they cannot achieve visual monitoring of speed and cannot reproduce the accident development process afterwards. On the other hand, different speed measurement methods have varying degrees of coupling with the rotor, reducing the robustness of the measurement. The advantages and disadvantages of the commonly used classical rotor speed measurement methods are shown in Table 1.

Table 1. Comparison of advantages and disadvantages of rotor speed measurement methods.

RSM Method	Advantages	Disadvantages
Toothed disc [10]	High measurement accuracy, Strong real-time performance, Strong anti-interference ability	The need to fix the processed toothed disc will change the spindle structure
PT residual pressure [11]	Able to obtain the voltage of the generator outlet PT, commonly used for electromagnetic measurement	Electrical faults and abnormal residual voltage of the primary equipment can cause inaccurate speed measurement
Photoelectric encoder [12]	High measurement accuracy, fast response	Susceptible to signal noise, contact type speed measurement requires coaxial installation
Laser Doppler [13]	Noncontact speed measurement without changing the spindle structure	High price, poor immunity
Machine Vision [14]	Noncontact speed measurement without changing the spindle structure, visualization of accident process	Limited usage scenarios

In recent years, machine vision, as a hot technology, has provided effective technical support for promoting the construction of smart grids. Research based on machine vision is constantly emerging and has been validated in practical power engineering applications [15–17]. Ref. [18] adopts an intelligent detection method for transmission line defects based on reparameterized YOLOv5, which solves the problem of slow edge reasoning caused by low computing power and low memory of power patrol edge equipment. Ref. [19] applies YOLOv5, which combines the weighted bidirectional feature pyramid (BiFPN) structure to the identification of power switch cabinet state lights, assigns different weights to the feature layer to transmit more effective feature information, and solves the problem of small target recognition caused by the high-density layout of state lights. Ref. [20] proposes a method of fan blade detection and spatial positioning based on the lightweight YOLOv5. ShuffleNetv2 is used as the feature extraction backbone network to achieve accurate positioning of fan blade tip. Ref. [21] uses a method of making fused image data set to solve the problem of the small number of defective insulator samples in the insulator image data set taken by UAV aerial photography. To sum up, machine vision has been widely used in power systems, but there is relatively little research on it in hydropower plants [22,23].

The RSM method for HGUs based on image processing is a type of indirect RSM method. This method mainly tracks the target in the measured structure video captured by the camera to obtain the motion trajectory of the measurement point in the image, and then determines the motion information of the structure through the geometric relationship between the image and the real world. Unlike the contact displacement monitoring method, which requires the installation of fixed support points on the structure, the camera is installed at a fixed point far from the measured object and does not have a coupling relationship with other electrical equipment, contributing to the non-interference of the RSM process and unit operation, greatly improving the anti-interference ability of RSM.

This paper applies the YOLOv5 algorithm to the state monitoring scenario of HGU, and proposes a method for rotor speed measurement and operation state recognition of HGU based on YOLOv5 combined with the period measurement method. A monitoring system for the rotor speed of HGU is also developed. This system not only achieves precise measurement of rotor speed, makes accurate judgments on the operating state of HGU, predicts rotor operating trends and alerts abnormal operating states, but also records real-time operating images of HGU, providing data sources for post analysis. The system developed in this paper can effectively ensure the safe and stable operation of HGUs, laying a theoretical foundation for the future development of more functional monitoring systems.

2. YOLOv5 Model Analysis

Object detection is a machine vision technology that can recognize semantic objects in images and provide their positions and categories. Traditional object detection methods typically include three steps: region selection, feature extraction, and feature classification. After the emergence of deep learning, object detection methods have enhanced the accuracy of feature classification and improved the efficiency of region selection, becoming a common method at present. There are two types of object detection methods: single-stage etection and two-stage detection. Single-stage detection integrates target classification, boundary localization, and feature extraction into a network, constructs end-to-end training methods, and uses regression to obtain the position of the target, reducing the repetitive calculation of image feature extraction steps. The main algorithms include SSD, YOLO, etc. Two-stage detection is based on constructing a deep convolutional neural network to extract target features, and then achieving target detection through image segmentation and positioning. The main algorithms include Faster RCNN, Fast R-CNN, SIFT, etc. [24].

The YOLO series algorithm is a target detection method based on regression thinking, which can directly predict the category and position of the target from the image without the need for candidate boxes or other intermediate steps. The advantage of this series of algorithms is its fast speed, which is suitable for real-time scenes. The YOLOv5 model has the advantages of fast reasoning speed, high precision and small model size, which makes it highly popular in the field of target detection. Figure 1 shows the overall block diagram of the YOLOv5 target detection algorithm. For a target detection algorithm, we can usually divide it into four general modules, specifically including Input, Backbone, Neck and Head, corresponding to the four red modules in Figure 1.

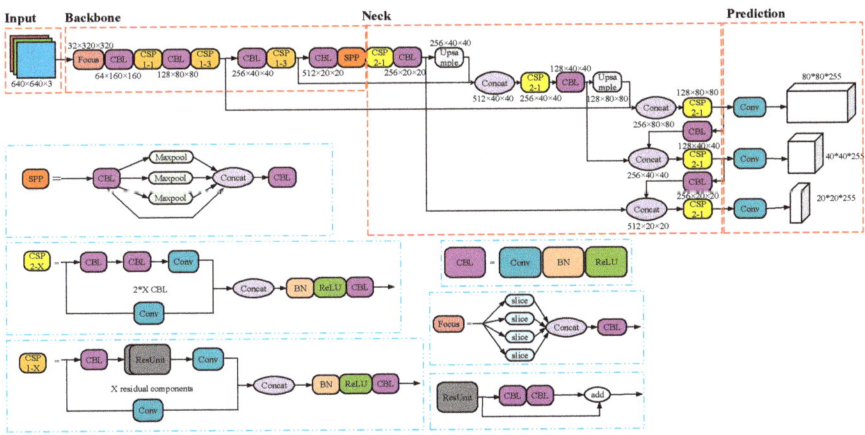

Figure 1. YOLOv5 model architecture diagram.

The Input represents the input image, which typically includes an image preprocessing stage, which scales the input image to the input size of the network and performs operations such as normalization. In the network training phase, YOLOv5 uses Mosaic data enhancement operations to improve the training speed of the model and the accuracy of the network. An adaptive anchor frame calculation and adaptive image zooming method are proposed. The Backbone is usually a network of classifiers with excellent performance, and this module is used to extract some common feature representations. YOLOv5 not only uses the CSPMarket53 structure, but also uses the Focus structure as the Backbone. The Neck network is usually located in the middle of the reference network and the header network, and it can be used to further enhance the diversity and robustness of features. Although YOLOv5 also uses the SPP module and FPN + PAN module, the implementation details are somewhat different. The Head is used to complete the output of target detection results. For different detection algorithms, the number of branches at the output end varies, usually including a classification branch and a regression branch. YOLOv5 leverages GIOU_Loss replaces the Smooth L1 Loss function to further improve the detection accuracy of the algorithm.

In order to evaluate the effectiveness and feasibility of the YOLOv5 model test results, in practical applications, Precision, Recall, Average Precision (AP) and mean Average Precision (mAP) are usually used as evaluation indicators [25]. The formula of the above indicators is as follows. The Precision represents the true correct proportion in the correct classification, and the Recall represents the proportion of the correct samples in the given correct samples.

$$\text{Precision} = \frac{\text{TP}}{\text{TP} + \text{FP}} \tag{1}$$

$$\text{Recall} = \frac{\text{TP}}{\text{TP} + \text{FN}} \tag{2}$$

$$\text{AP[Class]} = \sum_{i \in \text{confidence}} \text{Precision}_i [\text{Recall, Class, IOU}] \tag{3}$$

$$\text{mAP} = \frac{1}{N} \sum \text{AP}_i \tag{4}$$

In the formula, TP represents the number of tags that are positive samples and classified as positive samples. FP indicates the number of negative samples but classified as positive samples. FN indicates the number of positive samples but classified as negative samples.

Under a fixed Intersection Over Union (IOU), a given target will obtain different Precision and Recall values according to different confidence levels. Through interpolation of Precision and Recall, the continuous curve generated is the Precision–Recall (PR) curve. The AP represents the comprehensive performance of the model under different confidence levels by the area enclosed by the PR curve of a given target category and the horizontal and vertical coordinates [26]. The higher the AP value, the better the detection performance of the model. Each IOU corresponds to a different AP. The AP@.5 represents the AP when the IOU is taken as 0.5, and the AP@.5:.95 represents the average AP when the IOU is taken as 0.5 to 0.95, in steps of 0.05.

3. Analysis of RSM Principle

3.1. RSM by Period Method

The RSM principle based on image recognition technology is similar to that of digital circuit speed measurement. The digital circuit speed measurement is to count the known frequency high-frequency clock pulse with a counter within the interval of two adjacent output pulses, and then calculate the speed, which is called T-method speed measurement [27]. The schematic diagram of the period measurement method based on image recognition technology is shown in Figure 2. The time of one revolution of the HGU can be obtained by multiplying the frame rate of the camera by the number of frames taken during one revolution of the HGU. The key is to calculate the time difference of one revolution. In the continuous frames with markers (that is, within the irradiable range of the camera), the appearance and disappearance of markers are the two time points we focus on (the judgment of key frames in the program is shown in Figure 3). The time difference corresponding to the adjacent frames that appear or disappear is the time required for the rotor to rotate for one cycle, that is, the speed of the unit. Therefore, when the unit rotates for one and a half periods, we can obtain the real-time rotor speed of the two units.

$$n = \frac{60}{f_0 \times M} = \frac{60}{f_p} = \frac{60}{f_q} \tag{5}$$

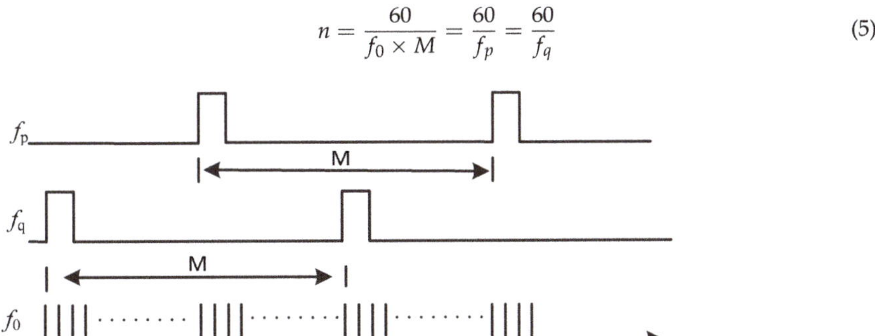

Figure 2. Waveform of velocity measurement principle of video measurement period method.

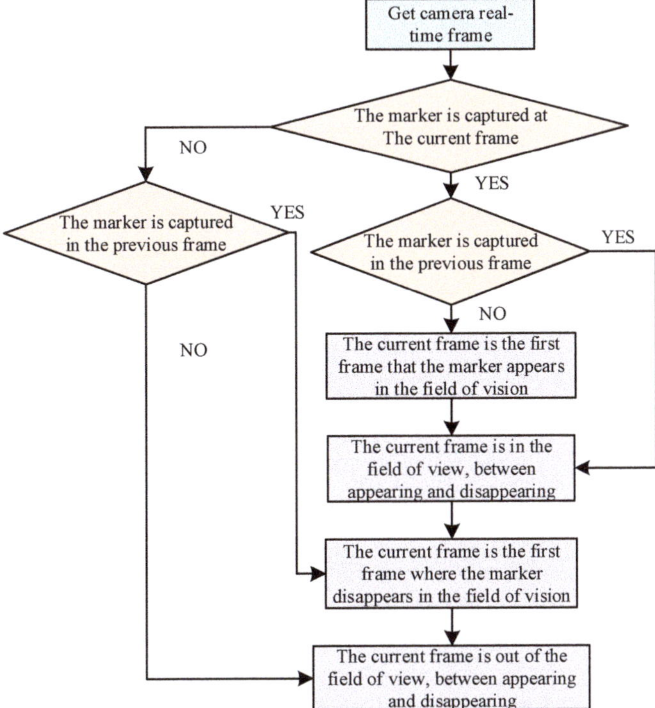

Figure 3. The judgment of the appearance and disappearance of markers.

In the formula, n is the rotor speed, rpm. f_0 is the camera frame rate, fps. M is the number of pictures collected by the camera in one rotation, frame. f_p is the frequency of the first frame of the marker in the field of vision, fps. f_q is the frequency of the last frame of the marker in the field of vision, fps.

The judgment process for key frames is shown in Figure 3. Firstly, by determining whether the current frame has captured the marker, it is determined whether the current frame is within the visible range of the camera. Secondly, if the current frame captures the marker and the previous frame also captures the marker, then the current frame is a continuous frame within the field of view. If the marker was not captured in the previous frame, the current frame is the first frame where the marker was captured. Finally, if the marker is not captured in the current frame and was captured in the previous frame, then the current frame is the first frame where the marker disappears. If the marker is not captured in the previous frame, then the current frame is a continuous frame outside the field of view.

The period measurement method calculates the time of adjacent pulses, but this part of the time calculated by video speed measurement is easily affected by the frame rate. Only when the time of HGU rotor rotation is a multiple of the camera sampling period (the reciprocal of frame rate), can it be ensured that the time corresponding to the frame of adjacent markers appearing (or disappearing) is exactly the time required for rotor rotation. In order to reduce such errors, an improvement link is set up: when calculating the time corresponding to the adjacent frame of the marker, plus the time difference corresponding to the adjacent frame of the marker disappearing, the two calculation results are put into the speed list, which is helpful to reduce the calculation error caused by accidental factors.

In order to avoid the impact of random noise or camera frame leakage on the real-time rotor speed, the data will be further digitally filtered after the speed measurement by the period measurement method. In this paper, the median average filtering method is adopted, that is, the maximum and minimum values are removed from a set of numerical lists and the average value is taken, which is equivalent to "median filtering method" + "arithmetic average filtering method".

The speed measurement method based on image recognition technology is similar to the digital circuit speed measurement method, which is applicable to low speed measurement. Since the rotor speed for HGU under normal or abnormal states is not more than 150 r/min, it belongs to low-speed rotation, so the speed measurement method of the period measurement is more suitable for the RSM of HGU software.

3.2. Algorithm Steps and Processes

The process of rotor speed calculation in this paper can be divided into five parts: obtaining the HGU rotor video, dynamic capture of markers, rotor speed calculation, HGU operation state analysis, and returning the data to the background. The four parts are the main algorithms, and the principle is shown in Figure 4.

(1) HGU rotor video: the real-time video of the HGU operation site is collected by the ip camera set around the water turbine. After the collection, the video is uploaded to the server for the next RSM preparation.

(2) Dynamic capture of markers: the RSM of HGU depends on the setting of the markers. The real-time rotor speed of HGU can be calculated by dynamically capturing the markers in each frame of the video captured in step (1).

(3) The rotor speed calculation of HGU: through the markers captured in step (2), find the key nodes, and the nodes where the markers appear and disappear, and measure the rotor speed by the period method.

(4) Judge the operation state of the HGU: through step (3), the two final speed values of the two cameras are obtained, and the mean value of the two values is calculated, which is the real-time rotor speed of the HGU calculated by the program, so that the operation state of the HGU can be judged.

(5) Return the data to the background: the server obtains the rotor speed of HGU, and finally sends the rotor speed information to the database according to the TCP com-

munication protocol and the video information according to the DUP communication protocol for storage.

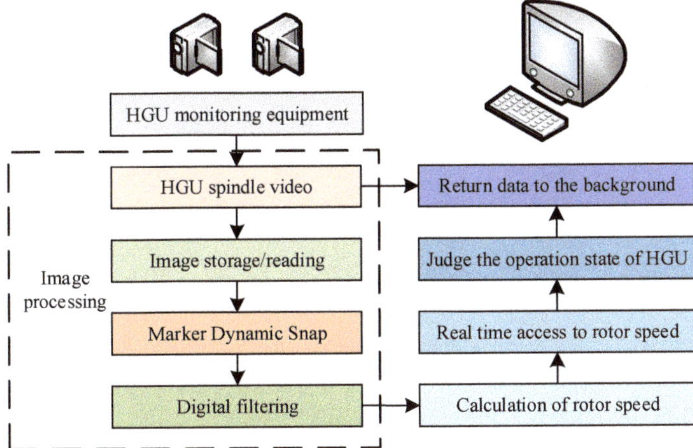

Figure 4. Speed measurement process.

4. YOLOv5 Test Result Analysis

4.1. Experimental Environment and Training Process

The parameters in the deep neural network mainly include the parameters that are automatically adjusted through learning and the hyperparameter that needs to be manually set. The adjustment of the hyperparameter is an important link between the theoretical knowledge of deep learning and the actual situation at present. The hyperparameter configuration of this training is as follows: the initial learning rate is 1×10^{-4}, the momentum is 0.0005, the batch size is 16, and the epoch is 200. The specific configuration of the computer is shown in Table 2.

Table 2. Computer configuration table.

Device	GPU	NVIDIA GeForce GTX 2080
Operating system	Operating system	Windows10
	Computer language	Python3.6.12
	Deep learning framework	Pytorch1.7.1

4.2. Data Set

YOLOv5, as a supervised learning algorithm, cannot be separated from the support of a large amount of data. The quality and distribution of data sets are important factors affecting the performance of the algorithm. The dataset used in this article originated from an actual hydropower plant and ultimately collected 2500 image data with a size of 640×480 in jpg format. The image data are divided into training set, validation set, and testing set in a ratio of 8:1:1.

In order to create a PASCAL VOC format dataset, LabelImg software was used to visually annotate each image. When annotating, the mouse was used to accurately and meticulously draw the border of the target as much as possible, which helps with the training and segmentation effect of the model. At the same time, the category name of the target is written on the border. As shown in Figure 5, it is the detection category "Mark" where the marker appears.

The result of image annotation is a Json format file, which is converted into a structured XML language based document, namely an xml format label file, which can better describe the target category, position, size and other attributes in the image.

Figure 5. YOLOv5 picture annotation diagram.

4.3. Evaluating Indicator

This paper uses labeled turbine rotor images as training and validation sets and conducts 200 epochs of training in the YOLOv5s model to obtain the optimal model weight file. Figure 6 shows the training results of the turbine rotor markers in this model.

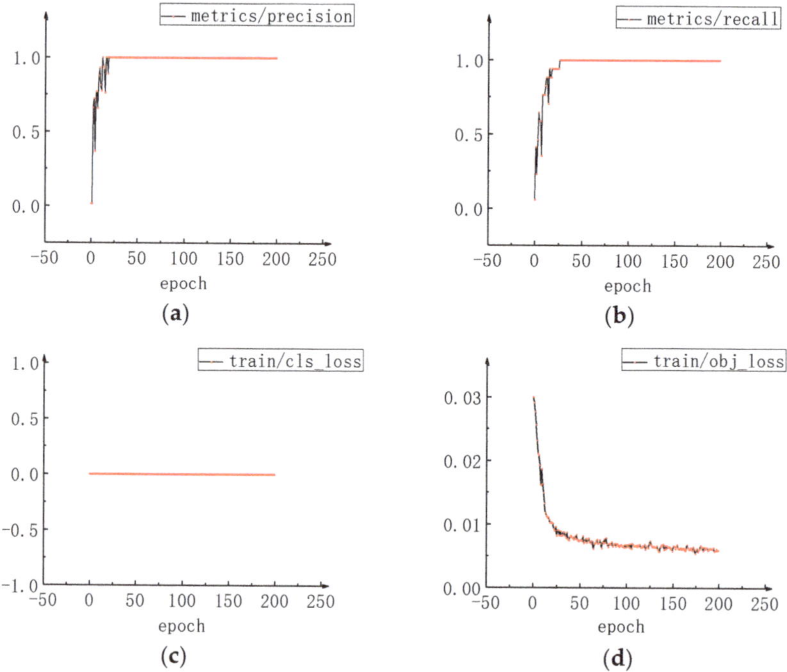

Figure 6. *Cont.*

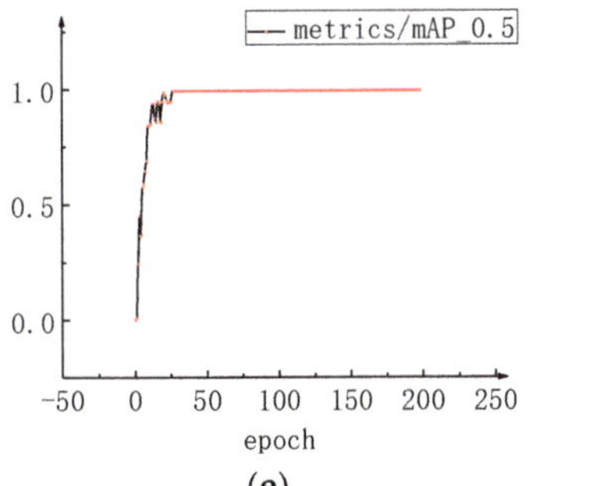
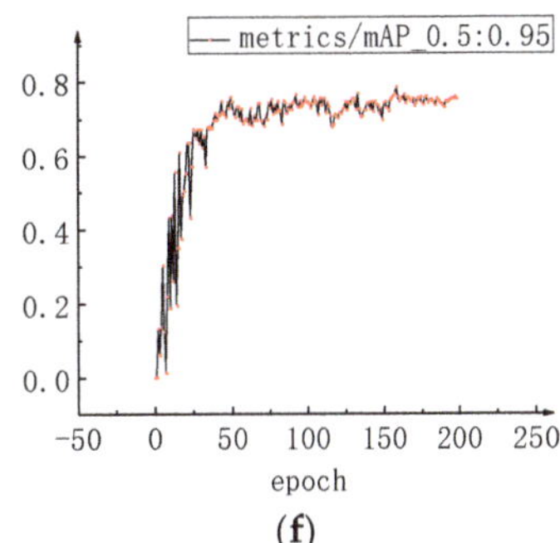

Figure 6. Analysis chart of training results. (**a**) Precision. (**b**) Recall. (**c**) Classified loss. (**d**) Confidence loss. (**e**) IOU = 0.5, mAP. (**f**) IOU $\in$ [0.5:0.05:0.95], mAP.

From the figure, it can be seen that the detection Precision and Recall of YOLOv5 for rotor images have reached 99.5% and 100%, respectively, indicating that the model has strong generalization ability and robustness, can adapt to various complex scenes and environments, and can accurately locate and recognize markers for rotor images. At the same time, observing the changes in Classified loss and Confidence loss with the training process, it was found that after approximately 100 epochs, both tended to stabilize, indicating that the model had converged to a better state. From the fact that the Classified loss in the training set and the test set is zero, it can be concluded that the model can identify all target categories in the training set without classification errors, and the Classified loss of the model in the test set is also zero, indicating that the model has no overfitting or under fitting problems.

In addition, through the mAP indicators under different IOU thresholds, it was found that when IOU is 0.5, mAP approaches one, and when mAP@.5:.95, it is also close to 80%, indicating that the model maintains high detection performance even for smaller or more difficult to recognize targets. These indicators all demonstrate that the model has good detection performance for markers and meets the precision requirements for target recognition in RSM.

4.4. Training Result Analysis

In order to verify the effectiveness of the YOLOv5 algorithm for marker capture, YOLOv5 algorithm was compared with the traditional object capture algorithm based on histogram reverse projection. The detection results are shown in Figure 7. As can be seen from the left column of Figure 7, when using histogram reverse projection, due to the single projection sample that can be selected, it is difficult to regularly frame and select markers during the operation of the HGU. Moreover, when the color of the surrounding environment is similar to the color of the marker, noise is prone to occur, as shown in the red box selected area in the last group of comparison images, and cannot be used for subsequent RSM of HGU.

The comparison of object capture results based on YOLOv5 is shown in the right column of Figure 7. During the operation of the HGU, the rotor markers have undergone a certain degree of deformation, but since the markers with different degrees of deformation have been labeled during the labeling phase, the model can still accurately identify various forms of markers during the detection phase, that is, the Recall of the model for the markers is 100%. As the marker area decreases, the confidence level of target detection decreases slightly, laying a good foundation for subsequent velocity measurement.

Figure 7. Comparison of YOLOv5 and histogram back projection detection results.

5. Field Measurement and Analysis

5.1. HGU State Definition

The test site is a hydropower plant in China, which is equipped with four units with a unit capacity of 200 MW and a total installed capacity of 850 MW. Based on the installed capacity and actual operation of the water turbine, different states of the HGU can be identified based on the RSM, as shown in Figure 8.

Firstly, the speed can be divided into a normal operating state and abnormal operating state. Secondly, during the startup state in normal operation, if the speed is greater than 95% of the rated speed, it is in the excitation state. During the shutdown state in normal operation, if the speed is less than 2% of the rated speed, it is considered a creeping state, and if the speed is less than 25% of the rated speed, it is considered an air brake state. Finally, abnormal operating states include electrical overspeed when the speed exceeds 1.15% of the rated speed, and mechanical overspeed when the speed exceeds 1.35% of the rated speed.

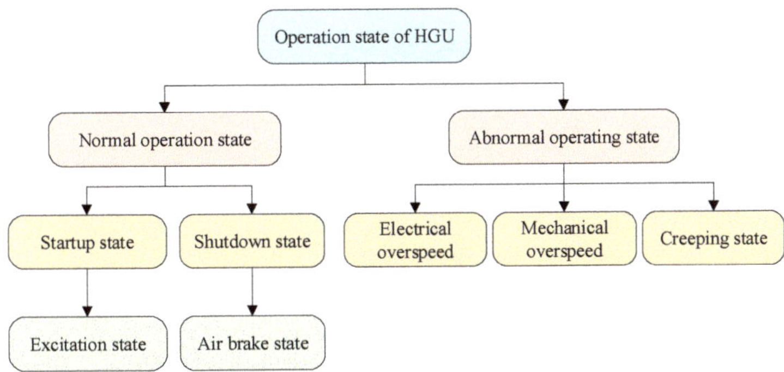

Figure 8. State of the HGU.

5.2. HGU State Judgment

The judgment of HGU state is an important content of HGU state monitoring. The HGU state not only depends on the numerical value obtained from the speed calculation, but also on the trend of the speed over time. For example, during the shutdown state, the unit may experience creep, and at this time, an air brake needs to be added to control the speed. During the startup state, the unit needs to input excitation to increase output power. In order to accurately determine the trend of rotational speed, a time window based method was adopted in the program. Specifically, a thread has been added to the program to record the speed V_q one second ago. When calculating the real-time speed V_p at the current moment, first compare it with V_q. If V_p is smaller than V_q and less than a preset fixed value (set to avoid data fluctuations causing misjudgment), then it can be considered that the unit is in a shutdown process, that is, the speed is gradually decreasing. On the contrary, if V_p is greater or equal than V_q, then it can be considered that the unit is in the start-up process, that is, the speed is gradually increasing or maintaining stability.

The method proposed in this paper determines whether the HGU has stopped operation by measuring the area changes of the markers in consecutive frames of the video. As shown in Figure 9, firstly, a certain marker on the HGU is captured and tracked in real-time, and each video image is processed to extract the contour of the marker and calculate its pixel area S. Then, compare the area difference of the markers in two adjacent video images ($\Delta S = |S_1 - S_2|$). If ΔS is less than a given threshold, it is considered that the HGU is in a stationary state. Finally, when using binocular cameras, it is necessary to synchronize the video images collected by the two cameras and use an "OR" statement to perform logical operations on the static state collected by the two cameras. As long as one camera detects that the water turbine has stopped running, it is considered that the speed is zero.

The on-site test is carried out during the startup and shutdown test after the unit maintenance, which is helpful for the judgment of multiple states in a short time. Since there are no two states of overspeed in the field test, this test only records the following state judgments (both of them have been judged in the laboratory), as shown in Figure 10 below. It can be seen from the figure that when the HGU transits from the shutdown state to the speed-up phase, the rotor speed increases from 0 rpm to 104.778 rpm, and the unit is in the excitation stage. The rotor speed continues to rise until the rotor speed reaches the rated rotor speed of 107.067 rpm. After maintaining operation for a period of time, the HGU starts to decelerate, and the rotor speed drops to 68.873 rpm. The unit entered the stage of air brake to speed-up the shutdown process, and finally the rotor speed continued to drop to zero. However, due to the large number of guide vanes, it is impossible for the guide vanes to be completely sealed under the closed state, and the water leakage of guide vanes is objective. When the water leakage increases to a certain extent, the water will impact the HGU runner, causing the rotating parts of the unit to produce slow rotating motion, and the rotor speed will reach 1.58 rpm from zero, which means the unit will appear in peristalsis.

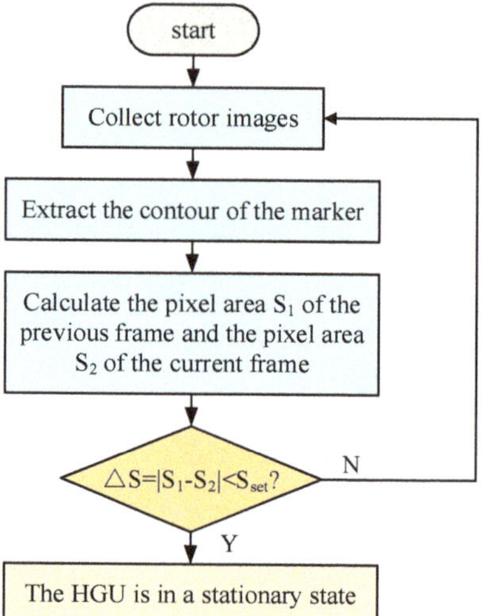

Figure 9. Flow chart of turbine shutdown state judgment.

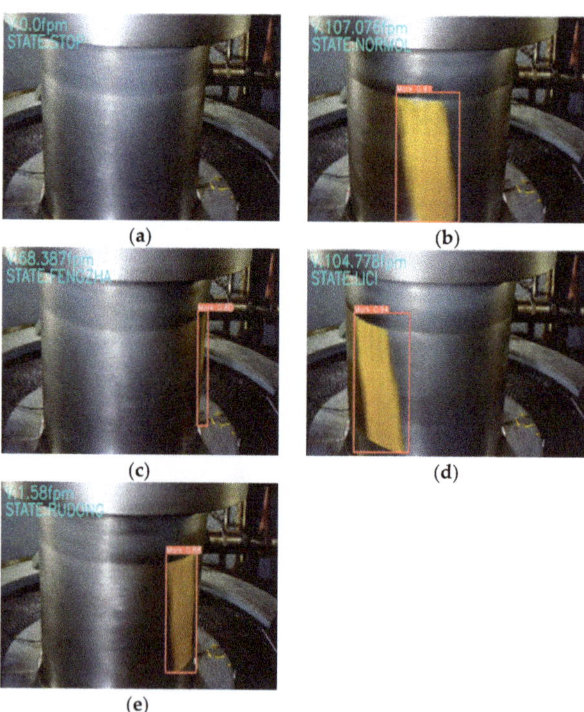

Figure 10. Unit state judgment diagram. (**a**) Shutdown state. (**b**) Excitation state. (**c**) Rated rotor speed state. (**d**) Air brake state. (**e**) Creeping state.

5.3. Analysis of the Change of HGU Start and Stop Rotor Speed

The rotor speed increased from 0 rpm to 107.1 rpm in the process of starting, and the starting time of this field test was about 180 s. In order to test the precision and tracking of the algorithm proposed in this paper, the real value of the rotor speed is recorded every five seconds and compared with the calculated rotor speed at the same time.

The overall change trend of rotor speed is shown in Figure 11a, with an average relative error of 2.45%. It can be seen from the figure that the overall following and precision of the calculated rotor speed are relatively high, but because the relative error is small, it is not convenient to analyze the error at different stages, so the low-speed stage, speed-up stage and stable stage of the rotor speed in Figure 11a are enlarged. It can be seen from the figure that, as shown in Figure 11c, due to the stable acceleration, the rotor speed in the speed-up stage is approximately linear, and the corresponding curve of the real value and the calculated value almost coincide, with high speed precision. In the low-speed stage (Figure 11b) and stable stage (Figure 11d), the calculated values are relatively small due to the slow speed change and the rotor speed is easily affected by historical data. It can be seen from the figure that the calculated value curve is below the true value curve. Finally, when the rotor speed is stabilized to the rated rotor speed, the calculated rotor speed will also tend to be stable.

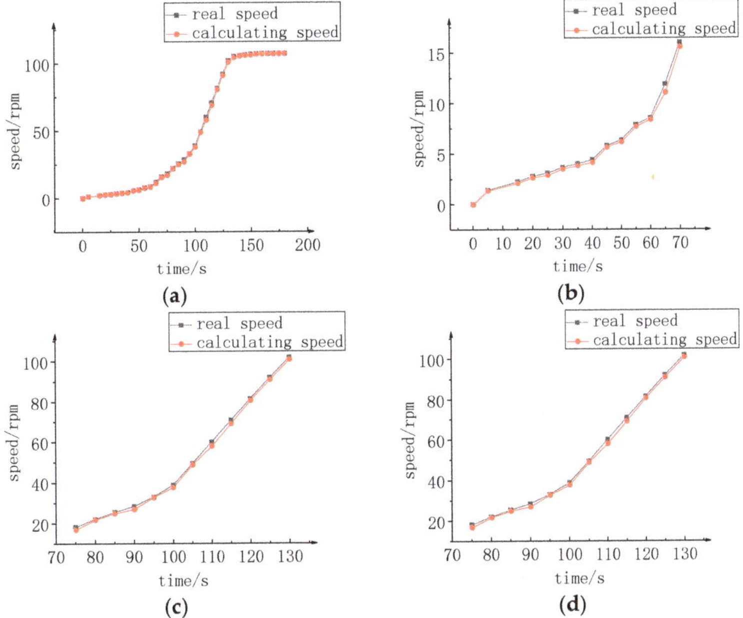

Figure 11. Rotor speed change diagram of HGU startup process. (**a**) Rotor speed change trend during startup process. (**b**) Rotor speed at low-speed stage. (**c**) Rotor speed at speed-up stage. (**d**) Rotor speed at stable stage.

During the shutdown process of the HGU, the rotor speed decreased from 107.1 rpm to 0 rpm. The shutdown test time was about 410 s, and the recording method was the same as that of the appeal startup process, which was recorded every 10 s. The overall trend diagram of the shutdown process is shown in Figure 12a. It can be seen from the figure that the precision of the calculated rotor speed is high and the tracking is good, with an average relative error of 2.44%. After the HGU is shut down, the guide vane is fully closed and the rotor speed decreases. However, because the HGU will damage the oil film of the bearing at low speed, resulting in excessive bearing friction, temperature rise and burning loss, the air brake must be put into operation to shorten the time of the unit at low speed, quickly stop rotating, and protect the HGU. From the Figure 12b–d, it can be seen that the calculated rotor speed are mostly above the true value, that is, slightly higher than the true value. This is mainly because the historical rotor speed during the shutdown process is relatively large, which affects the true value in the digital filtering stage. This law is just opposite to the startup process.

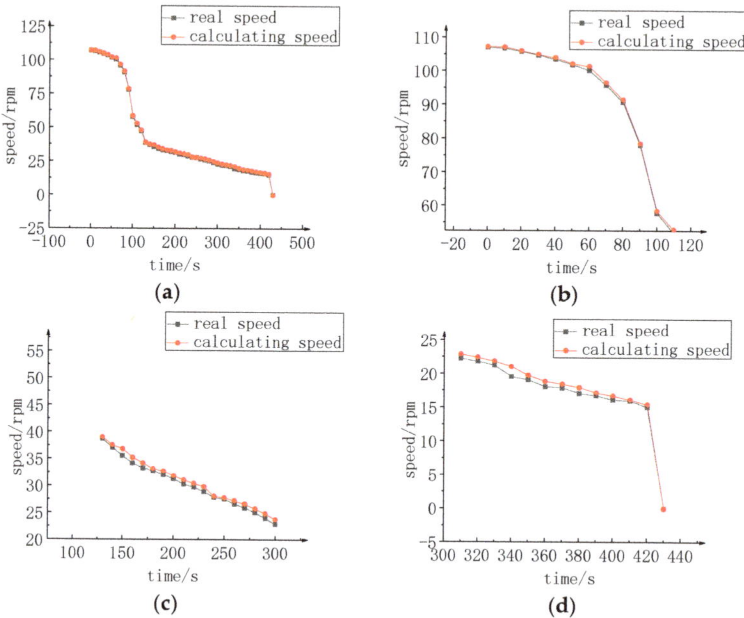

Figure 12. Rotor speed change diagram of HGU shutdown process. (**a**) Rotor speed change trend during shutdown process. (**b**) Rotor speed at low-speed stage. (**c**) Rotor speed at speed-up stage. (**d**) Rotor speed at stable stage.

5.4. Display of Monitoring System

This paper unifies the deployment of hardware devices and detection algorithms, and develops a monitoring system for the rotor speed of HGU based on mobile applications. The application of the HGU rotor speed monitoring system for continuous operation monitoring of a HGU rotor in operation, and the experimental results are shown in Figure 13.

The main functions of the HGU rotor speed monitoring system include calling a camera for real-time state monitoring, outputting algorithm recognition results for each stage of the rotor, judging the current state of the rotor, and summarizing and recording dimensional information such as time, state, and unit number, for users to view the historical state of the unit at any time. Based on the algorithm proposed in this paper, the system can effectively determine the rotor speed and current state of HGU, and the specific detection results are shown in Figure 14.

Figure 14 shows the judgment results of the system, where Figure 14a–f show the six operating states of the HGU, namely normal state, excitation state, acceleration state, air brake state, deceleration state, and shutdown state. After testing, the monitoring system can accurately identify the different operating states of the HGU.

The judgment results of the HGU rotor speed monitoring system are shown in Table 3. From the table, it can be seen that the speed accuracy measured by the system is relatively high, and the judgment of the unit's operating status is accurate.

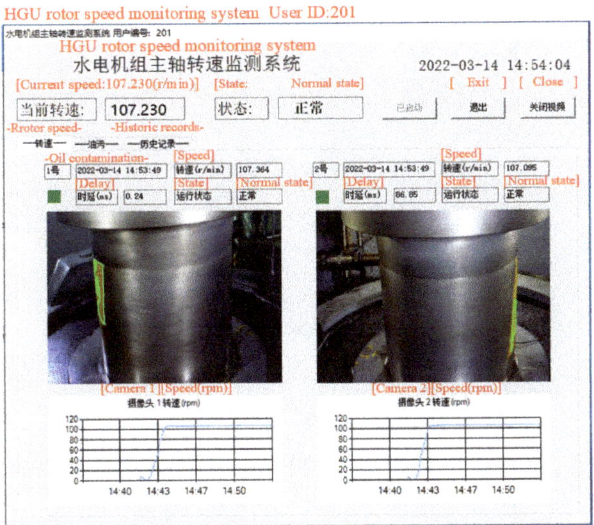

Figure 13. The experimental results of HGU rotor speed monitoring system.

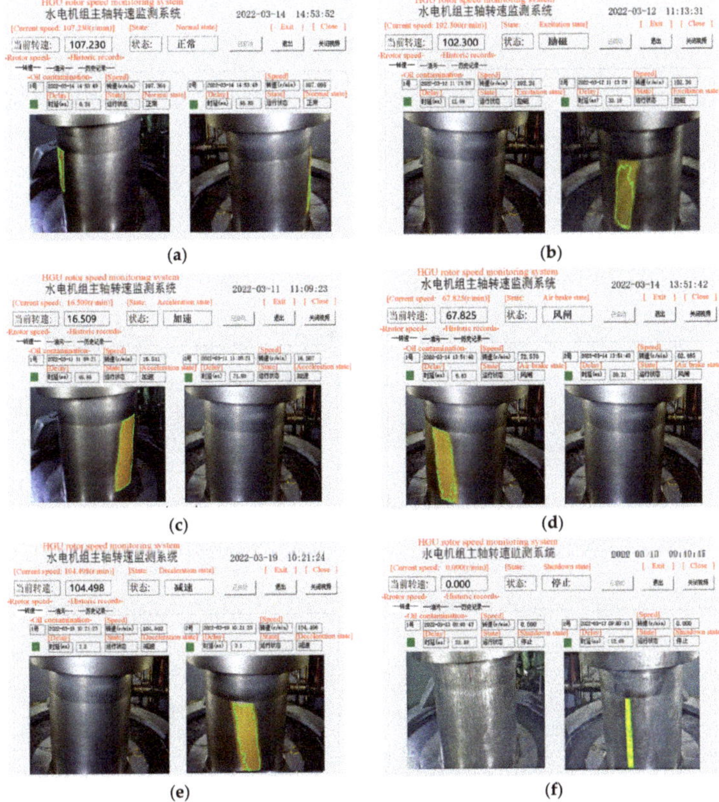

Figure 14. State determination results. (**a**) Normal state. (**b**) Excitation state. (**c**) Acceleration state. (**d**) Air brake state. (**e**) Deceleration state. (**f**) Shutdown state.

Table 3. Judgment results of the rotor speed monitoring system for HGU.

Rotor Speed	Judged State	Actual State	Is It Accurate?
107.230	Normal state	Normal state	Yes
102.300	Excitation state	Excitation state	Yes
16.509	Acceleration state	Acceleration state	Yes
67.825	Air brake state	Air brake state	Yes
104.498	Deceleration state	Deceleration state	Yes
0.000	Shutdown state	Shutdown state	Yes

The monitoring system can export the rotor speed change curve of the HGU starting and stopping, as shown in Figure 15. The horizontal axis represents time, and the vertical axis represents the rotor speed measured by the system.

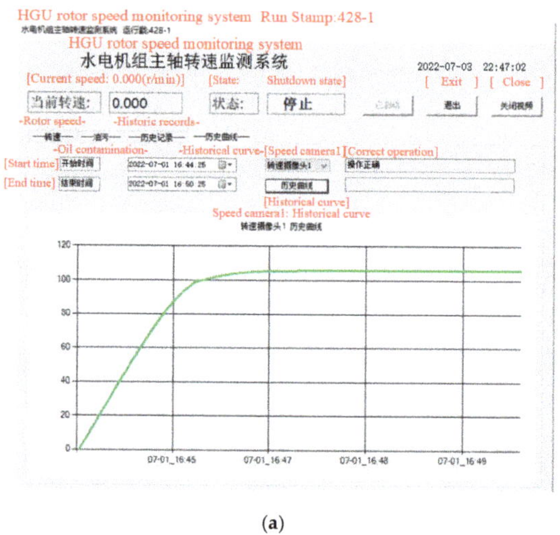

(**a**)

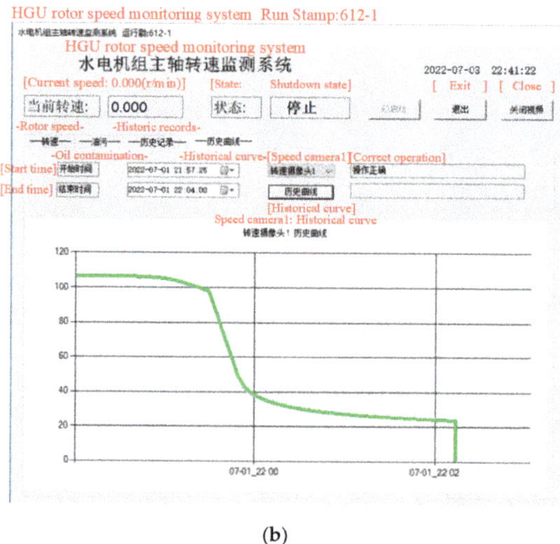

(**b**)

Figure 15. Startup and shutdown process curve of HGU. (**a**) startup process curve of HGU. (**b**) shutdown process curve of HGU.

The monitoring system can store daily data in detail and accurately and display the weekly or monthly data change trend in the data analysis module through the line chart.

The early warning module in the monitoring system can obtain abnormal data and establish a dataset to be transmitted to the database. The fault diagnosis module can identify abnormal data and diagnose faults with an accuracy rate of over 98%, and can provide accurate diagnostic reports. The effectiveness of the system has been verified through the application of the HGU rotor speed monitoring system for hydropower plants.

The performance comparison between the proposed method and the classical method in this article is shown in Table 4. From the table, it can be seen that the method proposed in this paper can achieve low-cost, noncontact, and high anti-interference RSM while ensuring accuracy, which has greatly improved performance compared to classical RSM methods.

Table 4. Performance comparison of different rotor speed measurement methods.

RSM Method	Precision	Additional Equipment Configuration	Contact Measurement	Anti-Interference
Toothed disc	0.001	Yes	Yes	High
PT residual pressure	0.01	Yes	Yes	Low
Photoelectric encoder	0.001	Yes	Yes	Low
Laser Doppler	0.01	Yes	No	Low
The method proposed in this paper	0.001	No	No	High

6. Conclusions

This paper proposes a method for RSM and OSI of HGU based on the YOLOv5. First, the YOLOv5 model is used to accurately capture the HGU rotor, then the period method is used to calculate the speed, and finally the operation state of the HGU is judged by the calculated speed. The method in this paper used to measure the speed of the HGU, which realizes the long-distance, noncontact and high-precision RSM of HGU, helps the staff better grasp the real-time operation of the HGU and makes the hydropower plant more intelligent and efficient. The specific performance is as follows:

(1) Using binocular camera to photograph the rotor can ensure real-time monitoring of the rotor.
(2) Accurately judging several different states of the HGU can help the staff find potential safety hazards quickly and on time according to the alarm prompt information, improve work efficiency and reduce labor costs.
(3) When the unit speed changes, the software can quickly follow its changing trend while maintaining the precision.

Author Contributions: Conceptualization, J.S. and J.L.; methodology, L.X., H.L. and J.L.; software, Python; validation, J.S., Y.L., R.Z. and J.L.; formal analysis, J.S. and L.X.; data curation, J.S., Y.L., R.Z. and L.X.; writing—original draft preparation, J.S., L.X., R.Z. and J.L.; writing—review and editing, J.S. and J.L.; project administration, J.L.; funding acquisition, J.L. All authors have read and agreed to the published version of the manuscript.

Funding: This research was funded by the Key R & D Program of State Grid Shaanxi Electric Power Company (SGTYHT/21-JS-223) and National Natural Science Foundation of China (52077176).

Data Availability Statement: Not applicable.

Acknowledgments: At the point of finishing this paper, I'd like to express my sincere thanks to all those who have lent me hands in the course of my writing this paper. First and foremost, I would like to thank my mentor Liu for his careful guidance at every stage of thesis writing. Secondly, I would like to thank my research group, which not only gives me economic support, but also gives me a perfect simulation platform. Last but not the least, I'd like to thank those leaders, teachers and working staff especially those in the Xi'an University of Technology. Without their help, it would be much harder for me to finish my study and this paper.

Conflicts of Interest: We would like to submit the enclosed manuscript titled "A Method for Rotor Speed Measurement and Operating State Identification of Hydro-generator Unit Based on YOLOv5", which we wish to be considered for publication in Machines. No conflicts of interest exit in the submission of this manuscript, and manuscript is approved by all authors for publication. I would like to declare on behalf of my co-authors that the work described was original research that has not been published previously, and not under consideration for publication elsewhere, in whole or in part.

References

1. De Santis, R.B.; Gontijo, T.S.; Costa, M.A. A Data-Driven Framework for Small Hydroelectric Plant Prognosis Using Tsfresh and Machine Learning Survival Models. *Sensors* **2023**, *23*, 12. [CrossRef] [PubMed]
2. Geng, Q.; Liang, C. Turbine health evaluation based on degradation degree. *Energy Rep.* **2022**, *8*, 435–444. [CrossRef]
3. Duan, R.; Liu, J.; Zhou, J.; Wang, P.; Liu, W. An Ensemble Prognostic Method of Francis Turbine Units Using Low-Quality Data under Variable Operating Conditions. *Sensors* **2022**, *22*, 525. [CrossRef] [PubMed]
4. Trojan, M.; Taler, J.; Smaza, K.; Wróbel, W.; Dzierwa, P.; Taler, D.; Kaczmarski, K. A new software program for monitoring the energy distribution in a thermal waste treatment plant system. *Renew. Energy* **2022**, *184*, 1055–1073. [CrossRef]
5. Vashishtha, G.; Kumar, R. Autocorrelation energy and aquila optimizer for MED filtering of sound signal to detect bearing defect in Francis turbine. *Meas. Sci. Technol.* **2022**, *33*, 015006. [CrossRef]
6. Zemouri, R.; Ibrahim, R.; Tahan, A. Hydrogenerator early fault detection: Sparse Dictionary Learning jointly with the Variational Autoencoder. *Eng. Appl. Artif. Intell.* **2023**, *120*, 105859. [CrossRef]
7. Guerrero, J.M.; Lumbreras, C.; Reigosa, D.; Fernandez, D.; Briz, F.; Charro, C.B. Accurate Rotor Speed Estimation for Low-Power Wind Turbines. *IEEE Trans. Power Electron.* **2020**, *35*, 373–381. [CrossRef]
8. Shang, L.; Cao, J.; Jia, X.; Yang, S.; Li, S.; Wang, L.; Wang, Z.; Liu, X. Effect of Rotational Speed on Pressure Pulsation Characteristics of Variable-Speed Pump Turbine Unit in Turbine Mode. *Water* **2023**, *15*, 609. [CrossRef]
9. Fang, H.Q.; Chen, L.; Li, X.M. Comparisons of Optimal Tuning Hydro Turbine Governor PID Gains Based on Linear and Nonlinear Mathematical Models. *Proc. CSEE* **2010**, *30*, 100–106.
10. Zhang, F.; Geng, Z.; Fu, J.; Pan, L. Flywheel Moment On-Line Determination Method and Application in Hydraulic-Turbine Unit. *J. Vib. Meas. Diagn.* **2015**, *35*, 927–931, 994.
11. Li, X.; Wang, X.; Zhang, Z.; Wang, L. Three redundant intelligent speed measuring devices improve the safety of hydroelectric power generation. *Mech. Electr. Tech. Hydropower Stn.* **2011**, *34*, 30+56.
12. Jia, X.; Wan, Q.; Zhao, C.; Du, Y.; Yu, H. Status and Prospect of Velocity Measurement Method with Optical Encoder. *Instrum. Tech. Sens.* **2018**, *3*, 102–107.
13. Zhou, J.; Wei, G.; Long, X. Research on direction discrimination and low-speed measurement for laser Doppler velocimeter. *Infrared Laser Eng.* **2012**, *41*, 632–638.
14. Xu, Z.; Shao, J.; Zhao, C.; Cao, S.; Xu, Z.; Jiang, Q. Intelligent Calibration System for Motor Vehicle Engine Speed Measuring Instrument Based on LabVIEW and Machine Vision. *Instrum. Tech. Sens.* **2022**, *6*, 71–74+79.
15. Başmaji, T.; Yaghi, M.; Alhalabi, M.; Rashed, A.; Zia, H.; Mahmoud, M.; Palavar, P.; Alkhadhar, S.; Alhmoudi, H.; Alkhedher, M.; et al. AI-powered health monitoring of anode baking furnace pits in aluminum production using autonomous drones. *Eng. Appl. Artif. Intell.* **2023**, *122*, 106143. [CrossRef]
16. Fang, J.; Liu, C.; Zheng, L.; Su, C. A data-driven method for online transient stability monitoring with vision-transformer networks. *Int. J. Electr. Power Energy Syst.* **2023**, *149*, 109020. [CrossRef]
17. Kong, Y.; Liu, Y.; Geng, J.; Huang, Z. Pixel-Level Assessment Model of Contamination Conditions of Composite Insulators Based on Hyperspectral Imaging Technology and a Semi-Supervised Ladder Network. *IEEE Trans. Dielectr. Electr. Insul.* **2023**, *30*, 326–335. [CrossRef]
18. Liu, M.; Li, Z.; Li, Y.C.; Liu, Y.D.; Jiang, X.C. A Method for Transmission Line Defect Edge Intelligent Inspection Based on Re-parameterized YOLOv5. *High Volt. Eng.* **2023**, 1–11. [CrossRef]
19. Cui, H.Y.; Yang, K.X.; Ge, H.H.; Xu, Y.P.; Wang, H.R.; Yang, C.; Dai, Y.Y. Lightweight GB-YOLOv5m State Detection Method for Power Switchgear. *J. Electron. Inf. Technol.* **2022**, *44*, 3777–3787.
20. Bai, J.P.; Wang, W.; Chen, Y.X.; Jiao, S.M. Detection and spatial location of wind turbine blades based on lightweight YOLOv5. *CAAI Trans. Intell. Syst.* **2022**, *17*, 1173–1181.
21. Wang, N.T.; Wang, S.Q.; Huang, J.F.; Yao, R.T.; Liu, Y.F. Insulator defect detection method based on improved YOLOv5 neural network. *Laser J.* **2022**, *43*, 60–65.
22. Duan, R.; Liu, J.; Zhou, J.; Liu, Y.; Wang, P.; Niu, X. Study on Performance Evaluation and Prediction of Francis Turbine Units Considering Low-Quality Data and Variable Operating Conditions. *Appl. Sci* **2022**, *12*, 4866. [CrossRef]
23. Wang, H.; Hou, Y.C.; Ma, G.F.; Wu, G.K.; Wang, D.; Huang, B.; Wu, P.; Wu, D.Z. Identification on Vortex Rope in Francis Turbine Draft Tube Based on Convkurgram. *IEEE Trans. Instrum. Meas.* **2022**, *71*, 7504014. [CrossRef]
24. Hao, S.; Yang, L.; Ma, X.; Ma, R.Z.; Wen, H. YOLOv5 Transmission Line Fault Detection Based on Attention Mechanism and Cross-scale Feature Fusion. *Proc. CSEE* **2023**, *43*, 2319–2331.

25. Li, X.B.; Li, Y.G.; Guo, N.; Fan, Z. Mask detection algorithm based on YOLOv5 integrating attention mechanism. *J. Graph.* **2023**, *44*, 16–25.
26. Pi, J.; Liu, Y.H.; Li, J.H. Research on lightweight forest fire detection algorithm based on YOLOv5. *J. Graph.* **2023**, *44*, 26–32.
27. Zhang, H.H.; Jia, Z.F.; Zuo, Y.D. Research on Range and Velocity Measurement Technology Based on Digital Processing. *Inf. Technol. Informatiz.* **2019**, *232*, 215–217.

Article

Ascertainment of Hydropower Potential Sites Using Location Search Algorithm in Hunza River Basin, Pakistan

Asim Qayyum Butt [1,2,3], Donghui Shangguan [1,2,3,*], Muhammad Waseem [4], Faraz ul Haq [4], Yongjian Ding [1,2,3], Muhammad Ahsan Mukhtar [1,2,3], Muhammad Afzal [2,5] and Ali Muhammad [1,2,3]

1   Northwest Institute of Eco-Environment and Resources, University of Chinese Academy of Sciences, Lanzhou 730000, China; asimbutt7891@mails.ucas.ac.cn (A.Q.B.); dyj@lzb.ac.cn (Y.D.); amscewre@gmail.com (M.A.M.); unahar.shigri@gmail.com (A.M.)
2   University of Chinese Academy of Sciences, Beijing 100049, China; iamafzal49@gmail.com
3   China-Pakistan Joint Research Center on Earth Sciences, CAS-HEC, Islamabad 45320, Pakistan
4   Centre of Excellence in Water Resources Engineering, University of Engineering and Technology Lahore, Lahore 54890, Pakistan; dr.waseem@uet.edu.pk (M.W.); engraraz@uet.edu.pk (F.u.H.)
5   Xinjiang Institute of Ecology and Geography, University of Chinese Academy of Sciences, Urumqi 830000, China
*   Correspondence: dhguan@izb.ac.cn

Abstract: The recent energy shortfall in Pakistan has prompted the need for the development of hydropower projects to cope with the energy and monetary crisis. Hydropower in the northern areas is available yet has not been explored too much. Focusing on the sustainable development goal (SDG) "Ensure access to affordable, reliable, sustainable and modern energy", thirteen proposed sites were selected from upstream to downstream of the Hunza River for analysis. The head on all the proposed sites was determined by taking the elevation difference between the proposed turbine and the intake at all sites. The discharge on all proposed ungauged sites was determined using ArcGIS for watershed delineation and the area ratio method along with Soil Conservation Service–Curve Number (SCS-CN) by using gauged data of Hunza River provided by Water and Power Development Authority (WAPDA) Pakistan at Daniyor bridge Gilgit, Shimshal and the Passo tributaries of Hunza River. The Location Search Algorithm (LSA) approach used a multi-criteria decision-making tool (MDM) to make a decision matrix considering the location and constraint criteria and then normalizing the decision matrix using benefit and cost criteria, the relative weights were assigned to all criteria using a rank sum weighted method and the sites were ranked on the basis of the final score. The results revealed that Hunza River has a significant hydropower potential and based on the final score in the LSA approach, proposed site 13, site 4 and site 9 were concluded as the most promising sites among proposed alternatives. The proposed methodology could be an encouraging step for decision makers for future hydropower development in Pakistan.

Keywords: Hunza River Basin; hydropower; GIS; watershed delineation; MDM; location search algorithm

Citation: Butt, A.Q.; Shangguan, D.; Waseem, M.; Haq, F.u.; Ding, Y.; Mukhtar, M.A.; Afzal, M.; Muhammad, A. Ascertainment of Hydropower Potential Sites Using Location Search Algorithm in Hunza River Basin, Pakistan. *Water* **2023**, *15*, 2929. https://doi.org/10.3390/w15162929

Academic Editors: Zhengwei Wang and Yongguang Cheng

Received: 30 May 2023
Revised: 29 July 2023
Accepted: 7 August 2023
Published: 14 August 2023

1. Introduction

As a renewable energy source, hydropower is a versatile, dependable, and economical source of electricity generation and sustainable water-resource management [1]. Modern and advanced hydropower plants are accelerating our transition to clean energy, providing adequate power, and storage for irrigation purposes, and helping in climate-mitigation services. Hydropower is the most environmentally clean kind of renewable energy, with its emissions of greenhouse gases among the minimum. It contributes 16% of the world's total energy consumption [2]. In the near future, the installed energy capacity of hydropower could reach more than 1064 GW, which would still be far less than the projected economically feasible hydropower generation. Economic constraints, urbanization, a revolution in hydrological regimes, and less available resources are the main issues affecting the exploitation and allocation of potential hydropower sites in different parts of the world, especially in developing countries, and about 70% of economically feasible hydroelectric energy is yet to be developed due to these constraints. Hence, sustainable hydropower could help us to meet the Paris agreement on climate change and the United Nations (UN) agenda of sustainable development [3].

Pakistan, the sixth largest population in the world, is one of the developing countries with an abundant potential for renewable energy resources; however, it is still facing a crisis of power production and there is a shortfall of the energy of about 5000 MW to 8000 MW [4]. I should be noted that the Ministry of Water and Power in Pakistan documented the first-ever energy policy in 2006 to mainstream hydropower development into the economic plan of the country. The policy consisted of three main phases: long-term, medium-term, and short-term plans. The short-term plans were designed to embellish the scheme to provide an attractive incentive for renewable energy development; however, the objectives could not be met due to financial and political constraints. A similar kind of situation was observed in other phases of energy policy in Pakistan. In energy policy, the exploration of different possible sites was also emphasized to put Pakistan on the renewable energy map of the world, especially the northern areas of Pakistan, which have significant hydropower potential which had still not been explored, mainly because of underinvestment by the government in the energy sector, partly implemented policies along with structural reforms, and a lack of proper identification of the hydropower potential of different rivers [5]. For example, only the Indus River has a potential of 60 GW; however, only 19% of it has been identified or used [6]. Moreover, despite a severe energy shortfall in the country, the progress in sustainable energy development has been too slow to meet the demand for energy in Pakistan, whereas hydropower installation has hastily increased worldwide. Today, the recent energy crises have provoked the Ministry to identify the hydropower potential of different rivers in-country and explore cheap, sustainable potential hydropower sites to meet the country's energy demands.

In the past, significant numbers of studies have been conducted for the estimation of the hydropower potential of the river and suitable sites using different approaches and tools. Most of these studies have used the classical approach or field investigation (one of the traditional methods for hydropower potential), whereas recently, advanced tools (e.g., Geographic Information Systems (GIS) and remote sensing (RS)) have been used in many research studies [7–13]. The conventional method (paper mapping) consumes a significant amount of money and time, which limits the scope of these methods, especially in remote regions with complex terrains. Moreover, rational approximation simply uses the river flow and elevation for power potential approximation through site survey at a single point, or two points, or at a very small area [14]. However, high power potentials can be found in inaccessible remote regions with rough terrain, making this technique impossible for the reliable assessment of power potential and significant sites. In addition to that, the chances of human error and the effect of subjective judgment hinder the reliability of these methods [15].

To overcome these difficulties, GIS and RS techniques and Geo-Spatial Information Systems (GSIS) have recently been adopted for location analysis because of their time- and cost-effectiveness. For example, [14] used GIS for assessing hydropower potential using the spatial decision support system. Other than this, numerous studies have been conducted to analyze GIS data and search for suitable locations [16,17], also recommended a location-analysis method for searching hydropower potential sites using GSIS. In addition, several studies have also used the spatial process-based development of location analysis systems by managing and analyzing RS data using GIS [18]. The mentioned studies comprehensibly considered various aspects of location analysis (e.g., environmental, economic, and social); however, they did not utilize precise methodology for the selection of a site based on multivariable, rather than the single-head, difference between the powerhouse and the weir of the site.

For location search analysis, techniques based on the multivariate, multi-criteria decision-making tool (MDM) are convenient methods that allow us to rank other options and to choose a suitable one among them by assessing up to a few measures. These methods are considered a useful technique to analyze the issues related to decision-making processes. More specifically, MDM is a process to find the optimum alternative from all the available practicable options [19,20]. In recent years, the use of MDM approaches (e.g., a technique for order preference by the simulation of the ideal solution (TOPSIS), the analytic hierarchy process (AHP) and analytic network process (ANP) method) for the complicated problem have increased, especially in water resource engineering and management [21–24]. Similarly, a study by [25] focused on the review of reservoir operation optimizations using a Metaheuristic Algorithm.

Hence, focusing on the sustainable development goal (SDG) "Ensure access to affordable, reliable, sustainable and modern energy" [26], and keeping in view the utilization of MDM and GIS, a location-analysis methodology was adopted to search for hydropower potential sites in the Hunza River basin and select the most promising hydel potential site at Hunza River. Details about the study area, methodology, results, and conclusion are represented in the following sections, respectively.

2. Materials and Methods

The methodology was adopted in this study to identify the best suitable location for hydropower generation using GIS and MDM approaches as represented in the flowchart (Figure 1). The detailed description is provided below.

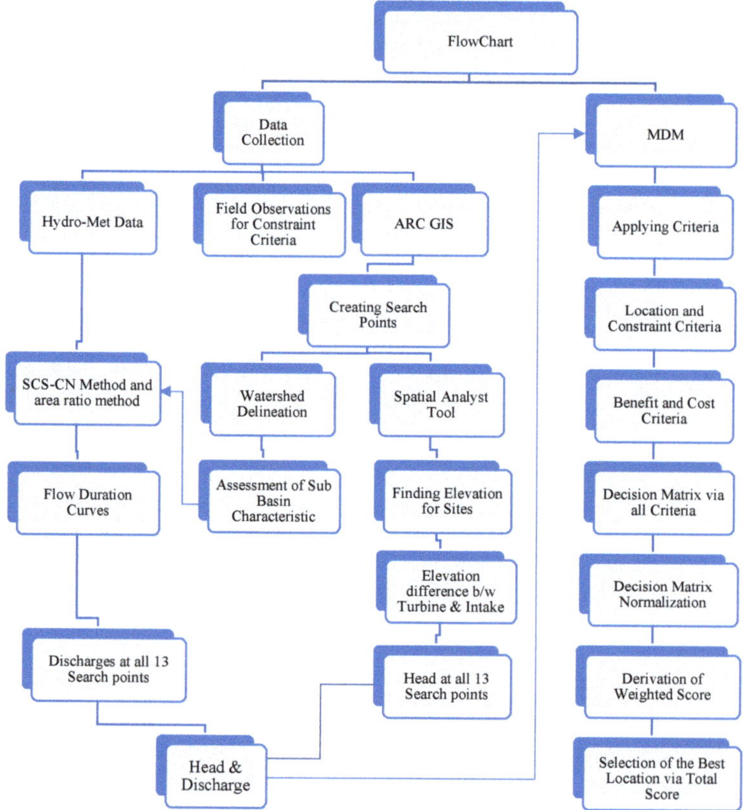

Figure 1. Flow chart of the adopted methodology.

2.1. Study Area

The study area, i.e., Hunza River basin (36.31° N, 74.64° E), as shown in Figure 2, is located in the mountainous valley in Gilgit-Baltistan, Pakistan, and has a basin area of 13,733 km^2 with a 231.76 km total length of the river. Several medium- and small-sized perennials as well as non-perennial tributaries cum rivers, and streams, such as Chalt, Daniyor, Chupurson, Khyber, Khuda_abad, Khunjeraab, Hisper, Hassanabad, Misgar, Hoper, Nalter, Rakaaposhi, Verjerab and Shimmshal, contribute to the flow of Hunza River, making its mean annual flow 323 m^3/s. The Hunza River basin joins the Nalter River and the Gilgit River before it falls into the mighty Indus River [4].

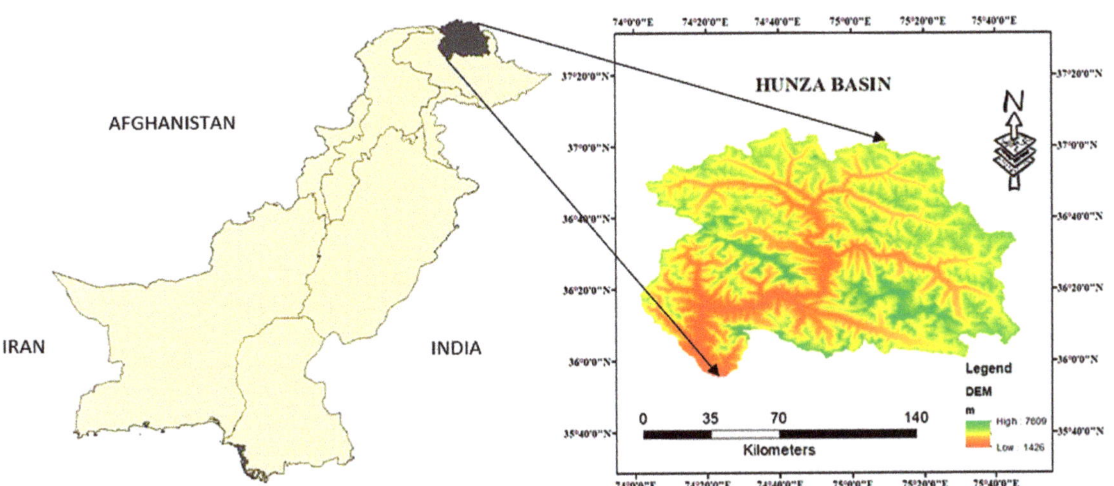

Figure 2. Study Area, i.e., Hunza River Basin.

Hence, keeping in mind the possibility of hydropower potential in the Hunza river basin, the current study focuses on the identification of suitable sites in Hunza district, Gilgit-Baltistan, Pakistan. To conduct the identification of potential sites, the flow data were collected from the Surface-water Hydrology Project of Water and Power Development Authority (SWHP-WAPDA) of Pakistan for a period of 55 years (1960–2015) at Daniyor Bridge and for a period of 20 years (1995–2015) at Shimshal and Passo tributaries. The meteorological data of four available stations, i.e., Khunjerab, Naltar, Passo, and Khunjerab were obtained for the assessment of flow at different sub-basins. Moreover, the open-access Advanced Spaceborne Thermal Emission (ASTER)'s 30 m Digital Elevation Model (http://srtm.csi.cgiar.org) (accessed on 28 July 2018) was used for GIS-based DEM processing and the estimation of elevation, area and available head and different proposed sites.

2.2. Multi-Criteria Decision Making (MDM)

Multi-criteria decision-making (MDM) is used for screening, prioritizing, selecting, or ranking the alternatives, and relies on the human assessment of a finite number of options according to a number of competing standards. The steps involved in MDM follow the sequence of (1) the selection of criteria; (2) the estimation of location criteria; (3) the normalization of location and constraint criteria; (4) assigning the weights to individual criteria; (5) the final Ranking of suitable sites based on the weighted average score; and (6) the estimation of hydropower potential and plant factor and conclusion.

Step 1: Selection of Criteria.

Considering conditions of the study area, different topographic, hydrologic, water resource-related, and socioeconomic criteria were used based on extensive literature, i.e., Refs [12,15,27,28], etc. The criteria were subdivided into location and constraint criteria, as represented in Table 1. For power calculations, head and discharge were used. Head and discharge were included in the location criteria and were estimated using GIS and a mathematical equation, respectively. The constraint criteria (as used by [11]) consist of criteria which can affect the sustainability of hydropower sites rather than the discharge and head. Furthermore, these constraint criteria were also divided based on the benefit (higher value represents positive aspect) or cost (higher value represents negative) aspect. Table 1 depicts the power criteria and constraint criteria along with their order number for MDM processing.

Table 1. Criteria for selection of best alternative.

Criterion	Location Criteria (Power)			Constraint Criteria (Environmental)		
Definition	Head	Discharge	Site Access	Agriculture Area	Residential Area	Interaction with other HPP *
Benefit or Cost aspect	Benefit	Benefit	Benefit	Cost	Cost	Cost

Note(s): * HPP = Hydropower Project.

Based on benefit and cost criteria analysis, the intangible criteria were converted to numerical values (from zero to ten) to create a decision matrix for each site, as shown in Table 2. If the proposed site contains a large agriculture area, the environmental impact assessment is considered, and resettlement issues would occur in case of construction of a hydropower project at that location. So, a value of 1 will be assigned to it according to the scale. Similarly, if the selected proposed location has a much smaller agriculture area, it means that no issues of environmental impact assessment would rise. So, a value of 9 is assigned to it. Similarly, a site with a particularly good access gets a value 9 on the scale while a site with difficult or no access is assigned a value up to 1.

Table 2. Values have been arbitrarily assigned to the constraint criteria. (Benefit criteria and cost criteria).

Benefit Criteria	Values	Cost Criteria
	00	
Very Low	01	Very High
Low	03	High
Average	05	Average
High	07	Low
Very High	09	Very Low
	10	

Step 2: Estimation of Location Criteria.

(a)   Head Determination at Proposed Sites

Thirteen search points were proposed as hydropower potential sites along the centerline of the river for marking the proposed locations of the plant. These search points were taken as locations of the powerhouse where turbines were to be installed. It was known that the Hunza River is completely supported by glacier-fed tributaries. So, from upstream to downstream, all thirteen recommended locations were chosen by selecting search points right before a new tributary to receive a varied discharge at each site to further examine its appropriateness. At all these search points, the longitude and latitude were marked on the map, as shown in Figure 3. Sites were selected on a GIS map along the river centerline to mark proposed plant/turbine locations and intake locations. Proposed sites were marked using ArcGIS Editor Toolbar. The elevation at each proposed site was extracted from DEM using a spatial analyst tool. Then, head was calculated for all thirteen search points/sites using the difference between the elevation head and tail water level. The head was calculated using the difference in elevation of the intake point and the turbine position. Three factors were examined while determining the natural head at each search site. First, upstream of each search location, a suitable thin cross-section of the river was discovered. Second, there was a curve to achieve a better fall or head at the powerhouse locations, and third, there was a cup-shaped valley behind the sites where a reservoir could be built.

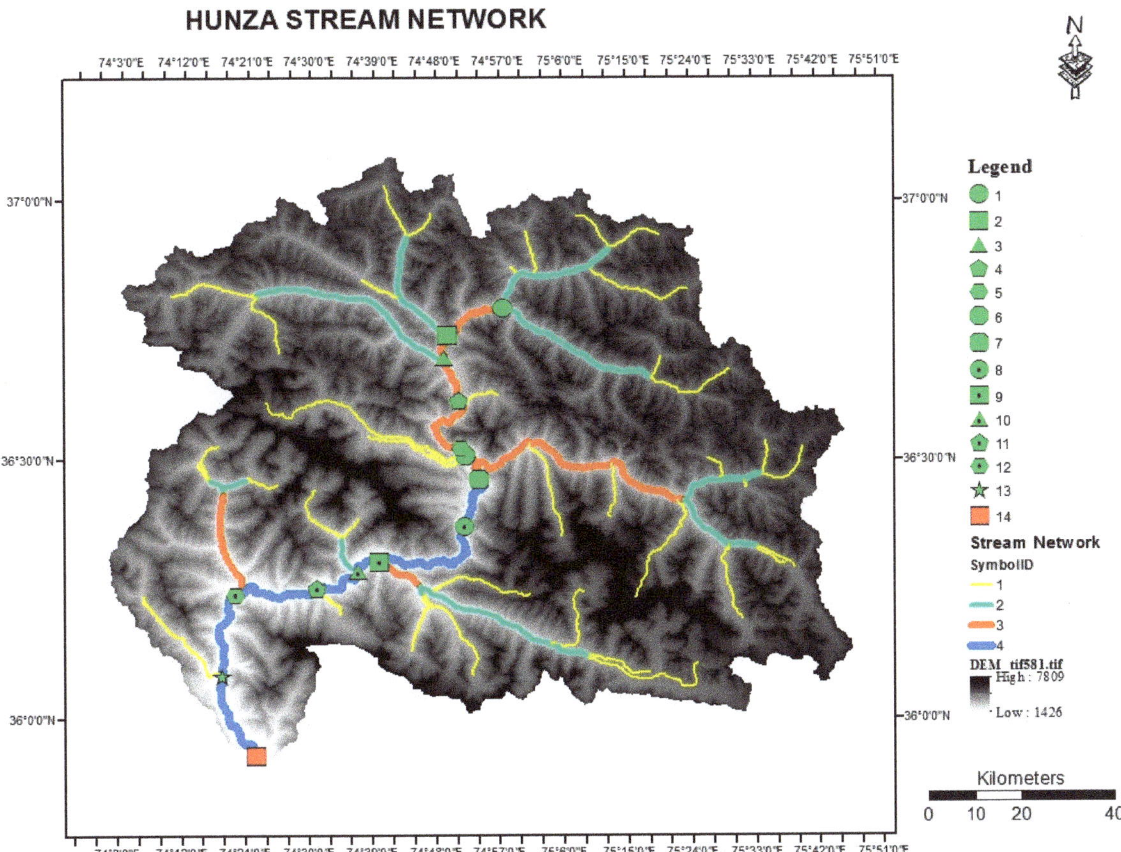

Figure 3. Proposed Alternatives/Sites in Hunza River Basin.

(b)   Estimation of Flow Duration Curve at Ungauged Sites

The data were obtained using the Soil Conservation Service Curve Number (SCS-CN) approach and the area ratio method for ungauged stations due to the lack of discharge data at various places. To estimate the discharge of sub-basins with meteorological stations, the SCS-CN approach was used, while the area ratio method was used to synthesize the data of remaining points. Moreover, watershed delineation at all thirteen points was carried out to divide the Hunza basin into thirteen sub-basins and the catchment characteristics (e.g., area, length, and CN) were estimated using GIS. Details of the SCS-CN method and area ratio method are given below.

The SCS-CN is the most widely used method for runoff estimation at the ungauged basin. SCS-CN is developed by the United States Department of Agriculture, keeping in view the hydrologic abstraction and land represented by CN. The curve number (CN) mainly depends on land use, soil type, and antecedent moisture conditions (AMCs) [29–31]. The SCS-CN is the most popular method for the hydrology of small catchment due to its simplicity, the incorporation of many factors, and well [30]. Using the SCS-CN method, the runoff (Q) is estimated using rainfall (P), curve number (CN), and initial abstraction (Ia), as given below:

$$\frac{P - Ia - Q}{S} = \frac{Q}{P - Ia} \tag{1}$$

$$Q = \frac{(P - \lambda S)^2}{P + (1 - \lambda)S} \qquad P \succ \lambda S \text{ else } Q = 0 \tag{2}$$

where S is maximum potential retention and calculated $S = \left(\frac{1000}{CN} - 10\right)$ if P in inches and $S = \left(\frac{2540}{CN} - 25.4\right)$ if P in cm, $\lambda = 0.2$, and CN is the tabulated dimensionless index and ranges from 0–100. For the current study, CN was taken as 80 for the area of Hunza, as Hunza is very vegetative and grassy.

In cases where no flow data are available, then the area ratio method can be used for the estimation of flow for sites using existing data from one or more nearby homogenous streamflow-gauging stations [32]. In the area ratio method, it is assumed that the runoff per unit area in the target ungauged basin is equal to that in the donor basin. To estimate the runoff of the target ungauged basin (Q_u), this method is quite simple and requires the runoff of the donor (Q_g), area of the donor (A_g), and area of the target ungauged basin (A_u). Several studies have used the area ratio method and found it quite effective in the region with a scarce gauging network [33–35]. Moreover, it is widely used in the region, where still no rainfall–runoff relation has been developed.

$$Q_u = K \left(\frac{A_u}{A_g}\right)^{\phi} \times Q_g \tag{3}$$

where ϕ is an exponent parameter estimated by the regression analysis of flow and area of the gauged basin, and K is a bias correction factor = 1.0 [32,35].

Following the estimation and collecting of flow data at each of the thirteen locations, the flow duration curve (FDC) for each of the points that were chosen was generated. To produce FDC, the runoff from each suggested site was separately arranged in descending order. Following this step, the associated percentiles were determined using the Weibull plotting algorithm. Furthermore, for the current study, Q_{100}, Q_{90}, Q_{80}, Q_{70}, Q_{60}, and Q_{50}, were extracted from the FDC of each proposed location.

Q_{100} represents the flow that is equaled or exceeded 100% of the time. Q_{50} represents the flow that is equaled or exceeded 50% of the time.

Step 3: Normalization of location and constraint criteria.

After the estimation of location criteria and conversion of constraint criteria into numerical values, a combined decision matrix for each proposed site was developed. The decision matrix was then normalized based on benefit and cost criteria using Equations (4) and (5), respectively.

$$\text{For Benefit Criteria } n_{ij} = \frac{b_{ij} - b_{max}}{bmin_{max}} \tag{4}$$

$$\text{For Cost Criteria } n_{ij} = \frac{cij_{max}}{cmin_{max}} \tag{5}$$

where b_{ij} represents the i_{th} value of j_{th} benefit criteria, and c_{ij} indicating the i_{th} value of j_{th} cost criteria, max, and min in subscript represents the respective maximum and minimum value of cost and benefit criteria, and n_{ij} is the normalized value.

Step 4: Assigning the weights to individual criterion.

The rank-sum weighted approach was used after the decision matrix was normalized and weights were assigned to each criterion. The weights in the rank-sum (RS) method are the normalized rankings of the individuals, which are determined by dividing their combined ranks by the sum of all ranks [36].

$$W_j = \frac{2(n+1-r_j)}{n(n+1)} \tag{6}$$

where r_j is the j_{th} criterion's ranking and $j = 1, 2, 3\ldots\ldots.n$.

Step 5: Final Ranking of suitable sites based on the weighted average score.

After the determination of relative weights for each criterion, each relative weight was multiplied by its associated criterion, and the sum of these six weights was used to rank the recommended sites.

Step 6: Estimation of hydropower potential and plant factor.

After evaluating the feasible locations using the MDM approach, the available hydropower potential was calculated for all recommended sites in the Hunza river basin using a worldwide efficiency of 81% [27].

$$P = \frac{\eta \rho g H Q}{1000} \tag{7}$$

where η = global efficiency, typically from 75 to 88 (%), g = gravitational acceleration, 9.81 (m/s^2), ρ = density of water 1000 (kg/m^3), P = mean annual electric power (kW), H = head (m), Q = discharge (m^3/s), i.e., Q_{50} is used for the discharge for the calculation of power as Q_{50} represents the flow that is equaled or exceeded 50% of the time [37]. According to [38], the flow rate in Q_{50} is almost constant throughout the year, despite being lower than in Q_{75} and Q_{90}. As a result, this discharge has been used in this study's Q_{50} estimation of rivers' potential for producing power.

Furthermore, for the calculation of the plant factor, using the average discharge equation, Qavg, the energy (E) in GWh was calculated using Equations (8) and (9) and then the plant factor using Equation (10).

$$Qavg = \frac{[Q_{100} + Q_{90} + Q_{80} + Q_{70} + Q_{60} + (5 * Q_{50})]}{10} \tag{8}$$

where Q_{100}, Q_{90}, Q_{80}, Q_{70}, Q_{60} and Q_{50} are the percentage exceedances of discharges for each site from flow duration curves at 100%, 90%, 80%, 70%, 60% and 50% probability.

$$E = \frac{8.76 * \eta * Qavg * g * H}{1000} \tag{9}$$

where E is energy in GWh, H = head (m), g = gravitational acceleration, 9.81 (m/s^2), η = global efficiency, typically from 75 to 88 (%) and Qavg is derived from Equation (7).

$$\text{Plant factor} = \frac{E}{8.76P} \times 100 \tag{10}$$

where E is energy in GWh, and P is power in MW.

3. Results

3.1. Head Determination

For assessment of the head at the marked thirteen sites, the elevations of both the proposed dam/weir and powerhouse locations were derived using DEM processing with the spatial analyst tools of GIS. The derived profile of the Hunza River is graphically represented in Figure 4 by using the derived elevation of each site against the reduced distance (RD) along with the derived river profile of Hunza River with respect to the elevation of the intake point of the proposed sites. The RD was taken from the upstream to downstream of the river in such a way that site one (1) was considered RD 0 km and so on.

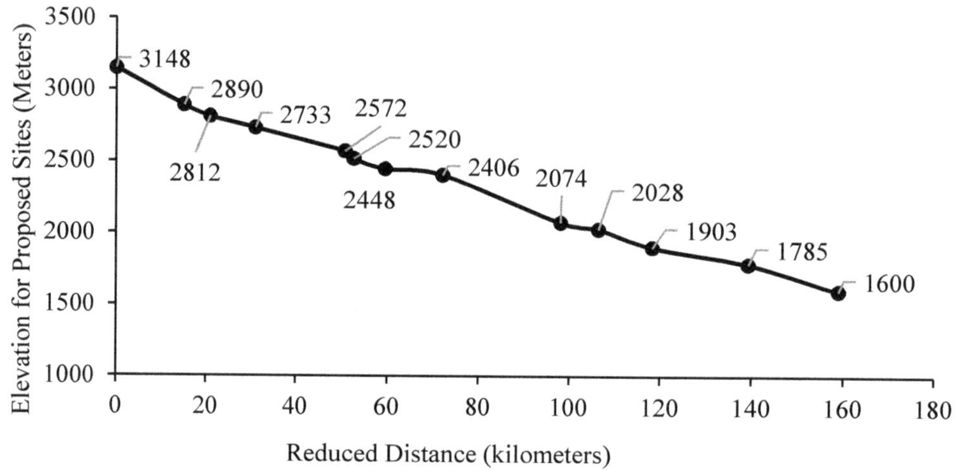

Figure 4. The derived elevation profile of the Hunza river.

For the purpose of head determination, locations were chosen on a GIS map of the Hunza River basin along the river's centerline to indicate potential plant/turbine locations and intake locations. ArcGIS Editor Toolbar was used to mark proposed locations. Using the spatial analyst tool, the elevation at each proposed site was extracted from the DEM. The difference between the elevations of the head and tail waters is then used to calculate the head for each of the thirteen search points/sites. The first point was used as the intake point and the second as the powerhouse/turbine position to calculate the elevation difference/head, as shown in Table 3. Based on the results, it was observed that sites 5, 4, and 1 have the top three highest heads, i.e., 117, 94.5 and 80.79, respectively.

Table 3. Sites/alternatives in Hunza River basin along with their heads.

Site No.	1	2	3	4	5	6	7	8	9	10	11	12	13
Head (m)	80.79	75.23	38.41	94.5	117.0	80.48	46.34	0.91	30.18	21.0	29.80	28.0	81.70

3.2. Watershed Delineation for All Sites

The catchment characteristics, including the watershed area of each site and CN, were calculated using GIS and CN-Tables to carry out the estimation of discharge with SCS-CN and AR methods. Figure 5 shows the watershed delineation from upstream to downstream using the arc hydro tool in GIS.

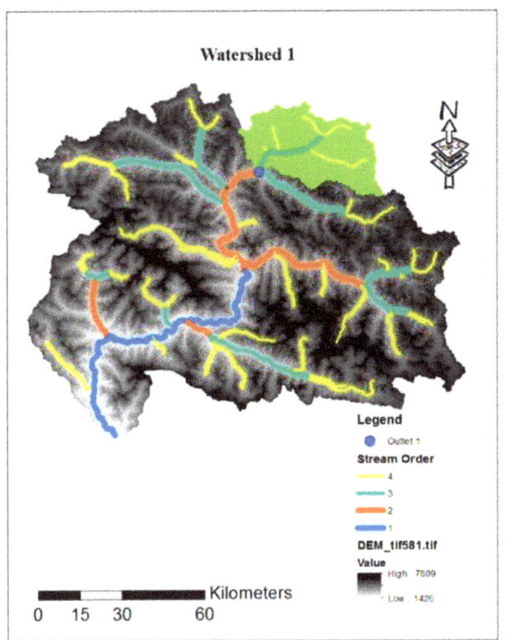

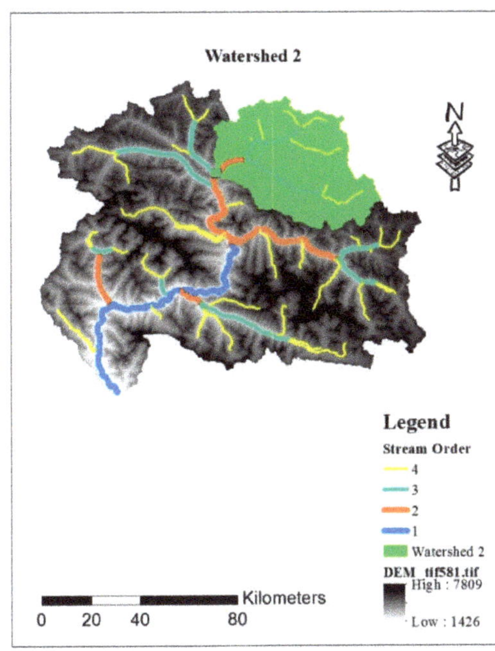

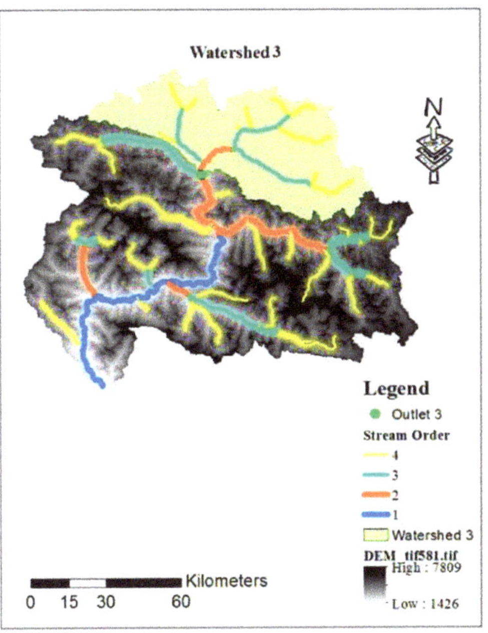

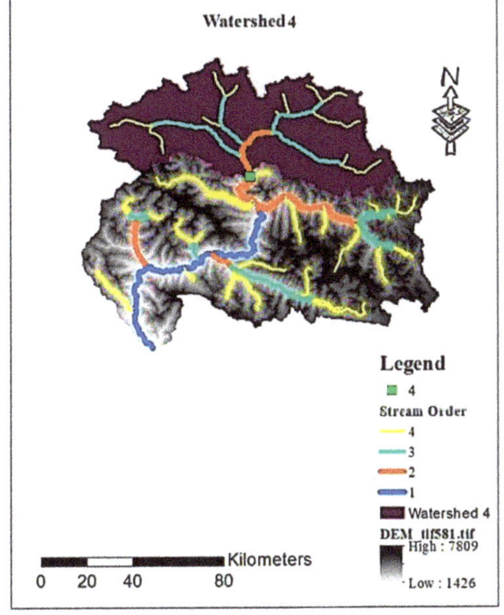

Figure 5. *Cont.*

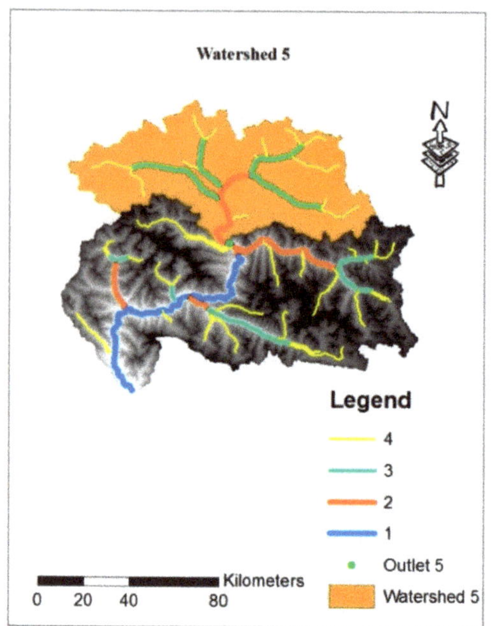

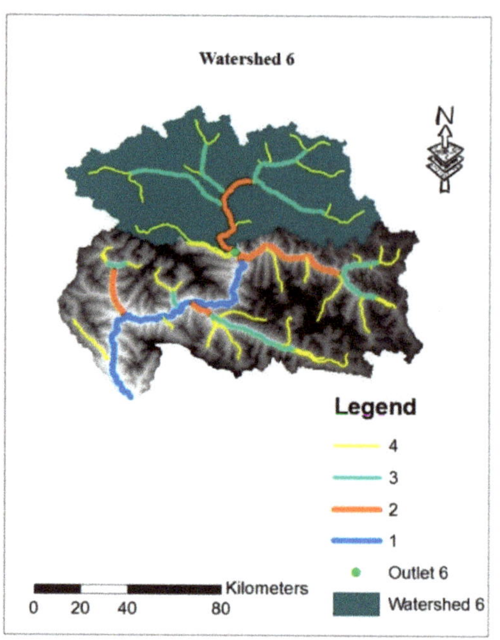

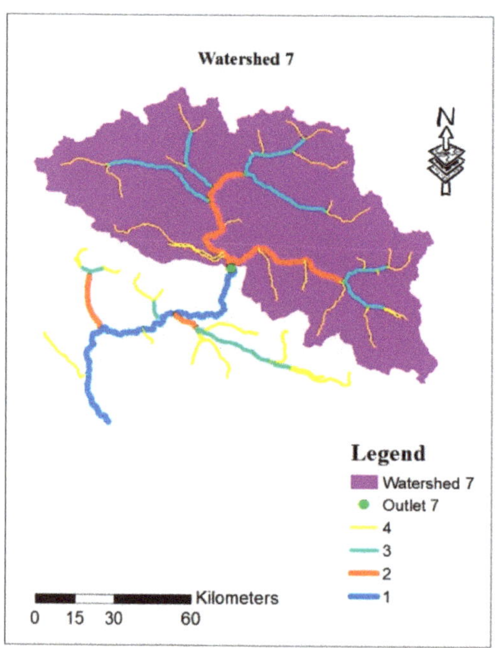

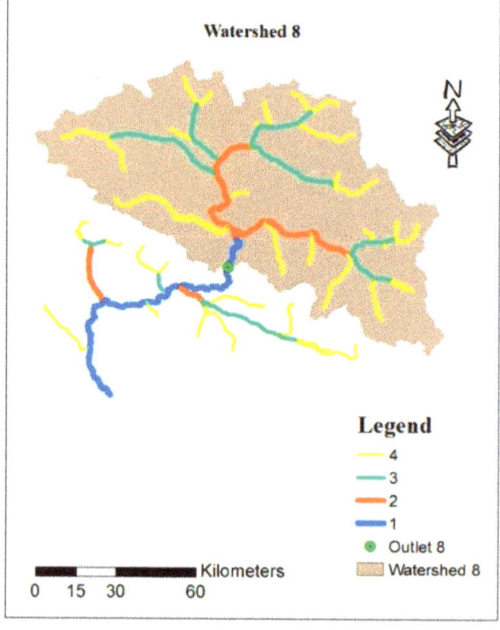

Figure 5. *Cont.*

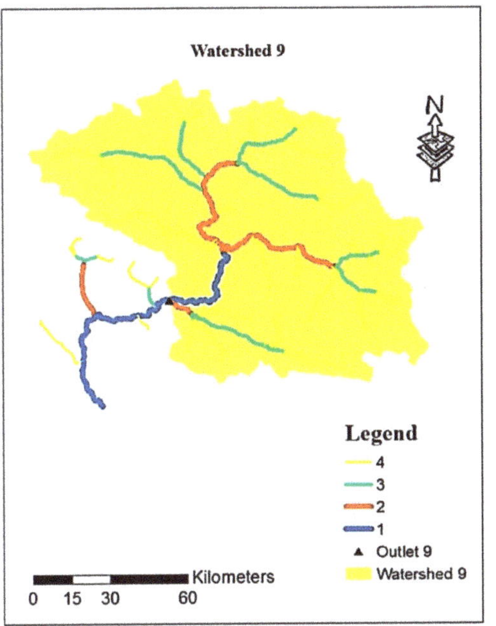

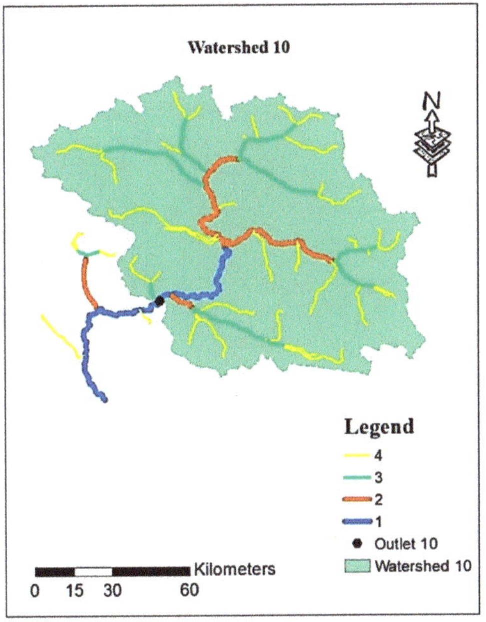

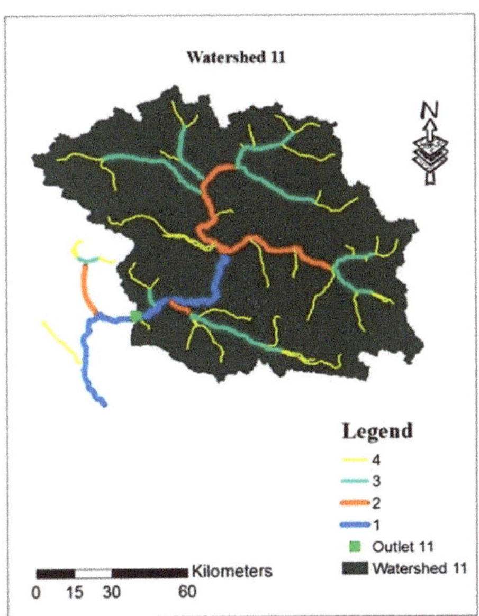

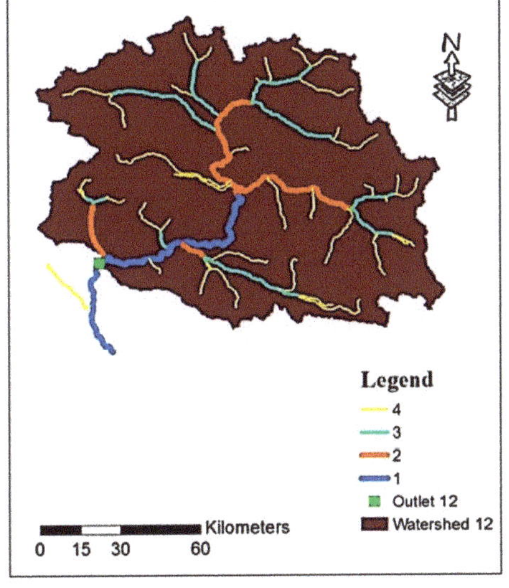

Figure 5. *Cont.*

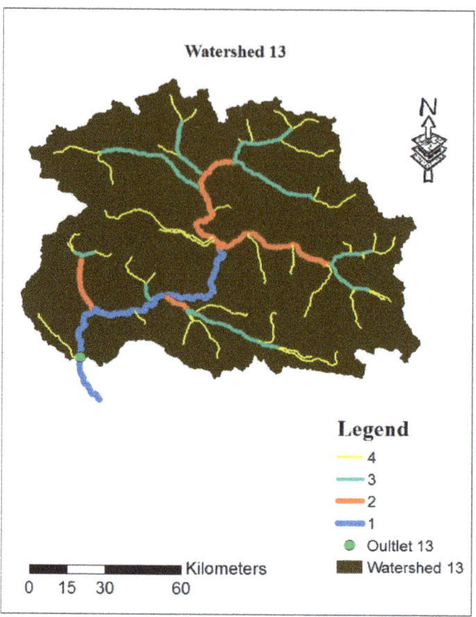

Figure 5. Watershed delineation of Hunza River at all 13 proposed alternatives.

3.3. Estimation of Flow Duration Curve at Each Site

As the accumulation of flow increases from upstream to downstream, site 13 has the highest value of discharge, while site one (1) has the lowest value of discharge. Furthermore, the discharge values at each site were used to develop flow-duration curves, as shown in Figure 6.

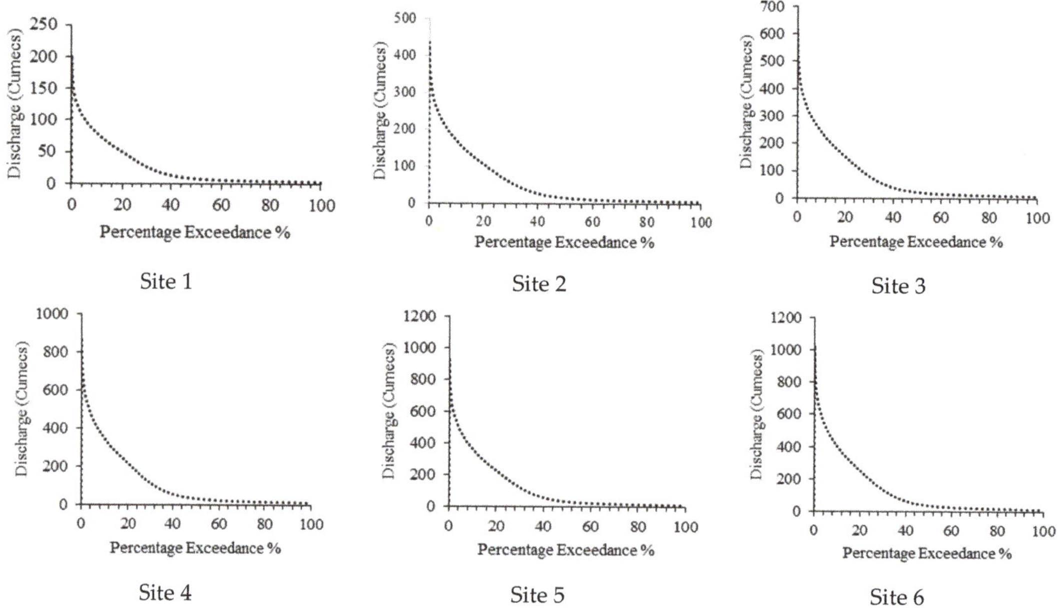

Figure 6. *Cont.*

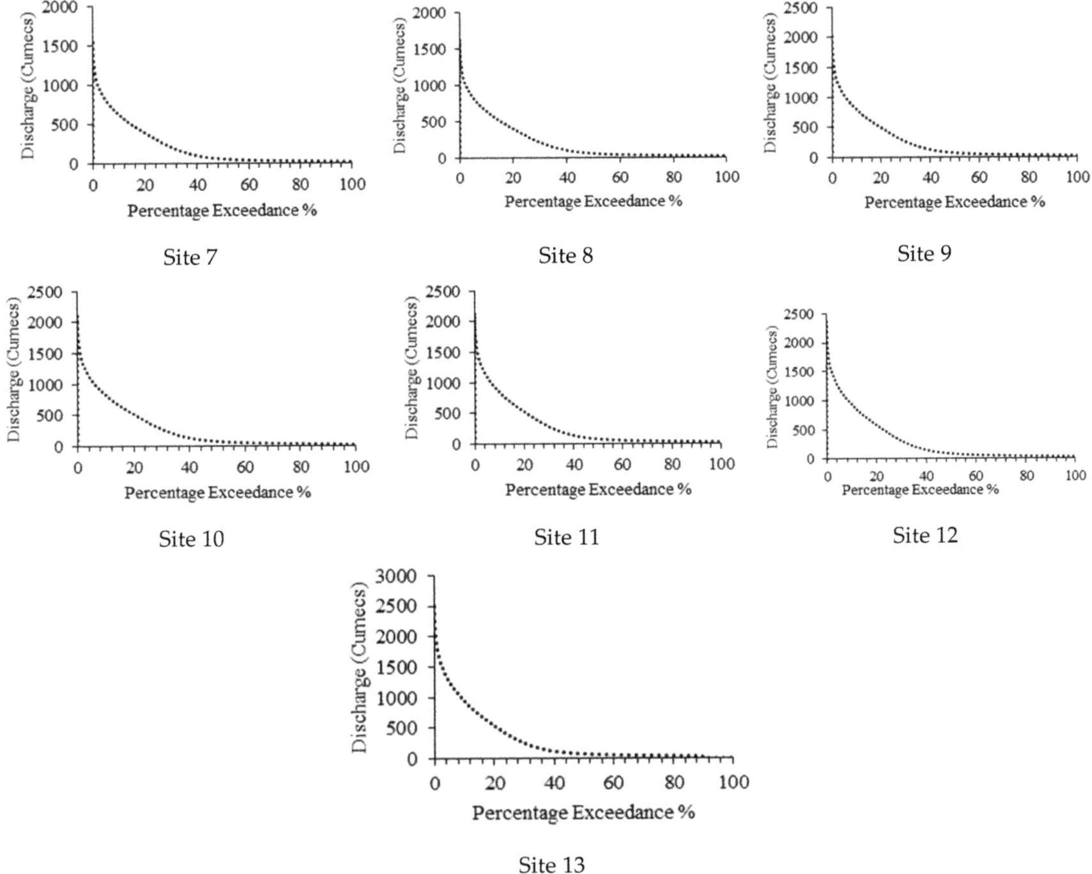

Figure 6. Flow-duration curves for discharges at all 13 proposed alternatives/sites.

3.4. Hydropower Potential

Calculated using Equation (6), the hydropower availability for all thirteen proposed sites is depicted in Table 4.

Table 4. Hydropower availability for all thirteen proposed sites.

Site No.	Q_{50} (Cumecs)	H (meters)	g (m/s^2)	η	Density (kg/m^3)	P (MW)
1	7.512	80.79	9.81	0.81	1000	04.82
2	16.210	75.23	9.81	0.81	1000	09.69
3	22.967	38.41	9.81	0.81	1000	07.01
4	32.566	94.50	9.81	0.81	1000	24.45
5	34.409	117.0	9.81	0.81	1000	31.99
6	38.124	80.48	9.81	0.81	1000	24.38
7	58.687	46.34	9.81	0.81	1000	21.61
8	60.648	0.91	9.81	0.81	1000	0.44
9	75.091	30.18	9.81	0.81	1000	18.00
10	78.383	21.0	9.81	0.81	1000	13.08
11	79.804	29.80	9.81	0.81	1000	18.90
12	88.079	28.0	9.81	0.81	1000	19.60
13	92.805	81.70	9.81	0.81	1000	60.25

3.5. Selection of Suitable Sites Using MDM Method

The numerical values were assigned to constraint criteria, as shown in Table 5, based on expert opinions and field analysis. Table 5 also shows the complete decision matrix, having both location and constraint criteria in numerical form for each proposed site. This decision matrix was further normalized based on benefit and cost criteria using Equations (3) and (4).

Table 5. Decision matrix with numerical values assigned to intangible criteria (Cost-wise and benefit-wise).

Site No.	Power (MW)	Site Access	Agriculture Area	Residential Area	Interaction with Other HPP
1	04.82	5	9	9	9
2	09.69	7	9	9	9
3	07.01	5	9	9	9
4	24.45	7	3	3	9
5	31.99	7	9	9	9
6	24.38	7	9	9	9
7	21.61	5	5	3	9
8	0.44	7	5	3	9
9	18.00	7	3	3	9
10	13.08	7	3	3	9
11	18.90	5	5	5	9
12	19.60	7	5	3	9
13	60.25	7	5	3	9

After the decision matrix was obtained in a numerical valued form, it was normalized on the basis of the benefit and cost criteria. As shown in Table 6, for the benefit criteria, the largest value is 60.25 for site 13. Then, all values of power were divided by 60.25., i.e., for site 13, 60.25/60.25 = 1.0. Similarly, for site 1, 4.823/60.25 = 0.08005.

Table 6. Normalized decision matrix after using the numerical scale on intangible criteria (benefit and cost wise).

Site No.	Power (MW)	Site Access	Agriculture Area	Residential Area	Interaction with Other HPP
1	0.080	1.0	0.33	0.33	1.0
2	0.161	1.0	0.33	0.33	1.0
3	0.116	1.0	0.33	0.33	1.0
4	0.406	1.0	1.0	1.0	1.0
5	0.531	1.0	0.33	0.33	1.0
6	0.405	1.0	0.33	0.33	1.0
7	0.359	1.0	0.60	1.0	1.0
8	0.007	1.0	0.60	1.0	1.0
9	0.299	1.0	1.0	1.0	1.0
10	0.217	1.0	1.0	1.0	1.0
11	0.314	1.0	0.60	0.60	1.0
12	0.325	1.0	0.60	1.0	1.0
13	1.0	1.0	0.60	1.0	1.0

For the cost criteria, the smallest value is 3 in the case of the agriculture area, then for all values of this, criterion 3 was divided by all the values of the proposed sites., i.e., for site 13, agriculture area would be 3/5 = 0.6, and the least value, i.e., 3/3, would obtain a value of 1.0.

Furthermore, the relative weights were estimated based on Equation (6) as depicted in Table 7.

Table 7. Relative weights of all criteria by their preferences.

Criterion	Preference	Relative Weight
Power	1	0.333
Site Access	2	0.267
Agriculture Area	3	0.200
Residential Area	4	0.133
Interaction with other HPP	5	0.067

As indicated in Table 8, the relative weights were multiplied by each appropriate criteria for each site, and the sum of all criteria was used as the final score for the recommended sites.

Table 8. Weighted decision matrix (normalized decision matrix * relative weights).

Site No.	Results Obtained by Multiplying Criteria with Relative Weights					Sum of All Criteria Weightage
	Power	Site Access	Agriculture Area	Residential Area	Interaction with Other HPP	
1	0.0173	0.2667	0.0667	0.0444	0.0667	0.4618
2	0.0348	0.2667	0.0667	0.0444	0.0667	0.4793
3	0.0251	0.2667	0.0667	0.0444	0.0667	0.4696
4	0.0877	0.2667	0.20	0.1333	0.0667	0.7544
5	0.1148	0.2667	0.0667	0.0444	0.0667	0.5593
6	0.0875	0.2667	0.0667	0.0444	0.0667	0.5319
7	0.0775	0.2667	0.120	0.1333	0.0667	0.6642
8	0.0016	0.2667	0.120	0.1333	0.0667	0.5883
9	0.0646	0.2667	0.20	0.1333	0.0667	0.7313
10	0.0469	0.2667	0.20	0.1333	0.0667	0.7136
11	0.0678	0.2667	0.120	0.0799	0.0667	0.6012
12	0.0703	0.2667	0.120	0.1333	0.0667	0.6570
13	0.2162	0.2667	0.120	0.1333	0.0667	0.8029

Table 9 shows the final rankings of the proposed sites based on sum of all criteria weightages and Table 10 depicts the energy and plant factors for the proposed sites along with their hydropower potentials.

Table 9. Ranking of each proposed site based on the final score.

Site No.	Longitude	Latitude	Sum of All Criteria Weightage (from Table 8)	Final Rankings
1	74.95 E	36.79 N	0.4618	13
2	74.82 E	36.74 N	0.4793	11
3	74.81 E	36.69 N	0.4696	12
4	74.85 E	36.61 N	0.7544	2
5	74.86 E	36.52 N	0.5593	9
6	74.87 E	36.51 N	0.5319	10
7	74.90 E	36.46 N	0.6642	5
8	74.86 E	36.37 N	0.5883	8
9	74.66 E	36.30 N	0.7313	3
10	74.61 E	36.28 N	0.7136	4
11	74.35 E	36.25 N	0.6012	7
12	74.35 E	36.25 N	0.6570	6
13	74.29 E	36.08 N	0.8029	1

Table 10. Hydropower potential of the proposed sites in Hunza River basin.

Site No.	1	2	3	4	5	6	7	8	9	10	11	12	13
Power (MW)	4.82	9.69	7.01	24.45	31.99	24.38	21.61	0.44	18.00	13.08	18.90	19.60	60.25
Energy (GWh)	31.41	63.13	45.66	159.3	208.4	158.9	140.8	02.85	117.3	85.22	123.1	127.7	392.6
Plant factor (%)	74.39	74.37	74.39	74.39	74.39	74.39	74.39	74.39	74.39	74.39	74.39	74.39	74.39

4. Discussion

Based on the analysis and final score in the LSA approach, proposed Site 13 (36 05′05″ N, 74 17′34″ E), Site 4 (36 37′02″ N, 74 51′10″ E), and Site 9 (36 18′28″ N, 74 39′57″ E) were concluded as the best locations among all proposed thirteen alternatives (Table 9).

Hydropower potential assessment of Hunza River basin resulted that this basin has huge potentials (Table 10) for development of hydropower. Moreover, analysis based on individual site resulted that the most suitable site, i.e., site 13 have 392.6 GWh energy potential, whereas site 4 have 159.3 and Site 9 has 117.3 GWh energy potential. On the other hand, site 5 has a comparatively high energy potential of 208.4 GWh but ranked 6 based on the MDM approach. It was mainly because of comparatively low benefit and high-cost constraints. Therefore, it was concluded that the only availability of significant favorable location criteria does not rank the site as the most suitable location, whereas other factors like constraint criteria also play a vital role in overall suitability. Hence, this study will provide an initial assessment and thorough analysis for the development of hydropower projects in the study area.

5. Conclusions

Hydropower is one of the clean, abundant, and renewable sources of energy that could help the country to meet the increasing energy demand and integrate intermittent renewables into the energy grid. Hunza River basin, situated in northern area Pakistan, was taken as the study area, and suitable hydropower locations were marked using GIS, Multi-Criteria Decision Making (MDM), and final score. Based on the location search algorithm and final scores, it is concluded that the Hunza River basin has an enormous potential for hydropower and that site 13, site 4, and site 9 could be considered as the three top probable sites among all alternatives, with potential hydropower availability estimations of 60.25 MW, 24.25 MW, and 18.01 MW, respectively.

It is important to note that the proposed methodology cannot replace fieldwork, but it can save time and give an accurate initial idea of potential sites. As a result, this study will help policymakers in the energy sector analyze current energy resources and locate areas with the potential for hydropower. Additionally, creating indices for those elements that could be used as constraint and location criteria will be necessary in the future to strengthen and improve the discriminating ability between searching spots, even in tiny regions. Therefore, results can be repeated in scenarios where sizable amounts of water and potential heads are present in Pakistan's northern regions, including Azad Kashmir. This study could not cover the geological studies, tectonic movements, sliding and riverbed investigations in the study area. All the understudy sites were conceived with a dam and powerhouse at the toe without any headrace tunnel. Future work can be carried out covering other aspects of ascertainment.

Author Contributions: The first author, A.Q.B., performed all the calculations and prepared the graphs and tables by development of the analyses with input and guidance from D.S., M.W. and Y.D. M.A., F.u.H., A.M. and M.A.M. contributed to the discussion of results and shared the writing of the paper with the first author. Funding acquisition was conducted by D.S. and Y.D. All authors have read and agreed to the published version of the manuscript.

Funding: This research and APC was funded by international partnership of the Chinese Academy of Sciences (Grant No. 131C11KYSB20200022), and the Ministry of Science and Technology (Grant No. 2018FY100502).

Institutional Review Board Statement: Not applicable.

Informed Consent Statement: Not applicable.

Data Availability Statement: The data used to substantiate the findings of this research are accessible upon request from the corresponding author.

Acknowledgments: The authors would like to thank every person who has collaborated throughout this research work, and mainly the Water and Power Development Authority (WAPDA) Pakistan for providing the Discharge data. The complexities of this research work could not have been possible to achieve without their cooperation. The constructive contribution of WAPDA and the careful review and sincere suggestions by the anonymous referees are gratefully acknowledged. The authors would also like to thank Haseeb Akbar, Muhammad Kaleem Sarwar and Abid Sheikh (WAPDA) for their help and guidance in completion of the research.

Conflicts of Interest: The authors declare no conflict of interest.

References

1. Bartle, A. Hydropower potential and development activities. *Energy Policy* **2002**, *30*, 1231–1239. [CrossRef]
2. World Energy Council, World Energy Resources Hydropower. 2016. Available online: https://www.worldenergy.org/assets/images/imported/2016/10/World-Energy-Resources-Full-report-2016.10.03.pdf (accessed on 1 May 2023).
3. UNCC. How Hydropower Can Help Climate Action. 2018. Available online: https://unfccc.int/news/how-hydropower-can-help-climate-action (accessed on 1 May 2023).
4. Tahir, A.A.; Chevallier, P.; Arnaud, Y.; Neppel, L.; Ahmad, B. Modeling snowmelt-runoff under climate scenarios in the Hunza River basin, Karakoram Range, Northern Pakistan. *J. Hydrol.* **2011**, *409*, 104–117. [CrossRef]
5. Mirjat, N.H.; Uqaili, M.A.; Harijan, K.; Das Valasai, G.; Shaikh, F.; Waris, M. A review of energy and power planning and policies of Pakistan. *Renew. Sustain. Energy Rev.* **2017**, *79*, 110–127. [CrossRef]
6. Siddiqi, A.; Wescoat, J.L.; Humair, S.; Afridi, K. An empirical analysis of the hydropower portfolio in Pakistan. *Energy Policy* **2012**, *50*, 228–241. [CrossRef]
7. Dudhani, S.; Sinha, A.; Inamdar, S. Assessment of small hydropower potential using remote sensing data for sustainable development in India. *Energy Policy* **2006**, *34*, 3195–3205. [CrossRef]
8. Rojanamon, P.; Chaisomphob, T.; Bureekul, T. Application of geographical information system to site selection of small run-of-river hydropower project by considering engineering/economic/environmental criteria and social impact. *Renew. Sustain. Energy Rev.* **2009**, *13*, 2336–2348. [CrossRef]
9. Kusre, B.; Baruah, D.; Bordoloi, P.; Patra, S. Assessment of hydropower potential using GIS and hydrological modeling technique in Kopili River basin in Assam (India). *Appl. Energy* **2010**, *87*, 298–309. [CrossRef]
10. Larentis, D.G.; Collischonn, W.; Olivera, F.; Tucci, C.E. Gis-based procedures for hydropower potential spotting. *Energy* **2010**, *35*, 4237–4243. [CrossRef]
11. Yi, C.-S.; Lee, J.-H.; Shim, M.-P. Site location analysis for small hydropower using geo-spatial information system. *Renew. Energy* **2010**, *35*, 852–861. [CrossRef]
12. Cuya, D.G.P.; Brandimarte, L.; Popescu, I.; Alterach, J.; Peviani, M. A GIS-based assessment of maximum potential hydropower production in La Plata basin under global changes. *Renew. Energy* **2013**, *50*, 103–114. [CrossRef]
13. Akbar, H.; Gheewala, S.H. Impact of Climate and Land Use Changes on flowrate in the Kunhar River Basin, Pakistan, for the Period (1992–2014). *Arab. J. Geosci.* **2021**, *14*, 707. [CrossRef]
14. Alashan, S.; Şen, Z.; Toprak, Z.F. Hydroelectric Energy Potential of Turkey: A Refined Calculation Method. *Arab. J. Sci. Eng.* **2016**, *41*, 1511–1520. [CrossRef]
15. Zaidi, A.Z.; Khan, M. Identifying high potential locations for run-of-the-river hydroelectric power plants using GIS and digital elevation models. *Renew. Sustain. Energy Rev.* **2018**, *89*, 106–116. [CrossRef]
16. Bayazıt, Y.; Bakış, R.; Koç, C. An investigation of small-scale hydropower plants using the geographic information system. *Renew. Sustain. Energy Rev.* **2017**, *67*, 289–294. [CrossRef]
17. Belmonte, S.; Núñez, V.; Viramonte, J.; Franco, J. Potential renewable energy resources of the Lerma Valley, Salta, Argentina for its strategic territorial planning. *Renew. Sustain. Energy Rev.* **2009**, *13*, 1475–1484. [CrossRef]
18. Punys, P.; Dumbrauskas, A.; Kvaraciejus, A.; Vyciene, G. Tools for Small Hydropower Plant Resource Planning and Development: A Review of Technology and Applications. *Energies* **2011**, *4*, 1258–1277. [CrossRef]
19. Tecle, A.; Fogel, M.; Duckstein, L. Multicriterion Selection of Wastewater Management Alternatives. *J. Water Resour. Plan. Manag.* **1988**, *114*, 383–398. [CrossRef]
20. Weng, S.; Huang, G.; Li, Y. An integrated scenario-based multi-criteria decision support system for water resources management and planning—A case study in the Haihe River Basin. *Expert Syst. Appl.* **2010**, *37*, 8242–8254. [CrossRef]
21. Atieh, M.; Gharabaghi, B.; Rudra, R. Entropy-based neural networks model for flow duration curves at ungauged sites. *J. Hydrol.* **2015**, *529*, 1007–1020. [CrossRef]

22. Gazendam, E.; Gharabaghi, B.; McBean, E.; Whiteley, H.; Kostaschuk, R. Ranking of Waterways Susceptible to Adverse Stormwater Effects. *Can. Water Resour. J.* **2009**, *34*, 205–228. [CrossRef]
23. Kodikara, P.; Perera, B.; Kularathna, M. Stakeholder preference elicitation and modelling in multi-criteria decision analysis—A case study on urban water supply. *Eur. J. Oper. Res.* **2010**, *206*, 209–220. [CrossRef]
24. Noori, A.; Bonakdari, H.; Morovati, K.; Gharabaghi, B. The optimal dam site selection using a group decision-making method through fuzzy TOPSIS model. *Environ. Syst. Decis.* **2018**, *38*, 471–488. [CrossRef]
25. Lai, V.; Huang, Y.F.; Koo, C.H.; Ahmed, A.N.; El-Shafie, A. A Review of Reservoir Operation Optimisations: From Traditional Models to Metaheuristic Algorithms. *Arch. Comput. Methods Eng.* **2022**, *29*, 3435–3457. [CrossRef] [PubMed]
26. Sustainable Development Goals. 2022. Available online: https://www.un.org/sustainabledevelopment/energy/ (accessed on 1 May 2023).
27. Jasso, A.T. *Evaluation of Small Hydro Power Potential in Three River Basins of Mexico*; Mexican Institute of Water Technology (IMTA): Jiutepec, Mexico, 2009; pp. 1–14.
28. Wali, U.G. Estimating Hydropower Potential of an Ungauged Stream. *Int. J. Emerg. Technol. Adv. Eng.* **2013**, *3*, 592–600.
29. Ajmal, M.; Kim, T.-W.; Ahn, J.-H. Stability assessment of the curve number methodology used to estimate excess rainfall in forest-dominated watersheds. *Arab. J. Geosci.* **2016**, *9*, 402. [CrossRef]
30. Soulis, K.X.; Valiantzas, J.D.; Dercas, N.; Londra, P.A. Investigation of the direct runoff generation mechanism for the analysis of the SCS-CN method applicability to a partial area experimental watershed. *Hydrol. Earth Syst. Sci.* **2009**, *13*, 605–615. [CrossRef]
31. Soulis, K.X.; Valiantzas, J.D. SCS-CN parameter determination using rainfall-runoff data in heterogeneous watersheds—The two-CN system approach. *Hydrol. Earth Syst. Sci.* **2012**, *16*, 1001–1015. [CrossRef]
32. Gianfagna, C.C.; Johnson, C.E.; Chandler, D.G.; Hofmann, C. Watershed area ratio accurately predicts daily streamflow in nested catchments in the Catskills, New York. *J. Hydrol. Reg. Stud.* **2015**, *4*, 583–594. [CrossRef]
33. Ergen, K.; Kentel, E. An integrated map correlation method and multiple-source sites drainage-area ratio method for estimating streamflows at ungauged catchments: A case study of the Western Black Sea Region, Turkey. *J. Environ. Manag.* **2016**, *166*, 309–320. [CrossRef]
34. Farmer, W.H.; Vogel, R.M. Performance-weighted methods for estimating monthly streamflow at ungauged sites. *J. Hydrol.* **2013**, *477*, 240–250. [CrossRef]
35. Shu, C.; Ouarda, T.B.M.J. Improved methods for daily streamflow estimates at ungauged sites. *Water Resour. Res.* **2012**, *48*, 1–15. [CrossRef]
36. Roszkowska, E.; Białymstoku, U.W. Rank Ordering Criteria Weighting Methods—A Comparative Overview. *Optimum. Econ. Stud.* **2013**, *5*, 14–33. [CrossRef]
37. Khan, M.; Zaidi, A.Z. Run-of-River Hydropower Potential of Kunhar River, Pakistan. *Pak. J. Meteorol.* **2015**, *12*, 25–32.
38. Cai, X.; Ye, F.; Gholinia, F. Application of artificial neural network and Soil and Water Assessment Tools in evaluating power generation of small hydropower stations. *Energy Rep.* **2020**, *6*, 2106–2118. [CrossRef]

energies

Article

Mid-Term Optimal Scheduling of Low-Head Cascaded Hydropower Stations Considering Inflow Unevenness

Shuo Huang [1], Xinyu Wu [1,*], Yiyang Wu [1] and Zheng Zhang [2]

[1] Institute of Hydropower and Hydroinformatics, Dalian University of Technology, Dalian 116024, China; huangshuo@mail.dlut.edu.cn (S.H.); wuyiyang@mail.dlut.edu.cn (Y.W.)

[2] China Yangtze Power Co., Ltd., Yichang 443000, China; zhangzhengno1@163.com

* Correspondence: wuxinyu@dlut.edu.cn

Abstract: China has a vast scale of hydropower, and the small hydropower stations account for a large proportion. In flood season, the excessive inflow keeps these stations at a high reservoir level, leading to a worse condition of hindered power output and a great error in the calculation of power generation. Therefore, this paper proposes a mid-term optimal scheduling model for low-head cascaded hydropower stations considering inflow unevenness, in which the power output is controlled by the expected power output curve and daily inflow–maximum power output curve. A case study of nine hydropower stations on the Guangxi power grid shows that, regardless of considering the fitted curve or not, there are different degrees of error between the planned and actual situations. However, the error and power generation are decreased when considering the fitted curve, which reflects the impact of hindered power output. Meanwhile, according to the comparison, the weekly plan is more in line with the real condition when using this model to solve the problem. The results indicate that this model improves the accuracy of power output calculation for low-head hydropower stations with uneven inflow, playing a key role in the process of scheduling.

Keywords: inflow unevenness; low-head hydropower stations; hindered power output; mid-term optimal scheduling

Citation: Huang, S.; Wu, X.; Wu, Y.; Zhang, Z. Mid-Term Optimal Scheduling of Low-Head Cascaded Hydropower Stations Considering Inflow Unevenness. *Energies* **2023**, *16*, 6368. https://doi.org/10.3390/en16176368

Academic Editor: Helena M. Ramos

Received: 23 July 2023
Revised: 26 August 2023
Accepted: 27 August 2023
Published: 2 September 2023

1. Introduction

In recent years, more and more studies have focused on the mid-term optimal scheduling of hydropower stations for its comprehensive function [1]. On the one hand, it has the ability to meet the needs of the power market in time and maintain the balance between supply and demand [2]. Therefore, in power market transactions and dispatching, it can optimize the economic benefits and improve the market competitiveness of hydropower stations based on market prices and competition conditions [3]. On the other hand, mid-term scheduling considers future water inflow, seasonal changes in hydropower's power output, and the probable trend of the power market's development. Over a long period, it can reasonably dispatch water resources to make full use of water inflow and reduce spilled water. Thus, it appears to be significant to derive the best plan for mid-term scheduling.

For cascade hydropower stations, there are commonly three methods for addressing hydropower scheduling problems [4]—linear programming (LP), heuristic algorithms, and dynamic programming (DP)—and each method has its advantages and shortcomings. The biggest advantage of linear programming is that it is a feasible solution that is easier to derive compared with other methods [5], but the restrictions are equally obvious because this method is only suitable for the condition that the constraints and the objective function are all linear. When it comes to nonlinear problems, there is no solution using this method [6], and the other two methods appear to be functional. Heuristic algorithms, such as genetic algorithms (GA) [7], particle swarm optimization (PSO) [8], ant colony optimization (ACO) [9], and simulated annealing (SA) [10], are more suitable to solve the problem of large-scale cascaded hydropower stations on account of their powerful computing ability. However, what hinders their development is that heuristic algorithms easily prematurely converge to a locally optimal solution. As for dynamic programming [11], it is widely adopted in cascaded hydropower station scheduling. It divides a complex problem into several easier subproblems and obtains the globally optimal solution by solving the subproblems. Nevertheless, the curse of dimensionality appears as the scale of the problem increases, making it difficult to figure out a high-dimensional problem. On this issue, a great deal of improved dynamic programming algorithms [12], like progressive optimality algorithms (POA) [13,14], dynamic programming with successive approximation (DPSA) [15,16], and discrete differential dynamic programming (DDDP) [17,18], come into use. For example, Shen et al. [19] proposed a method combined with random sampling and a progressive optimality algorithm to address the problem of the curse of dimensionality caused by the large scale of cascaded hydropower stations, and the results proved that the amount of computation is greatly reduced. Li et al. [20] proposed an improved decomposition–coordination and discrete differential dynamic programming (IDC-DDDP) method to solve the problem of optimizing large-scale hydropower systems. He et al. [21] proposed an improved DPSA with a relaxation strategy (named DPSARS) based on the above mathematical derivations to solve the long-term joint power generation scheduling (LJPGS) of a large-scale hydropower station group (LHSG) and improved the calculation accuracy.

However, there is a large error for low-head hydropower stations if we derive solutions through these methods directly. Compared with high-head hydropower stations with robust adjustment ability, like the Three Gorges hydropower station, which plays a vital role in the optimal scheduling of reservoirs [22], the practical plan of low-head hydropower stations is more difficult to derive [23]. On the one hand, these low-head hydropower stations have small storage capacity, making it hard for them to change the power output in time based on inflow, especially when it comes to extreme rainfall. Meanwhile, due to the inflow being uneven every day, if we ignore the variation in inflow within a day and calculate power output with mean daily inflow, there is much error for power generation and water spillage. Although in many studies inflow unevenness is not considered—for example, Fredo et al. [24] related output with head, turbine outflow, and efficiency of the generating units to derive the operation rules in long-term optimal scheduling without considering inflow unevenness—in reality, inflow unevenness seriously affects the accuracy of optimal scheduling. Over the past 30 years, optimal scheduling considering inflow unevenness has gradually been studied. Ge et al. [25] proposed a novel stochastic optimization algorithm using Latin hypercube sampling and Cholesky decomposition combined with scenario bundling and sensitivity analysis (LC-SB-SA) to address the problem of inflow unevenness. Akbari et al. [26] integrated fuzzy-state SDP and multicriteria decision making to address inflow unevenness for a single reservoir. Zhang et al. [27] estimated the unevenness of reservoir operating rules through the Bayesian model averaging model, which combined three popular operating rules to reduce the unevenness of operating rules. Wang et al. [28] developed a hedging model for hydropower operation, which determined more rational carryover storage decisions when considering inflow unevenness. On the other hand, these low-head hydropower stations have small storage capacity, so it is possible for them to experience significant fluctuations in water level. In flood season, the inflow surges, and if the reservoir is already near or at its water level limit, measures including flood discharge and water release should be taken to control the reservoir's water level to prevent safety issues. As a result, frequent fluctuations in water level can result in unstable generating flow, which reduces the efficiency of the turbine, and, finally, leads to hindered power output. This is why, with the increase in inflow, low-head hydropower stations' power output increases first and then decreases. Additionally, when it comes to extreme rainfall, the tailwater level of hydropower stations continues to rise and the water head for generating becomes lower, leading to the result that power output is seriously hindered [29] and estimating low-head hydropower stations' power output becomes more difficult. In the past, the expected output curve was used to compute power output, but the results were quite different from the actual ones for the reason that hindered power output was not considered. Consequently, optimal scheduling of low-head hydropower stations is more complex, especially in the mid-term scheduling with a day as the time interval and a week or month as the scheduling period.

This paper puts forward a mid-term optimal operation model for low-head cascaded hydropower stations considering inflow unevenness. In this model, the power output is controlled by the expected power output curve and inflow–maximum power output curve which reflects the influence of daily inflow on the maximum power output. Compared to the traditional method that only uses the expected power output curve to limit power output, this model considers the influence of water head and daily inflow on power output simultaneously to reduce the error brought by hindered output and inflow unevenness. The characteristic of this model is that it first calculates the maximum power output corresponding to the two curves, and then takes the minimum value as the maximum power output limit. Therefore, we can use the most suitable curve to limit power output in different conditions. In flood season, inflow unevenness has a greater impact on the power output due to a surge in inflow, so the inflow–maximum power output curve is mainly used to control the power output. In the dry season, the impact of inflow is small, so the calculation of power output is mainly based on the expected power output curve. In order to solve this mid-term optimal scheduling model of low-head cascaded hydropower stations, it is necessary to consider the relationship between upstream and downstream inflow and then compute the hindered output. Through the process, this model significantly improves the accuracy of optimal scheduling of low-head hydropower stations and overcomes the difficulty of model solution caused by the dual factors—uneven inflow and hindered power output. Meanwhile, it enhances the practicability of low-head cascaded hydropower stations' mid-term optimal scheduling, enabling us to allocate water resources rationally and improving low-head hydropower stations' utilization efficiency.

The remainder of this paper is organized as follows. Section 2 provides details of the methodology, including optimization model construction and its solution method. Section 3 describes a case study of nine hydropower stations on the Guangxi power grid and presents the results and analysis. Finally, Section 4 states the conclusion of the study.

2. Materials and Methods

As Figure 1 shows, in flood season, the inflow seriously fluctuates in a day. The most significant impact of uneven inflow is the frequent fluctuation of reservoir level, which prevents the hydropower stations from running at a stable generating head. Hydropower stations' units thereby require constant adjustments to keep them functioning reasonably. This not only results in increased equipment wear and tear but also reduces the efficiency and stability of power generation. Furthermore, most small hydropower stations are considered runoff hydropower stations for lacking stability [30]. On this occasion, neglecting the intra-day variation of inflow and calculating power output based on the average daily inflow may lead to great errors. This can result in unnecessary water spilling while meeting the constantly changing demands of the power grid. Therefore, the drastic fluctuations in a short time caused by uneven inflow lead to longer system response times and complicate the scheduling optimization.

In view of the particular situation, the constraints of inflow and generating heads are considered simultaneously in this model to restrict power output, making the calculation more accurate. First, fit the relationship between daily inflow and the maximum power output of each hydropower station in each month. Second, build the mid-term optimal operation model of low-head cascaded hydropower stations. In this model, the power output of the hydropower stations in each period is controlled by the curve of head–expected power output and the curve of inflow–maximum power output. At last, solve this optimal model in the condition that the relationship between upstream and downstream is considered, and meanwhile, calculate the hindered power output.

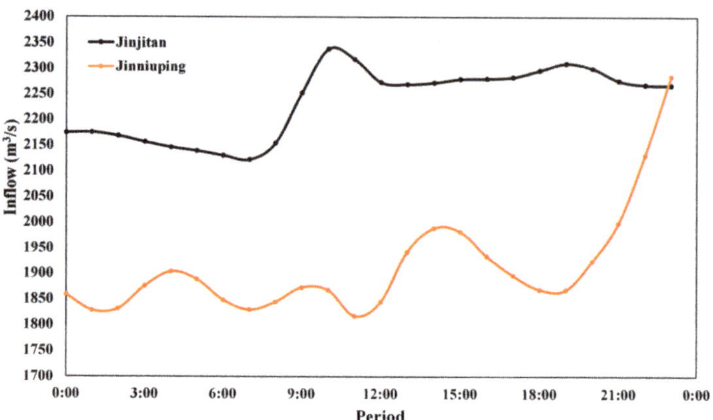

Figure 1. Inflow process on a day in July at Jinjitan and Jinniuping hydropower station.

2.1. Optimal Model

2.1.1. Daily Inflow–Maximum Power Output Curve

The steps to obtain the daily inflow–maximum power output curve are as follows. First, record the data of daily inflow $Q_m^{i,j,k}$ and the corresponding power output $P_m^{i,j,k}$ in month j and store them in sets U_m^j and V_m^j, respectively, where i, j, k, m represents year, month, day, and the sequence number of hydropower station, respectively; $K_{I,j}$ is the total number of days in month j of the year i, and i ranges from 1 to N which is the number of years the hydropower station has been in operation. Then, draw the upper envelope of the scatterplot of $P_m^{i,j,k}$ and $Q_m^{i,j,k}$, and fit the upper envelope into a piecewise linear function $pp_m^j = f_m^j\left(qq_m^j\right)$, where pp_m^j and qq_m^j represents daily inflow and the maximum power output of hydropower station m in month j, respectively. The fitted curve $pp_m^j = f_m^j\left(qq_m^j\right)$ is the daily inflow–maximum power output curve of hydropower station m in month j. According to these steps, the fitted curve, also called the daily inflow–maximum power output curve, can be obtained for different hydropower stations in different periods.

$$U_m^j = \left(Q_m^{1,j,1}, Q_m^{1,j,2}, \ldots, Q_m^{1,j,K_{1,j}}, Q_m^{2,j,1}, Q_m^{2,j,2}, \ldots, Q_m^{2,j,K_{2,j}}, \ldots, Q_m^{N,j,1} Q_m^{N,j,2}, \ldots, Q_m^{N,j,K_{N,j}}\right)$$

$$V_m^j = \left(P_m^{1,j,1}, P_m^{1,j,2}, \cdots, P_m^{1,j,K_{1,j}}, P_m^{2,j,1}, P_m^{2,j,2}, \cdots, P_m^{2,j,K_{2,j}}, \cdots, P_m^{N,j,1}, P_m^{N,j,2}, \cdots, P_m^{N,j,K_{N,j}}\right)$$

2.1.2. Objective Function

For hydropower stations, under the condition of ensuring enough water supply, how to maximize their power generation income is the priority issue. Therefore, in this paper, we choose the most power generation as the objective function.

$$\text{Max}F = \sum_{t=1}^{T} \sum_{m=1}^{M} p_m^t \Delta^t \tag{1}$$

where F is the power generation objective function; T is the number of periods; M is the number of hydropower stations; p_m^t is the average power output of hydropower station m in period t; Δ^t is the period hour number.

2.1.3. Constraints

Constraints in the model include water balance (Equation (2)), maximum and minimum storage constraints (Equation (3)), maximum and minimum power output constraints, and minimum outflow constraints. In particular, the maximum power output is controlled by the expected power output curve (Equation (4)) and inflow–maximum output curve (Equation (5)). For low-head hydropower stations with poor adjusting ability, in flood season their power output is primarily decided by inflow. So, in this condition Equation (5) plays the major role compared to Equation (4). Conversely, in the dry season, their power output is mainly decided by Equation (4).

$$s_m^t + Q_m^t - r_m^t = s_m^{t+1} \tag{2}$$

$$\underline{s}_m^t \leq s_m^t \leq \overline{s}_m^t \tag{3}$$

$$p_m^t \leq g\left(H_m^t\right) \tag{4}$$

$$p_m^t \leq f_m^{j(t)}\left(Q_m^t\right) \tag{5}$$

$$H_m^t = \left(\frac{Z_m^t + Z_m^{t+1}}{2}\right) - zd_m^t - Hl(q_m^t) \tag{6}$$

where m, t are sequence numbers of hydropower stations and periods; s_m^t is initial reservoir storage; $\underline{s}_m^t$ and $\overline{s}_m^t$ are lower and upper reservoir storage bounds; H_m^t is mean water head, and it is equal to the mean reservoir water level in the period t minus the downstream water level and the head loss; Z_m^t and Z_m^{t+1} are average water level upstream of hydropower station m in period t and $t + 1$, respectively; zd_m^t is average water level downstream, q_m^t is average generating flow; $Hl\left(q_m^t\right)$ is head loss; $g\left(H_m^t\right)$ is the maximum power output of hydropower station m when the mean water head is H_m^t; Q_m^t is inflow; $j(t)$ is the month in which period t is located; $f_m^{j(i)}$ is the maximum power output determined by inflow.

As mentioned above, the relationship between the inflow upstream and downstream should be considered to solve this model, so there are different ways to compute the mean water head when hydropower stations downstream exist or not.

If there are no hydropower stations downstream:

$$zd_m^t = \varphi\left(r_m^t\right) \tag{7}$$

where r_m^t is average outflow; $\varphi\left(r_m^t\right)$ is the function of outflow and downstream reservoir level.

If there are some hydropower stations downstream:

$$zd_m^t = \max\left[\varphi\left(r_m^t\right), \frac{Z_{Lm}^t + Z_{Lm}^{t+1}}{2}\right] \tag{8}$$

where Z_{Lm}^t is the average reservoir water level of the downstream hydropower station of hydropower station m in period t.

For minimum outflow constraints and minimum output constraints, we use a penalty function to describe:

$$F' = \sum_{t=1}^{T}\sum_{m=1}^{M} p_m^t \Delta^t + a\sum_{t=1}^{T}\sum_{m=1}^{M}\left[\min(p_m^t - \underline{p}_m^t, 0)\right]^2 + b\sum_{t=1}^{T}\sum_{m=1}^{M}\left[\min(r_m^t - \underline{r}_m^t, 0)\right]^2 + c\sum_{t=1}^{T}\sum_{m=1}^{M}\left[\min(r_m^t, 0)\right]^2 \tag{9}$$

where $\underline{p}_m^t$, $\underline{r}_m^t$ are the minimum power output and outflow offline; a, b, and c are penalty coefficients. The values of a and b are determined based on experience, with a relatively wide range of possible values. In this case, a value of 10 is suitable. The value of c is effectively positive infinity, and in this case, it is represented as 10^{15}.

2.2. Solution Method

Considering that there are a lot of hydropower stations in a cascade, traditional dynamic programming (DP) is not suitable for solving this kind of problem with high dimensions. Therefore, a method combined with a progressive optimality algorithm (POA) and successive approximation method of dynamic programming (DPSA) is used to address this problem.

2.2.1. POA and DPSA

POA reduces dimensionality by dividing the optimization problem of T periods into T-1 two-stage problems and then figures out these two-stage problems when a feasible solution is already known. DPSA reduces dimensionality by controlling the number of variables. Each time only one variable is considered in the model, greatly decreasing the difficulty of deriving the solution. As a consequence, in this method, what is needed to address are these subproblems which consider two periods and one variable every time. The solution of these subproblems only concludes the optimal value of the variable in the middle period, which is easy enough that the traditional DP can work out.

2.2.2. Detailed Solution Process

For low-head hydropower stations whose upstream and downstream have a close connection, it is necessary to optimize the hydropower station's power output's change process in the condition that all the upstream and downstream stations next to it fix their reservoir levels because the water storage of the downstream reservoir has a supporting effect on the upstream. First, obtain the initial solution for the reservoirs by following the rules that the generating flow is the same for each period. Second, for hydropower station m, the search step and search direction are determined to get Z_1, Z_2, and Z_3. Third, one of the water levels is selected as the initial water level of hydropower station m in the $t + 1$ period, equal to the final water level in the t period. If there are hydropower stations upstream of hydropower station m, the initial and final water levels of one of the upstream hydropower stations at time t and $t + 1$ are fixed to calculate its power output and outflow. After completing the calculation for the upstream hydropower stations in sequence, the power output and outflow of hydropower station m and its downstream hydropower stations are calculated according to the same method. Then, the value of F' of the whole cascade can be derived, which is the criteria to judge the quality of optimization results. Similarly, the final results when the water level of hydropower station m in the $t + 1$ period is Z_2 or Z_3 can be obtained by the same method. Therefore, we can choose the best results among Z_1, Z_2, and Z_3 and the corresponding water level is the optimized water level of hydropower station m in the $t + 1$ period. This is the process of hydropower station m to find the optimal water level in the $t + 1$ period. The steps for other hydropower stations to optimize their water levels in different periods are the same. At last, if the water level of all hydropower stations is equal to the initial water level at the beginning of this optimization cycle, and the search step is less than the specified value, the cycle ends. The flowchart of the calculation process is shown in Figure 2:

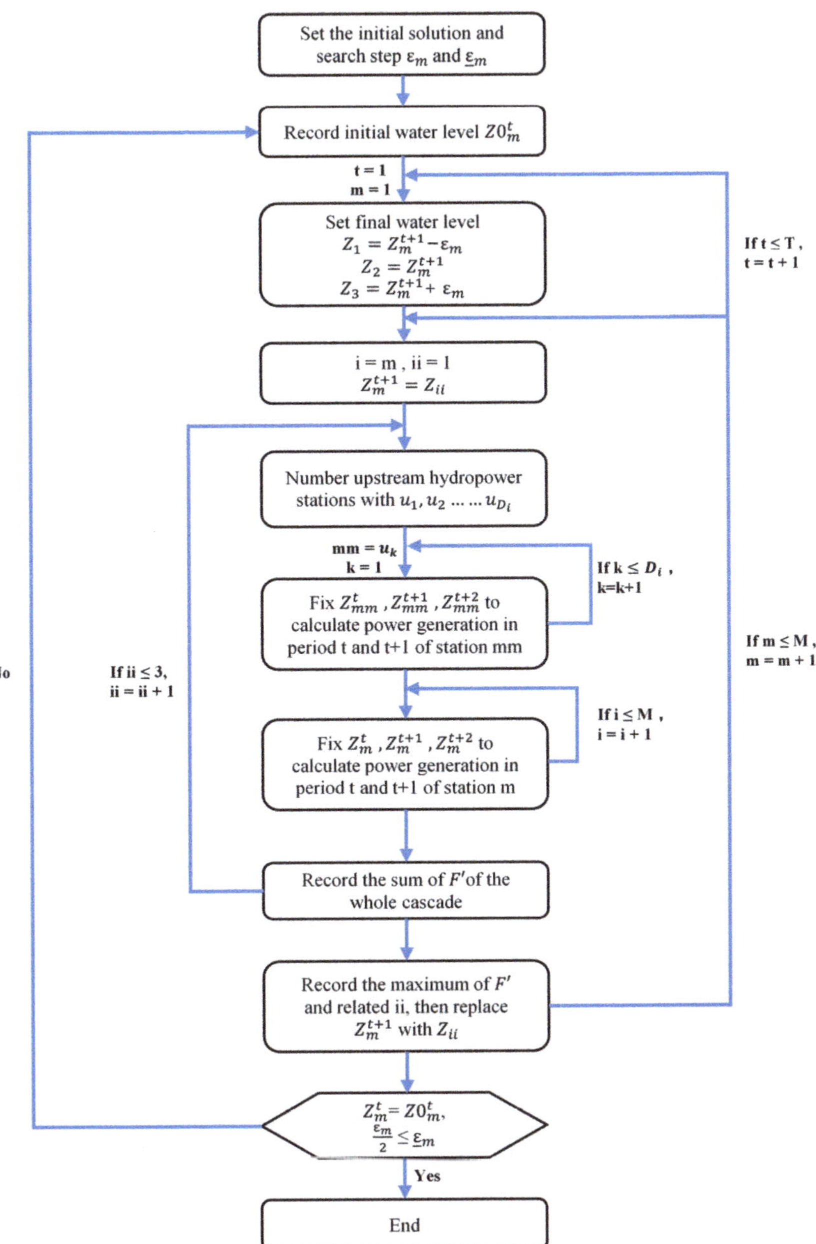

Figure 2. Flowchart of the calculation process.

3. Case Study

This paper uses the weekly plan formulation of the Guangxi power grid with many low-head hydropower stations as the research object. The installed capacity of hydropower dispatched by the Guangxi power grid is about 9000 MW, among which there are more than 30 hydropower stations in provincial scheduling, with an installed capacity of more than 7000 MW. What is more, the region has over 800 small hydropower stations whose total installed capacity is more than 1800 MW. Under the current situation, how to coordinate the operating rules between adjustable hydropower stations, large and small hydropower stations, and hydropower and wind power stations, so as to obtain the maximum benefit of the power grid is a very challenging issue. However, among the hydropower stations with provincial scheduling, only the Youjiang hydropower station (with an installed capacity of 540 MW) has annual adjustment capability. The Yantan hydropower station (with an installed capacity of 1800 MW) with the largest installed capacity on the power grid only has seasonal adjustment capability. Hence, in the flood season, other small installed capacity stations are equal to runoff hydropower stations. When the inflow comes concentratedly from various river basins, due to the insufficient overall regulation capacity of hydropower stations, it is very difficult to adjust the peak of the power grid in low valleys. For example, the installed capacity of cascaded hydropower stations below Yantan, the mainstream of the Hongshui River, accounts for more than half of the installed capacity of the provincial dispatched hydropower stations, but their inflow is controlled by Longtan hydropower station which has poor adjusting ability. In this situation, their optimal scheduling must be coordinated with the upstream hydropower stations in order to regulate the daily peak value of power output and seasonal adjustment, definitely increasing the difficulty of decisions. Therefore, the problem that it needs to discard water for the balance between demand and supply in the low valleys matters for Guangxi's hydropower scheduling. Unlike other hydropower stations on China Southern Power Grid (CSG), which have high water heads of more than 100 m or even more than 200 m, the water head for generating in Guangxi is generally below 50 m, and the Youjiang Power Station with the highest designed water head is less than 90 m. Due to the low water head, the power output is severely hindered when extreme rainfall appears in the flood season, so the condition that although the inflow increases significantly, the power output decreases instead often occurs. In actual operation, it is important to decrease the hindered output and precisely calculate power generation according to the inflow.

This paper takes the weekly plan of 34 cascaded hydropower stations under the jurisdiction of the Guangxi power grid as an example. Figure 3 shows the relation of the studied hydropower stations. A week in July during the flood season was selected to study, and the maximum power generation model was used to make the weekly plan. Most of the cascaded hydropower stations in Yujiang River, Liujiang River, and Guijiang River on the Guangxi power grid are low-head hydropower stations. After calculation and analysis, we found that apart from Naji, Jinjitan, Yemao, Luodong, Mashi, Guding, Dabu, Honghua, and Jinniuping Hydropower Stations, the differences in power output are relatively insignificant for the other hydropower stations regardless of considering the fitted curve or not. Excluding these nine hydropower stations, the remaining ones have a total power generation ranging from 56,400 MWh to 62,448 MWh over the course of these seven days. Therefore, in order to study the impact of fitted curves on power output, a total of 9 low-head hydropower stations including Naji, Jinjitan, Yemao, Luodong, Mashi, Guding, Dabu, Honghua, and Jinniuping were selected for analysis.

The daily inflow–maximum power output curve is obtained by fitting the historical data according to the method in Section 2, so it is also called the fitted curve. According to the inflow–maximum power out curve demonstrated in Figure 4, which perfectly reflects the serious impact of hindered power output, it can be easily seen that they have the same variation trend. With the increase in inflow, their maximum power output increases first and then decreases. What is more, when a hydropower station's inflow is more than others, its decline is even greater. Considering this factor, the final results are shown in Figures 5–7:

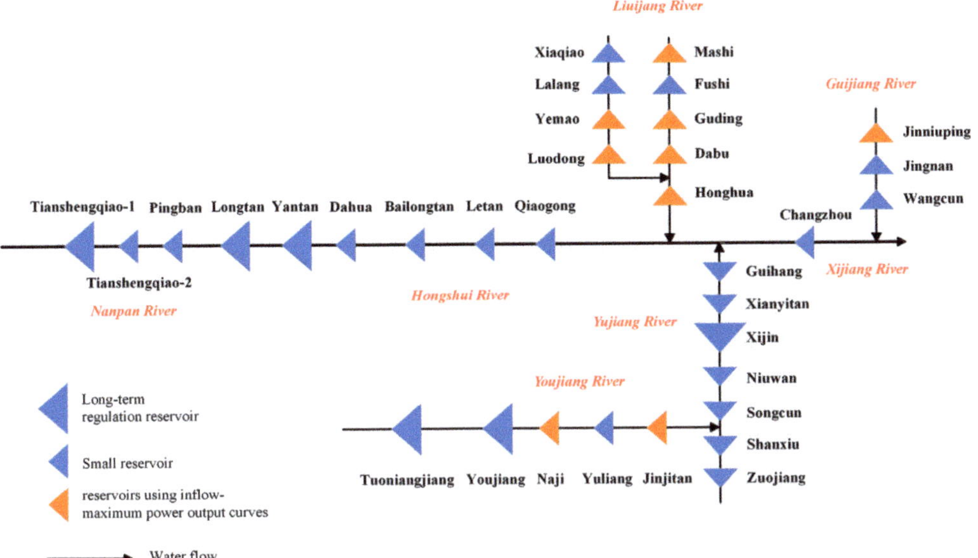

Figure 3. The structure diagram of the studied hydropower stations.

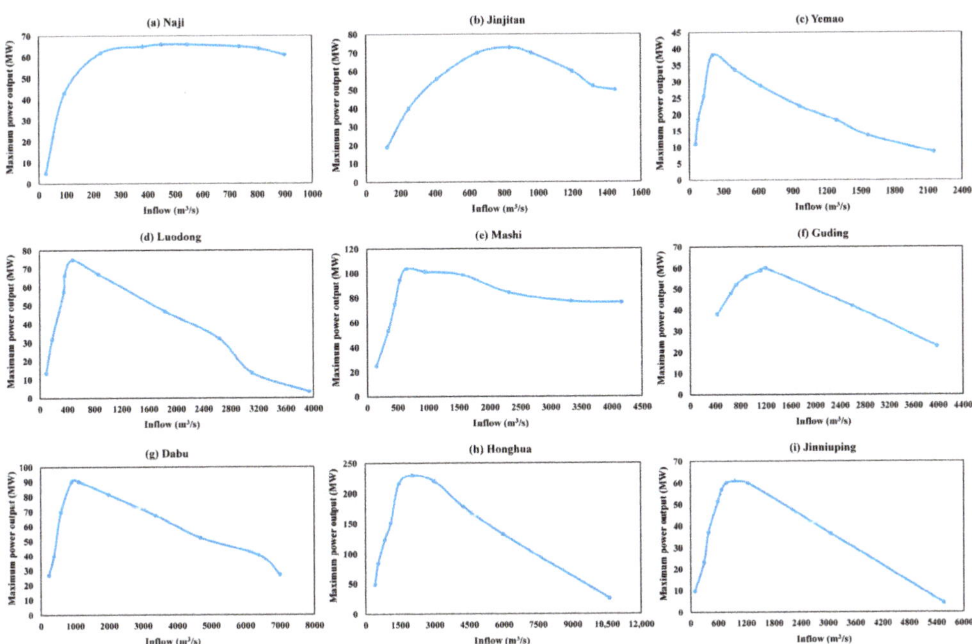

Figure 4. The inflow–maximum power output curve of the nine studied hydropower stations.

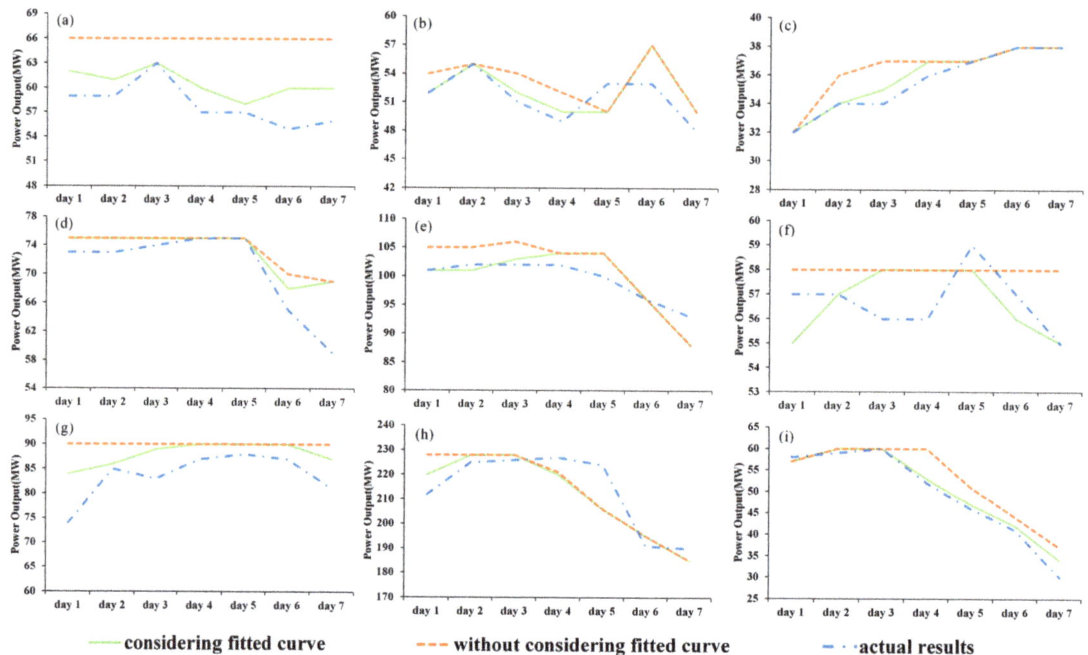

Figure 5. Comparison of the weekly plan in the three conditions: (**a**) Naji; (**b**) Jinjitan; (**c**) Yemao; (**d**) Luodong; (**e**) Mashi; (**f**) Guding; (**g**) Dabu; (**h**) Honghua; (**i**) Jinniuping.

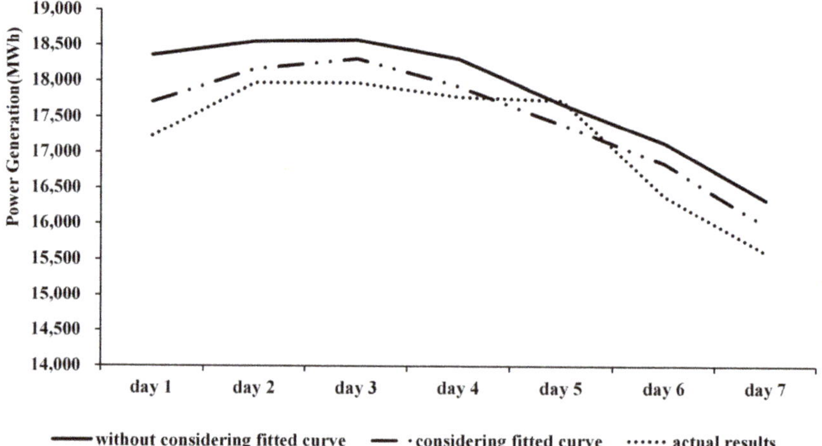

Figure 6. Comparison of the daily total power generation of the nine hydropower stations in the three conditions.

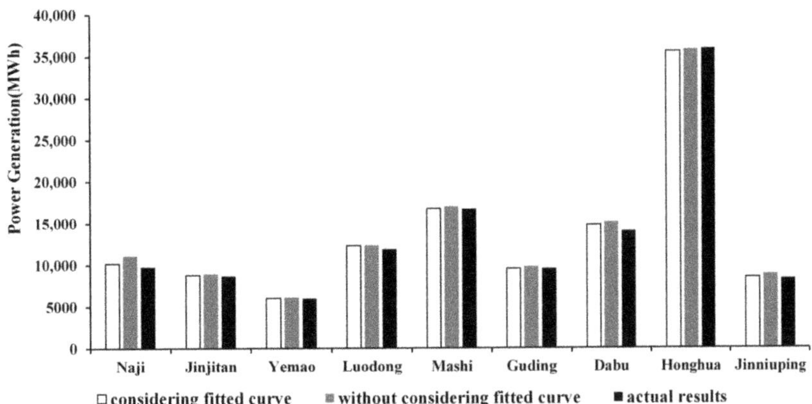

Figure 7. Comparison of the total power generation of each station in the three conditions.

Figure 5 shows the comparison of the weekly plan of the above nine power stations considering the fitted curve, the weekly plan without considering the fitted curve, and the actual results. Considering the fitted curve is to consider the daily inflow–maximum power output curve, then use the expected power output curve and the daily inflow–maximum power output curve to limit the power output according to the method in Section 2. Since the selected time period for the case study falls within the flood season, the fitted curve plays a major role compared to the expected power output curve. Not considering the fitted curve means directly using the expected power output curve as the constraint of the power output. Therefore, according to Figure 5, through comparative analysis of the power output process of hydropower stations in various river basins, it can be concluded that whether or not the fitted curve is considered in the optimal scheduling, there are different degrees of error between the planned and actual results. The main reason for the error is that the weekly plan is usually made this Thursday for the next week, so the actual inflow and the network loads may have a huge error from the prediction. This leads to a continuous adjustment of power output decisions in the actual process for the scheduling plan may greatly differ from the previous plan. However, the errors between considering the fitted curve and not considering the fitted curve are different. It can be easily seen that on most days, the power output when considering the fitted curve is 2–8 MW less than that when not considering the fitted curve, which is closer to the real power output and has less error. So, when considering the fitted curve and the expected power output curve simultaneously, the scheduling plan we derived is more similar to the real ones and more feasible than the results under the circumstance that only the expected power output curve is considered. Especially for Dabu and Jinniuping hydropower stations, the maximum value of error is 5 MW and 4 MW, respectively, when considering the fitted curve while that reaches up to 11 MW and 8 MW without considering the fitted curve. Meanwhile, the larger the power output, the greater the error. The actual power output of Honghua hydropower station on the first day is 8 MW less than that considering the fitted curve, and 16 MW less than that not considering the fitted curve. The cause of these phenomena is the uneven inflow, leading to power output hindered. In some periods of these stations, the expected power output curve is used to calculate power output, so in flood season they can work with the maximum power output. In fact, there is power output hindered when calculating the daily average flow because inflow in a day is uneven, so the power output of some stations like Naji and Dabu cannot reach the maximum value all the time.

Figures 6 and 7 also effectively illustrate the same phenomenon since the total power generation is derived from the power output, so the results are similar to what Figure 5 reflects. Figure 6 shows the comparison of daily total power generation in three conditions of all nine hydropower stations. It can be seen that the average error when considering the fitted curve is 236.57 MW, while that is 610.29 MW when not considering the fitted curve. What is more, when we kick off the relation between inflow and maximum power output, the power generation in six of seven days is more, reflecting the impact of hindered power output and proving that this model is practical. Figure 7 shows the comparison between the total power generation in three conditions of each station. By comparing, it can be observed that the total power generation of the nine hydropower stations when considering the fitted curve is smaller than that when not considering the fitted curve. Except for the Honghua hydropower station, the total power generation of the other eight hydropower stations is much closer to the actual situation when considering the fitted curve. A possible reason for this may be that on the fifth day, the inflow of Honghua hydropower station is even enough that the fitted curve on that day is unsuitable. Therefore, considering the limitations of the fitted curve on power output may result in the theoretically total power generation being smaller than the actual results. These reasons may have caused the planned power output of the Honghua hydropower station to be smaller than the actual power output on that day, resulting in a smaller total power generation than the real situation. In conclusion, the proposed model in this paper can decrease the impact of hindered power output, making the weekly plan more similar to real conditions.

4. Conclusions

In this paper, a mid-term optimization scheduling model is proposed for low-head cascaded hydropower stations considering the unevenness of water inflow and the hindered power output. This model controls power output by using the expected power output curve and the daily inflow–maximum power output curve together. Nine hydropower stations on the Guangxi power grid are studied, and the results whether the fitted curve is considered or not are compared with the actual results. The results show that (1) considering the fitted curve, the scheduling plan is more in line with the actual conditions, resulting in an error reduction of 2–8 MW compared to that when not considering the fitted curve. (2) When considering the fitted curve, the total daily power output of the nine studied stations has a smaller error of 61% compared to that when not considering it, and the total power output for each station over seven days is also smaller.

It can be observed that the unevenness of inflow has a significant impact on the optimal scheduling of hydropower stations, especially for stations with small installed capacities. In addition, the hindered power output further reduces the accuracy of calculations and complicates the optimal scheduling problem. Therefore, the model proposed in this paper partially addresses this problem and improves the accuracy of power output calculation for low-head hydropower stations with uneven inflow.

Author Contributions: Conceptualization, X.W. and Z.Z.; methodology, X.W.; software, X.W.; validation, X.W. and S.H.; formal analysis, X.W.; investigation, S.H.; resources, X.W. and Z.Z.; data curation, X.W.; writing—original draft preparation, S.H.; writing—review and editing, S.H. and Y.W.; visualization, S.H.; supervision, X.W. and Y.W.; project administration, X.W.; funding acquisition, X.W. All authors have read and agreed to the published version of the manuscript.

Funding: This research was funded by The National Natural Science Foundation of China, Grant/Award Number: 52179005.

Data Availability Statement: Not applicable.

Conflicts of Interest: The authors declare no conflict of interest.

References

1. Schaffer, L.E.; Helseth, A.; Korpas, M. A stochastic dynamic programming model for hydropower scheduling with state-dependent maximum discharge constraints. *Renew. Energy* **2022**, *194*, 571–581. [CrossRef]
2. Feng, S.Z.; Zheng, H.; Qiao, Y.F.; Yang, Z.T.; Wang, J.W.; Liu, S.Q. Weekly hydropower scheduling of cascaded reservoirs with hourly power and capacity balances. *Appl. Energy* **2022**, *311*, 118620. [CrossRef]
3. Perez-Diaz, J.I.; Guisandez, I.; Chazarra, M.; Helseth, A. Medium-term scheduling of a hydropower plant participating as a price-maker in the automatic frequency restoration reserve market. *Electr. Ppwer Syst. Res.* **2020**, *185*, 106339. [CrossRef]

4. Fang, Z.; Liao, S.L.; Cheng, C.T.; Zhao, H.Y.; Liu, B.X.; Su, H.Y. Parallel improved DPSA algorithm for medium-term optimal scheduling of large-scale cascade hydropower plants. *Renew. Energy* **2023**, *210*, 134–147. [CrossRef]

5. Wu, X.Y.; Wu, Y.Y.; Cheng, X.L.; Cheng, C.T.; Li, Z.H.; Wu, Y.Q. A mixed-integer linear programming model for hydro unit commitment considering operation constraint priorities. *Renew. Energy* **2023**, *204*, 507–520. [CrossRef]

6. Piekutowski, M.R.; Litwinowicz, T.; Frowd, R.J. Optimal short-term scheduling for a large-scale cascaded hydro system. *IEEE Trans. Power Syst.* **1994**, *9*, 805–811. [CrossRef]

7. Rashid, M.U.; Abid, I.; Latif, A. Optimization of hydropower and related benefits through Cascade Reservoirs for sustainable economic growth. *Renew. Energy* **2022**, *185*, 241–254. [CrossRef]

8. Hatamkhani, A.; Moridi, A.; Yazdi, J. A simulation—Optimization models for multi-reservoir hydropower systems design at watershed scale. *Renew. Energy* **2020**, *149*, 253–263. [CrossRef]

9. Wang, C.; Zhou, J.Z.; Lu, P.; Yuan, L. Long-term scheduling of large cascade hydropower stations in Jinsha River, China. *Energy Convers. Manag.* **2015**, *90*, 476–487. [CrossRef]

10. Teegavarapu, R.S.V.; Simonovic, S.P. Optimal operation of reservoir systems using simulated annealing. *Water Resour. Manag.* **2002**, *16*, 401–428. [CrossRef]

11. Bellman, R. On the Theory of Dynamic Programming. *Proc. Natl. Acad. Sci. USA* **1952**, *38*, 716–719. [CrossRef] [PubMed]

12. Feng, Z.K.; Niu, W.J.; Cheng, C.T.; Wu, X.Y. Optimization of hydropower system operation by uniform dynamic programming for dimensionality reduction. *Energy* **2017**, *134*, 718–730. [CrossRef]

13. Heidari, M.; Chow, V.T.; Kokotovic, P.V.; Meredith, D.D. Discrete Differential Dynamic Programing Approach to Water Resources Systems Optimization. *Water Resour. Res.* **1971**, *7*, 273–282. [CrossRef]

14. Zhang, C.; Zhou, J.; Wang, C.; Zhang, Y.; Mo, L. Variable period progressive optimality algorithm for optimal dispatch of cascade reservoirs. *J. Hydroelectr. Eng.* **2016**, *35*, 12–21. [CrossRef]

15. Giles, J.E.; Wunderlich, W.O. Weekly Multipurpose Planning Model for TVA Reservoir System. *J. Water Resour. Plan. Manag. Div.* **1981**, *107*, 495–511. [CrossRef]

16. Yi, J.; Labadie, J.W.; Stitt, S. Dynamic optimal unit commitment and loading in hydropower systems. *J. Water Resour. Plan. Manag.* **2003**, *129*, 388–398. [CrossRef]

17. Tospornsampan, J.; Kita, I.; Ishii, M.; Kitamura, Y. Optimization of a multiple reservoir system operation using a combination of genetic algorithm and discrete differential dynamic programming: A case study in Mae Klong system, Thailand. *Paddy Water Environ.* **2005**, *3*, 29–38. [CrossRef]

18. Chow, V.T.; Maidment, D.R.; Tauxe, G.W. Computer Time and Memory Requirements for DP and DDDP in Water Resource Systems Analysis. *Water Resour. Res.* **1975**, *11*, 621–628. [CrossRef]

19. Shen, J.J.; Zhu, W.L.; Cheng, C.T.; Zhong, H.; Jiang, Y.; Li, X.F. Method for high-dimensional hydropower system operations coupling random sampling with feasible region identification. *J. Hydrol.* **2021**, *599*, 126357. [CrossRef]

20. Li, C.L.; Zhou, J.Z.; Ouyang, S.; Ding, X.L.; Chen, L. Improved decomposition-coordination and discrete differential dynamic programming for optimization of large-scale hydropower system. *Energy Convers. Manag.* **2014**, *84*, 363–373. [CrossRef]

21. He, Z.Z.; Wang, C.; Wang, Y.Q.; Wei, B.W.; Zhou, J.Z.; Zhang, H.R.; Qin, H. Dynamic programming with successive approximation and relaxation strategy for long-term joint power generation scheduling of large-scale hydropower station group. *Energy* **2021**, *222*, 119960. [CrossRef]

22. Liu, X.Y.; Chen, L.; Zhu, Y.H.; Singh, V.P.; Qu, G.; Guo, X.H. Multi-objective reservoir operation during flood season considering spillway optimization. *J. Hydrol.* **2017**, *552*, 554–563. [CrossRef]

23. Ceran, B.; Jurasz, J.; Wroblewski, R.; Guderski, A.; Zlotecka, D.; Kazmierczak, L. Impact of the Minimum Head on Low-Head Hydropower Plants Energy Production and Profitability. *Energies* **2020**, *13*, 6728. [CrossRef]

24. Fredo, G.L.M.; Finardi, E.C.; de Matos, V.L. Assessing solution quality and computational performance in the long-term generation scheduling problem considering different hydro production function approaches. *Renew. Energy* **2019**, *131*, 45–54. [CrossRef]

25. Ge, X.L.; Xia, S.; Lee, W.J. An efficient stochastic algorithm for mid-term scheduling of cascaded hydro systems. *J. Mod. Power Syst. Clean Energy* **2019**, *7*, 163–173. [CrossRef]

26. Akbari, M.; Afshar, A.; Mousavi, S.J. Stochastic multiobjective reservoir operation under imprecise objectives: Multicriteria decision-making approach. *J. Hydroinform.* **2011**, *13*, 110–120. [CrossRef]

27. Zhang, J.W.; Liu, P.; Wang, H.; Lei, X.H.; Zhou, Y.L. A Bayesian model averaging method for the derivation of reservoir operating rules. *J. Hydrol.* **2015**, *528*, 276–285. [CrossRef]

28. Wang, J.; Cheng, C.T.; Wu, X.Y.; Shen, J.J.; Cao, R. Optimal Hedging for Hydropower Operation and End-of-Year Carryover Storage Values. *J. Water Resour. Plan. Manag.* **2019**, *145*, 04019003. [CrossRef]

29. Zhao, Z.; Liu, J.; Cheng, C.; Liao, S.; Jin, X. A MILP model for day-ahead peak operation of cascade hydropower stations considering backwater. *J. Hydraul. Eng.* **2019**, *50*, 925–935. [CrossRef]

30. Molénat, J.; Dagès, C.; Bouteffeha, M.; Mekki, I. Can small reservoirs be used to gauge stream runoff? *J. Hydrol.* **2021**, *603*, 127087. [CrossRef]

Article

Testing and Numerical Analysis of Abnormal Pressure Pulsations in Francis Turbines

Lu Jia [1,*], **Yongzhong Zeng** [1], **Xiaobing Liu** [1], **Wanting Huang** [2] and **Wenzhuo Xiao** [2]

[1]   Key Laboratory of Fluid and Power Machinery, Ministry of Education, Xihua University, Chengdu 610097, China; zyzzyzhome@163.com (Y.Z.); liuxb@mail.xhu.edu.cn (X.L.)
[2]   Sichuan Southwest Vocational College of Civil Aviation, Chengdu 610039, China; huangwanting2005@163.com (W.H.); flipped_bx@163.com (W.X.)
*   Correspondence: kasangjialu7788@163.com

Abstract: During the flood season, Francis turbines often operate under low-head and full-load conditions, frequently experiencing significant pressure pulsations, posing potential threats to the safe and stable operation of the units. However, the factors contributing to substantial pressure pulsations in Francis turbines are multifaceted. This paper focuses on a mixed-flow hydroelectric generating unit at a specific hydropower station. Field tests were conducted to investigate abnormal vibrations and hydraulic pressure pulsations under low-head and full-load conditions. Utilizing the Navier–Stokes equations and the RNG k-ε turbulence model, the unsteady flow field within the turbine under these conditions was calculated. The results indicate that the abnormal pressure pulsations detected in the bladeless zone between the wicket gates and the turbine inlet are due to operational points deviating from the normal operating range of the turbine. When water flows at a large inflow angle, striking the turbine blade heads, it leads to significant flow separation and vortex formation at the back of the blade inlet edges, causing severe vibrations in the hydroelectric generating unit. These findings provide a basis and assurance for the safe and stable operation of the power station.

Keywords: Francis turbines; turbine vibration; stability test; pressure pulsation; numerical analysis; separation vortex

Citation: Jia, L.; Zeng, Y.; Liu, X.; Huang, W.; Xiao, W. Testing and Numerical Analysis of Abnormal Pressure Pulsations in Francis Turbines. *Energies* **2024**, *17*, 237. https://doi.org/10.3390/en17010237

Academic Editor: Massimiliano Renzi

Received: 19 November 2023
Revised: 9 December 2023
Accepted: 27 December 2023
Published: 2 January 2024

1. Introduction

Stability has always been a key research focus in the operation of hydroelectric generating units, particularly the stability of large hydropower stations, which draws significant attention from researchers. In China, hydropower stations are often required to operate at full or even overload capacity during the flood season, in accordance with grid regulation demands, especially those equipped with Francis turbines. However, many of these stations experience abnormal pressure pulsations during full-load operation. For instance, domestic projects such as the Yantan and Xiaolangdi Hydropower Stations have exhibited abnormal vibrations under full-load conditions, with some stations even unable to sustain long-term operation under these conditions [1–3]. Prolonged self-excited or excessive vibrations in hydroelectric units can lead to component deformation, loosening, or even fracturing, thereby posing a threat to the safety of the entire unit and the power station [4–6].

Recently, the occurrence of abnormal vibrations during the operation of Francis turbines has become increasingly frequent, and the stability of these units has turned into a pressing concern in the industry. However, the factors inducing abnormal hydraulic vibrations in turbines are diverse [7,8]. Tian Shutang, based on the low-head operation experience of the Longyangxia Hydropower Station, conducted a series of model and prototype turbine tests and analyzed the impact of head variations on different types of hydraulic vibrations [9]. Qiao Liangliang and others studied the pressure pulsations in the cover area caused by flow separation and the formation of Karman vortices and blade passage vortices, which is due to the worsening flow conditions around the blades when the operation area of Francis turbine units deviates from the optimal operating range [10]. Feng Jinhai and others, aiming to study the stability of Francis turbines under off-design conditions, performed static and modal analyses on turbines under three typical off-design conditions [11]. Furthermore, Feng Yanmin and others researched the stability characteristics of units with large head variations, suggesting that stability tests should be conducted under multiple characteristic heads to further determine the vibration zones of the units at different heads, thereby improving the safety and stability of unit operations [12]. Additionally, Wu Daoping analyzed the problem of large vibrations under certain conditions. To ensure the stable operation of the units, the Francis turbine units at the Youshui River Hydropower Station were taken as the test subjects, and field tests on the vibration, noise, and hydraulic pressure pulsation of these units were conducted [13].

While the above research outcomes have provided some explanation for the causes of abnormal vibrations in turbines, the factors leading to these vibrations are complex. The results of on-site tests of turbines often result from a combination of multiple factors, and relying solely on field test methods to investigate the root causes of abnormal turbine vibrations clearly has its limitations.

In the past decade or so, fruitful research outcomes have been achieved in analyzing the internal flow field of turbines using CFD software 2021. Numerical simulations enable the dissection of local details in the turbine flow field, further revealing the causes of abnormal vibrations. Lu Lei and others applied numerical simulation methods to analyze the intense vibrations and noise emitted by turbines under full and overload conditions, concluding that the hydraulic instability at the station was caused by blade passage vortices in the turbine [14]. Dekterev et al. further investigated the evolution of vortex cores in turbine blade passages based on hydrodynamic models of hydraulic units [15]. Luo Xingqi and others summarized that the root cause of turbine vibrations lies in the vortex structures of unsteady flows within the turbine, which can lead to large pressure fluctuations, significant unit vibrations, and fatigue of turbine components [16]. Sun L and others conducted unsteady numerical calculations and visualization experiments of turbine flow fields, researching the evolution and pressure pulsation characteristics of blade passage cavitation vortices under partial-load conditions. The results indicated that changes in the volume of cavitation vortices primarily occur near the junction of the trailing edge of the turbine blade back and the lower ring, subsequently inducing local amplification of pressure pulsation magnitudes [17].

The phenomenon of abnormal vibrations in Francis turbines during full-load operation is a widespread issue in the industry. However, the factors causing abnormal hydraulic vibrations in turbines are complex, and particularly, the relationship between the unsteady flow field structure and unstable pressure pulsations under low-head, full-load conditions is still unclear.

At a specific hydropower station in Guizhou, Unit 4 operated at the historically lowest gross head of 110.19 m due to the rise in downstream water levels during the flood season. The turbine of this station was stable and operated normally with the wicket gates at a relative opening of 78% and an active power of 215.57 MW. However, when the relative opening of the wicket gates further increased to 80.449% and the active power rose to 218.863 MW, operating at full load, the unit experienced severe abnormal vibrations accompanied by noticeable noise.

This paper systematically investigates the full-load abnormal vibrations of Unit 4's turbine at a specific hydropower station. Using numerical simulation methods, the unsteady internal flow field of the turbine was calculated for the full-load operating conditions (a wicket gate relative opening of 80.449%, an active power of 218.863 MW, and a gross head of 110.19 m). The study initially involved solving three-dimensional turbulent flow equations to perform unsteady calculations of the flow field in the bladeless zone between the wicket gates and the turbine inlet, obtaining the pressure pulsation curves at computational monitoring points. Additionally, by combining field test data, the root causes of the unit's severe vibrations were analyzed, providing references to improve the operating conditions with abnormal pressure pulsations and to ensure the stable operation of the unit.

2. Unsteady Calculation and Testing of the Internal Flow Field in the Turbine

To dissect the flow conditions within the internal flow field of Unit 4's turbine at a specific hydropower station under full-load conditions, this paper established a geometric model of the turbine's internal flow passages using the three-dimensional geometric modeling software Unigraphics (UG) 2021. This was followed by mesh division, and then, numerical analysis and the calculation of the unsteady flow within the turbine unit were conducted using numerical simulation software.

2.1. Control Equations [18–20]

When applying the principle of mass conservation to incompressible fluids, the continuity equation is:

$$\frac{\partial u}{\partial x} + \frac{\partial v}{\partial y} + \frac{\partial w}{\partial z} = 0 \tag{1}$$

In the equation: t—time, s; u, v, w—components of fluid velocity in different directions; ρ—fluid density.

Motion Equation (Navier–Stokes Equation) [21]:

$$\nabla \cdot \vec{u} = 0$$
$$\frac{\partial \vec{u}}{\partial t} + \left(\vec{u} \cdot \nabla \right) \vec{u} = -\frac{\nabla p}{\rho} + \nabla^2 \cdot \vec{u} + \vec{F} \tag{2}$$

where $\vec{u}$ is the flow velocity, a vector field; ρ is the fluid density, p is the pressure, v is the kinematic viscosity, and $\vec{F}$ represents external force (per unit of mass in a volume) acting on the fluid. Let us also choose the Ox axis that coincides with the main direction of flow propagation. Moreover, we assume here the external force $\vec{F}$ above to be the force, which has a potential Φ represented by $\vec{F} = -\nabla\Phi$.

The Re-Normalization Group (RNG) k-ε turbulence model is an advanced version of the standard k-ε model used in computational fluid dynamics (CFD) for simulating turbulence. The RNG model modifies the standard k-ε model by including additional terms derived from statistical mechanics. These modifications enhance the model's accuracy for a wider range of flows, especially those involving rapid strain rates, swirling, and complex flows. The equations for the RNG k-ε model are as follows [22]:

Turbulent Kinetic Energy (k) Equation:

$$\frac{\partial(\rho k)}{\partial t} + \frac{\partial(\rho u_i k)}{\partial x_i} = \frac{\partial}{\partial x_j}\left[(\mu + \frac{\mu_t}{\sigma_k})\frac{\partial k}{\partial x_j}\right] + P_k + P_b - \rho\varepsilon - Y_M + S_k \tag{3}$$

where P_k is the production of turbulence kinetic energy due to mean velocity gradients, P_b is the production of turbulence kinetic energy due to buoyancy, Y_M is the contribution of the fluctuating dilatation in compressible turbulence to the overall dissipation rate, and S_k is a user-defined source term.

Turbulent Dissipation Rate (ε) Equation:

$$\frac{\partial(\rho\varepsilon)}{\partial t} + \frac{\partial(\rho u_i \varepsilon)}{\partial x_i} = \frac{\partial}{\partial x_j}\left[(\mu + \frac{\mu_t}{\sigma_\varepsilon})\frac{\partial \varepsilon}{\partial x_j}\right] + C_{1\varepsilon}\frac{\varepsilon}{k}(P_k + C_{3\varepsilon}P_b) - C_{2\varepsilon}\rho\frac{\varepsilon^2}{k} + S_\varepsilon \tag{4}$$

where $C_{1\varepsilon}$, $C_{2\varepsilon}$, and $C_{3\varepsilon}$ are constants; σ_k and σ_ε are the turbulent Prandtl numbers for k and ε, respectively; and S_ε is a user-defined source term.

The RNG k-ε model introduces additional terms and constants into the ε equation that improve its accuracy for complex flows. The model also uses an analytical formula for turbulent Prandtl numbers, which enhances its performance over the standard k-ε model under certain flow conditions.

2.2. Hydroelectric Turbine 3D Modeling and Grid Generation

2.2.1. 3D Modeling

After making the corresponding spatial adjustments in the 3D modeling software UG, the guide vanes of the hydroelectric turbine were reconstructed using the spline surface method, and a mathematical model for surface fitting of the Francis turbine blades was provided. Geometric models for other flow components, such as the volute casing and tailpipe, were also established. The geometric modeling of the entire flow passage of the hydroelectric turbine, as shown in Figure 1, is divided into four parts: the volute casing, guide vanes (including fixed guide vanes and adjustable guide vanes), runner, and tailpipe. Figure 1a represents the overall 3D model of the hydroelectric turbine, and Figure 1b shows an exploded view of the complete 3D model.

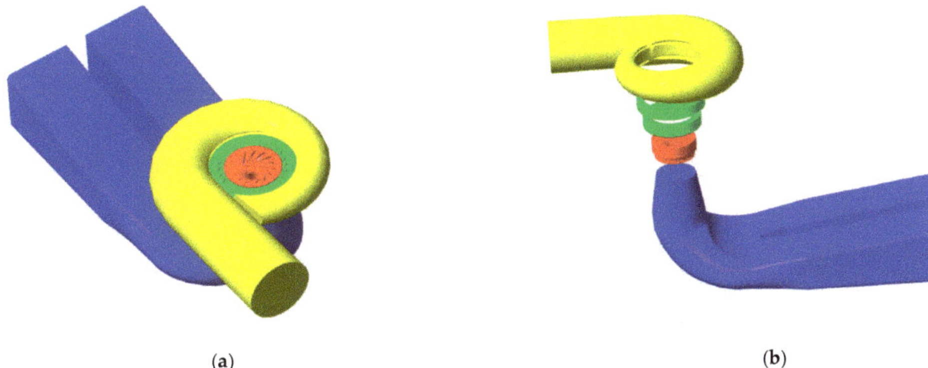

(a)   (b)

Figure 1. Overall 3D schematic of the hydroelectric turbine. (**a**) Overall 3D model of the hydroelectric turbine. (**b**) Exploded 3D model of the hydroelectric turbine.

2.2.2. Grid Partitioning

In CFD simulations, to achieve higher accuracy, it often becomes necessary to refine the computational grid and, thus, determine the appropriate size of the near-wall mesh. For rapid solutions near the wall, it is essential to use a turbulence model in conjunction with wall function methods that incorporate accurate empirical data. For example, when using the standard k-ε turbulence model with wall function modeling, the closest internal node to the wall must be within the logarithmic layer of turbulence, meaning y+ must be greater than 11.63 (ideally between 30 and 500). This sets a range for the distance from the closest grid to the wall and the grid size. Using NUMECA's IGG software 2021 [23], the geometric model was partitioned into grids. After grid independence testing, a final grid partitioning scheme was selected, with a total number of grid elements of approximately 150 million. The grid model is shown in Figure 2.

2.3. Boundary Conditions

We set the volume flow rate as the inlet boundary condition and absolute static pressure at the tailrace outlet as the outlet boundary condition, and choose the no-slip wall condition for the wall. The specific numerical values are given in Table 1.

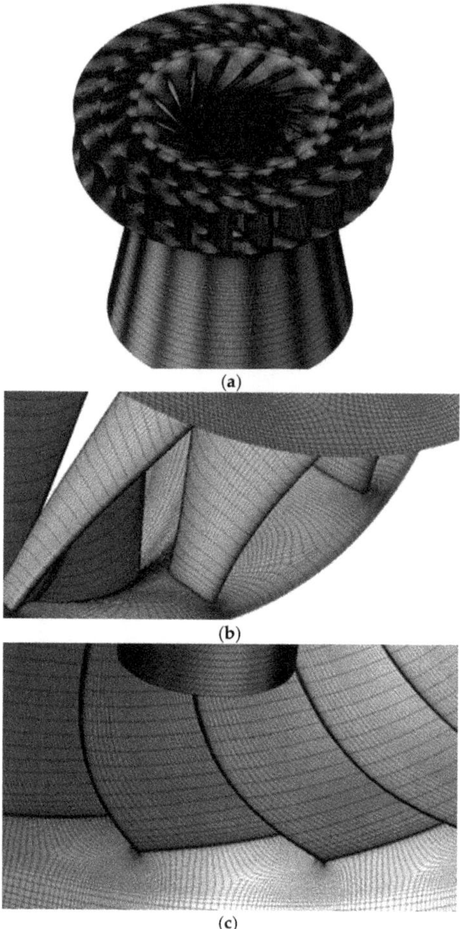

Figure 2. Grid Diagram of the Hydraulic Turbine. (**a**) Grid diagram of the fluid domain for the guide vanes, runner, draft tube cone vanes, runner, and draft tube cone. (**b**) Leading edge of the impeller blades. (**c**) Trailing edge of the runner blades.

Table 1. Boundary condition parameters.

Boundary Conditions for Calculation	Parameters
Flow rate (m^3/s)	232.77
Outlet absolute pressure At the tailrace (kPa)	307.26
Rotation speed (r/min)	166.7

2.4. Configuration of Monitoring Points for Pressure Pulsation Calculation

To facilitate comparison with measured data, three computational monitoring points (A1, A2, and A3) were strategically placed within the bladeless region between the active guide vane and the runner inlet of the turbine. Point A3 was positioned at the top cover of the stay ring, which coincided with the sensor location. Additionally, two more computational monitoring points, A1 and A2, were introduced to further monitor pressure pulsations in the central region of the bladeless area. These points were located at 50% of the blade height and near the lower stay ring, as illustrated in Figure 3. This configuration allows us to observe and compare pressure variations within the entire bladeless region with greater precision, enabling a more accurate assessment of the turbine's operational state. Additionally, this level of attention to detail and careful setup underscores our high standards and expertise in equipment monitoring and maintenance.

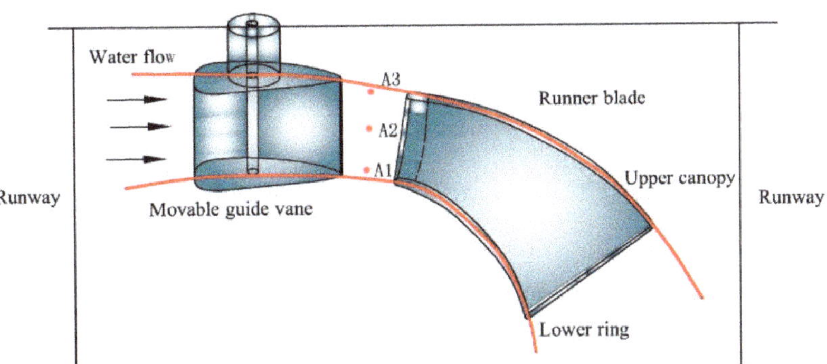

Figure 3. Illustration of the arrangement of computational monitoring points for pressure pulsations in the turbine's flow components.

2.5. Field Testing

2.5.1. Test Methodology

The vibration monitoring system consists of three main components, including sensors, a signal acquisition unit, and a data processing unit. The signal acquisition unit is composed of intelligent data acquisition devices and DC power supply devices, responsible for collecting sensor signals and other data inputs. The data processing unit is managed by the main control workstation, which handles the real-time monitoring, analysis, and diagnostic applications of the system. Vibration sensors installed at designated locations convert vibration signals generated during the turbine's operation into electrical signals. These signals are then transmitted to the intelligent data acquisition devices. The main control workstation retrieves the acquisition data from the intelligent data acquisition devices via Ethernet. After processing through software algorithms and data processing, various characteristic parameters reflecting the turbine's operational status are obtained.

2.5.2. Sensor Measurement Point Arrangement

The selection and arrangement of sensor measurement points during on-site testing are presented in Table 2.

Table 2. Selection and arrangement of sensor measurement point serial.

Serial Number	Signal Type	Measurement Point Name	Sensor Type
1		Upper frame X-axis horizontal vibration	Low-frequency velocity vibration sensor
2		Upper frame Y-axis horizontal vibration	Low-frequency velocity vibration sensor
3		Upper frame vertical vibration	Low-frequency velocity vibration sensor
4		Lower frame X-axis horizontal vibration	Low-frequency velocity vibration sensor
5	Vibration	Lower frame Y-axis horizontal vibration	Low-frequency velocity vibration sensor
6		Lower frame vertical vibration	Low-frequency velocity vibration sensor
7		Top cover X-axis horizontal vibration	Low-frequency velocity vibration sensor
8		Top cover Y-axis horizontal vibration	Low-frequency velocity vibration sensor
8		Top cover vertical vibration	Low-frequency velocity vibration sensor
10	Pressure	Guide vane outlet pressure pulsation	Pressure pulsation sensor
11	pulsation	Top cover bottom pressure pulsation	Pressure pulsation sensor

3. Results and Analysis

3.1. Numerical Calculation Results

The flow distribution within the runner and guide vanes at three different blade heights under the operating conditions of relative opening at 80.449%, an active power of 218.863 MW, and a gross head of 110.19 m was obtained through numerical simulations, as shown in Figures 4–6. At different time instants, the flow patterns within the flow field vary, and flow separation occurs as the water flows through the active guide vanes. From the figures, it can be observed that a Karman vortex appears at the rear of the active guide vanes, leading to turbine vibrations.

Under these operating conditions, the kinetic energy at the inlet region of the fixed guide vanes remains stable. However, significant flow separation occurs from the runner inlet to the outlet region. This is because the water jet impacts the leading edge of the runner blades at a large angle of attack, causing a noticeable flow separation vortex to form as the water flows from the runner blade inlet edge toward the blade's backside. This results in an increase in vortex energy between the blades, generating significant turbulence. It is this flow separation that induces unstable flow within the turbine, leading to severe vibrations in the hydroelectric generator unit.

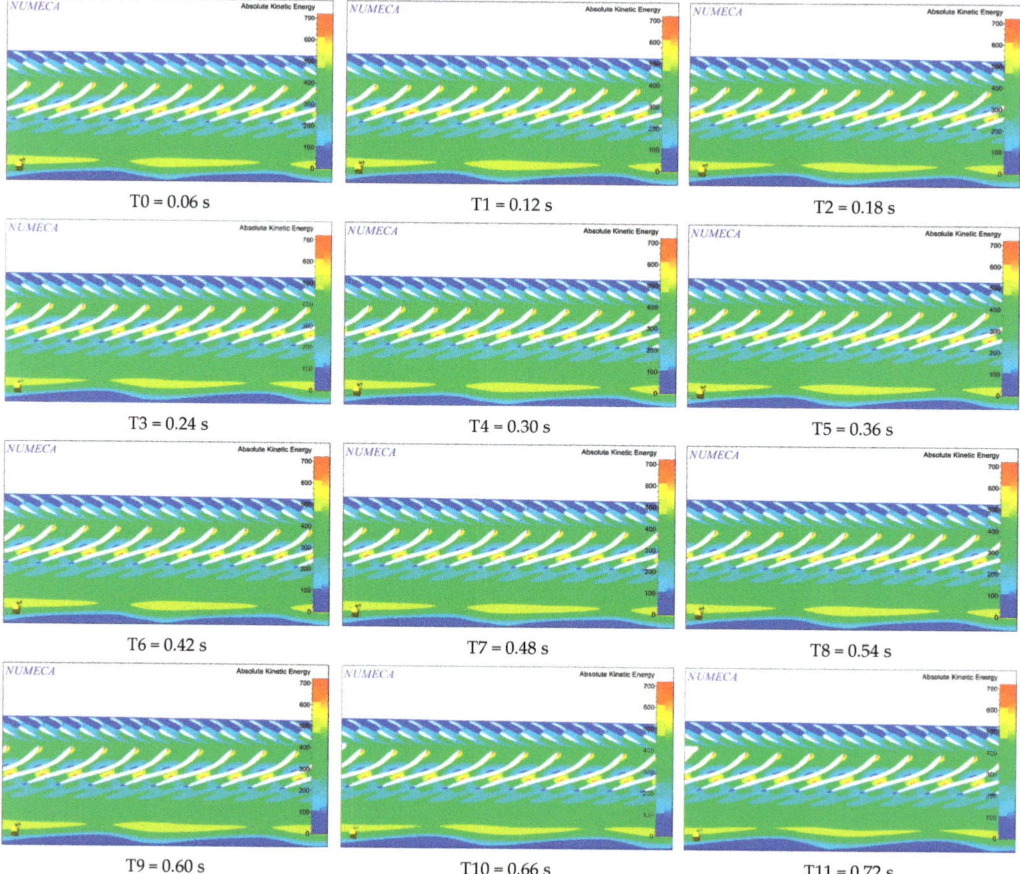

Figure 4. Flow distribution on the near upper crown flow surface (90% blade height) of the turbine.

3.2. Field Test Results

The specific test data for severe abnormal vibrations of the unit under this operating condition are shown in Table 3. From Table 3, it can be observed that during the abnormal operation of the unit, both the vertical vibration amplitude of the top cover and the vertical vibration amplitude of the lower frame have significantly exceeded the standard values. According to the standards [24], the vertical vibration displacement of the top cover should not exceed 90 μm, and the vertical vibration displacement of the lower frame should not exceed 70 μm.

Figures 7 and 8 show the waveform and order ratio plots of some monitoring points obtained from the on-site online monitoring system during the abnormal vibration of the unit. It can be seen from the figures that the reason for the excessive vibration of the unit is the very clear 0.56250× low-frequency component present in both the unit's vibration and pressure pulsation signals. Regarding the frequency components causing the excessive vibration amplitude, this frequency is mainly generated by hydraulic factors. Its source mainly originates from the instability of water flow in the bladeless area at the outlet of the unit's guide vane, leading to a significant increase in pressure pulsation. This, in turn, generates a huge hydraulic force in the axial direction of the unit, resulting in excessive vertical vibrations of the top cover and lower frame.

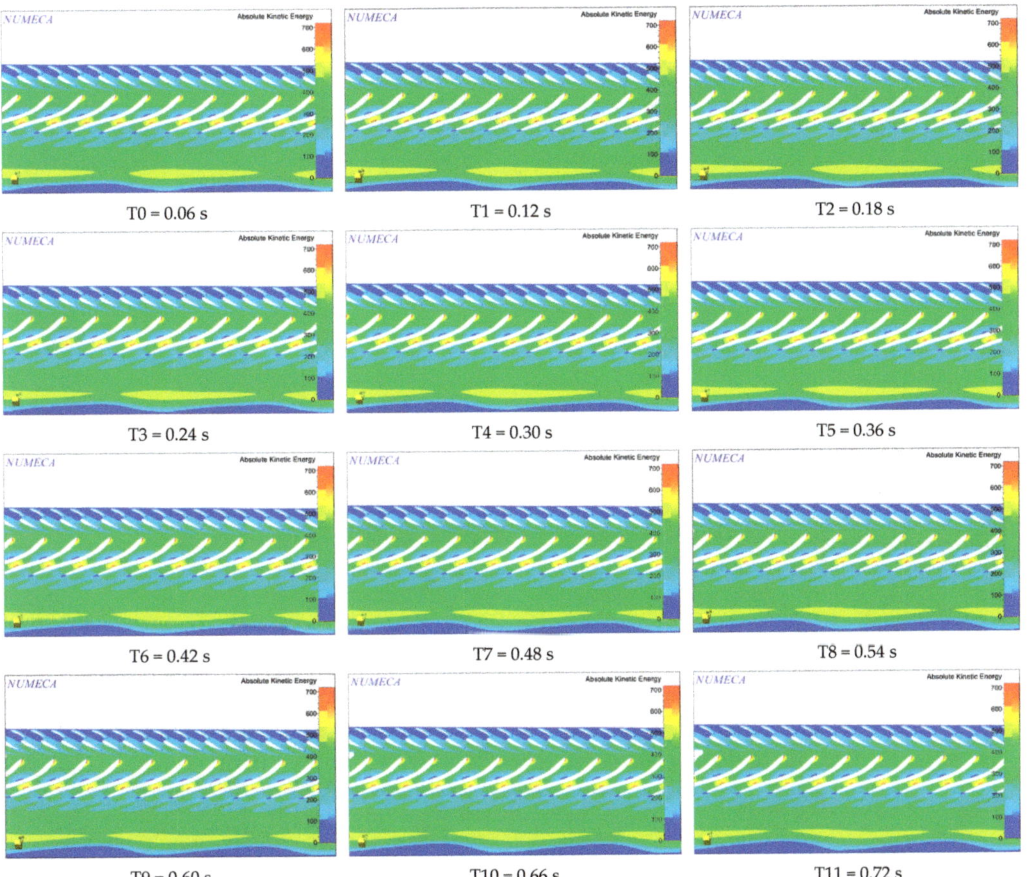

T0 = 0.06 s   T1 = 0.12 s   T2 = 0.18 s

T3 = 0.24 s   T4 = 0.30 s   T5 = 0.36 s

T6 = 0.42 s   T7 = 0.48 s   T8 = 0.54 s

T9 = 0.60 s   T10 = 0.66 s   T11 = 0.72 s

Figure 5. Flow distribution at the middle flow surface (50% blade height) of the turbine.

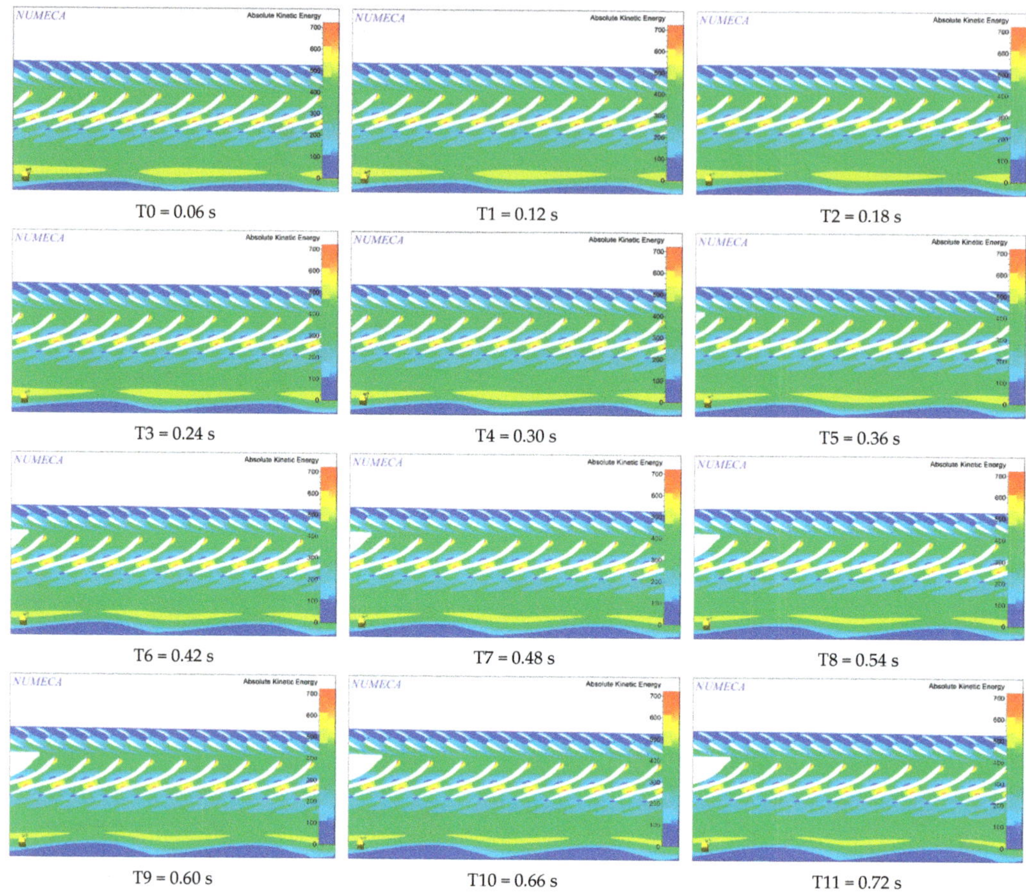

Figure 6. Flow distribution on the near lower circumferential flow surface (10% blade height) of the turbine.

Table 3. Comparison table between unit swinging alarms and normal data.

Serial Number	Guide Vane Opening (%)	Active Power (MW)	Top Cover Vertical Vibration (μm)	Lower Frame Vertical Vibration (μm)	Spiral Casing Inlet Pressure Pulsation (kPa)	Guide Vane Outlet Pressure Pulsation (kPa)	Remarks
1	80.449	218.863	378	139	93.2	131.04	Alarm value
2	78	215.57	19	18	6.0	9.3	Normal Operating value

3.3. Verification Calculation of Working Head under Operating Conditions

After calculating the working head of the No.4 unit of Dongqing Hydropower Station operating at a relative openness of 80.449%, with an active power of 218.863 MW, and a gross head of 110.19 m, it was found that the operating condition points of severe vibration in the hydro-turbine generator set deviate from the safe operating range specified by the turbine manufacturer. The following details the specific calculation process [25–27].

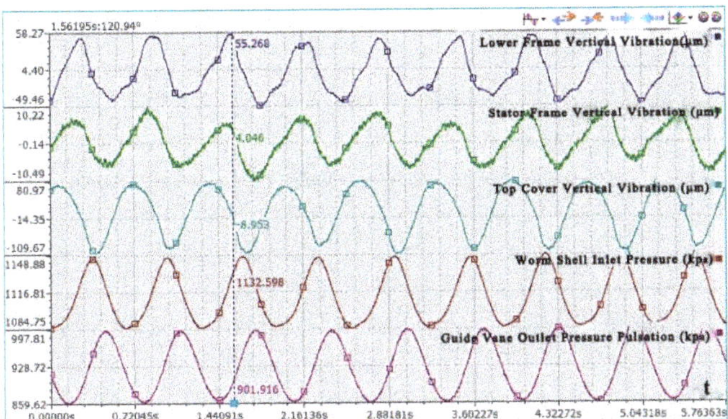

Figure 7. Vibration measurement pressure waveform curves for the unit operating at 80.449% openness.

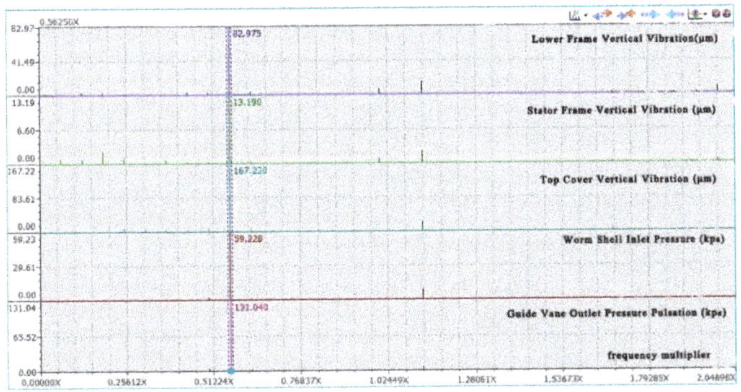

Figure 8. Harmonic ratio graph of vibration measurement pressure during unit operation at 80.449% openness.

Based on the field test data, including upstream and downstream water levels, gross head, active power, and generator efficiency, the output calculation formula for the hydro-turbine can be established as:

$$N = \rho \cdot g \cdot Q \cdot H \cdot \eta_w \cdot \eta_e \tag{5}$$

In the formula, ρ represents the density of water; g is the acceleration due to gravity, a variable parameter; Q is the flow rate of water in the pipeline; H is the working head of the hydropower station; η_e is the efficiency of the generator; and η_w is the efficiency of the turbine.

Using the trial calculation method, and initially assuming the working efficiency of the turbine at this condition based on the turbine test report, with $\eta_w = 92.3\%$ and $\eta_e = 98\%$, the flow rate of the turbine is calculated as:

$$Q = \frac{P}{\rho \cdot g \cdot H \cdot \eta_w \cdot \eta_e} \tag{6}$$

From this, the total head loss in the water diversion system of the hydropower station can be calculated as:

$$\Delta H = \left(\sum S_0 L + \sum T_0 \right) Q^2 \tag{7}$$

In the formula, $T_0 = \frac{8\zeta}{\pi^2 g d^4}$ represents the local specific resistance, S_0 is the pipe-specific resistance, L is the length of the pipe section, ζ is the coefficient of local head loss, and d is the hydraulic diameter.

Finally, $H = H_g - \Delta H$ is calculated as the working head of the turbine, where H_g represents the gross head.

Based on the pressure pipeline layout diagram of the power station, the working head for the No.4 unit of the hydropower station operating at a relative openness of 80.449%, with an active power of 218.863 MW, and a gross head of 110.19 m, was calculated. The results are presented in Table 4.

Table 4. Parameters of the turbine under full-load operating conditions.

Parameter Name	Parameter Values
Relative guide vane opening (%)	80.449
Flow rate (m^3/s)	232.77
Gross head (m)	110.19
Head loss (m)	4.00
Working head (m)	106.19
Active power (MW)	218.863

The operating condition values from Table 4 are marked on the comprehensive characteristic curve diagram of the turbine operation for the station, as shown by the red circle in Figure 9. It was observed that the operating condition point (a relative openness of 80.449%, an active power of 218.863 MW, and a gross head of 110.19 m, corresponding to a working head of 106.19 m) has deviated from the normal operating range provided by the turbine manufacturer.

3.4. Calculation Results of Pressure Pulsation and Frequency at Monitoring Points

Through unsteady numerical simulations of the flow field, pressure pulsation curves and frequency curves at monitoring points were obtained. The results are presented in Figures 10 and 11, which show the time-domain and frequency-domain diagrams of pressure pulsations at three computational monitoring points located at the exit of the movable guide vanes of the turbine under this abnormal operating condition. As shown in Figure 10, although there are minor differences in the pressure fluctuation amplitudes at the three different blade height positions at the monitoring points near the exit of the movable guide vanes, the pattern of pressure pulsation is essentially the same across these points, with similar main frequencies and amplitudes of pressure pulsation. Moreover, the amplitude of fluctuations is very large, exceeding 10 m of the water column, with the results at point A3 closely matching the field-measured data (as shown in Figure 7). Applying Fast Fourier Transform (FFT) spectral analysis, the frequency of pressure pulsations at the monitoring points was found to be 0.56250X (as shown in Figure 8), which is consistent with the calculated and measured values, as shown in Figure 11. This indicates that the numerical simulation method used in this paper can accurately predict the hydraulic performance of the turbine, with the amplitude and frequency of the main vibration sources obtained through numerical simulation corresponding well with the actual flow field test data under this operating condition.

Figure 12 shows the pressure distribution inside the turbine and guide vanes at a 50% blade height under 80% operating conditions of the hydro-turbine. The diagram reveals that under 80% operating conditions, the water pressure is relatively high around the guide vanes and lower in the turbine area. The pressure is stable at the inlet of the guide vanes but fluctuates significantly at the outlet. Additionally, the pressure inside the guide vanes exhibits good circumferential symmetry. At the turbine inlet, due to the impact of the water flow, the pressure increases noticeably compared to the surrounding water, and within the turbine flow field, the pressure first decreases, then increases, and decreases again from top to bottom. Due to the influence of the Karman vortex, significant flow separation occurs at the exit of the movable guide vanes, resulting in the boundary layer that was originally adhering to the wall at the front of the blade separating from the wall.

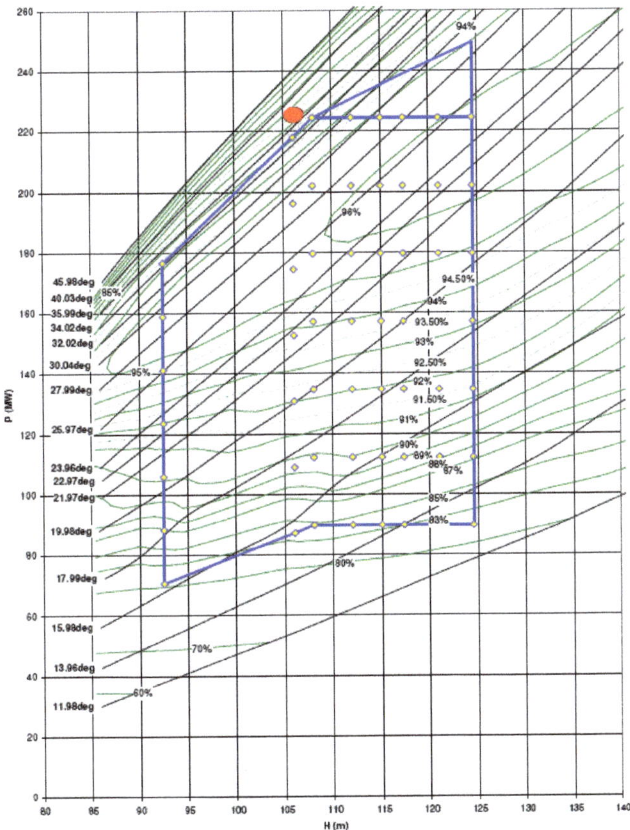

Figure 9. Position of the turbine operating point on the comprehensive operating characteristic curve.

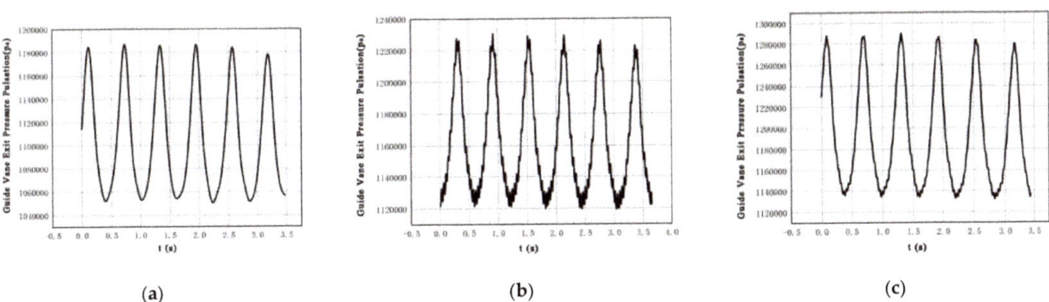

Figure 10. Time-domain diagram of pressure pulsation numerical simulation in the vaneless space. (**a**) Point A1 at the exit of the movable guide vane. (**b**) Point A2 at the exit of the movable guide vane. (**c**) Point A3 at the exit of the movable guide vane.

Pressure Amplitude (pa)
1040000
780000
520000
260000
0
A1
A2
A3
0
5
10
15
20
25
30
f/f_n

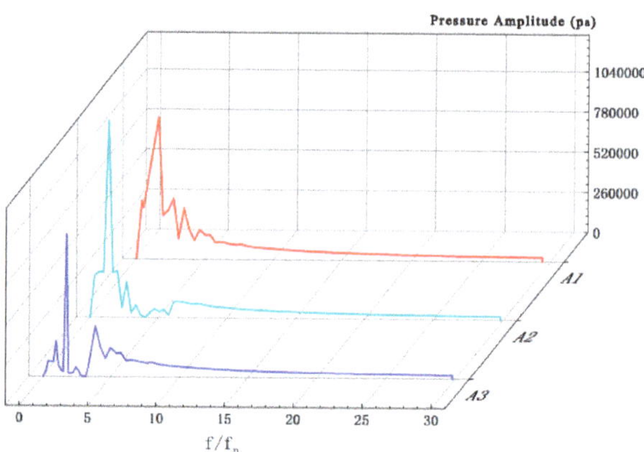

Figure 11. Frequency-domain diagram of pressure pulsation numerical simulation in the vaneless space.

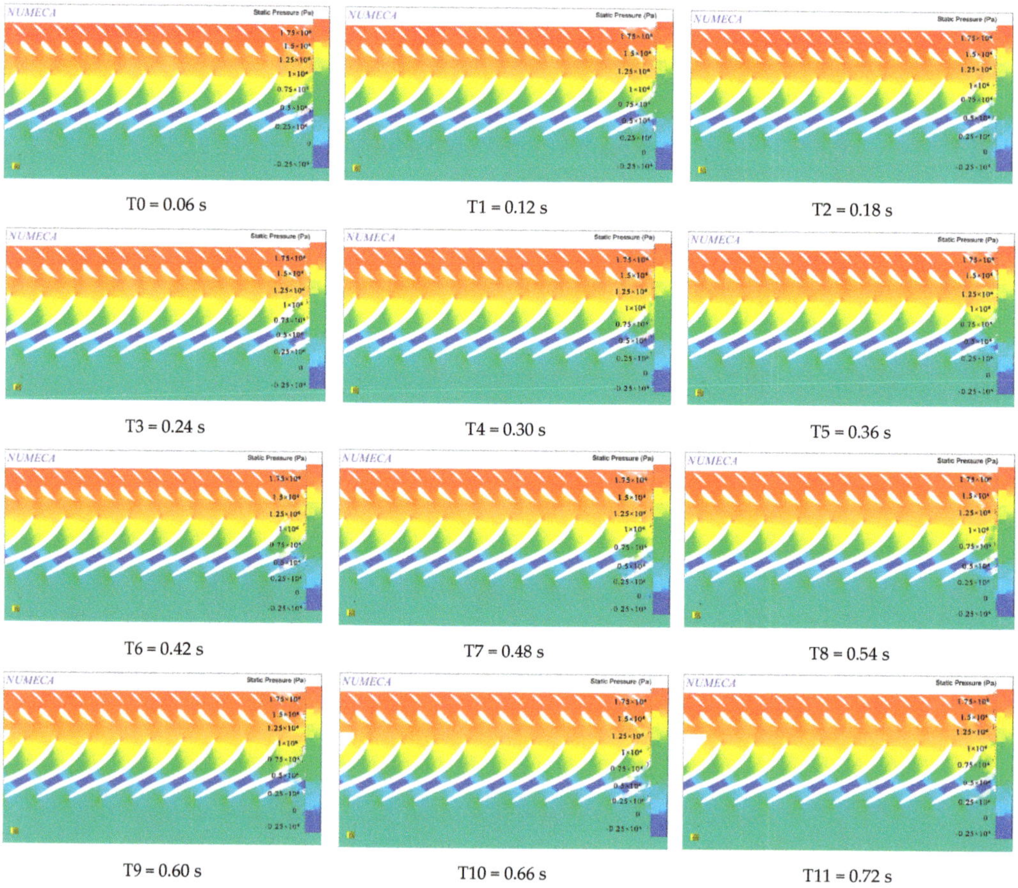
T0 = 0.06 s
T1 = 0.12 s
T2 = 0.18 s
T3 = 0.24 s
T4 = 0.30 s
T5 = 0.36 s
T6 = 0.42 s
T7 = 0.48 s
T8 = 0.54 s
T9 = 0.60 s
T10 = 0.66 s
T11 = 0.72 s

Figure 12. Pressure distribution on the flow surface inside the turbine and guide vanes at 50% blade height under 80% operating conditions of the hydro-turbine.

At a relative openness of 80%, the pattern of pressure pulsation amplitude changes at the three measurement points at the exit of the movable guide vanes is similar, with the pressure pulsation amplitudes being essentially the same and several times higher than the pressure pulsation values at the same points under normal operating conditions. With the increase in load, the flow rate of the unit increases, leading to higher head losses, resulting in a mismatch between the water flow characteristics and the guide vane profiles. This results in increased pressure pulsation at the exit of the movable guide vanes as the water flows around them, ultimately reducing the efficiency of the turbine.

3.5. Analysis of Abnormal Pressure Pulsation Formation

The experimental tests and computational results indicate that the high-amplitude pressure pulsations occurring under full-load conditions in the power station are directly related to the deviation of the turbine operating point from the operating region. The calculations and simulations have demonstrated that the hydraulic vibrations occurring in the Dongjing Hydropower Station's turbine under the condition of a 222.8 MW load, a gross head of 110.19 m, and a relative opening of 80.449% are due to the operating point being outside the turbine output limit line on the comprehensive characteristic curve provided by the manufacturer. This deviation of the operating point leads to a detached flow in the moving guide vanes within the internal flow field, resulting in the generation of low-frequency Kármán vortex streets.

The coupling of the Kármán vortices at the exit of the moving guide vanes with the detached vortices occurring in the backside region of the turbine blade inlet leads to an abnormal increase in the amplitude of pressure pulsations in the vaneless space.

4. Conclusions

This paper, utilizing CFD software, conducted unsteady flow computations for the internal flow field of the No.4 unit of a hydropower station in Guizhou, and compared these with on-site experimental test data. Applying FFT spectrum analysis, the study investigated the pressure pulsation frequencies at monitoring points and arrived at the following conclusions:

1. The generation of low-frequency, large-amplitude pressure pulsations at 0.56250X in the vaneless space of the turbine at the Guizhou Hydropower Station is due to the operating point deviating from the operational range. This deviation led to significant detached vortices in the backside area of the turbine blade inlets, causing low-frequency vibrations in the vaneless space and inducing unstable flows in the turbine.
2. The coupling of the detached vortices from the trailing edge of the moving guide vanes with those formed in the backside area of the turbine blade inlets led to a dramatic increase in pressure pulsation amplitudes in the vaneless space, reaching 1.29 MPa, thus causing excessive vertical vibration amplitudes in the top cover and lower frame.
3. These findings suggest that employing fluid simulation technology combined with experimental testing to analyze hydraulic vibrations in turbines is an effective approach. Simultaneously, by reducing the opening of the movable guide vanes and lowering the load, the abnormal vibrations of the turbine were eliminated.

Author Contributions: Investigation, resources, writing—original draft, L.J.; conceptualization, X.L.; methodology, project administration, Y.Z.; investigation, W.H.; supervision, W.X. All authors have read and agreed to the published version of the manuscript.

Funding: This research was supported by the National Natural Science Foundation of China (approval No.: 51279172), the National Key Research and Development Program (No.: 2018YFB0905200), and the Key Laboratory of Fluid and Power Machinery, Ministry of Education, Xihua University (szjj2016-002).

Data Availability Statement: Data are contained within the article.

Conflicts of Interest: The authors declare that they have no known competing financial interests or personal relationships that could have appeared to influence the work reported in this paper.

Abbreviations

RNG	Re-normalization group
CFD	Computational fluid dynamics
UG	Unigraphics
t	Time variable (s)
v	Viscosity (kg m^{-1} s^{-1})
g	Gravitational acceleration (m s^{-2})
DC	Direct current
η_e	Engine efficiency
η_w	Hydro turbine efficiency
FFT	Fast Fourier transform

Greek letters

∂	The differentiation of a function

References

1. Shen, K.; Zhang, Z.; Liang, Z. Calculation of Hydraulic Vibrations in the Powerhouse of Yantan Hydropower Station. *Hydropower Energy Sci.* **2003**, *21*, 73–75.
2. Zheng, M.; Ma, X.; Li, W. Causes and Treatment Measures for Turbine Cracks at Xiaolangdi Power Station. *Hydropower Energy Sci.* **2008**, *26*, 153–155.
3. Liu, P.; Chen, X.; Wang, Q.; Li, D. Study on Dynamic and Static Interference and Vibration Issues in High Head Francis Turbines. *J. Hydroelectr. Eng.* **2016**, *35*, 8.
4. Xiao, R.; Wang, F.; Gui, Z. Analysis of Fatigue Cracks in Blades of Francis Turbines and Improvement Strategies. *J. Hydraul. Eng.* **2011**, *42*, 5.
5. Zhang, L. Research on Fatigue Crack Control of Francis Turbine Runner Blades. Ph.D. Thesis, Tsinghua University, Beijing, China, 2010.
6. Niu, L. Study and Control of Vortex Rope Characteristics in the Draft Tube of Francis Turbines. Master's Thesis, Zhejiang Sci-Tech University, Hangzhou, China, 2015.
7. Zhang, S. Research on the Hydraulic Stability of Large Francis Turbines. Ph.D. Thesis, Huazhong University of Science and Technology, Wuhan, China, 2008.
8. Gui, Z.; Chang, Y.; Chai, X.; Wang, Y. Research Progress on Pressure Pulsation and Vibration Stability of Francis Turbines. *Large Electr. Mach. Technol.* **2014**, 61–65.
9. Tian, S. Discussion on Early Power Generation at the Three Gorges Based on the Low Head Operation Experience at Longyangxia. *Yangtze River* **1994**, *25*, 5.
10. Qiao, L.; Chen, Q. Operation Stability Test and Analysis of a 200 MW Francis Turbine Unit. *Hydropower Energy Sci.* **2014**, *32*, 4.
11. Feng, J.; Ling, Z.; Zhao, Z.; Li, H.; Chen, D. Study on the Structural Stability of Francis Turbine Runner under Partial Load Conditions. *J. Hydroelectr. Eng.* **2021**, *40*, 107–114.
12. Feng, Y.; Chang, H.; Zhang, E.; Liu, C. Field Stability Test and Vibration Protection Strategy Research for a Francis Turbine Generator Unit. *Hydropower Energy Sci.* **2016**, *34*, 178–182.
13. Wu, D.; Wen, Q.; Dai, Y. Vibration and Noise Test Analysis and Treatment Measures for Francis Turbines: A Case Study of Shangyoujiang Hydropower Station. *Yangtze River* **2020**, *51*, 218–224.
14. Lu, L.; Zhang, L.; Yang, J.; Zhou, L. Cause Analysis and Research on Vibration and Noise of Low Head Francis Turbines. *Hydropower* **2014**, *40*, 47–49.
15. Dekterev, D.; Maslennikova, A.; Abramov, A. An experimental study of dependence of hydro turbine vibration parameters on pressure pulsations in the flow path. *J. Phys. Conf. Ser.* **2017**, *899*, 022003. [CrossRef]
16. Luo, X.; Zhu, G.; Feng, J. Technological Advancements and Development Trends in Turbine Technology. *J. Hydroelectr. Eng.* **2020**, *39*, 1–18.
17. Sun, L.; Guo, P.; Yan, J. Transient analysis of load rejection for a high-head Francis turbine based on structured overset mesh. *Renew. Energy* **2021**, *171*, 658–671. [CrossRef]
18. Wang, Z.; Sun, L.; Guo, P.; Wang, X. Analysis and Suppression of Vortex Formation in the Blade Passages of Francis Turbines. *J. Hydroelectr. Eng.* **2020**, *39*, 113–120.
19. Guo, P.; Sun, L.; Luo, X. Study on the Flow Characteristics of Vortices in the Blade Passages of Francis Turbines. *J. Agric. Eng.* **2019**, *35*, 43–51.

20. Wang, P.; Xu, D.; Zhang, B. Numerical Simulation of Pressure Pulsation and Flow Field Characteristics in Axial-Flow Turbines. *China Rural. Water Hydropower* **2018**, 208–211.

21. Ershkov, S.V. Non-stationary Riccati-type flows for incompressible 3D Navier-Stokes equations. *Comput. Math. Appl.* **2016**, *71*, 1392–1404. [CrossRef]

22. Ahn, S.-H.; Xiao, Y.; Wang, Z.; Luo, Y.; Fan, H. Unsteady prediction of cavitating flow around a three dimensional hydrofoil by using a modified RNG k-ε model. *Ocean. Eng.* **2018**, *158*, 275–285. [CrossRef]

23. Galerkin, Y.B.; Voinov, I.B.; Drozdov, A.A. Comparison of CFD-calculations of centrifugal compressor stages by NUMECA Fine Turbo and ANSYS CFX programs. *IOP Conf. Ser. Mater. Sci. Eng.* **2017**, *232*, 012044. [CrossRef]

24. *EN 60994-1992*; Guide for Field Measurement of Vibrations and Pulsations in Hydraulic Machinery (Turbines, Storage Pumps and Pump-Turbines) (IEC 994-1991). International Electrotechnical Commission (IEC): Geneva, Switzerland, 1992.

25. Li, Q. *Research on Hydraulic Stability of Francis Turbines*; China Water & Power Press: Beijing, China, 2014.

26. Grein, H.; Goede, E. Site experience with Francis turbines operating under very large head Variatio. In Proceedings of the 17th IAHR Symposium, St. Petersburg, Russia, 21–25 June 2004.

27. Wu, P. Characteristics and Hazards of Blade Passage Vortices. In Proceedings of the 15th Academic Symposium on Chinese Hydropower Equipment, Beijing, China, 15–19 September 1994; Chinese Society for Electrical Engineering: Beijing, China; Chinese Society for Hydropower Engineering: Beijing, China; Chinese Society for Power Engineering: Beijing, China; Chinese Hydraulic Engineering Society: Beijing, China, 2004.

Article

Representing Hourly Energy Prices in a Large-Scale Monthly Water System Model

Mustafa Sahin Dogan [1], Ellie White [2], Yiqing Yao [2] and Jay R. Lund [2,*]

[1] Department of Civil Engineering, Aksaray University, Aksaray 68100, Türkiye; msahindogan@aksaray.edu.tr
[2] Department of Civil and Environmental Engineering, University of California, Davis, One Shields Avenue, Davis, CA 95616, USA; white.elaheh@gmail.com (E.W.); yqyao@ucdavis.edu (Y.Y.)
* Correspondence: jrlund@ucdavis.edu

Abstract: Water system management models represent different purposes, such as water supply, flood control, recreation, and hydropower. When building large-scale system models to represent these diverse objectives, their most appropriate time steps for each purpose often do not coincide. A monthly time step is usually sufficient for water supply modeling, but it can be too coarse for flood control, hydropower, and energy operations, where hourly time steps are preferred. Large-scale water management and planning models mostly employ monthly time steps, but using monthly average energy prices underestimates hydropower revenue and overestimates pumping energy cost because these plants tend to operate during times with above- or below-average energy prices within any month. The approach developed here uses hourly varying prices depending on the percent of monthly operating hours. This paper examines an approach that approximately incorporates hourly energy price variations for hydropower and pumping into large-scale monthly time-step water system model operations without affecting water delivery results. Results from including hourly varying energy prices in a large-scale monthly water supply model of California (CALVIN) are presented. CALVIN is a hydroeconomic linear programming optimization model that allocates water to agricultural and urban users with an objective to minimize total scarcity costs, operating costs, and hydropower revenue loss. Thirteen hydropower plants are modeled with hourly varying prices, and their revenue increased by 25 to 58% compared to revenue calculated with monthly average constant energy prices. Hydropower revenue improvements are greater in critically dry years. For pumping plants modeled with hourly varying prices, the energy use cost decreased by 10 to 59%. This study improves system representation and results for large-scale modeling.

Keywords: energy prices; hydropower; pumping cost; revenue; system models

Citation: Dogan, M.S.; White, E.; Yao, Y.; Lund, J.R. Representing Hourly Energy Prices in a Large-Scale Monthly Water System Model. *Water* **2024**, *16*, 562. https://doi.org/10.3390/w16040562

Academic Editors: Zhengwei Wang and Yongguang Cheng

Received: 22 December 2023
Revised: 3 February 2024
Accepted: 6 February 2024
Published: 12 February 2024

1. Introduction

Computer models help operators and policymakers explore and compare management alternatives, better operate complex water resources systems, and predict the future performance of existing or proposed decisions and conditions [1]. Model use can increase the benefits or decrease the costs of managing water for wide-ranging purposes, such as agricultural, urban, and wildlife water supply; hydropower; flood control; recreation; and navigation [2–4]. Yeh [5], Wurbs [6], and Labadie [7] reviewed water resource system models, including simulation and optimization models, ranging from small- to large-scale. Integrated river basin management models provide a framework for more efficiently managing scarce water among users, at different locations and times, considering economic and hydrologic variables [8,9]. Davis [10] defines integrated water management as a facilitated stakeholder process to promote coordinated activities in pursuit of common objectives for better development and management. Marques and Tilmant [11] point out that operational coordination is critical to maintain and increase system-wide benefits under uncertain conditions in multireservoir water systems. Li et al. [12] reviewed uncertainties in integrated simulation–optimization modeling.

Many methods are available in water resource modeling. However, energy prices are often omitted or greatly simplified compared to other parameters. Several studies explicitly address energy price representation [13–15], which is especially important for hydropower and pumping operations. Models exist for simulating energy market operations [16–19], but they often are not integrated with water system models, or their integration with complex water models is computationally difficult, isolating market price models from water system models. Additionally, the availability of hourly input data often limits the explicit representation of energy prices in many places.

Water resource system models use different time scales, varying from fractions of an hour to annual. Short-term models have hourly to daily time steps. Time steps for long-term models range from weekly to annual. Long-term models can be useful for short-term models by providing preliminary promising solution regions [20,21]. Hourly or shorter-peak and off-peak times and energy value rates are important for hydropower decisions and pumping plant operations. Short-term models, with hourly time steps and time horizons of a few days, can directly represent hourly price variations in their operations. Long-term models rarely use hourly time steps due to computational demands; instead, monthly time steps are preferred. However, using monthly average energy prices for long-term planning and management often underestimates hydropower values [15] and overestimates pumping costs. This is because, with economically optimal operations, reservoirs with an afterbay operate during the most economically advantageous hours [13], the so-called peak hours. Similarly, pumping plants with a forebay or afterbay operate during off-peak hours, when energy prices are lower.

Hourly price variations can be represented with an implicit method that uses different prices depending on capacity use [15]. The method assumes that to maximize revenue, a hydropower plant with reservoir storage and an afterbay allocates hydropower releases preferentially to peak price times, when energy demand and prices are highest [22]. The reservoir operator is assumed to have good short-term foresight of energy prices, allowing revenue-maximizing releases during peak times. Energy prices are exogenous, so the operator cannot influence them. Tejada-Giubert et al. [13] used such a method to maximize the Central Valley Project energy revenue. A monthly capacity factor for each plant is assigned to the price duration curve factors, assuming each plant can be dispatched and operated during the most economically valued hours in a month.

Using variable energy prices, rather than constant prices, increases overall hydropower revenue. Olivares and Lund [15] studied the representation of energy prices in long- and medium-term hydropower operations. Their model, focusing on a single reservoir, relates hourly energy prices with the proportion of monthly hours available for energy generation. Hourly varying prices are estimated with a moving average method, a function of the percentage volume allocation. Madani and Lund [14] used energy prices that vary with the number of hours of operation in a suite of optimization models for California's high-elevation hydropower plants. The model assumes that a power plant will prioritize releasing water during high-valued times of the month. It only releases water during lower-valued times when water is abundant, if the objective is solely hydropower operation. These studies concluded that there is a great benefit of incorporating hourly energy price variations in long-term models. Former studies focused on regional hydropower modeling with small reservoir storage capacities or with a limited number of large-scale reservoirs with hourly price variations. The effects of hourly price variations on pumping operations have not been studied.

This study presents an implicit incorporation of hourly energy price variations in a large-scale long-term model of California's extensive water supply system, CALVIN, that uses monthly time steps over an 82-year hydrologic record. CALVIN is useful for planning, management, and policy studies [23,24]. It integrates and economically optimizes reservoir, power plant, pumping plant, and other water supply operations. The method used here is similar to earlier hydropower studies, except at a much larger scale. Also, hourly varying energy prices are applied to pumping operations with a few modifications to the method.

CALVIN is operated with both constant and hourly varying energy prices. Generation, revenue, and agricultural and urban water scarcity results are compared to assess the most effective representation of prices for long-term, multi-purpose modeling.

2. Materials and Methods

2.1. CALVIN Hydroeconomic Model

CALVIN (CALifornia Value Integrated Network) is a hydro-economic deterministic optimization model for California's inter-tied water supply and delivery system [23]. CALVIN represents California's water infrastructure with 49 surface reservoirs, 38 groundwater reservoirs, 600+ conveyance links, 1250+ nodes, and 36 agricultural, 41 urban and 8 wildlife refuge water demand areas, where the full-size model has more than 5 million decision variables. The CALVIN model covers about 88 percent of California's irrigated acreage and 92 percent of the state's urban population. Prescribed CALVIN operations are based on 82 years of historical data to represent hydrologic variability. CALVIN is a linear programming model that uses a generalized network-flow optimization, where operations are driven by convex cost-based piecewise linear penalty functions (for hydropower plants and agricultural and urban demands) or operating cost curves (for pumping and treatment plants). CALVIN operates and allocates surface and groundwater resources with deterministic hydrologic inflows under demand conditions of the year 2050, within physical and environmental constraints [23]. CALVIN is modeled with Pyomo, a high-level optimization modeling library in Python, and can connect to freely available solvers that provide a solution for the network-flow optimization problem [24]. CALVIN has an economics-driven objective, shown in Equation (1), to minimize system-wide operating costs (such as pumping and treatment) and scarcity costs to water users. Scarcity volume is defined as the amount of water that the user is willing to pay for but did not receive. Whenever a user's target demand is not met, scarcity occurs with a cost derived from the user's willingness to pay. The physical system is represented by a set of nodes (N) and links (or arcs) (A) in network-flow optimization, shown in Figure 1. Links are defined by $(i, j, k) \in A$, where i is the origin node, j is the terminal node, and k is the index of the piecewise linear component for links [24].

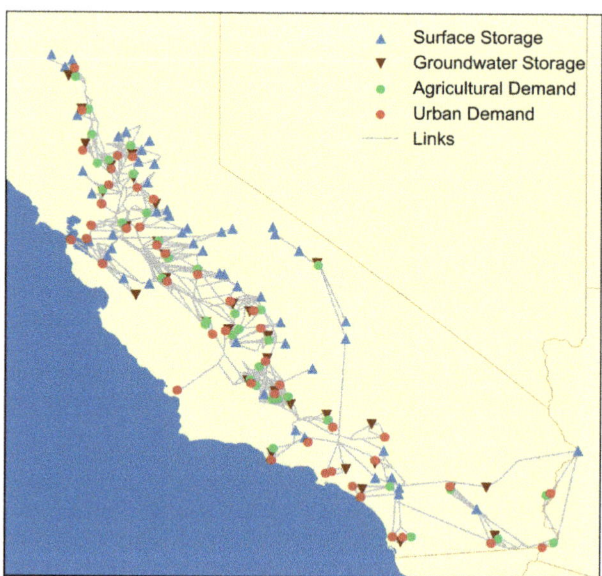

Figure 1. Network representation of California's water supply system in CALVIN [24].

CALVIN's objective function and constraints are

$$\mathrm{min}\, z = \sum_i \sum_j \sum_k c_{ijk} X_{ijk} \qquad (1)$$

subject to

$$X_{ijk} \le u_{ijk}, \ \forall (i,j,k) \in A \qquad (2)$$

$$X_{ijk} \ge l_{ijk}, \ \forall (i,j,k) \in A \qquad (3)$$

$$\sum_i \sum_k X_{jik} = \sum_i \sum_k a_{ijk} X_{ijk}, \ \forall j \in N \qquad (4)$$

where z is a summation over all links (i, j, k) and represents total scarcity and operating costs. X_{ijk} is the decision variable and represents the flow conveyed in a link. u_{ijk} and l_{ijk} are upper- and lower-bound constraints, respectively. a_{ijk} is the amplitude and represents losses (<1) or reuse (>1) within the system. Equations (2)–(4) enforce upper and lower bounds on each link and mass balance on each node, respectively.

CALVIN assumes a decentralized energy market in California with hydropower as a price-taker. Each facility allocates hydropower releases during the most valuable hours while considering water scarcity costs for agricultural and urban users, as well as downstream minimum in-stream flow requirements. Hydropower and water scarcity penalty curves dictate the economically optimal release time and volume. Seasonally, it is fortunate that California's energy and water use peaks generally coincide, reducing conflicts across these purposes.

CALVIN's limitations include its perfect hydrologic foresight (except for Arnold [25]); the linearization of nonlinear operations, such as hydropower; and simplified environmental regulations, water quality, and stream–aquifer interaction behavior. CALVIN currently provides a minimal representation of flood control and recreation operations [23]. Despite its limitations, CALVIN can simulate various water management scenarios and offer insights for statewide and regional water policy and planning decisions.

2.2. Representing Hourly Varying Energy Prices in Monthly Models

CALVIN operates explicitly to minimize statewide operating and scarcity costs. Hydropower in CALVIN is modeled using economic penalty curves, which represent the benefits lost from not generating hydropower. Hourly wholesale energy prices for the year of 2010 are obtained from the California Independent System Operator [26]. The moving average method proposed by Olivares and Lund [15] is used to represent hourly varying energy prices in the monthly model. This method relates the monthly percentage of hours of generation at turbine capacity with a price duration curve (Figure 2). Prices are averaged up to percent use. As seen in Figure 3, marginal variable energy prices decrease with increased hours of generation. This trend represents electricity market operations. Small releases have greater marginal benefits (energy prices), and as the hours of operation increase, the marginal hydropower revenue decreases, whereas the marginal revenue does not change with constant average monthly prices. For economically optimal hydropower operations with hourly varying prices, CALVIN makes small releases when the marginal energy price is high, and the lowest average price occurs when energy is generated at the monthly turbine capacity [14]. CALVIN is a deterministic model with fixed monthly price fluctuations, allocating water for maximum profitability. The total hydropower generation (G) in month i can be calculated as follows:

$$G(Q_i, h_i) = \mu \cdot \gamma \cdot Q_i \cdot h_i \cdot \Delta t \qquad (5)$$

where Q is the release obtained from CALVIN; h is the head, which is dynamically calculated from a polynomial function depending on reservoir storage; μ is efficiency; γ is the specific weight of water; and Δt is the time period.

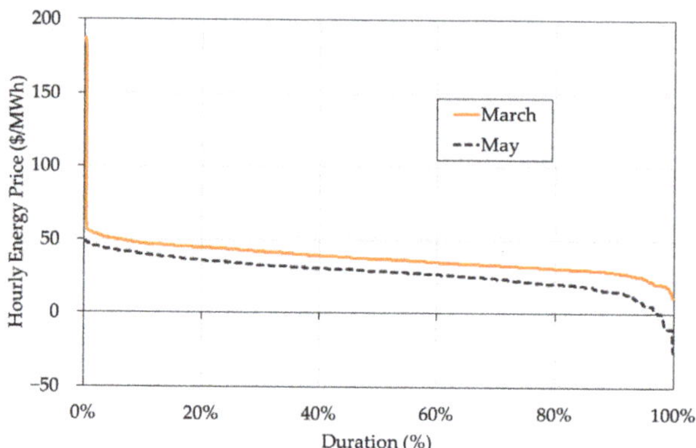

Figure 2. Price duration curves of hourly energy prices for months of March and May. Each month has a different hourly price variation pattern.

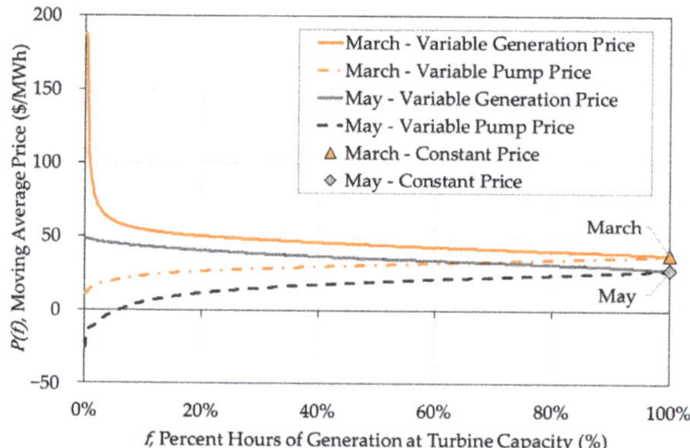

Figure 3. Moving average curves for months of March and May. The marginal variable energy price slope of the continuous and dashed price curves above decrease with increased hours of generation. Constant prices oversimplify plants that operate based on the marginal value.

Defining Q_{cap} as the plant's turbine release capacity and $G_{cap} = \mu \cdot \gamma \cdot Q_{cap} \cdot h_i \cdot \Delta t$ as the plant's generation capacity, the percentage of hours of generation at full capacity (f) can be obtained as

$$f = \frac{G_i}{G_{cap}} \tag{6}$$

Multiplying generation (G) with energy prices (P_i) yields the hydropower revenue (B) in month i, which can be written as

$$B_i = \overline{PG}(f)_i \cdot G_i \tag{7}$$

$$C_i = \overline{PP}(f)_i \cdot E_i \tag{8}$$

where $\overline{PG}(f)_i$ is the average of all prices exceeding $P(f)$, with the price obtained from the moving average price curve (Figure 3). For the energy cost of pumping plants (C_i), the unit energy price ($\overline{PP}(f)_i$), the average of all prices not exceeding $P(f)$, is multiplied with the total energy use (E_i).

Afterbays provide operational flexibility to large-scale reservoir operations. Several California hydropower facilities effectively use afterbays to regulate reservoir releases. These large-storage power plants are used to meet peak-time electricity demand. A total of 13 out of 33 CALVIN facilities with a large storage capacity and an afterbay are modeled with hourly varying energy prices. The remaining facilities, including run-of-river hydropower plants, use monthly constant average energy prices. Run-of-river plants are not modeled with hourly varying energy prices because they are operated continuously, depending on stream conditions. Meanwhile, pumping plants with a forebay or afterbay can also optimize pumping hours to minimize pumping costs. Six of twenty-four CALVIN pumping facilities have forebays and are modeled with hourly varying energy prices. CALVIN's hydropower and pumping plants are listed in Table 1 and Table 2, respectively.

Table 1. Modeled CALVIN hydropower facilities and plant characteristics.

Hydropower Facility	Power Plant Capacity (MW)	Reservoir Storage Capacity (10^6 m^3)	Turbine Release Capacity (m^3/s)
Shasta [a]	629	5612	561
Spring Creek [a]	180	296	143
Carr	154	296	97
Trinity [a]	140	3019	132
Keswick	117	29	494
Hyatt [a]	644	4367	523
Colgate [a]	325	1147	107
Folsom [a]	199	1205	245
New Narrows	49	86	121
Thermalito Fore/Afterbay	115	70	503
New Melones [a]	300	2953	270
Don Pedro [a]	203	2504	260
Holm [a]	157	372	23
Kirkwood [a]	122	444	21
New Exchequer	95	1263	89
Gianelli [a]	424	2518	376
Moccasin [a]	104	1	19
O'Neill [a]	25	70	42
Pine Flat	190	1233	231
Castaic	1247	401	123
Devil Canyon	280	90	46
Warne	78	7	47
San Francisquito 1 and 2	123	-	22
Gorges	112	-	23
Mojave	32	-	56
Others	92	-	359

Note: [a] Modeled with hourly varying prices.

Table 2. Modeled CALVIN pumping facilities and plant characteristics.

Pumping Facility	Pumping Plant Capacity (MW)	Flow Capacity (m^3/s)
Badger Hill	6	270
Banks [a]	212	245
Buena Vista	90	146
Contra Costa 1	3	9
Dos Amigos	99	340
Del Valle [a]	0	3
Eagle	72	52
Edmonston	622	127
Gianelli [a,b]	151	317
Iron	72	52
Julian Hinds	72	52
Los Vaqueros	23	6
Las Perillas	2	270
Mallard Slough	5	1
Old River	36	7
O'Neill [a,b]	19	121

Table 2. *Cont.*

Pumping Facility	Pumping Plant Capacity (MW)	Flow Capacity (m^3/s)
Oso [a]	37	90
Pearlblossom	95	83
South Bay [a]	4	9
Tracy	92	133
Walnut Creek	61	14
Wheeler Ridge	107	127
Chrisman	233	127

Notes: [a] Modeled with hourly varying prices; [b] operated for both pumping and power generation.

Figure 4 illustrates the run procedure of CALVIN with conventional constant average and proposed hourly varying energy prices. First, the input data are organized. These data include unit costs for moving water, economic demands for users (agricultural, urban, and hydropower), and hydrological time series, such as reservoir inflow, local inflow, and groundwater recharge. In the conventional model run, hydropower and pumping plant operating curves are modeled based on monthly constant average prices. CALVIN's objective is to minimize total costs. Therefore, a hydropower penalty curve, which is the inverse of its benefit curve, represents the loss for not generating hydropower, depending on the turbine discharge rate. For a model run with the proposed method to incorporate hourly energy price variations, moving average energy prices are used to generate piecewise linear operating curves. Third, the CALVIN model is run with defined energy prices, and optimized flow and reservoir storage decisions are obtained. Finally, hydropower generation and revenue, as well as pumping costs, are postprocessed and compared, as presented in Section 3.

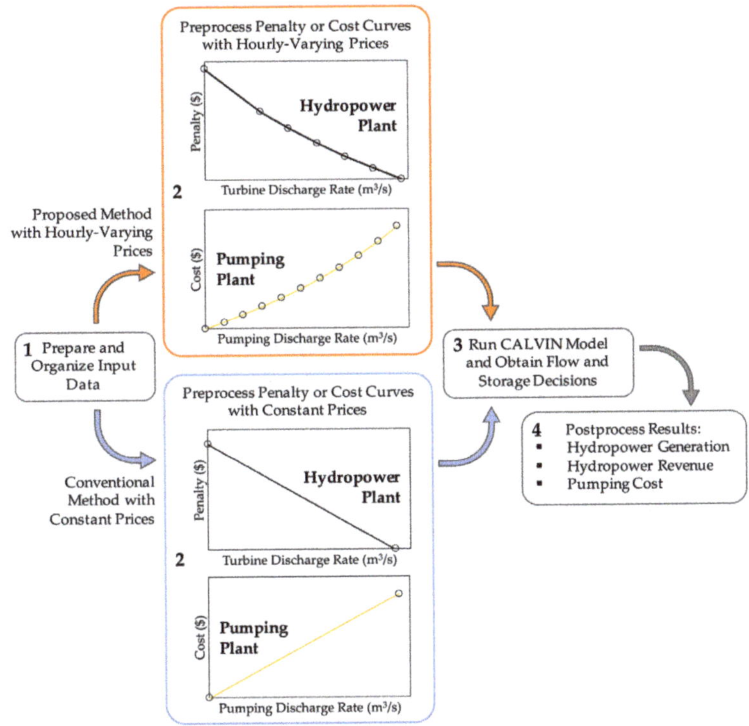

Figure 4. Modeling procedure with conventional constant and proposed hourly varying energy prices.

3. Results

3.1. Power Generation and Revenue

CALVIN is run with both monthly constant average prices and hourly varying prices. With hourly varying prices, the model tends to allocate small releases with higher hydropower benefits. As the operating time at turbine capacity increases, the marginal benefits decrease. With hourly varying prices, statewide total hydropower generation is slightly less than with constant monthly prices, but the energy revenue is significantly greater (Table 3). With constant monthly average prices, the annual total modeled hydropower revenue is about USD 550 million per year, which increases by 24% to USD 681 million per year when modeled with hourly variable prices. The hydropower generation of plants modeled with variable prices does not notably change, except for Holm and O'Neill, where annual average hydropower generation and revenue decrease. Holm is upstream of Don Pedro and O'Neill is between the California Aqueduct and Delta–Mendota Canal. California's intertied water system allows other supply routes to be utilized in the model using hourly varying prices, without impacting overall agricultural and urban deliveries. With hourly varying prices, less water is pumped from O'Neill (discussed later), reducing the hydropower generation of this pumped-storage plant. For other plants modeled with variable prices, the average annual revenues increase by 25 to 58%.

Table 3. Modeled annual average hydropower operations with constant and hourly varying prices. Facilities modeled with variable energy prices show a greater change (mostly positive) in the revenue reported from the model.

Facility	Turbine Release (m^3/s)		Generation (GWh/Year)		Revenue (USD M/Year)		Revenue
	Constant Price	Variable Price	Constant Price	Variable Price	Constant Price	Variable Price	Change (%)
Shasta [a]	217	216	2286	2285	75.5	110.6	46.5
Spring Creek [a]	32	31	381	381	13.5	19.4	43.4
Carr	25	25	354	354	12.2	12.4	1.5
Trinity [a]	48	48	408	409	13.0	19.0	46.7
Keswick	244	243	465	464	15.5	15.4	−0.8
Hyatt [a]	154	152	2190	2183	76.7	111.6	45.5
Colgate [a]	47	46	1356	1361	45.8	64.0	39.9
Folsom [a]	100	99	655	653	22.1	32.0	44.6
New Narrows	60	60	277	277	9.5	9.4	−1.0
Thermalito	137	136	293	290	10.3	10.2	−0.9
New Melones [a]	40	40	495	496	15.9	25.2	58.5
Don Pedro [a]	56	55	611	610	18.6	29.5	58.2
Holm [a]	17	13	869	673	30.6	27.8	−9.2
Kirkwood [a]	13	13	364	355	12.6	15.8	25.1
New Exchequer	36	36	275	277	8.5	8.6	1.0
Gianelli [a]	0	0	0	0	0	0	0
Moccasin [a]	13	13	167	167	5.8	7.4	27.4
O'Neill [a]	37	10	33	9	1.2	0.4	−68.0
Pine Flat	60	60	445	444	13.2	13.2	−0.1
Castaic	35	35	872	872	33.2	33.2	0.0
Devil's Canyon	37	37	1051	1051	37.2	37.2	0.0
Warne	36	36	522	522	19.2	19.2	0.0
San Francisquito 1 and 2	14	14	629	629	23.1	23.1	0.1
Gorges	7	7	366	366	12.5	12.5	0.0
Mojave	37	37	90	90	3.2	3.2	0.0
Others	244	242	642	642	20.5	20.5	−0.3
Statewide	1745	1704	16,094	15,857	550	681	24

Note: [a] Modeled with hourly varying prices.

Figure 5 shows the monthly statewide modeled hydropower generation and revenue with constant and variable prices. Overall, the monthly generation and revenue patterns do not differ greatly. Generation and revenue have a similar monthly trend, higher in the summer. Hydropower revenue with variable prices (VPs) is greater than with constant prices (CPs) in all months. Hydropower generation is greater with variable prices in the spring (April–June) when average energy prices are the lowest, but the hourly price fluctuations are the highest. In these months, with price fluctuations, the differences between constant and variable price revenues are much greater. In other months, constant prices lead to higher hydropower generation but lower revenue (underestimating with constant prices). As the hourly price fluctuations decrease, the moving average of variable prices gradually approaches constant average prices. Months with identical power generation or revenue can be compared for analysis. For example, for March, the power generation in both price schemes is quite close (1.38 TWh and 1.39 TWh for constant and variable prices, respectively) but with a big difference in revenue. In March, with constant prices, the revenue is about USD 52 million, while the revenue with variable prices is USD 62 million. In October, January, and February, the revenue difference between variable and constant prices is fairly small, although generation with constant prices is higher. Using constant monthly average prices rather than hourly varying prices underestimates hydropower revenue.

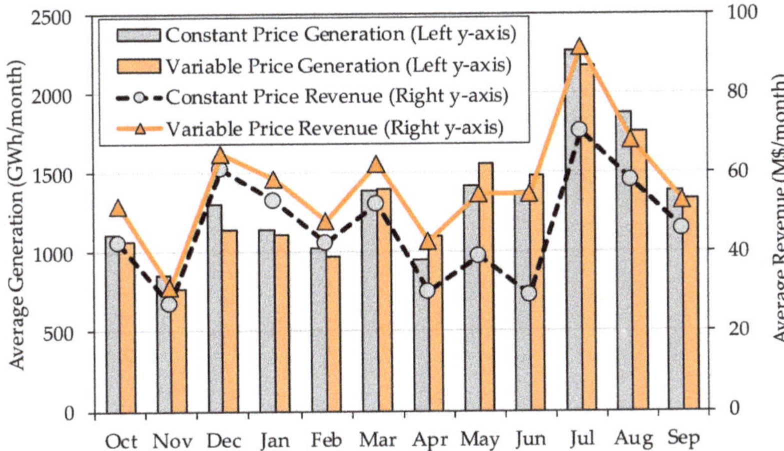

Figure 5. Modeled statewide monthly hydropower generation and power revenue with constant and variable energy prices. Revenues tend to diverge more in the months with greater energy price variations between the two methods.

Figure 6 shows monthly revenue from the 82-year (October 1921 to September 2003) modeling period by the percent of hours of generation at turbine capacity for selected CALVIN hydropower facilities with integrated hydropower and water supply operations. The proposed hourly varying energy price scheme increases the revenue per capacity use, as indicated by the slopes of the linear regression lines. The divergent regression lines also imply that these hydropower plants are mostly operated for shorter durations in a given month. Hyatt is operated more at turbine capacity compared to other modeled plants. When water availability is greater, such as in wet months and years, plants are operated during most hours and generate more revenue.

CALVIN covers Sacramento, the San Joaquin Valley, and southern California. Water year types (WYTs) are runoff indices used in Sacramento and the San Joaquin Valley, the northern and southern parts of the Central Valley [27]. Since a WYT index is unavailable for the much drier southern California region, plants in this area are excluded from the WYT analysis. Hydropower generation increases during wetter years. The average hydropower generation in the Central Valley CALVIN facilities differs by water year types, as shown in Table 4. In wet and above-normal years, generation with constant monthly prices is higher, whereas generation in other year types is fairly similar. During periods of abundant water, hydropower operations become more prominent. As water becomes scarce, water supply operations for agricultural and urban users dominate over hydropower operations, so more hydropower production becomes incidental. As water availability decreases from wet to critically dry years, discrepancies in generation and revenue between the two pricing types increase. In wet years, revenues are highest, while the average annual generation with constant prices, 16.3 TWh per year, exceeds generation with variable prices, 15.7 TWh per year. When the WYT is below normal, the average annual generations are equal, although revenue with variable prices, USD 501 million per year, exceeds the annual average revenue of USD 367 million per year with constant prices. Only in dry years does variable price generation exceed constant price generation. Because water year types are not evenly distributed, the overall average generation and revenue with constant and variable prices do not align with historical averages. Hydropower revenue is underestimated for all year types with constant average prices. With hourly varying prices, hydropower revenue increases by 28% in wet years to 40% in critically dry years, with an overall average increase of 33% for all Central Valley facilities.

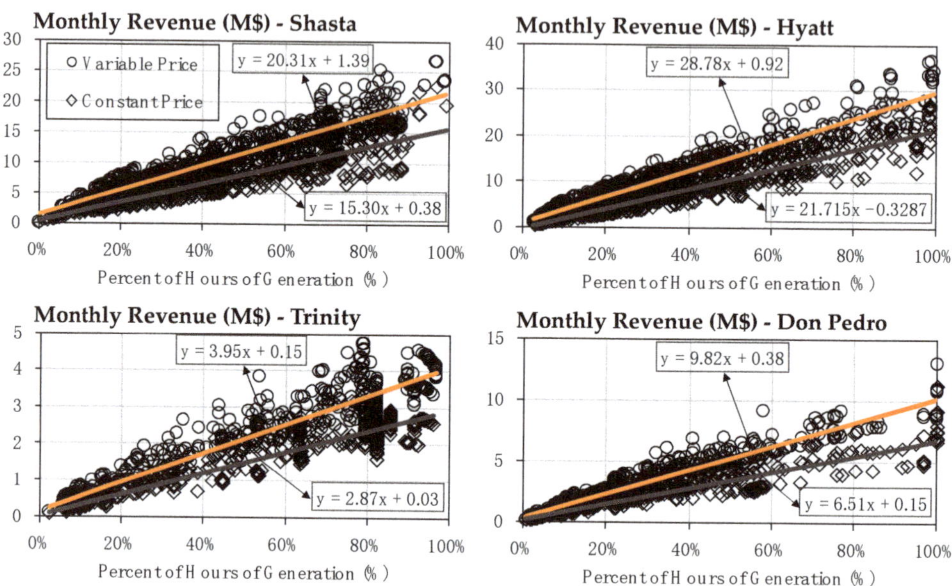

Figure 6. Modeled monthly generation and percent hours of generation in a given month for selected hydropower plants. Variable energy prices increase revenue by percent of generation hours.

Table 4. Modeled generation and power revenue in different water year types for the Central Valley facilities. The drier the year, the more the revenue changes and the more impactful having variable pricing schemes are.

Year Type	Average Generation (TWh/Year)		Average Revenue (USD M/Year)		Revenue Change (%)
	Constant Price	Variable Price	Constant Price	Variable Price	
Wet	16.3	15.7	548	701	28
Above Normal	12.8	12.4	436	569	31
Below Normal	10.9	10.9	367	501	36
Dry	9.4	9.5	313	435	39
Critical	7.3	7.2	242	338	40
Central Valley Average	12.0	11.8	403	535	33

3.2. Water Supply Operations

Scarcity occurs when a user's demand is not completely fulfilled (when scarcity costs are positive). CALVIN has water delivery targets for development, land use, and population estimates, for each agricultural and urban demand area, for the year 2050. Water is allocated from the statewide water system to agricultural, urban, and environmental users to meet these target demands. The difference between target demand and actual delivery is defined as the water shortage or scarcity amount. The objective of CALVIN is to minimize statewide scarcity and operating costs.

The proposed hourly varying price scheme has little overall effect on water supply operations from surface reservoirs (Figure 7). Shasta's storage is slightly greater in March and Don Pedro's storage is greater in June through November with hourly variable energy prices. Consistent water supply operations alongside higher hydropower revenue suggest an improved representation of hydropower with hourly varying energy prices.

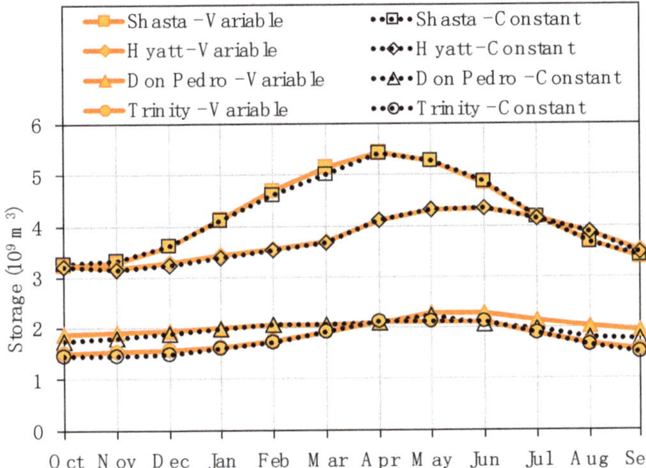

Figure 7. Modeled monthly average reservoir storage of selected facilities. Reservoir operations are little affected when variable energy prices are used.

3.3. Pumping Plant Operations

A similar variable energy pricing method is also applied to CALVIN's pumping plant operations. However, in this case, moving average curves of hourly energy prices are calculated in descending order of prices, meaning that for economically optimal pumping operations, plants should operate when prices are lowest (Figure 3). This is easier achieved when pumping plants have forebays and afterbays. Plants with a forebay can store water during high-energy-price periods (peak times) and commence operations during low-energy-price periods (off-peak times). Six of twenty-four pumping plants have major afterbays and are modeled with variable average prices in CALVIN. The remaining pumping plants use constant monthly average energy prices.

When hourly varying prices are used, the annual average pumping energy costs decrease by 10 to 59% (Table 5). The Banks pumping plant's energy use increases by 30 GWh, while O'Neill's energy use decreases by 6 GWh per year with variable prices. However, the total statewide energy usage remains constant. With constant monthly average energy prices, CALVIN's pumping plants use 12,574 GWh of energy with an annual cost of USD 442 million per year. However, without causing significant changes to agricultural and urban deliveries, the statewide optimized annual average pumping cost was reduced to USD 430 million by simply using hourly varying energy prices in modeling pumping operations, resulting in a savings of about USD 12 million per year (a 3% reduction). The difference in pumping costs between constant and hourly varying price schemes may be small compared to the annual total pumping cost. This could be partly because either the plants modeled with hourly varying prices have small capacities or their pumping capacity is not fully utilized. Additionally, only 6 of 24 pumping plants are modeled with varying energy prices due to data availability and the applicability of the method. However, this shows that plants with a forebay have more flexibility for operations and reduced operating costs.

Figure 8 shows the monthly energy cost by percent of hours of operation for the Banks and Oso pumping plants. Overall, the energy costs are lower with hourly varying prices for both plants. But the Banks pumping plant is operated in most hours of a given month, so its regression line has greater slope with variable prices. As discussed earlier, when a plant operates during most hours, the constant average and hourly varying energy price schemes converge. Operations at the Oso pumping plant do not exceed 50% of hours in a given month. So, using constant average monthly prices greatly overestimates pumping energy costs.

Table 5. Modeled annual average pumping plant operations with constant and hourly varying energy prices. Variable energy price scheme results in less pumping costs.

Facility	Discharge Rate (m³/s)		Energy Use (GWh/Year)		Energy Cost (USD M/Year)		Energy Cost Change (%)
	Constant Price	Variable Price	Constant Price	Variable Price	Constant Price	Variable Price	
Banks [a]	148.71	152.69	1130	1160	39	36	−10
Del Valle [a]	0.06	0.06	0.10	0.10	0.0044	0.0018	−59
Gianelli [a,b]	31.83	31.81	198	198	9	5	−41
O'Neill [a,b]	24.49	20.54	37	31	2	1	−40
Oso [a]	31.37	31.37	225	225	8	5	−41
South Bay [a]	0.06	0.06	1.28	1.28	0.0555	0.0242	−56
Others	845.95	841.90	10,983	10,959	384	383	0
Statewide	1928	1920	12,574	12,574	442	430	−3

Notes: [a] Modeled with hourly varying prices; [b] operated for both pumping and power generation.

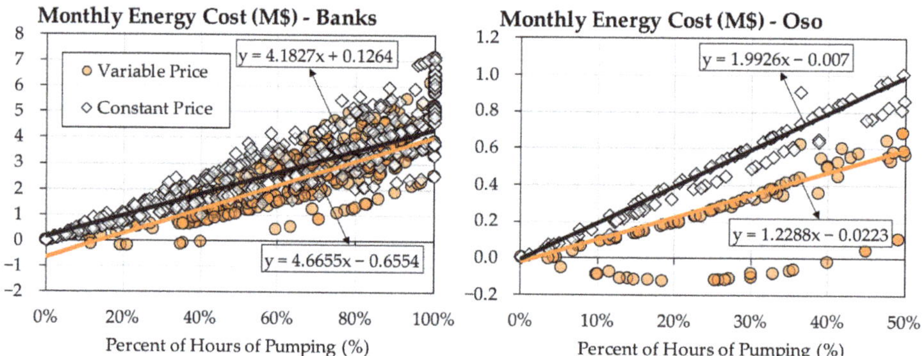

Figure 8. Modeled monthly energy and percent of hours of operation in a given month for pumping selected facilities. Pumping energy costs are smaller with variable energy prices, but as hours of operation increase, cost calculations from two methods become closer.

4. Discussion

Hydropower plants with large water storage capacities are operated mostly to meet peak energy demands, especially when they have afterbays, while run-of-river plants run continuously, depending on water availability. For plants with large storage capabilities, energy releases in peak hours have greater profits because energy prices during these hours are much higher. So, in a given month, these peak hours are targeted to maximize hydropower revenue. For plants with small storage capacities or run-of-river plants, the effectiveness of the proposed hourly varying energy prices method is limited, as they operate continuously throughout most of a given month. In addition, when there is less energy price variation, constant monthly average and variable prices converge and variable energy prices does not improve revenue estimates. However, in energy markets with significant variations, assuming constant energy prices for monthly models greatly underestimates hydropower revenue and overestimates pumping energy costs.

5. Conclusions

A simple method was used to implicitly represent hourly varying energy prices in monthly time-step, large-scale model operations, without affecting model runtime and water supply operations. Instead of having to run in hourly time steps, the method uses price duration curves and estimates hourly varying prices as a function of hours of operation at turbine or pumping plant capacity to capture hourly price variations in models with large time steps and better represent hydropower and pumping plant operations. The proposed method was applied to CALVIN, a hydroeconomic optimization model for California's intertied water supply system. Thirteen of thirty-three hydropower plants with afterbays and 6 of 24 pumping plants with forebays are modeled with hourly varying prices in CALVIN.

The model results were improved with variable prices, leading to increased hydropower revenue with a slight reduction in generation and decreased pumping costs. Overall, the use of constant energy prices tends to underestimate hydropower benefits and overestimate pumping costs.

The representation of hourly energy price variability had only a small, but sometimes significant, impact on overall water operations, particularly in drier years when water availability is limited. The proposed method did not increase water shortages, and water supply operations from reservoirs remained unchanged. The existence of afterbays and forebays in hydropower and pumping facilities provides operational flexibility and supports the utilization of hourly energy price variability to increase benefits and reduce costs. Forebays and afterbays also facilitate the implicit representation of variable energy prices in monthly models. This technique and its application demonstrate that it is sometimes possible to adequately represent short-period (hourly) phenomena and performance in long-period (monthly) models. This method can be applied to any model representing a hydropower facility with an effective afterbay or a pumping facility with a forebay, thereby improving the integration of hydropower and pumping operations into longer-term water resource system models.

Author Contributions: Conceptualization, M.S.D. and J.R.L.; analysis, M.S.D., E.W. and Y.Y.; writing—original draft preparation, M.S.D.; writing—review and editing, M.S.D., E.W., Y.Y. and J.R.L.; visualization, M.S.D.; supervision, J.R.L. All authors have read and agreed to the published version of the manuscript.

Funding: This research received no external funding.

Data Availability Statement: The data that support the findings of this study, the model's documentation, the supplementary material, and the source code are openly available at https://github.com/ucd-cws/calvin (accessed on 4 December 2023) and https://github.com/ucd-cws/calvin-network-data (accessed on 4 December 2023).

Conflicts of Interest: The authors declare no conflicts of interest.

References

1. Loucks, D.P. Water resource systems models: Their role in planning. *J. Water Resour. Plan. Manag.* **1992**, *118*, 214–223. [CrossRef]
2. Yeh, W.W.-G.; Becker, L. Multiobjective analysis of multireservoir operations. *Water Resour. Res.* **1982**, *18*, 1326–1336. [CrossRef]
3. Savic, D.A.; Simonovic, S.P. An interactive approach to selection and use of single multipurpose reservoir models. *Water Resour. Res.* **1991**, *27*, 2509–2521. [CrossRef]
4. Leta, M.K.; Demissie, T.A.; Tränckner, J. Optimal operation of Nashe hydropower reservoir under land use land cover change in Blue Nile River Basin. *Water* **2022**, *14*, 1606. [CrossRef]
5. Yeh, W.W.-G. Reservoir management and operations models: A state-of-the-art review. *Water Resour. Res.* **1985**, *21*, 1797–1818. [CrossRef]
6. Wurbs, R.A. Reservoir-system simulation and optimization models. *J. Water Resour. Plan. Manag.* **1993**, *119*, 455–472. [CrossRef]
7. Labadie, J.W. Optimal operation of multireservoir systems: State-of-the-art review. *J. Water Resour. Plan. Manag.* **2004**, *130*, 93–111. [CrossRef]
8. Ward, F.A.; Lynch, T.P. Integrated river basin optimization: Modeling economic and hydrologic interdependence. *J. Am. Water Resour. Assoc.* **1996**, *32*, 1127–1138. [CrossRef]
9. Vichete, W.D.; Méllo Júnior, A.V.; Soares, G.A.d.S. A water allocation model for multiple uses based on a proposed hydro-economic method. *Water* **2023**, *15*, 1170. [CrossRef]
10. Davis, M.D. Integrated water resource management and water sharing. *J. Water Resour. Plan. Manag.* **2007**, *133*, 427–445. [CrossRef]

11. Marques, G.F.; Tilmant, A. The economic value of coordination in large-scale multireservoir systems: The Parana River case. *Water Resour. Res.* **2013**, *49*, 7546–7557. [CrossRef]

12. Li, C.; He, L.; Liu, D.; Feng, Z. A Scientometric review for uncertainties in integrated simulation–optimization modeling system. *Water* **2024**, *16*, 285. [CrossRef]

13. Tejada-Guibert, J.A.; Stedinger, J.R.; Staschus, K. Optimization of value of CVP's hydropower production. *J. Water Resour. Plan. Manag.* **1990**, *116*, 52–70. [CrossRef]

14. Madani, K.; Lund, J.R. Modeling California's high-elevation hydropower systems in energy units. *Water Resour. Res.* **2009**, *45*. [CrossRef]

15. Olivares, M.A.; Lund, J.R. Representing energy price variability in long-and medium-term hydropower optimization. *J. Water Resour. Plan. Manag.* **2012**, *138*, 606–613. [CrossRef]

16. Ring, B.J. Dispatch Based Pricing in Decentralised Power Systems. Ph.D. Dissertation, University of Canterbury, Christchurch, New Zealand, 1995.

17. Madrigal Martinez, M. Optimization Models and Techniques for Implementation and Pricing of Electricity Markets. Ph.D. Dissertation, University of Waterloo, Waterloo, ON, Canada, 2001.

18. Pritchard, G.; Philbot, A.; Neame, P. Hydroelectric reservoir optimization in a pool market. *Math. Program.* **2004**, *103*, 445–461. [CrossRef]

19. Wu, Y.; Su, C.; Liu, S.; Guo, H.; Sun, Y.; Jiang, Y.; Shao, Q. Optimal decomposition for the monthly contracted electricity of cascade hydropower plants considering the bidding space in the day-ahead spot market. *Water* **2022**, *14*, 2347. [CrossRef]

20. Liu, S.; Luo, J.; Chen, H.; Wang, Y.; Li, X.; Zhang, J.; Wang, J. Third-monthly hydropower scheduling of cascaded reservoirs using successive quadratic programming in trust corridor. *Water* **2023**, *15*, 716. [CrossRef]

21. Li, J.; Saw, M.M.M.; Chen, S.; Yu, H. Short-term optimal operation of Baluchaung II hydropower plant in Myanmar. *Water* **2020**, *12*, 504. [CrossRef]

22. Dogan, M.S. Integrated Water Operations in California: Hydropower, Overdraft, and Climate Change. Master's Thesis, University of California, Davis, CA, USA, 2015.

23. Draper, A.J.; Jenkins, M.W.; Kirby, K.W.; Lund, J.R.; Howitt, R.E. Economic-engineering optimization for California water management. *J. Water Resour. Plan. Manag.* **2003**, *129*, 155–164. [CrossRef]

24. Dogan, M.S.; Fefer, M.A.; Herman, J.D.; Hart, Q.J.; Merz, J.R.; Medellin-Azuara, J.; Lund, J.R. An open-source Python implementation of California's hydro-economic optimization model. *Environ. Model. Softw.* **2018**, *108*, 8–13. [CrossRef]

25. Arnold, W. The Economic Value of Carryover Storage in California's Water Supply System with Limited Hydrologic Foresight. Master's Thesis, University of California, Davis, CA, USA, 2021.

26. Locational Marginal Energy Price Data. Available online: http://www.caiso.com (accessed on 6 December 2023).

27. Chronological reconstructed Sacramento and San Joaquin Valley Water Year Hydrologic Classification Indices. Available online: https://cdec.water.ca.gov/reportapp/javareports?name=WSIHIST (accessed on 22 January 2024).

MDPI

St. Alban-Anlage 66

4052 Basel

Switzerland

www.mdpi.com

MDPI Books Editorial Office

E-mail: books@mdpi.com

www.mdpi.com/books